VISUAL COMPLEX ANALYSIS

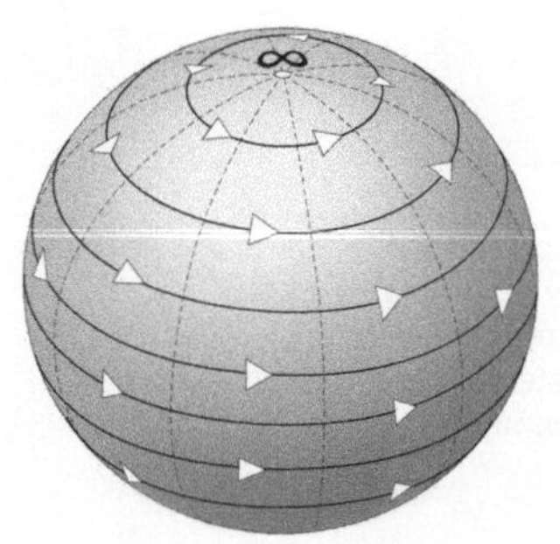

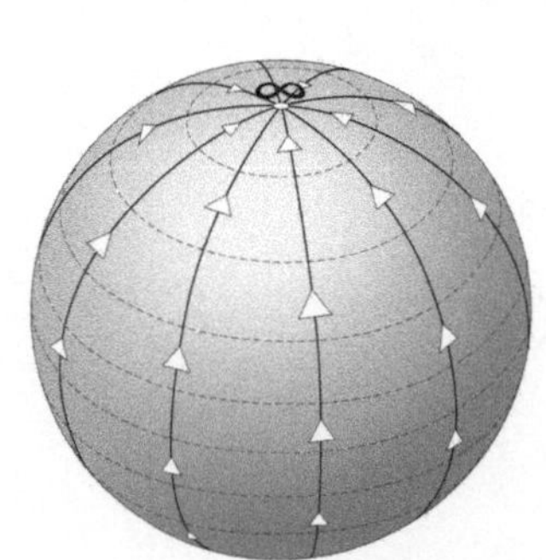

VISUAL COMPLEX ANALYSIS

TRISTAN NEEDHAM

UNIVERSITY OF SAN FRANCISCO

Foreword by
ROGER PENROSE

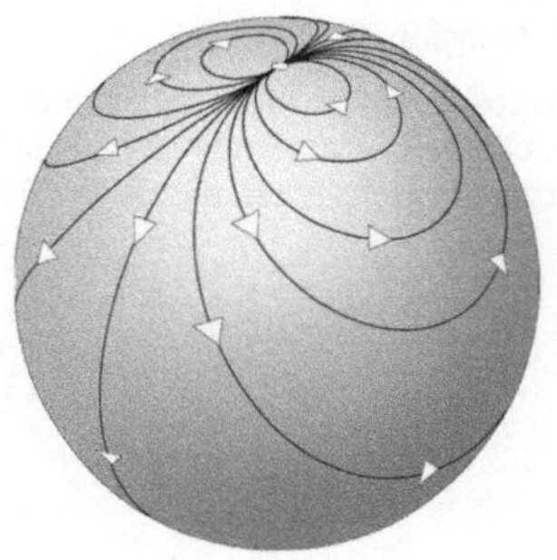

25TH ANNIVERSARY EDITION

OXFORD
UNIVERSITY PRESS

OXFORD
UNIVERSITY PRESS

Great Clarendon Street, Oxford, OX2 6DP,
United Kingdom

Oxford University Press is a department of the University of Oxford.
It furthers the University's objective of excellence in research, scholarship,
and education by publishing worldwide. Oxford is a registered trade mark of
Oxford University Press in the UK and in certain other countries

First edition published in 1997

25th Anniversary Edition 2023

Published in the United States of America by Oxford University Press
198 Madison Avenue, New York, NY 10016, United States of America

British Library Cataloguing in Publication Data
Data available

Library of Congress Control Number: 2022948019

ISBN 978–0–19–286891–6 (hbk.)
ISBN 978–0–19–286892–3 (pbk.)

DOI: 10.1093/oso/9780192868916.001.0001

Printed and bound by
CPI Group (UK) Ltd, Croydon, CR0 4YY

For
Roger Penrose
and
George Burnett-Stuart

FOREWORD

Roger Penrose

Complex analysis is the theory of functions of complex numbers or, more specifically, *holomorphic* functions of such numbers. This theory is both profoundly beautiful and vastly influential, both in pure mathematics and in many areas of application, particularly in physics, indeed being central to the underlying formalism of quantum mechanics. However, the very concept of a complex number is an essentially abstract one, depending upon the seemingly absurd notion of a square root of -1, the square of any ordinary real number being, unlike -1, necessarily non-negative.

Yet, it should be borne in mind that even the notion of a so-called "real" number is also an abstraction, and we must move far beyond the immediate notion of "counting numbers" 0, 1, 2, 3, 4, . . ., and beyond even the fractions $\frac{1}{2}$, $\frac{3}{5}$, $-\frac{7}{4}$, etc., if we are to express even the square root of 2. But here we are helped by a *visual image*, and can perceive a straight line extended indefinitely in both directions to give us a good intuitive impression of the full array of real numbers. The slightly misleading term "real" for this imagined array is excusable, as we can indeed imagine a ruler, or a line of ink drawn on a piece of paper, as providing us with some sort of conceptual image of this array. This greatly helps our understandings of what the mathematician's precise notion of a "real number" is intended to idealize. We are not concerned with whatever might be the nature of the physics of the particles or fields that might compose our ruler or ink-line; nor, indeed, do we require any concept of the cosmology that may be relevant to the extension out to infinity of our imagined ruler or ink-line. Our abstract mathematical notion of a "real number" remains aloof from any such realities of the actual world. Yet, in a curious reversal of roles, it is this very mathematical idealization that underlies most of our theories of the actual world.

So, what can be the driving force behind a need to go beyond these seemingly ubiquitous "real numbers"? What purpose might there be for the introduction of a "square root of -1"? Such a number fits nowhere within the span of the real numbers, and it would appear that we have no reason to demand that the equation $x^2 + 1 = 0$ have any kind of "solution". The answer to this desire for such entities lies in the *magic* that lies hidden within them, but it is not a magic that immediately reveals itself. In fact, when the first hint of this magic was actually perceived, in the mid 16th century, by Girolamo Cardano, and then more completely by Rafael Bombelli, these strange numbers were dismissed, even by them, as being as useless as they were mysterious.

It is of some interest to note that it was not in the equation $x^2 + 1 = 0$ that this hint of magic was first perceived, but in cubic equations like $x^3 = 15x + 4$, which has the perfectly sensible solution $x = 4$. Yet, as Bombelli had noted, Cardano's general expression for the solution of such cubic equations necessarily involves a detour into a mysterious world of numbers of a sort where the equation $x^2 + 1 = 0$ is deemed to have the two solutions, now referred to as the *imaginary units* $x = i$ and $x = -i$, of an algebra—now called *complex-number* algebra—that had appeared to be consistent, but not what had been regarded as "*real*".

This dismissive attitude did not change much until the mid to late 18th century, with Leonhard Euler's remarkable formula $e^{i\theta} = \cos\theta + i\sin\theta$ and, even more importantly, the geometrical representation of the entire family of complex numbers as points in a Euclidean plane, as initially proposed by Caspar Wessel, where the algebraic operations on complex numbers are readily understood in geometrical terms. This provided a kind of 2-dimensional "visual reality" to the array of complex numbers that could be combined with topological notions, such as employed initially by Carl Friedrich Gauss, and soon followed by others. The early to mid 19th century saw many important advances, many of these being due to Augustin-Louis Cauchy, especially with the beauty and the power of contour integration, and, perhaps most profoundly, with ideas due to Bernhard Riemann. The very notion of "complex smoothness" of complex functions was expressed by use of the Cauchy–Riemann equations, and this provided the powerful concept of a *holomorphic function* that implies that a power-series expansion always locally exists, this leading to a vast and powerful theory with numerous magical properties.

The two revolutions of early 20th century physics both owe a profound debt to complex-number mathematics. This is most manifest with quantum mechanics, since the basic formalism of that theory depends fundamentally on complex numbers and holomorphic functions. We see a remarkable interplay between quantum spin and the geometry of complex numbers. The basic equations of Schrödinger and Dirac are both complex equations. In relativity theory, the transformations relating the visual field of two observers passing close by each other at different relativistic speeds is most easily understood in terms of simple holomorphic functions. Moreover, many solutions of Einstein's equations for general relativity benefit greatly from properties of holomorphic functions, as does the description of gravitational waves.

In view of the undoubted importance of complex analysis in so much of mathematics and physics, it is clearly important that there are basic accounts of these topics available to those unfamiliar (or only partly familiar) with the basic ideas of complex analysis. In this foreword I have very much stressed how the visual or geometric viewpoint has been of vital importance, not only to the historical development of complex analysis, but also to the proper understanding of the subject. Tristan Needham's *Visual Complex Analysis* as originally published in 1997 was, to my knowledge, unique in the extent to which it was able to cover these fundamental ideas with such thoroughness, visual elegance, and clarity.

With this *25th Anniversary Edition* there have been some significant improvements, most particularly in the incorporation of captions to the diagrams. This makes it easier for the reader to dip into the arguments, as illustrated so elegantly in such wonderfully expressive pictures, without necessarily having to look through to find the relevant portion of the text. In any case, I am sure that readers, over a broad range of relevant knowledge—from those with no prior experience of complex analysis to those already experts—will gain greatly from the charm, distinct originality, and visual clarity of the arguments presented here.

of complex analysis. In this foreword I have very much stressed how the visual or geometric viewpoint has been of vital importance, not only to the historical development of complex analysis, but also to the proper understanding of the subject. Tristan Needham's *Visual Complex Analysis*, as originally published in 1997, was, to my knowledge, unique in the extent to which it was able to cover these fundamental ideas with such thoroughness, visual elegance, and clarity.

With this *25th Anniversary Edition* there have been some significant improvements, most particularly in the incorporation of captions to the diagrams. This makes it easier for the reader to dip into the arguments, as illustrated so elegantly in these wonderfully expressive pictures, without necessarily having to look through the detailed text portions of the book. In any case, I am sure that readers, over a broad range of different knowledge [illegible] will [illegible] and [illegible] clarity of this remarkable book.

PREFACE TO THE 25TH ANNIVERSARY EDITION

Introduction

Mathematical reality exists outside the confines of space and time, but books about mathematics do not. A quarter century after its publication, I am grateful that an entirely new generation of mathematicians and scientists has continued to embrace VCA's unorthodox, intuitive, and, above all, *geometrical* approach to Complex Analysis. To mark the occasion, Oxford University Press has graciously permitted me to revisit the work, resulting in the creation of this significantly improved *25th Anniversary Edition*.

Principal Changes in the New Edition

Perhaps I should begin by noting what has *not* changed. As I shall elaborate below, back in 1997, one of my key mathematical innovations was the application of Newton's geometrical methods from the *Principia* (now 335 years old) to the Complex Analysis of Cauchy and Riemann (now more than 200 years old). Therefore, I have not seen any need to update the main body of the text, which has remained almost unchanged, save for a few additional corrections.

All of the changes to this new edition of VCA—both in appearance and substance—came about as a direct result of the knowledge I gained creating my second and final book,[1] *Visual Differential Geometry and Forms: A Mathematical Drama in Five Acts*, Princeton University Press, 2021 (hereafter abbreviated to VDGF). In particular, I thank Wanda España of PUP for her beautiful work on the design of VDGF, for this in turn inspired some design elements that I requested of OUP for this new edition of VCA. Of course I also thank Oxford University Press for *accepting* some of these suggestions, while also imprinting their own distinctive style on the work.

- The most obvious change is in the physical dimensions of the book: it has expanded from 6″×9″ to 7″×10″. There are two significant advantages to this change: (1) the book is more comfortable to hold and read; (2) the 503[2]

[1] Needham (2021).

[2] The original edition only contained 501 figures: The figure in this Preface and [6.41] are both new.

figures—the beating heart of the work!—are now 36% larger, and, one may dare to hope, 36% clearer!

- Another obvious change is the introduction of the standard numbering system for sections, subsections, equations, and figures. In hindsight, my deliberate avoidance of the standard scheme was perhaps a rather childish, tantrum-like expression of my disgust with the prevailing, life-sucking reduction of wildly exciting mathematical ideas to the arid structure enshrined in traditional treatises: "Lemma 12.7.2 implies Theorem 14.3.8". Let me take this opportunity to apologize—better 25 years late than never?—to every professor who was ever brave enough to adopt VCA as the text for their course, only to discover that they had to struggle to refer their students to any given figure or result!
- The *Bibliography* has been updated. Not only have previously cited works been updated to their latest editions, but I have also added a number of *new* works, which were not cited in the first edition, for the simple reason that 25 years ago they did not *exist*!
- The *Index* has been improved significantly. Many new entries have been added, and, perhaps more importantly, wherever a single main entry formerly listed a long, frustratingly unhelpful string of undifferentiated page numbers, I have now split it into many individual and *helpfully specific subentries*. By way of proof, consider the entry for "Jacobian matrix", then and now!
- *Giving credit to Eugenio Beltrami*. In the original edition I pointed out that in 1868 Beltrami discovered[3] (and published) the conformal models of the hyperbolic plane, which Poincaré then rediscovered 14 years later, in 1882. Now, following the example I set myself in VDGF, I have gone one step further, attempting to put the record straight by *renaming* these models as the *Beltrami–Poincaré* disc and half-plane. Beltrami also discovered the projective model, and I have renamed that, too, as the *Beltrami–Klein* model. Correspondingly, my fancifully named inhabitants of these hyperbolic worlds have been renamed from "Poincarites" to *Beltrami–Poincarites*!
- Twenty five years ago, VCA was on the bleeding edge of what was *typographically* possible. Indeed, my editor told me that VCA was the *first* mathematics book published by Oxford University Press to be composed in LaTeX and yet *not* to be typeset using the standard Computer Modern fonts designed by Donald Knuth, the creator of TeX. I was able to achieve this feat by virtue of Michael Spivak's[4] then newly created *MathTime* typeface, and by virtue of the wonderful

[3] See Milnor (1982), Stillwell (1996), and Stillwell (2010).

[4] Yes, the famous differential geometer!

Y&Y TEX System for Windows.[5] I was thereby able to typeset the book in Times text and (mainly) MathTime mathematics—a *vast* aesthetic improvement over Computer Modern, in my humble but strong opinion!

Needless to say, the TEX world has moved on considerably in the past 25 years. But in order to take advantage of these advances, I was forced to grapple with the task of updating my original TEX files (including my countless macros) from the ancient (dead?) language of LATEX2.09 to the modern LATEX 2ε. This effort was rewarded with the ability to typeset this new edition of VCA using the same typefaces that I very carefully selected for VDGF, all three of which sprang from the genius mind of Hermann Zapf (1918–2015): Optima for the headings, Palatino for the text, and, crucially, the remarkable Euler fonts for the *mathematics*.

Let me pause for a moment to pay tribute to Zapf's Euler mathematical fonts. They would appear to me to be at home onboard the starship of an advanced alien civilization, yet they also evoke the time-worn stone engravings within an ancient Greek ruin—their beauty transcends space and time: Picture an inscription on the *Guardian of Forever*![6]

But my fatal attraction to these fonts brought about conflict between the Euler mathematics in the text of the new edition and the *MathTime* mathematics in all the *figures* of the original edition. Having been born cursed (and I suppose blessed) with a compulsion to strive for perfection, I had no choice but to undertake the self-inflicted, Herculean task of hand-editing (within CorelDRAW) all 501 of my original, hand-drawn figures! I then output new versions in which each MathTime symbol is here replaced with its matching Euler counterpart, thereby bringing the figures into perfect alignment with the new text.

I am the impossibly proud father of remarkable twin daughters, Faith and Hope. I am also the father of VCA and of VDGF. It is therefore a source of deep, resonating joy that *both* sets of twins now *look* like twins!

- Now let me explain the most fundamental change, the one that took me the greatest time and effort to accomplish, and the one that I believe transforms this *25th Anniversary Edition* into a truly *new* edition, one that may even be of value to owners of the original edition.

The 503 figures are the mathematical *soul* of the work. They crystallize all the geometrical insights I was able to glean from my many years spent struggling to *understand* Complex Analysis. Yet, in the entire original edition, *you will not find a single* ***caption***. Why?!

Sadly, the answer is simple: *cowardice*. As a newly minted DPhil student of Penrose, with no track record or reputation, I feared that the mathematical community

[5] Sadly, this system (designed by Professor Berthold Horn of MIT) ceased to exist in 2004.
[6] The space–time portal featured in the *Star Trek* episode, "The City on the Edge of Forever".

would reject or even ridicule[7] my Newtonian arguments. I therefore sought to control the narrative (literally!) by only revealing the arguments within the text proper, where I could spell out my reasoning in full.

Well, 25 years later, I am certainly older, perhaps a tad wiser, and I have also gained a modicum of confidence by virtue of the enthusiastic reviews that VCA has since received in all the major journals. Perhaps more emotionally significant to me has been the large number of *individual* notes of appreciation I have received, which I continue to receive to this day, from readers of all stripes—graduate students, professors, and working scientists—from around the globe.

So, finally, I have done something that I should have had the courage to do at the outset:

> ***In this* 25th Anniversary Edition*, every figure now has a* caption *that fully explains its mathematical content. Additionally, many of these new captions include a* title *(in bold print) that further distils the figure's meaning down to its essence.***

This approach[8] is directly inspired by the works of my teacher, friend, and mentor, Sir Roger Penrose—to whom this book is dedicated, and to whom I now offer sincere thanks for his generous Foreword!—most remarkably in his *Road to Reality*,[9] where a single figure's caption can take up a quarter of a page!

This innovation now makes it possible to read VCA in an entirely new way—as a highbrow comic book! *Much* of the geometrical reasoning of the work can now be grasped simply by studying a figure and its accompanying caption, only turning to the main text for the complete explanation as needed. Furthermore, instead of undertaking a systematic, linear reading of the work, you are now invited to skip and hop about, lighting upon whichever figure happens to catch your eye.

- Newton's concept of *ultimate equality* underlies many of the arguments in this book, but its use was not made *explicit* in the original edition, and this led some to suppose that the arguments were less rigorous than they actually were (and remain). In this new edition I have only occasionally modified the main text, but I *have* felt at liberty to explicitly introduce ultimate equalities into some of the newly minted *captions*. As in my earlier works,[10] I have used the symbol $\asymp$ to denote this concept.

Let me now expand upon this vitally important point ...

[7] In my defence, such fears were not entirely groundless: When Princeton University Press sent out draft chapters of VDGF for review, one of the three anonymous reviewers bluntly declared, "This is not even mathematics!"

[8] This is also my approach in VDGF.

[9] Penrose (2005).

[10] Needham (1993), Needham (2014).

More on Newton's *Principia* and His Concept of *Ultimate Equality*

As I explain in the original Preface (immediately following this new Preface), VCA (and VDGF) could never have come to be had I not undertaken a careful study of Newton's *Principia* in 1982, while I was still a DPhil student of Penrose. As I struggled to penetrate Newton's ancient diagrams, I intuitively sensed the power and beauty of his geometrical reasoning, long before I fully understood it.

But one aspect of Newton's thinking did make sense to me *immediately*: why should Newton settle for studying an *equation* describing the orbit of a planet around the Sun when he possessed a geometrical tool of enormous power, capable of instead analyzing the orbit *itself*?! This in turn provoked within me a life-changing *mathematical crisis of conscience*: how many other mathematical phenomena had I fooled myself into *pretending* I had understood, when all I had done was grasp an equation that *described* the phenomenon from afar, rather than daring to stare mathematical reality squarely and *geometrically* in the face?

I felt excited and strangely fearful, like one of the apes in the opening of Stanley Kubrick's, *2001: A Space Odyssey*, shrieking and wildly leaping about in the stark silent presence of the black monolith. I wanted to be that one ape that was brave enough to *touch* the monolith, now taking the form of a different ancient reservoir of deeply alien knowledge: Newton's mind!

So I did. And once I had done so, I stared down at Newton's geometrical concept of *ultimate equality*, as the ape had stared down at the large femur bone on the ground before him, picked it up, gingerly at first, examined it closely, and then spent the next 35 years of my life wielding it with all the force and ingenuity I could muster—a tool to crack open and lay bare to our visual intuition the secrets of Complex Analysis, and, 25 years later, the secrets of Differential Geometry and Forms.

The presentation of the geometrical reasoning in VDGF *appears* to be more rigorous than that in VCA because there I *explicitly* make use of Newton's concept of ultimate equality. Let me now steal an excerpt from the Prologue of VDGF in order to explain what this means.

As I have discussed elsewhere,[11] Newtonian scholars have painstakingly dismantled the pernicious myth[12] that the results in the 1687 *Principia* were first derived by Newton using his original 1665 version of the calculus, and only later recast into the geometrical form that we find in the finished work.

Instead, it is now understood that by the mid-1670s, having studied Apollonius, Pappus, and Huygens, in particular, the mature Newton became disenchanted with

[11] See Needham (1993), the original 1997 Preface to VCA, and Needham (2014).

[12] Sadly, this myth originated with Newton himself, in the heat of his bitter priority battle with Leibniz over the discovery of the calculus. See Arnol'd (1990), Bloye and Huggett (2011), de Gandt (1995), Guicciardini (1999), Newton (1687, p.123), and Westfall (1980).

the form in which he had originally discovered the calculus in his youth—which is different again from the Leibnizian form we all learn in college today—and had instead embraced purely geometrical methods.

Thus it came to pass that by the 1680s Newton's algebraic infatuation with power series gave way to a new form of calculus—what he called the "synthetic method of fluxions"[13]—in which the geometry of the Ancients was transmogrified and reanimated by its application to shrinking geometrical figures in their moment of vanishing. *This* is the potent but non-algorithmic form of calculus that we find in full flower in his great *Principia* of 1687.

Let me now immediately spell out Newton's approach, and in significantly greater detail than I did in the first edition of VCA, in the vain hope that this new edition may bolster my efforts in VDGF to inspire more mathematicians and physicists to adopt Newton's intuitive (yet rigorous[14]) methods than did the first edition.

If two quantities A and B depend on a small quantity ϵ, and their ratio approaches unity as ϵ approaches zero, then I shall avoid the more cumbersome language of limits by following Newton's precedent in the *Principia*, saying simply that, "A is ultimately equal to B". Also, as I did in earlier works [(Needham, 1993), (Needham, 2014)], I shall employ the symbol $\asymp$ to denote this concept of *ultimate equality*.[15] In short,

$$\text{"A is ultimately equal to B"} \iff \mathrm{A} \asymp \mathrm{B} \iff \lim_{\epsilon \to 0} \frac{\mathrm{A}}{\mathrm{B}} = 1.$$

It follows [exercise] from the theorems on limits that ultimate equality is an equivalence relation, and that it also inherits additional properties of ordinary equality, e.g.,

$$\mathrm{X} \asymp \mathrm{Y} \ \& \ \mathrm{P} \asymp \mathrm{Q} \implies \mathrm{X} \cdot \mathrm{P} \asymp \mathrm{Y} \cdot \mathrm{Q}, \quad \text{and} \quad \mathrm{A} \asymp \mathrm{B} \cdot \mathrm{C} \iff (\mathrm{A}/\mathrm{B}) \asymp \mathrm{C}.$$

Before I begin to apply this idea in earnest, I also note that the jurisdiction of ultimate equality can be extended naturally to things other than numbers, enabling one to say, for example, that two triangles are "ultimately similar", meaning that their angles are ultimately equal.

As I explain in the original Preface, having grasped Newton's method, I immediately tried my own hand at using it to simplify my teaching of introductory calculus, only later realizing how I might apply it to Complex Analysis (in VCA), and later still to Differential Geometry (in VDGF). Though I might choose any number of simple, illustrative examples [see Needham (1993) for more], I will reuse the specific one I gave in the original Preface to VCA, and for one simple reason: *this*

[13] See Guicciardini (2009, Ch. 9).

[14] Fine print to follow!

[15] This notation was subsequently adopted (with attribution) by the Nobel physicist, Subrahmanyan Chandrasekhar (see Chandrasekhar, 1995, p. 44).

time I will use the "$\asymp$"–notation to *present* the argument rigorously, whereas in the first edition I did not. Indeed, this example may be viewed as a recipe for transforming many of VCA's "explanations" into "proofs",[16] merely by sprinkling on the requisite $\asymp$'s. With the addition of figure captions, I am now able to do some of this sprinkling myself, but some must still be left to the reader.

Before looking at this example, I suggest you first read the original presentation of the argument in the original Preface (immediately following this one) and *then* return to this point.

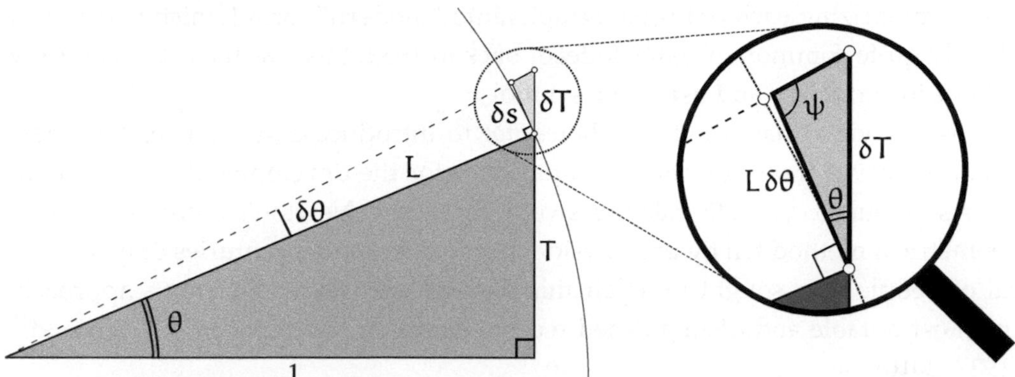

Let us show that if $T = \tan\theta$, then $\frac{dT}{d\theta} = 1 + T^2$ (see figure). If we increase θ by a small (ultimately vanishing) amount $\delta\theta$, then T will increase by the length of the vertical hypotenuse δT of the small triangle, in which the other two sides of this triangle have been constructed to lie in the directions $(\theta + \delta\theta)$ and $(\theta + \frac{\pi}{2})$, as illustrated. To obtain the result, we first observe that in the limit that $\delta\theta$ vanishes, the small triangle with hypotenuse δT is ultimately similar to the large triangle with hypotenuse L, because $\psi \asymp \frac{\pi}{2}$. Next, as we see in the magnifying glass, the side δs adjacent to θ in the small triangle is ultimately equal to the illustrated arc of the circle with radius L, so $\delta s \asymp L\,\delta\theta$; note that we have moved the dot from the corner of the triangle to the arc, to stress this point. Thus,

$$\frac{dT}{Ld\theta} \asymp \frac{\delta T}{L\delta\theta} \asymp \frac{\delta T}{\delta s} \asymp \frac{L}{1} \quad \Longrightarrow \quad \frac{dT}{d\theta} = L^2 = 1 + T^2.$$

So far as I know, Newton never wrote down this specific example, but compare the illuminating directness of his *style* of geometrical reasoning with the unilluminating computations we teach our students today, more than three centuries later! As Newton himself put it,[17] the geometric method is to be preferred by virtue of the "clarity and brevity of the reasoning involved and because of the simplicity of the conclusions and the illustrations required." Indeed, Newton went even further, resolving that *only* the synthetic method was "worthy of public utterance".

[16] I was already using the $\asymp$ notation (both privately and in print) at the time of writing VCA, and, in hindsight, it was a mistake that I did not employ it throughout the original edition of VCA.

[17] See Guicciardini (2009, p. 231)

Newton himself did not employ *any* symbol to represent his concept of "ultimate equality". Instead, his devotion to the geometrical *method* of the Ancients spilled over into emulating their *mode* of expression, causing him to write out the words "ultimately have the ratio of equality" every single time the concept was invoked in a proof. As Newton (1687, p. 124) explained, the *Principia* is "written in words at length in the manner of the Ancients". Even when Newton claimed that two ratios were ultimately equal, he insisted on expressing *each ratio* in words. As a result, I myself was quite unable to follow Newton's reasoning without first transcribing and summarizing each of his paragraphs into "modern" form (which was in fact already quite common in 1687). Indeed, back in 1982, this was the catalyst for my private introduction and use of the symbol $\asymp$.

It is my view that Newton's choice *not* to introduce a symbol for "ultimate equality" was a tragically consequential error for the development of mathematics. As Leibniz's symbolic calculus swept the world, Newton's more penetrating geometrical method fell by the wayside. In the intervening centuries only a handful of people ever sought to repair this damage and revive Newton's approach, the most notable and distinguished recent champion having been V. I. Arnol'd[18] [1937–2010].

Had Newton shed the trappings of this ancient mode of exposition and instead employed some symbol (*any* symbol!) in place of the words "ultimately have the ratio of equality", his dense, paragraph-length proofs in the *Principia* might have been reduced to a few succinct lines, and his mode of thought might still be widely employed today. Both VCA and VDGF are attempts to demonstrate, very concretely, the continuing relevance and vitality of Newton's geometrical approach, in areas of mathematics whose discovery lay a century in the future at the time of his death in 1727.

Allow me to insert some fine print concerning my use of the words "rigour" and "proof". Yes, my occasional explicit use of Newtonian ultimate equalities in this new edition represents a quantum jump in rigour, as compared to my original exposition in VCA, but there will be some mathematicians who will object (with justification!) that even this increase in rigour is insufficient, and that *none* of the "proofs" in this work are worthy of that title, including the one just given: I did not actually prove that the side of the triangle is ultimately equal to the arc of the circle.

I can offer no *logical* defence, but will merely repeat the words I wrote in the original Preface to VCA, 25 years ago: "... suppose one believes, as I do, that our mathematical theories are attempting to capture aspects of a robust Platonic world that is not of our making. I would then contend that an initial lack of rigour is a small price to pay if it allows the reader to see into this world more directly and

[18] See, for example, Arnol'd (1990).

pleasurably than would otherwise be possible." So, to preemptively address my critics, let me therefore concede, from the outset, that when I claim that an assertion is "proved", it may be read as, *"proved beyond a reasonable doubt"*![19]

The Continued Relevance of the Unorthodox *Contents* of VCA

Leaving aside its unorthodox geometrical *methods*, VCA is also distinguished by its unorthodox *contents*. The concept of the complex derivative (or *amplitwist*, as I call it) is not even introduced until Chapter 4, and many of the most interesting, unusual, and important parts of the book make little or no use of analysis.

For many decades before VCA, and now for decades after VCA, almost all[20] introductions to Complex Analysis have seemed to follow almost exactly in the footsteps of the ones that came before—the same topics, explained in the same order, and in the same manner, like a single-file procession of monks through the snow, quietly and sombrely intent on stepping into the footprints of the monk who went before them: Do not disturb the pristine snow lying all about you! But when I embarked upon VCA, 35 years ago, I was like a happily disobedient child, jumping about wildly, rolling my whole mind in the snow!

I started from scratch, asking myself, *which ideas connected to Complex Analysis have become the most vital to modern mathematics and physics?* A pair of closely related answers to this question immediately presented themselves, and while I make no claim to prescience, it is certainly true that the following have only become *more* vital to mathematics and physics over the past 25 years:

- **Hyperbolic Geometry** violates the normal rules of Euclidean geometry, and it is therefore also called *Non-Euclidean Geometry*. In the hyperbolic plane there are infinitely many different lines through a given point that are parallel to a given line, and the angles in a triangle always add up to *less* than π, the departure from π being proportional to the *area* of the triangle.

Yet, as Poincaré was the first to recognize, this strange geometry arises naturally across many parts of mathematics and physics. For example, the final figure of this book reveals how hyperbolic geometry unifies *all* the methods of solving the two-dimensional Dirichlet Problem. Furthermore, the visionary insights of Thurston[21]

[19] Upon reading these words, a strongly supportive member of the Editorial Board of Princeton University Press suggested to my editor that in place of "Q.E.D.", I conclude each of my proofs in VDGF with the letters, "P.B.R.D."!

[20] I freely admit that I have not undertaken a systematic study of all the Complex Analysis textbooks that have been published during the last 25 years, and I am aware that there exist excellent *exceptions* to the following generalization, my favourite ones being Shaw (2006) and Stewart and Tall (2018), which I highly recommend precisely *because* of their very unusual contents. Incidentally, Shaw just happens to have been a fellow student of Penrose!

[21] For details of Thurston's *Geometrization Conjecture*—now *Theorem*!—see Thurston (1997).

(1946–2012)—subsequently vindicated by Perelman in 2003, six years after the publication of VCA—have established that hyperbolic geometry is in some sense more *fundamental* than Euclidean geometry.

This fascinating and important geometry is intimately entwined with the *complex numbers*, via the Möbius transformations that we shall discuss next. Yet, after a quarter century, VCA's long Chapter 6 remains the most complete and the most *geometrical* treatment of hyperbolic geometry that I have seen in any introduction to Complex Analysis.

- **Möbius Transformations** are mappings of the form

$$z \mapsto M(z) = \frac{az+b}{cz+d},$$

where a, b, c, d are complex constants. Whereas hyperbolic geometry is rarely even mentioned in introductory texts on Complex Analysis, these transformations are discussed in *all* of them, but usually in a perfunctory and superficial manner. In stark contrast, my long Chapter 3 remains the most complete and the most *geometrical* treatment of these transformations that I have seen in any introductory text.

The reason I lavished such extravagant attention upon these deceptively simple transformations is that they are possessed of *magical* powers, manifesting themselves in multifaceted guises across mathematics and physics. Again, it was Poincaré who was the first to recognize this. They are the isometries of both the *two*-dimensional hyperbolic plane (introduced above) and of *three*-dimensional hyperbolic space; they are the famous *Lorentz transformations* (isometries) of Minkowski and Einstein's *four*-dimensional spacetime; and when the constants are all integers, and $ad - bc = 1$, they form the *modular group*, describing the symmetries of the *modular functions* so important in modern number theory; and the list goes on ...

The second pair of innovations with respect to the contents of VCA is centred on the use of *vector fields* as an alternative means of visualizing complex mappings. Instead of picturing a point z in $\mathbb{C}$ as being mapped to another point $w = f(z)$ in another copy of $\mathbb{C}$—which is the paradigm in force throughout the first nine chapters—we instead *picture* $f(z)$ *as a vector emanating from* z.

- **The Topology of Vector Fields** is the subject of Chapter 10. Here we shed new light on the vital topological concept of the *winding number* (the subject of Chapter 7) by instead viewing it through the prism of the *index* of a singular point of a vector field. The climax of the topological analysis is a proof of the glorious Poincaré–Hopf Theorem, (10.4), which relates the indices of a flow on a closed surface to the surface's topological *genus*, which counts how many holes the surface has.

So far as I can tell, my final[22] major innovative topic in VCA has *still* not had the impact I believe it deserves, despite my best efforts, and despite the even earlier, independent efforts of its first principal champion, Professor Bart Braden, who should also be credited[23] with having *named* the concept in honour of its first proponent:

- **The Pólya Vector Field** of a complex mapping $z \mapsto f(z)$ was introduced by George Pólya (1887–1985) in his 1974 textbook with Gordon Latta, (Pólya and Latta, 1974), and it is the subject of VCA's Chapter 11: At each point z in $\mathbb{C}$ we draw a vector $P(z)$ that is defined to be the *complex conjugate* of $f(z)$, so $P(z) \equiv \overline{f(z)}$. This *Pólya vector field* then provides a wonderfully vivid means of *visualizing* the real and imaginary parts of the integral[24] of $f(z)$ along a directed path (aka a *contour*) C connecting two points in $\mathbb{C}$, namely, (11.1) on page 549:

$$\int_C f(z)\,dz = [\text{flow of P } \textit{along } C] + i\,[\text{flow of P } \textit{across } C]\,.$$

Here, the imaginary flux component is *positive* if P flows across C *from our left to our right* as we face forward in the direction of travel along C.

Next, it follows immediately from the *Cauchy–Riemann equations*, (4.7), that there is a wonderfully vivid *physical interpretation* of the existence of the complex derivative (what I call the *amplitwist*), namely,

> *If and only if the amplitwist* $f'(z)$ *exists, then the Pólya vector field* $P(z) \equiv \overline{f(z)}$ *is* divergence-free *and* curl-free.

This in turn immediately provides a marvellously *physical explanation* of one the central results of Complex Analysis, namely, *Cauchy's Theorem*, (8.20), the simplest version of which states that

> *If the amplitwist* f' *exists everywhere on and inside a simple closed loop* L, *then* $\oint_L f(z)\,dz = 0$.

But if L contains a singularity of $f(z)$, then the integral need not vanish. An example of central importance is $f(z) = (1/z)$, for which the Pólya vector field is the radial outward flow from a *source* at the origin. Choosing L to be an origin-centred circle

[22] In fact there are many other innovations scattered across the work, most notably my introduction of a concept I christened the *complex curvature*, which I apply to central force fields in Chapter 5, and to the *geometry* of harmonic functions in Chapter 12; there's also the *Topological Argument Principle* in Chapter 7, and more besides, but I cannot attempt to catalogue *all* these ideas and discoveries here. Some of the observations in Chapter 12 were previously published in Needham (1994), which won the MAA's *Carl B. Allendoerfer Award* in 1995.

[23] See Braden (1985), Braden (1987)—which won the MAA's *Carl B. Allendoerfer Award* in 1988—and Braden (1991).

[24] Rest assured that the following ideas will all be explained *ab initio* in the main text; this section of the Preface is addressed to experienced readers who are *already familiar with* the fundamentals of standard Complex Analysis.

K of radius r, traced counterclockwise, we obtain an *immediate visual and physical explanation* of the iconic fact that

$$\oint_K \frac{1}{z}\,dz = 2\pi i.$$

For clearly there is no flow along K, and since $P(z) = (1/\bar{z})$ flows orthogonally across K (from left to right) with speed $(1/r)$, its flux across K is $(2\pi r)(1/r) = 2\pi$. Furthermore, this physical interpretation explains why the value of the integral will *remain* $2\pi i$ if K is continuously deformed into a general loop encircling the source at the origin, so long as K does not *cross* that point as it is deformed.

Lastly, the ability of modern computers to quickly and easily draw the Pólya vector field of any explicit formula for $f(z)$ makes the concept all the more powerful as a means of *visualizing* Complex Analysis.

While the evangelical work of VCA (and of Braden's earlier papers) failed to have its desired effect in the 20th century, my fervent hope now is that by explicitly singling out this concept for praise in this new Preface, more people may take notice of it, and the gospel of the Pólya vector field may thereby be spread amongst multitudes of new believers in the 21st century!

The Quiet Revolution

In hindsight, something was clearly in the air. No sooner had VCA been published than I began to notice the emergence of a wave of kindred works that likewise challenged the prevailing dominance of an arid, purely formal approach to mathematics, and that instead embraced intuitive explanations, the meaning of results, and, crucially, the revelatory power of *geometry*.

It appeared that I had unwittingly been enlisted into a resistance movement of sorts, one in which cell members were not permitted to know each others identities—perhaps for their own safety?! But, unlike Bourbaki, this quiet revolution had no name and no leaders; it was, to coin a phrase, "By the people, for the people." Long may this healthy embrace of *meaning, intuition, and* ***geometry*** continue!

Cheers!
I raise my glass to the *next* 25 years!

T. N.
Mill Valley, California
June, 2022

PREFACE

Theories of the known, which are described by different physical ideas, may be equivalent in all their predictions and hence scientifically indistinguishable. However, they are not psychologically identical when trying to move from that base into the unknown. For different views suggest different kinds of modifications which might be made and hence are not equivalent in the hypotheses one generates from them in one's attempt to understand what is not yet understood.

Feynman (1966)

A Parable

Imagine a society in which the citizens are encouraged, indeed compelled up to a certain age, to read (and sometimes write) musical scores. All quite admirable. However, this society also has a very curious—few remember how it all started—and disturbing law: *Music must never be listened to or performed!*

Though its importance is universally acknowledged, for some reason music is not widely appreciated in this society. To be sure, professors still excitedly pore over the great works of Bach, Wagner, and the rest, and they do their utmost to communicate to their students the beautiful meaning of what they find there, but they still become tongue-tied when brashly asked the question, "What's the point of all this?!"

In this parable, it was patently unfair and irrational to have a law forbidding would-be music students from experiencing and understanding the subject directly through "sonic intuition." But in our society of mathematicians we *have* such a law. It is not a written law, and those who flout it may yet prosper, but it says, *Mathematics must not be visualized!*

More likely than not, when one opens a random modern mathematics text on a random subject, one is confronted by abstract symbolic reasoning that is divorced from one's sensory experience of the world, *despite* the fact that the very phenomena one is studying were often discovered by appealing to geometric (and perhaps physical) intuition.

This reflects the fact that steadily over the last hundred years the honour of visual reasoning in mathematics has been besmirched. Although the great mathematicians have always been oblivious to such fashions, it is only recently that the "mathematician in the street" has picked up the gauntlet on behalf of geometry.

The present book openly challenges the current dominance of purely symbolic logical reasoning by using new, visually accessible arguments to explain the truths of elementary complex analysis.

Computers

In part, the resurgence of interest in geometry can be traced to the mass-availability of computers to draw mathematical objects, and perhaps also to the related, somewhat breathless, popular interest in chaos theory and in fractals. This book instead advocates the more sober use of computers as an aid to geometric *reasoning*.

I have tried to encourage the reader to think of the computer as a physicist would his laboratory—it may be used to check existing ideas about the construction of the world, or as a tool for discovering new phenomena which then demand new ideas for their explanation. Throughout the text I have suggested such uses of the computer, but I have deliberately avoided giving *detailed* instructions. The reason is simple: whereas a mathematical idea is a timeless thing, few things are more ephemeral than computer hardware and software.

Having said this, the program "f(z)" is currently[25] the best tool for visually exploring the ideas in this book; a free demonstration version can be downloaded directly from Lascaux Graphics. On occasion it would also be helpful if one had access to an all-purpose mathematical engine such as *Maple* or *Mathematica*. However, I would like to stress that none of the above software is essential: the entire book can be fully understood without *any* use of a computer.

Finally, some readers may be interested in knowing how computers were used to produce this book. Perhaps five of the 501 diagrams were drawn using output from *Mathematica*; the remainder I drew by hand (or rather "by mouse") using CorelDRAW, occasionally guided by output from "f(z)". I typeset the book in LaTeX using the wonderful *Y&Y TeX System for Windows*[26], the figures being included as EPS files. The text is Times[27], with Helvetica heads, and the mathematics is principally MathTime, though nine other mathematical fonts make cameo appearances. All of these Adobe Type 1 fonts were obtained from Y&Y, Inc., with the exception of Adobe's *MathematicalPi-Six* font, which I used to represent quaternions. Finally, OUP printed the book directly from my PostScript file.

The Book's Newtonian Genesis

It is fairly well known that Newton's original 1665 version of the calculus was different from the one we learn today: its essence was the manipulation of power

[25] Sadly, this no longer exists.
[26] Sadly, this no longer exists.
[27] In this new edition, the headings are Optima, the text is Palatino, and the mathematics is Euler.

series, which Newton likened to the manipulation of decimal expansions in arithmetic. The symbolic calculus—the one in every standard textbook, and the one now associated with the name of Leibniz—was also perfectly familiar to Newton, but apparently it was of only incidental interest to him. After all, armed with his power series, Newton could evaluate an integral like $\int e^{-x^2}\,dx$ just as easily as $\int \sin x\,dx$. Let Leibniz try *that!*

It is less well known that around 1680 Newton became disenchanted with both these approaches, whereupon he proceeded to develop a *third* version of calculus, based on *geometry*. This "geometric calculus" is the mathematical engine that propels the brilliant physics of Newton's *Principia*.

Having grasped Newton's method, I immediately tried my own hand at using it to simplify my teaching of introductory calculus. An example will help to explain what I mean by this. Let us show that if $T = \tan\theta$, then $\frac{dT}{d\theta} = 1 + T^2$. If we increase θ by a small amount $d\theta$ then T will increase by the amount dT in the figure below. To obtain the result, we need only observe that in the limit as $d\theta$ tends to zero, the black triangle is ultimately similar [exercise] to the shaded triangle. Thus, in this limit,

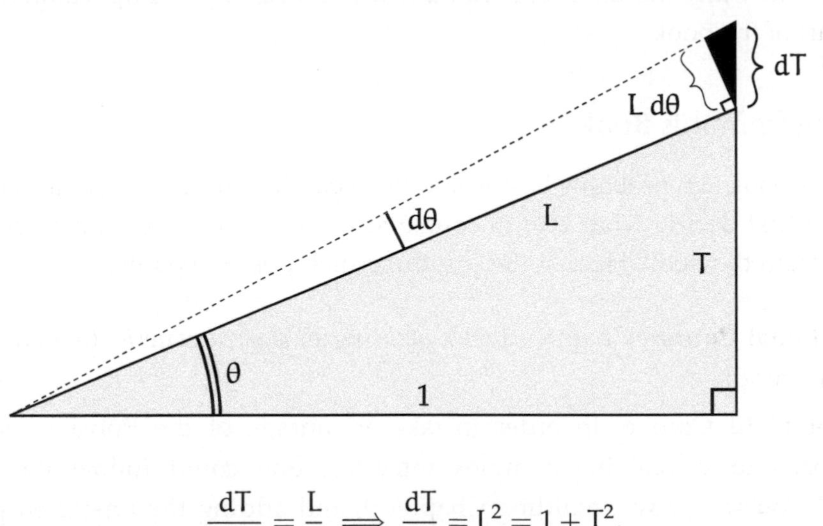

$$\frac{dT}{L\,d\theta} = \frac{L}{1} \implies \frac{dT}{d\theta} = L^2 = 1 + T^2.$$

Only gradually did I come to realize how naturally this mode of thought could be applied—almost exactly 300 years later!—to the geometry of the complex plane.

Reading This Book

In the hope of making the book fun to read, I have attempted to write as though I were explaining the ideas directly to a friend. Correspondingly, I have tried to make you, the reader, into an active participant in developing the ideas. For example, as an argument progresses, I have frequently and deliberately placed a pair of logical stepping stones sufficiently far apart that you may need to pause and stretch slightly to pass from one to the next. Such places are marked "[exercise]"; they often require nothing more than a simple calculation or a moment of reflection.

This brings me to the exercises proper, which may be found at the end of each chapter. In the belief that the essential prerequisite for finding the answer to a question is the *desire* to find it, I have made every effort to provide exercises that provoke curiosity. They are considerably more wide-ranging than is common, and they often establish important facts which are then used freely in the text itself. While problems whose be all and end all is routine calculation are thereby avoided, I believe that readers will automatically develop considerable computational skill *in the process* of seeking solutions to these problems. On the other hand, my intention in a large number of the exercises is to illustrate how geometric thinking can often *replace* lengthy calculation.

Any part of the book marked with a star ("*") may be omitted on a first reading. If you do elect to read a starred section, you may in turn choose to omit any starred *sub*sections. Please note, however, that a part of the book that is starred is not necessarily any more difficult, nor any less interesting or important, than any other part of the book.

Teaching from this Book

The entire book can probably be covered in a year, but in a single semester course one must first decide what *kind* of course to teach, then choose a corresponding path through the book. Here I offer just three such possible paths:

- **Traditional Course.** Chapters 1 to 9, *omitting all starred material* (e.g., the whole of Chapter 6).
- **Vector Field Course.** In order to take advantage of the Pólya vector field approach to visualizing complex integrals, one could follow the "Traditional Course" above, omitting Chapter 9, and adding the unstarred parts of Chapters 10 and 11.
- **Non-Euclidean Course.** At the expense of teaching any integration, one could give a course focused on Möbius transformations and non-Euclidean geometry. These two related parts of complex analysis are probably the most important ones for contemporary mathematics and physics, and yet they are also the ones that are almost entirely neglected in undergraduate-level texts. On the other hand, graduate-level works tend to assume that you have already encountered the main ideas as an undergraduate: Catch 22!

Such a course might go as follows: All of Chapter 1; the unstarred parts of Chapter 2; all of Chapter 3, including the starred sections but (possibly) omitting the starred *sub*sections; all of Chapter 4; all of Chapter 6, including the starred sections but (possibly) omitting the starred *sub*sections.

Omissions and Apologies

If one believes in the ultimate unity of mathematics and physics, as I do, then a very strong case for the necessity of complex numbers can be built on their apparently fundamental role in the quantum mechanical laws governing *matter*. Also, the work of Sir Roger Penrose has shown (with increasing force) that complex numbers play an equally central role in the relativistic laws governing the structure of *spacetime*. Indeed, if the laws of matter and of spacetime are ever to be reconciled, then it seems very likely that it will be through the auspices of the complex numbers. This book cannot explore these matters; instead, we refer the interested reader to Feynman (1963, Vol. III), Feynman (1985), Penrose and Rindler (1984), and especially Penrose (2005).

A more serious omission is the lack of discussion of *Riemann surfaces*, which I had originally intended to treat in a final chapter. This plan was aborted once it became clear that a serious treatment would entail expanding the book beyond reason. By this time, however, I had already erected much of the necessary scaffolding, and this material remains in the finished book. In particular, I hope that the interested reader will find the last three chapters helpful in understanding Riemann's original physical insights, as expounded by Klein (1881). See also Springer (1981, Ch. 1), which essentially reproduces Klein's monograph, but with additional helpful commentary.

I consider the history of mathematics to be a vital tool in understanding both the current state of mathematics, and its trajectory into the future. Sadly, however, I can do no more than touch on historical matters in the present work; instead I refer you to the remarkable book, *Mathematics and Its History*, by John Stillwell (2010). Indeed, I strongly encourage you to think of his book as a companion to mine: not only does it trace and explain the development of complex analysis, but it also explores and illuminates the connections with other areas of mathematics.

To the expert reader I would like to apologize for having invented the word "amplitwist" [Chapter 4] as a synonym (more or less) for "derivative", as well the component terms "amplification" and "twist". I can only say that the need for *some* such terminology was forced on me in the classroom: if you try teaching the ideas in this book *without* using such language, I think you will quickly discover what I mean! Incidentally, a precedence argument in defence of "amplitwist" might be that a similar term was coined by the older German school of Klein, Bieberbach, *et al*. They spoke of "eine Drehstreckung", from "drehen" (to twist) and "strecken" (to stretch).

A significant proportion of the geometric observations and arguments contained in this book are, to the best of my knowledge, new. I have not drawn attention to this in the text itself as this would have served no useful purpose: students don't need to know, and experts will know without being told. However, in cases where an idea is clearly unusual but I am aware of it having been published by someone else, I have tried to give credit where credit is due.

In attempting to rethink so much classical mathematics, I have no doubt made mistakes; the blame for these is mine alone. Corrections will be gratefully received at VCA.correction@gmail.com.

My book will no doubt be flawed in many ways of which I am not yet aware, but there is one "sin" that I have intentionally committed, and for which I shall not repent: many of the arguments are not rigorous, at least as they stand. This is a serious crime if one believes that our mathematical theories are merely elaborate mental constructs, precariously hoisted aloft. Then rigour becomes the nerve-racking balancing act that prevents the entire structure from crashing down around us. But suppose one believes, as I do, that our mathematical theories are attempting to capture aspects of a robust Platonic world that is not of our making. I would then contend that an initial lack of rigour is a small price to pay if it allows the reader to see into this world more directly and pleasurably than would otherwise be possible.

T. N.
San Francisco, California
June, 1996

ACKNOWLEDGEMENTS

First and foremost I wish to express my indebtedness to Dr. Stanley Nel. He is my friend, my colleague, and my Dean, and in all three of these capacities he has helped me to complete this book. As a friend he offered support when progress was slow and my spirits were low; as a mathematical colleague he read much of the book and offered helpful criticisms; as Dean he granted me a succession of increasingly powerful computers, and when the US Immigration Service sought to have my position filled by an "equally qualified" American, he successfully fought them on my behalf. For all this, and much else besides, I offer him my deep gratitude.

Next I would like to thank Prof. John Stillwell of Monash University. The great value I place on his writings should be clear from the frequency with which I refer to them in the pages that follow. Also, though I lack his gift for conciseness, I have sought to emulate elements of his approach in an attempt to give back *meaning* to mathematical concepts. Finally, my greatest and most concrete debt arises from the fact that he read each draft chapter as it was written, and this despite the fact that we had never even met! The book owes a great deal to his numerous helpful suggestions and corrections.

I consider myself very fortunate that the mathematics department here at the University of San Francisco is completely free of political intrigue, rivalry, and other assorted academic blights. I am grateful to *all* my colleagues for creating such a friendly and supportive atmosphere in which to work. In particular, however, I should like to single out the following people for thanks:

- Nancy Campagna for her diligent proof-reading of half the book;
- Allan Cruse and Millianne Lehmann, not only for granting all my software requests during their respective tenures as Department Chair, but also for all their kind and sage advice since my arrival in the United States;
- James Finch for his patience and expertise in helping me overcome various problems associated with my typesetting of the book in LaTeX;
- Robert Wolf for having built up a superb mathematics collection in our library;
- Paul Zeitz for his great faith in me and in the value of what I was trying to accomplish, for his concrete suggestions and corrections, and for his courage in being the first person (other than myself) to teach complex analysis using chapters of the book.

Prof. Gerald Alexanderson of Santa Clara University has my sincere thanks for the encouragement he offered me upon reading some of the earliest chapters, as well as for his many subsequent acts of kindness.

I will always be grateful for the education I received at Merton College, Oxford. It is therefore especially pleasing and fitting to have this book published by OUP, and I would particularly like to thank Dr. Martin Gilchrist, the former Senior Mathematics Editor, for his enthusiastic encouragement when I first approached him with the idea of the book.

When I first arrived at USF from England in 1989 I had barely seen a computer. The fact that OUP printed this book directly from my Internet-transmitted PostScript files is an indication of how far I have come since then. I owe all this to James Kabage. A mere graduate student at the time we met, Jim quickly rose through the ranks to become Director of Network Services. Despite this fact, he never hesitated to spend *hours* with me in my office resolving my latest hardware or software crisis. He always took the extra time to clearly explain to me the reasoning leading to his solution, and in this way I became his student.

I also thank Dr. Benjamin Baab, the Executive Director of Information Technology Services at USF. Despite his lofty position, he too was always willing to roll up his sleeves in order to help me resolve my latest Microsoft conundrum.

Eric Scheide (our multitalented Webmaster) has my sincere gratitude for writing an extremely nifty *Perl* program that greatly speeded my creation of the index.

I thank Prof. Berthold Horn of MIT for creating the magnificent *Y&Y TEX System for Windows*[28], for his generous help with assorted TEXnical problems, and for his willingness to adopt my few suggestions for improving what I consider to be the Mercedes-Benz of the TEX world.

Similarly, I thank Martin Lapidus of Lascaux Graphics for incorporating many of my suggestions into his "f(z)"[29] program, thereby making it into an even better tool for doing "visual complex analysis".

This new printing of the book incorporates a great many corrections. Most of these were reported by readers, and I very much appreciate their efforts. While I cannot thank each one of these readers by name, I must acknowledge Dr. R. von Randow for single-handedly having reported more than 30 errors.

As a student of Roger Penrose I had the privilege of watching him think out loud by means of his beautiful blackboard drawings. In the process, I became convinced that if only one tried hard enough—or were clever enough!—every mathematical mystery could be resolved through geometric reasoning. George Burnett-Stuart and I became firm friends while students of Penrose. In the course of our endless discussions of music, physics, and mathematics, George helped me to refine both

[28] While *Y & Y* no longer exists, my gratitude lives on.
[29] While "f(z)" no longer exists, my gratitude lives on.

my conception of the nature of mathematics, and of what constitutes an acceptable explanation within that subject. My dedication of this book to these two friends scarcely repays the great debt I owe them.

The care of several friends helped me to cope with depression following the death of my beloved mother Claudia. In addition to my brother Guy and my father Rodney, I wish to express my appreciation to Peter and Ginny Pacheco, and to Amy Miller. I don't know what I would have done without their healing affection.

Lastly, I thank my dearest wife Mary. During the writing of this book she allowed me to pretend that science was the most important thing in life; now that the book is over, she is my daily proof that there is something even more important.

my conception of the nature of mathematics and of what constitutes an acceptable explanation within that subject. My dedication of this book to these two friends scarcely repays the great debt I owe them.

The care of several friends helped me to cope with depression following the death of my beloved mother Claudine in addition to my brother Guy and my father Rodney. I wish to express my appreciation to Esther and Gerry Pedersen and to Ann Miller; I don't know what I would have done without their healing affection.

Lastly, I thank my beloved wife Mary. During the writing of this book she allowed me to pretend that mathematics was the most important thing in life. Now that the book is over, she is my daily proof that there is something even more important.

CONTENTS

CHAPTER 1

Geometry and Complex Arithmetic

1.1 Introduction

1.1.1 *Historical Sketch*

Half a millennium has elapsed since complex numbers were first discovered. Here, as the reader is probably already aware, the term *complex number* refers to an entity of the form $a + ib$, where a and b are ordinary real numbers and, unlike any ordinary number, i has the property that $i^2 = -1$. This discovery would ultimately have a profound impact on the whole of mathematics, unifying much that had previously seemed disparate, and explaining much that had previously seemed inexplicable. Despite this happy ending—in reality the story continues to unfold to this day—progress following the initial discovery of complex numbers was painfully slow. Indeed, relative to the advances made in the nineteenth century, *little was achieved during the first 250 years of the life of the complex numbers.*

How is it possible that complex numbers lay dormant through ages that saw the coming and the passing of such great minds as Descartes, Fermat, Leibniz, and even the visionary genius of Newton? The answer appears to lie in the fact that, far from being embraced, complex numbers were initially greeted with suspicion, confusion, and even hostility.

Girolamo Cardano's *Ars Magna*, which appeared in 1545, is conventionally taken to be the birth certificate of the complex numbers. Yet in that work Cardano introduced such numbers only to immediately dismiss them as "subtle as they are useless". As we shall discuss, the first substantial *calculations* with complex numbers were carried out by Rafael Bombelli, appearing in his *L'Algebra* of 1572. Yet here too we find the innovator seemingly disowning his discoveries (at least initially), saying that "the whole matter seems to rest on sophistry rather than truth". As late as 1702, Leibniz described i, the square root of -1, as "that amphibian between existence and nonexistence". Such sentiments were echoed in the terminology of the period. To the extent that they were discussed at all, complex

Visual Complex Analysis. 25th Anniversary Edition. Tristan Needham, Oxford University Press.
 DOI: 10.1093/oso/9780192868916.003.0001

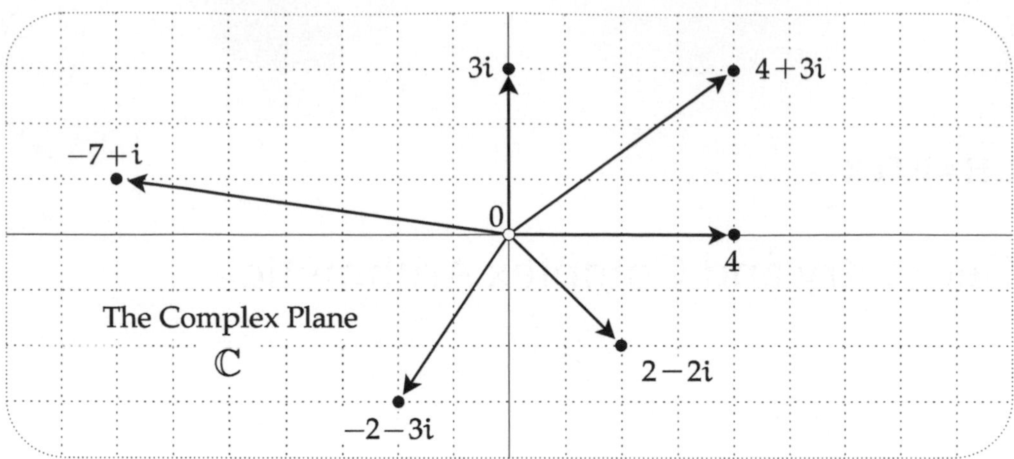

[1.1] The complex plane *enables us to visualize the abstract complex number* $a + ib$ *as a* simple point in the plane, *with Cartesian coordinates* (a, b).

numbers were called "impossible" or "imaginary", the latter term having (unfortunately) lingered to the present day[1]. Even in 1770 the situation was still sufficiently confused that it was possible for so great a mathematician as Euler to mistakenly argue that $\sqrt{-2}\sqrt{-3} = \sqrt{6}$.

The root cause of all this trouble seems to have been a psychological or philosophical block. How could one investigate these matters with enthusiasm or confidence when nobody felt they knew the answer to the question, "What *is* a complex number?"

A satisfactory answer to this question was only found at the end of the eighteenth century[2]. Independently, and in rapid succession, Wessel, Argand, and Gauss all recognized that complex numbers could be given a simple, concrete, *geometric interpretation* as points (or vectors) in the plane: The mystical quantity $a + ib$ should be viewed simply as the point in the xy-plane having Cartesian coordinates (a, b), or equivalently as the vector connecting the origin to that point. See [1.1]. When thought of in this way, the plane is denoted $\mathbb{C}$ and is called the *complex plane*[3].

The operations of adding or multiplying two complex numbers could now be given equally definite meanings as geometric operations on the two

[1] However, an "imaginary number" now refers to a real multiple of i, rather than to a general complex number. Incidentally, the term "real number" was introduced precisely to distinguish such a number from an "imaginary number".

[2] Wallis almost hit on the answer in 1673; see Stillwell (2010, §14.4) for a detailed account of this interesting near miss.

[3] Also known as the "Gauss plane" or the "Argand plane".

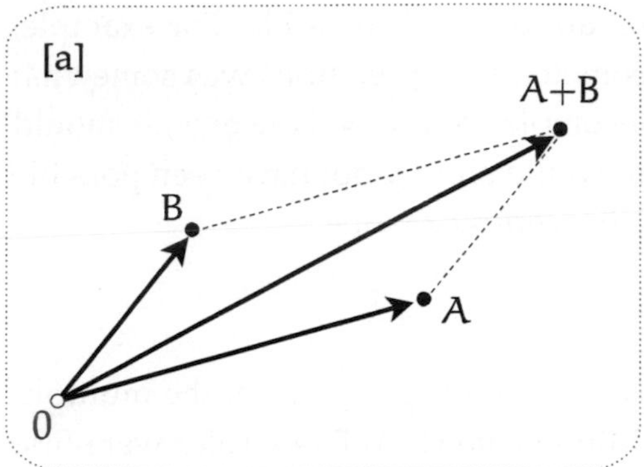

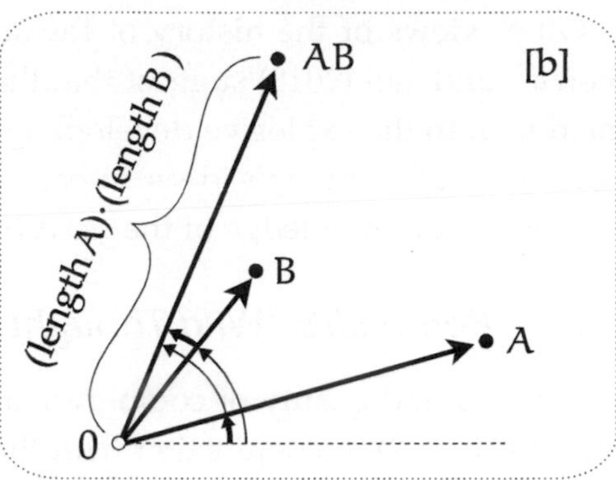

[1.2] [a] *Geometrically, the* **sum** *of the complex numbers* A *and* B *is the diagonal of the parallelogram with sides* A *and* B. **[b]** *Geometrically, the length of the* **product** *of* A *and* B *is the product of their separate lengths, and its angle is the sum of their separate angles.*

corresponding points (or vectors) in the plane. The rule for addition is illustrated in [1.2a]:

> *The sum* $A + B$ *of two complex numbers is given by the parallelogram rule of ordinary vector addition.* (1.1)

Note that this is consistent with [1.1], in the sense that $4+3i$ (for example) is indeed the sum of 4 and $3i$.

Figure [1.2b] illustrates the much less obvious rule for multiplication:

> *The length of* AB *is the product of the lengths of* A *and* B, *and the angle of* AB *is the sum of the angles of* A *and* B. (1.2)

This rule is not forced on us in any obvious way by [1.1], but note that it is at least consistent with it, in the sense that $3i$ (for example) is indeed the product of 3 and i. Check this for yourself. As a more exciting example, consider the product of i with itself. Since i has unit length and angle $(\pi/2)$, i^2 has unit length and angle π. Thus $i^2 = -1$.

The publication of the geometric interpretation by Wessel and by Argand went all but unnoticed, but the reputation of Gauss (as great then as it is now) ensured wide dissemination and acceptance of complex numbers as points in the plane. Perhaps less important than the details of this new interpretation (at least initially) was the mere fact that there now existed *some* way of making sense of these numbers—that they were now *legitimate* objects of investigation. In any event, the floodgates of invention were about to open.

It had taken more than two and a half centuries to come to terms with complex numbers, but the development of a beautiful new theory of how to do *calculus* with such numbers (what we now call *complex analysis*) was astonishingly rapid. Most of the fundamental results were obtained (by Cauchy, Riemann, and others) between 1814 and 1851—a span of less than forty years!

Other views of the history of the subject are certainly possible. For example, Stewart and Tall (2018) suggest that the geometric interpretation[4] was somewhat incidental to the explosive development of complex analysis. However, it should be noted that Riemann's ideas, in particular, would simply not have been possible without prior knowledge of the *geometry* of the complex plane.

1.1.2 Bombelli's "Wild Thought"

The power and beauty of complex analysis ultimately springs from the multiplication rule (1.2) in conjunction with the addition rule (1.1). These rules were first discovered by Bombelli in *symbolic* form; more than two centuries passed before the complex plane revealed figure [1.2]. Since we merely plucked the rules out of thin air, let us return to the sixteenth century in order to understand their algebraic origins.

Many texts seek to introduce complex numbers with a convenient historical fiction based on solving quadratic equations,

$$x^2 = mx + c. \tag{1.3}$$

Two thousand years BCE, it was already known that such equations could be solved using a method that is equivalent to the modern formula,

$$x = \tfrac{1}{2}\left[m \pm \sqrt{m^2 + 4c}\right].$$

But what if $m^2 + 4c$ is negative? This was the very problem that led Cardano to consider square roots of negative numbers. Thus far the textbook is being historically accurate, but next we read that the need for (1.3) to always have a solution *forces* us to take complex numbers seriously. This argument carries almost as little weight now as it did in the sixteenth century. Indeed, we have already pointed out that Cardano did not hesitate to discard such "solutions" as useless.

It was not that Cardano lacked the imagination to pursue the matter further, rather he had a fairly compelling reason *not to*. For the ancient Greeks mathematics was synonymous with geometry, and this conception still held sway in the sixteenth century. Thus an algebraic relation such as (1.3) was not so much thought of as a problem in its own right, but rather as a mere vehicle for solving a genuine problem of geometry. For example, (1.3) may be considered to represent the problem of finding the intersection points of the parabola $y = x^2$ and the line $y = mx + c$. See [1.3a].

In the case of L_1 the problem has a solution; algebraically, $(m^2 + 4c) > 0$ and the two intersection points are given by the formula above. In the case of L_2 the problem clearly does *not* have a solution; algebraically, $(m^2 + 4c) < 0$ and the absence of

[4] We must protest one piece of their evidence: Although Wallis did attempt a geometric interpretation of complex numbers in 1673, he did *not* hit upon the "correct" picture; see footnote 2.

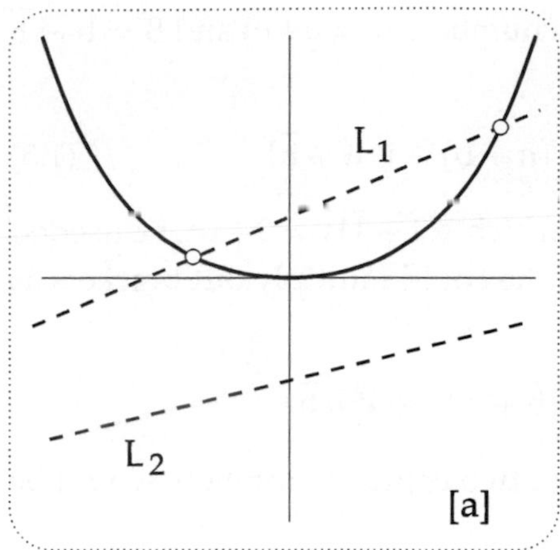

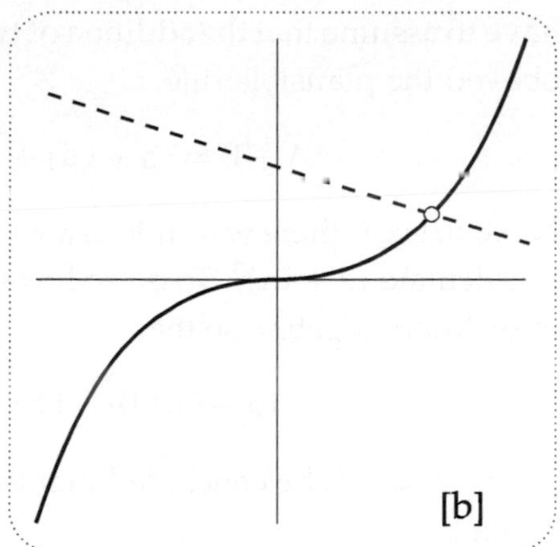

[1.3] **[a]** *While some lines intersect a parabola, some do not, so the quadratic equation* $x^2 = mx + c$ *may or may not have solutions.* **[b]** *But* $x^3 = mx + c$ *must* always *have a solution.*

solutions is correctly manifested by the occurrence of "impossible" numbers in the formula.

It was not the quadratic that forced complex numbers to be taken seriously, it was the *cubic*,

$$x^3 = 3px + 2q.$$

[Ex. 1 shows that a general cubic can always be reduced to this form.] This equation represents the problem of finding the intersection points of the cubic curve $y = x^3$ and the line $y = 3px+2q$. See [1.3b]. Building on the work of del Ferro and Tartaglia, Cardano's *Ars Magna* showed that this equation could be solved by means of a remarkable formula [see Ex. 2]:

$$x = \sqrt[3]{q + \sqrt{q^2 - p^3}} + \sqrt[3]{q - \sqrt{q^2 - p^3}}. \tag{1.4}$$

Try it yourself on $x^3 = 6x + 6$.

Some thirty years after this formula appeared, Bombelli recognized that there was something strange and paradoxical about it. First note that if the line $y = 3px + 2q$ is such that $p^3 > q^2$ then the formula involves complex numbers. For example, Bombelli considered $x^3 = 15x + 4$, which yields

$$x = \sqrt[3]{2 + 11i} + \sqrt[3]{2 - 11i}.$$

In the previous case of [1.3a] this merely signalled that the geometric problem had no solution, but in [1.3b] it is clear that the line will *always* hit the curve! In fact inspection of Bombelli's example yields the solution $x = 4$.

As he struggled to resolve this paradox, Bombelli had what he called a "wild thought": perhaps the solution $x = 4$ could be recovered from the above expression if $\sqrt[3]{2 + 11i} = 2 + ni$ and $\sqrt[3]{2 - 11i} = 2 - ni$. Of course for this to work he would have to assume that the addition of two complex numbers $A = a + i\tilde{a}$ and $B = b + i\tilde{b}$ obeyed the plausible rule,

$$A + B = (a + i\tilde{a}) + (b + i\tilde{b}) = (a + b) + i(\tilde{a} + \tilde{b}). \tag{1.5}$$

Next, to see if there was indeed a value of n for which $\sqrt[3]{2 + 11i} = 2 + in$, he needed to calculate $(2 + in)^3$. To do so he assumed that he could multiply out brackets as in ordinary algebra, so that

$$(a + i\tilde{a})\,(b + i\tilde{b}) = ab + i(a\tilde{b} + \tilde{a}b) + i^2\,\tilde{a}\tilde{b}.$$

Using $i^2 = -1$, he concluded that the product of two complex numbers would be given by

$$AB = (a + i\tilde{a})\,(b + i\tilde{b}) = (ab - \tilde{a}\tilde{b}) + i(a\tilde{b} + \tilde{a}b). \tag{1.6}$$

This rule vindicated his "wild thought", for he was now able to show that $(2 \pm i)^3 = 2 \pm 11i$. Check this for yourself.

While complex numbers themselves remained mysterious, Bombelli's work on cubic equations thus established that perfectly real problems required complex arithmetic for their solution.

Just as with its birth, the subsequent development of the theory of complex numbers was inextricably bound up with progress in other areas of mathematics (and also physics). Sadly, we can only touch on these matters in this book; for a full and fascinating account of these interconnections, the reader is instead referred to Stillwell (2010). Repeating what was said in the Preface, we cannot overstate the value of reading Stillwell's book alongside this one.

1.1.3 Some Terminology and Notation

Leaving history behind us, we now introduce the modern terminology and notation used to describe complex numbers. The information is summarized in the table below, and is illustrated in [1.4].

Name	Meaning	Notation
modulus of z	length r of z	$\lvert z\rvert$
argument of z	angle θ of z	$\arg(z)$
real part of z	x coordinate of z	$\mathrm{Re}(z)$
imaginary part of z	y coordinate of z	$\mathrm{Im}(z)$
imaginary number	real multiple of i	
real axis	set of real numbers	
imaginary axis	set of imaginary numbers	
complex conjugate of z	reflection of z in the real axis	$\overline{z}$

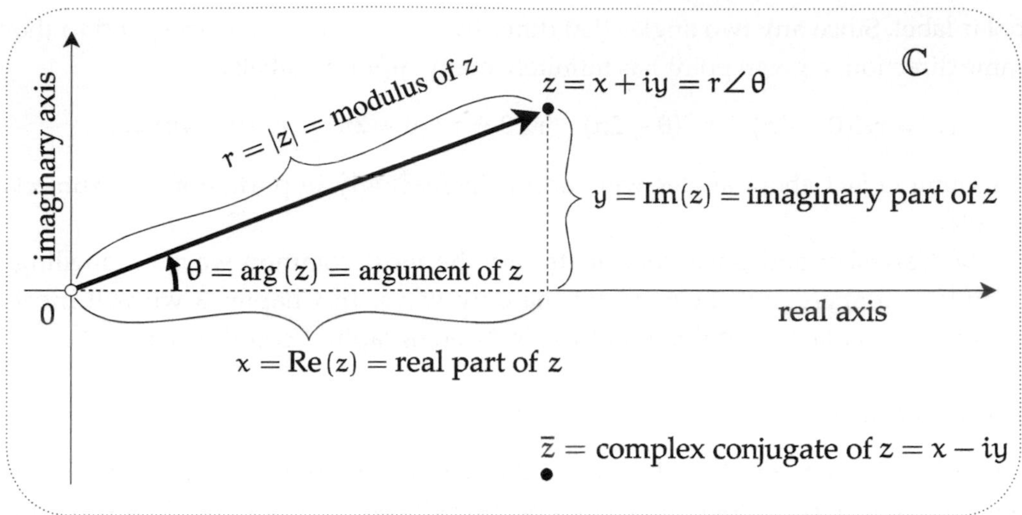

[1.4] *Visual summary of the terminology and notation used to describe complex numbers.*

It is valuable to grasp from the outset that (according to the geometric view) a complex number is a single, indivisible entity—a point in the plane. Only when we choose to describe such a point with numerical coordinates does a complex number appear to be compound or "complex". More precisely, $\mathbb{C}$ is said to be *two dimensional*, meaning that *two* real numbers (coordinates) are needed to label a point within it, but exactly *how* the labelling is done is entirely up to us.

One way to label the points is with Cartesian coordinates (the real part x and the imaginary part y), the complex number being written as $z = x + iy$. This is the natural labelling when we are dealing with the addition of two complex numbers, because (1.5) says that the real and imaginary parts of $A+B$ are obtained by adding the real and imaginary parts of A and B.

In the case of multiplication, the Cartesian labelling no longer appears natural, for it leads to the messy and unenlightening rule (1.6). The much simpler geometric rule (1.2) makes it clear that we should instead label a typical point z with its *polar* coordinates, $r = |z|$ and $\theta = \arg z$. In place of $z = x+iy$ we may now write $z = r\angle\theta$, where the symbol $\angle$ serves to remind us that θ is the *angle* of z. [Although this notation is still used by some, we shall only employ it briefly; later in this chapter we will discover a much better notation (the standard one) which will then be used throughout the remainder of the book.] The geometric multiplication rule (1.2) now takes the simple form,

$$(R\angle\phi)\,(r\angle\theta) = (Rr)\angle(\phi+\theta). \tag{1.7}$$

In common with the Cartesian label $x + iy$, a given polar label $r\angle\theta$ specifies a unique point, but (unlike the Cartesian case) a given point does not have a unique polar label. Since any two angles that differ by a multiple of 2π correspond to the same direction, a given point has infinitely many different labels:

$$\ldots = r\angle(\theta - 4\pi) = r\angle(\theta - 2\pi) = r\angle\theta = r\angle(\theta + 2\pi) = r\angle(\theta + 4\pi) = \ldots$$

This simple fact about angles will become increasingly important as our subject unfolds.

The Cartesian and polar coordinates are the most common ways of labelling complex numbers, but they are not the only ways. In Chapter 3 we will meet another particularly useful method, called "stereographic" coordinates.

1.1.4 Practice

Before continuing, we strongly suggest that you make yourself comfortable with the concepts, terminology, and notation introduced thus far. To do so, try to convince yourself geometrically (*and/or* algebraically) of each of the following facts:

$$\operatorname{Re}(z) = \tfrac{1}{2}[z + \overline{z}] \qquad \operatorname{Im}(z) = \tfrac{1}{2i}[z - \overline{z}] \qquad |z| = \sqrt{x^2 + y^2}$$

$$\tan[\arg z] = \tfrac{\operatorname{Im}(z)}{\operatorname{Re}(z)} \qquad z\overline{z} = |z|^2 \qquad r\angle\theta = r\,(\cos\theta + i\sin\theta)$$

$$\text{Defining } \tfrac{1}{z} \text{ by } (1/z)\,z = 1, \text{ it follows that } \tfrac{1}{z} = \tfrac{1}{r\angle\theta} = \tfrac{1}{r}\angle(-\theta).$$

$$\frac{R\angle\phi}{r\angle\theta} = \frac{R}{r}\angle(\phi - \theta) \qquad \frac{1}{(x + iy)} = \frac{x}{x^2 + y^2} - i\frac{y}{x^2 + y^2}$$

$$(1 + i)^4 = -4 \qquad (1 + i)^{13} = -2^6(1 + i) \qquad (1 + i\sqrt{3})^6 = 2^6$$

$$\frac{(1 + i\sqrt{3})^3}{(1 - i)^2} = -4i \qquad \frac{(1 + i)^5}{(\sqrt{3} + i)^2} = -\sqrt{2}\angle - (\pi/12) \qquad \overline{r\angle\theta} = r\angle(-\theta)$$

$$\overline{z_1 + z_2} = \overline{z_1} + \overline{z_2} \qquad \overline{z_1\,z_2} = \overline{z_1}\,\overline{z_2} \qquad \overline{z_1/z_2} = \overline{z_1}/\overline{z_2}.$$

Lastly, establish the so-called *generalized triangle inequality*:

$$|z_1 + z_2 + \cdots + z_n| \leqslant |z_1| + |z_2| + \cdots + |z_n|. \tag{1.8}$$

When does equality hold?

1.1.5 Equivalence of Symbolic and Geometric Arithmetic

We have been using the symbolic rules (1.5) and (1.6) interchangeably with the geometric rules (1.1) and (1.2), and we now justify this by showing that they are indeed equivalent. The equivalence of the addition rules (1.1) and (1.5) will be familiar to those who have studied vectors; in any event, the verification is sufficiently straightforward that we may safely leave it to the reader. We therefore only address the equivalence of the multiplication rules (1.2) and (1.6).

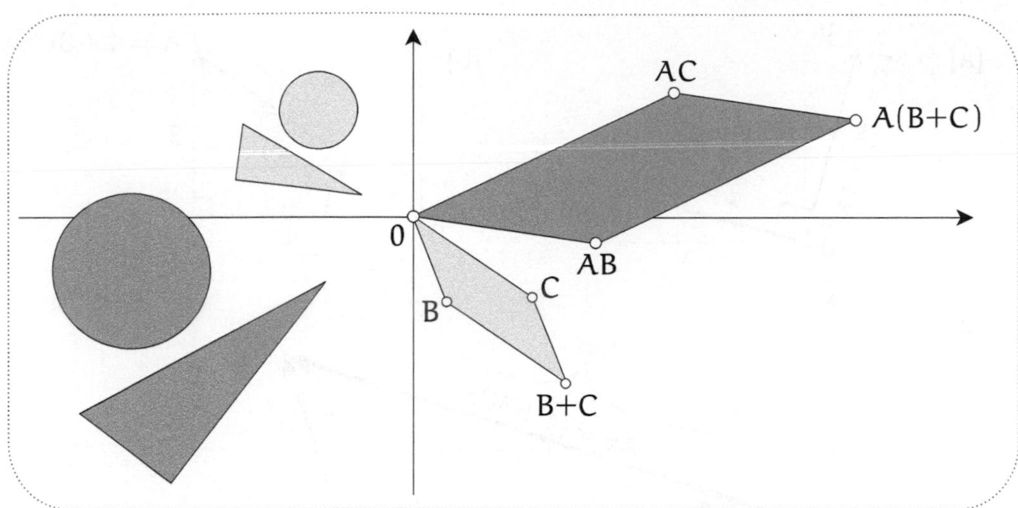

[1.5] A complex number can be interpreted as a rotation and expansion of the plane. *Geometrically, multiplication by a complex number* $A = R\angle\phi$ *rotates the plane through angle* ϕ*, and expands it by* R*, so, in particular, parallelograms are transformed into similar parallelograms. It follows that* $A(B + C) = AB + AC$.

First we will show how the symbolic rule may be derived from the geometric rule. To do so we shall rephrase the geometric rule (1.7) in a particularly useful and important way. Let z denote a general point in $\mathbb{C}$, and consider what happens to it—where it moves to—when it is multiplied by a fixed complex number $A = R\angle\phi$. According to (1.7), the length of z is magnified by R, while the angle of z is increased by ϕ. Now imagine that this is done simultaneously to *every* point of the plane:

> *Geometrically, multiplication by a complex number* $A = R\angle\phi$ *is a rotation of the plane through angle* ϕ*, and an expansion of the plane by factor* R. (1.9)

A few comments are in order:

- Both the rotation and the expansion are centred at the origin.
- It makes no difference whether we do the rotation followed by the expansion, or the expansion followed by the rotation.
- If $R < 1$ then the "expansion" is in reality a contraction.

Figure [1.5] illustrates the effect of such a transformation, the lightly shaded shapes being transformed into the darkly shaded shapes. Check for yourself that in this example $A = 1 + i\sqrt{3} = 2\angle\frac{\pi}{3}$.

It is now a simple matter to deduce the symbolic rule from the geometric rule. Recall the essential steps taken by Bombelli in deriving (1.6): (i) $i^2 = -1$; (ii) brackets can be multiplied out, i.e., if A, B, C, are complex numbers then

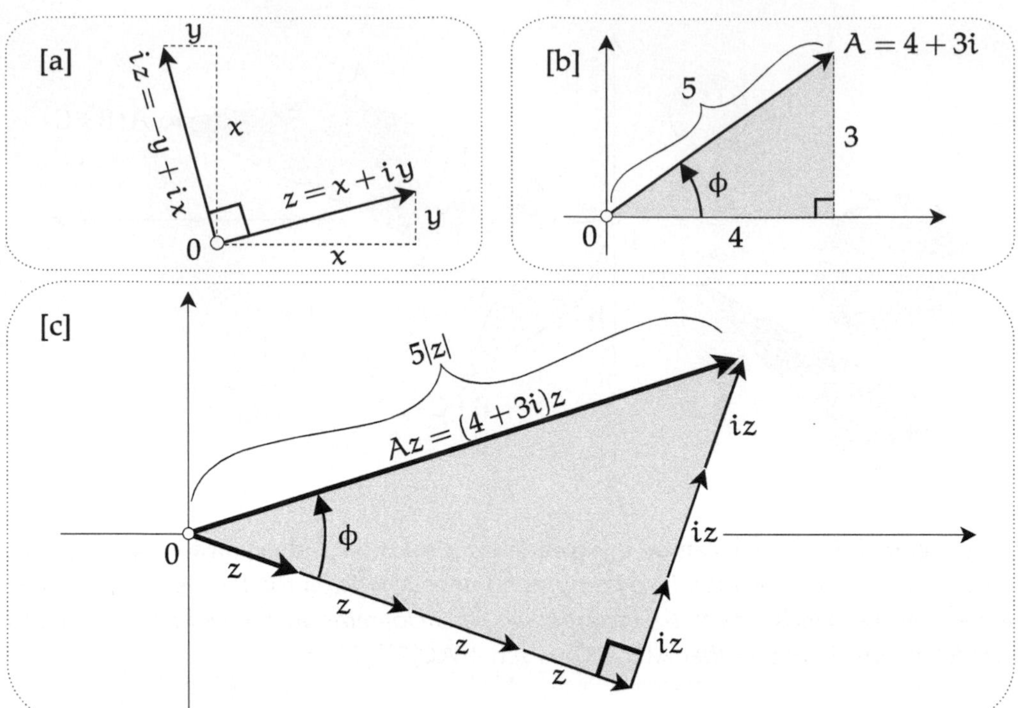

[1.6] Geometric *explanation* of the geometry of multiplication. [a] $iz = (z \text{ rotated by } \frac{\pi}{2})$. **[b]** $A = 4 + 3i = 5\angle\phi$. **[c]** *Since this shaded triangle is similar to the one in* **[b]**, *we see that multiplication by* $5\angle\phi$ *does indeed rotate the plane by* ϕ, *and expand it by* 5.

$A(B + C) = AB + AC$. We have already seen that the geometric rule gives us (i), and figure [1.5] now reveals that (ii) is also true, for the simple reason that *rotations and expansions preserve parallelograms*. By the geometric definition of addition, $B + C$ is the fourth vertex of the parallelogram with vertices 0, B, C. To establish (ii), we merely observe that multiplication by A rotates and expands this parallelogram into another parallelogram with vertices 0, AB, AC and $A(B + C)$. This completes the derivation of (1.6).

Conversely, we now show how the geometric rule may be derived from the symbolic rule[5]. We begin by considering the transformation $z \mapsto iz$. According to the symbolic rule, this means that $(x + iy) \mapsto (-y + ix)$, and [1.6a] reveals that iz *is z rotated through a right angle.* We now use this fact to interpret the transformation $z \mapsto Az$, where A is a general complex number. How this is done may be grasped sufficiently well using the example $A = 4 + 3i = 5\angle\phi$, where $\phi = \tan^{-1}(3/4)$.

[5] In every text we have examined this is done using trigonometric identities. We believe that the present argument supports the view that such identities are merely complicated manifestations of the simple rule for complex multiplication.

See [1.6b]. The symbolic rule says that brackets can be multiplied out, so our transformation may be rewritten as follows:

$$\begin{aligned} z \mapsto Az &= (4+3i)z \\ &= 4z + 3(iz) \\ &= 4z + 3\left(z \text{ rotated by } \tfrac{\pi}{2}\right). \end{aligned}$$

This is visualized in [1.6c]. We can now see that the shaded triangles in [1.6c] and [1.6b] are *similar*, so multiplication by $5\angle\phi$ does indeed rotate the plane by ϕ, and expand it by 5. Done.

1.2 Euler's Formula

1.2.1 Introduction

It is time to replace the $r\angle\theta$ notation with a much better one that depends on the following miraculous fact:

$$\boxed{e^{i\theta} = \cos\theta + i\sin\theta}\,! \tag{1.10}$$

This result was discovered by Leonhard Euler around 1740, and it is called *Euler's formula* in his honour.

Before attempting to explain this result, let us say something of its meaning and utility. As illustrated in [1.7a], the formula says that $e^{i\theta}$ is the point on the unit circle at angle θ. Instead of writing a general complex number as $z = r\angle\theta$, we can now write $z = r\,e^{i\theta}$. Concretely, this says that to reach z we must take the unit

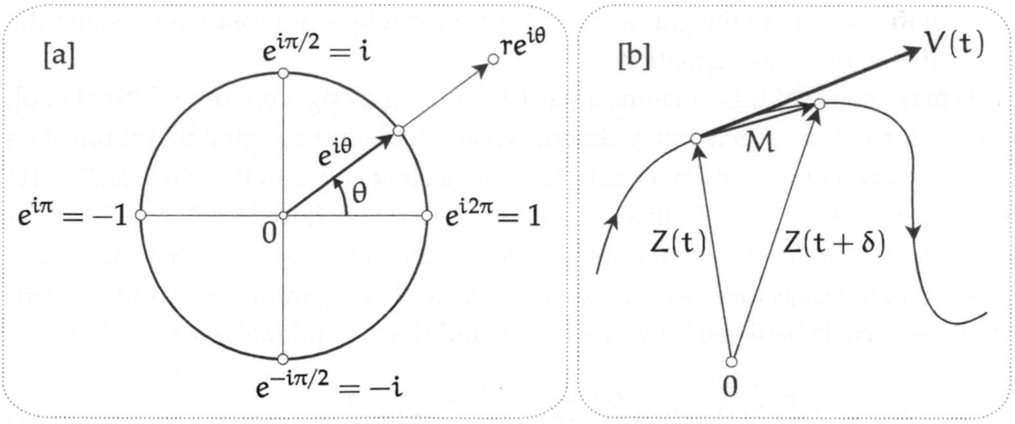

[1.7] [a] Euler's formula *states that* $e^{i\theta}$ *has unit length and points at angle* θ*; expanding this by* r*, we may reach any complex number.* **[b]** *The complex velocity* $V(t)$ *(tangent to the complex orbit* $Z(t)$*) is ultimately equal to* M/δ*, as* $\delta \to 0$.

vector $e^{i\theta}$ that points at z, then stretch it by the length of z. Part of the beauty of this representation is that the geometric rule (1.7) for multiplying complex numbers now looks almost obvious:

$$\left(R\,e^{i\phi}\right)\left(r\,e^{i\theta}\right) = Rr\,e^{i(\phi+\theta)}.$$

Put differently, algebraically manipulating $e^{i\theta}$ in the same way as the real function e^x yields true facts about complex numbers.

In order to explain Euler's formula we must first address the more basic question, "What does $e^{i\theta}$ *mean*?" Surprisingly, many authors answer this by defining $e^{i\theta}$, out of the blue, *to be* $(\cos\theta + i\sin\theta)$! This gambit is logically unimpeachable, but it is also a low blow to Euler, reducing one of his greatest achievements to a mere tautology. We will therefore give two heuristic arguments in support of (1.10); deeper arguments will emerge in later chapters.

1.2.2 Moving Particle Argument

Recall the basic fact that e^x is its own derivative: $\frac{d}{dx}e^x = e^x$. This is actually a *defining* property, that is, if $\frac{d}{dx}f(x) = f(x)$, and $f(0) = 1$, then $f(x) = e^x$. Similarly, if k is a real constant, then e^{kx} may be defined by the property $\frac{d}{dx}f(x) = k\,f(x)$. To extend the action of the ordinary exponential function e^x from real values of x to imaginary ones, let us cling to this property by insisting that it remain true if $k = i$, so that

$$\frac{d}{dt}e^{it} = ie^{it}. \tag{1.11}$$

We have used the letter t instead of x because we will now think of the variable as being *time*. We are used to thinking of the derivative of a real function as the slope of the tangent to the graph of the function, but how are we to understand the derivative in the above equation?

To make sense of this, imagine a particle moving along a curve in $\mathbb{C}$. See [1.7b]. The motion of the particle can be described *parametrically* by saying that at time t its position is the complex number $Z(t)$. Next, recall from physics that the *velocity* $V(t)$ is the vector—now thought of as a complex number—whose length and direction are given by the instantaneous speed, and the instantaneous direction of motion (tangent to the trajectory), of the moving particle. The figure shows the movement M of the particle between time t and $t + \delta$, and this should make it clear that

$$\frac{d}{dt}Z(t) = \lim_{\delta\to 0}\frac{Z(t+\delta) - Z(t)}{\delta} = \lim_{\delta\to 0}\frac{M}{\delta} = V(t).$$

Thus, given a complex function $Z(t)$ of a real variable t, we can always visualize Z as the position of a moving particle, and $\frac{dZ}{dt}$ as its velocity.

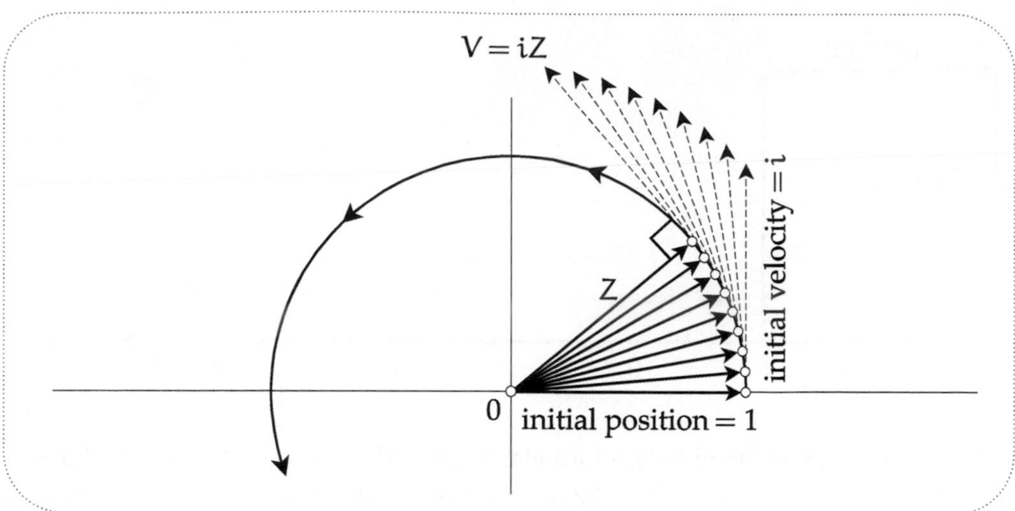

[1.8] First explanation of Euler's formula. *The orbit* $Z(t) = e^{it}$ *is characterized by the fact that its velocity* $V = iZ =$ (its position rotated by $\frac{\pi}{2}$). *Thus, the particle must always move at right angles to its current position, and at unit speed. Therefore, after time* $t = \theta$ *it will have travelled distance* θ *round the unit circle, subtending angle* θ *at the origin, so* $e^{i\theta} = \cos\theta + i\sin\theta$.

We can now use this idea to find the trajectory in the case $Z(t) = e^{it}$. See [1.8]. According to (1.11),

$$\textit{velocity} = V = iZ = \textit{position, rotated through a right angle.}$$

Since the initial position of the particle is $Z(0) = e^0 = 1$, its initial velocity is i, and so it is moving vertically upwards. A split second later the particle will have moved very slightly in this direction, and its *new* velocity will be at right angles to its *new* position vector. Continuing to construct the motion in this way, it is clear that the particle will travel round the unit circle.

Since we now know that $|Z(t)|$ remains equal to 1 throughout the motion, it follows that the particle's speed $|V(t)|$ also remains equal to 1. Thus after time $t = \theta$ the particle will have travelled a distance θ round the unit circle, and so the angle of $Z(\theta) = e^{i\theta}$ will be θ. This is the geometric statement of Euler's formula.

1.2.3 Power Series Argument

For our second argument, we begin by re-expressing the defining property $\frac{d}{dx}f(x) = f(x)$ in terms of power series. Assuming that $f(x)$ can be expressed in the form $a_0 + a_1x + a_2x^2 + \cdots$, a simple calculation shows that

$$e^x = f(x) = 1 + x + \frac{x^2}{2!} + \frac{x^3}{3!} + \cdots,$$

and further investigation shows that this series converges for all (real) values of x.

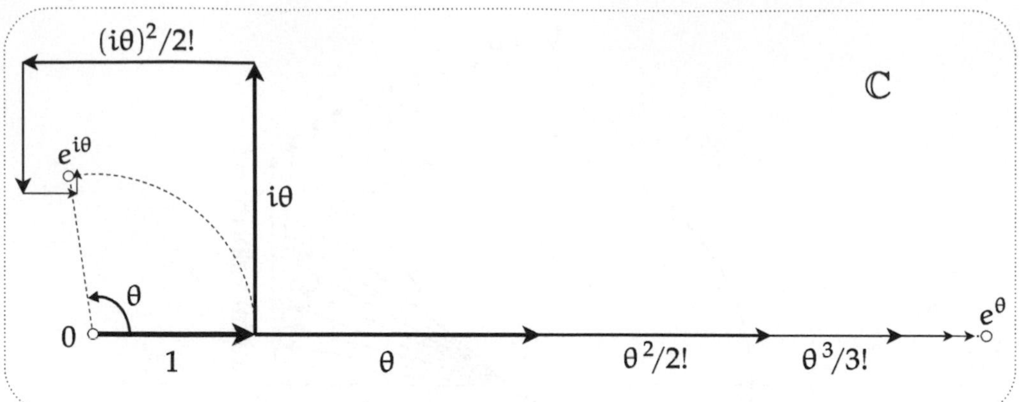

[1.9] Second explanation of Euler's formula. *Imagine the terms of the horizontal power series for e^θ to be rods connected by hinges. Now wrap the series into the illustrated spiral, with each rod making a right angle with the one before it. The argument in the text confirms Euler's miraculous conclusion: the spiral ends at the point on the unit circle at angle θ. So, imagining the rigid, vertical $i\theta$-rod transformed into flexible string, we may wrap it onto the unit circle, and its end will* coincide *with the end of the spiral!*

Putting x equal to a real value θ, this infinite sum of horizontal real numbers is visualized in [1.9]. To make sense of $e^{i\theta}$, we now cling to the power series and put $x = i\theta$:

$$e^{i\theta} = 1 + i\theta + \frac{(i\theta)^2}{2!} + \frac{(i\theta)^3}{3!} + \cdots.$$

As illustrated in [1.9], this series is just as meaningful as the series for e^θ, but instead of the terms all having the same direction, here each term makes a right angle with the previous one, producing a kind of spiral.

This picture makes it clear that the known convergence of the series for e^θ guarantees that the spiral series for $e^{i\theta}$ converges to a definite point in $\mathbb{C}$. However, it is certainly *not* clear that it will converge to the point on the unit circle at angle θ. To see this, we split the spiral into its real and imaginary parts:

$$e^{i\theta} = C(\theta) + iS(\theta),$$

where

$$C(\theta) = 1 - \frac{\theta^2}{2!} + \frac{\theta^4}{4!} - \cdots, \qquad \text{and} \qquad S(\theta) = \theta - \frac{\theta^3}{3!} + \frac{\theta^5}{5!} - \cdots.$$

At this point we could obtain Euler's formula by appealing to Taylor's Theorem, which shows that $C(\theta)$ and $S(\theta)$ are the power series for $\cos\theta$ and $\sin\theta$. However, we can also get the result by means of the following elementary argument that does not require Taylor's Theorem.

We wish to show two things about $e^{i\theta} = C(\theta) + iS(\theta)$: (i) it has unit length, and (ii) it has angle θ. To do this, first note that differentiation of the power series C and S yields

$$C' = -S \qquad \text{and} \qquad S' = C,$$

where a prime denotes differentiation with respect to θ.

To establish (i), observe that

$$\frac{d}{d\theta}|e^{i\theta}|^2 = (C^2 + S^2)' = 2(CC' + SS') = 0,$$

which means that the length of $e^{i\theta}$ is independent of θ. Since $e^{i0} = 1$, we deduce that $|e^{i\theta}| = 1$ for all θ.

To establish (ii) we must show that $\Theta(\theta) = \theta$, where $\Theta(\theta)$ denotes the angle of $e^{i\theta}$, so that

$$\tan\Theta(\theta) = \frac{S(\theta)}{C(\theta)}.$$

Since we already know that $C^2 + S^2 = 1$, we find that the derivative of the LHS of the above equation is

$$[\tan\Theta(\theta)]' = (1 + \tan^2\Theta)\,\Theta' = \left(1 + \frac{S^2}{C^2}\right)\Theta' = \frac{\Theta'}{C^2},$$

and that the derivative of the RHS is

$$\left[\frac{S}{C}\right]' = \frac{S'C - C'S}{C^2} = \frac{1}{C^2}.$$

Thus

$$\frac{d\Theta}{d\theta} = \Theta' = 1,$$

which implies that $\Theta(\theta) = \theta + \mathtt{const}$. Taking the angle of $e^{i0} = 1$ to be 0 [would it make any geometric difference if we took it to be 2π?], we find that $\Theta = \theta$.

Although it is incidental to our purpose, note that we can now conclude (without Taylor's Theorem) that $C(\theta)$ and $S(\theta)$ are the power series of $\cos\theta$ and $\sin\theta$.

1.2.4 Sine and Cosine in Terms of Euler's Formula

A simple but important consequence of Euler's formula is that sine and cosine can be constructed from the exponential function. More precisely, inspection of [1.10] yields

$$e^{i\theta} + e^{-i\theta} = 2\cos\theta \qquad \text{and} \qquad e^{i\theta} - e^{-i\theta} = 2i\sin\theta,$$

or equivalently,

$$\cos\theta = \frac{e^{i\theta} + e^{-i\theta}}{2} \qquad \text{and} \qquad \sin\theta = \frac{e^{i\theta} - e^{-i\theta}}{2i}. \tag{1.12}$$

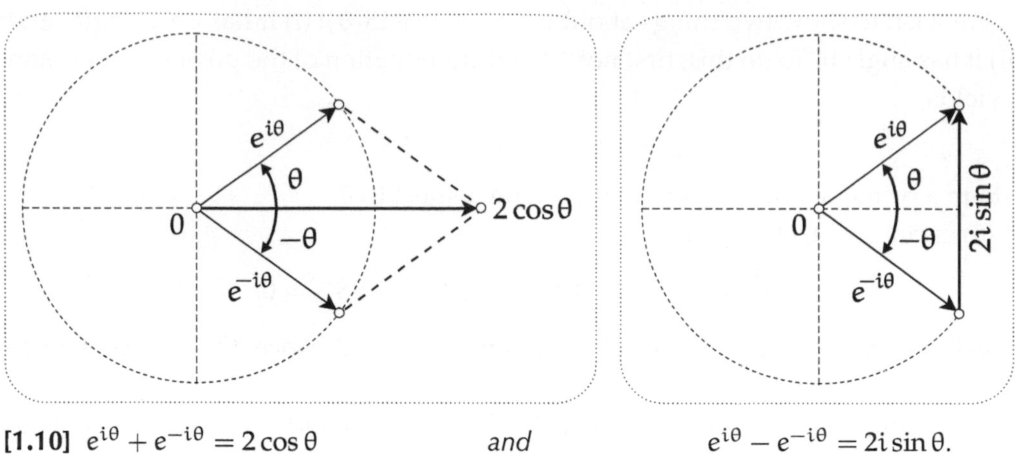

[1.10] $e^{i\theta} + e^{-i\theta} = 2\cos\theta$ *and* $e^{i\theta} - e^{-i\theta} = 2i\sin\theta$.

1.3 Some Applications

1.3.1 Introduction

Often problems that do not appear to involve complex numbers are nevertheless solved most elegantly by viewing them through complex spectacles. In this section we will illustrate this point with a variety of examples taken from diverse areas of mathematics. Further examples may be found in the exercises at the end of the chapter.

The first example [trigonometry] merely illustrates the power of the concepts already developed, but the remaining examples develop important new ideas.

1.3.2 Trigonometry

All trigonometric identities may be viewed as arising from the rule for complex multiplication. In the following examples we will reduce clutter by using the following shorthand: $C \equiv \cos\theta$, $S \equiv \sin\theta$, and similarly, $c \equiv \cos\phi$, $s \equiv \sin\phi$.

To find an identity for $\cos(\theta+\phi)$, view it as a component of $e^{i(\theta+\phi)}$. See [1.11a]. Since

$$\begin{aligned}\cos(\theta+\phi) + i\sin(\theta+\phi) &= e^{i(\theta+\phi)}\\ &= e^{i\theta}e^{i\phi}\\ &= (C+iS)(c+is)\\ &= [Cc - Ss] + i[Sc + Cs],\end{aligned}$$

we obtain not only an identity for $\cos(\theta+\phi)$, but also one for $\sin(\theta+\phi)$:

$$\cos(\theta+\phi) = Cc - Ss \qquad \text{and} \qquad \sin(\theta+\phi) = Sc + Cs.$$

This illustrates another powerful feature of using complex numbers: every complex equation says two things at once.

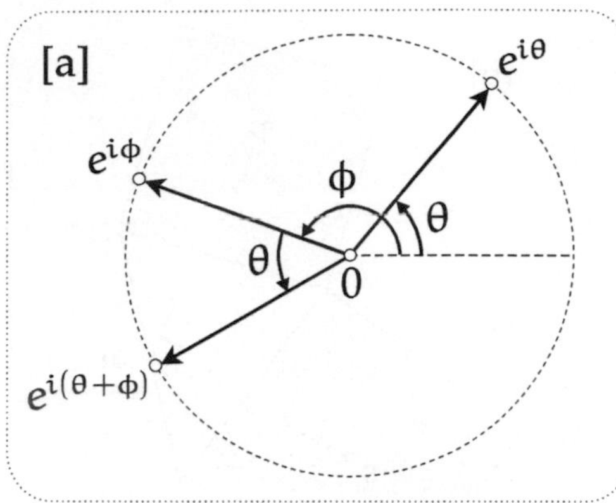

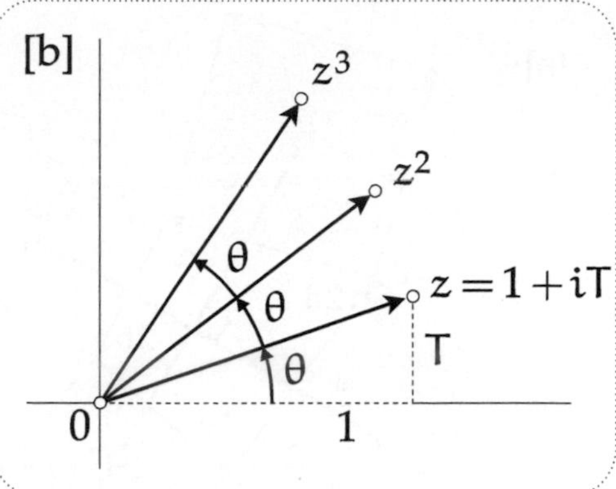

[1.11] [a] Trigonometric identities for $\cos(\theta+\phi)$ and $\sin(\theta+\phi)$ *are easily obtained by taking the real and imaginary parts of* $e^{i(\theta+\phi)} = e^{i\theta}e^{i\phi}$. **[b] An identity for $\tan 3\theta$** *in terms of* $T = \tan\theta$ *follows easily from the illustrated fact that* $(1+iT)^3$ *has angle* 3θ.

To simultaneously find identities for $\cos 3\theta$ and $\sin 3\theta$, consider $e^{i3\theta}$:

$$\cos 3\theta + i \sin 3\theta = e^{i3\theta} = (e^{i\theta})^3 = (C + iS)^3 = \left[C^3 - 3CS^2\right] + i\left[3C^2S - S^3\right].$$

Using $C^2 + S^2 = 1$, these identities may be rewritten in the more familiar forms,

$$\cos 3\theta = 4C^3 - 3C \qquad \text{and} \qquad \sin 3\theta = -4S^3 + 3S.$$

We have just seen how to express trig functions of multiples of θ in terms of powers of trig functions of θ, but we can also go in the opposite direction. For example, suppose we want an identity for $\cos^4\theta$ in terms of multiples of θ. Since $2\cos\theta = e^{i\theta} + e^{-i\theta}$,

$$\begin{aligned} 2^4\cos^4\theta &= \left(e^{i\theta} + e^{-i\theta}\right)^4 \\ &= \left(e^{i4\theta} + e^{-i4\theta}\right) + 4\left(e^{i2\theta} + e^{-i2\theta}\right) + 6 \\ &= 2\cos 4\theta + 8\cos 2\theta + 6 \\ \Longrightarrow \cos^4\theta &= \tfrac{1}{8}\left[\cos 4\theta + 4\cos 2\theta + 3\right]. \end{aligned}$$

Although Euler's formula is extremely convenient for doing such calculations, it is not essential: all we are really using is the equivalence of the geometric and symbolic forms of complex multiplication. To stress this point, let us do an example without Euler's formula.

To find an identity for $\tan 3\theta$ in terms of $T = \tan\theta$, consider $z = 1 + iT$. See [1.11b]. Since z is at angle θ, z^3 will be at angle 3θ, so $\tan 3\theta = \text{Im}(z^3)/\text{Re}(z^3)$. Thus,

$$z^3 = (1 + iT)^3 = (1 - 3T^2) + i(3T - T^3) \qquad \Longrightarrow \qquad \tan 3\theta = \frac{3T - T^3}{1 - 3T^2}.$$

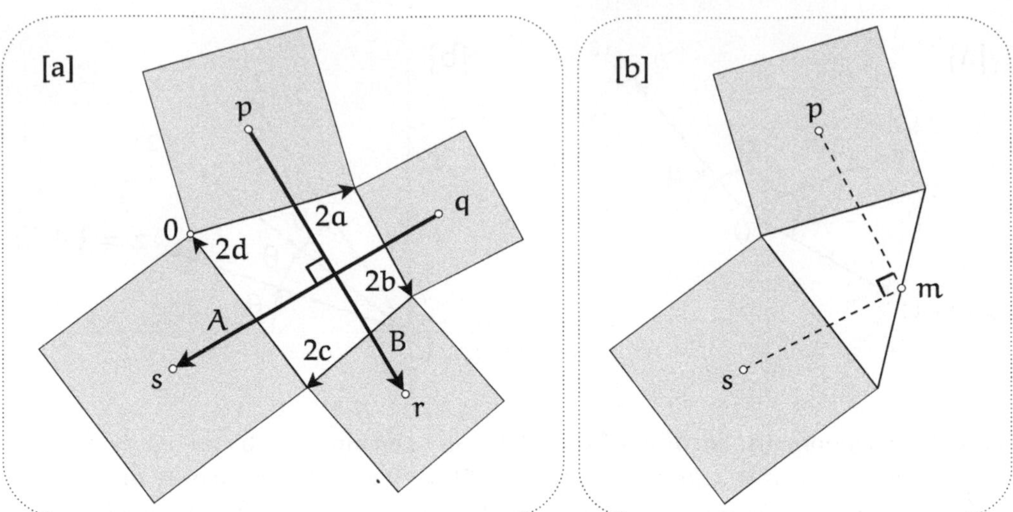

[1.12] A surprising property of quadrilaterals. [a] If squares are constructed on the sides of a quadrilateral, the line-segments joining the centres of opposite squares are perpendicular and of equal length! *A short algebraic proof is made possible by taking the edges of the quadrilateral to be* complex numbers *that sum to zero.* **[b]** *An alternative geometric proof is based on the illustrated fact:* If squares are constructed on two sides of an arbitrary triangle, then the line-segments from their centres to the midpoint m of the remaining side are perpendicular and of equal length.

1.3.3 Geometry

We shall base our discussion of geometric applications on a single example. In [1.12a] we have constructed squares on the sides of an arbitrary quadrilateral. Let us prove what this picture strongly suggests: *the line-segments joining the centres of opposite squares are perpendicular and of equal length.* It would require a great deal of ingenuity to find a purely geometric proof of this surprising result, so instead of relying on our own intelligence, let us invoke the intelligence of the complex numbers!

Introducing a factor of 2 for convenience, let $2a$, $2b$, $2c$, and $2d$ represent complex numbers running along the edges of the quadrilateral. The only condition is that the quadrilateral close up, i.e.,

$$a + b + c + d = 0.$$

As illustrated, choose the origin of $\mathbb{C}$ to be at the vertex where $2a$ begins. To reach the centre p of the square constructed on that side, we go along a, then an equal distance at right angles to a. Thus, since ia is a rotated through a right angle, $p = a + ia = (1 + i)a$. Likewise,

$$q = 2a + (1 + i)b, \quad r = 2a + 2b + (1 + i)c, \quad s = 2a + 2b + 2c + (1 + i)d.$$

The complex numbers $A = s - q$ (from q to s) and $B = r - p$ (from p to r) are therefore given by

$$A = (b + 2c + d) + i(d - b) \qquad \text{and} \qquad B = (a + 2b + c) + i(c - a).$$

We wish to show that A and B are perpendicular and of equal length. These two statements can be combined into the single complex statement $B = iA$, which says that B is A rotated by $(\pi/2)$. To finish the proof, note that this is the same thing as $A + iB = 0$, the verification of which is a routine calculation:

$$A + iB = (a + b + c + d) + i\,(a + b + c + d) = 0.$$

As a first step towards a purely geometric explanation of the result in [1.12a], consider [1.12b]. Here squares have been constructed on two sides of an arbitrary triangle, and, as the picture suggests, *the line-segments from their centres to the mid-point* m *of the remaining side are perpendicular and of equal length.* As is shown in Ex. 21, [1.12a] can be quickly deduced[6] from [1.12b]. The latter result can, of course, be proved in the same manner as above, but let us instead try to find a purely geometric argument.

To do so we will take an interesting detour, investigating translations and rotations of the plane in terms of complex functions. In reality, this "detour" is much more important than the geometric puzzle to which our results will be applied.

Let $\mathcal{T}_v$ denote a translation of the plane by v, so that a general point z is mapped to $\mathcal{T}_v(z) = z + v$. See [1.13a], which also illustrates the effect of the translation on a triangle. The *inverse* of $\mathcal{T}_v$, written $\mathcal{T}_v^{-1}$, is the transformation that undoes it; more formally, $\mathcal{T}_v^{-1}$ is defined by $\mathcal{T}_v^{-1} \circ \mathcal{T}_v = \mathcal{E} = \mathcal{T}_v \circ \mathcal{T}_v^{-1}$, where $\mathcal{E}$ is the "do nothing" transformation (called the *identity*) that maps each point to itself: $\mathcal{E}(z) = z$. Clearly, $\mathcal{T}_v^{-1} = \mathcal{T}_{-v}$.

If we perform $\mathcal{T}_v$, followed by another translation $\mathcal{T}_w$, then the composite mapping $\mathcal{T}_w \circ \mathcal{T}_v$ of the plane is another translation:

$$\mathcal{T}_w \circ \mathcal{T}_v(z) = \mathcal{T}_w(z + v) = z + (w + v) = \mathcal{T}_{w+v}(z).$$

This gives us an interesting way of motivating addition itself. If we had introduced a complex number v *as being* the translation $\mathcal{T}_v$, then we could have defined the "sum" of two complex numbers $\mathcal{T}_v$ and $\mathcal{T}_w$ to be the net effect of performing these translations in succession (in either order). Of course this would have been equivalent to the definition of addition that we actually gave.

Let $\mathcal{R}_a^\theta$ denote a rotation of the plane through angle θ about the point a. For example, $\mathcal{R}_a^\phi \circ \mathcal{R}_a^\theta = \mathcal{R}_a^{\theta+\phi}$, and $\left(\mathcal{R}_a^\theta\right)^{-1} = \mathcal{R}_a^{-\theta}$. As a first step towards expressing rotations as complex functions, note that (1.9) says that a rotation about the origin can be written as $\mathcal{R}_0^\theta(z) = e^{i\theta} z$.

[6] This approach is based on a paper of Finney (1970).

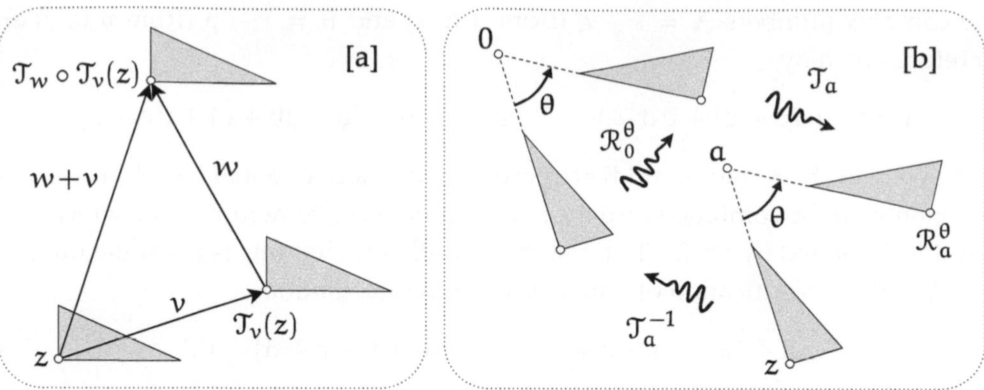

[1.13] Translations and rotations as complex functions. [a] *Composition of translations is equivalent to addition:* $\mathcal{T}_w \circ \mathcal{T}_v(z) = \mathcal{T}_{w+v}(z)$. **[b]** *The general rotation* $\mathcal{R}_a^\theta$ *can be performed by translating* a *to* 0*, rotating by* θ *about* 0*, then translating* 0 *back to* a.

As illustrated in [1.13b], the general rotation $\mathcal{R}_a^\theta$ can be performed by translating a to 0, rotating θ about 0, then translating 0 back to a:

$$\mathcal{R}_a^\theta(z) = \left(\mathcal{T}_a \circ \mathcal{R}_0^\theta \circ \mathcal{T}_a^{-1}\right)(z) = e^{i\theta}(z-a) + a = e^{i\theta}z + k,$$

where $k = a(1-e^{i\theta})$. Thus we find that a rotation about any point can instead be expressed as an equal rotation about the origin, followed by a translation: $\mathcal{R}_a^\theta = (\mathcal{T}_k \circ \mathcal{R}_0^\theta)$. Conversely, a rotation of α about the origin followed by a translation of v can always be reduced to a single rotation:

$$\mathcal{T}_v \circ \mathcal{R}_0^\alpha = \mathcal{R}_c^\alpha, \quad \text{where} \quad c = v/(1-e^{i\alpha}).$$

In the same way, you can easily check that if we perform the translation before the rotation, the net transformation can again be accomplished with a single rotation: $\mathcal{R}_0^\theta \circ \mathcal{T}_v = \mathcal{R}_p^\theta$. What is p?

The results just obtained are certainly not obvious geometrically [try them], and they serve to illustrate the power of thinking of translations and rotations as complex functions. As a further illustration, consider the net effect of performing two rotations about different points. Representing the rotations as complex functions, an easy calculation [exercise] yields

$$\left(\mathcal{R}_b^\phi \circ \mathcal{R}_a^\theta\right)(z) = e^{i(\theta+\phi)}z + v, \qquad \text{where} \qquad v = ae^{i\phi}(1-e^{i\theta}) + b(1-e^{i\phi}).$$

Unless $(\theta+\phi)$ is a multiple of 2π, the previous paragraph therefore tells us that

$$\mathcal{R}_b^\phi \circ \mathcal{R}_a^\theta = \mathcal{R}_c^{(\theta+\phi)}, \qquad \text{where} \qquad c = \frac{v}{1-e^{i(\theta+\phi)}} = \frac{ae^{i\phi}(1-e^{i\theta}) + b(1-e^{i\phi})}{1-e^{i(\theta+\phi)}}.$$

[What should c equal if $b=a$ or $\phi=0$? Check the formula.] This result is illustrated in [1.14a]. Later we shall find a purely geometric explanation of this result, and, in the process, a very simple geometric construction of the point c given by the complicated formula above.

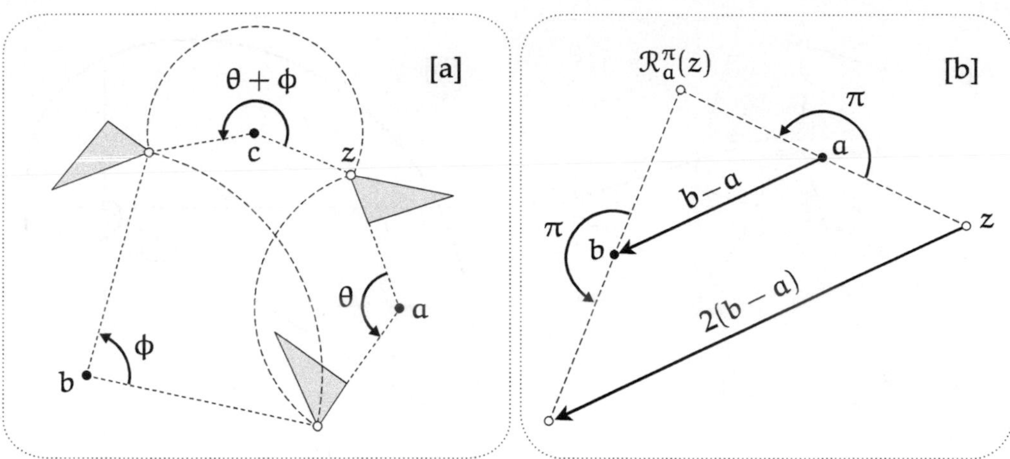

[1.14] Composition of rotations. [a] *In general, two successive rotations (of* θ *and* φ*) about different points (*a *and* b*) are equivalent to a single rotation of* $(\theta + \phi)$ *about a special point* c. **[b]** *But if* $(\theta + \phi) = 2\pi$ *then the net result is instead a translation, as can easily be seen here in the special case* $\theta = \phi = \pi$.

If, on the other hand, $(\theta + \phi)$ is a multiple of 2π, then $e^{i(\theta+\phi)} = 1$, and

$$\mathcal{R}_b^\phi \circ \mathcal{R}_a^\theta = \mathcal{T}_v, \qquad \text{where} \qquad v = (1 - e^{i\phi})(b - a).$$

For example, putting $\theta = \phi = \pi$, this predicts that $\mathcal{R}_b^\pi \circ \mathcal{R}_a^\pi = \mathcal{T}_{2(b-a)}$ is a translation by twice the complex number connecting the first centre of rotation to the second. That this is indeed true can be deduced directly from [1.14b].

The above result on the composition of two rotations implies [exercise] the following:

> *Let* $\mathcal{M} = \mathcal{R}_{a_n}^{\theta_n} \circ \cdots \circ \mathcal{R}_{a_2}^{\theta_2} \circ \mathcal{R}_{a_1}^{\theta_1}$ *be the composition of* n *rotations, and let* $\Theta = \theta_1 + \theta_2 + \cdots + \theta_n$ *be the total amount of rotation. In general,* $\mathcal{M} = \mathcal{R}_c^\Theta$ *(for some* c*), but if* Θ *is a multiple of* 2π *then* $\mathcal{M} = \mathcal{T}_v$*, for some* v.

Returning to our original problem, we can now give an elegant geometric explanation of the result in [1.12b]. Referring to [1.15a], let $\mathcal{M} = \mathcal{R}_m^\pi \circ \mathcal{R}_p^{(\pi/2)} \circ \mathcal{R}_s^{(\pi/2)}$. According to the result just obtained, $\mathcal{M}$ is a translation. To find out *what* translation, we need only discover the effect of $\mathcal{M}$ on a single point. Clearly, $\mathcal{M}(k) = k$, so $\mathcal{M}$ is the zero translation, i.e., the identity transformation $\mathcal{E}$. Thus

$$\mathcal{R}_p^{(\pi/2)} \circ \mathcal{R}_s^{(\pi/2)} = (\mathcal{R}_m^\pi)^{-1} \circ \mathcal{M} = \mathcal{R}_m^\pi.$$

If we define $s' = \mathcal{R}_m^\pi(s)$ then m is the midpoint of ss'. But, on the other hand,

$$s' = \left(\mathcal{R}_p^{(\pi/2)} \circ \mathcal{R}_s^{(\pi/2)}\right)(s) = \mathcal{R}_p^{(\pi/2)}(s).$$

Thus the triangle sps' is isosceles and has a right angle at p, so sm and pm are perpendicular and of equal length. Done.

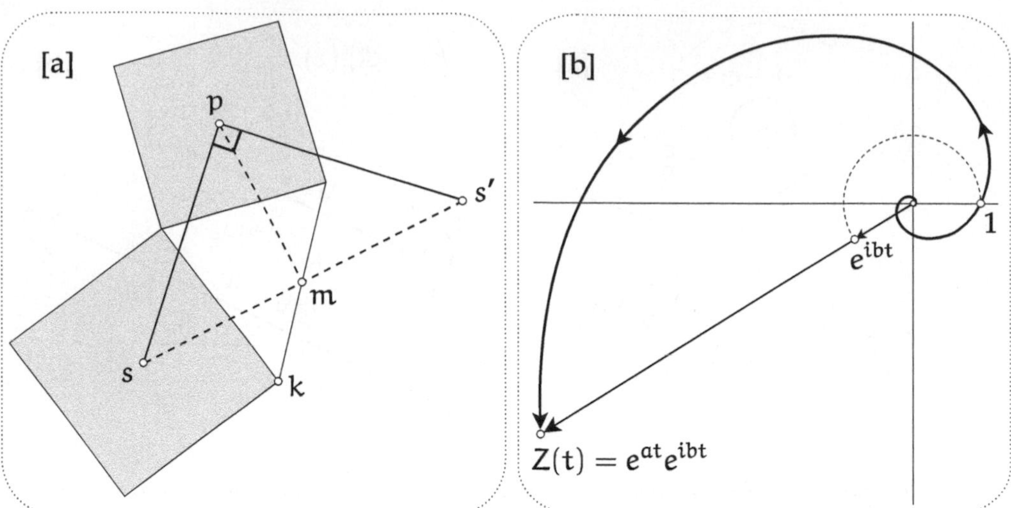

[1.15] [a] *The total angle of rotation of* $\mathcal{M} = \mathcal{R}_m^{\pi} \circ \mathcal{R}_p^{(\pi/2)} \circ \mathcal{R}_s^{(\pi/2)}$ *is* 2π*, so it is a translation. But we see that* $\mathcal{M}(k) = k$*, so the translation vanishes—*$\mathcal{M}$ *is the identity. It follows [see text] that* sm *and* pm *are perpendicular and of equal length.* **[b]** *The orbit* $Z(t) = e^{at}e^{ibt}$ *is a spiral: as* e^{ibt} *rotates round the unit circle with angular speed* b*, it is stretched by the rapidly increasing factor* e^{at}*.*

1.3.4 Calculus

For our calculus example, consider the problem of finding the 100th derivative of $e^x \sin x$. More generally, we will show how complex numbers may be used to find the nth derivative of $e^{ax} \sin bx$.

In discussing Euler's formula we saw that e^{it} may be thought of as the location at time t of a particle travelling around the unit circle at unit speed. In the same way, e^{ibt} may be thought of as a unit complex number rotating about the origin with (angular) speed b. If we stretch this unit complex number by e^{at} as it turns, then its tip describes the motion of a particle that is spiralling away from the origin. See [1.15b].

The relevance of this to the opening problem is that the location of the particle at time t is

$$Z(t) = e^{at}e^{ibt} = e^{at}\cos bt + i\,e^{at}\sin bt.$$

Thus the derivative of $e^{at}\sin bt$ is simply the vertical (imaginary) component of the velocity V of Z.

We could find V simply by differentiating the components of Z in the above expression, but we shall instead use this example to introduce the geometric approach that will be used throughout this book. In [1.16], consider the movement $M = Z(t+\delta) - Z(t)$ of the particle between time t and $(t+\delta)$.

Recall that V is defined to be the limit of (M/δ) as δ tends to zero. Thus V and (M/δ) are very nearly equal if δ is very small. This suggests two intuitive ways of

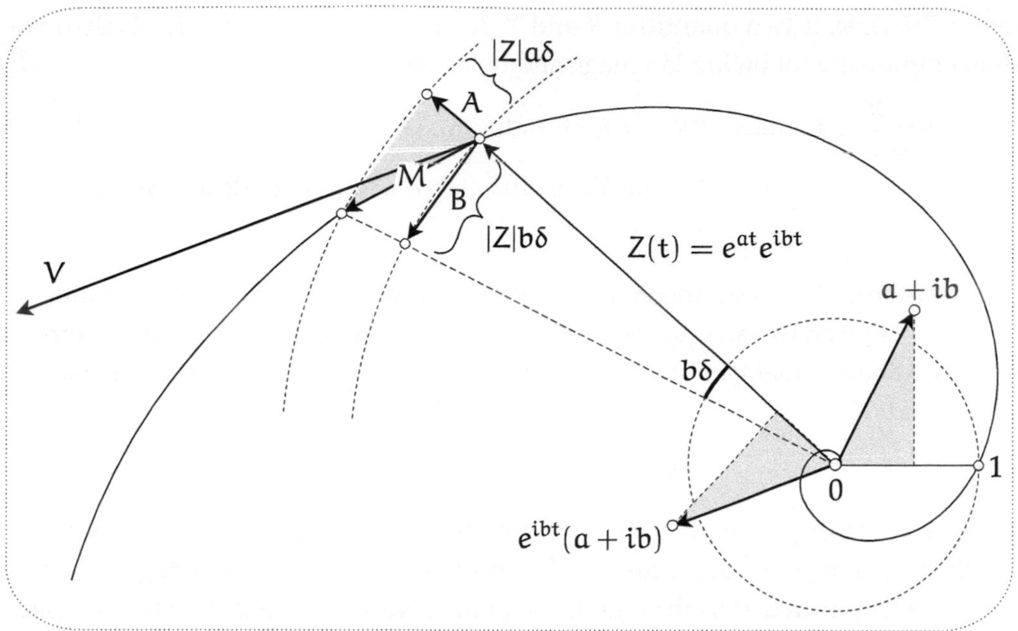

[1.16] Geometric evaluation of the velocity: $V\delta \asymp M \asymp A + B \asymp e^{ibt}|Z|a\delta + ie^{ibt}|Z|b\delta$, *so* $V = (a + ib)Z$.

speaking, both of which will be used in this book: (i) we shall say that "$V = (M/\delta)$ when δ is *infinitesimal*" or (ii) that "V and (M/δ) are *ultimately equal*" (as δ tends to zero).

As explained in the new Preface to this *25th Anniversary Edition*, both the concept and the language of *ultimate equality* were introduced by Newton in his great *Principia* of 1687, which is filled with inspired geometrical constructions, *not* tedious calculus computations. Sadly, although it was the mathematical engine that propelled his extraordinary discoveries, Newton chose to emulate the expository style of the ancient Greek geometers, writing out his proofs in *words*: he did not introduce any *symbol* to represent his critical concept of ultimate equality. We, however, shall employ the symbol "$\asymp$" to denote Newton's concept, enabling us to write, for example, $V \asymp (M/\delta)$.

We stress that here the words "ultimately equal" and "infinitesimal" are being used in definite, technical senses; in particular, "infinitesimal" does not refer to some mystical, infinitely small quantity.[7] More precisely, following Newton's lead

[7] For more on this distinction, see the discussion in Chandrasekhar (1995).

in the *Principia,* if two quantities X and Y depend on a third quantity δ, then we shall employ the following language and notation:

$$\lim_{\delta\to 0}\frac{X}{Y}=1 \iff \text{"}X=Y \text{ for infinitesimal } \delta\text{".}$$

$$\iff \text{"X and Y are \textit{ultimately equal} (as } \delta \text{ tends to zero)".}$$

$$\iff X \asymp Y.$$

It follows from the basic theorems on limits that Newton's concept of "ultimate equality" is [exercise] an *equivalence relation,* and that it also inherits other properties of ordinary equality. For example, since V and (M/δ) are ultimately equal, so are $V\delta$ and M:

$$V \asymp \frac{M}{\delta} \iff V\delta \asymp M.$$

We now return to the problem of finding the velocity of the spiralling particle. As illustrated in [1.16], draw rays from 0 through $Z(t)$ and $Z(t+\delta)$, together with circular arcs (centred at 0) through those points. Now let A and B be the complex numbers connecting $Z(t)$ to the illustrated intersection points of these rays and arcs. If δ is infinitesimal, then B is at right angles to A and Z, and $M = A + B$.

Let us find the ultimate lengths of A and B. During the time interval δ, the angle of Z increases by $b\delta$, so the two rays cut off an arc of length $b\delta$ on the unit circle, and an arc of length $|Z|b\delta$ on the circle through Z. Thus $|B|$ is ultimately equal to $|Z|b\delta$. Next, note that $|A|$ is the increase in $|Z(t)|$ occurring in the time interval δ. Thus, since

$$\frac{d}{dt}|Z(t)| = \frac{d}{dt}e^{at} = a|Z|,$$

$|A|$ is ultimately equal to $|Z|a\delta$.

The shaded triangle at Z is therefore ultimately similar to the shaded right triangle with hypotenuse $a + ib$. Rotating the latter triangle by the angle of Z, you should now be able to see that if δ is infinitesimal then

$$\begin{aligned} M &= (a+ib) \text{ rotated by the angle of } Z, \text{ and expanded by } |Z|\,\delta \\ &= (a+ib)Z\delta \\ \Longrightarrow V &= \frac{d}{dt}Z = (a+ib)Z. \end{aligned} \tag{1.13}$$

Thus all rays from the origin cut the spiral at the same angle [the angle of $(a+ib)$], and the speed of the particle is proportional to its distance from the origin.

Note that although we have not yet given meaning to e^z (where z is a general complex number), it is certainly tempting to write $Z(t) = e^{at}e^{ibt} = e^{(a+ib)t}$. This makes the result (1.13) look very natural. Conversely, this suggests that we should *define* $e^z = e^{(x+iy)}$ to be $e^x e^{iy}$; another justification for this step will emerge in the next chapter.

Using (1.13), it is now easy to take further derivatives. For example, the acceleration of the particle is

$$\frac{d^2}{dt^2}Z = \frac{d}{dt}V = (a+ib)^2\,Z = (a+ib)\,V.$$

Continuing in this way, each new derivative is obtained by multiplying the previous one by $(a+ib)$. [Try sketching these successive derivatives in [1.16].] Writing $(a+ib) = R\,e^{i\phi}$, where $R = \sqrt{a^2+b^2}$ and ϕ is the appropriate value of $\tan^{-1}(b/a)$, we therefore find that

$$\frac{d^n}{dt^n}Z = (a+ib)^n\,Z = R^n\,e^{in\phi}\,e^{at}e^{ibt} = R^n e^{at} e^{i(bt+n\phi)}.$$

Thus

$$\frac{d^n}{dt^n}\left[e^{at}\sin bt\right] = (a^2+b^2)^{\frac{n}{2}}\,e^{at}\sin\left[bt + n\,\tan^{-1}(b/a)\right]. \tag{1.14}$$

1.3.5 Algebra

In the final year of his life (1716) Roger Cotes made a remarkable discovery that enabled him (in principle) to evaluate the family of integrals,

$$\int \frac{dx}{x^n - 1},$$

where $n = 1, 2, 3, \ldots$. To see the connection with algebra, consider the case $n=2$. The key observations are that the denominator (x^2-1) can be *factorized* into $(x-1)(x+1)$, and that the integrand can then be split into *partial fractions*:

$$\int \frac{dx}{x^2-1} = \tfrac{1}{2}\int\left[\frac{1}{x-1} - \frac{1}{x+1}\right]dx = \tfrac{1}{2}\ln\left[\frac{x-1}{x+1}\right].$$

As we shall see, for higher values of n one cannot completely factorize (x^n-1) into linear factors without employing complex numbers—a scarce and dubious commodity in 1716! However, Cotes was aware that if he could break down (x^n-1) into real *linear and quadratic* factors, then he would be able to evaluate the integral. Here, a "real quadratic" refers to a quadratic whose coefficients are all real numbers.

For example, (x^4-1) can be broken down into $(x-1)(x+1)(x^2+1)$, yielding a partial fraction expression of the form

$$\frac{1}{x^4-1} = \frac{A}{x-1} + \frac{B}{x+1} + \frac{Cx}{x^2+1} + \frac{D}{x^2+1},$$

and hence an integral that can be evaluated in terms of ln and $\tan^{-1}$. More generally, even if the factorization involves more complicated quadratics than (x^2+1), it is easy to show that only ln and $\tan^{-1}$ are needed to evaluate the resulting integrals.

In order to set Cotes' work on $(x^n - 1)$ in a wider context, we shall investigate the general connection between the roots of a polynomial and its factorization. This connection can be explained by considering the geometric series,

$$G_{m-1} = c^{m-1} + c^{m-2}z + c^{m-3}z^2 + \cdots + cz^{m-2} + z^{m-1},$$

in which c and z are complex. Just as in real algebra, this series may be summed by noting that zG_{m-1} and cG_{m-1} contain almost the same terms—try an example, say $m = 4$, if you have trouble seeing this. Subtracting these two expressions yields

$$(z - c)G_{m-1} = z^m - c^m, \tag{1.15}$$

and thus

$$G_{m-1} = \frac{z^m - c^m}{z - c}.$$

If we think of c as fixed and z as variable, then $(z^m - c^m)$ is an mth-degree polynomial in z, and $z = c$ is a root. The result (1.15) says that this m^{th}-degree polynomial can be factored into the product of the linear term $(z - c)$ and the $(m-1)^{\text{th}}$-degree polynomial G_{m-1}.

In 1637 Descartes published an important generalization of this result. Let $P_n(z)$ denote a general polynomial of degree n:

$$P_n(z) = z^n + Az^{n-1} + \cdots + Dz + E,$$

where the coefficients $A, \ldots, E$ may be complex. Since (1.15) implies

$$P_n(z) - P_n(c) = (z - c)\left[G_{n-1} + AG_{n-2} + \cdots + D\right],$$

we obtain *Descartes' Factor Theorem* linking the existence of roots to factorizability:

> *If* c *is a solution of* $P_n(z) = 0$ *then* $P_n(z) = (z - c)\,P_{n-1}$, *where* P_{n-1} *is of degree* $(n - 1)$.

If we could in turn find a root c' of P_{n-1}, then the same reasoning would yield $P_n = (z - c)(z - c')\,P_{n-2}$. Continuing in this way, Descartes' theorem therefore holds out the promise of factoring P_n into precisely n linear factors:

$$P_n(z) = (z - c_1)(z - c_2)\cdots(z - c_n). \tag{1.16}$$

If we do not acknowledge the existence of complex roots (as in the early 18th century) then this factorization will be possible in some cases (e.g., $z^2 - 1$), and impossible in others (e.g., $z^2 + 1$). But, in splendid contrast to this, if one admits complex numbers then it can be shown that P_n *always has* n *roots in* $\mathbb{C}$, *and the factorization (1.16) is always possible*. This is called the *Fundamental Theorem of Algebra*, and we shall explain its truth in Chapter 7.

Each factor $(z - c_k)$ in (1.16) represents a complex number connecting the root c_k to the variable point z. Figure [1.17a] illustrates this for a general cubic polynomial.

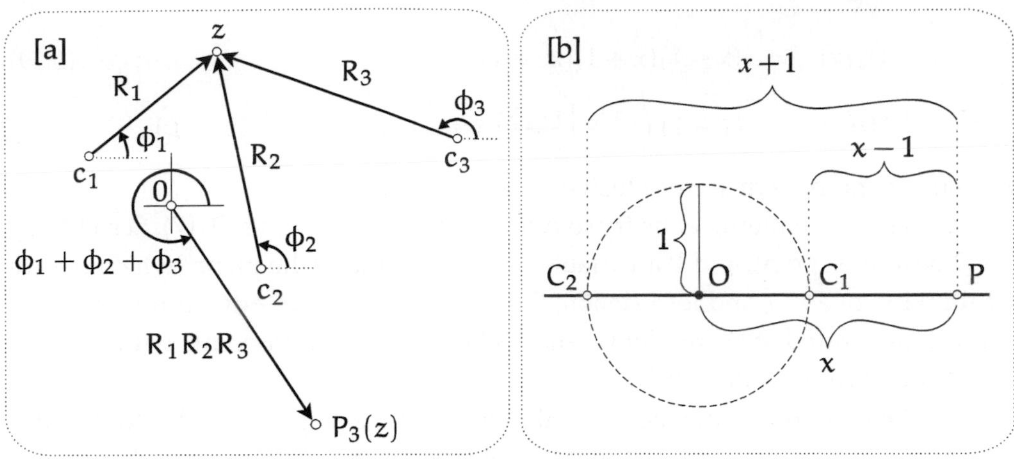

[1.17] Geometric evaluation of polynomials. [a] *If the roots of polynomial* P *are known, then its value at* z *can easily be visualized: Connect the roots to* z*, then add the angles and multiply the lengths to obtain the direction and length of* P(z). **[b]** *First example of Cotes's geometrical factorization of* $x^n - 1$*: Here,* $x^2 - 1 = PC_1 \cdot PC_2$.

Writing each of these complex numbers in the form $R_k\, e^{i\phi_k}$, (1.16) takes the more vivid form

$$P_n(z) = R_1 R_2 \cdots R_n\, e^{i(\phi_1+\phi_2+\cdots+\phi_n)}.$$

Although the Fundamental Theorem of Algebra was not available to Cotes, let us see how it guarantees that he would succeed in his quest to decompose $x^n - 1$ into real linear and quadratic factors. Cotes' polynomial has real coefficients, and, quite generally, we can show that

> *If a polynomial has real coefficients then its complex roots occur in complex conjugate pairs, and it can be factorized into real linear and quadratic factors.*

For if the coefficients $A, \ldots, E$ of $P_n(z)$ are all real then $P_n(c) = 0$ implies [exercise] $P_n(\overline{c}) = 0$, and the factorization (1.16) contains

$$(z - c)(z - \overline{c}) = z^2 - (c + \overline{c})\, z + c\overline{c} = z^2 - 2\,\mathrm{Re}(c)\, z + |c|^2,$$

which is a real quadratic.

Let us now discuss how Cotes was able to factorize $x^n - 1$ into real linear and quadratic factors *by appealing to the geometry of the regular* n*-gon*. [An "n-gon" is an n-sided polygon.] To appreciate the following, place yourself in his 18th century shoes and forget all you have just learnt concerning the Fundamental Theorem of Algebra; even forget about complex numbers and the complex plane!

For the first few values of n, the desired factorizations of $U_n(x) = x^n - 1$ are not too hard to find:

$$U_2(x) = (x-1)(x+1), \tag{1.17}$$

$$U_3(x) = (x-1)(x^2+x+1), \tag{1.18}$$

$$\begin{aligned} U_4(x) &= (x-1)(x+1)(x^2+1), \\ U_5(x) &= (x-1)\left(x^2+\left[\tfrac{1+\sqrt{5}}{2}\right]x+1\right)\left(x^2+\left[\tfrac{1-\sqrt{5}}{2}\right]x+1\right), \end{aligned} \tag{1.19}$$

but the general pattern seems elusive.

To find such a pattern, let us try to *visualize* the simplest case, (1.17). See [1.17b]. Let O be a fixed point, and P a variable point, on a line in the plane (which we are *not* thinking of as $\mathbb{C}$), and let x denote the distance OP. If we now draw a circle of unit radius centred at O, and let C_1 and C_2 be its intersection points with the line, then clearly[8] $U_2(x) = PC_1 \cdot PC_2$.

To understand quadratic factors in this spirit, let us skip over (1.18) to the simpler quadratic in (1.19). This factorization of $U_4(x)$ is the best we could do without complex numbers, but ideally we would have liked to have decomposed $U_4(x)$ into *four* linear factors. This suggests that we rewrite (1.19) as

$$U_4(x) = (x-1)(x+1)\,\sqrt{x^2+1}\,\sqrt{x^2+1},$$

the last two "factors" being analogous to genuine linear factors. If we are to interpret this expression (by analogy with the previous case) as the product of the distances of P from four fixed points, then the points corresponding to the last two "factors" must be *off the line*. More precisely, Pythagoras' Theorem tells us that a point whose distance from P is $\sqrt{x^2+1^2}$ must lie at unit distance from O in a direction at right angles to the line OP.

Referring to [1.18a], we can now see that $U_4(x) = PC_1 \cdot PC_2 \cdot PC_3 \cdot PC_4$, where $C_1C_2C_3C_4$ is the illustrated square inscribed in the circle.

Since we have factorized $U_4(x)$ with the regular 4-gon (the square), perhaps we can factorize $U_3(x)$ with the regular 3-gon (the equilateral triangle). See [1.18b].

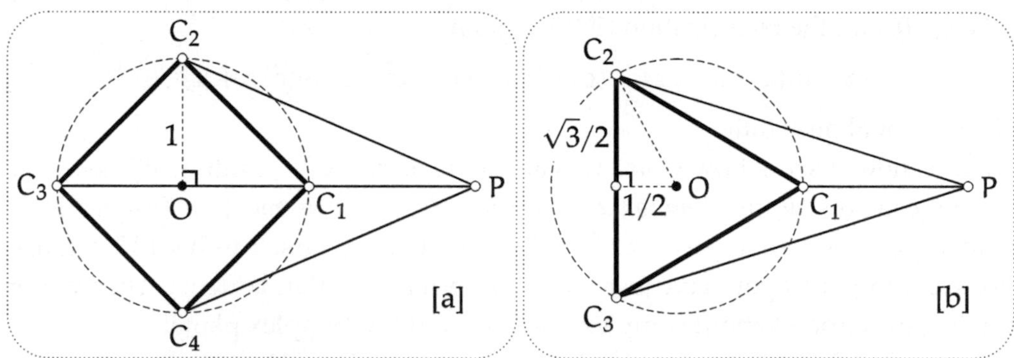

[1.18] Cotes's geometrical factorization of $(x^n - 1)$ using regular n-gons: If OP = x, then [a] $x^4 - 1 = PC_1 \cdot PC_2 \cdot PC_3 \cdot PC_4$; [b] $x^3 - 1 = PC_1 \cdot PC_2 \cdot PC_3$.

[8] Here, and in what follows, we shall suppose for convenience that $x > 1$, so that $U_n(x)$ is positive.

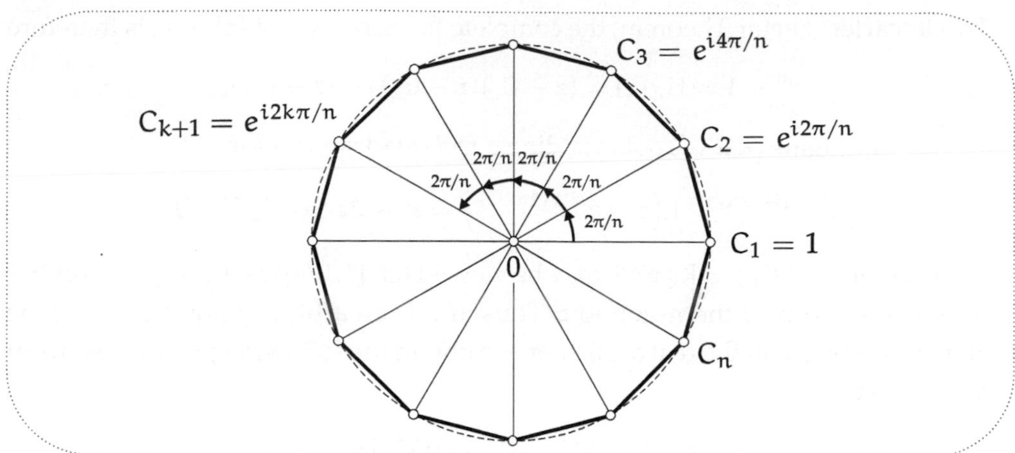

[1.19] The nth roots of unity: *Each vertex of the regular n-gon is a solution of $z^n = 1$.*

Applying Pythagoras' Theorem to this figure,

$$\begin{aligned} PC_1 \cdot PC_2 \cdot PC_3 &= PC_1 \cdot (PC_2)^2 = (x-1)\left(\left[x+\tfrac{1}{2}\right]^2 + \left[\tfrac{\sqrt{3}}{2}\right]^2\right) \\ &= (x-1)(x^2+x+1), \end{aligned}$$

which is indeed the desired factorization (1.18) of $U_3(x)$!

A plausible generalization for U_n now presents itself:

> *If $C_1C_2C_3 \cdots C_n$ is a regular n-gon inscribed in a circle of unit radius centred at O, and P is the point on OC_1 at distance x from O, then $U_n(x) = PC_1 \cdot PC_2 \cdots PC_n$.*

This is Cotes' result. Unfortunately, he stated it without proof, and he left no clue as to how he discovered it. Thus we can only speculate that he may have been guided by an argument like the one we have just supplied[9].

Since the vertices of the regular n-gon will always come in symmetric pairs that are equidistant from P, the examples in [1.18] make it clear that Cotes' result is indeed equivalent to factorizing $U_n(x)$ into real linear and quadratic factors.

Recovering from our feigned bout of amnesia concerning complex numbers and their geometric interpretation, Cotes' result becomes simple to understand and to prove. Taking O to be the origin of the complex plane, and C_1 to be 1, the vertices of Cotes' n-gon are given by $C_{k+1} = e^{ik(2\pi/n)}$. See [1.19], which illustrates the case $n = 12$. Since $(C_{k+1})^n = e^{ik2\pi} = 1$, all is suddenly clear: *The vertices of the regular n-gon are the n complex roots of $U_n(z) = z^n - 1$.* Because the solutions of $z^n - 1 = 0$ may be written formally as $z = \sqrt[n]{1}$, the vertices of the n-gon are called the *n^{th} roots of unity*.

[9] Stillwell (2010) has instead speculated that Cotes used complex numbers (as we are about to), but then deliberately stated his findings in a form that did not require them.

By Descartes' Factor Theorem, the complete factorization of $(z^n - 1)$ is therefore

$$z^n - 1 = U_n(z) = (z - C_1)(z - C_2)\cdots(z - C_n),$$

with each conjugate pair of roots yielding a *real* quadratic factor,

$$\left(z - e^{ik(2\pi/n)}\right)\left(z - e^{-ik(2\pi/n)}\right) = z^2 - 2z\cos\left[\tfrac{2k\pi}{n}\right] + 1.$$

Each factor $(z - C_k) = R_k\, e^{i\phi_k}$ may be viewed (cf. [1.17a]) as a complex number connecting a vertex of the n-gon to z. Thus, if P is an arbitrary point in the plane (not merely a point on the real axis), then we obtain the following generalized form of Cotes' result:

$$U_n(P) = [PC_1 \cdot PC_2 \cdots PC_n]\, e^{i\Phi},$$

where $\Phi = (\phi_1 + \phi_2 + \cdots + \phi_n)$. If P happens to be a real number (again supposed greater than 1) then $\Phi = 0$ [make sure you see this], and we recover Cotes' result.

We did not immediately state and prove Cotes' result in terms of complex numbers because we felt there was something rather fascinating about our first, direct approach. Viewed in hindsight, it shows that even if we attempt to avoid complex *numbers*, we cannot avoid the geometry of the complex *plane!*

1.3.6 Vectorial Operations

Not only is complex addition the same as vector addition, but we will now show that the familiar vectorial operations of dot and cross products (also called scalar and vector products) are both subsumed by complex multiplication. Since these vectorial operations are extremely important in physics—they were discovered by physicists!—their connection with complex multiplication will prove valuable both in applying complex analysis to the physical world, and in using physics to understand complex analysis.

When a complex number $z = x + iy$ is being thought of merely as a vector, we shall write it in bold type, with its components in a column:

$$z = x + iy \iff \mathbf{z} = \begin{pmatrix} x \\ y \end{pmatrix}.$$

Although the dot and cross product are meaningful for arbitrary vectors in space, we shall assume in the following that our vectors all lie in a single plane—the complex plane.

Given two vectors **a** and **b**, figure [1.20a] recalls the definition of the dot product as the length of one vector, times the projection onto that vector of the other vector:

$$\mathbf{a} \cdot \mathbf{b} = |a|\,|b| \cos\theta = \mathbf{b} \cdot \mathbf{a},$$

where θ is the angle between **a** and **b**.

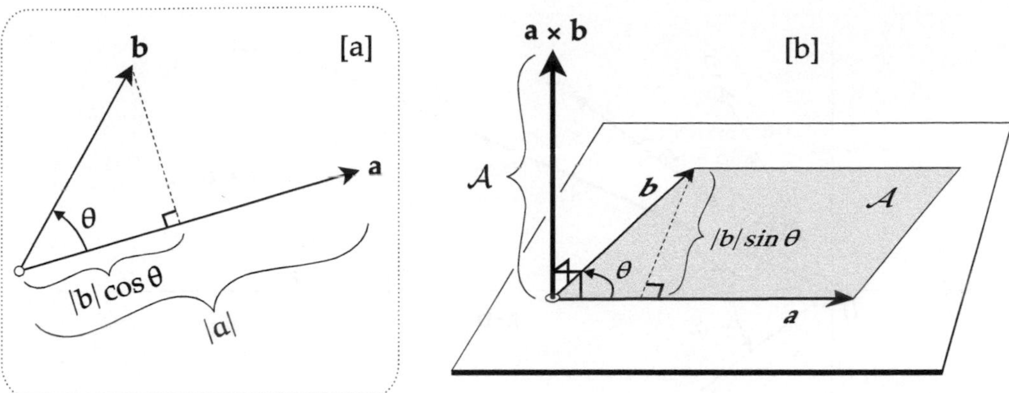

[1.20] [a] *The* **dot product** *is defined to be the orthogonal projection of one vector onto the other, times the length of the other vector.* **[b]** *The* **cross product** *is defined to be the vector orthogonal to both (oriented according to the "right hand rule"), with length equal to the area of the parallelogram they span.*

Figure [1.20b] recalls the definition of the cross product: $\mathbf{a} \times \mathbf{b}$ is the vector perpendicular to the plane of **a** and **b** whose length is equal to the area $\mathcal{A}$ of the parallelogram spanned by **a** and **b**. But wait, there are *two* (opposite) directions perpendicular to $\mathbb{C}$; which should we choose?

Writing $\mathcal{A} = |a|\,|b| \sin\theta$, the area $\mathcal{A}$ has a *sign* attached to it. An easy way to see this sign is to think of the angle θ *from* ***a*** *to* ***b*** as lying in range $-\pi$ to π; the sign of $\mathcal{A}$ is then the same as θ. If $\mathcal{A} > 0$, as in [1.20b], then we define $\mathbf{a} \times \mathbf{b}$ to point upwards from the plane, and if $\mathcal{A} < 0$ we define it to point downwards. It follows that $\mathbf{a} \times \mathbf{b} = -(\mathbf{b} \times \mathbf{a})$.

This conventional definition of $\mathbf{a} \times \mathbf{b}$ is intrinsically three-dimensional, and it therefore presents a problem: if **a** and **b** are thought of as complex numbers, $\mathbf{a} \times \mathbf{b}$ *cannot be*, for it does not lie in the (complex) plane of **a** and **b**. No such problem exists with the dot product because $\mathbf{a} \cdot \mathbf{b}$ is simply a real number, and this suggests a way out.

Since all our vectors will be lying in the same plane, their cross products will all have equal (or opposite) directions, so the only distinction between one cross product and another will be the value of $\mathcal{A}$. For the purposes of this book we will therefore *redefine the cross product to be the (signed) area $\mathcal{A}$ of the parallelogram spanned by* **a** *and* **b**:

$$\mathbf{a} \times \mathbf{b} = |a|\,|b| \sin\theta = -(\mathbf{b} \times \mathbf{a}).$$

Figure [1.21] shows two complex numbers $a = |a|\, e^{i\alpha}$ and $b = |b|\, e^{i\beta}$, the angle from a to b being $\theta = (\beta - \alpha)$. To see how their dot and cross products are related to complex multiplication, consider the effect of multiplying each point in $\mathbb{C}$ by $\overline{a}$. This is a rotation of $-\alpha$ and an expansion of $|a|$, and if we look at the image

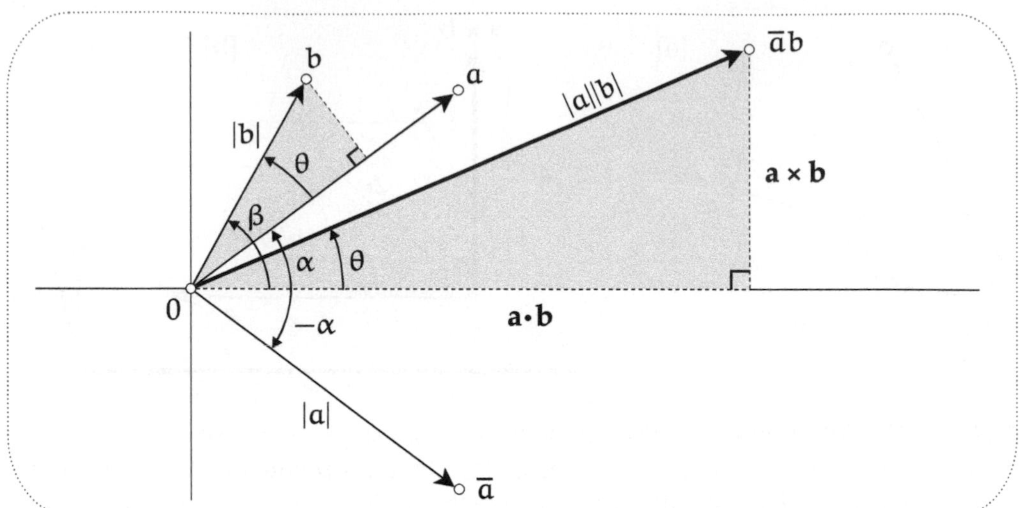

[1.21] The dot and cross products in terms of complex multiplication. *Multiplication of $\mathbb{C}$ by $\overline{a}$ is a rotation of $-\alpha$ and an expansion of $|a|$. It follows that the dot and cross products can be expressed as the real and imaginary parts of $\overline{a}\,b$.*

under this transformation of the shaded right triangle with hypotenuse b, then we immediately see that

$$\overline{a}\,b = \mathbf{a}\cdot\mathbf{b} + i\,(\mathbf{a}\times\mathbf{b}). \tag{1.20}$$

Of course we could also have obtained this by simple calculation:

$$\overline{a}\,b = (|a|\,e^{-i\alpha})(|b|\,e^{i\beta}) = |a|\,|b|\,e^{i(\beta-\alpha)} = |a|\,|b|\,e^{i\theta} = |a|\,|b|(\cos\theta + i\,\sin\theta).$$

When we refer to the dot and cross products as "vectorial operations" we mean that they are defined *geometrically*, independently of any particular choice of coordinate axes. However, once such a choice has been made, (1.20) makes it easy to *express* these operations in terms of Cartesian coordinates. Writing $a = x + iy$ and $b = x' + iy'$,

$$\overline{a}\,b = (x - iy)(x' + iy') = (xx' + yy') + i\,(xy' - yx'),$$

so

$$\begin{pmatrix} x \\ y \end{pmatrix} \cdot \begin{pmatrix} x' \\ y' \end{pmatrix} = xx' + yy' \qquad \text{and} \qquad \begin{pmatrix} x \\ y \end{pmatrix} \times \begin{pmatrix} x' \\ y' \end{pmatrix} = xy' - yx'.$$

We end with an example that illustrates the importance of the sign of the area $(\mathbf{a}\times\mathbf{b})$. Consider the problem of finding the area $\mathcal{A}$ of the quadrilateral in [1.22a], whose vertices are, *in counterclockwise order*, a, b, c, and d. Clearly this is just the sum of the ordinary, unsigned areas of the four triangles formed by joining the

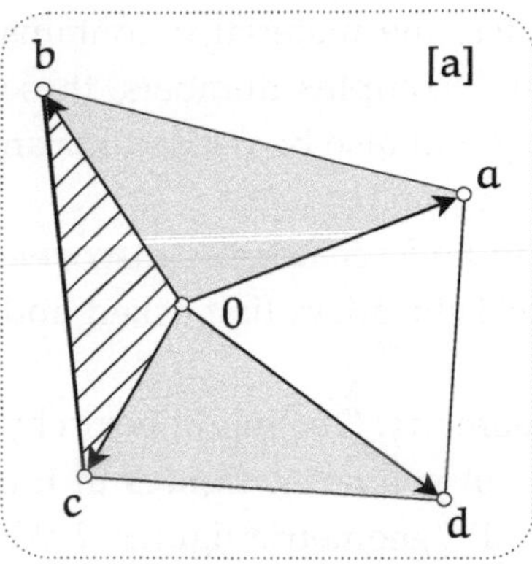

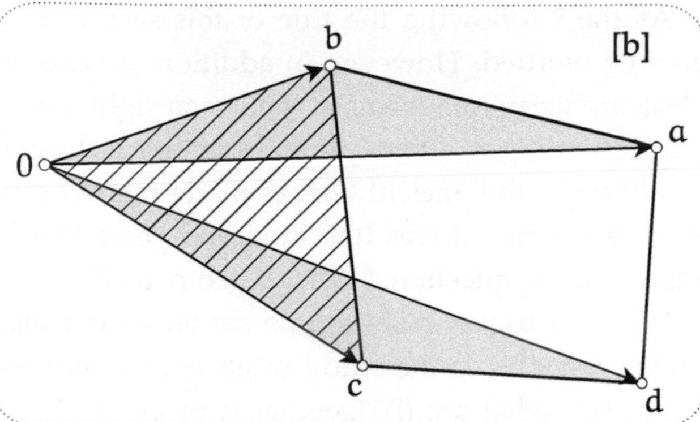

[1.22] Area of a quadrilateral in terms of signed areas of triangles. [a] *The area of the quadrilateral is the sum of the areas of the four triangles, and is therefore given by (1.21).* **[b]** *The area of the quadrilateral is* still *the sum of the areas of the four triangles, because the striped area is automatically* subtracted *from the areas of the other three triangles.*

vertices of the quadrilateral to the origin. Thus, since the area of each triangle is simply half the area of the corresponding parallelogram,

$$\begin{aligned} \mathcal{A} &= \tfrac{1}{2}\left[(\mathbf{a}\times\mathbf{b}) + (\mathbf{b}\times\mathbf{c}) + (\mathbf{c}\times\mathbf{d}) + (\mathbf{d}\times\mathbf{a})\right] \\ &= \tfrac{1}{2}\operatorname{Im}\left[\overline{a}\,b + \overline{b}\,c + \overline{c}\,d + \overline{d}\,a\right]. \end{aligned} \tag{1.21}$$

Obviously this formula could easily be generalized to polygons with more than four sides.

But what if 0 is outside the quadrilateral? In [1.22b], $\mathcal{A}$ is clearly the sum of the ordinary areas of three of the triangles, *minus* the ordinary area of the striped triangle. Since the angle from **b** to **c** is negative, $\frac{1}{2}(\mathbf{b}\times\mathbf{c})$ is automatically the negative of the striped area, and $\mathcal{A}$ *is therefore given by exactly the same formula as before!*

Can you find a location for 0 that makes two of the signed areas negative? Check that the formula still works. Exercise 35 shows that (1.21) *always* works.

1.4 Transformations and Euclidean Geometry*

1.4.1 Geometry Through the Eyes of Felix Klein

Even with the benefit of enormous hindsight, it is hard to introduce complex numbers in a compelling manner. Historically, we have seen how cubic equations forced them upon us algebraically, and in discussing Cotes' work we saw something of the inevitability of their geometric interpretation. In this section we will attempt to show how complex numbers arise very naturally, almost inevitably, from a careful re-examination of *plane Euclidean geometry*[10].

[10] The excellent book by Nikulin and Shafarevich (1987) is the only other work we know of in which a similar attempt is made.

As the * following the title of this section indicates, the material it contains may be omitted. However, in addition to "explaining" complex numbers, these ideas are very interesting in their own right, and they will also be needed for an understanding of other optional sections of the book.

Although the ancient Greeks made many beautiful and remarkable discoveries in geometry, it was two thousand years later that Felix Klein first asked and answered the question, "What *is* geometry?"

Let us restrict ourselves from the outset to *plane* geometry. One might begin by saying that this is the study of geometric properties of geometric figures in the plane, but what are (i) "geometric properties", and (ii) "geometric figures"? We will concentrate on (i), swiftly passing over (ii) by interpreting "geometric figure" as anything we might choose to draw on an infinitely large piece of flat paper with an infinitely fine pen.

As for (i), we begin by noting that if two figures (e.g., two triangles) have the same geometric properties, then (from the point of view of geometry) they must be the "same", "equal", or, as one usually says, *congruent*. Thus if we had a clear definition of congruence ("geometric equality") then we could reverse this observation and *define geometric properties as those properties that are common to all congruent figures.* How, then, can we tell if two figures are geometrically equal?

Consider the triangles in [1.23], and imagine that they are pieces of paper that you could pick up in your hand. To see if T is congruent to T′, you could pick up T and check whether it could be placed on top of T′. Note that it is essential that we be allowed to move T in space: in order to place T on top of $\widetilde{\mathrm{T}}$ we must first flip it over; we can't just slide T around within the plane. Tentatively generalizing, this suggests that *a figure* F *is congruent to another figure* F′ *if there exists a motion of* F *through space that makes it coincide with* F′. Note that the discussion suggests that there are two fundamentally different types of motion: those that involve flipping the figure over, and those that do not. Later, we shall return to this important point.

It is clearly somewhat unsatisfactory that in attempting to define geometry in the *plane* we have appealed to the idea of motion through *space*. We now rectify this. Returning to [1.23], imagine that T and T′ are drawn on separate, transparent sheets of plastic. Instead of picking up just the triangle T, we now pick up the *entire sheet* on which it is drawn, then try to place it on the second sheet so as to make T coincide with T′. At the end of this motion, each point A on T's sheet lies over a point A′ of T′'s sheet, and we can now define the motion $\mathcal{M}$ to be this mapping $A \mapsto A' = \mathcal{M}(A)$ of the plane to itself.

However, not any old mapping qualifies as a motion, for we must also capture the (previously implicit) idea of the sheet remaining *rigid* while it moves, so that

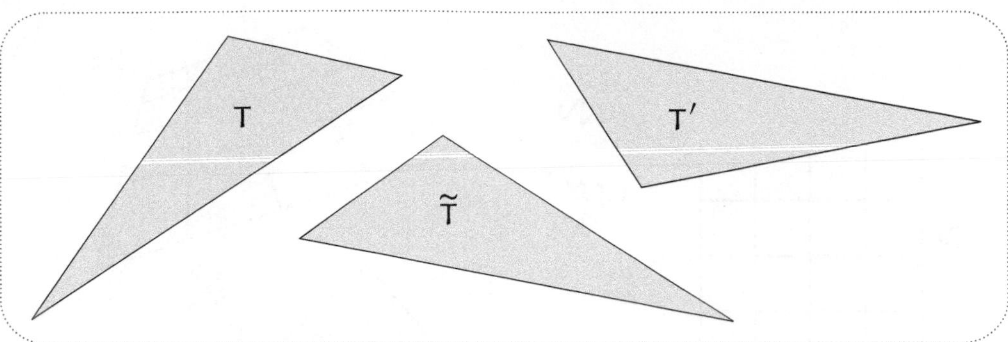

[1.23] *The triangles* T' *and* $\widetilde{\mathsf{T}}$ *are both "geometrically equal" ("congruent") to* T *because there exists a "motion" that carries* T *to each of these. But the motion that carries* T *to* $\widetilde{\mathsf{T}}$ *requires a flip.*

distances between points remain constant during the motion. Here, then, is our definition:

> *A "motion"* $\mathcal{M}$ *is a mapping of the plane to itself such that the distance between any two points* A *and* B *is equal to the distance between their images* $\mathsf{A}' = \mathcal{M}(\mathsf{A})$ *and* $\mathsf{B}' = \mathcal{M}(\mathsf{B})$. (1.22)

Note that what we have called a motion is often termed a "rigid motion", or an "isometry".

Armed with this precise concept of a motion, our final definition of geometric equality becomes

> F *is congruent to* F', *written* $\mathsf{F} \cong \mathsf{F}'$, *if there exists a motion* $\mathcal{M}$ *such that* $\mathsf{F}' = \mathcal{M}(\mathsf{F})$. (1.23)

Next, as a consequence of our earlier discussion, *a geometric property of a figure is one that is unaltered by all possible motions of the figure.* Finally, in answer to the opening question of "What is geometry?", Klein would answer that it is the study of these so-called *invariants* of the set of motions.

One of the most remarkable discoveries of the 19th century was that Euclidean geometry is not the *only* possible geometry. Two of these so-called *non-Euclidean* geometries will be studied in Chapter 6, but for the moment we wish only to explain how Klein was able to generalize the above ideas so as to embrace such new geometries.

The aim in (1.23) was to use a family of transformations to introduce a concept of geometric equality. But will this $\cong$-type of equality behave in the way we would like and expect? To answer this we must first make these expectations explicit. So as not to confuse this general discussion with the particular concept of congruence in (1.23), let us denote geometric equality by $\sim$.

(i) A figure should equal itself: $\mathsf{F} \sim \mathsf{F}$, for all F.

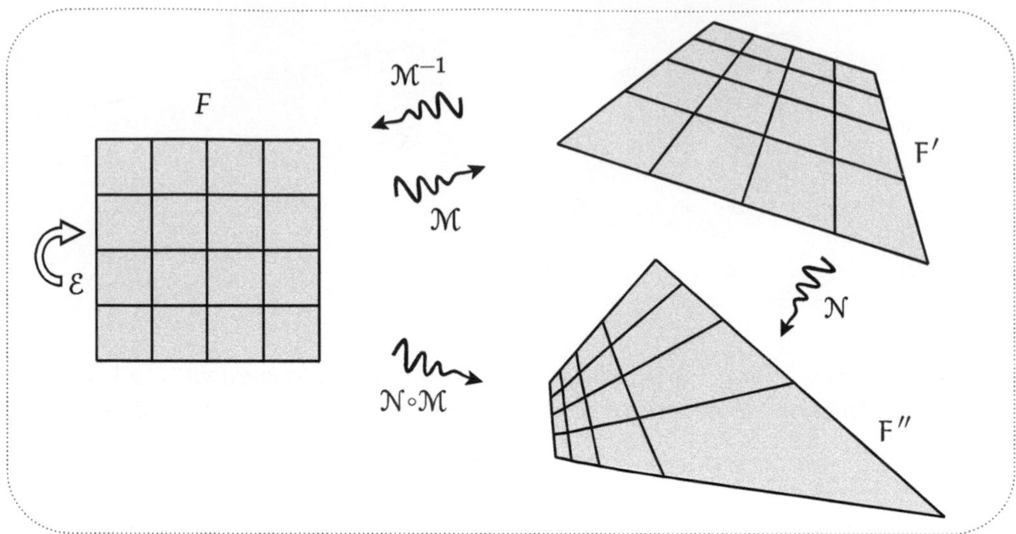

[1.24] Klein's vision of a geometry in terms of a group of transformations. *In* projective geometry *"geometric equality" is defined using more general transformations than rigid Euclidean motions. But geometric equality only makes sense if it is an* equivalence relation, *and this in turn implies that the generalized transformations satisfy the three requirements of a* group.

(ii) If F equals F′, then F′ should equal F: $F \sim F' \Rightarrow F' \sim F$.

(iii) If F and F′ are equal, and F′ and F″ are equal, then F and F″ should also be equal: $F \sim F'$ & $F' \sim F'' \Rightarrow F \sim F''$.

Any relation satisfying these expectations is called an *equivalence relation*.

Now suppose that we retain the definition (1.23) of geometric equality, but that we generalize the definition of "motion" given in (1.22) by replacing the family of distance-preserving transformations with some other family $\mathcal{G}$ of transformations. It should be clear that not any old $\mathcal{G}$ will be compatible with our aim of defining geometric equality. Indeed, (i), (ii), and (iii) imply that $\mathcal{G}$ must have the following very special structure, which is illustrated[11] in [1.24].

(i) The family $\mathcal{G}$ must contain a transformation $\mathcal{E}$ (called the *identity*) that maps each point to itself.

(ii) If $\mathcal{G}$ contains a transformation $\mathcal{M}$, then it must also contain a transformation $\mathcal{M}^{-1}$ (called the *inverse*) that undoes $\mathcal{M}$. [Check for yourself that for $\mathcal{M}^{-1}$ to exist (let alone be a member of $\mathcal{G}$) $\mathcal{M}$ must have the special properties of being

[11] Here G is the group of *projections*. If we do a perspective drawing of figures in the plane, then the mapping from that plane to the "canvas" plane is called a *perspectivity*. A projection is then defined to be any sequence of perspectivities. Can you see why the set of projections should form a group?

(a) *onto* and (b) *one-to-one*, i.e., (a) every point must be the image of some point, and (b) distinct points must have distinct images.]

(iii) If $\mathcal{M}$ and $\mathcal{N}$ are members of $\mathcal{G}$ then so is the composite transformation $\mathcal{N} \circ \mathcal{M} =$ ($\mathcal{M}$ followed by $\mathcal{N}$). This property of $\mathcal{G}$ is called *closure*.

We have thus arrived, very naturally, at a concept of fundamental importance in the whole of mathematics: a family $\mathcal{G}$ of transformations that satisfies these three[12] requirements is called a *group*.

Let us check that the motions defined in (1.22) do indeed form a group: (i) Since the identity transformation preserves distances, it is a motion. (ii) Provided it exists, the inverse of a motion will preserve distances and hence will be a motion itself. As for existence, (a) it is certainly *plausible* that when we apply a motion to the entire plane then the image is the entire plane—we will prove this later—and (b) the non-zero distance between distinct points is preserved by a motion, so their images are again distinct. (iii) If two transformations do not alter distances, then applying them in succession will not alter distances either, so the composition of two motions is another motion.

Klein's idea was that we could first select a group $\mathcal{G}$ at will, then define a corresponding "geometry" as the study of the invariants of that $\mathcal{G}$. [Klein first announced this idea in 1872—when he was 23 years old!—at the University of Erlangen, and it has thus come to be known as his *Erlangen Program*.] For example, if we choose $\mathcal{G}$ to be the group of rigid motions, we recover the familiar Euclidean geometry of the plane. But this is far from being the *only* geometry of the plane, as the so-called *projective geometry* of [1.24] illustrates.

Klein's vision of geometry was broader still. We have been concerned with what geometries are possible *when figures are drawn anywhere in the plane*, but suppose for example that we are only allowed to draw within some disc D. It should be clear that we can construct "geometries of D" in exactly the same way that we constructed geometries of the plane: given a group $\mathcal{H}$ of transformations of D to itself, the corresponding geometry is the study of the invariants of $\mathcal{H}$. If you doubt that any such groups exist, consider the set of all rotations around the centre of D.

The reader may well feel that the above discussion is a chronic case of mathematical generalization running amuck—that the resulting conception of geometry is (to coin a phrase) "as subtle as it is useless". Nothing could be further from the truth! In Chapter 3 we shall be led, very naturally, to consider a particularly interesting group of transformations of a disc to itself. The resulting non-Euclidean geometry is called *hyperbolic* or *Lobachevskian* geometry, and it is the subject of Chapter 6. Far from being useless, this geometry has proved to be an immensely powerful tool in

[12] In more abstract settings it is necessary to add a fourth requirement of *associativity*, namely, $\mathcal{A} \circ (\mathcal{B} \circ \mathcal{C}) = (\mathcal{A} \circ \mathcal{B}) \circ \mathcal{C}$. Of course for transformations this is automatically true.

diverse areas of mathematics and physics, and the insights it continues to provide lie on the cutting edge of contemporary research.

1.4.2 Classifying Motions

To understand the foundations of Euclidean geometry, it seems we must study its group of motions. At the moment, this group is defined rather abstractly as the set of distance-preserving mappings of the plane to itself. However, it is easy enough to think of concrete examples of motions: a rotation of the plane about an arbitrary point, a translation of the plane, or a reflection of the plane in some line. Our aim is to understand the most general possible motions in equally vivid terms.

We begin by stating a key fact:

> *A motion is uniquely determined by its effect on any triangle (i.e., on any three non-collinear points).* (1.24)

By this we mean that knowing what happens to the three points tells us what must happen to *every* point in the plane. To see this, first look at [1.25]. This shows that each point P is uniquely determined by its distances from the vertices A, B, C of such a triangle[13]. The distances from A and B yield two circles which (in general) intersect in two points, P and Q. The third distance (from C) then picks out P.

To obtain the result (1.24), now look at [1.26]. This illustrates a motion $\mathcal{M}$ mapping A, B, C to A′, B′, C′. By the very definition of a motion, $\mathcal{M}$ must map an arbitrary point P to a point P′ whose distances from A′, B′, C′ are equal to the original distances of P from A, B, C. Thus, as shown, P′ is uniquely determined. Done.

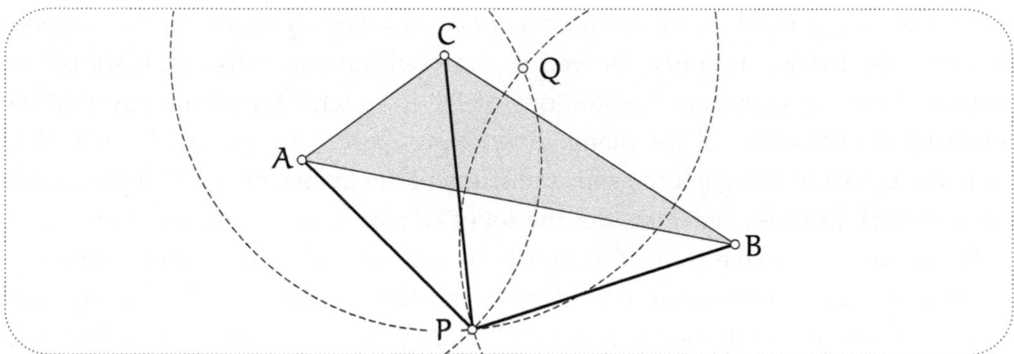

[1.25] *Each point* P *is uniquely determined by its distances from the vertices* A, B, C *of a triangle.*

[13] This is how earthquakes are located. Two types of wave are emitted by the quake as it begins: fast-moving "P-waves" of compression, and slower-moving "S-waves" of destructive shear. Thus the P-waves will arrive at a seismic station before the S-waves, and the time-lag between these events may be used to calculate the distance of the quake from that station. Repeating this calculation at two more seismic stations, the quake may be located.

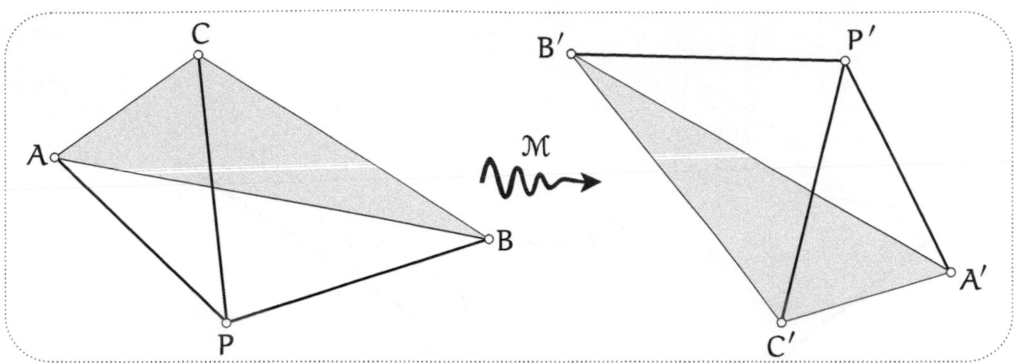

[1.26] A rigid motion is uniquely determined by its effect on any triangle: *If* $\mathcal{M}$ *maps* ABC *to* A′B′C′, *then an arbitrary point* P *must be mapped to the illustrated point* $P' = \mathcal{M}(P)$.

A big step towards classification is the realization that there are two fundamentally different kinds of motions. In terms of our earlier conception of motion through space, the distinction is whether or not a figure must be flipped over before it can be placed on top of a congruent figure. To see how this dichotomy arises in terms of the new definition (1.22), suppose that a motion sends two points A and B to A′ and B′. See [1.27]. According to (1.24), the motion is not yet determined: we need to know the image of any (non-collinear) third point C, such as the one shown in [1.27]. Since motions preserve the distances of C from A and B, there are just two possibilities for the image of C, namely, C′ and its reflection $\widetilde{C}$ in the line L through A′ and B′. Thus there are precisely two motions ($\mathcal{M}$ and $\widetilde{\mathcal{M}}$, say) that map A, B to A′, B′: $\mathcal{M}$ sends C to C′, and $\widetilde{\mathcal{M}}$ sends C to $\widetilde{C}$.

A distinction can be made between $\mathcal{M}$ and $\widetilde{\mathcal{M}}$ by looking at how they affect angles. All motions preserve the magnitude of angles, but we see that $\mathcal{M}$ also preserves the *sense* of the angle θ, while $\widetilde{\mathcal{M}}$ reverses it. The fundamental nature of this distinction can be seen from the fact that $\mathcal{M}$ must in fact preserve *all* angles, while $\widetilde{\mathcal{M}}$ must reverse all angles.

To see this, consider the fate of the angle ϕ in the triangle T. If C goes to C′ (i.e., if the motion is $\mathcal{M}$) then, carrying out the construction indicated in [1.26], the image of T is T′, and the angle is preserved. If, on the other hand, C goes to $\widetilde{C}$ (i.e., if the motion is $\widetilde{\mathcal{M}}$) then the image of T is the reflection $\widetilde{T}$ of T′ in L, and the angle is reversed. Motions that preserve angles are called *direct*, and those that reverse angles are called *opposite*. Thus rotations and translations are direct, while reflections are opposite. Summarizing what we have found,

> *There is exactly one direct motion* $\mathcal{M}$ *(and exactly one opposite motion* $\widetilde{\mathcal{M}}$*) that maps a given line-segment* AB *to another line-segment* A′B′ *of equal length. Furthermore,* $\widetilde{\mathcal{M}} = (\mathcal{M}$ *followed by reflection in the line* A′B′). (1.25)

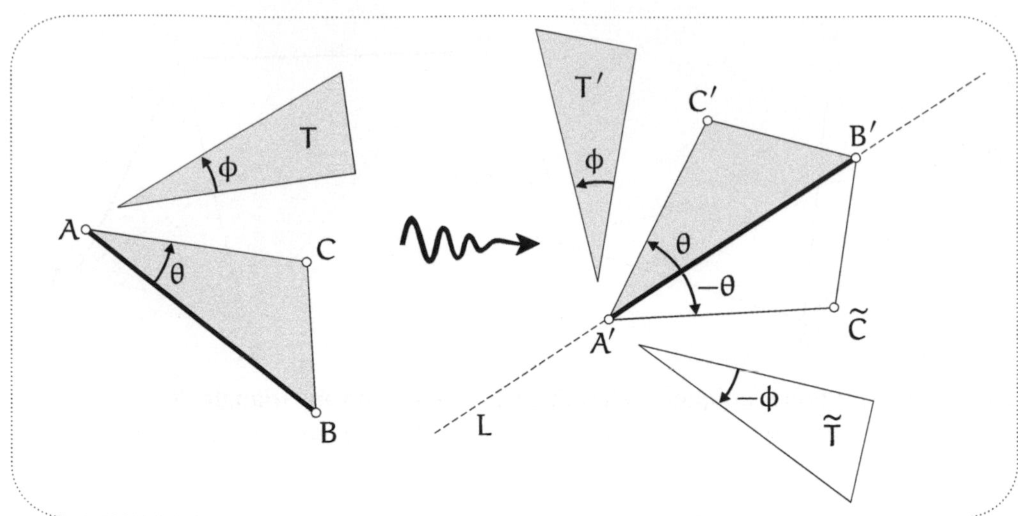

[1.27] *If a motion* $\mathcal{M}$ *maps a given line-segment* AB *to another line-segment* A′B′ *of equal length, then a third point* C *must be mapped to either* C′ *or its reflection* $\widetilde{C}$ *in* A′B′. *Thus,* **there is exactly one direct motion** $\mathcal{M}$ **(and exactly one opposite motion** $\widetilde{\mathcal{M}}$**) that maps a given line-segment AB to another line-segment A′B′ of equal length**. *Furthermore,* $\widetilde{\mathcal{M}}$ = ($\mathcal{M}$ followed by reflection in the line A′B′).

To understand motions we may thus consider two randomly drawn segments AB and A′B′ of equal length, then find *the* direct motion (and *the* opposite motion) that maps one to the other. It is now easy to show that

Every direct motion is a rotation, or else (exceptionally) a translation. (1.26)

Note that this result gives us greater insight into our earlier calculations on the composition of rotations and translations: since the composition of any two direct motions is another direct motion [why?], it can only be a rotation or a translation. Conversely, those calculations allow us to restate (1.26) in a very neat way:

Every direct motion can be expressed as a complex function of the form $\mathcal{M}(z) = e^{i\theta}z + v$. (1.27)

We now establish (1.26). If the line-segment A′B′ is parallel to AB then the vectors $\overrightarrow{AB}$ and $\overrightarrow{A'B'}$ are either equal or opposite. If they are equal, as in [1.28a], the motion is a translation; if they are opposite, as in [1.28b], the motion is a rotation of π about the intersection point of the lines AA′ and BB′.

If the segments are not parallel, produce them (if necessary) till they meet at M, and let θ be the angle between the directions of $\overrightarrow{AB}$ and $\overrightarrow{A'B'}$. See [1.28c]. First recall an elementary property of circles: the chord AA′ subtends the same angle θ at every point of the circular arc AMA′. Next, let O denote the intersection point of this arc with the perpendicular bisector of AA′. We now see that the direct motion carrying AB to A′B′ is a rotation of θ about O, for clearly A is rotated to A′, and the direction of $\overrightarrow{AB}$ is rotated into the direction of $\overrightarrow{A'B'}$. Done.

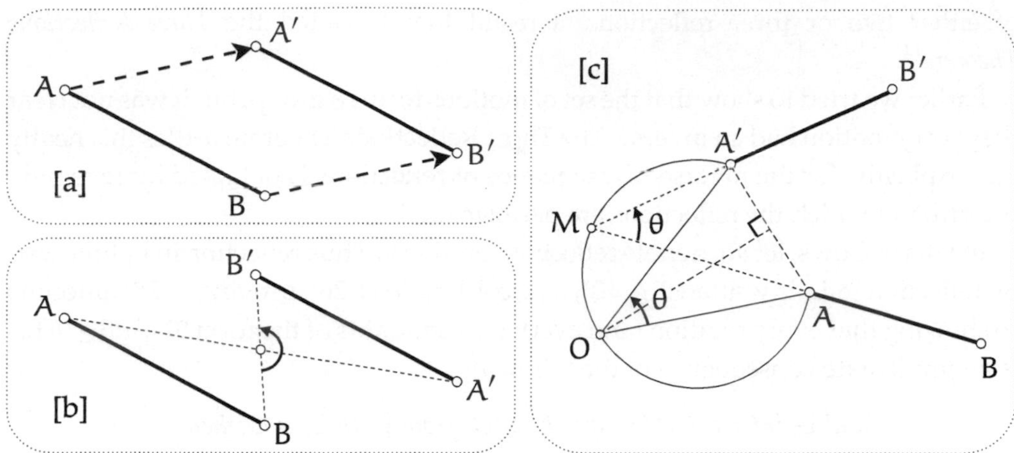

[1.28] Every direct motion is a rotation, or else (exceptionally) a translation. *In the exceptional case that the motion* $\mathcal{M}$ *maps a line segment* AB *to a* parallel *segment* A′B′, $\mathcal{M}$ *is either* **[a]** *a translation, or* **[b]** *a rotation of* π *about the intersection point of the lines* AA′ *and* BB′. *However, in the general case* **[c]**, $\mathcal{M}$ *is a rotation of* θ *about* O, *where* θ *and* O *are constructed as shown.*

The sense in which translations are "exceptional" is that if the two segments are drawn at random then it is very unlikely that they will be parallel. Indeed, given AB, a translation is only needed for one possible direction of A′B′ out of infinitely many, so the mathematical probability that a random direct motion is a translation is actually *zero!*

Direct transformations will be more important to us than opposite ones, so we relegate the investigation of opposite motions to Exs. 39, 40, 41. The reason for the greater emphasis on direct motions stems from the fact that they form a group (a *subgroup* of the full group of motions), while the opposite motions do not. Can you see why?

1.4.3 Three Reflections Theorem

In chemistry one is concerned with the interactions of atoms, but to gain deeper insights one must study the electrons, protons, and neutrons from which atoms are built. Likewise, though our concern is with direct motions, we will gain deeper insights by studying the opposite motions from which direct motions are built. More precisely,

$$\textit{Every direct motion is the composition of two reflections.} \tag{1.28}$$

Note that the second sentence of (1.25) then implies that *every opposite motion is the composition of three reflections*. See Ex. 39. In brief, every motion is the composition of either two or three reflections, a result that is called the *Three Reflections Theorem*[14].

Earlier we tried to show that the set of motions forms a group, but it was not clear that every motion had an inverse. The Three Reflections Theorem settles this neatly and explicitly, for the inverse of a sequence of reflections is obtained by reversing the order in which the reflections are performed.

In what follows, let $\mathfrak{R}_L$ denote reflection in a line L. Thus reflection in L_1 followed by reflection in L_2 is written $\mathfrak{R}_{L_2} \circ \mathfrak{R}_{L_1}$. According to (1.26), proving (1.28) amounts to showing that every rotation (and every translation) is of the form $\mathfrak{R}_{L_2} \circ \mathfrak{R}_{L_1}$. This is an immediate consequence of the following:

> *If* L_1 *and* L_2 *intersect at* O, *and the angle from* L_1 *to* L_2 *is* ϕ, *then* $\mathfrak{R}_{L_2} \circ \mathfrak{R}_{L_1}$ *is a rotation of* 2ϕ *about* O,

and

> *If* L_1 *and* L_2 *are parallel, and* **V** *is the perpendicular connecting vector from* L_1 *to* L_2, *then* $\mathfrak{R}_{L_2} \circ \mathfrak{R}_{L_1}$ *is a translation of* $2\mathbf{V}$.

Both these results are easy enough to prove directly [try it!], but the following is perhaps more elegant.

First, since $\mathfrak{R}_{L_2} \circ \mathfrak{R}_{L_1}$ is a direct motion (because it reverses angles twice), it is either a rotation or a translation. Second, note that rotations and translations may be distinguished by their *invariant curves*, that is, curves that are mapped into themselves. For a rotation about a point O, the invariant curves are circles centred at O, while for a translation they are lines parallel to the translation.

Now look at [1.29a]. Clearly $\mathfrak{R}_{L_2} \circ \mathfrak{R}_{L_1}$ leaves invariant any circle centred at O, so it is a rotation about O. To see that the angle of the rotation is 2ϕ, consider the image P' of any point P on L_1. Done.

Now look at [1.29b]. Clearly $\mathfrak{R}_{L_2} \circ \mathfrak{R}_{L_1}$ leaves invariant any line perpendicular to L_1 and L_2, so it is a translation parallel to such lines. To see that the translation is $2\mathbf{V}$, consider the image P' of any point P on L_1. Done.

Note that a rotation of θ can be represented as $\mathfrak{R}_{L_2} \circ \mathfrak{R}_{L_1}$, where L_1, L_2 is *any* pair of lines that pass through the centre of the rotation and that contain an angle $(\theta/2)$. Likewise, a translation of **T** corresponds to *any* pair of parallel lines separated by $\mathbf{T}/2$. This circumstance yields a very elegant method for composing rotations and translations.

For example, see [1.30a]. Here a rotation $\mathcal{R}_a^{\theta}$ about a through θ is being represented as $\mathfrak{R}_{L_2} \circ \mathfrak{R}_{L_1}$, and a rotation $\mathcal{R}_b^{\phi}$ about b through ϕ is being represented as $\mathfrak{R}_{L'_2} \circ \mathfrak{R}_{L'_1}$. To find the net effect of rotating about a and then about b, choose $L_2 = L'_1$ to be the line through a and b. If $\theta + \phi \neq 2\pi$, then L_1 and L'_2 will intersect at some

[14] Results such as (1.26) may instead be viewed as consequences of this theorem; see Stillwell (1992) for an elegant and elementary exposition of this approach.

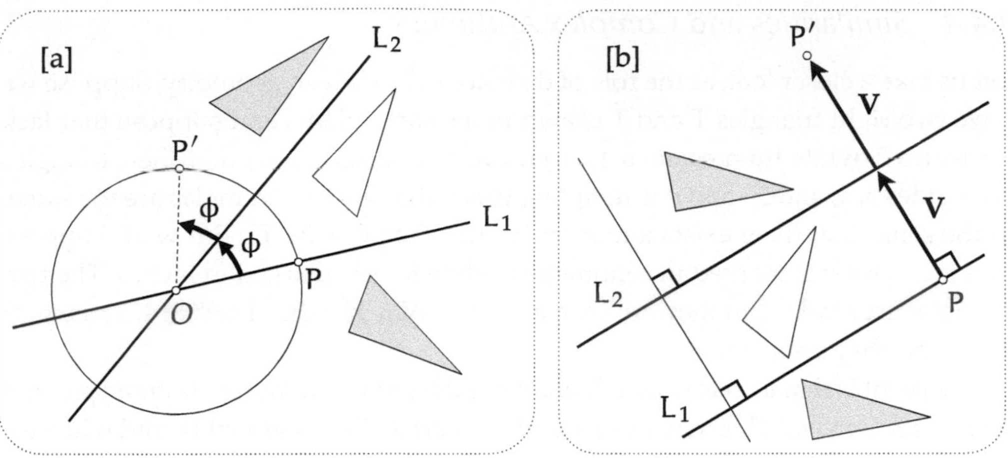

[1.29] The composition of two reflections is a rotation or a translation. [a] *If* L_1 *and* L_2 *intersect at* O, *and the angle from* L_1 *to* L_2 *is* ϕ, *then* $\mathfrak{R}_{L_2} \circ \mathfrak{R}_{L_1}$ *is a rotation of* 2ϕ *about* O. **[b]** *If* L_1 *and* L_2 *are parallel, and* $\boldsymbol{V}$ *is the perpendicular connecting vector from* L_1 *to* L_2, *then* $\mathfrak{R}_{L_2} \circ \mathfrak{R}_{L_1}$ *is a translation of* $2\boldsymbol{V}$.

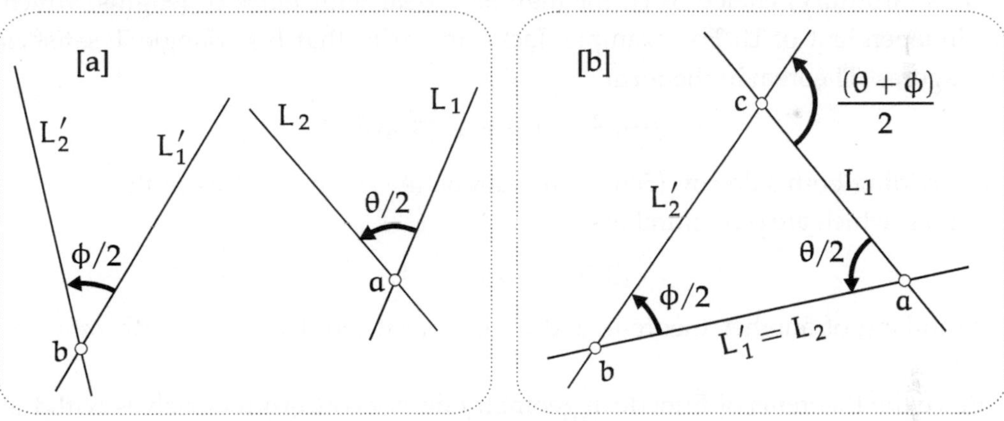

[1.30] Geometric reduction of two rotations to a single rotation. [a] *The composition of two rotations* $\mathcal{R}_b^{\phi} \circ \mathcal{R}_a^{\theta}$ *can be expressed as the composition of four reflections:* $(\mathfrak{R}_{L_2'} \circ \mathfrak{R}_{L_1'}) \circ (\mathfrak{R}_{L_2} \circ \mathfrak{R}_{L_1})$. **[b]** *But if we choose* $L_2 = L_1'$ *then the middle two reflections* cancel, *and we discover that* $\mathcal{R}_b^{\phi} \circ \mathcal{R}_a^{\theta} = \mathcal{R}_c^{\theta+\phi}$.

point c, as in [1.30b]. Thus the middle two reflections *cancel*, and the composition of the two rotations is given by

$$(\mathfrak{R}_{L_2'} \circ \mathfrak{R}_{L_1'}) \circ (\mathfrak{R}_{L_2} \circ \mathfrak{R}_{L_1}) = \mathfrak{R}_{L_2'} \circ \mathfrak{R}_{L_1},$$

which is a rotation $\mathcal{R}_c^{\theta+\phi}$ about c through $(\theta + \phi)$! That this construction agrees with our calculation on p. 20 is demonstrated in Ex. 36.

Further examples of this method may be found in Ex. 42 and Ex. 43.

1.4.4 Similarities and Complex Arithmetic

Let us take a closer look at the role of distance in Euclidean geometry. Suppose we have two right triangles T and $\widetilde{\mathrm{T}}$ drawn in the same plane, and suppose that Jack measures T while Jill measures $\widetilde{\mathrm{T}}$. If Jack and Jill both report that their triangles have sides 3, 4, and 5, then it is tempting to say that the two triangles are the same, in the sense that there exists a motion $\mathcal{M}$ such that $\widetilde{\mathrm{T}} = \mathcal{M}(\mathrm{T})$. But wait! Suppose that Jack's ruler is marked in centimetres, while Jill's is marked in inches. The two triangles are *similar*, but they are *not* congruent. Which is the "true" 3, 4, 5 triangle? Of course they both are.

The point is that *whenever we talk about distances numerically, we are presupposing a unit of measurement.* This may be pictured as a certain line-segment U, and when we say that some other segment has a length of 5, for example, we mean that precisely 5 copies of U can be fitted into it. But on our flat[15] plane any choice of U is as good as any other—there is no *absolute* unit of measurement, and our geometric theorems should reflect that fact.

Meditating on this, we recognize that Euclidean theorems do not in fact depend on this (arbitrary) choice of U, for they only deal with *ratios* of lengths, which are independent of U. For example, Jack can verify that his triangle T satisfies Pythagoras' Theorem in the form

$$(3\text{cm})^2 + (4\text{cm})^2 = (5\text{cm})^2,$$

but, dividing both sides by $(5\text{cm})^2$, this can be rewritten in terms of the ratios of the sides, which are pure numbers:

$$(3/5)^2 + (4/5)^2 = 1.$$

Try thinking of another theorem, and check that it too deals only with ratios of lengths.

Since the theorems of Euclidean geometry do not concern themselves with the actual sizes of figures, our earlier definition of geometric equality in terms of motions is clearly too restrictive: two figures should be considered the same if they are *similar*. More precisely, we now consider two figures to be the same if there exists a *similarity* mapping one to the other, where

A similarity $\mathcal{S}$ is a mapping of the plane to itself that preserves ratios of distances.

It is easy to see [exercise] that a given similarity $\mathcal{S}$ expands every distance by the same (non-zero) factor r, which we will call the *expansion* of $\mathcal{S}$. We can therefore refine our notation by including the expansion as a superscript, so that a general similarity of expansion r is written $\mathcal{S}^r$. Clearly, the identity transformation is a similarity, $\mathcal{S}^k \circ \mathcal{S}^r = \mathcal{S}^{kr}$, and $(\mathcal{S}^r)^{-1} = \mathcal{S}^{(1/r)}$, so it is fairly clear that the set of all

[15] In the non-Euclidean geometries of Chapter 6 we will be drawing on *curved* surfaces, and the amount of curvature in the surface *will* dictate an absolute unit of length.

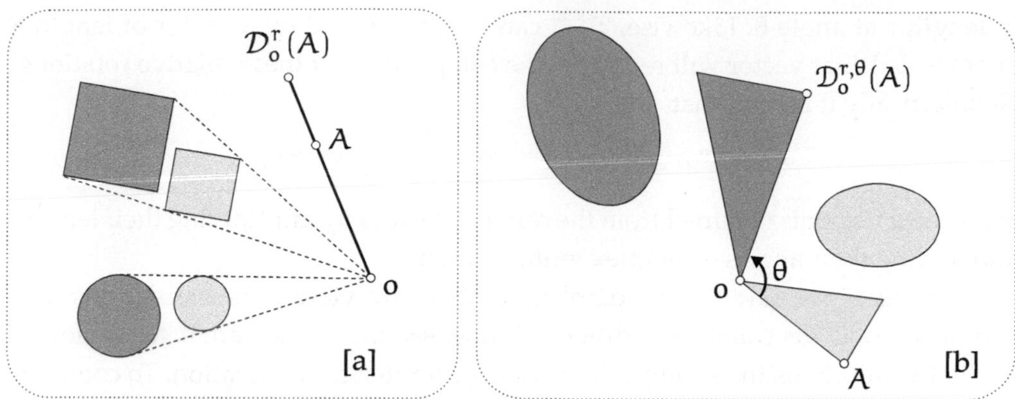

[1.31] Dilative rotations. [a] *A* central dilation $\mathcal{D}_o^r$ *leaves* o *fixed and radially stretches each segment* oA *by* r. **[b]** *If this central dilation is followed by (or preceded by) a rotation* $\mathcal{R}_o^\theta$ *with the same centre, then we obtain the* dilative rotation $\mathcal{D}_o^{r,\theta} \equiv \mathcal{R}_o^\theta \circ \mathcal{D}_o^r = \mathcal{D}_o^r \circ \mathcal{R}_o^\theta$.

similarities forms a group. We thus arrive at the definition of Euclidean geometry that Klein gave in his Erlangen address:

Euclidean geometry is the study of those properties of geometric figures that are invariant under the group of similarities. (1.29)

Since the motions are just the similarities $\mathcal{S}^1$ of unit expansion, the group of motions is a subgroup of the group of similarities; our previous attempt at defining Euclidean geometry therefore yields a "subgeometry" of (1.29).

A simple example of an $\mathcal{S}^r$ is a *central dilation* $\mathcal{D}_o^r$. As illustrated in [1.31a], this leaves o fixed and radially stretches each segment oA by r. Note that the inverse of a central dilation is another central dilation with the same centre: $(\mathcal{D}_o^r)^{-1} = \mathcal{D}_o^{(1/r)}$. If this central dilation is followed by (or preceded by) a rotation $\mathcal{R}_o^\theta$ with the same centre, then we obtain the *dilative rotation*

$$\mathcal{D}_o^{r,\theta} \equiv \mathcal{R}_o^\theta \circ \mathcal{D}_o^r = \mathcal{D}_o^r \circ \mathcal{R}_o^\theta,$$

shown in [1.31b]. Note that a central dilation may be viewed as a special case of a dilative rotation: $\mathcal{D}_o^r = \mathcal{D}_o^{r,0}$.

This figure should be ringing loud bells. Taking o to be the origin of $\mathbb{C}$, (1.9) says that $\mathcal{D}_o^{r,\theta}$ corresponds to multiplication by $r\,e^{i\theta}$:

$$\mathcal{D}_o^{r,\theta}(z) = \left(r\,e^{i\theta}\right) z.$$

Conversely, and this is the key point, *the rule for complex multiplication may be viewed as a consequence of the behaviour of dilative rotations.*

Concentrate on the set of dilative rotations with a common, fixed centre o, which will be thought of as the origin of the complex plane. Each $\mathcal{D}_o^{r,\theta}$ is uniquely determined by its expansion r and rotation θ, and so it can be represented by a vector of length r at angle θ. Likewise, $\mathcal{D}_o^{R,\phi}$ can be represented by a vector of length R at angle ϕ. What vector will represent the composition of these dilative rotations? Geometrically it is clear that

$$\mathcal{D}_o^{R,\phi} \circ \mathcal{D}_o^{r,\theta} = \mathcal{D}_o^{r,\theta} \circ \mathcal{D}_o^{R,\phi} = \mathcal{D}_o^{Rr,(\theta+\phi)},$$

so the new vector is obtained from the original vectors by multiplying their lengths and adding their angles—complex multiplication!

On page 19 we saw that if complex numbers are viewed as translations then composition yields complex addition. We now see that if they are instead viewed as dilative rotations then composition yields complex multiplication. To complete our "explanation" of complex numbers in terms of geometry, we will show that these translations and dilative rotations are fundamental to Euclidean geometry as defined in (1.29).

To understand the general similarity $\mathcal{S}^r$ involved in (1.29), note that if p is an arbitrary point, $\mathcal{M} \equiv \mathcal{S}^r \circ \mathcal{D}_p^{(1/r)}$ is a *motion*. Thus *any similarity is the composition of a dilation and a motion:*

$$\mathcal{S}^r = \mathcal{M} \circ \mathcal{D}_p^r. \tag{1.30}$$

Our classification of motions therefore implies that similarities come in two kinds: if $\mathcal{M}$ preserves angles then so will $\mathcal{S}^r$ [a *direct similarity*]; if $\mathcal{M}$ reverses angles then so will $\mathcal{S}^r$ [an *opposite similarity*].

Just as we concentrated on the group of direct motions, so we will now concentrate on the group of direct similarities. The fundamental role of translations and dilative rotations in Euclidean geometry finally emerges in the following surprising theorem:

Every direct similarity is a dilative rotation or (exceptionally) a translation. (1.31)

For us, at least, this fact constitutes one satisfying "explanation" of complex numbers; as mentioned in the Preface, other equally compelling explanations may be found in the laws of physics.

To begin to understand (1.31), observe that (1.25) and (1.30) imply that a direct similarity is determined by the image A′B′ of any line-segment AB. First consider the exceptional case in which A′B′ are of equal length AB. We then have the three cases in [1.28], all of which are consistent with (1.31). If A′B′ and AB are parallel but *not* of equal length, then we have the two cases shown in [1.32a] and [1.32b], in both of which we have drawn the lines AA′ and BB′ intersecting in p. By appealing to the similar triangles in these figures, we see that in [1.32a] the similarity is $\mathcal{D}_p^{r,0}$, while in [1.32b] it is $\mathcal{D}_p^{r,\pi}$, where in both cases $r = (pA'/pA) = (pB'/pB)$.

Now consider the much more interesting general case where A′B′ and AB are neither the same length, nor parallel. Take a peek at [1.32d], which illustrates this. Here n is the intersection point of the two segments (produced if necessary), and

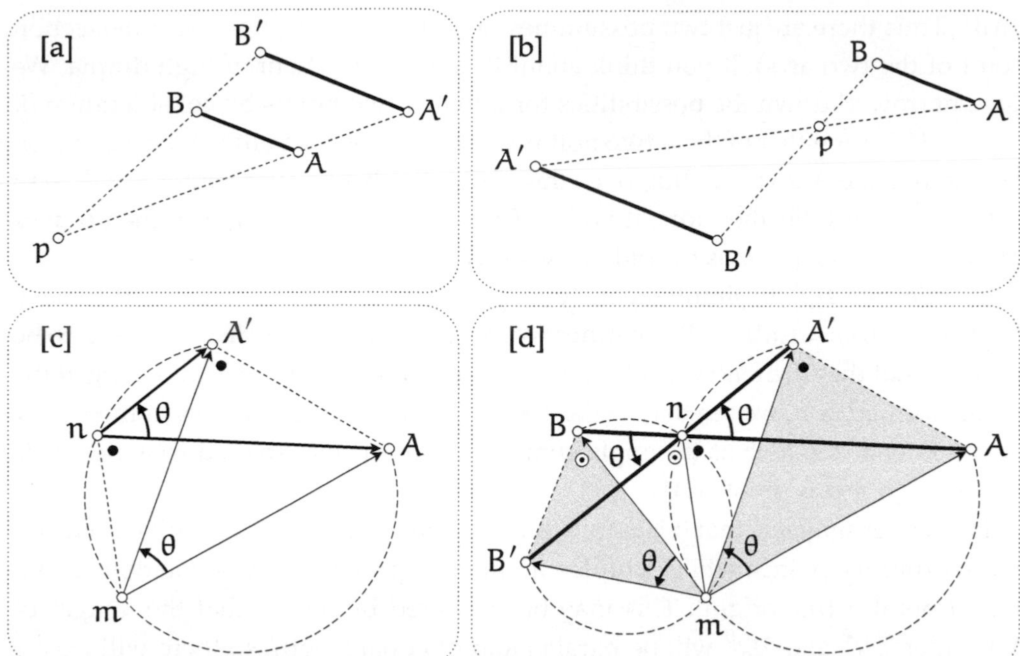

[1.32] Every direct similarity is a dilative rotation *(with the sole exception of an isometric translation). If a direct similarity* $\mathcal{S}$ *carries a line segment* AB *to a* parallel *segment* A′B′ *of different length, then either* **[a]** $\mathcal{S} = \mathcal{D}_p^{r,0}$, *or* **[b]** $\mathcal{S} = \mathcal{D}_p^{r,\pi}$, *where in both cases* $r = (pA'/pA) = (pB'/pB)$. **[c]** *For a given* θ, *there exists a* $\mathcal{D}_q^{r,\theta}$ mapping A to A′ if and only if q lies on the circular arc AnA′. **[d]** *If* $\mathcal{D}_q^{r,\theta}$ *is to simultaneously map* A *to* A′ and B *to* B′ *then* θ *must be the angle between* AB *and* A′B′, *so* q *must lie on the circular arc* AnA′ and *on the circular arc* BnB′, *so* q = n *or* q = m. *But clearly* q = n *fails, so* q = m *is our only hope. Miraculously, thanks to the similarity of the two shaded triangles,* $\mathcal{D}_m^{r,\theta}$ succeeds (!) *in simultaneously mapping* A *to* A′ *and* B *to* B′, *thereby completing the proof.*

θ is the angle between them. To establish (1.31), we must show that we can carry AB to A′B′ with a single dilative rotation. For the time being, simply note that if AB is to end up having the same direction as A′B′ then it must be rotated by θ, so the claim is really this: *There exists a point* q, *and an expansion factor* r, *such that* $\mathcal{D}_q^{r,\theta}$ *carries* A *to* A′ *and* B *to* B′.

Consider the part of [1.32d] that is reproduced in [1.32c]. Clearly, by choosing $r = (nA'/nA)$, $\mathcal{D}_n^{r,\theta}$ will map A to A′. More generally, you see that we can map A to A′ with $\mathcal{D}_q^{r,\theta}$ if and only if AA′ subtends angle θ at q. Thus, with the appropriate value of r, $\mathcal{D}_q^{r,\theta}$ *maps* A *to* A′ *if and only if* q *lies on the circular arc* AnA′. The figure illustrates one such position, q = m. Before returning to [1.32d], we need to notice one more thing: mA subtends the same angle (marked •) at n and A′.

Let us return to [1.32d]. We want $\mathcal{D}_q^{r,\theta}$ to map A to A′ and B to B′. According to the argument above, q must lie on the circular arc AnA′ *and* on the circular arc

BnB′. Thus there are just two possibilities: $q = n$ or $q = m$ (the other intersection point of the two arcs). If you think about it, this is a moment of high drama. We have narrowed down the possibilities for q to just two points by consideration of angles *alone*; for either of these two points we can choose the value of the expansion r so as to make A go to A′, but, once this choice has been made, either B will map to B′ or it won't! Furthermore, it is clear from the figure that if $q = n$ then B does *not* map to B′, so $q = m$ is the only possibility remaining.

In order for $\mathcal{D}_m^{r,\theta}$ to simultaneously map A to A′ and B to B′, we need to have $r = (mA'/mA) = (mB'/mB)$; in other words, the two shaded triangles need to be similar. That they are *indeed* similar is surely something of a miracle. Looking at the angles formed at n, we see that $\theta + \odot + \bullet = \pi$, and the result follows immediately by thinking of the RHS as the angle-sum of each of the two shaded triangles. This completes our proof[16] of (1.31).

The reader may feel that it is unsatisfactory that (1.31) calls for dilative rotations about arbitrary points, while complex numbers represent dilative rotations about a *fixed* point o (the origin). This may be answered by noting that the images of AB under $\mathcal{D}_q^{r,\theta}$ and $\mathcal{D}_o^{r,\theta}$ will be parallel and of equal length, so there will exist a translation [see Ex. 44 for details] $\mathcal{T}_v$ mapping one onto the other. In other words, a general dilative rotation differs from an origin-centred dilative rotation by a mere translation: $\mathcal{D}_q^{r,\theta} = \mathcal{T}_v \circ \mathcal{D}_o^{r,\theta}$. To sum up,

> *Every direct similarity $\mathcal{S}^r$ can be expressed as a complex function of the form* $\mathcal{S}^r(z) = re^{i\theta}z + v$.

1.4.5 Spatial Complex Numbers?

Let us briefly attempt to generalize the above ideas to *three*-dimensional space. Firstly, a central dilation of space (centred at O) is defined exactly as before, and a dilative rotation with the same centre is then the composition of such a dilation with a rotation of space about an axis passing through O. Once again taking (1.29) as the definition of Euclidean geometry, we get off to a flying start, because the key result (1.31) generalizes: *Every direct similarity of space is a dilative rotation, a translation, or the composition of a dilative rotation and a translation along its rotation axis.* See Coxeter (1969, p. 103) for details.

It is therefore natural to ask if there might exist "spatial complex numbers" for which addition would be composition of translations, and for which multiplication would be composition of dilative rotations. With addition all goes well: the position vector of each point in space may be viewed as a translation, and composition of

[16] The present argument has the advantage of proceeding in steps, rather than having to be discovered all at once. For other proofs, see Coxeter and Greitzer (1967, p. 97), Coxeter (1969, p. 73), and Eves (1992, p. 71). Also, see Ex. 45 for a simple proof using complex functions.

these translations yields ordinary vector addition in space. Note that this vector addition makes equally good sense in *four*-dimensional space, or n-dimensional space for that matter.

Now consider the set $\mathcal{Q}$ of dilative rotations with a common, fixed centre O. Initially, the definition of multiplication goes smoothly, for the "product" $Q_1 \circ Q_2$ of two such dilative rotations is easily seen to be another dilative rotation (Q_3, say) of the same kind. This follows from the above classification of direct similarities by noting that $Q_1 \circ Q_2$ leaves O fixed. If the expansions of Q_1 and Q_2 are r_1 and r_2 then the expansion of Q_3 is clearly $r_3 = r_1 r_2$, and in Chapter 6 we shall give a simple geometric construction for the rotation of Q_3 from the rotations of Q_1 and Q_2. However, unlike rotations in the plane, it makes a difference in what order we perform two rotations in space, so our multiplication rule is *not commutative*:

$$Q_1 \circ Q_2 \neq Q_2 \circ Q_1. \tag{1.32}$$

We are certainly *accustomed* to multiplication being commutative, but there is nothing inconsistent about (1.32), so this cannot be considered a decisive obstacle to an algebra of "spatial complex numbers".

However, a fundamental problem does arise when we try to represent these dilative rotations as points (or vectors) in space. By analogy with complex multiplication, we wish to interpret the equation $Q_1 \circ Q_2 = Q_3$ as saying that the dilative rotation Q_1 maps the point Q_2 to the point Q_3. But this interpretation is impossible! The specification of a point in space requires *three* numbers, but the specification of a dilative rotation requires *four:* one for the expansion, one for the angle of rotation, and two[17] for the direction of the axis of the rotation.

Although we have failed to find a three-dimensional analogue of complex numbers, we have discovered the *four*-dimensional space $\mathcal{Q}$ of dilative rotations (centred at O) of three-dimensional space. Members of $\mathcal{Q}$ are called *quaternions,* and they may be pictured as points or vectors in four dimensions, but the details of how to do this will have to wait till Chapter 6. Quaternions can be added by ordinary vector addition, and they can be multiplied using the non-commutative rule above (composition of the corresponding dilative rotations).

The discoveries of the rules for multiplying complex numbers and for multiplying quaternions have some interesting parallels. As is well known, the quaternion rule was discovered in *algebraic* form by Sir William Rowan Hamilton in 1843. It is less well known that three years earlier Olinde Rodrigues had published an elegant geometric investigation of the composition of rotations in space that contained

[17] To see this, imagine a sphere centred at O. The direction of the axis can be specified by its intersection with the sphere, and this point can be specified with two coordinates, e.g., longitude and latitude.

essentially the same result; only much later[18] was it recognized that Rodrigues' geometry was equivalent to Hamilton's algebra.

Hamilton and Rodrigues are just two examples of hapless mathematicians who would have been dismayed to examine the unpublished notebooks of the great Carl Friedrich Gauss. There, like just another log entry in the chronicle of his private mathematical voyages, Gauss recorded his discovery of the quaternion rule in 1819.

In Chapter 6 we shall investigate quaternion multiplication in detail and find that it has elegant applications. However, the immediate benefit of this discussion is that we can now see what a remarkable property it is of *two*-dimensional space that it is possible to interpret points *within it* as the fundamental Euclidean transformations *acting on it*.

[18] See Altmann (1989) for the intriguing details of how this was unravelled.

1.5 Exercises

1 The roots of a general cubic equation in X may be viewed (in the XY-plane) as the intersections of the X-axis with the graph of a cubic of the form,

$$Y = X^3 + AX^2 + BX + C.$$

(i) Show that the point of inflection of the graph occurs at $X = -\frac{A}{3}$.

(ii) Deduce (geometrically) that the substitution $X = \left(x - \frac{A}{3}\right)$ will reduce the above equation to the form $Y = x^3 + bx + c$.

(iii) Verify this by calculation.

2 In order to solve the cubic equation $x^3 = 3px + 2q$, do the following:

(i) Make the inspired substitution $x = s + t$, and deduce that x solves the cubic if $st = p$ and $s^3 + t^3 = 2q$.

(ii) Eliminate t between these two equations, thereby obtaining a quadratic equation in s^3.

(iii) Solve this quadratic to obtain the two possible values of s^3. By symmetry, what are the possible values of t^3?

(iv) Given that we know that $s^3 + t^3 = 2q$, deduce the formula (1.4).

3 In 1591, more than forty years after the appearance of (1.4), François Viète published another method of solving cubics. The method is based on the identity (see p. 17) $\cos 3\theta = 4C^3 - 3C$, where $C = \cos\theta$.

(i) Substitute $x = 2\sqrt{p}\,C$ into the (reduced) general cubic $x^3 = 3px + 2q$ to obtain $4C^3 - 3C = \frac{q}{p\sqrt{p}}$.

(ii) Provided that $q^2 \leqslant p^3$, deduce that the solutions of the original equation are

$$x = 2\sqrt{p}\cos\left[\tfrac{1}{3}(\phi + 2m\pi)\right],$$

where m is an integer and $\phi = \cos^{-1}(q/p\sqrt{p})$.

(iii) Check that this formula gives the correct solutions of $x^3 = 3x$, namely, $x = 0, \pm\sqrt{3}$.

4 Here is a basic fact about integers that has many uses in number theory: *If two integers can be expressed as the sum of two squares, then so can their product.* With the understanding that each symbol denotes an integer, this says that if $M = a^2 + b^2$ and $N = c^2 + d^2$, then $MN = p^2 + q^2$. Prove this result by considering $|(a + ib)(c + id)|^2$.

5 The figure below shows how two similar triangles may be used to construct the product of two complex numbers. Explain this.

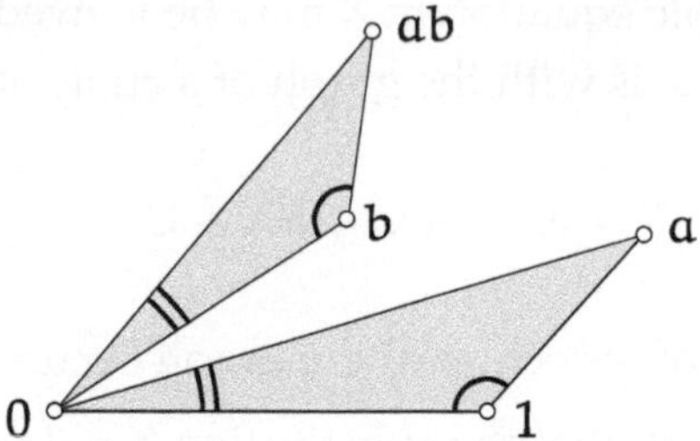

6 (i) If c is a fixed complex number, and R is a fixed real number, explain with a picture why $|z - c| = R$ is the equation of a circle.

(ii) Given that z satisfies the equation $|z + 3 - 4i| = 2$, find the minimum and maximum values of $|z|$, and the corresponding positions of z.

7 Use a picture to show that if a and b are fixed complex numbers then $|z - a| = |z - b|$ is the equation of a line.

8 Let L be a straight line in $\mathbb{C}$ making an angle ϕ with the real axis, and let d be its distance from the origin. Show geometrically that if z is any point on L then

$$d = \left|\text{Im}[e^{-i\phi} z]\right|.$$

[*Hint:* Interpret $e^{-i\phi}$ using (1.9).]

9 Let A, B, C, D be four points on the unit circle. If $A + B + C + D = 0$, show that the points must form a rectangle.

10 Show geometrically that if $|z| = 1$ then

$$\text{Im}\left[\frac{z}{(z+1)^2}\right] = 0.$$

Apart from the unit circle, what other points satisfy this equation?

11 Explain geometrically why the locus of z such that

$$\arg\left(\frac{z-a}{z-b}\right) = \text{const.}$$

is an arc of a certain circle passing through the fixed points a and b.

12 By using pictures, find the locus of z for each of the following equations:

$$\text{Re}\left(\frac{z-1-i}{z+1+i}\right) = 0, \quad \text{and} \quad \text{Im}\left(\frac{z-1-i}{z+1+i}\right) = 0.$$

[*Hints:* What does $\text{Re}(W) = 0$ imply about the angle of W? Now use the previous exercise.]

13 Find the geometric configuration of the points a, b, and c if

$$\left(\frac{b-a}{c-a}\right) = \left(\frac{a-c}{b-c}\right).$$

[*Hint:* Separately equate the lengths and angles of the two sides.]

14 By considering the product $(2+i)(3+i)$, show that

$$\frac{\pi}{4} = \tan^{-1}\frac{1}{2} + \tan^{-1}\frac{1}{3}.$$

15 Draw $e^{i\pi/4}$, $e^{i\pi/2}$, and their sum. By expressing each of these numbers in the form $(x+iy)$, deduce that

$$\tan\frac{3\pi}{8} = 1+\sqrt{2}.$$

16 Starting from the origin, go one unit east, then the same length north, then (1/2) of the previous length west, then (1/3) of the previous length south, then (1/4) of the previous length east, and so on. What point does this "spiral" converge to?

17 If $z = e^{i\theta} \neq -1$, then $(z-1) = (i\tan\frac{\theta}{2})(z+1)$. Prove this (i) by calculation, (ii) with a picture.

18 Prove that

$$e^{i\theta} + e^{i\phi} = 2\cos\left[\tfrac{\theta-\phi}{2}\right] e^{\frac{i(\theta+\phi)}{2}} \quad \text{and} \quad e^{i\theta} - e^{i\phi} = 2i\sin\left[\tfrac{\theta-\phi}{2}\right] e^{\frac{i(\theta+\phi)}{2}}$$

(i) by calculation, and (ii) with a picture.

19 The "centroid" G of a triangle T is the intersection of its medians. If the vertices are the complex numbers a, b, and c, then you may assume that

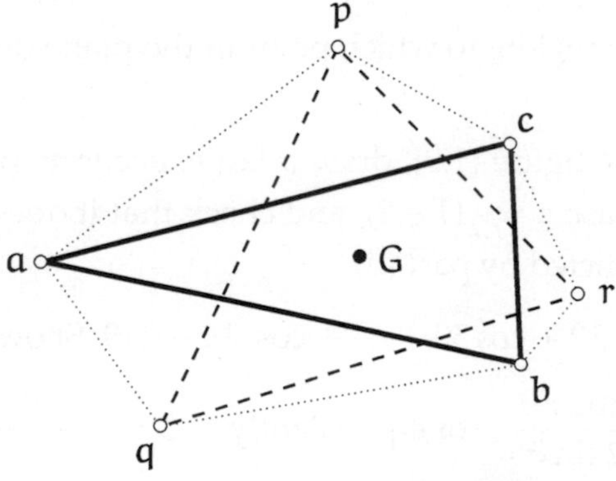

$$G = \tfrac{1}{3}(a+b+c).$$

On the sides of T we have constructed three *similar* triangles [dotted] of arbitrary shape, so producing a new triangle [dashed] with vertices p, q, r. Using complex algebra, show that the centroid of the new triangle is in exactly the same place as the centroid of the old triangle!

20 *Gaussian integers* are complex numbers of the form $m + in$, where m and n are integers—they are the grid points in [1.1]. Show that it is impossible to draw an equilateral triangle such that all three vertices are Gaussian integers. [*Hints*: You may assume that one of the vertices is at the origin; try a proof by contradiction; if a triangle is equilateral, you can rotate one side into another; remember that $\sqrt{3}$ is irrational.]

21 Make a copy of [1.12a], draw in the diagonal of the quadrilateral shown in [1.12b], and mark its midpoint m. As in [1.12b], draw the line-segments connecting m to p, q, r, and s. According to the result in [1.12b], what happens to p and to r under a rotation of $(\pi/2)$ about m? So what happens to the line-segment pr? Deduce the result shown in [1.12a].

22 Will the result in [1.12a] survive if the squares are instead constructed on the inside of the quadrilateral?

23 Draw an arbitrary triangle, and on each side draw an equilateral triangle lying outside the given triangle. What do you suspect is special about the new triangle formed by joining the centroids (cf. Ex. 19) of the equilateral triangles? Use complex algebra to prove that you are right. What happens if the equilateral triangles are instead drawn on the inside of the given triangle?

24 From (1.15), we know that

$$1 + z + z^2 + \cdots + z^{n-1} = \frac{z^n - 1}{z - 1}.$$

(i) In what region of $\mathbb{C}$ must z lie in order that the *infinite* series $1+z+z^2+\cdots$ converges?

(ii) If z lies in this region, to which point in the plane does the infinite series converge?

(iii) In the spirit of figure [1.9], draw a large, accurate picture of the infinite series in the case $z = \frac{1}{2}(1 + i)$, and check that it does indeed converge to the point predicted by part (ii).

25 Let $S = \cos\theta + \cos 3\theta + \cos 5\theta + \cdots + \cos(2n-1)\theta$. Show that

$$S = \frac{\sin 2n\theta}{2\sin\theta} \quad \text{or equivalently} \quad S = \frac{\sin n\theta \cos n\theta}{\sin\theta}.$$

[*Hint:* Use Ex. 24, then Ex. 18 to simplify the result.]

26 (i) By considering $(a + ib)(\cos\theta + i\sin\theta)$, show that

$$b\cos\theta + a\sin\theta = \sqrt{a^2 + b^2}\,\sin\left[\theta + \tan^{-1}(b/a)\right].$$

(ii) Use this result to prove (1.14) by the method of induction.

27 Show that the polar equation of the spiral $Z(t) = e^{at}e^{ibt}$ in [1.15b] is $r = e^{(a/b)\theta}$.

28 Reconsider the spiral $Z(t) = e^{at}e^{ibt}$ in [1.15b], where a and b are fixed real numbers. Let τ be a variable real number. According to (1.9), $z \mapsto \mathcal{F}_\tau(z) = (e^{a\tau}e^{ib\tau})\, z$ is an expansion of the plane by factor $e^{a\tau}$, combined with a rotation of the plane through angle $b\tau$.

(i) Show that $\mathcal{F}_\tau[Z(t)] = Z(t+\tau)$, and deduce that the spiral is an *invariant curve* (cf. p. 42) of the transformations $\mathcal{F}_\tau$.

(ii) Use this to give a calculus-free demonstration that all rays from the origin cut the spiral at the same angle.

(iii) Show that if the spiral is rotated about the origin through an arbitrary angle, the new spiral is again an invariant curve of each $\mathcal{F}_\tau$.

(iv) Argue that the spirals in the previous part are the *only* invariant curves of $\mathcal{F}_\tau$.

29 (i) If $V(t)$ is the complex velocity of a particle whose orbit is $Z(t)$, and dt is an infinitesimal moment of time, then $V(t)\, dt$ is a complex number along the orbit. Thinking of the integral as the (vector) sum of these movements, what is the geometric interpretation of $\int_{t_1}^{t_2} V(t)\, dt$?

(ii) Referring to [1.15b], sketch the curve $Z(t) = \frac{1}{a+ib}\, e^{at}e^{ibt}$.

(iii) Given the result (1.13), what is the velocity of the particle in the previous part.

(iv) Combine the previous parts to deduce that $\int_0^1 e^{at}e^{ibt}dt = \left[\frac{1}{a+ib}\, e^{at}e^{ibt}\right]_0^1$, and draw in this complex number in your sketch for part (ii).

(v) Use this to deduce that

$$\int_0^1 e^{at} \cos bt\, dt = \frac{a\,(e^a \cos b - 1) + b\, e^a \sin b}{a^2 + b^2},$$

and

$$\int_0^1 e^{at} \sin bt\, dt = \frac{b\,(1 - e^a \cos b) + a\, e^a \sin b}{a^2 + b^2}.$$

30 Given two starting numbers S_1, S_2, let us build up an infinite sequence $S_1, S_2, S_3, S_4, \ldots$ with this rule: *each new number is twice the difference of the previous two.* For example, if $S_1 = 1$ and $S_2 = 4$, we obtain $1, 4, 6, 4, -4, -16, -24, \ldots$. Our aim is to find a formula for the nth number S_n.

(i) Our generating rule can be written succinctly as $S_{n+2} = 2(S_{n+1} - S_n)$. Show that $S_n = z^n$ will solve this *recurrence relation* if $z^2 - 2z + 2 = 0$.

(ii) Use the quadratic formula to obtain $z = 1 \pm i$, and show that if A and B are arbitrary complex numbers, $S_n = A(1+i)^n + B(1-i)^n$ is a solution of the recurrence relation.

(iii) If we want only real solutions of the recurrence relation, show that $B = \overline{A}$, and deduce that $S_n = 2\,\text{Re}[A(1+i)^n]$.

(iv) Show that for the above example $A = -(1/2) - i$, and by writing this in polar form deduce that $S_n = 2^{n/2}\sqrt{5}\cos\left[\frac{(n+4)\pi}{4} + \tan^{-1} 2\right]$.

(v) Check that this formula predicts $S_{34} = 262144$, and use a computer to verify this.

[Note that this method can be applied to any recurrence relation of the form $S_{n+2} = pS_{n+1} + qS_n$.]

31 With the same recurrence relation as in the previous exercise, use a computer to generate the first 30 members of the sequence given by $S_1 = 2$ and $S_2 = 4$. Note the repeating pattern of zeros.

(i) With the same notation as before, show that this sequence corresponds to $A = -i$, so that $S_n = 2\,\text{Re}[-i(1+i)^n]$.

(ii) Draw a sketch showing the locations of $-i(1+i)^n$ for $n = 1$ to $n = 8$, and hence explain the pattern of zeros.

(iii) Writing $A = a + ib$, our example corresponds to $a = 0$. More generally, explain geometrically why such a repeating pattern of zeros will occur if and only if $(a/b) = 0, \pm 1$ or $b = 0$.

(iv) Show that $\frac{S_1}{S_2} = \frac{1}{2}\left[1 - \frac{a}{b}\right]$, and deduce that a repeating pattern of zeros will occur if and only if $S_2 = 2S_1$ (as in our example), $S_1 = S_2$, $S_1 = 0$, or $S_2 = 0$.

(v) Use a computer to verify these predictions.

32 The Binomial Theorem says that if n is a positive integer,

$$(a+b)^n = \sum_{r=0}^{n} \binom{n}{r} a^{n-r} b^r, \quad \text{where} \quad \binom{n}{r} = \frac{n!}{(n-r)!\,r!}$$

are the binomial coefficients [*not vectors!*]. The algebraic reasoning leading to this result is equally valid if a and b are *complex* numbers. Use this fact to show that if $n = 2m$ is even then

$$\binom{2m}{1} - \binom{2m}{3} + \binom{2m}{5} - \cdots + (-1)^{m+1}\binom{2m}{2m-1} = 2^m \sin\left(\frac{m\pi}{2}\right).$$

33 Consider the equation $(z-1)^{10} = z^{10}$.

(i) Without attempting to solve the equation, show geometrically that all 9 solutions [why not 10?] must lie on the vertical line, $\text{Re}\,(z) = \frac{1}{2}$. [*Hint:* Ex. 7.]

(ii) Dividing both sides by z^{10}, the equation takes the form $w^{10} = 1$, where $w = (z-1)/z$. Hence solve the original equation.

(iii) Express these solutions in the form $z = x+iy$, and thereby verify the result in (i). [*Hint:* To do this neatly, use Ex. 18.]

34 Let S denote the set of 12th roots of unity shown in [1.19], one of which is $\xi = e^{i(\pi/6)}$. Note that ξ is a *primitive* 12th root of unity, meaning that its powers yield *all* the 12th roots of unity: $S = \{\xi, \xi^2, \xi^3, \ldots, \xi^{12}\}$.

(i) Find all the primitive 12th roots of unity, and mark them on a copy of [1.19].

(ii) Write down, in the form of (1.16), the factorization of the polynomial $\Phi_{12}(z)$ whose roots are the primitive 12th roots of unity. [In general, $\Phi_n(z)$ is the polynomial (with the coefficient of the highest power of z equal to 1) whose roots are the primitive n^{th} roots of unity; it is called the nth *cyclotomic polynomial*.]

(iii) By first multiplying out pairs of factors corresponding to conjugate roots, show that $\Phi_{12}(z) = z^4 - z^2 + 1$.

(iv) By repeating the above steps, show that $\Phi_8(z) = z^4 + 1$.

(v) For a general value of n, explain the fact that if ζ is a primitive n^{th} root of unity, then so is $\overline{\zeta}$. Deduce that if $n > 2$ then $\Phi_n(z)$ always has even degree and real coefficients.

(vi) Show that if p is a prime number then $\Phi_p(z) = 1 + z + z^2 + \cdots + z^{p-1}$. [*Hint:* Ex. 24.]

[In these examples it is striking that $\Phi_n(z)$ has integer coefficients. In fact it can be shown that this is true for *every* $\Phi_n(z)$! For more on these fascinating polynomials, see Stillwell (1994).]

35 Show algebraically that the formula (1.21) is invariant under a translation by k, i.e., its value does not change if a becomes $a+k$, b becomes $b+k$, etc. Deduce from [1.22a] that the formula always gives the area of the quadrilateral. [*Hint:* Remember, $(z+\overline{z})$ is always real.]

36 According to the calculation on p. 20, $\mathcal{R}_b^{\phi} \circ \mathcal{R}_a^{\theta} = \mathcal{R}_c^{(\theta+\phi)}$, where

$$c = \frac{ae^{i\phi}(1-e^{i\theta}) + b(1-e^{i\phi})}{1-e^{i(\theta+\phi)}}.$$

Let us check that this c is the same as the one given by the geometric construction in [1.30b].

(i) Explain why the geometric construction is equivalent to saying that c satisfies the two conditions

$$\arg\left[\frac{c-b}{a-b}\right] = \tfrac{1}{2}\phi \quad \text{and} \quad \arg\left[\frac{c-a}{b-a}\right] = -\tfrac{1}{2}\theta.$$

(ii) Verify that the calculated value of c (given above) satisfies the first of these conditions by showing that

$$\frac{c-b}{a-b}=\left[\frac{\sin\frac{\theta}{2}}{\sin\frac{(\theta+\phi)}{2}}\right]e^{i\phi/2}. \tag{1.33}$$

[*Hint*: Use $(1-e^{i\alpha})=-2i\sin(\alpha/2)\,e^{i\alpha/2}$.]

(iii) In the same way, verify that the second condition is also satisfied.

37 Deduce (1.33) directly from [1.30b]. [*Hint*: Draw in the altitude through b of the triangle abc, and express its length first in terms of $\sin\frac{\theta}{2}$, then in terms of $\sin\frac{(\theta+\phi)}{2}$.]

38 On page 20 we calculated that for any non-zero α, $\mathcal{T}_v\circ\mathcal{R}_0^\alpha$ is a rotation:

$$\mathcal{T}_v\circ\mathcal{R}_0^\alpha=\mathcal{R}_c^\alpha,\quad\text{where}\quad c=v/(1-e^{i\alpha}).$$

However, if $\alpha=0$ then $\mathcal{T}_v\circ\mathcal{R}_0^\alpha=\mathcal{T}_v$ is a translation. Try to reconcile these facts by considering the behaviour of $\mathcal{R}_c^\alpha$ in the limit that α tends to zero.

39 A *glide reflection* is the composition $\mathcal{T}_v\circ\mathfrak{R}_L=\mathfrak{R}_L\circ\mathcal{T}_v$ of reflection in a line L and a translation v in the direction of L. For example, if you walk at a steady pace in the snow, your tracks can be obtained by repeatedly applying the same glide reflection to a single footprint. Clearly, a glide reflection is an opposite motion.

(i) Draw a line L, a line-segment AB, the image $\widetilde{A}\widetilde{B}$ of the segment under $\mathfrak{R}_L$, and the image A′B′ of AB under the glide reflection $\mathcal{T}_v\circ\mathfrak{R}_L$.

(ii) Suppose you erased L from your picture; by considering the line-segments AA′ and BB′, show that you can reconstruct L.

(iii) Given any two segments AB and A′B′ of equal length, use the previous part to construct the glide reflection that maps the former to the latter.

(iv) Deduce that every opposite motion is a glide reflection.

(v) Express a glide reflection as the composition of three reflections.

40 Let L be a line making angle ϕ (or $\phi+\pi$) with the real axis, and let p be the point on L that is closest to the origin, so that $|p|$ is the distance to the line. Consider the glide reflection [cf. previous exercise] $G=\mathcal{T}_v\circ\mathfrak{R}_L$, where the translation is through distance r parallel to L. Let us fix the value of ϕ by writing $v=+r\,e^{i\phi}$.

(i) Use a picture to show that $p=\pm i|p|\,e^{i\phi}$, and explain the geometric significance of the $\pm$.

(ii) What transformation is represented by the complex function $H(z)=\overline{z}+r$?

(iii) Use pictures to explain why $G=\mathcal{T}_p\circ\mathcal{R}_0^\phi\circ H\circ\mathcal{R}_0^{-\phi}\circ\mathcal{T}_{-p}$.

(iv) Deduce that $G(z)=e^{i2\phi}\,\overline{z}+e^{i\phi}\,(r\pm 2i|p|)$.

(v) Hence describe (in geometric terms) the glide reflection represented by $G(z) = i\bar{z} + 4i$. Check your answer by looking at the images of -2, $2i$, and 0.

41 Let $\widetilde{\mathcal{M}}(z)$ be the representation of a general opposite motion as a complex function.

(i) Explain why $\overline{\widetilde{\mathcal{M}}(z)}$ is a direct motion, and deduce from (1.27) that $\widetilde{\mathcal{M}}(z) = e^{i\alpha}\,\bar{z} + w$, for some α and w.

(ii) Using the previous exercise, deduce that every opposite motion is a glide reflection.

42 On p. 20 we calculated that if $(\theta + \phi) = 2\pi$ then

$$\mathcal{R}_b^{\phi} \circ \mathcal{R}_a^{\theta} = \mathcal{T}_v, \quad \text{where} \quad v = (1 - e^{i\phi})(b - a).$$

(i) Let $Q = (b - a)$ be the complex number from the first centre of rotation to the second. Show algebraically that v has length $2\sin(\theta/2)\,|Q|$, and that its direction makes an angle of $\left(\frac{\pi-\theta}{2}\right)$ with Q.

(ii) Give direct geometric proofs of these results by redrawing figure [1.30b] in the case $(\theta + \phi) = 2\pi$.

43 On p. 20 we calculated that

$$\mathcal{T}_v \circ \mathcal{R}_0^{\alpha} = \mathcal{R}_c^{\alpha}, \quad \text{where} \quad c = v/(1 - e^{i\alpha}).$$

(i) Show algebraically that the complex number from the old centre of rotation (the origin) to the new centre of rotation (c) has length $\frac{|v|}{2\sin(\alpha/2)}$, and that its direction makes an angle of $\left(\frac{\pi-\alpha}{2}\right)$ with v.

(ii) Representing both $\mathcal{R}_0^{\alpha}$ and $\mathcal{T}_v$ as the composition of two reflections, use the idea in [1.30b] to give direct, geometric proofs of these results.

44 Just as in [1.13b], a dilative rotation $\mathcal{D}_p^{r,\theta}$ centred at an arbitrary point p may be performed by translating p to the origin, doing $\mathcal{D}_o^{r,\theta}$, then translating o back to p. Representing these transformations as complex functions, show that

$$\mathcal{D}_p^{r,\theta}(z) = r\,e^{i\theta}z + v, \quad \text{where} \quad v = p(1 - r\,e^{i\theta}).$$

Conversely, if v is given, deduce that

$$\mathcal{T}_v \circ \mathcal{D}_o^{r,\theta} = \mathcal{D}_p^{r,\theta}, \quad \text{where} \quad p = v/(1 - r\,e^{i\theta}).$$

45 In the previous exercise you showed that an arbitrary dilative rotation or translation can be written as a complex function of the form $f(z) = az + b$, and, conversely, that every such function represents a unique dilative rotation or translation.

(i) Given two pairs of distinct points $\{A, B\}$ and $\{A', B'\}$, show [by finding them explicitly] that a and b exist such that $f(A) = A'$ and $f(B) = B'$.

(ii) Deduce the result (1.31).

CHAPTER 2

Complex Functions as Transformations

2.1 Introduction

A complex function f is a rule that assigns to a complex number z an image complex number $w = f(z)$. In order to investigate such functions it is essential that we be able to visualize them. Several methods exist for doing this, but (until Chapter 10) we shall focus almost exclusively on the method introduced in the previous chapter. That is, we shall view z and its image w as points in the complex plane, so that f becomes a *transformation of the plane*.

Conventionally, the image points w are drawn on a fresh copy of $\mathbb{C}$, called the *image plane* or the *w-plane*. This convention is illustrated in [2.1], which depicts the transformation $z \mapsto w = f(z) = (1 + i\sqrt{3})z = 2\,e^{i\pi/3}\,z$ (cf. figure [1.5], p. 9).

Usually, the real and imaginary parts of z are denoted x and y, and those of the image point w are denoted u and v, so that $w = f(z) = u(z) + iv(z)$, where $u(z)$ and $v(z)$ are real functions of z. The precise forms of these functions will depend on whether we describe z with Cartesian or polar coordinates. For instance, writing $z = x + iy$ in the above example yields

$$u(x+iy) = x - \sqrt{3}\,y \qquad \text{and} \qquad v(x+iy) = \sqrt{3}\,x + y,$$

while writing $z = r\,e^{i\theta}$ and $(1 + i\sqrt{3}) = 2e^{i\pi/3}$ yields

$$u(r\,e^{i\theta}) = 2r\cos\left[\theta + \tfrac{\pi}{3}\right] \qquad \text{and} \qquad v(r\,e^{i\theta}) = 2r\sin\left[\theta + \tfrac{\pi}{3}\right].$$

Of course we may also describe the w-plane with polar coordinates so that $w = f(z) = R\,e^{i\phi}$, where $R(z)$ and $\phi(z)$ are real functions of z. With the same example as before, the transformation becomes

$$R(r\,e^{i\theta}) = 2r \qquad \text{and} \qquad \phi(r\,e^{i\theta}) = \theta + \tfrac{\pi}{3}.$$

Visual Complex Analysis. 25th Anniversary Edition. Tristan Needham, Oxford University Press.
 DOI: 10.1093/oso/9780192868916.003.0002

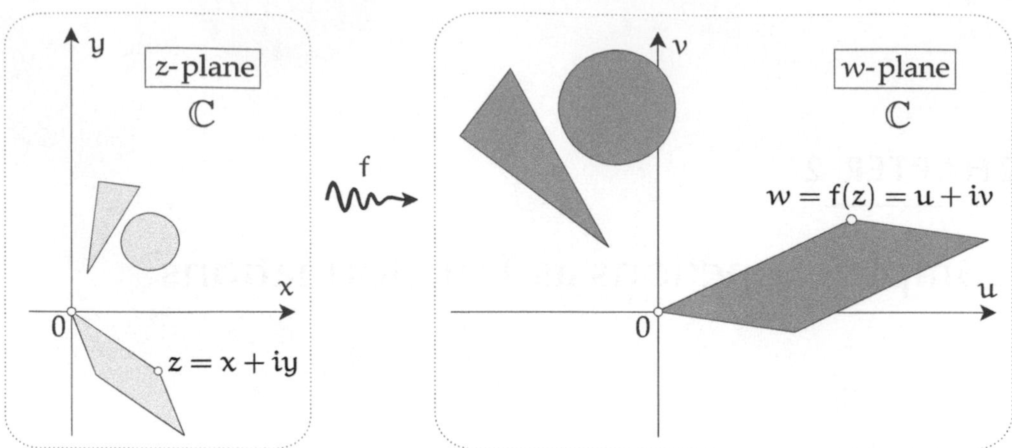

[2.1] *A complex function* $f(z)$ *is most commonly viewed as a mapping of points in one copy of* $\mathbb{C}$ *(the "z-plane") to points in another copy of* $\mathbb{C}$ *(the "w-plane"):* $z \mapsto w = f(z)$. *Here,* $f(z) = (1 + i\sqrt{3})z = 2\,e^{i\pi/3}\,z$ *is the combined effect of an expansion by factor* 2 *and a rotation by angle* $(\pi/3)$. *Most commonly, we write* $z = x + iy$ *and* $w = u + iv$.

We shall find that we can gain considerable insight into a given f by drawing pictures showing its effect on points, curves, and shapes. However, it would be nice if we could simultaneously grasp the behaviour of f for *all* values of z. One such method is to instead represent f as a *vector field*, whereby $f(z)$ is depicted as a vector emanating from the point z; for more detail, the reader is invited to read the beginning of Chapter 10.

Yet other methods are based on the idea of a graph. In the case of a real function $f(x)$ of a real variable x we are accustomed to the convenience of visualizing the overall behaviour of f by means of its graph, i.e., the curve in the two-dimensional xy-plane made up of the points $(x, f(x))$. In the case of a complex function this approach does not seem viable because to depict the pair of complex numbers $(z, f(z))$ we would need *four* dimensions: two for $z = x+iy$ and two for $f(z) = u+iv$.

Actually, the situation is not quite as hopeless as it seems. First, note that although two-dimensional space is needed to draw the graph of a real function f, the graph itself [the set of points $(x, f(x))$] is only a one-dimensional curve, meaning that only one real number (namely x) is needed to identify each point within it. Likewise, although four-dimensional space is needed to draw the set of points with coordinates $(x, y, u, v) = (z, f(z))$, the graph itself is *two*-dimensional, meaning that only two real numbers (namely x and y) are needed to identify each point within it. Thus, intrinsically, the graph of a complex function is merely a two-dimensional surface (a so-called *Riemann surface*), and it is thus susceptible to visualization in ordinary three-dimensional space. This approach will not be explored in this book, though the last three chapters in particular should prove helpful in understanding Riemann's original physical insights, as expounded by Klein (1881). See also

Springer (1981, Ch. 1), which essentially reproduces Klein's monograph, but with additional helpful commentary.

There is another type of graph of a complex function that is sometimes useful. The image $f(z)$ of a point z may be described by its distance $|f(z)|$ from the origin, and the angle $\arg[f(z)]$ it makes with the real axis. Let us discard half of this information (the angle) and try to depict how the modulus $|f(z)|$ varies with z. To do so, imagine the complex z-plane lying horizontally in space, and construct a point at height $|f(z)|$ vertically above each point z in the plane, thereby producing a surface called the *modular surface of f*. Figure [2.2] illustrates the conical modular surface of $f(z) = z$, while [2.3] illustrates the paraboloid modular surface of $f(z) = z^2$.

A note on computers. Beginning in this chapter, we will often suggest that you use a computer to expand your understanding of the mathematical phenomenon

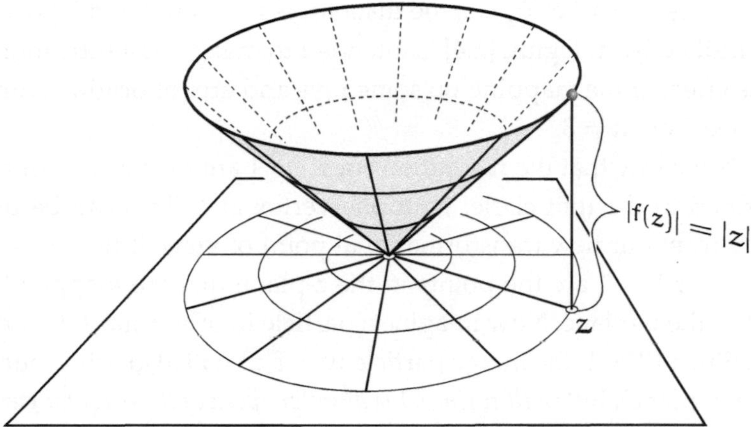

[2.2] *The height above* $\mathbb{C}$ *of the* **modular surface** *of a complex function* $f(z)$ *is its modulus,* $|f(z)|$. *Here,* $f(z) = z$*, producing a cone.*

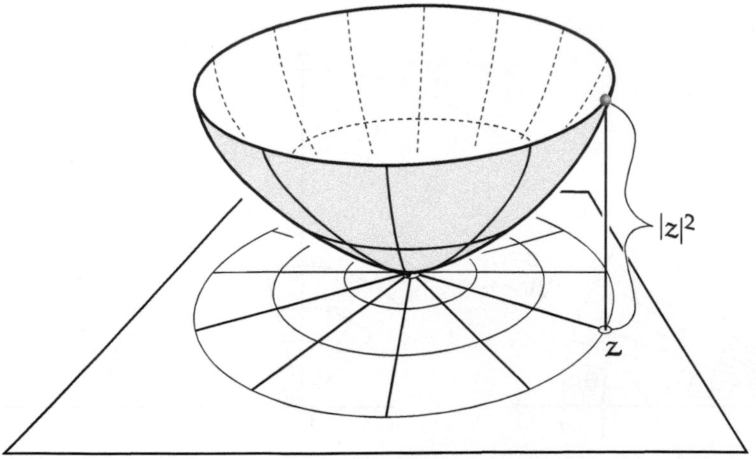

[2.3] *The paraboloid modular surface of* $f(z) = z^2$.

under discussion. However, we wish to stress that the *specific* uses of the computer that we have suggested in the text are only a beginning. Think of the computer as a physicist would his laboratory—you may use it to check your existing ideas about the construction of the world, or as a tool for discovering new phenomena which then demand new ideas for their explanation. In the Preface we make concrete suggestions (probably of only fleeting relevance) as to how your laboratory should be equipped.

2.2 Polynomials

2.2.1 Positive Integer Powers

Consider the mapping $z \mapsto w = z^n$, where n is a positive integer. Writing $z = r\,e^{i\theta}$ this becomes $w = r^n\,e^{in\theta}$, i.e., the distance is raised to the nth power and the angle is multiplied by n. Figure [2.4] is intended to make this a little more vivid by showing the effect of the mapping on some rays and arcs of origin-centred circles. As you can see, here $n = 3$.

On page 29 we saw that the n solutions of $z^n = 1$ are the vertices of the regular n-gon inscribed in the unit circle, with one vertex at 1. This can be understood more vividly from our new transformational point of view. If $w = f(z) = z^n$ then the solutions of $z^n = 1$ are the points in the z-plane that are mapped by f to the point $w = 1$ in the w-plane. Now imagine a particle in orbit round the unit circle in the z-plane. Since $1^n = 1$, the image particle $w = f(z)$ will also orbit round the unit circle (in the w-plane), but *with n times the angular speed of the original particle*. Thus each time z executes $(1/n)$ of a revolution, w will execute a complete revolution and return to the same image point. The preimages of any given w on the unit

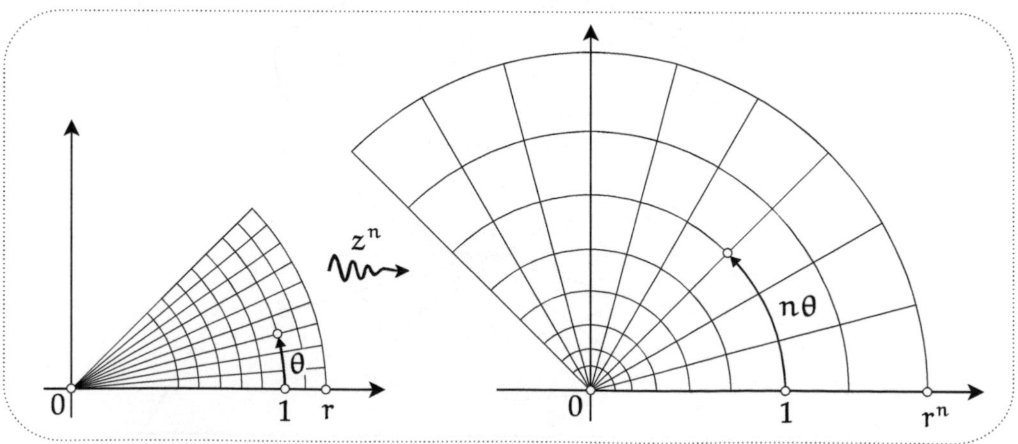

[2.4] *The geometric effect of the power mapping, $z \mapsto z^n$.*

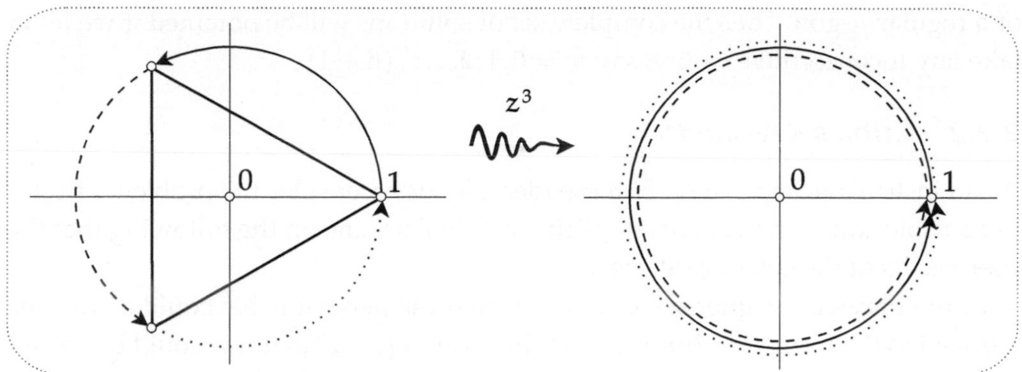

[2.5] *As z rotates around the unit circle, starting at 1, its image under $z \mapsto z^3$ rotates three times as fast, executing a complete revolution every time z executes one third of a revolution. Thus the cube roots of 1 form an equilateral triangle. More generally, the same reasoning explains why the nth roots of unity form a regular n-gon.*

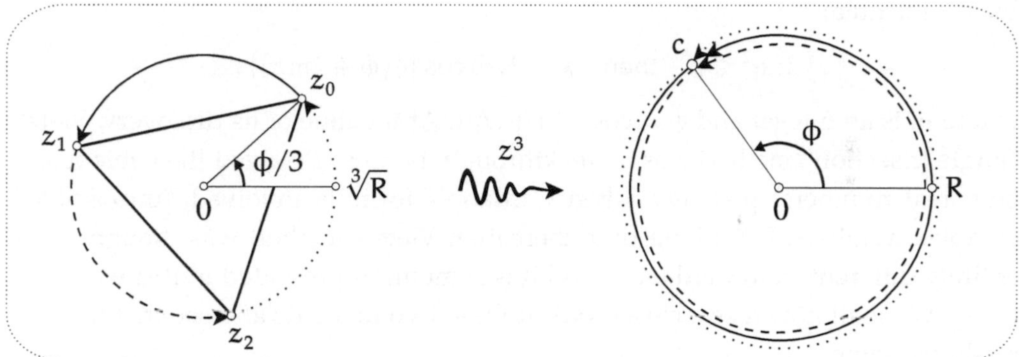

[2.6] *Extending the reasoning of the previous figure, the cube roots of an arbitrary complex number c will form an equilateral triangle, and the nth roots will form a regular n-gon.*

circle will therefore be successive positions of z as it repeatedly executes $(1/n)$ of a revolution, i.e., they will be the vertices of a regular n-gon. With $w = 1$, figure [2.5] illustrates this idea for the mapping $w = f(z) = z^3$.

More generally, [2.6] shows how to solve $z^3 = c = R\,e^{i\phi}$ by inscribing an equilateral triangle in the circle $|z| = \sqrt[3]{R}$. By the same reasoning, it is clear that the solutions of $z^n = c$ are the vertices of the regular n-gon inscribed in the circle $|z| = \sqrt[n]{R}$, with one vertex at angle (ϕ/n).

To arrive at the same result symbolically, first note that if ϕ is one value of arg c, then the complete set of possible angles is $(\phi + 2m\pi)$, where m is an arbitrary integer. Setting $z = r\,e^{i\theta}$,

$$r^n\,e^{in\theta} = z^n = c = R\,e^{i(\phi+2m\pi)} \quad\Longrightarrow\quad r^n = R \;\text{ and }\; n\theta = \phi + 2m\pi,$$

so the solutions are $z_m = \sqrt[n]{R}\,e^{i(\phi+2m\pi)/n}$. Each time we increase m by 1, z_m is rotated by $(1/n)$ of a revolution (because $z_{m+1} = e^{\frac{2\pi i}{n}}\,z_m$), producing the vertices of a regular n-gon. Thus the complete set of solutions will be obtained if we let m take any n consecutive values, say $m = 0, 1, 2, \ldots, (n-1)$.

2.2.2 Cubics Revisited*

As an instructive application of these ideas, let us reconsider the problem of solving a cubic equation in x. For simplicity, we shall assume in the following that the coefficients of the cubic are all real.

In the previous chapter we saw [Ex. 1] that the general cubic could always be reduced to the form $x^3 = 3px + 2q$. We then found [Ex. 2] that this could be solved using Cardano's formula,

$$x = s + t, \quad \text{where} \quad s^3 = q + \sqrt{q^2 - p^3}, \quad t^3 = q - \sqrt{q^2 - p^3}, \quad \text{and} \quad st = p.$$

Once again, observe that if $q^2 < p^3$ then this formula involves complex numbers.

On the other hand, we also saw [Ex. 3] that the cubic could be solved using Viète's formula:

$$\text{if } q^2 \leqslant p^3, \text{ then } \quad x = 2\sqrt{p}\cos\left[\tfrac{1}{3}(\phi + 2m\pi)\right],$$

where m is an integer and $\phi = \cos^{-1}(q/p\sqrt{p})$. At the time of its discovery, Viète's "angle trisection" method was a breakthrough, because it solved the cubic (using only real numbers) precisely when Cardano's formula involved "impossible", complex numbers. For a long time thereafter, Viète's method was thought to be entirely different from Cardano's, and it is sometimes presented in this way even today. We shall now take a closer look at these two methods and see that they are really the same.

If $q^2 \leqslant p^3$, then in Cardano's formula s^3 and t^3 are *complex conjugates*:

$$s^3 = q + i\sqrt{p^3 - q^2} \qquad \text{and} \qquad t^3 = \overline{s^3} = q - i\sqrt{p^3 - q^2}.$$

These complex numbers are illustrated on the RHS of [2.7]. By Pythagoras' Theorem, they both have length $|s^3| = p\sqrt{p}$, and so the angle ϕ occurring in Viète's formula is simply the angle of s^3.

Since s^3 and t^3 lie on the circle of radius $(\sqrt{p})^3$, their preimages under the mapping $z \mapsto z^3$ will lie on the circle of radius $\sqrt{p}$. The LHS of [2.7] shows these preimages; note that the three values of t are the complex conjugates of the three values of s.

According to the Fundamental Theorem of Algebra, the original cubic should have three solutions. However, by combining each of the three values of s with each of the three values of t, it would seem that Cardano's formula $x = s + t$ yields *nine* solutions.

The resolution lies in the fact that we also require $st = p$. Since p is real, this means s and t must have equal and opposite angles. In the formula $x = s + t$, each

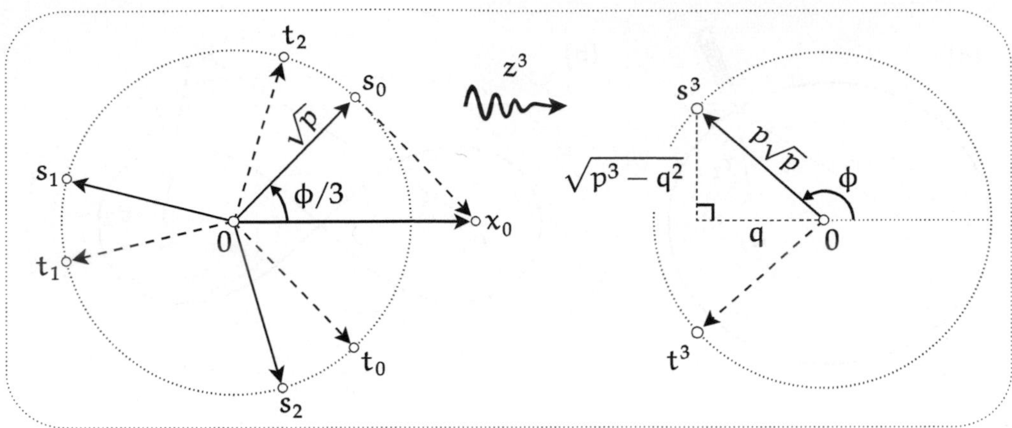

[2.7] *The solution of the cubic* $x^3 = 3px + 2q$ *was long thought to have two completely different solutions. If* $q^2 \leqslant p^3$*, Cardano's 1545 formula says that* $x = s + \bar{s}$*, where* $s^3 = q + i\sqrt{p^3 - q^2}$*. On the other hand, Viète's 1591 "angle trisection method" yields* $x = 2\sqrt{p}\cos\left[\frac{1}{3}(\phi + 2m\pi)\right]$*, where* $\phi = \cos^{-1}(q/p\sqrt{p})$*. The figure reveals that the two methods are in fact the same.*

of the three values of s must therefore be paired with the conjugate value of t. We can now see how Cardano's formula becomes Viète's formula:

$$x_m = s_m + t_m = s_m + \overline{s_m} = 2\sqrt{p}\cos\left[\tfrac{1}{3}(\phi + 2m\pi)\right].$$

In Ex. 4 the reader is invited to consider the case $q^2 > p^3$.

2.2.3 Cassinian Curves*

Consider [2.8a]. The ends of a piece of string of length l are attached to two fixed points a_1 and a_2 in $\mathbb{C}$, and, with its tip at z, a pencil holds the string taut. The figure illustrates the well known fact that if we move the pencil (continuing to keep the string taut) it traces out an *ellipse*, with foci a_1 and a_2. Writing $r_{1,2} = |z - a_{1,2}|$, the equation of the ellipse is thus

$$r_1 + r_2 = l.$$

By choosing different values of l we obtain the illustrated family of confocal ellipses.

In 1687 Newton published his great *Principia*, in which he demonstrated that the planets orbit in such ellipses, with the sun at one of the foci. Seven years earlier, however, Giovanni Cassini had instead proposed that the orbits were curves for which the *product* of the distances is constant:

$$r_1 \cdot r_2 = \text{const.} = k^2. \tag{2.1}$$

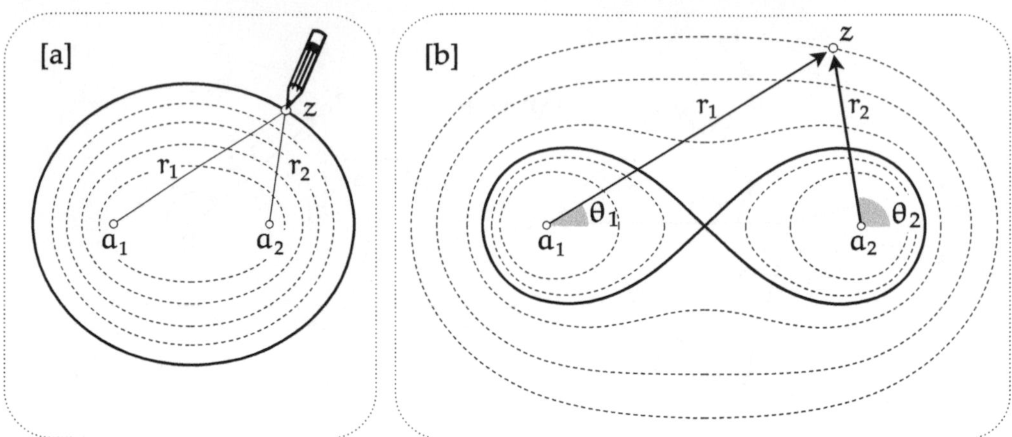

[2.8] **[a]** *Ellipses with* foci a_1 *and* a_2 *have equation* $r_1 + r_2 = \text{const}$. *In 1687 Newton proved that the planets orbit the sun in such ellipses, with the sun at one focus.* **[b]** *Seven years earlier, however, Giovanni Cassini had instead proposed that they orbit the sun in what are now called* Cassinian curves, *with equation* $r_1 r_2 = \text{const}$. *The figure eight curve is called the* lemniscate, *and it played a pivotal role in understanding* elliptic integrals *and* elliptic functions.

These curves are illustrated in [2.8b]; they are called *Cassinian curves*, and the points a_1 and a_2 are again called *foci*.

The following facts will become clearer in a moment, but you might like to think about them for yourself. If k is small then the curve consists of two separate pieces, resembling small circles centred at a_1 and a_2. As k increases, these two components of the curve become more egg shaped. When k reaches a value equal to half the distance between the foci then the pointed ends of the egg shapes meet at the midpoint of the foci, producing a figure eight [shown solid]. Increasing the value of k still further, the curve first resembles an hourglass, then an ellipse, and finally a circle.

Although Cassinian curves turned out to be useless as a description of planetary motion, the figure eight curve proved extremely valuable in quite another context. In 1694 it was rediscovered by James Bernoulli and christened the *lemniscate*—it then became the catalyst in unravelling the behaviour of the so-called *elliptic integrals* and *elliptic functions*. See Stillwell (2010) and Siegel (1969) for more on this fascinating story.

Cassinian curves arise naturally in the context of complex polynomials. A general quadratic $Q(z) = z^2 + pz + q$ will have two roots (say, a_1 and a_2) and so can be factorized as $Q(z) = (z - a_1)(z - a_2)$. In terms of [2.8b], this becomes

$$Q(z) = r_1 r_2 \, e^{i(\theta_1 + \theta_2)}.$$

Therefore, by virtue of (2.1), $z \mapsto w = Q(z)$ will map each curve in [2.8b] to an origin-centred circle, $|w| = k^2$, and it will map the foci to the origin.

If we follow this transformation by a translation of c, i.e., if we change $z \mapsto Q(z)$ to $z \mapsto Q(z) + c$, then the images will instead be concentric circles centred at $c =$ (image of foci). Conversely, given any quadratic mapping $z \mapsto w = Q(z)$, the preimages of a family of concentric circles in the w-plane centred at c will be the Cassinian curves whose foci are the preimages of c.

In particular, consider the case $c = 1$ and $w = Q(z) = z^2$. The preimages of $w = 1$ are $z = \pm 1$, so these are the foci, and the Cassinian curves are thus centred at the origin. See [2.9]. Since Q leaves the origin fixed, the lemniscate must be mapped (as illustrated) to the circle of radius 1 passing through the origin. Writing $z = r\,e^{i\theta}$, $w = r^2\,e^{i2\theta}$, and so we see from the figure that the polar equation of the lemniscate is

$$r^2 = 2\cos 2\theta. \tag{2.2}$$

Returning to [2.8b], the form of the Cassinian curves may be grasped more intuitively by sketching the modular surface of $Q(z) = (z-a_1)(z-a_2)$. First observe that as z moves further and further away from the origin, $Q(z)$ behaves more and more like z^2. Indeed, since the ratio $[Q(z)/z^2]$ is easily seen [exercise] to tend to unity as $|z|$ tends to infinity, we may say that $Q(z)$ is ultimately equal to z^2 in this limit. Thus, for large values of $|z|$, the modular surface of Q will look like the paraboloid in [2.3].

Next, consider the behaviour of the surface near a_1. Writing $D = |a_1 - a_2|$ for the distance between the foci, we see [exercise] that $|Q(z)|$ is ultimately equal to Dr_1 as z tends to a_1. Thus the surface meets the plane at a_1 in a cone like that shown in [2.2]. Of course the same thing happens at a_2.

Combining these facts, we obtain the surface shown in [2.10]. Since a Cassinian curve satisfies $|Q(z)| = r_1 r_2 = k^2$, it is the intersection of this surface with a plane

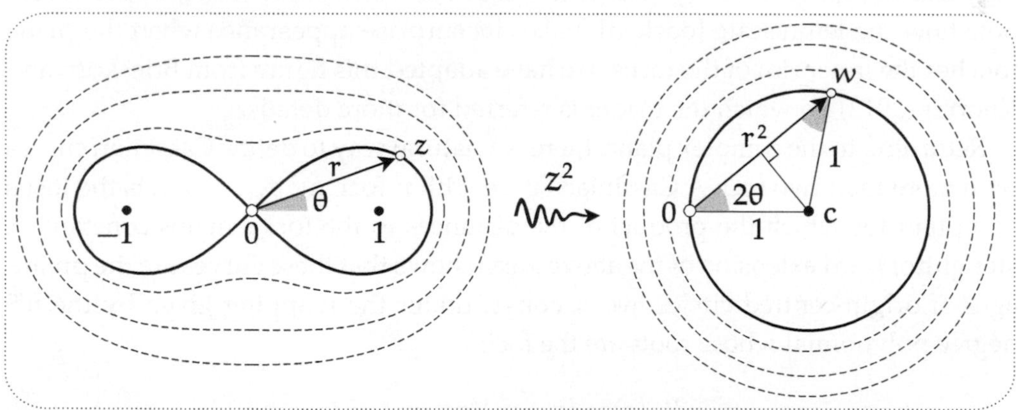

[2.9] *On the left, the Cassinian curves with foci* ± 1 *have equation* $|(z-1)(z+1)| = \text{const}$. *It follows that their images under* $z \mapsto z^2$ *are the circles illustrated on the right, centred at* 1. *It then follows immediately that the equation of the lemniscate is* $r^2 = 2\cos 2\theta$.

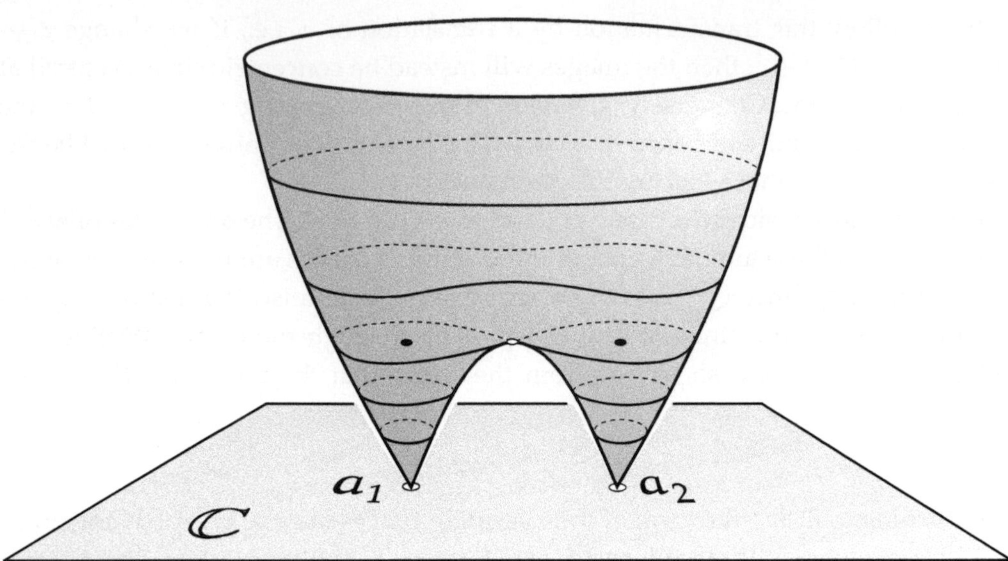

[2.10] *The form of the Cassinian curves in [2.8b] may be grasped intuitively as horizontal sections of the modular surface of* $Q(z) = (z - a_1)(z - a_2)$.

parallel to $\mathbb{C}$, and at height k^2 above it. As k increases from 0 to a large value, it is now easy to follow the evolution of the curves in [2.8b] by looking at how this intersection varies as the plane moves upward in [2.10]. Thus the Cassinian curves may be viewed as a geographical contour map of the modular surface of the quadratic.

Interestingly, Cassinian curves were already known to the ancient Greeks. Around 150 BCE, Perseus considered the intersection curves of a *torus* [obtained by rotating a circle C about an exterior line l in its plane] with planes parallel to l. It turns out that if the distance of the plane from l equals the radius of C then the resulting *spiric section of Perseus* is a Cassinian curve. See [2.11]; in particular, note how the lemniscate [dashed] makes its surprise appearance when the plane touches the inner rim of the torus. We have adapted this figure from Brieskorn and Knörrer (2012), to which the reader is referred for more details.

Returning to the complex plane, there is a natural way to define Cassinian curves with more than two foci: A Cassinian curve with n foci, $a_1, a_2, \ldots, a_n$, is the locus of a point for which the product of the distances to the foci remains constant. A straightforward extension of the above ideas shows that these curves are the preimages of origin-centred circles $|w| = \text{const.}$ under the mapping given by the nth degree polynomial whose roots are the foci:

$$z \mapsto w = P_n(z) = (z - a_1)(z - a_2) \cdots (z - a_n).$$

Equivalently, the Cassinian curves are the cross-sections of the modular surface of $P_n(z)$. This surface has n cone-like legs resting on $\mathbb{C}$ at $a_1, a_2, \ldots, a_n$, and for large values of $|z|$ it resembles the axially symmetric modular surface of z^n.

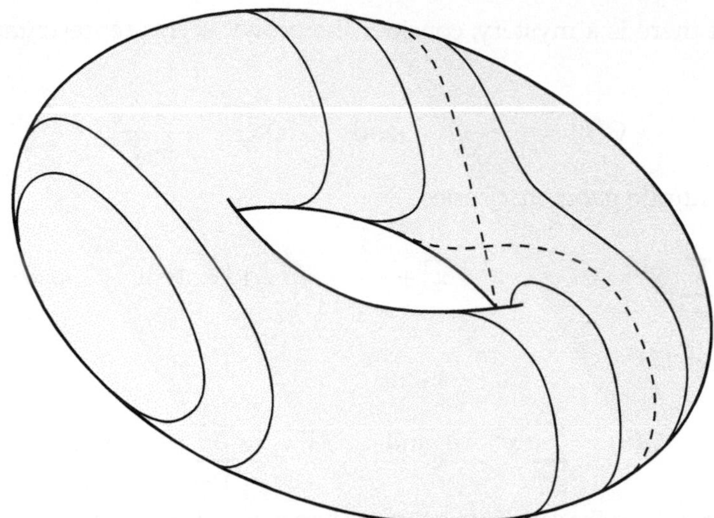

[2.11] The Spiric Sections of Perseus (150 BCE). *The intersection of a torus with a plane whose distance from the axis of symmetry equals the radius of the generating circle turns out to be a Cassinian curve. If the plane touches the inner rim, the* **lemniscate** *makes a surprise appearance!*

2.3 Power Series

2.3.1 The Mystery of Real Power Series

Many real functions $F(x)$ can be expressed (e.g., via Taylor's Theorem) as power series:

$$F(x) = \sum_{j=0}^{\infty} c_j\, x^j = c_0 + c_1\, x + c_2\, x^2 + c_3\, x^3 + \cdots,$$

where the c_j's are real constants. Of course, this infinite series will normally only converge to $F(x)$ in some origin-centred *interval of convergence* $-R < x < R$. But how is R (the *radius of convergence*) determined by $F(x)$?

It turns out that this question has a beautifully simple answer, *but only if we investigate it in the complex plane.* If we instead restrict ourselves to the real line—as mathematicians were forced to in the era in which such series were first employed—then the relationship between R and $F(x)$ is utterly mysterious. Historically, it was precisely this mystery[1] that led Cauchy to several of his breakthroughs in complex analysis.

[1] Cauchy was investigating the convergence of series solutions to *Kepler's equation*, which describes where a planet is in its orbit at any given time.

To see that there is a mystery, consider the power series representations of the functions

$$G(x) = \frac{1}{1-x^2} \qquad \text{and} \qquad H(x) = \frac{1}{1+x^2}.$$

The familiar infinite geometric series,

$$\frac{1}{1-x} = \sum_{j=0}^{\infty} x^j = 1 + x + x^2 + x^3 + \cdots \qquad \text{if and only if } -1 < x < 1, \tag{2.3}$$

immediately yields

$$G(x) = \sum_{j=0}^{\infty} x^{2j} \qquad \text{and} \qquad H(x) = \sum_{j=0}^{\infty} (-1)^j x^{2j},$$

where *both series have the same interval of convergence*, $-1 < x < 1$.

It is easy to understand the interval of convergence of the series for $G(x)$ if we look at the graph [2.12a]. The series becomes divergent at $x = \pm 1$ because these points are *singularities* of the function itself, i.e., they are places where $|G(x)|$ becomes infinite. But if we look at $y = |H(x)|$ in [2.12b], there seems to be no reason for the series to break down at $x = \pm 1$. Yet break down it does.

To begin to understand this, let us expand these functions into power series centred at $x = k$ (instead of $x = 0$), i.e., into series of the form $\sum_{j=0}^{\infty} c_j X^j$, where $X = (x-k)$ measures the displacement of x from the centre k. To expand G we first generalize (2.3) by expanding $1/(a-x)$ about k:

$$\frac{1}{a-x} = \frac{1}{a-(X+k)} = \frac{1}{(a-k)}\,\frac{1}{\left[1-\left(\frac{X}{a-k}\right)\right]},$$

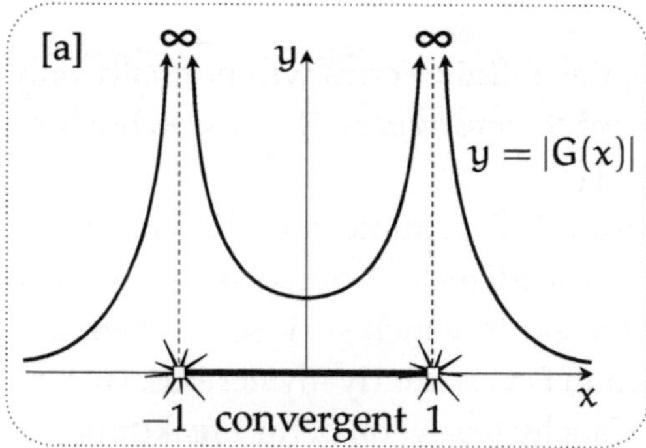

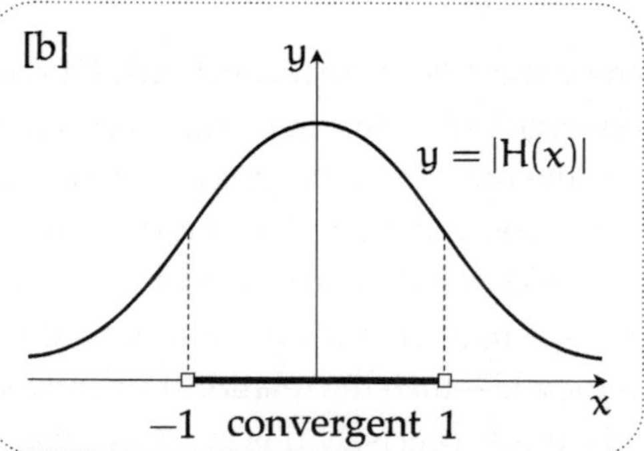

[2.12] [a] *The convergence of the power series for* $G(x) = \frac{1}{1-x^2}$ *is readily understood: clearly it must end when we arrive at the singularities* $x = \pm 1$. **[b]** *In contrast to this, the fact that the power series for* $H(x) = \frac{1}{1+x^2}$ also *stops converging at* ± 1 *is utterly mysterious!*

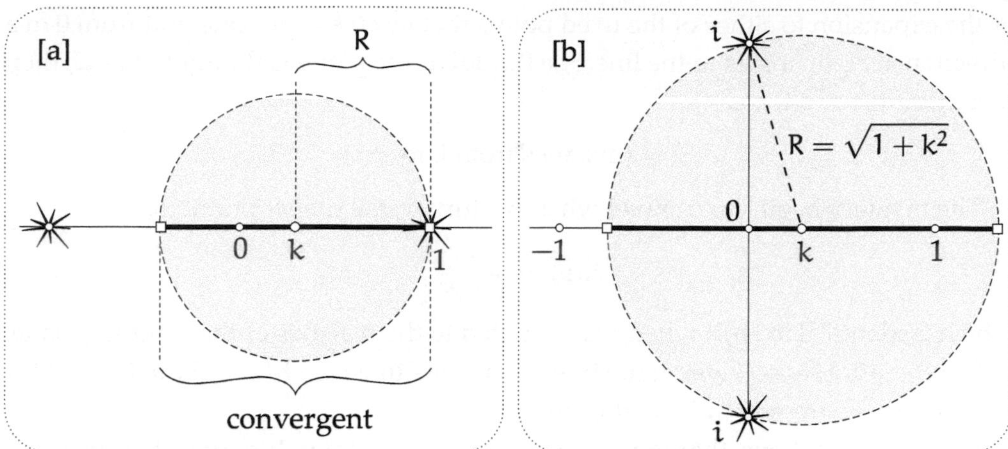

[2.13] The radius of convergence is the distance to the nearest singularity. [a] *It is intuitively clear that the radius of convergence* R *of the power series for* $G(x) = \frac{1}{1-x^2}$ *centred at* k *is the distance from* k *to the nearest singularity.* **[b]** *The previously mysterious formula* $R = \sqrt{1^2 + k^2}$ *for the convergence of the real power series for* $H(x) = \frac{1}{1+x^2}$ *is explained by its* imaginary *singularities at* $\pm i$.

and so

$$\frac{1}{a-x} = \sum_{j=0}^{\infty} \frac{X^j}{(a-k)^{j+1}}, \quad \text{if and only if } |X| < |a-k|. \tag{2.4}$$

To apply this result to G, we factorize $(1-x^2) = (1-x)(1+x)$ and then decompose G into partial fractions:

$$\frac{1}{1-x^2} = \frac{1}{2}\left[\frac{1}{1-x} - \frac{1}{-1-x}\right] = \frac{1}{2}\sum_{j=0}^{\infty}\left[\frac{1}{(1-k)^{j+1}} - \frac{1}{(-1-k)^{j+1}}\right] X^j,$$

where $|X| < |1-k|$ and $|X| < |1+k|$. Thus the interval of convergence $|X| < R$ is given by

$$R = \min\{|1-k|, |1+k|\} = \text{(distance from k to the nearest singularity of G)}.$$

This readily comprehensible result is illustrated in [2.13a]; ignore the shaded disc for the time being.

In the case of $H(x)$, I cannot think of an elegant method of finding the expansion using only real numbers, but see Ex. 9 for an attempt. Be that as it may, it can be shown that the radius of convergence of the series in X is given by the strange formula $R = \sqrt{1+k^2}$. As with Cotes' work in the previous chapter, we have here a result about real functions that is trying to tell us about the existence of the complex plane.

If we picture the real line as embedded in a plane then Pythagoras' Theorem tells us that $R = \sqrt{1^2 + k^2}$ should be interpreted as the distance from the centre k of the expansion to either of the fixed points that lie *off the line*, one unit from 0 in a direction at right angles to the line. See [2.13b]. If the plane is thought of as $\mathbb{C}$, then these points are $\pm i$, and

$$R = (\text{distance from } k \text{ to } \pm i).$$

The mystery begins to unravel when we turn to the *complex* function

$$h(z) = \frac{1}{1+z^2},$$

which is identical to $H(x)$ when z is restricted to the real axis of the complex plane. In fact there is a sense—we cannot be explicit yet—in which $h(z)$ is the *only* complex function that agrees with H on this line.

While [2.12b] shows that $h(z)$ is well-behaved for real values of z, it is clear that $h(z)$ *has two singularities in the complex plane*, one at $z = i$ and the other at $z = -i$; these are shown as little explosions in [2.13b]. Figure [2.14] tries to make this more vivid by showing the modular surface of $h(z)$, the singularities at $\pm i$ appearing as "volcanoes" erupting above these points. We will sort through the details in a moment, but the mystery has all but disappeared: *in both [2.13a] and [2.13b], the radius of convergence is the distance to the nearest singularity.*

If we intersect the surface in [2.14] with a vertical plane through the real axis then we recover the deceptively tranquil graph in [2.12b], but if we instead slice

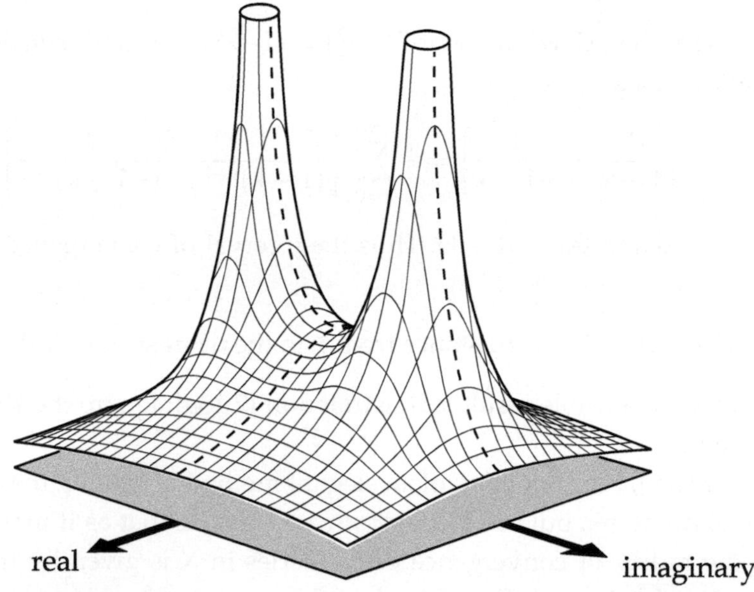

[2.14] *Twin volcanoes erupt in the modular surface of* $h(z) = \frac{1}{1+z^2}$, *directly above its singularities at* $z = \pm i$. *The graphs in [2.12] can now be understood in a unified way, as the vertical sections of this surface in the imaginary and real directions. The mysterious radius of convergence of the real power series for* $h(x)$ *is thereby explained!*

the surface along the imaginary axis then we obtain the graph in [2.12a]. That this is no accident may be seen by first noting that $G(x)$ is just the restriction to the real axis of the complex function $g(z) = 1/(1 - z^2)$. Since $g(z) = h(iz)$, h and g are essentially the same: if we rotate the plane by $(\pi/2)$ and then do h, we obtain g. In particular the modular surface of g is simply [2.14] rotated by $(\pi/2)$, the volcanoes at $\pm i$ being rotated to ± 1.

2.3.2 *The Disc of Convergence*

Let us consider the convergence of complex power series, leaving aside for the moment the question of whether a given complex function can be expressed as such a series.

A complex power series $P(z)$ (centred at the origin) is an expression of the form

$$P(z) = \sum_{j=0}^{\infty} c_j \, z^j = c_0 + c_1 \, z + c_2 \, z^2 + c_3 \, z^3 + \cdots, \tag{2.5}$$

where the c_j's are complex constants, and z is a complex variable. The partial sums of this infinite series are just the ordinary polynomials,

$$P_n(z) = \sum_{j=0}^{n} c_j \, z^j = c_0 + c_1 \, z + c_2 \, z^2 + c_3 \, z^3 + \cdots + c_n \, z^n.$$

For a given value of $z = a$, the sequence of points $P_1(a), P_2(a), P_3(a), \ldots$ is said to *converge to the point* A if for any given positive number ϵ, *no matter how small*, there exists a positive integer N such that $|A - P_n(a)| < \epsilon$ for every value of n greater than N. Figure [2.15a] illustrates that this is much simpler than it sounds: all it says is that once we reach a certain point $P_N(a)$ in the sequence $P_1(a), P_2(a), P_3(a), \ldots$, all of the subsequent points lie within an arbitrarily small disc of radius ϵ centred at A.

In this case we say that the power series $P(z)$ *converges* to A at $z = a$, and we write $P(a) = A$. If the sequence $P_1(a), P_2(a), P_3(a), \ldots$ does not converge to a particular point, then the power series $P(z)$ is said to *diverge* at $z = a$. Thus for each point z, $P(z)$ will either converge or diverge.

Figure [2.15b] shows a magnified view of the disc in [2.15a]. If $n > m > N$ then $P_m(a)$ and $P_n(a)$ both lie within this disc, and consequently the distance between them must be less than the diameter of the disc:

$$|c_{m+1} a^{m+1} + c_{m+2} a^{m+2} + \cdots + c_n a^n| = |P_n(a) - P_m(a)| < 2\epsilon. \tag{2.6}$$

Conversely, it can be shown [exercise] that if this condition is met then $P(a)$ converges. Thus we have a new way of phrasing the definition of convergence: $P(a)$ *converges if and only if there exists an* N *such that inequality* (2.6) *holds (for arbitrarily small* ϵ*) whenever* m *and* n *are both greater than* N.

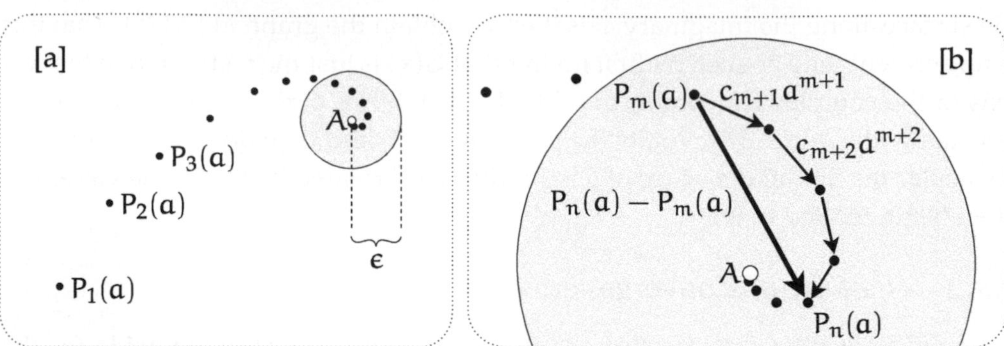

[2.15] [a] *The sequence* $P_n(a)$ *is said to* converge to A *if it inevitably enters every disc centred at A, no matter how small we make its radius* ϵ*, and the disc then acts like a black hole: having entered the disc, the sequence can never escape it!* **[b]** *The direct route from* $P_m(a)$ *to* $P_n(a)$ *has length* $|P_n(a) - P_m(a)|$ *and is shorter than the indirect route that goes via* $P_{m+1}(a)$*,* $P_{m+2}(a)$*, etc., which has length* $\widetilde{P}_n(a) - \widetilde{P}_m(a)$*. But if* $P(a)$ *is* absolutely convergent *then the indirect route must have a length that tends to zero for sufficiently large* m *and* n*, and therefore the direct route must, too. Therefore,* absolute convergence implies convergence.

The complex power series $P(z)$ is said to be *absolutely convergent* at $z = a$ if the real series

$$\widetilde{P}(z) \equiv \sum_{j=0}^{\infty} |c_j\, z^j| = |c_0| + |c_1\, z| + |c_2\, z^2| + |c_3\, z^3| + \cdots ,$$

converges there. Absolute convergence is certainly different from ordinary convergence. For example, [exercise] $P(z) = \sum z^j/j$ is convergent at $z = -1$, but it is not absolutely convergent there. On the other hand,

> *If* $P(z)$ *is absolutely convergent at some point, then it will also be convergent at that point.* (2.7)

Thus absolute convergence is a stronger requirement than convergence.

To establish (2.7), suppose that $P(z)$ is absolutely convergent at $z = a$, so that (by definition) $\widetilde{P}(a)$ is convergent. In terms of the partial sums $\widetilde{P}_n(z) = \sum_{j=0}^{n} |c_j\, z^j|$ of the real series $\widetilde{P}(z)$, this says that for sufficiently large values of m and n we can make $[\widetilde{P}_n(a) - \widetilde{P}_m(a)]$ as small as we please. But, referring to [2.15b], we see that

$$\widetilde{P}_n(a) - \widetilde{P}_m(a) = |c_{m+1} a^{m+1}| + |c_{m+2} a^{m+2}| + \cdots + |c_n a^n|$$

is the total length of the roundabout journey from $P_m(a)$ to $P_n(a)$ that goes via $P_{m+1}(a)$, $P_{m+2}(a)$, etc. Since $|P_n(a) - P_m(a)|$ is the length of the shortest journey from $P_m(a)$ to $P_n(a)$,

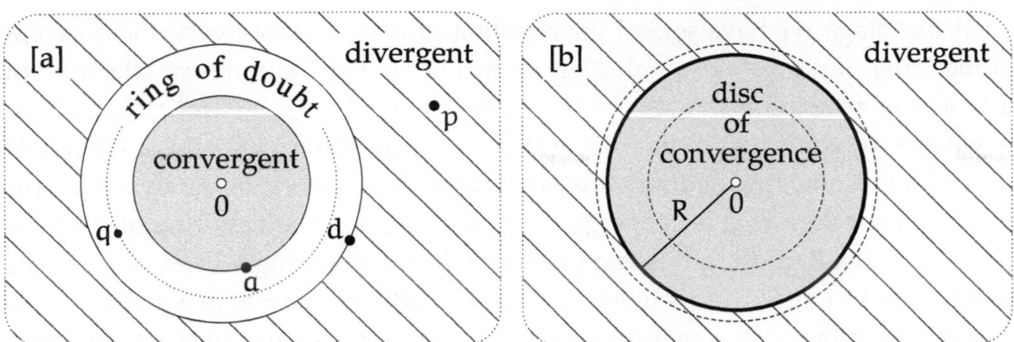

[2.16] [a] *If* $P(z)$ *converges at* a *then it also converges everywhere inside the shaded disc* $|z| < |a|$. *And if it diverges at* d *then it also diverges in the striped region* $|z| > |d|$. *Its behaviour is therefore unknown only in "the ring of doubt",* $|a| \leqslant |z| \leqslant |d|$. *Testing a point* q *half way across this ring then halves the width of the ring.* **[b]** *Iterating this* q*-algorithm, the ring of doubt shrinks and converges to a definite circle, the* circle of convergence, *the interior of which is the* disc of convergence, *and the radius of which,* R, *is the* radius of convergence.

$$|P_n(a) - P_m(a)| \leqslant \widetilde{P}_n(a) - \widetilde{P}_m(a).$$

Thus $|P_n(a) - P_m(a)|$ must also become arbitrarily small for sufficiently large m and n. Done.

We can now establish the following fundamental fact:

> *If* $P(z)$ *converges at* $z = a$, *then it will also converge everywhere inside the disc* $|z| < |a|$. (2.8)

See [2.16a]. In fact we will show that $P(z)$ is *absolutely* convergent in this disc; the result then follows directly from (2.7).

If $P(a)$ converges then the length $|c_n\, a^n|$ of each term must die away to zero as n goes to infinity [why?]. In particular, there must be a number M such that $|c_n\, a^n| < M$ for all n. If $|z| < |a|$ then $\rho = |z|/|a| < 1$ and so $|c_n\, z^n| < M\rho^n$. Thus,

$$\widetilde{P}_n(z) - \widetilde{P}_m(z) \leqslant M(\rho^{m+1} + \rho^{m+2} + \cdots + \rho^n) = \frac{M}{1-\rho}\,[\rho^{m+1} - \rho^{n+1}], \tag{2.9}$$

where the RHS is as small as we please for sufficiently large m and n. Done.

If $P(z)$ does not converge everywhere in the plane then there must be at least one point d where it diverges. Now suppose that $P(z)$ were to converge at some point p further away from the origin than d. See [2.16a]. By (2.8) it would then converge everywhere inside the disc $|z| < |p|$, and in particular it would converge at d, contradicting our initial hypothesis. Thus,

> *If* $P(z)$ *diverges at* $z = d$, *then it will also diverge everywhere outside the circle* $|z| = |d|$. (2.10)

At this stage we have settled the question of convergence everywhere except in the "ring of doubt", $|a| \leqslant |z| \leqslant |d|$, shown in [2.16a]. Suppose we take a point q half way across the ring of doubt (i.e., on the circle $|z| = \frac{|a|+|d|}{2}$), then check whether $P(q)$ converges or not. Regardless of the outcome, (2.8) and (2.10) enable us to obtain a new ring of doubt that is half as wide as before. For example, if $P(q)$ is convergent then $P(z)$ is convergent for $|z| < |q|$, and the new ring of doubt is $|q| \leqslant |z| \leqslant |d|$. Repeating this test procedure in the new ring will again halve its width. Continuing in this manner, the ring of doubt will narrow to a definite circle $|z| = R$ (called the *circle of convergence*) such that $P(z)$ converges everywhere inside the circle, and diverges everywhere outside the circle. See [2.16b]. The radius R is called the *radius of convergence*—at last we see where this name comes from!—and the interior of the circle is called the *disc of convergence*.

Note that this argument tells us nothing about the convergence of $P(z)$ *on* the circle of convergence. In principle, we can imagine convergence at all, some, or none of the points on this circle, and one can actually find examples of power series that realize each of these three possibilities. See Ex. 11.

All of the above results immediately generalize to a power series centred at an arbitrary point k, that is to a series of the form $P(z) = \sum c_j Z^j$, where $Z = (z - k)$ is the complex number from the centre k to the point z. Thus, restating our main conclusion (due to Niels Abel) in general form,

Given a complex power series $P(z)$ *centred at* k, *there exists a circle* $|z - k| = R$ *centred at* k *such that* $P(z)$ *converges everywhere inside the circle, and* $P(z)$ *diverges everywhere outside the circle.* (2.11)

Of course one can also have a series that converges everywhere, but this may be thought of as the limiting case in which the circle of convergence is infinitely large.

Returning to figures [2.13a] and [2.13b], we now recognize the illustrated discs as the discs of convergence of the series for $1/(1 \mp z^2)$.

2.3.3 Approximating a Power Series with a Polynomial

Implicit in the definition of convergence is a simple but very important fact: if $P(a)$ converges, then its value can be *approximated* by the partial sum $P_m(a)$, and by choosing a sufficiently large value of m we can make the approximation as accurate as we wish. Combining this observation with (2.11),

At each point z in the disc of convergence, $P(z)$ *can be approximated with arbitrarily high precision by a polynomial* $P_m(z)$ *of sufficiently high degree.*

For simplicity's sake, let us investigate this further in the case that $P(z)$ is centred at the origin. The error $E_m(z)$ at z associated with the approximation $P_m(z)$ can be defined as the distance $E_m(z) = |P(z) - P_m(z)|$ between the exact answer and the approximation. For a fixed value of m, the error $E_m(z)$ will vary as z moves around in the disc of convergence. Clearly, since $E_m(0) = 0$, the error will be extremely small if z is close to the origin, but what if z approaches the circle of convergence? The answer depends on the particular power series, but *it can happen that the error becomes enormous!* [See Ex. 12.] This does not contradict the above result: for any fixed z, no matter how close to the circle of convergence, the error $E_m(z)$ will become arbitrarily small as m tends to infinity.

This problem is avoided if we restrict z to the disc $|z| \leqslant r$, where $r < R$, because this prevents z from getting *arbitrarily* close to the circle of convergence, $|z| = R$. In attempting to approximate $P(z)$ within this disc, it turns out that we can do the following. We first decide on the maximum error (say ϵ) that we are willing to put up with, then choose (once and for all) an approximating polynomial $P_m(z)$ of sufficiently high degree that the error is smaller than ϵ throughout the disc. That is, throughout the disc, the approximating point $P_m(z)$ lies less than ϵ away from the true point, $P(z)$. One describes this by saying that $P(z)$ is *uniformly convergent* on this disc:

If $P(z)$ has disc of convergence $|z| < R$, then $P(z)$ is uniformly convergent on the closed disc $|z| \leqslant r$, where $r < R$. (2.12)

Although we may not have uniform convergence on the whole disc of convergence, the above result shows that this is really a technicality: we do have uniform convergence on a disc that *almost* fills the complete disc of convergence, say $r = (0.999999999)\,R$.

To verify (2.12), first do Ex. 12, then have a good look at (2.9).

2.3.4 Uniqueness

If a complex function can be expressed as a power series, then it can only be done so in one way—the power series is *unique*. This is an immediate consequence of the *Identity Theorem:*

If

$$c_0 + c_1 z + c_2 z^2 + c_3 z^3 + \cdots = d_0 + d_1 z + d_2 z^2 + d_3 z^3 + \cdots$$

for all z in a neighbourhood (no matter how small) of 0, then the power series are identical: $c_j = d_j$.

Putting $z = 0$ yields $c_0 = d_0$, so they may be cancelled from both sides. Dividing by z and again putting $z = 0$ then yields $c_1 = d_1$, and so on. [Although this was easy, Ex. 13 shows that it is actually rather remarkable.] The result can be strengthened considerably: *If the power series merely agree along a segment of curve (no matter how small) through 0, or if they agree at every point of an infinite sequence of points that converges to 0, then the series are identical.* The verification is essentially the same, only instead of putting $z = 0$, we now take the limit as z *approaches* 0, either along the segment of curve or through the sequence of points.

We can perhaps make greater intuitive sense of these results if we first recall that a power series can be approximated with arbitrarily high precision by a *polynomial* of sufficiently high degree. Given two points in the plane (no matter how close together) there is a unique line passing through them. Thinking in terms of a graph $y = f(x)$, this says that a polynomial of degree 1, say $f(x) = c_0 + c_1 x$, is uniquely determined by the images of any two points, no matter how close together. Likewise, in the case of degree 2, if we are given three points (no matter how close together), there is only one parabolic graph $y = f(x) = c_0 + c_1 x + c_2 x^2$ that can be threaded through them. This idea easily extends to complex functions: *there is one, and only one, complex polynomial of degree* n *that maps a given set of* $(n+1)$ *points to a given set of* $(n+1)$ *image points.* The above result may therefore be thought of as the limiting case in which the number of known points (together with their known image points) tends to infinity.

Earlier we alluded to a sense in which $h(z) = 1/(1+z^2)$ is the *only* complex function that agrees with the real function $H(x) = 1/(1+x^2)$ on the real line. Yet clearly we can easily write down infinitely many complex functions that agree with $H(x)$ in this way. For example,

$$g(z) = g(x+iy) = \frac{\cos[x^2 y] + i \sin[y^2]}{e^y + x^2 \ln(e + y^4)}.$$

Then in what sense can $h(z)$ be considered the unique generalization of $H(x)$?

We already know that $h(z)$ can be expressed as the power series $\sum_{j=0}^{\infty}(-1)^j z^{2j}$, and this fact yields [exercise] a provisional answer: $h(z)$ is the only complex function that (i) agrees with $H(x)$ on the real axis, and (ii) can be expressed as a power series in z. This still does not completely capture the sense in which $h(z)$ is unique, but it's a start.

More generally, suppose we are given a real function $F(x)$ that can be expressed as a power series in x on a (necessarily origin-centred) segment of the real line: $F(x) = \sum_{j=0}^{\infty} c_j x^j$. Then the complex power series $f(z) = \sum_{j=0}^{\infty} c_j z^j$ with the same coefficients can be used to define the unique complex function $f(z)$ that
(i) agrees with F on the given segment of the real axis, and (ii) can be expressed as a power series in z.

For example, consider the *complex exponential function*, written e^z, the geometry of which we will discuss in the next section. Since $e^x = \sum_{j=0}^{\infty} x^j/j!$,

$$e^z = 1 + z + \tfrac{1}{2!}z^2 + \tfrac{1}{3!}z^3 + \tfrac{1}{4!}z^4 + \cdots.$$

Note that our heuristic, power-series approach to Euler's formula [Chapter 1] is starting to look more respectable!

2.3.5 Manipulating Power Series

The fact that power series can be approximated with arbitrarily high precision by polynomials implies [see Ex. 14] that

Two power series with the same centre can be added, multiplied, and divided in the same way as polynomials. (2.13)

If the two series $P(z)$ and $Q(z)$ have discs of convergence D_1 and D_2, then the resulting series for $[P+Q]$ and PQ will both converge in the smaller of D_1 and D_2, though they may in fact converge within a still larger disc. No such general statement is possible in the case of $(P/Q) = P(1/Q)$, because the convergence of the series for $(1/Q)$ is limited not only by the boundary circle of D_2, but also by any points inside D_2 where $Q(z) = 0$.

Let us illustrate (2.13) with a few examples. Earlier we actually assumed this result in order to find the series for $1/(1-z^2)$ centred at k. Using the partial fraction decomposition

$$\frac{1}{1-z^2} = \frac{(1/2)}{1-z} + \frac{(1/2)}{1+z},$$

we obtained two power series for the functions on the RHS, and then *assumed* that these power series could be added like two polynomials, by adding the coefficients.

In the special case $k = 0$ we can check that this procedure works, because we already *know* the correct answer for the series centred at the origin:

$$\frac{1}{1-z^2} = 1 + z^2 + z^4 + z^6 + \cdots .$$

Since

$$\frac{1}{1-z} = 1 + z + z^2 + z^3 + z^4 + z^5 + \cdots$$

$$\text{and} \quad \frac{1}{1+z} = 1 - z + z^2 - z^3 + z^4 - z^5 + \cdots ,$$

we see that adding the coefficients of these series does indeed yield the correct series for $1/(1-z^2)$.

Since

$$\frac{1}{1-z^2} = \left[\frac{1}{1-z}\right]\left[\frac{1}{1+z}\right],$$

we can recycle this example to illustrate the correctness of multiplying power series as if they were polynomials:

$$[1 + z + z^2 + z^3 + z^4 + z^5 + \cdots][1 - z + z^2 - z^3 + z^4 - z^5 + \cdots]$$
$$= 1 + (1\text{–}1)\,z + (1\text{–}1\text{+}1)\,z^2 + (1\text{–}1\text{+}1\text{–}1)\,z^3 + (1\text{–}1\text{+}1\text{–}1\text{+}1)\,z^4 + \cdots ,$$

which is again the correct series for $1/(1-z^2)$.

Next, let's use (2.13) to find the series for $1/(1-z)^2$:

$$\begin{aligned} & [1+z+z^2+z^3+z^4+z^5+\cdots][1+z+z^2+z^3+z^4+z^5+\cdots] \\ = \; & 1+(1{+}1)\,z+(1{+}1{+}1)\,z^2+\;(1{+}1{+}1{+}1)\,z^3+\;(1{+}1{+}1{+}1{+}1)\,z^4+\cdots, \end{aligned}$$

and so $(1-z)^{-2} = \sum_{j=0}^{\infty}(j+1)\,z^j$.

You may check for yourself that the above series for $(1-z)^{-1}$ and $(1-z)^{-2}$ are both special cases of the general *Binomial Theorem*, which states that if n is any real number (not just a positive integer), then within the unit disc,

$$(1+z)^n = 1+nz+\tfrac{n(n-1)}{2!}z^2+\tfrac{n(n-1)(n-2)}{3!}z^3+\tfrac{n(n-1)(n-2)(n-3)}{4!}z^4+\cdots. \quad (2.14)$$

Historically, this result was one of Newton's key weapons in developing calculus, and later it played an equally central role in the work of Euler.

In Exs. 16, 17, 18, we show how manipulation of power series may be used to demonstrate the Binomial Theorem, first for all negative integers, then for all rational powers. Although we shall not discuss it further, the case of an irrational power ρ may be treated by taking an infinite sequence of rational numbers that converges to ρ. Later we shall use other methods to establish a still more general version of (2.14) in which the power n is allowed to be a *complex* number!

Next we describe how to divide two power series $P(z)$ and $Q(z)$. In order to find the series $P(z)/Q(z) = \sum c_j z^j$, one multiplies both sides by $Q(z)$ to obtain $P(z) = Q(z)\sum c_j z^j$, and then multiplies the two power series on the right. By the uniqueness result, the coefficients of this series must equal the known coefficients of $P(z)$, and this enables one to calculate the c_j's. An example will make this process much clearer.

In order to find the coefficients c_j in the series $1/e^z = \sum c_j z^j$, we multiply both sides by e^z to obtain

$$\begin{aligned} 1 &= [1+z+\frac{z^2}{2!}+\frac{z^3}{3!}+\frac{z^4}{4!}+\cdots][c_0+c_1 z+c_2 z^2+c_3 z^3+c_4 z^4+\cdots] \\ &= c_0+(c_0+c_1)\,z+\left(\frac{c_0}{2!}+\frac{c_1}{1!}+\frac{c_2}{0!}\right)z^2+\left(\frac{c_0}{3!}+\frac{c_1}{2!}+\frac{c_2}{1!}+\frac{c_3}{0!}\right)z^3\cdots. \end{aligned}$$

By the uniqueness result, we may equate coefficients on both sides to obtain an infinite set of linear equations:

$$\begin{aligned} 1 &= c_0, \\ 0 &= c_0+c_1, \\ 0 &= c_0/2!+c_1/1!+c_2/0!, \\ 0 &= c_0/3!+c_1/2!+c_2/1!+c_3/0!, \text{etc.} \end{aligned}$$

Successively solving the first few of these equations [exercise] quickly leads to the guess $c_n = (-1)^n/n!$, which is then easily verified [exercise] by considering the binomial expansion of $(1-1)^m$, where m is a positive integer. Thus we find that

$$1/e^z = 1 - z + \tfrac{1}{2!}z^2 - \tfrac{1}{3!}z^3 + \tfrac{1}{4!}z^4 - \tfrac{1}{5!}z^5 + \cdots = e^{-z},$$

just as with the real function e^x.

2.3.6 *Finding the Radius of Convergence*

Given a complex power series $P(z) = \sum c_j z^j$, there are several ways of determining its radius of convergence directly from its coefficients. Since they are formally identical to the methods used on real series, we merely state them, leaving it to you to generalize the standard real proofs.

The *ratio test* says that

$$R = \lim_{n\to\infty} \left|\frac{c_n}{c_{n+1}}\right|,$$

provided this limit exists. For example, if

$$P(z) = 1 + z + \frac{z^2}{2^2} + \frac{z^3}{3^2} + \frac{z^4}{4^2} + \cdots,$$

then

$$R = \lim_{n\to\infty} \frac{1/n^2}{1/(n+1)^2} = \lim_{n\to\infty} \left(1 + \tfrac{1}{n}\right)^2 = 1.$$

If $|c_n/c_{n+1}|$ tends to infinity then (formally) $R = \infty$, corresponding to convergence everywhere in the plane. For example, $e^z = \sum_{j=0}^{\infty} z^j/j!$ converges everywhere, because

$$R = \lim_{n\to\infty} \frac{1/n!}{1/(n+1)!} = \lim_{n\to\infty} (n+1) = \infty.$$

When the ratio test fails, or becomes difficult to apply, we can often use the *root test*, which says that

$$R = \lim_{n\to\infty} \frac{1}{\sqrt[n]{|c_n|}},$$

provided this limit exists. For example, if we first recall [we will discuss this later] that the real function e^x may be written as

$$e^x = \lim_{n\to\infty} \left(1 + \frac{x}{n}\right)^n,$$

then applying the root test to the series

$$P(z) = \sum_{j=1}^{\infty} \left(\frac{j-3}{j}\right)^{j^2} z^j,$$

yields [exercise] $R = e^3$.

On occasion both the ratio and root tests will fail, but there exists a slightly refined version of the latter which can be shown to work in *all* cases. It is called the *Cauchy-Hadamard Theorem*, and it says that

$$R = \frac{1}{\limsup \sqrt[n]{|c_n|}}.$$

We will not discuss this further since it is not needed in this book.

The above examples of power series were plucked out of thin air, but often our starting point is a known complex function $f(z)$ which is then *expressed* as a power series. The problem of determining R then has a conceptually much more satisfying answer. Roughly[2],

If $f(z)$ *can be expressed as a power series centred at* k, *then the radius of convergence is the distance from* k *to the nearest singularity of* $f(z)$. (2.15)

Figure [2.17a] illustrates this, the singularities of $f(z)$ being represented as explosions. To understand which functions can be expanded into power series we need deep results from later in the book, but we are already in a position to verify that a rational function [the ratio of two polynomials] can be, and that the radius of convergence for its expansion is given by (2.15).

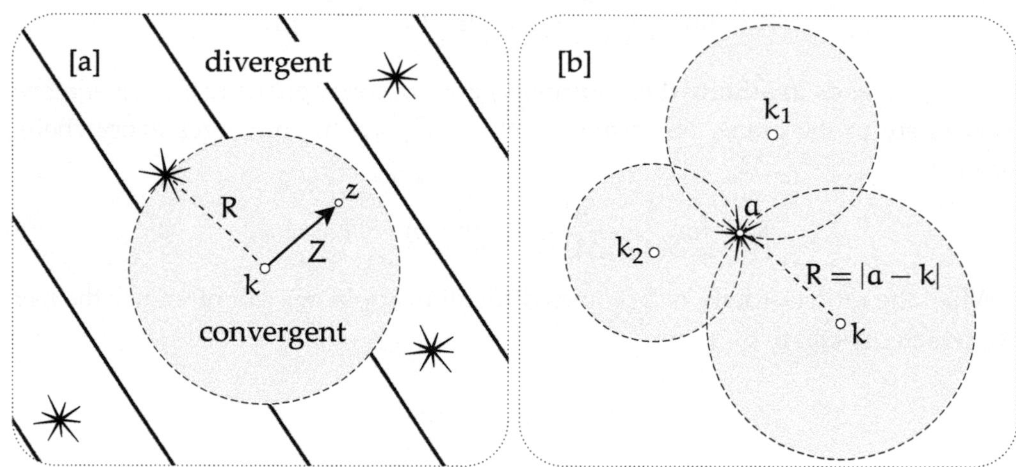

[2.17] [a] *Here, explosions mark the locations of singularities of a function* $f(z)$ *that we wish to expand into a power series centred at* k, *i.e., a series in powers of* $Z \equiv (z - k)$. *Then the radius of convergence* R = (the distance from k to the nearest singularity.) **[b]** *Here we see the three discs of convergence that arise from three different choices of the centre* k, *assuming that* a *is the nearest singularity for all three centres.*

[2] Later [p. 107] we shall have to modify the statement in the case that $f(z)$ is a "multifunction", having more than one value for a given value of z.

To begin with, reconsider [2.13a] and [2.13b], both of which are examples of (2.15). Recall that in [2.13b] we merely *claimed* that $R = \sqrt{1+k^2}$ for the series expansion of $h(z) = 1/(1+z^2)$ centred at the real point k. We now verify this and explicitly find the series.

To do so, first note that (2.4) easily generalizes to

$$\frac{1}{a-z} = \sum_{j=0}^{\infty} \frac{Z^j}{(a-k)^{j+1}}, \quad \text{if and only if } |Z| < |a-k|, \tag{2.16}$$

where a and k are now arbitrary *complex* numbers, and $Z = (z-k)$ is the complex number connecting the centre of the expansion to z. The condition $|z-k| < |a-k|$ for convergence is that z lie in the interior of the circle centred at k and passing through a. See [2.17b], which also shows the discs of convergence when we instead choose to expand $1/(a-z)$ about k_1 or k_2. Since the function $1/(a-z)$ has just one singularity at $z = a$, we have verified (2.15) for this particular function.

Earlier we found the expansion of $1/(1-x^2)$ by factorizing the denominator and using partial fractions. We are now in a position to use exactly the same approach to find the expansion of $h(z) = 1/(1+z^2)$ centred at an arbitrary complex number k:

$$\frac{1}{1+z^2} = \frac{1}{(z-i)(z+i)} = \frac{1}{2i}\left[\frac{1}{-i-z} - \frac{1}{i-z}\right].$$

Applying (2.16) to both terms then yields

$$\frac{1}{1+z^2} = \sum_{j=0}^{\infty} \frac{1}{2i}\left[\frac{1}{(-i-k)^{j+1}} - \frac{1}{(i-k)^{j+1}}\right] Z^j. \tag{2.17}$$

The series for $1/(\pm i - z)$ converge inside the concentric circles $|z-k| = |\pm i - k|$ centred at k and passing through the points $\mp i$, which are the singularities of $h(z)$. But (2.17) will only converge when *both* these series converge, i.e., in the disc $|z-k| < R$ where R is the distance from the centre k to the nearest singularity of $h(z)$. Thus we have confirmed (2.15) for $h(z)$.

In particular, if k is real then (2.17) converges in the disc shown in [2.13b]. If z is restricted to the real axis then $h(z)$ reduces to the real function $1/(1+x^2)$, and the expansion of this function into powers of $X = (x-k)$ can be deduced easily from (2.17). Since k is now real, $|i-k| = \sqrt{1+k^2}$, and we may write $(i-k) = \sqrt{1+k^2}\, e^{i\phi}$, where $\phi = \arg(i-k)$ is the appropriate value of $\tan^{-1}(-1/k)$. Thus [exercise]

$$\frac{1}{1+x^2} = \sum_{j=0}^{\infty} \left[\frac{\sin(j+1)\phi}{(\sqrt{1+k^2})^{j+1}}\right] X^j. \tag{2.18}$$

Again, we have here a result concerning real functions that would be very difficult to obtain using only real numbers.

The above analysis of $1/(1+z^2)$ can easily be generalized [exercise] to show that any rational function can be expressed as a power series, with radius of convergence given by (2.15).

2.3.7 Fourier Series*

On the 21st of December 1807, Joseph Fourier announced to the French Academy a discovery so remarkable that his distinguished audience found it *literally* incredible. His claim was that any[3] real periodic function $F(\theta)$, no matter how capricious its graph, may be decomposed into a sum of sinusoidal waves of higher and higher frequency. For simplicity's sake, let the period be 2π; then the *Fourier series* is

$$F(\theta) = \tfrac{1}{2}a_0 + \sum_{n=1}^{\infty} \left[a_n \cos n\theta + b_n \sin n\theta\right],$$

where [see Ex. 20]

$$a_n = \frac{1}{\pi}\int_0^{2\pi} F(\theta)\cos n\theta \, d\theta \qquad \text{and} \qquad b_n = \frac{1}{\pi}\int_0^{2\pi} F(\theta)\sin n\theta \, d\theta. \qquad (2.19)$$

This optional section is addressed primarily to readers who have already encountered such series. For those who have not, we hope that this brief discussion (together with the exercises at the end of the chapter) may whet your appetite for more on this fascinating subject[4].

In the world of the real numbers there appears to be no possible connection between the concepts of Fourier series and Taylor series, but when we pass into the complex realm a beautiful and remarkable fact emerges:

> *Taylor series and Fourier series of real functions are merely two different ways of viewing complex power series.*

We will explain this cryptic pronouncement by means of an example.

Consider the complex function $f(z) = 1/(1-z)$. Writing $z = r\,e^{i\theta}$, one finds [exercise] that the real and imaginary parts of $f(r\,e^{i\theta})$ are given by

$$f(r\,e^{i\theta}) = u(r\,e^{i\theta}) + i\,v(r\,e^{i\theta}) = \left[\frac{1 - r\cos\theta}{1 + r^2 - 2r\cos\theta}\right] + i\left[\frac{r\sin\theta}{1 + r^2 - 2r\cos\theta}\right].$$

Let's concentrate on just one of these real functions, say v.

[3] Later it was found that *some* restrictions must be placed on F, but they are astonishingly weak.

[4] In many areas of mathematics it is hard to find even one really enlightening book, but Fourier analysis has been blessed with at least two: Lanczos (1966), and Körner (1988).

If z moves outward from the origin along a ray $\theta = \text{const.}$ then $v(r\,e^{i\theta})$ becomes a function of r alone, say $V_\theta(r)$. For example,

$$V_{\frac{\pi}{4}}(r) = \frac{r}{\sqrt{2}(1+r^2) - 2r}.$$

If z instead travels round and round a circle $r = \text{const.}$ then v becomes a function of θ alone, say $\widetilde{V}_r(\theta)$. For example,

$$\widetilde{V}_{\frac{1}{2}}(\theta) = \frac{2\sin\theta}{5 - 4\cos\theta}.$$

Note that this is a periodic function of θ, with period 2π. The reason is simple and applies to any $\widetilde{V}_r(\theta)$ arising from a (single-valued) function $f(z)$: each time z makes a complete revolution and returns to its original position, $f(z)$ travels along a closed loop and returns to its original position.

Now, to see the unity of Taylor and Fourier series, recall that (within the unit disc) $f(z) = 1/(1-z)$ can be expressed as a convergent complex power series:

$$\begin{aligned} f(r\,e^{i\theta}) &= 1 + (r\,e^{i\theta}) + (r\,e^{i\theta})^2 + (r\,e^{i\theta})^3 + (r\,e^{i\theta})^4 + \cdots \\ &= 1 + r(\cos\theta + i\sin\theta) + r^2(\cos 2\theta + i\sin 2\theta) + r^3(\cos 3\theta + i\sin 3\theta) + \cdots. \end{aligned}$$

In particular,

$$v(r\,e^{i\theta}) = r\sin\theta + r^2\sin 2\theta + r^3\sin 3\theta + r^4\sin 4\theta + r^5\sin 5\theta + \cdots.$$

If we put $\theta = (\pi/4)$, we immediately obtain the Taylor series for $V_{\frac{\pi}{4}}(r)$:

$$\frac{r}{\sqrt{2}(1+r^2) - 2r} = V_{\frac{\pi}{4}}(r) = \tfrac{1}{\sqrt{2}}\,r + r^2 + \tfrac{1}{\sqrt{2}}\,r^3 - \tfrac{1}{\sqrt{2}}\,r^5 - r^6 - \tfrac{1}{\sqrt{2}}\,r^7 + \tfrac{1}{\sqrt{2}}\,r^9 + \cdots.$$

Once again, consider how difficult this would be to obtain using only real numbers. From this we find, for example, that

$$\frac{d^{98}}{dr^{98}}\left[\frac{r}{\sqrt{2}(1+r^2) - 2r}\right]\Bigg|_{r=0} = 98!$$

If we instead put $r = (1/2)$, we immediately obtain the Fourier series for $\widetilde{V}_{\frac{1}{2}}(\theta)$:

$$\frac{2\sin\theta}{5 - 4\cos\theta} = \widetilde{V}_{\frac{1}{2}}(\theta) = \tfrac{1}{2}\sin\theta + \tfrac{1}{2^2}\sin 2\theta + \tfrac{1}{2^3}\sin 3\theta + \tfrac{1}{2^4}\sin 4\theta + \cdots.$$

The absence of cosine waves in this series correctly reflects the fact that $\widetilde{V}_{\frac{1}{2}}(\theta)$ is an *odd* function of θ.

This connection between complex power series and Fourier series is not merely aesthetically satisfying, it can also be very practical. The conventional derivation of the Fourier series of $\widetilde{V}_{\frac{1}{2}}(\theta)$ requires that we evaluate the tricky integrals in (2.19), whereas we have obtained the result using only simple algebra! Indeed, we can now use our Fourier series to do integration:

$$\int_0^{2\pi} \left[\frac{2\sin\theta\sin n\theta}{5-4\cos\theta}\right] d\theta = \frac{\pi}{2^n}.$$

Further examples may be found in Exs. 21, 37, 38.

We end with a premonition of things to come. The coefficients in a Taylor series may be calculated by differentiation, while those in a Fourier series may be calculated by integration. Since these two types of series are really the same in the complex plane, this suggests that there exists some hidden connection between differentiation and integration that only complex numbers can reveal. Later we shall see how Cauchy confirmed this idea in spectacular fashion.

2.4 The Exponential Function

2.4.1 Power Series Approach

We have seen that the only complex function expressible as a power series that generalizes the real function e^x to complex values is

$$e^z = 1 + z + \tfrac{1}{2!}z^2 + \tfrac{1}{3!}z^3 + \tfrac{1}{4!}z^4 + \cdots,$$

which converges everywhere in $\mathbb{C}$. We now investigate the geometric nature of this function.

Figure [2.18] visualizes the above series as a spiral journey, the angle between successive legs of the journey being fixed and equal to • $= \arg(z) = \arg(x+iy)$. In the special case where this angle is a right angle, we saw in Chapter 1 that the spiral converges to a point on the unit circle given by Euler's formula, $e^{iy} = \cos y + i \sin y$. In fact this special spiral enables us to figure out what happens in the case of the general spiral in [2.18]: for an arbitrary value of $z = x + iy$, *the spiral converges to the illustrated point at distance* e^x *and at angle* y. In other words,

$$e^{x+iy} = e^x\, e^{iy}.$$

This is a consequence of the fact that if a and b are arbitrary complex numbers, then $e^a\, e^b = e^{a+b}$. To verify this we simply multiply the two series:

$$\begin{aligned}
e^a\, e^b &= \left[1 + a + \tfrac{1}{2!}a^2 + \tfrac{1}{3!}a^3 + \cdots\right]\left[1 + b + \tfrac{1}{2!}b^2 + \tfrac{1}{3!}b^3 + \cdots\right] \\
&= 1 + (a+b) + \left[\frac{a^2 + 2ab + b^2}{2!}\right] + \left[\frac{a^3 + 3a^2b + 3ab^2 + b^3}{3!}\right] + \cdots \\
&= 1 + (a+b) + \tfrac{1}{2!}(a+b)^2 + \tfrac{1}{3!}(a+b)^3 + \cdots \\
&= e^{a+b}.
\end{aligned}$$

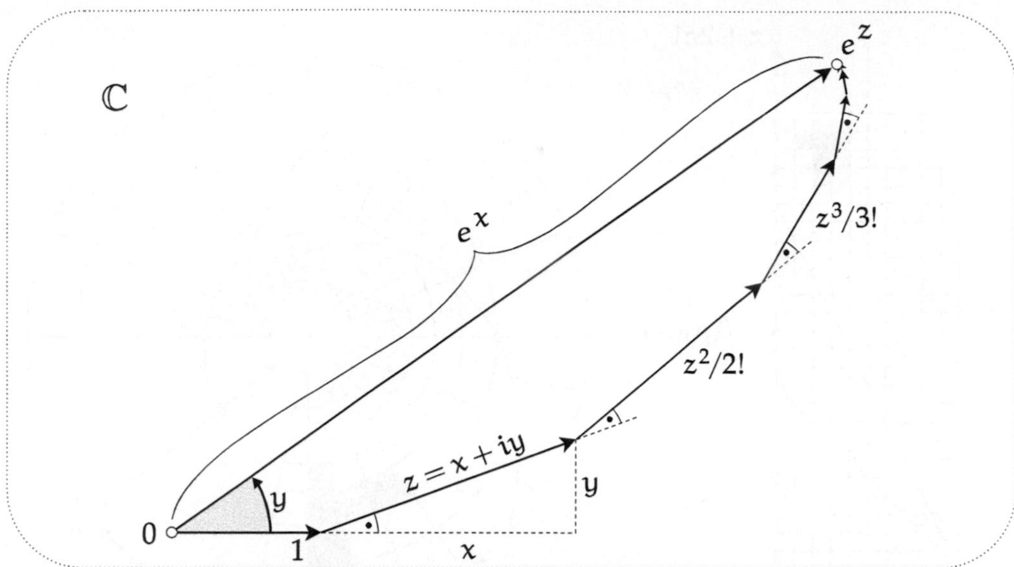

[2.18] *The power series for* e^z *can be visualized as a spiral journey, the angle between successive legs of the journey being fixed and equal to* $\bullet = \arg(z) = \arg(x+iy)$. *In the text we show that the power series has the property that* $e^a e^b = e^{a+b}$, *so that* $e^{x+iy} = e^x e^{iy}$. *Thus the spiral ends at the point that has length* e^x *and angle* y, *as illustrated.*

Here we have left it to you to show that the *general* term in the penultimate line is indeed $(a+b)^n/n!$.

2.4.2 The Geometry of the Mapping

Figure [2.19] illustrates the essential features of the mapping $z \mapsto w = e^z$. Study it carefully, noting the following facts:

- If z travels upward at a steady speed s, then w rotates about the origin at angular speed s. After z has travelled a distance of 2π, w returns to its starting position. Thus the mapping is *periodic*, with period $2\pi i$.
- If z travels westward at a steady speed, w travels towards the origin, with ever decreasing speed. Conversely, if z travels eastward at a steady speed, w travels away from the origin with ever increasing speed.
- Combining the previous two facts, the entire w-plane (with the exception of $w = 0$) will be filled by the image of any horizontal strip in the z-plane of height 2π.
- A line in general position is mapped to a spiral of the type discussed in the previous chapter.
- Euler's formula $e^{iy} = \cos y + i \sin y$ can be interpreted as saying that e^z wraps the imaginary axis round and round the unit circle like a piece of string.

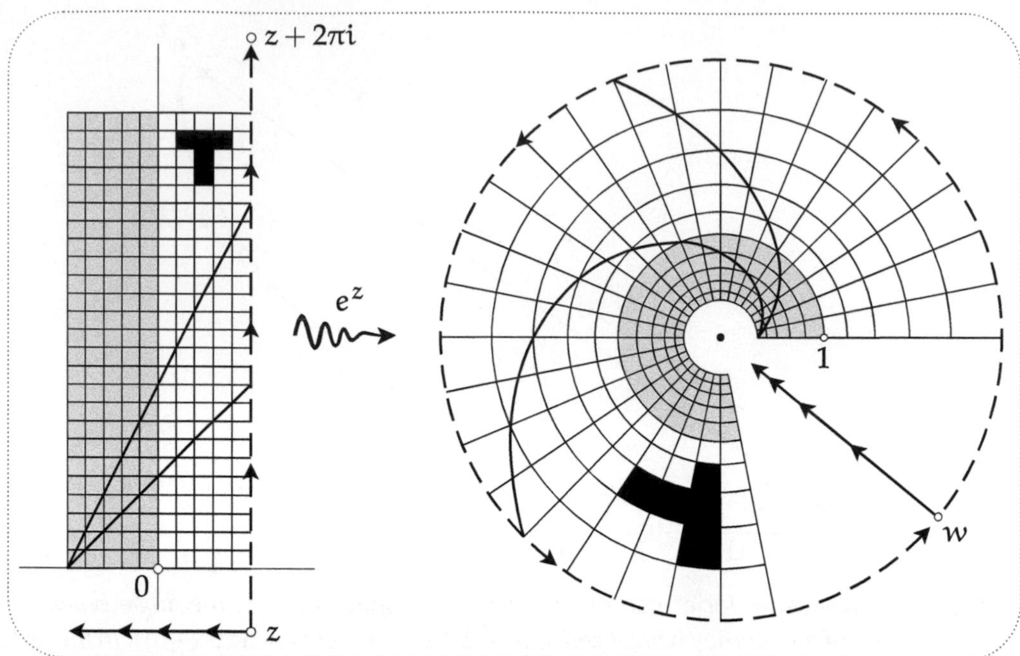

[2.19] The geometry of the complex exponential *follows from the fact that* e^{x+iy} *has length* e^x *and angle* y. *Thus horizontal lines* $y = \text{const.}$ *map to rays, and vertical lines* $x = \text{const.}$ *map to circles. Mysteriously, for now, a computer can be used to empirically confirm that small squares are ultimately mapped to small squares, in the limit that they shrink and vanish. This is a* central *mystery of complex analysis, and it will begin to unravel in Chapter 4.*

- The half-plane to the left of the imaginary axis is mapped to the interior of the unit circle, and the half-plane to the right of the imaginary axis is mapped to the exterior of the unit circle.
- The images of the small squares closely resemble squares, and (related to this) any two intersecting lines map to curves that intersect at the same angle as the lines themselves.

The last of these observations is not intended to be self-evident—in Chapter 4 we will begin to explore this fundamental property and to see that it is shared by many other important complex mappings.

2.4.3 *Another Approach*

The advantage of the power series approach to e^z is that it suggests that there is something unique about this generalization of e^x to complex values. The disadvantage is the amount of unilluminating algebra needed to decipher the geometric

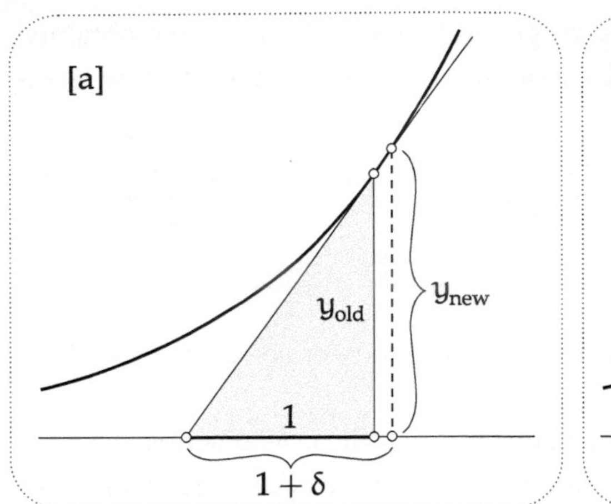

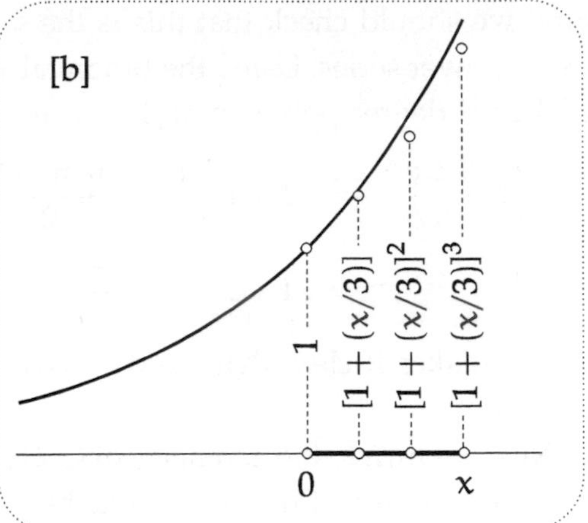

[2.20] [a] *The fact that* $(e^x)' = e^x$ *implies that the shaded triangle has unit base. It follows that* $y_{\text{new}} \asymp (1+\delta)\, y_{\text{old}}$ *as* δ *vanishes.* **[b]** *Now start with* $x = 0$ *and* $y_{\text{old}} = 1$. *Taking* $\delta = (x/n)$ *and repeatedly moving* δ *to the right, the height is ultimately multiplied by* $[1+\frac{x}{n}]$ *each time, so* $e^x \asymp [1+\frac{x}{n}]^n$.

meaning of the series. We now describe a different approach in which the geometry lies much closer to the surface. The idea is to generalize the real result,

$$e^x = \lim_{n\to\infty} \left(1 + \frac{x}{n}\right)^n. \tag{2.20}$$

Here is one way of understanding (2.20). As we discussed in Chapter 1, $f(x) = e^x$ may be defined by the property $f'(x) = f(x)$. Figure [2.20a] interprets this in terms of the graph of $y = f(x)$. Drawing a tangent at an arbitrary point, the base of the shaded triangle is always equal to 1. As you see from the figure, it follows that if the height is y_{old} at some point x, then moving x an infinitesimal distance δ to the right yields a new height given by

$$y_{\text{new}} = (1+\delta)\, y_{\text{old}}.$$

To find the height e^x at x, we divide the interval $[0, x]$ into a large number n of very short intervals of length (x/n). Since the height at $x = 0$ is 1, the height at (x/n) will be approximately $[1 + (x/n)] \cdot 1$, and so the height at $2(x/n)$ will be approximately $[1 + (x/n)] \cdot [1 + (x/n)] \cdot 1$, and so........, and so the height at $x = n(x/n)$ will be approximately $[1 + (x/n)]^n$. [For clarity's sake, [2.20b] illustrates this geometric progression with the small (hence inaccurate) value $n = 3$.] It is now plausible that the approximation $[1+(x/n)]^n$ becomes more and more accurate as n tends to infinity, thereby yielding (2.20). Try using a computer to verify empirically that the accuracy does indeed increase with n.

Generalizing (2.20) to complex values, we may define e^z as

$$e^z = \lim_{n\to\infty} \left(1 + \frac{z}{n}\right)^n. \tag{2.21}$$

First we should check that this is the *same* generalization of e^x that we obtained using power series. Using the Binomial Theorem to write down the first few terms of the nth degree polynomial $[1+(z/n)]^n$, we get

$$\begin{aligned}\left(1+\frac{z}{n}\right)^n &= 1+n\left[\frac{z}{n}\right]+\frac{n(n-1)}{2!}\left[\frac{z}{n}\right]^2+\frac{n(n-1)(n-2)}{3!}\left[\frac{z}{n}\right]^3+\cdots\\ &= 1+z+\frac{\left(1-\frac{1}{n}\right)}{2!}z^2+\frac{\left(1-\frac{1}{n}\right)\left(1-\frac{2}{n}\right)}{3!}z^3+\cdots,\end{aligned}$$

which makes it clear that we do recover the original power series as n tends to infinity.

Next we turn to the geometry of (2.21). In deciphering the power series for e^z we felt free to assume Euler's formula, because in Chapter 1 we used the power series to derive that result. However, it would smack of circular reasoning if we were to assume Euler's formula while following our new approach to e^z, based on (2.21). Temporarily, we shall therefore revert to our earlier notation and write $r\angle\theta$ instead of $r\,e^{i\theta}$; the fact we wish to understand is therefore written $e^{x+iy} = e^x\angle y$.

With $n = 6$, figure [2.21] uses Ex. 1.5, p. 52, to geometrically construct the successive powers of $a \equiv [1+(z/n)]$ for a specific value of z. [All six shaded triangles are similar; the two kinds of shading merely help to distinguish one triangle from the next.] Even with this small value of n, we see empirically that in this particular case $[1+(z/n)]^n$ is close to $e^x\angle y$. To understand this mathematically, we will try to approximate $a = [1+(z/n)]$ when n is large.

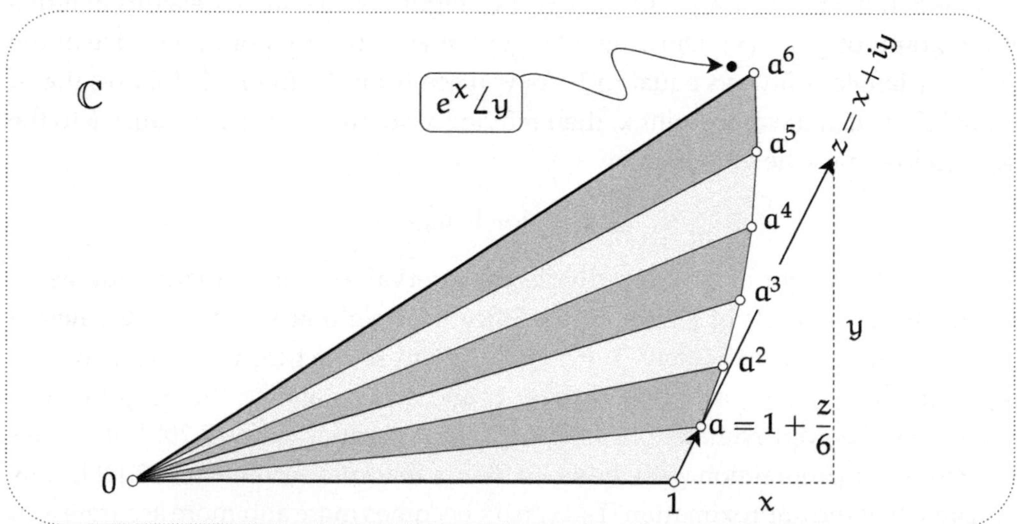

[2.21] *Begin with the lightly shaded triangle with vertices 0, 1, and* $a \equiv [1+(z/n)]$*, here illustrated with* $n = 6$*. Now successively construct triangles that are all similar to the original, as illustrated. We have thereby constructed the successive powers of* a*, culminating in* $a^n = [1+(z/n)]^n$*. Our earlier computations demonstrate that this must ultimately be equal to* $e^x\angle y$*, and this concrete example seems to confirm this. But what is the geometrical explanation?!*

Let ϵ be a small, ultimately vanishing, complex number. Consider the length r and angle θ of the number $(1+\epsilon) = r\angle\theta$ shown in [2.22]. The origin-centred circular arc [not shown] connecting $(1+\epsilon)$ to the point r on the real axis almost coincides with the illustrated perpendicular from $(1+\epsilon)$ to the real axis. Thus r is approximately equal to $[1+\mathrm{Re}(\epsilon)]$, and is ultimately equal[5] to it as ϵ tends to zero. Similarly, we see that the angle θ (the illustrated arc of the unit circle) is ultimately equal to $\mathrm{Im}(\epsilon)$. Thus

$$(1+\epsilon) \asymp [1+\mathrm{Re}(\epsilon)]\angle\mathrm{Im}(\epsilon), \qquad \text{as } \epsilon \text{ vanishes.}$$

Now set $\epsilon = (z/n) = (x+iy)/n$. With the same values of z and n as in [2.21], figure [2.23] shows the approximation $b \equiv \left(1+\frac{x}{n}\right) \angle \left(\frac{y}{n}\right)$ to a, together with its successive powers.

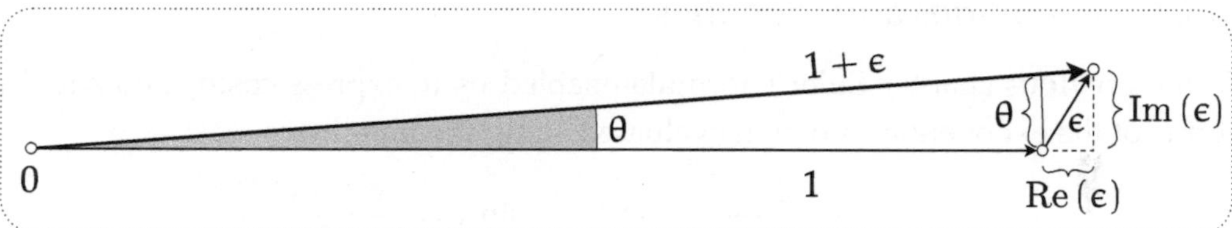

[2.22] *As ϵ vanishes, we see that* $(1+\epsilon) \asymp [1+\mathrm{Re}(\epsilon)]\angle\mathrm{Im}(\epsilon)$.

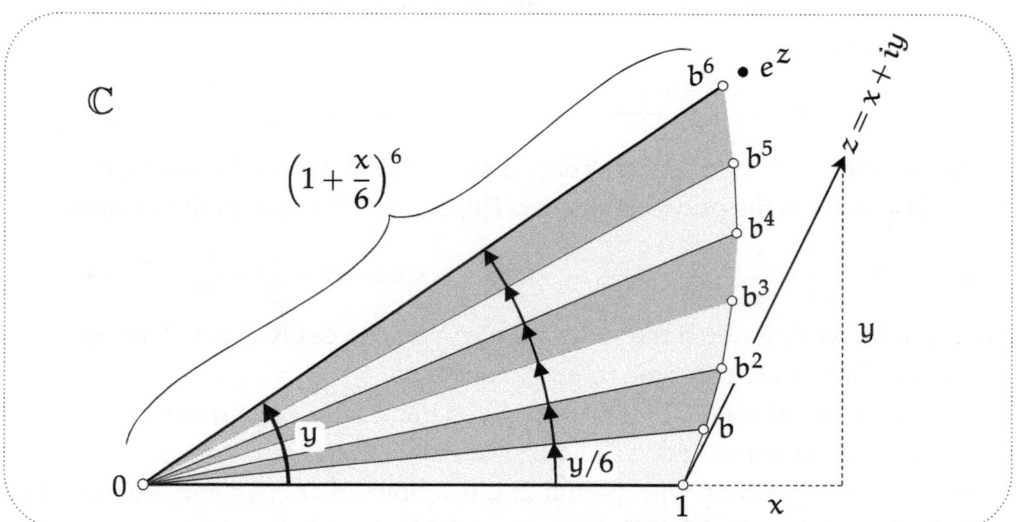

[2.23] Geometrical explanation of [2.21]. *Taking* $\epsilon = (z/n) = (x+iy)/n$ *in the previous figure, we may now approximate* $a \equiv [1+(z/n)]$ *as* $b \equiv \left(1+\frac{x}{n}\right) \angle \left(\frac{y}{n}\right)$, *because* $a \asymp b$ *as* n *goes to infinity. Here we once again illustrate the case* $n = 6$. *Thus,* $\left[1+\frac{z}{n}\right]^n \asymp \left[\left(1+\frac{x}{n}\right) \angle \left(\frac{y}{n}\right)\right]^n = \left(1+\frac{x}{n}\right)^n \angle y \asymp e^x \angle y$, *and the mystery is solved!*

[5] Once again, "$\asymp$" denotes Newton's concept of *ultimate equality*; see the new Preface.

Returning to the general case, the geometry of (2.21) should now be clear. As n goes to infinity,

$$\left[1+\frac{z}{n}\right]^n \asymp \left[\left(1+\frac{x}{n}\right) \angle \left(\frac{y}{n}\right)\right]^n = \left(1+\frac{x}{n}\right)^n \angle y.$$

Thus, in this limit as n goes to infinity, and using (2.20), we deduce that

$$e^{x+iy} = e^x \angle y,$$

as was to be shown. In particular, if we put $x = 0$ then we recover Euler's formula, $e^{iy} = 1\angle y$, and so we are entitled to write $e^{x+iy} = e^x\, e^{iy}$.

For a slightly different way of looking at (2.21), see Ex. 22.

2.5 Cosine and Sine

2.5.1 *Definitions and Identities*

In the previous chapter Euler's formula enabled us to express cosine and sine in terms of the exponential function evaluated along the imaginary axis:

$$\cos x = \frac{e^{ix}+e^{-ix}}{2} \quad \text{and} \quad \sin x = \frac{e^{ix}-e^{-ix}}{2i}.$$

Now that we understand the effect of e^z on arbitrary points (not merely points on the imaginary axis), it is natural to extend the definitions of cosine and sine to the complex functions

$$\cos z \equiv \frac{e^{iz}+e^{-iz}}{2} \quad \text{and} \quad \sin z \equiv \frac{e^{iz}-e^{-iz}}{2i}. \tag{2.22}$$

Of course another way of generalizing $\cos x$ and $\sin x$ would be via their power series, discussed in the previous chapter. This leads to the alternative definitions,

$$\cos z = 1 - \frac{z^2}{2!} + \frac{z^4}{4!} - \frac{z^6}{6!} + \cdots, \quad \text{and} \quad \sin z = z - \frac{z^3}{3!} + \frac{z^5}{5!} - \frac{z^7}{7!} + \cdots.$$

However, by writing down the series for $e^{\pm iz}$ you can easily check that these two approaches both yield the same complex functions.

From the definitions (2.22) we see that $\cos z$ and $\sin z$ have much in common with their real ancestors. For example, $\cos(-z) = \cos z$, and $\sin(-z) = -\sin z$. Also, since e^z is periodic with period $2\pi i$, it follows that $\cos z$ and $\sin z$ are also periodic, but with period 2π. The meaning of this periodicity will become clearer when we examine the geometry of the mappings.

Other immediate consequences of (2.22) are the following important generalizations of Euler's formula:

$$e^{iz} = \cos z + i \sin z \quad \text{and} \quad e^{-iz} = \cos z - i \sin z.$$

WARNING: $\cos z$ and $\sin z$ are now complex numbers—they are *not* the real and imaginary parts of e^{iz}.

It is not hard to show that all the familiar identities for $\cos x$ and $\sin x$ continue to hold for our new complex functions. For example, we still have

$$\cos^2 z + \sin^2 z = (\cos z + i \sin z)(\cos z - i \sin z) = e^{iz}e^{-iz} = e^0 = 1,$$

despite the fact that this identity no longer expresses Pythagoras' Theorem. Similarly, we will show that if a and b are arbitrary complex numbers then

$$\cos(a+b) = \cos a \cos b - \sin a \sin b \tag{2.23}$$

$$\sin(a+b) = \sin a \cos b + \cos a \sin b, \tag{2.24}$$

despite the fact that these identities no longer express the geometric rule for multiplying points on the unit circle. First,

$$\begin{aligned}\cos(a+b) + i\sin(a+b) &= e^{i(a+b)}\\ &= e^{ia}e^{ib}\\ &= (\cos a + i \sin a)(\cos b + i \sin b)\\ &= (\cos a \cos b - \sin a \sin b) + i(\sin a \cos b + \cos a \sin b),\end{aligned}$$

exactly as in the previous chapter. However, in view of the warning above, we *do not* obtain (2.23) and (2.24) simply by equating real and imaginary parts. Instead [exercise] one first finds the analogous identity for $\cos(a+b) - i \sin(a+b)$, then adds it to (or subtracts it from) the one above.

2.5.2 *Relation to Hyperbolic Functions*

Recall the definitions of the hyperbolic cosine and sine functions:

$$\cosh x \equiv \frac{e^x + e^{-x}}{2} \qquad \text{and} \qquad \sinh x \equiv \frac{e^x - e^{-x}}{2}.$$

By interpreting each of these as the average (i.e., midpoint) of e^x and $\pm e^{-x}$, it is easy to obtain the graphs $y = \cosh x$ and $y = \sinh x$ shown in [2.24a] and [2.24b].

As you probably know, $\cosh x$ and $\sinh x$ satisfy identities that are remarkably similar to those satisfied by $\cos x$ and $\sin x$, respectively. For example, if r_1 and r_2 are arbitrary real numbers, then [exercise]

$$\cosh(r_1 + r_2) = \cosh r_1 \cosh r_2 + \sinh r_1 \sinh r_2 \tag{2.25}$$

$$\sinh(r_1 + r_2) = \sinh r_1 \cosh r_2 + \cosh r_1 \sinh r_2. \tag{2.26}$$

Nevertheless, [2.24] shows that the actual behaviour of the hyperbolic functions is quite unlike the circular functions: they are not periodic, and they become arbitrarily large as x tends to infinity. It is therefore surprising and pleasing that the introduction of complex numbers brings about a unification of these two types of functions.

We begin to see this if we restrict $z = iy$ to the imaginary axis, for then

$$\cos(iy) = \cosh y \qquad \text{and} \qquad \sin(iy) = i \sinh y.$$

This connection becomes particularly vivid if we consider the modular surface of $\sin z$. Since $|\sin z|$ is ultimately equal to $|z|$ as z approaches the origin, it follows that

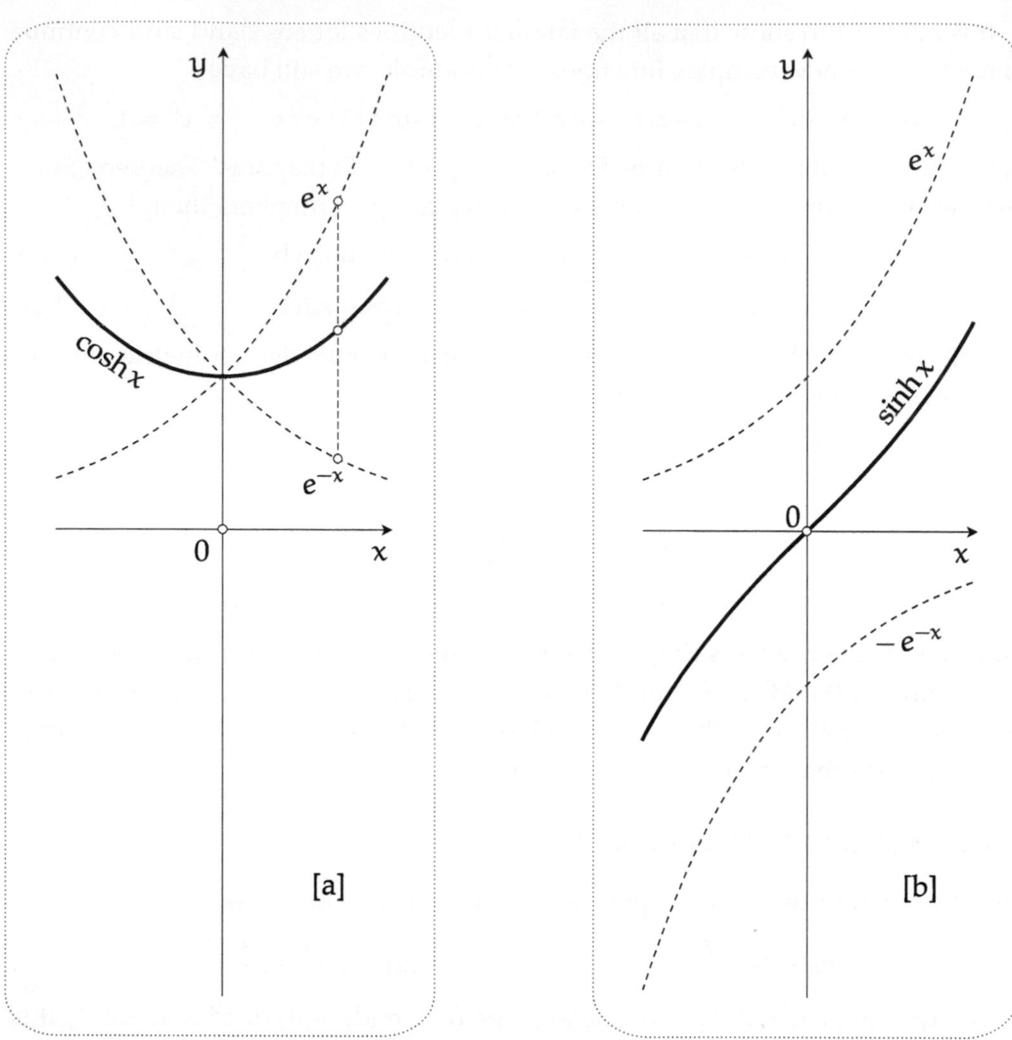

[2.24] [a] *The graph of* $\cosh x$ *may be visualized as the average (i.e., midpoint) of the graphs of* e^x *and* e^{-x}. **[b]** *The graph of* $\sinh x$ *may be visualized in the same way, as the average of* e^x *and* $-e^{-x}$.

the surface rises above the origin in the form of a cone. Also, $|\sin(z + \pi)| = |\sin z|$, so there is an identical cone at each multiple of π along the real axis. These are the only points [exercise] at which the surface hits the plane. Figure [2.25]—which we have adapted from Markushevich (2005)—shows a portion of the surface. Notice that this surface also yields the cosh graph, for if we restrict $z = (3\pi/2) + iy$ to the line $x = (3\pi/2)$, for example, then $|\sin z| = \cosh y$.

A practical benefit of this unification is that if you can remember (or quickly derive using Euler's formula) a trig identity involving cosine and sine, then you can immediately write down the corresponding identity for the hyperbolic functions. For example, if we substitute $a = ir_1$ and $b = ir_2$ into (2.23) and (2.24), then we obtain (2.25) and (2.26).

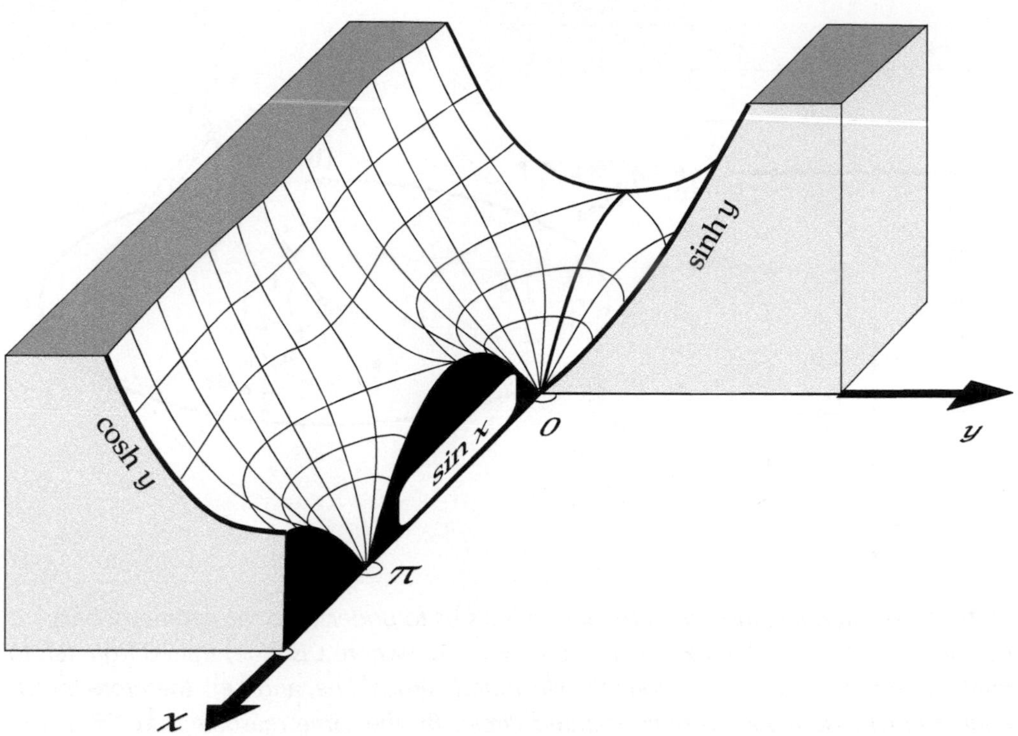

[2.25] *Cutting open the modular surface of the complex function* sin z *reveals the previously hidden unity of the circular and hyperbolic functions.*

The connection between the circular and hyperbolic functions becomes stronger still if we generalize the latter to complex functions in the obvious way:

$$\cosh z \equiv \frac{e^z + e^{-z}}{2} \quad \text{and} \quad \sinh z \equiv \frac{e^z - e^{-z}}{2}.$$

Since we now have

$$\cosh z = \cos(iz) \quad \text{and} \quad \sinh z = -i\,\sin(iz),$$

the distinction between the two kinds of function has all but evaporated: cosh is the composition of a rotation through $(\pi/2)$, followed by cos; also, sinh is the composition of a rotation through $(\pi/2)$, followed by sin, followed by a rotation through $-(\pi/2)$.

2.5.3 *The Geometry of the Mapping*

Just as in the real case, $\sin z = \cos(z - \frac{\pi}{2})$, which means that we may obtain sin from cos by first translating the plane by $-(\pi/2)$. It follows from the preceding remarks that it is sufficient to study just $\cos z$ in order to understand all four functions, $\cos z$, $\sin z$, $\cosh z$, and $\sinh z$. We now consider the geometric nature of the mapping $z \mapsto w = \cos z$.

We begin by finding the image of a horizontal line $y = -c$ lying below the real axis. It is psychologically helpful to picture the line as the orbit of a particle moving

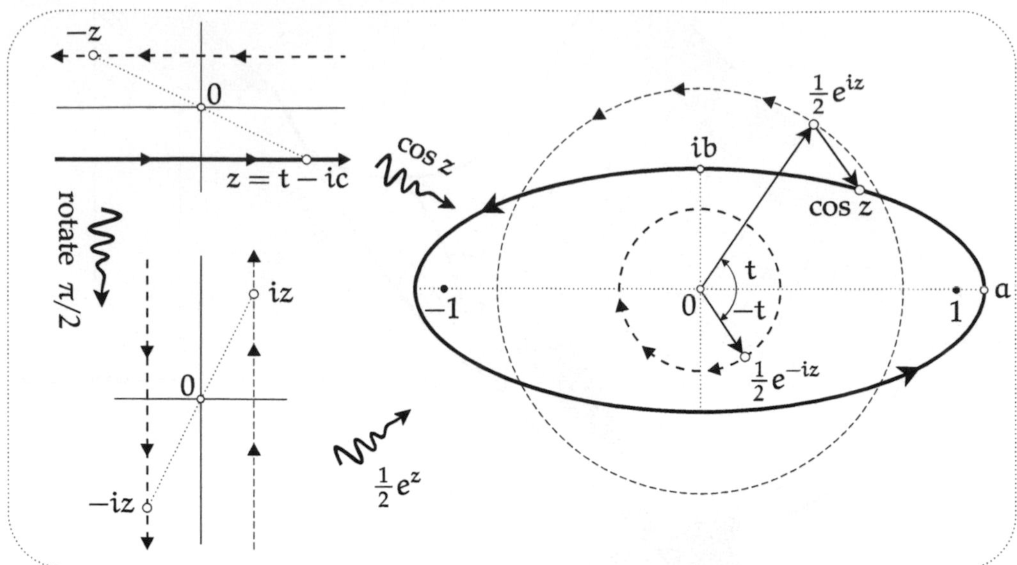

[2.26] *Understanding the geometry of* e^z *allows us to understand the geometry of* $\cos z$. *Consider the illustrated horizontal line* $z(t) = t - ic$ *(where* t *is time) traced from left to right. Then* iz *travels upward along the illustrated vertical line, and* $\frac{1}{2}e^{iz}$ *therefore travels counterclockwise along an origin-centred circle. By the same reasoning,* $\frac{1}{2}e^{-iz}$ *travels clockwise along the smaller circle. Finally,* $\cos z$ *is the sum of these two counter-rotating circular motions, and it traces an* ellipse. *It turns out that all horizontal lines are mapped to* confocal *ellipses, with foci* ± 1!

eastward at unit speed, whose position at time t is $z = t - ic$. See [2.26], in which the line is shown heavy and unbroken. As z traces this line, $-z$ traces the line $y = c$, but in the opposite direction. Applying the mapping $z \mapsto iz$ (which is a rotation of $\frac{\pi}{2}$), the image particles trace the vertical lines $x = \pm c$, again with unit speed and in opposite directions. Finally applying $z \mapsto \frac{1}{2}e^z$, the image particles orbit with equal and opposite angular speeds in origin-centred circles of radii $\frac{1}{2}e^{\pm c}$.

The image orbit under $z \mapsto w = \cos z$ of the original particle travelling on the line $y = -c$ is just the sum of these counter-rotating circular motions. This is clearly some kind of symmetrical oval hitting the real and imaginary axes at $a = \cosh c$ and $ib = i \sinh c$. It is also clear that $\cos z$ executes a complete orbit of this oval with each movement of 2π by z; this is the geometric meaning of the periodicity of $\cos z$.

I haven't found a simple geometric explanation, but it's easy to show symbolically that the oval traced by $\cos z$ is a perfect ellipse. Writing $w = u + iv$, we find from the figure [exercise] that $u = a \cos t$ and $v = b \sin t$, which is the familiar parametric representation of the ellipse $(u/a)^2 + (v/b)^2 = 1$. Furthermore,

$$\sqrt{a^2 - b^2} = \sqrt{\cosh^2 c - \sinh^2 c} = 1,$$

so the foci are at ± 1, independent of which particular horizontal line z travels along.

Try mulling this over. How does the shape of the ellipse change as we vary c? How do we recover the real cosine function as c tends to zero? What is the orbit of $\cos z$ as z travels eastward along the line $y = c$, above the real axis? What is the image of the vertical line $x = c$ under $z \mapsto \cosh z$? What is the orbit of $\sin z$ as z travels eastward along the line $y = c$; how does it differ from the orbit of $\cos z$; and is the resulting variation of $|\sin z|$ consistent with the modular surface shown in [2.25]?

Before reading on, try using the idea in [2.26] to sketch for yourself the image under $z \mapsto \cos z$ of a *vertical* line.

As illustrated in [2.27], the answer is a *hyperbola*. We can show this using the addition rule (2.23), which yields

$$u + iv = \cos(x + iy) = \cos x \cosh y - i \sin x \sinh y.$$

On a horizontal line, y is constant, so $(u/\cosh y)^2 + (v/\sinh y)^2 = 1$, as before. On a vertical line, x is constant, so $(u/\cos x)^2 - (v/\sin x)^2 = 1$, which is the equation of a hyperbola. Furthermore, since $\cos^2 x + \sin^2 x = 1$, it follows that the foci of the hyperbola are always ± 1, independent of which vertical line is being mapped.

Figure [2.27] tries to make these results more vivid by showing the image of a grid of horizontal and vertical lines. Note the empirical fact that *each small square in*

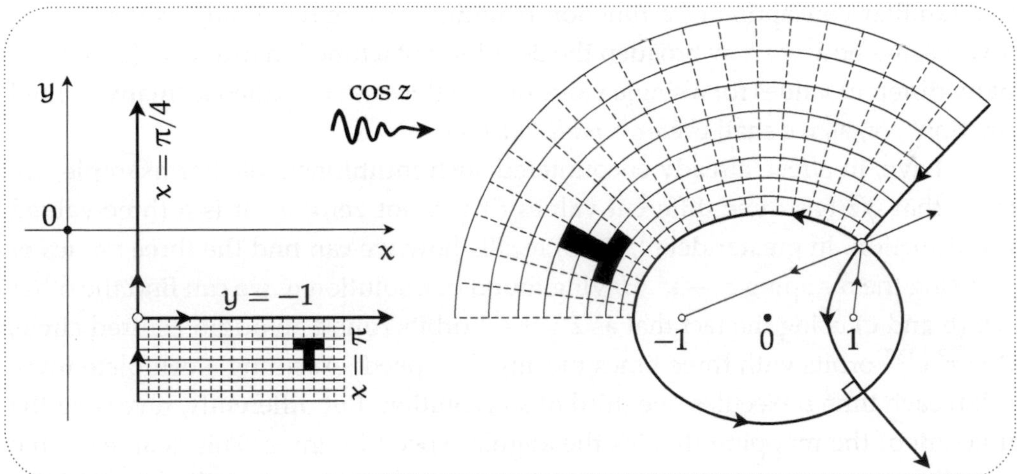

[2.27] *Using the same reasoning as the previous figure, vertical lines are mapped by* $\cos z$ *to hyperbolas, and calculation reveals that these are also* confocal, *with the* same *foci as the ellipses,* ± 1*! Next, recall the* reflection property *of the conic sections: a ray of light emitted from a focus is reflected directly towards the other focus by the ellipse, and directly away from the other focus by the hyperbola, as illustrated. It follows that* these two sets of conic sections are the orthogonal trajectories of each other. *Furthermore—and this is an additional, surprising, and presently* mysterious *fact—as the size of the squares in the grid on the left tends to zero, their images under* $\cos z$ *are ultimately* squares, *too, as illustrated!*

the grid is mapped by $\cos z$ *to an image shape that is again approximately square.* This is the same surprising (and visually pleasing) phenomenon that we observed in the case of $z \mapsto e^z$.

We hope your curiosity is piqued—later chapters are devoted to probing this phenomenon in depth. In the present case of $z \mapsto \cos z$ we can at least give a mathematical explanation of part of the result, namely, that the sides of the image "squares" do indeed meet at right angles; in other words, each ellipse cuts each hyperbola at right angles.

This hinges on the fact that these ellipses and hyperbolas are *confocal*. To prove the desired result [exercise], think of each curve as a mirror, then appeal to the familiar *reflection property* of the conic sections: a ray of light emitted from a focus is reflected directly towards the other focus by the ellipse, and directly away from the other focus by the hyperbola, as illustrated in [2.27].

2.6 Multifunctions

2.6.1 Example: Fractional Powers

Thus far we have considered a complex function f to be a rule that assigns to each point z (perhaps restricted to lie in some region) a single complex number $f(z)$. This familiar conception of a function is unduly restrictive. Using examples, we now discuss how we may broaden the definition of a function to allow $f(z)$ to have many different values for a single value of z. In this case f is called a "many-valued function", or, as we shall prefer, a *multifunction*.

We have, in effect, already encountered such multifunctions. For example, we know that $\sqrt[3]{z}$ has three different values (if z is not zero), so it is a three-valued multifunction. In greater detail, [2.28] recalls how we can find the three values of $\sqrt[3]{p}$ using the mapping $z \mapsto z^3$. Having found one solution a, we can find the other two (b and c) using the fact that as $z = r\,e^{i\theta}$ orbits round an origin-centred circle, $z^3 = r^3\,e^{i3\theta}$ orbits with three times the angular speed, executing a complete revolution each time z executes one third of a revolution. Put differently, reversing the direction of the mapping divides the angular speed by three. This is an essential ingredient in understanding the mapping $z \mapsto \sqrt[3]{z}$, which we will now study in detail.

Writing $z = r\,e^{i\theta}$, we have $\sqrt[3]{z} = \sqrt[3]{r}\,e^{i(\theta/3)}$. Here $\sqrt[3]{r}$ is uniquely defined as the real cube root of the length of z; the sole source of the three-fold ambiguity in the formula is the fact that there are infinitely many different choices for the angle θ of a given point z.

Think of z as a moving point that is initially at $z = p$. If we arbitrarily choose θ to be the angle ϕ shown in [2.28], then $\sqrt[3]{p} = a$. As z gradually moves away from p, θ gradually changes from its initial value ϕ, and $\sqrt[3]{z} = \sqrt[3]{r}\,e^{i(\theta/3)}$ gradually moves

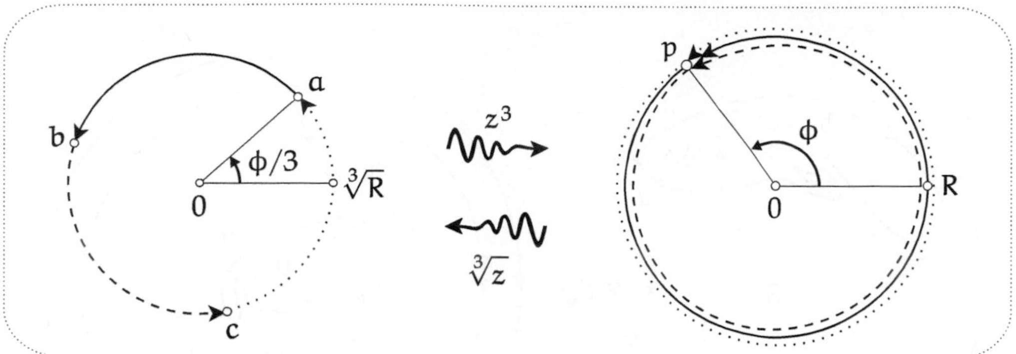

[2.28] The cube root is a three-valued multifunction. *Since $z \mapsto z^3$ multiplies angles by 3, $z \mapsto \sqrt[3]{z}$ divides angles by 3. Now suppose we start at the illustrated point p at angle ϕ on the circle of radius R. Then it would seem clear that its cube root is the illustrated point $\sqrt[3]{p} = a$ at angle $(\phi/3)$ on the circle of radius $\sqrt[3]{R}$. But now suppose p executes a complete revolution and returns to its starting point. Then $\sqrt[3]{z}$ rotates at one third the angular speed and arrives at b: a second equally good value of $\sqrt[3]{p}$. If p keeps rotating, executing a second revolution, we arrive at $\sqrt[3]{p} = c$, and a third revolution returns us to the original value $\sqrt[3]{p} = a$. Thus the three values of $\sqrt[3]{p}$ are a, b, and c, and they form an equilateral triangle.*

away from its initial position a, *but in a completely determined way*—its distance from the origin is the cube root of the distance of z, and its angular speed is one third that of z.

Figure [2.29] illustrates this. Usually we draw mappings going from left to right, but here we have reversed this convention to facilitate comparison with [2.28].

As z travels along the closed loop A (finally returning to p), $\sqrt[3]{z}$ travels along the illustrated closed loop and returns to its original value a. However, if z instead travels along the closed loop B, which goes round the origin once, then $\sqrt[3]{z}$ does *not* return to its original value but instead ends up at a different cube root of p, namely b. Note that the detailed shape of B is irrelevant, all that matters is that it encircles the origin once. Similarly, if z travels along C, encircling the origin twice, then $\sqrt[3]{z}$ ends up at c, the third and final cube root of p. Clearly, if z were to travel along a loop [not shown] that encircled the origin three times, then $\sqrt[3]{z}$ would return to its original value a.

The premise for this picture of $z \mapsto \sqrt[3]{z}$ was the arbitrary choice of $\sqrt[3]{p} = a$, rather than b or c. If we instead chose $\sqrt[3]{p} = b$, then the orbits on the left of [2.29] would simply be rotated by $(2\pi/3)$. Similarly, if we chose $\sqrt[3]{p} = c$, then the orbits would be rotated by $(4\pi/3)$.

The point $z = 0$ is called a *branch point* of $\sqrt[3]{z}$. More generally, let $f(z)$ be a multifunction and let $a = f(p)$ be one of its values at some point $z = p$. Arbitrarily choosing the initial position of $f(z)$ to be a, we may follow the movement of $f(z)$ as z travels along a closed loop beginning and ending at p. When z returns to p, $f(z)$

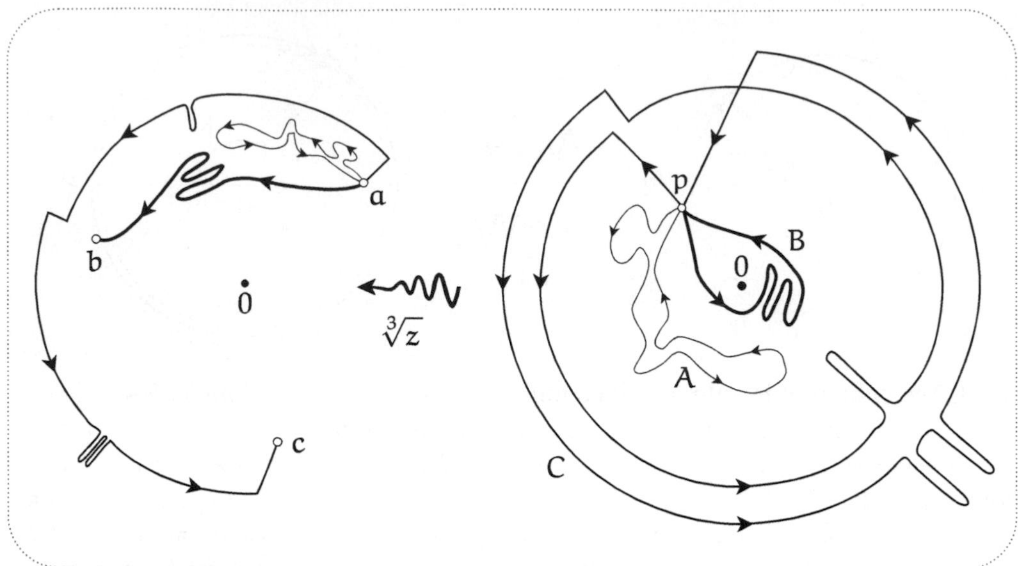

[2.29] *Writing* $z = r\,e^{i\theta}$, *we have* $\sqrt[3]{z} = \sqrt[3]{r}\,e^{i(\theta/3)}$. *Once again, let us arbitrarily start at* $\sqrt[3]{p} = a$, *and then let* p *move along the three illustrated paths, A, B, and C. As you see, which cube root we end up at does not depend on the detailed shape of the path, it* only *depends on the net change in* θ*—how many complete revolutions have been executed when* p *returns home. The cube root rotates one third of a revolution each time* p *loops around the* branch point *at the origin.*

will either return to a or it will not. A branch point $z = q$ of f is a point such that $f(z)$ fails to return to a as z travels along any loop that encircles q once.

Returning to the specific example $f(z) = \sqrt[3]{z}$, we have seen that if z executes three revolutions round the branch point at $z = 0$ then $f(z)$ returns to its original value. If $f(z)$ were an ordinary, single-valued function then it would return to its original value after only one revolution. Thus, relative to an ordinary function, *two extra* revolutions are needed to restore the original value of $f(z)$. We summarize this by saying that 0 is a branch point of $\sqrt[3]{z}$ of *order* two.

More generally, if q is a branch point of some multifunction $f(z)$, and $f(z)$ first returns to its original value after N revolutions round q, then q is called an *algebraic branch point* of order $(N - 1)$; an algebraic branch point of order 1 is called a *simple branch point*. We should stress that it is perfectly possible that $f(z)$ never returns to its original value, no matter how many times z travels round q. In this case q is called a *logarithmic branch point*—the name will be explained in the next section.

By extending the above discussion of $\sqrt[3]{z}$, check for yourself that if n is an integer then $z^{(1/n)}$ is an n-valued multifunction whose only (finite) branch point is at $z = 0$, the order of this branch point being $(n - 1)$. More generally, the same is true for any fractional power $z^{(m/n)}$, where (m/n) is a fraction reduced to lowest terms.

2.6.2 *Single-Valued Branches of a Multifunction*

Next we will show how we may extract three ordinary, single-valued functions from the three-valued multifunction $\sqrt[3]{z}$. First, [2.30] introduces some terminology which we need for describing sets of points in $\mathbb{C}$.

A set S is said to be *connected* (see [2.30a]) if any two points in S can be connected by an unbroken curve lying entirely within S. Conversely, if there exist pairs of points that cannot be connected in this way (see [2.30b]), then the set is *disconnected*. Amongst connected sets we may single out the *simply connected* sets (see [2.30c]) as those that do not have holes in them. More precisely, if we picture the path connecting two points in the set as an elastic string, then this string may be continuously deformed into any other path connecting the points, without any part of the string ever leaving the set. Conversely, if the set does have holes in it then it is *multiply connected* (see [2.30d]) and there exist two paths connecting two points such that one path cannot be deformed into the other.

Now let us return to [2.29]. By arbitrarily picking one of the three values of $\sqrt[3]{p}$ at $z = p$, and then allowing z to move, we see that we obtain a unique value of $\sqrt[3]{Z}$ associated with any particular path from p to Z. However, we are still dealing with a multifunction: by going round the branch point at 0 we can end up at any one of the three possible values of $\sqrt[3]{Z}$.

On the other hand, the value of $\sqrt[3]{Z}$ does not depend on the detailed shape of the path: *if we continuously deform the path without crossing the branch point then we obtain the same value of* $\sqrt[3]{Z}$. This shows us how we may obtain a single-valued function. If we restrict z to any simply connected set S that contains p but does not contain the branch point, then every path in S from p to Z will yield the same value of $\sqrt[3]{Z}$,

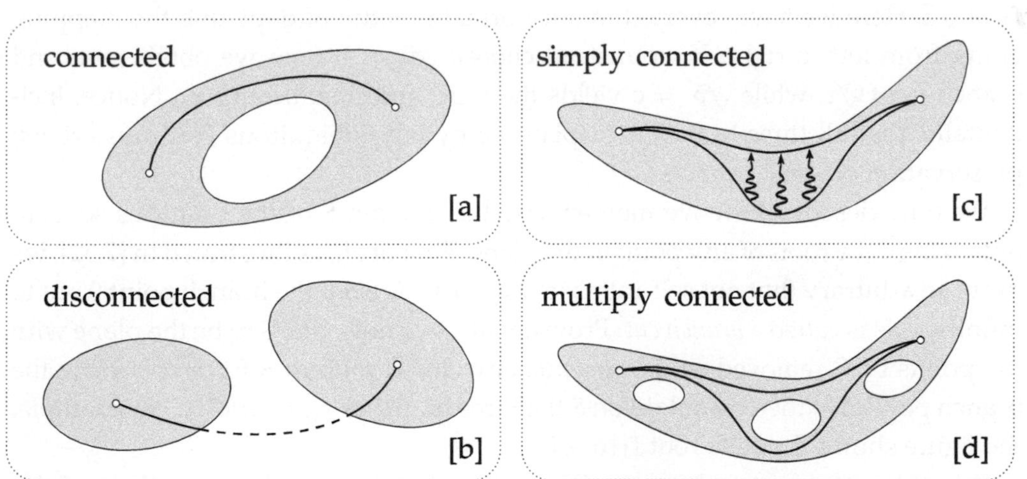

[2.30] *Here is the* terminology *that is used to describe different types of regions in the plane.*

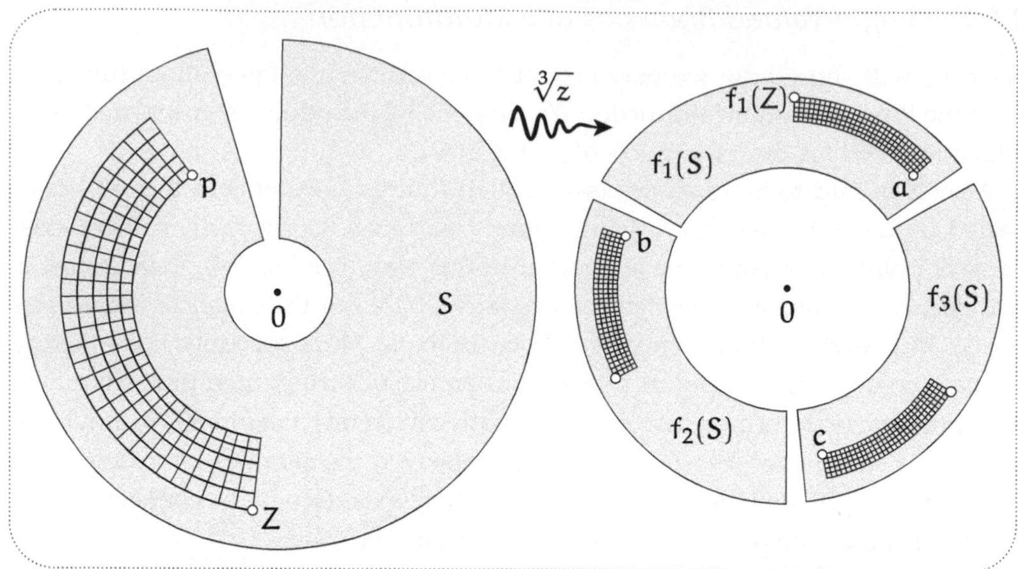

[2.31] *If we chose a simply connected region S that does not contain the branch point at 0, then no path within S can encircle the branch point at 0, and therefore it is possible to define a single-valued branch of $\sqrt[3]{z}$ within S. For example, if we arbitrarily take $\sqrt[3]{p} = a$, then let $z = p$ move about within S and arrive at $z = Z$, say, then $\sqrt[3]{z}$ arrives at a unique value $f_1(Z) = \sqrt[3]{Z}$ that is independent of the path. But since there are three values of $\sqrt[3]{p}$, there are three distinct branches of $\sqrt[3]{z}$. Incidentally, note that these three branches display the ubiquitous (yet mysterious) preservation of small squares.*

which we will call $f_1(Z)$. Since the path is irrelevant, f_1 is an ordinary, single-valued function of position on S; it is called a *branch* of the original multifunction $\sqrt[3]{z}$.

Figure [2.31] illustrates such a set S, together with its image under the branch f_1 of $\sqrt[3]{z}$. Here we have reverted to our normal practice of depicting the mapping going from left to right. If we instead choose $\sqrt[3]{p} = b$ then we obtain a second branch f_2 of $\sqrt[3]{z}$, while $\sqrt[3]{p} = c$ yields the third and final branch f_3. Notice, incidentally, that all three branches display the by now ubiquitous (yet mysterious) preservation of small squares.

We now describe how we may enlarge the domain S of the branches so as to obtain the cube roots of any point in the plane. First of all, as illustrated in [2.32], we draw an arbitrary (but not self-intersecting) curve C from the branch point 0 out to infinity; this is called a *branch cut*. Provisionally, we now take S to be the plane with the points of C removed—this prevents any closed path in S from encircling the branch point. We thereby obtain on S the three branches f_1, f_2, and f_3. For example, the figure shows the cube root $f_1(d)$ of d.

What about a point such as e *on* C? Imagine that z is travelling round an origin-centred circle through e. The figure illustrates the fact that $f_1(z)$ approaches two different values according as z arrives at e with positive or negative angular speed.

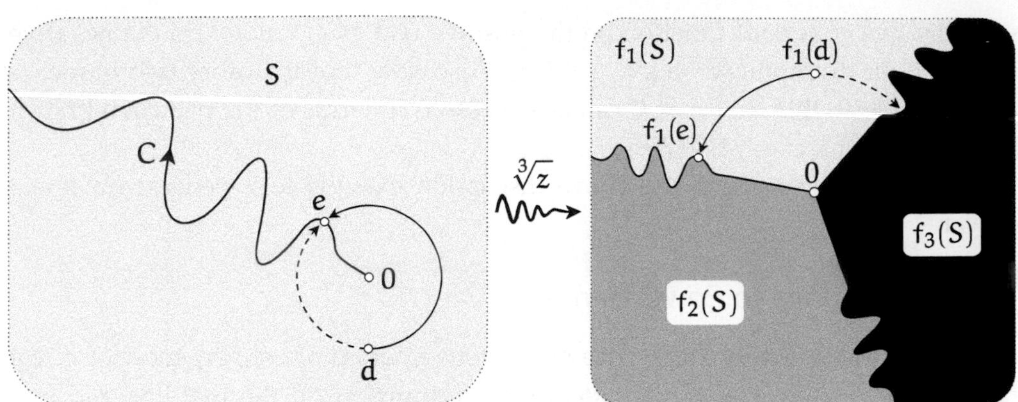

[2.32] *In order to define* **a single-valued branch of** $f(z) = \sqrt[3]{z}$ *throughout the* entire *complex plane, we begin by drawing an arbitrary curve* C *that starts at the branch point* 0 *and extends out to infinity, as illustrated, and define* S *to be* $\mathbb{C}$ *with the points of* C removed. *The curve* C *is called a* branch cut. *Out of the three possible choices for* $\sqrt[3]{d}$, *let us arbitrarily choose the illustrated point* $f_1(d)$. *Then* $f_1(Z)$ *is single-valued throughout* S, *because it is impossible for* z *to encircle the branch point at* 0 *as it travels from* d *to* Z. *But we still need to define* $f_1(e)$ *at a point such as* e *that lies* on C. *By convention, it is defined to be the value of* $f_1(z)$ *when* z *arrives at* e *travelling* counterclockwise, *i.e., arriving from the* right *of* C. *The two other branches are defined in exactly the same way.*

If we (arbitrarily) define $f_1(e)$ to be the value of $f(z)$ when z travels *counterclockwise* round the circle, then $f_1(z)$ is well defined on the whole plane. Similarly for the other two branches.

Of course the branch cut C is the work of man—the multifunction $\sqrt[3]{z}$ is oblivious to our desire to dissect it into three single-valued functions. As we have just seen, this shows up in the fact that the resulting branches are *discontinuous* on C, despite the fact that the three values of $\sqrt[3]{z}$ *always* move continuously as z moves continuously. As z crosses C travelling counterclockwise then we must switch from one branch to the next in order to maintain continuous motion of $\sqrt[3]{z}$: for example, f_1 switches to f_2. If z executes three counterclockwise revolutions round the branch point, then the branches permute cyclically, each finally returning to itself: using an arrow to denote a crossing of C,

$$\left\{\begin{array}{c} f_1 \\ f_2 \\ f_3 \end{array}\right\} \to \left\{\begin{array}{c} f_2 \\ f_3 \\ f_1 \end{array}\right\} \to \left\{\begin{array}{c} f_3 \\ f_1 \\ f_2 \end{array}\right\} \to \left\{\begin{array}{c} f_1 \\ f_2 \\ f_3 \end{array}\right\}.$$

A common choice for C is the negative real axis. If we do not allow z to cross the cut then we may restrict the angle $\theta = \arg(z)$ to lie in the range $-\pi < \theta \leqslant \pi$. This is called the *principal value of the argument*, written Arg (z); note the capital first letter. With this choice of θ, the single-valued function $\sqrt[3]{r}\, e^{i(\theta/3)}$ is called the *principal branch* of the cube root; let us write it as $[\sqrt[3]{z}\,]$. Note that the principal branch agrees with the real cube root function on the positive real axis, but *not* on the negative real axis; for example, $[\sqrt[3]{-8}\,] = 2\,e^{i(\pi/3)}$. Also note that the other two branches associated with this choice of C can be expressed in terms of the principal branch as $e^{i(2\pi/3)}[\sqrt[3]{z}\,]$ and $e^{i(4\pi/3)}[\sqrt[3]{z}\,]$.

It should be clear how the above discussion extends to a general fractional power.

2.6.3 Relevance to Power Series

Earlier we explained the otherwise mysterious interval of convergence for a real function such as $1/(1+x^2)$ by extending the function off the real line and into the complex plane: the obstruction to convergence was the existence of points at which the complex function became infinite (singularities). We now discuss the more subtle fact that branch points also act as obstacles to the convergence of power series.

The real Binomial Theorem says that if n is any real number (not just a positive integer), then

$$(1+x)^n = 1 + nx + \tfrac{n(n-1)}{2!}x^2 + \tfrac{n(n-1)(n-2)}{3!}x^3 + \tfrac{n(n-1)(n-2)(n-3)}{4!}x^4 + \cdots .$$

If n is a positive integer then the series terminates at x^n and the issue of convergence does not arise. If n is not a positive integer then the ratio test tells us that the interval of convergence of the power series is $-1 < x < 1$. This interval is easily understood when n is negative, because the function then has a singularity at $x = -1$. But how, for example, are we to explain this interval of convergence in the case $n = (1/3)$?

Figure [2.33a] shows the graph $y = (1+x)^{\frac{1}{3}}$ of the real function $f(x) = (1+x)^{\frac{1}{3}}$, which is well defined for all x since every real number has a unique real cube root. Looking at this graph, there seems to be no good reason for the series to break down at ± 1, yet break down it does. This is illustrated rather vividly by the dashed curve, which is the graph of the 30th degree polynomial obtained by truncating the binomial series at x^{30}. As you can see, this curve follows $y = f(x)$ very closely (actually more closely than illustrated) between ± 1, but just beyond this interval it suddenly starts to deviate wildly.

Unlike the case of $1/(1+x^2)$, observe that the mystery does not disappear when we extend the real function $f(x)$ to the complex function $f(z) = (1+z)^{\frac{1}{3}}$, because *$f(z)$ does not have any singularities.*

We have already discussed the fact [see (2.14) and Exs. 16, 17, 18] that the Binomial Theorem extends to the complex plane. In the present case it says that

$$f(z) = (1+z)^{\frac{1}{3}} = 1 + \tfrac{1}{3}z - \tfrac{1}{9}z^2 + \tfrac{5}{81}z^3 - \tfrac{10}{243}z^4 + \tfrac{22}{729}z^5 - \cdots ,$$

with convergence inside the unit disc shown in [2.33b]. In common with all power series, the RHS of the above equation is a single-valued function. For example, at

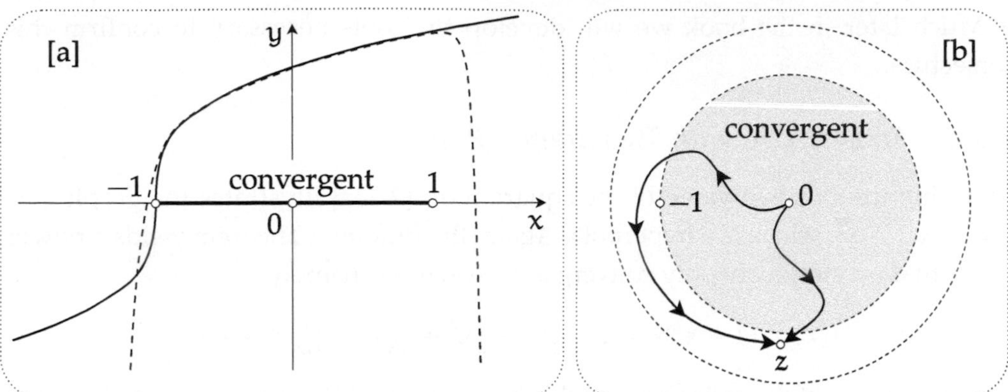

[2.33] [a] *The real function* $f(x) = (1+x)^{\frac{1}{3}}$ *can be expressed as a power series, but only for* $-1 < x < 1$*; the dashed curve shows the* 30th *degree polynomial obtained by truncating the binomial series at* x^{30}*. But* why *does convergence break down here?! There are no singularities in sight, and, unlike the resolution of the convergence mystery of* $\frac{1}{1+x^2}$*, the complex generalization* $f(z) = (1+z)^{\frac{1}{3}}$ also *does not have any singularities!* **[b]** *But* $f(z)$ *does have a* branch point *at* -1*. Suppose the series were to converge at the illustrated point z outside the unit disc. Then the two illustrated paths could encircle the branch point, yielding two different values of* $f(z)$*. But the power series is single-valued, so this is impossible!*

$z = 0$ the series equals 1. But while $f(x)$ was an ordinary single-valued function of x, the LHS of the above equation is a three-valued multifunction of z, with a second order branch point at $z = -1$. For example, $f(0)$ takes three values: 1, $e^{i\frac{2\pi}{3}}$, and $e^{-i\frac{2\pi}{3}}$. We now recognize that the power series represents just one branch of $f(z)$, namely the one for which $f(0) = 1$.

This solves the mystery. For suppose that the series were to converge inside the larger circle in [2.33b], and in particular at the illustrated point z. Starting at $z = 0$ with the value $f(0) = 1$, then travelling along the two illustrated paths to z, we clearly end up with two *different* values of $f(z)$, because together the two paths enclose the branch point at -1. But the power series cannot mimic this behaviour since it is necessarily single-valued—its only way out is to cease converging outside the unit disc. We have demanded the impossible of the power series, and it has responded by committing suicide!

This example shows that a branch point is just as real an obstacle to convergence as a singularity. Quite generally, this argument shows that if a branch of a multifunction can be expressed as a power series, the disc of convergence cannot be large enough to contain any branch points of the multifunction. This strongly suggests a further generalization of the (unproven) statement (2.15):

> *If a complex function or a branch of a multifunction can be expressed as a power series, the radius of convergence is the distance to the nearest singularity or branch point.* (2.27)

Much later in the book we will develop the tools necessary to confirm this conjecture.

2.6.4 An Example with Two Branch Points

Choosing the positive value of the square root, [2.34a] illustrates the graph $y = f(x) = \sqrt{1+x^2}$, which is a hyperbola. Again, the Binomial Theorem yields a power series that mysteriously only converges between ± 1, namely,

$$f(x) = (1+x^2)^{\frac{1}{2}} = 1 + \tfrac{1}{2}x^2 - \tfrac{1}{8}x^4 + \tfrac{1}{16}x^6 - \tfrac{5}{128}x^8 + \cdots .$$

The divergence of the series beyond this interval is vividly conveyed by the dashed curve, which is the graph of the 20th degree polynomial obtained by truncating the binomial series at x^{20}.

As before, the explanation lies in $\mathbb{C}$, where $f(x)$ becomes the two-valued multifunction $f(z) = \sqrt{z^2+1}$. This can be rewritten as $f(z) = \sqrt{(z-i)(z+i)}$, which makes it clear that $f(z)$ has two simple branch points, one at i and the other at $-i$. These branch points obstruct the convergence of the corresponding complex series, limiting it to the unit disc shown in [2.34b].

In greater detail, the notation of [2.34b] enables us to write

$$f(z) = \sqrt{r_1 r_2}\, e^{i(\theta_1+\theta_2)/2}. \tag{2.28}$$

Here we must bear in mind that the figure illustrates only one possibility (out of infinitely many) for each of the angles θ_1 and θ_2. To see that i is indeed a branch

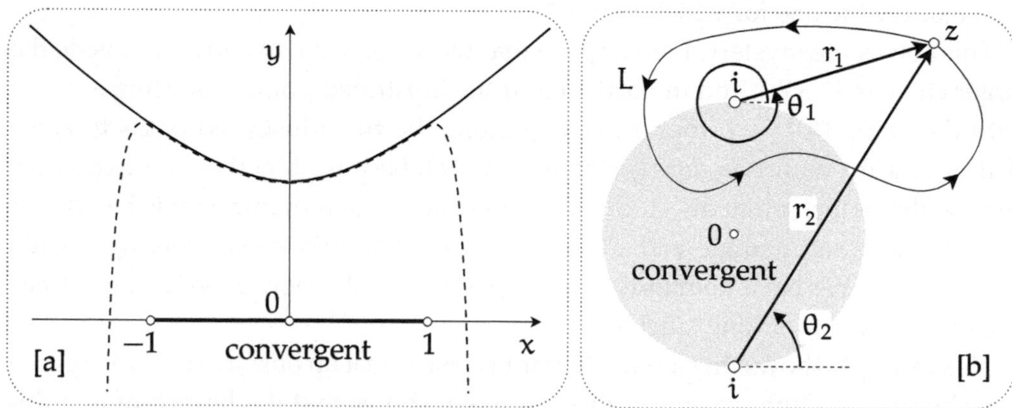

[2.34] [a] *The real function* $f(x) = (1+x^2)^{\frac{1}{2}}$ *can be expressed as a power series, but only for* $-1 < x < 1$*; the dashed curve shows the* 20th *degree polynomial obtained by truncating the binomial series at* x^{20}. **[b]** *In the complex plane,* $f(z) = \sqrt{(z-i)(z+i)} = \sqrt{r_1 r_2}\, e^{i(\theta_1+\theta_2)/2}$*, which makes it clear that* $f(z)$ *has branch points at* $\pm i$*, and these obstruct the convergence of the complex power series beyond the unit disc. If* z *loops around* i *along* L*, as illustrated, then the net changes in* θ_1 *and* θ_2 *are* 2π *and* 0*, respectively. So* $f(z) \rightsquigarrow e^{i\pi} f(z) = -f(z)$.

point, suppose we start with the value of $f(z)$ given by the illustrated values of θ_1 and θ_2. Now let z travel round the illustrated loop L. As it does so, $(z + i)$ rocks back and forth, so θ_2 merely oscillates, finally returning to its original value. But $(z - i)$ undergoes a complete revolution, and so θ_1 increases by 2π. Thus when z returns to its original position, (2.28) shows that $f(z)$ does not return to its original value, but rather to

$$f_{\text{new}}(z) = \sqrt{r_1 r_2}\, e^{i(\theta_1+2\pi+\theta_2)/2} = e^{i\pi} \sqrt{r_1 r_2}\, e^{i(\theta_1+\theta_2)/2} = -f_{\text{old}}(z).$$

Of course the same thing happens if z travels along a loop that goes once round $-i$, instead of round $+i$.

In order to dissect $f(z)$ into two single-valued branches, we appear to need two branch cuts: one cut C_1 from i to infinity (to prevent us encircling the branch point at i), and another cut C_2 from $-i$ to infinity, for the same reason. Figure [2.35a] illustrates a particularly common and important choice of these cuts, namely, rays going due west. If we do not allow z to cross the cuts then we may restrict the angle $\theta_1 = \arg(z - i)$ to its principal value, in the range $-\pi < \theta_1 \leqslant \pi$. For example, the angle in [2.34b] is not the principal value, while the one in [2.35a] is. If θ_2 is likewise restricted to its unique principal value then (2.28) becomes the single-valued principal branch of $f(z)$, say $F(z)$. The other branch of $f(z)$ is simply $-F(z)$.

Let us return to the previous situation in which we allowed θ_1 and θ_2 to take general values rather than their principal values. Figure [2.35b] illustrates the fact that it is possible to define two branches of $f(z)$ using only a single branch cut C that connects the two branch points. If z is restricted to the shaded, multiply-connected region S, then it cannot loop around either branch point singly. It can,

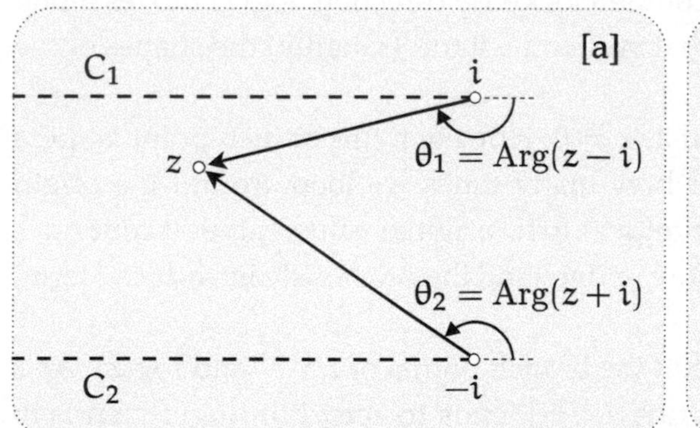

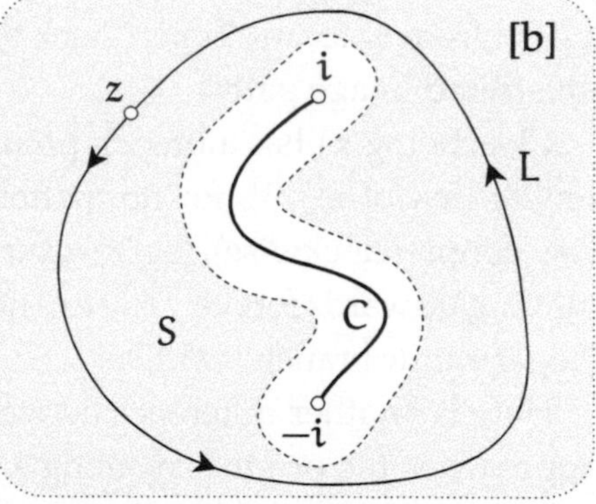

[2.35] [a] *We can define two single-valued branches of* $f(z) = \sqrt{z^2+1} = \sqrt{r_1 r_2}\, e^{i(\theta_1+\theta_2)/2}$ *by making branch cuts going due west from both branch points, and then taking* θ_1 *and* θ_2 *to be their principal values, as illustrated.* **[b]** *In fact we only need a single branch cut connecting the two branch points, such as* C. *For then a loop* L *cannot encircle either branch point singly, and if it encircles both, as illustrated, then* $f(z)$ *does not change its value.*

however, travel along a loop such as L that encircles both branch points together. But in this case both θ_1 and θ_2 increase by 2π, so (2.28) shows that $f(z)$ returns to its original value. Thus we can define two single-valued branches on S. Finally, we may expand S until it borders on C.

2.7 The Logarithm Function

2.7.1 *Inverse of the Exponential Function*

The complex logarithm function $\log(z)$ may be introduced as the "inverse" of e^z. More precisely, we define $\log(z)$ to be any complex number such that $e^{\log(z)} = z$. It follows [exercise] that

$$\log(z) = \ln|z| + i\,\arg(z).$$

Since $\arg(z)$ takes infinitely many values, differing from each other by multiples of 2π, we see that $\log(z)$ is a multifunction taking infinitely many values, differing from each other by multiples of $2\pi i$. For example,

$$\log(2+2i) = \ln 2\sqrt{2} + i(\pi/4) + 2n\pi i,$$

where n is an arbitrary integer.

The reason we get infinitely many values is clear if we go back to the exponential mapping shown in [2.19], p. 90: each time z travels straight upward by $2\pi i$, e^z executes a complete revolution and returns to its original value. Figure [2.36] rephrases this using the above example of $\log(2+2i)$. If we arbitrarily choose the initial value $w = \ln 2\sqrt{2} + i\,(\pi/4)$ for $\log(2+2i)$, then as z travels along a loop that encircles the origin ν times in the counterclockwise direction, $\log(z)$ moves along a path from w to $w + 2\nu\pi i$. Check that you understand (roughly) the shapes of the illustrated image paths.

Clearly $\log(z)$ has a branch point at $z = 0$. However, this branch point is quite unlike that of $z^{(1/n)}$, for no matter how many times we loop around the origin (say counterclockwise), $\log(z)$ *never* returns to its original value, rather it continues moving upwards forever. You can now understand the previously introduced term, "logarithmic branch point".

Here is another difference between the branch points of $z^{(1/n)}$ and $\log(z)$. As z approaches the origin, say along a ray, $|z^{(1/n)}|$ tends to zero, but $|\log(z)|$ tends to infinity, and in this sense the origin is a singularity as a well as a branch point. On the other hand, algebraic branch points can also be singularities; consider $(1/\sqrt{z})$.

To define single-valued branches of $\log(z)$ we make a branch cut from 0 out to infinity. The most common choice for this cut is the negative real axis. In this cut plane we may restrict $\arg(z)$ to its principal value $\text{Arg}\,(z)$; remember, this is

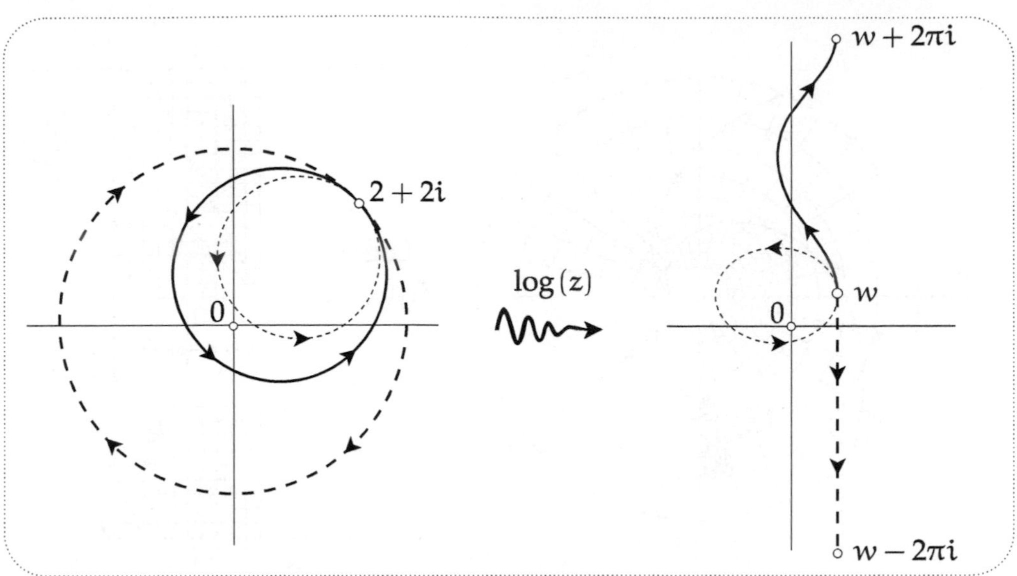

[2.36] *Here we see that* 0 *is a branch point of* $\log(z)$. *However, this branch point is quite unlike that of* $z^{(1/n)}$, *for no matter how many times we loop around the origin (say counterclockwise),* $\log(z)$ never *returns to its original value, rather it continues moving upwards forever. If we make a branch cut along the negative real axis, we can restrict the angle to be its principal value, in which case we obtain the* principal value *of* $\log(z)$, *written* $\text{Log}(z) \equiv \ln|z| + i\,\text{Arg}(z)$.

defined by $-\pi < \text{Arg}(z) \leqslant \pi$. This yields the *principal branch* or *principal value* of the logarithm, written $\text{Log}(z)$, and defined by

$$\text{Log}(z) \equiv \ln|z| + i\,\text{Arg}(z).$$

For example, $\text{Log}(-\sqrt{3}-i) = \ln 2 - i(5\pi/6)$, $\text{Log}(i) = i(\pi/2)$, and $\text{Log}(-1) = i\pi$. Note that if $z = x$ is on the positive real axis, $\text{Log}(x) = \ln(x)$.

Figure [2.37] illustrates how the mapping $z \mapsto w = \text{Log}(z)$ sends rays to horizontal lines, and circles to vertical line-segments connecting the horizontal lines at heights $\pm\pi$; the entire z-plane is mapped to the horizontal strip of the w-plane bounded by these lines. Study this figure until you are completely at peace with it. You can see the price we pay for forcing the logarithm to be single-valued: it becomes discontinuous at the cut. As z crosses the cut travelling counterclockwise, the height of w suddenly jumps from π to $-\pi$. If we wish w to instead move continuously, then we must switch to the branch $\text{Log}(z) + 2\pi i$ of the logarithm.

Another problem with restricting ourselves to the principal branch is that the familiar rules for the logarithm break down. For example, $\text{Log}(ab)$ is *not* always equal to $\text{Log}(a) + \text{Log}(b)$; try $a = -1$ and $b = i$, for example. However, if we keep all values of the logarithm in play then it *is* true [exercise] that

$$\log(ab) = \log(a) + \log(b) \quad \text{and} \quad \log(a/b) = \log(a) - \log(b),$$

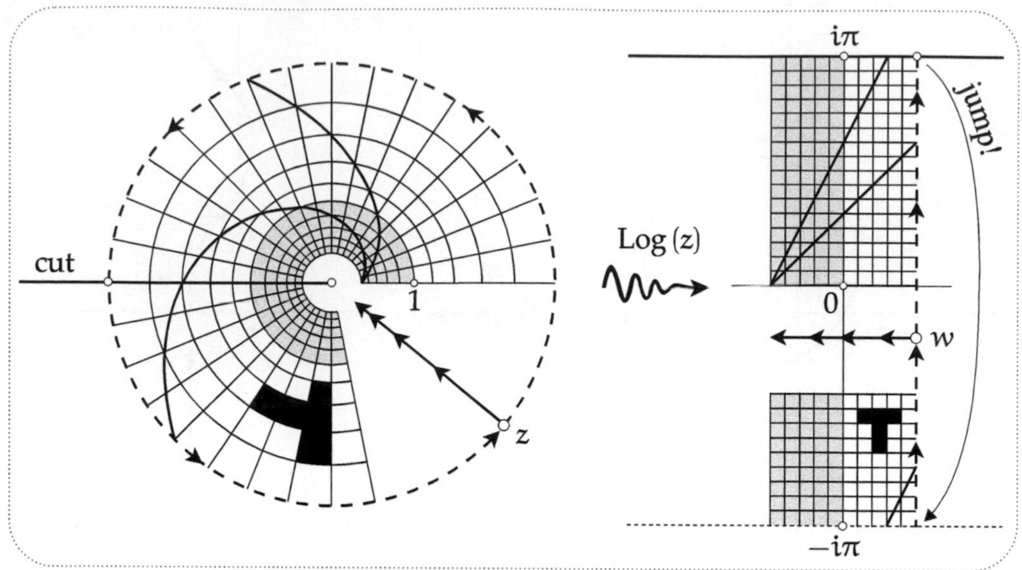

[2.37] The geometry of Log (z). *Note that the price we pay for having a single-valued logarithm is that it is* discontinuous*: as we cross the branch cut, its value jumps.*

in the sense that every value of the LHS is *contained amongst* the values of the RHS, and vice versa.

2.7.2 The Logarithmic Power Series

If we wish to find a power series for the complex logarithm, two problems immediately arise. First, since a power series is single-valued, the best we can hope for is to represent a single branch of $\log(z)$; let's choose the principal branch, $\operatorname{Log}(z)$. Second, the origin is both a singularity and a branch point of $\operatorname{Log}(z)$ so we cannot have a power series centred there (i.e., in powers of z); let us therefore try an expansion centred at $z = 1$, i.e., in powers of $(z-1)$. [Of course any other non-zero point would be equally suitable.] Writing $Z = (z-1)$, our problem, then, is to expand $\operatorname{Log}(1+Z)$ in powers of Z.

Let us use the abbreviation $L(z) = \operatorname{Log}(1+z)$. Since the branch point of $L(z)$ is $z = -1$, the largest disc of convergence we can have is the unit disc. To find the series we will use the fact that $e^{L(z)} = (1+z)$. Recall from (2.21) [on p. 91] that by taking n to be a sufficiently large positive integer, we can approximate e^L as precisely as we wish using $[1 + (L/n)]^n$. Thus, as $(1/n)$ vanishes,[6]

$$\left(1+\frac{L}{n}\right)^n \asymp e^L = (1+z) \implies 1+\frac{L}{n} \asymp (1+z)^{\frac{1}{n}}.$$

[6] Once again, "$\asymp$" denotes Newton's concept of *ultimate equality*; see the new Preface.

There are n branches of $(1+z)^{\frac{1}{n}}$ within the unit disc, but since $L(0) = 0$ we need the branch of $(1+z)^{\frac{1}{n}}$ that equals 1 when $z = 0$. Appealing to the Binomial series for this principal branch, we obtain

$$1 + \frac{L}{n} \asymp 1 + \frac{1}{n}z + \frac{\frac{1}{n}(\frac{1}{n}-1)}{2!}z^2 + \frac{\frac{1}{n}(\frac{1}{n}-1)(\frac{1}{n}-2)}{3!}z^3 + \cdots,$$

and hence

$$L(z) \asymp z + \frac{(\frac{1}{n}-1)z^2}{2} + \frac{(\frac{1}{n}-1)(\frac{1}{2n}-1)z^3}{3} + \frac{(\frac{1}{n}-1)(\frac{1}{2n}-1)(\frac{1}{3n}-1)z^4}{4} + \cdots.$$

Finally, since this ultimate equality becomes equality in the limit that $(1/n)$ tends to zero, we obtain the following *logarithmic power series*:

$$\text{Log}\,(1+z) = z - \frac{z^2}{2} + \frac{z^3}{3} - \frac{z^4}{4} + \frac{z^5}{5} - \frac{z^6}{6} + \cdots. \tag{2.29}$$

For other approaches to this series, see Exs. 31, 32.

Using the ratio test, you can check for yourself that this series does indeed converge inside the unit circle. In fact it can be shown [see Ex. 11] that the series also converges everywhere *on* the unit circle, except obviously at $z = -1$. This yields some very interesting special cases. For example, putting $z = i$ and then equating real and imaginary parts, we get

$$\ln\sqrt{2} = \frac{1}{2} - \frac{1}{4} + \frac{1}{6} - \frac{1}{8} + \frac{1}{10} - \frac{1}{12} + \cdots$$

$$\text{and} \quad \pi = 4\left[1 - \frac{1}{3} + \frac{1}{5} - \frac{1}{7} + \frac{1}{9} - \frac{1}{11} + \cdots\right].$$

Try checking the first series by noting that if $z = 1$, then $\ln\sqrt{2} = \frac{1}{2}\ln(1+z)$. The remarkable second series for π was first published by Leibniz in 1676, though it was already known to Newton and Gregory. It is usually obtained from the power series for $\tan^{-1} x$, evaluated at $x = 1$. Astonishingly, *that* power series was discovered *centuries* before Newton, Gregory, and Leibniz roamed the Earth, by the Indian mathematician, Mādhava of Sangamagrāma (c.1340 – c.1425); see (Stillwell, 2010, p. 166).

For other interesting applications of the logarithmic series, see Exs. 36, 37, 38.

2.7.3 General Powers

If x is a real variable then we are accustomed to being able to express x^3, for example, as $e^{3\ln x}$. Let's see whether we can do the same thing using the complex exponential and logarithm. That is, let us investigate the possibility of writing

$$z^k = e^{k\log(z)}. \tag{2.30}$$

Let $z = r\,e^{i\theta}$, where θ is chosen to be the principal value, Arg (z). Then

$$e^{3\,\text{Log}\,(z)} = e^{3(\ln r + i\theta)} = e^{3\ln r}\,e^{i3\theta} = r^3\,e^{i3\theta} = z^3.$$

But the most general branch of $\log(z)$ is simply Log $(z) + 2n\pi i$, where n is an integer, so

$$e^{3\log(z)} = e^{6n\pi i}\,e^{3\,\text{Log}\,(z)} = e^{6n\pi i}\,z^3 = z^3$$

is true irrespective of which branch of the logarithm is chosen. Clearly, by the same argument, (2.30) is true for all integer values of k.

Next, consider the three branches of $z^{\frac{1}{3}}$. Recalling that the principal branch $[z^{\frac{1}{3}}]$ of this function is $\sqrt[3]{r}\,e^{i(\theta/3)}$, where θ again represents the principal angle, you can easily check that $e^{\frac{1}{3}\,\text{Log}\,(z)} = [z^{\frac{1}{3}}]$. Thus the general branch of the logarithm yields

$$e^{\frac{1}{3}\log(z)} = e^{i\frac{2n\pi}{3}}\,[z^{\frac{1}{3}}].$$

Thus we have again confirmed (2.30), in the sense that the infinitely many branches of $\log(z)$ yield precisely the three branches of the cube root: $[z^{\frac{1}{3}}]$, $e^{i(2\pi/3)}[z^{\frac{1}{3}}]$, and $e^{i(4\pi/3)}[z^{\frac{1}{3}}]$. By the same reasoning, if (p/q) is a fraction reduced to lowest terms then $e^{\frac{p}{q}\log(z)}$ yields precisely the q branches of $z^{\frac{p}{q}}$.

Finally, note that the RHS of (2.30) is still meaningful if $k = (a + ib)$ is a *complex* number. Emboldened by the above successes, we now take (2.30) as the *definition* of a complex power. If we use Log (z) in (2.30) then we find that the principal branch of $z^{(a+ib)}$ is given by [exercise]

$$[z^{(a+ib)}] \equiv e^{(a+ib)\,\text{Log}\,(z)} = r^a e^{-b\theta}\,e^{i(a\theta + b\ln r)}.$$

If z now travels along a closed loop encircling the origin n times, then $\log(z)$ moves along a path from Log (z) to Log $(z) + 2n\pi i$, and $z^{(a+ib)}$ moves along a path from $[z^{(a+ib)}]$ to

$$z^{(a+ib)} = e^{i2\pi n a}e^{-2\pi n b}\,[z^{(a+ib)}].$$

If $b \neq 0$ then the factor $e^{-2\pi n b}$ makes it obvious that $z^{(a+ib)}$ never returns to its original value, no matter how many times we go round the origin. Thus $z = 0$ is a logarithmic branch point in this case. This is still true even if $b = 0$, provided [exercise] that the real power a is *irrational*. Only when a is a rational number does z^a return to its original value after a finite number of revolutions, and only when a is an integer does z^a become single-valued.

We end with an important observation on the use of "e^z" to denote the single-valued exponential mapping. Reversing the roles of the constant and variable in (2.30), we are forced to define $f(z) = k^z$ to be the "multifunction" [see Ex. 29] $f(z) = e^{z\log(k)}$. But if we now put $k = e = 2.718\ldots$ then we are suddenly in hot water: the exponential mapping "e^z" is merely one branch [what are the others?] of the newly defined multifunction $(2.718\ldots)^z$. To avoid this confusion, some authors always write the exponential mapping as $\exp(z)$. However, we shall retain the notation "e^z", which is both convenient and rooted in history, with the understanding that e^z *always refers to the single-valued exponential mapping, and never to the multifunction* $(2.718\ldots)^z$.

2.8 Averaging over Circles*

2.8.1 The Centroid

This entire section is optional because the chief result to which we shall be led (*Gauss's Mean Value Theorem*) will be derived again later, in fact more than once. It is nevertheless fun and instructive to attempt to understand the result using only the most elementary of methods.

Consider a set of n point particles in $\mathbb{C}$, located at $z_1, z_2, \ldots, z_n$. If the mass of the particle at z_j is m_j then the *centroid* Z of the set of particles (also called the "centre of mass") is defined to be

$$Z \equiv \frac{\sum_{j=1}^{n} m_j\, z_j}{\sum_{j=1}^{n} m_j}.$$

If we imagine the plane to be massless, Z is the point at which we could rest the plane on a pin so as to make it balance.

Throughout this section *we shall take the masses of the particles to be equal*, in which case the centroid becomes the average position of the particles:

$$Z = \frac{1}{n}\sum_{j=1}^{n} z_j\,.$$

This is the case depicted in [2.38a]. An immediate consequence of this definition is that $\sum (z_j - Z) = 0$. In other words, *the complex numbers from Z to the particles cancel.* This vanishing sum is illustrated in [2.38b]. Conversely, if some point Z has the property that the complex numbers connecting it to the particles cancel, then Z must be the centroid.

Another immediate result is that if we translate the set of points by b, then the centroid will translate with them, i.e., the new centroid will be $Z + b$. The same thing happens if we rotate the set of points about the origin—the centroid rotates with them. In general,

If Z is the centroid of $\{z_j\}$, then the centroid of $\{az_j + b\}$ is $aZ + b$. (2.31)

Given a second set of n points $\{\widetilde{z}_j\}$ (with centroid $\widetilde{Z}$), we may add pairs from the two sets to obtain the set $\{z_j + \widetilde{z}_j\}$, and it is easy to see that the centroid of the latter is $Z + \widetilde{Z}$. In particular, the centroid Z of $\{z_j = x_j + iy_j\}$ is the sum $X + iY$ of the centroid X of the points $\{x_j\}$ on the real axis and the centroid iY of the points $\{iy_j\}$ on the imaginary axis.

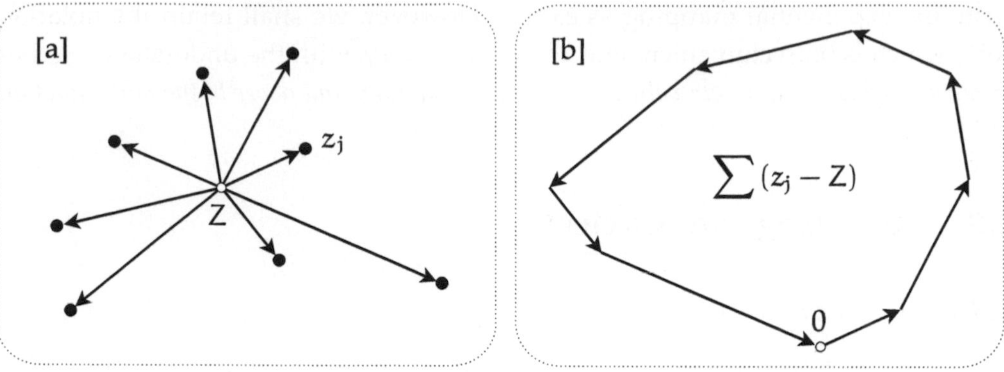

[2.38] [a] The *centroid* Z *of masses* m_j *located at* z_j *is* $Z \equiv \frac{\sum_{j=1}^{n} m_j z_j}{\sum_{j=1}^{n} m_j}$. *In the case that the masses are all* equal, *join* Z *to the locations of the masses, and sum these connecting complex numbers* $(z_j - Z)$ ... **[b]** ... *and discover that their sum vanishes! Furthermore, this vanishing sum* characterizes *the location of* Z.

Our next result will play a minor role at the end of this section, but later we shall see that it has other interesting consequences. The *convex hull* H of the set of particles $\{z_j\}$ is defined to be the smallest convex polygon such that each particle lies on H or inside it. More intuitively, first imagine pegs sticking out of the plane at each point z_j, then stretch an imaginary rubber band so as to enclose all the pegs. When released, the rubber band will contract into the desired polygon H shown in [2.39a]. We can now state the result:

The centroid Z *must lie in the interior of the convex hull* H. (2.32)

For if p is outside this set, we see that the complex numbers from p to the particles cannot possibly cancel, as they must do for Z. More formally, we take it as visually evident that through any exterior point p we may draw a line L such that H and its shaded interior lie entirely on one side of L. The impossibility of the complex numbers cancelling now follows from their lying entirely on this side of L, for they all must have positive components in the direction of the illustrated complex number N normal to L. Except when the particles are collinear (in which case H collapses to a line-segment), the same reasoning forbids Z from lying *on* H.

As illustrated in [2.39b], an immediate consequence of (2.32) is that

If all the particles lie within some circle then their centroid must also lie within that circle. (2.33)

The main result we wish to derive in this section is based on the following fact. Defining the *centre* of a regular n-gon to be the centre of the circumscribing circle,

The centre of a regular n*-gon is the centroid of its vertices.* (2.34)

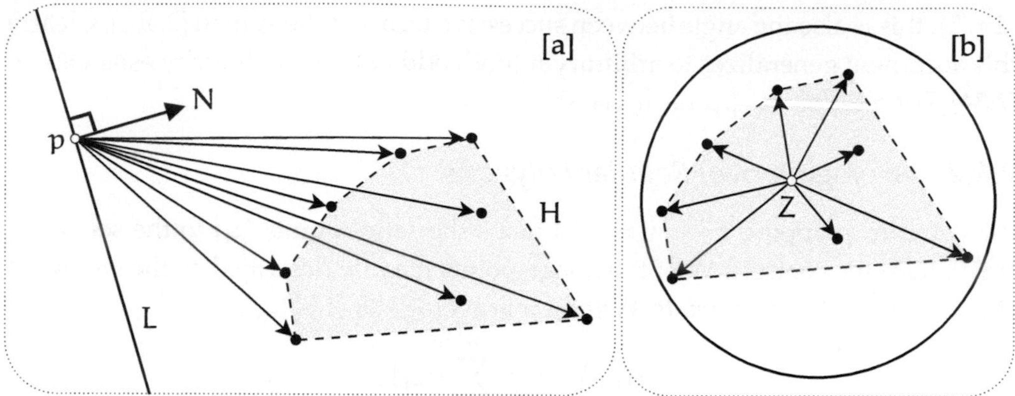

[2.39] The centroid must lie within the convex hull. [a] *If* p *lies outside the convex hull* H *of the masses, then the connecting complex numbers from* p *to the masses cannot vanish, therefore the centroid* Z *must lie within* H. **[b]** *Therefore, if all the particles lie within some circle then their centroid must also lie within that circle.*

By virtue of (2.31), we may as well choose the n-gon to be centred at the origin, in which case the claim is that the sum of the vertices vanishes. As illustrated in [2.40a], this is obvious if n is even since the vertices then occur in opposite pairs.

The explanation is not quite so obvious when n is odd; see [2.40b], which illustrates the case $n = 5$. However, if we draw $\sum z_j$ systematically, taking the vertices z_j in counterclockwise order, then we obtain [2.40c], and the answer is suddenly clear: *the sum of the vertices of the regular* 5*-gon forms another regular* 5*-gon*. The figure explains why this happens. Since the angle between successive vertices in [2.40b] is

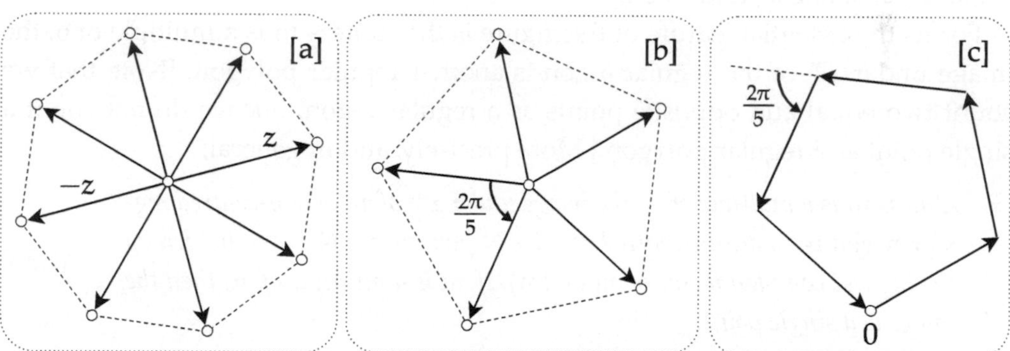

[2.40] The sum of the vertices of a regular n-gon vanishes. [a] *If* n *is even, it is obvious that the sum of the vertices of any regular, origin-centred* n*-gon must vanish, for the vertices occur in diametrically opposite pairs.* **[b]** *In fact the sum of the vertices must also vanish when* n *is odd, but this is no longer immediately obvious. Begin by noting that the angle between successive vertices is* $(2\pi/n)$. **[c]** *Drawing the sum of the vertices in counterclockwise order, all is suddenly clear: the sum* itself *forms a regular* n*-gon, and therefore it vanishes.*

$(2\pi/5)$, this is also the angle between successive terms of the sum in [2.40c]. Clearly this argument generalizes to arbitrary n (both odd and even), thereby establishing (2.34). For a different approach, see Ex. 40.

2.8.2 Averaging over Regular Polygons

If a complex mapping $z \mapsto w = f(z)$ maps the set of points $\{z_j\}$ to the set $\{w_j = f(z_j)\}$, then the centroid W of the image points may be described as the *average of* $f(z)$ *over the set* $\{z_j\}$ *of* n *points*. Writing this average as $\langle f(z)\rangle_n$,

$$\langle f(z)\rangle_n = \tfrac{1}{n}\sum_{j=1}^{n} f(z_j)\,.$$

Note that if $f(z) = c$ is constant, then its average over any set of points is equal to c.

Henceforth, we shall restrict ourselves to the case where $\{z_j\}$ are the vertices of a regular n-gon; correspondingly, $\langle f(z)\rangle_n$ will be understood as the average of $f(z)$ over the vertices of such a regular n-gon. Note that if we write $f(z) = u(z) + iv(z)$, then

$$\langle f(z)\rangle_n = \langle u(z)\rangle_n + i\,\langle v(z)\rangle_n. \tag{2.35}$$

Initially, we consider only *origin-centred* polygons.

Consider, then, the average of $f(z) = z^m$ over the vertices of such a regular n-gon. Figure [2.41] illustrates the case $n = 6$. In the centre of the figure is a shaded regular hexagon, and on the periphery are the images of its vertices under the mappings $z, z^2, \ldots, z^6$. Study this figure carefully, and see if you can understand what's going on. If we take still higher values of m, then this pattern repeats cyclically: z^7 is like z^1, z^8 is like z^2, and so on.

For us the essential feature of this figure is that unless m is a multiple of 6, the image under z^m of the regular 6-gon is another regular polygon. [Note that we count two equal and opposite points as a regular 2-gon, but we do not count a single point as a regular polygon.] More precisely, and in general,

> *Unless* m *is a multiple of* n, *the image under* z^m *of an origin-centred regular* n*-gon is an origin-centred regular* N*-gon, where* $N = ($$n$ *divided by the highest common factor of* m *and* n*). If* m *is a multiple of* n*, then the image is a single point.* (2.36)

Check that this agrees with [2.41]. Try to establish the result on your own, but see Ex. 41 if you get stuck.

Combining this result with (2.34), we obtain the following key fact: *If* $n > m$ *then* $\langle z^m\rangle_n = 0$. This is easy to generalize. If

$$P_m(z) = c_0 + c_1 z + c_2 z^2 + c_3 z^3 + \cdots + c_m z^m$$

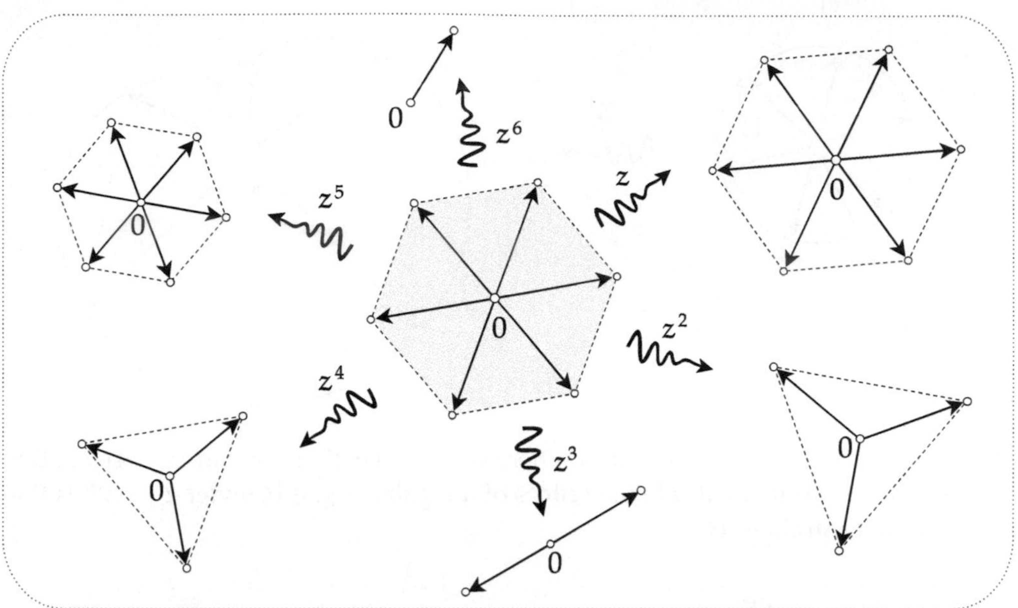

[2.41] *Unless* m *is a multiple of* n*, the image under* z^m *of an origin-centred regular* n*-gon is an origin-centred regular* N*-gon, where* $N = (n$ *divided by the highest common factor of* m *and* $n)$*. If* m *is a multiple of* n*, then the image is a single point.*

is a general polynomial of degree m, then its average over the vertices of the n-gon is

$$\langle P_m(z)\rangle_n = \langle c_0\rangle_n + c_1 \langle z\rangle_n + c_2 \langle z^2\rangle_n + c_3 \langle z^3\rangle_n + \cdots + c_m \langle z^m\rangle_n.$$

If the number n of vertices is greater than the degree m of the polynomial, we therefore obtain

$$\langle P_m(z)\rangle_n = \langle c_0\rangle_n = c_0 = P_m(0).$$

In other words, *the centroid of the image points is the image of the centroid.* Expressing this result in the language of averages,

> *If* $n > m$ *then the average of an* m^{th} *degree polynomial* $P_m(z)$ *over the vertices of an origin-centred regular* n*-gon is its value* $P_m(0)$ *at the centre of the* n*-gon.* (2.37)

Finally, let us generalize to regular n-gons that are centred at an arbitrary point k, instead of the origin. Of course when we apply z^m to the vertices of such a regular polygon, the image points do *not* form a regular polygon. See [2.42], which shows the effect of z^4 on the vertices of a regular hexagon H centred at k, together with the image of the entire circle on which these vertices lie. Nevertheless, the figure also illustrates the surprising and beautiful fact that, once again, *the centroid of the image points is the image of the centroid of* H. Figure [2.43a] confirms this empirically

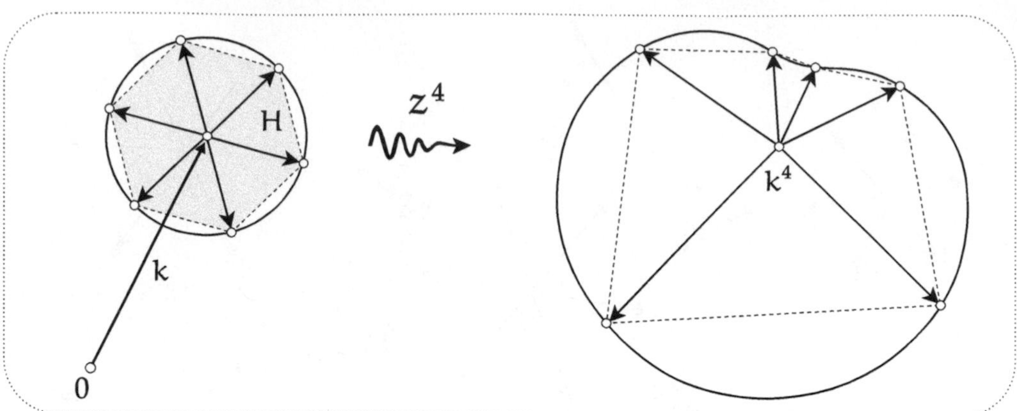

[2.42] *It is a surprising and beautiful fact (here illustrated with* $n = 6$ *and* $m = 4$*) that* **the centroid of the image points of the vertices of a regular** n**-gon H under** $z \mapsto z^m$ **is the image of the centroid of H.**

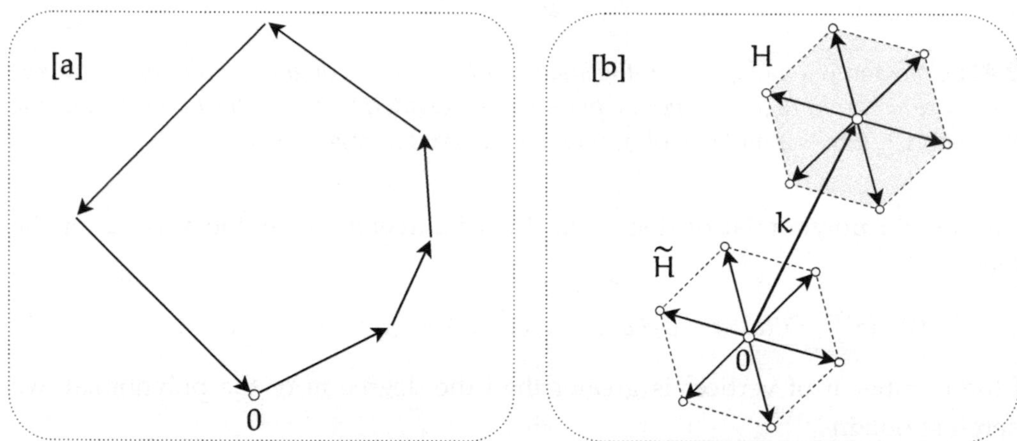

[2.43] [a] Empirical verification of the claim made in the previous figure: *the sum of the connecting complex numbers from* k^4 *to images of the vertices of* H *vanishes, so* k^4 *is indeed the centroid of these image points.* **[b]** *In order to mathematically explain this phenomenon, we view* H *to be the translation by* k *of the origin-centred* $\widetilde{H}$*; see main text for the argument.*

by showing that the sum of complex numbers connecting k^4 to the image points is indeed zero.

Extending our notation slightly, we may write the average of z^m over the vertices of H as $\langle z^m \rangle_H$, so what we must show is that $\langle z^4 \rangle_H = k^4$. It is no harder to treat the general case of z^m acting on the vertices of a regular n-gon H centred at k. First note that H can be obtained by translating an origin-centred n-gon $\widetilde{H}$ by k. See the example in [2.43b]. Since a vertex z_j of $\widetilde{H}$ translates to a vertex $z_j + k$ of H, it follows that

$$\langle z^m \rangle_H = \langle (z+k)^m \rangle_{\widetilde{H}}.$$

But $(z+k)^m = \sum_{j=0}^{m} \binom{m}{j} z^j k^{m-j}$ is just an mth degree polynomial which maps 0 to k^m. Using (2.37), we conclude that if $n > m$ then $\langle z^m \rangle_H = k^m$, as was to be shown.

Generalizing the argument that led to (2.37), we see that (2.37) is a special case of the following result:

If $n > m$ then the average of an mth degree polynomial $P_m(z)$ over the vertices of a regular n-gon centred at k is its value $P_m(k)$ at the centre of the n-gon. (2.38)

2.8.3 Averaging over Circles

Since at least the time of Archimedes, mathematicians have found it fruitful to think of a circle as the limit of a regular n-gon as n tends to infinity. We will now use this idea to investigate the average of a complex function over a circle.

Inscribing a regular n-gon in a given circle, and taking the limit as n tends to infinity, (2.38) shows that

The average over a circle C of a polynomial of arbitrarily high degree is equal to the value of the polynomial at the centre of C. (2.39)

By (2.35), the average $\langle f(z) \rangle_C$ of a complex function $f(z) = u(z) + iv(z)$ over a circle C may be expressed as $\langle f(z) \rangle_C = \langle u(z) \rangle_C + i\,\langle v(z) \rangle_C$. Using a familiar idea from ordinary calculus, the averages of the two real functions u and v may be expressed as integrals. If C has centre k and radius R, then as θ varies between 0 and 2π, $z = k + R\,e^{i\theta}$ traces out C. Thus,

$$\langle u(z) \rangle_C = \frac{1}{2\pi}\int_0^{2\pi} u(k + R\,e^{i\theta})\,d\theta \qquad \text{and} \qquad \langle v(z) \rangle_C = \frac{1}{2\pi}\int_0^{2\pi} v(k + R\,e^{i\theta})\,d\theta.$$

More compactly, we may write

$$\langle f(z) \rangle_C = \frac{1}{2\pi}\int_0^{2\pi} f(k + R\,e^{i\theta})\,d\theta,$$

in which it is understood that the complex integral may be evaluated in terms of the real integrals above.

Once again denoting a general m^{th} degree polynomial by $P_m(z)$, (2.39) can therefore be expressed as an integral formula:

$$\frac{1}{2\pi}\int_0^{2\pi} P_m(k + R\,e^{i\theta})\,d\theta = \langle P_m(z) \rangle_C = P_m(k). \tag{2.40}$$

For example, if C is centred at the origin and $P_m(z) = z^m$, then

$$\langle z^m \rangle_C = \frac{1}{2\pi}\int_0^{2\pi} R^m\,e^{im\theta}\,d\theta = \frac{R^m}{2\pi}\int_0^{2\pi} [\cos m\theta + i \sin m\theta]\,d\theta = 0,$$

in agreement with (2.40).

The fact that (2.40) holds for polynomials of arbitrarily high degree immediately suggests that it might also hold for *power series*. We shall show that it does.

As usual we will only give the details for origin-centred power series, the generalization to arbitrary centres being straightforward. Let $P(z) = \sum_{j=0}^{\infty} c_j z^j$ be the power series, so that $P_m(z) = \sum_{j=0}^{m} c_j z^j$ are its approximating polynomials. If the circle C lies inside the disc of convergence of $P(z)$, then (2.12) implies the following. No matter how small we choose a real number ϵ, we can find a sufficiently large m such that $P_m(z)$ approximates $P(z)$ with accuracy ϵ throughout C and its interior. If we write $\mathcal{E}(z)$ for the complex number from the approximation $P_m(z)$ to the exact answer $P(z)$, then

$$P(z) = P_m(z) + \mathcal{E}(z), \quad \text{where} \quad |\mathcal{E}(z)| < \epsilon$$

for all z on and inside C, and in particular at the centre k of C.

At this point we could immediately study $\langle P(z)\rangle_C$ in terms of its integral representation, but it is more instructive to first consider the average $\langle P(z)\rangle_n$ of $P(z)$ over a regular n-gon inscribed in C. Once this is done, we may let n tend to infinity to obtain $\langle P(z)\rangle_C$.

First note that $\mathcal{E}(z)$ maps the vertices of the n-gon to points lying inside an origin-centred disc of radius ϵ. By (2.33), or directly from the generalized triangle inequality (1.8) on p. 8, the centroid $\langle \mathcal{E}(z)\rangle_n$ of these points must also lie in this disc. Choosing n greater than m, say $n = (m+1)$, (2.38) yields

$$\begin{aligned}\langle P(z)\rangle_{m+1} &= \langle P_m(z)\rangle_{m+1} + \langle \mathcal{E}(z)\rangle_{m+1} \\ &= P_m(k) + \langle \mathcal{E}(z)\rangle_{m+1} \\ &= P(k) + [\langle \mathcal{E}(z)\rangle_{m+1} - \mathcal{E}(k)]\,.\end{aligned}$$

The term in square brackets is the connecting complex number from $\mathcal{E}(k)$ to $\langle \mathcal{E}(z)\rangle_{m+1}$, and since both these points lie within a disc of radius ϵ, their connecting complex number must be shorter than 2ϵ. Finally, since the term in square brackets may also be interpreted as the connecting complex number from $P(k)$ to $\langle P(z)\rangle_{m+1}$, we have the following result:

> *Let m be chosen so that $P_m(z)$ approximates the power series $P(z)$ with accuracy ϵ on and within a circle C centred at k. If a regular $(m+1)$-gon is inscribed in C, then the average $\langle P(z)\rangle_{m+1}$ of $P(z)$ over its vertices will approximate $P(k)$ with accuracy 2ϵ.* (2.41)

We have thus transformed an exact result concerning the approximation $P_m(z)$ into an approximation result concerning the exact mapping $P(z)$.

For example, let C be the unit circle, and let $P(z) = e^z$. If we desire an accuracy of $\epsilon = 0.004$ everywhere on the unit disc then it turns out that $m = 5$ is sufficient, i.e., the approximating polynomial of lowest degree that has this accuracy is

$$P_5(z) = 1 + z + \tfrac{1}{2!}z^2 + \tfrac{1}{3!}z^3 + \tfrac{1}{4!}z^4 + \tfrac{1}{5!}z^5.$$

Figure [2.44] shows the image under $z \mapsto e^z$ of C, and in particular it shows the images of the vertices of a regular hexagon inscribed in C. According to the result, the centroid of these image points should differ from $e^0 = 1$ by no more than 0.008—an indiscernible discrepancy in a drawing done to this scale. This prediction is convincingly borne out in [2.45], which shows the sum of the complex numbers connecting 1 to the images of the vertices of the hexagon. To within the accuracy of the drawing, the sum is indeed zero!

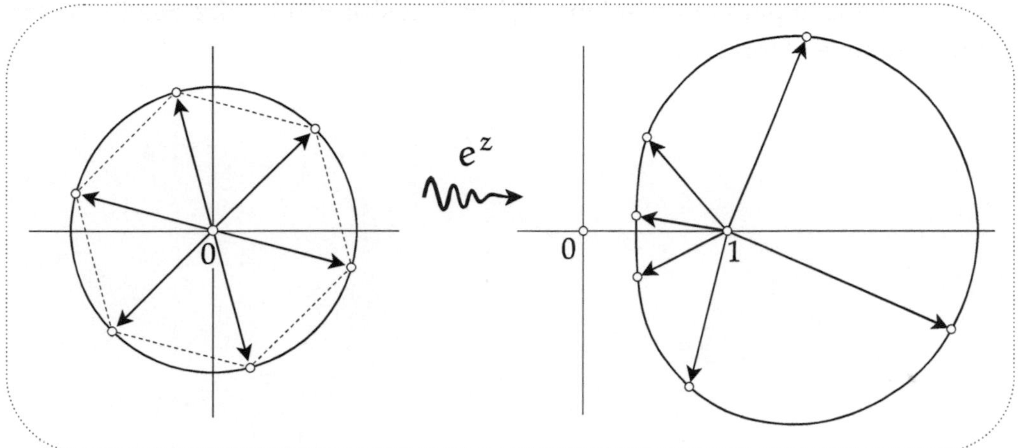

[2.44] *Applying $z \mapsto e^z$ to the unit circle and to the vertices of the inscribed regular hexagon on the left yields the figure on the right.* **The centroid of the six image points is *almost* the image of the centre of the original hexagon**, *namely, $e^0 = 1$. As the number of vertices goes to infinity, the result is ultimately exact.*

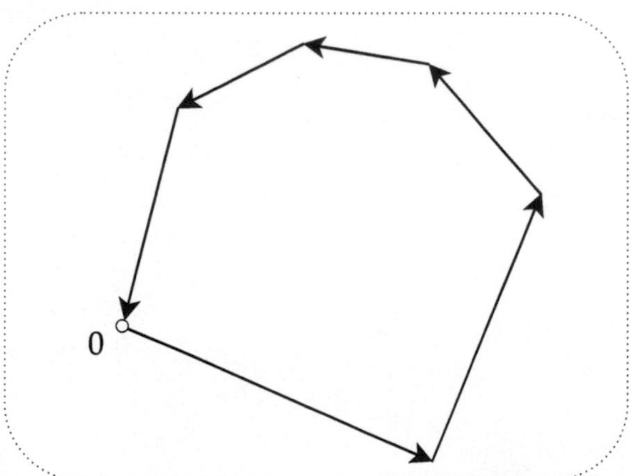

[2.45] Empirical confirmation of the claim made in the previous figure: *the connecting complex numbers from 1 to the images of the vertices of the hexagon do indeed sum to zero, at least to within the accuracy of this drawing.*

In the limit that ϵ tends to zero and m tends to infinity, (2.41) yields a form of *Gauss's Mean Value Theorem*:

> *If a complex function* $f(z)$ *can be expressed as a power series, and a circle* C *(radius* R *and centre* k*) lies within the disc of convergence of that power series, then*
>
> $$\langle f(z)\rangle_C = \frac{1}{2\pi}\int_0^{2\pi} f(k + R\,e^{i\theta})\,d\theta = f(k).$$

In addition to its theoretical importance, this formula can sometimes be used to evaluate difficult real integrals. For example, the exact version of [2.44] is $\langle e^z\rangle_C = e^0 = 1$, and this implies [exercise] that $\int_0^{2\pi} e^{\cos\theta}\,\cos[\sin\theta]\,d\theta = 2\pi$. See Ex. 43 for another example of this idea.

2.9 Exercises

1 Sketch the circle $|z-1| = 1$. Find (geometrically) the polar equation of the image of this circle under the mapping $z \mapsto z^2$. Sketch this image curve, which is called a *cardioid*.

2 Consider the complex mapping $z \mapsto w = (z-a)/(z-b)$. Show geometrically that if we apply this mapping to the perpendicular bisector of the line-segment joining a and b, then the image is the unit circle. In greater detail, describe the motion of w round this circle as z travels along the line at constant speed.

3 Consider the family of complex mappings

$$z \mapsto M_a(z) = \frac{z-a}{\overline{a}\,z-1} \qquad (a \text{ constant}).$$

[These mappings will turn out to be fundamental to non-Euclidean geometry.] Do the following problems algebraically; in the next chapter we will provide geometric explanations.

(i) Show that $M_a[M_a(z)] = z$. In other words, M_a is self-inverse.

(ii) Show that $M_a(z)$ maps the unit circle to itself.

(iii) Show that if a lies inside the unit disc then $M_a(z)$ maps the unit disc to itself.

Hint: Use $|q|^2 = q\,\overline{q}$ to verify that

$$|\overline{a}\,z-1|^2 - |z-a|^2 = (1-|a|^2)\,(1-|z|^2).$$

4 In figure [2.7] we saw that if $q^2 \leqslant p^3$ then the solutions of $x^3 = 3px + 2q$ are all real. Draw the corresponding picture in the case $q^2 > p^3$, and deduce that one solution is real, while the other two form a complex conjugate pair.

5 Show that the mapping $z \mapsto z^2$ doubles the angle between two rays coming out of the origin. Use this to deduce that the lemniscate (see [2.9] on p. 69) must self-intersect at right angles.

6 This question refers to the Cassinian curves in [2.9] on p. 69.

(i) On a copy of this figure, sketch the curves that intersect each Cassinian curve at right angles; these are called the *orthogonal trajectories* of the original family of curves.

(ii) Give an argument to show that each orthogonal trajectory hits one of the foci at ± 1.

(iii) If the Cassinian curves are thought of as a geographical contour map of the modular surface (cf. [2.10]) of (z^2-1), then what is the interpretation of the orthogonal trajectories in terms of the surface?

(iv) In Chapter 4 we will show that if two curves intersect at some point $p \neq 0$, and if the angle between them at p is ϕ, then the image curves under $z \mapsto w = z^2$ will *also* intersect at angle ϕ, at the point $w = p^2$. Use this to deduce that as z travels out from one of the foci along an orthogonal trajectory, $w = z^2$ travels along a ray out of $w = 1$.

(v) Check the result of the previous part by using a computer to draw the images under $w \mapsto \sqrt{w}$ of (A) circles centred at $w = 1$; (B) the radii of such circles.

(vi) Writing $z = x + iy$ and $w = u + iv$, find u and v as functions of x and y. By writing down the equation of a line in the w-plane through $w = 1$, show that the orthogonal trajectories of the Cassinian curves are actually segments of *hyperbolas*.

7 Sketch the modular surface of $C(z) = (z+1)(z-1)(z+1+i)$. Hence sketch the Cassinian curves $|C(z)| = \text{const.}$, then check your answer using a computer. To answer the following questions, recall that if $R(z)$ is a real function of position in the plane, then $R(p)$ is a *local minimum* of R if $R(p) < R(z)$ for all z in the immediate neighbourhood of p. A *local maximum* is defined similarly.

(i) Referring to the previous exercise, what is the significance of the orthogonal trajectories of the Cassinian curves you have just drawn?

(ii) Does $|C(z)|$ have any local maxima?

(iii) Does $|C(z)|$ have any non-zero local minima?

(iv) If D is a disc (or indeed a more arbitrary shape), can the maximum of $|C(z)|$ on D occur at a point inside D, or must the maximum occur at a boundary point of D? What about the minimum of $|C(z)|$ on D?

(v) Do you get the same answers to these questions if $C(z)$ is replaced by an arbitrary polynomial? What about a complex function that is merely known to be expressible as a power series?

8 On page 69 we saw that the polar equation of the lemniscate with foci at ± 1 is $r^2 = 2\cos 2\theta$. In fact James Bernoulli and his successors worked with a slightly different lemniscate having equation $r^2 = \cos 2\theta$. Let us call this the *standard lemniscate*.

(i) Where are the foci of the standard lemniscate?

(ii) What is the value of the product of the distances from the foci to a point on the standard lemniscate?

(iii) Show that the Cartesian equation of the standard lemniscate is

$$(x^2 + y^2)^2 = x^2 - y^2.$$

9 Here is an attempt [ultimately doomed] at using real methods to expand $H(x) = 1/(1+x^2)$ into a power series centred at $x = k$, i.e., into a series of the form $H(x) = \sum_{j=0}^{\infty} c_j X^j$, where $X = (x-k)$. According to Taylor's Theorem, $c_j = H^{(j)}(k)/j!$, where $H^{(j)}(k)$ is the jth derivative of H.

(i) Show that $c_0 = 1/(1+k^2)$ and $c_1 = -2k/(1+k^2)^2$, and find c_2. Note how it becomes increasingly difficult to calculate the successive derivatives.

(ii) Recall (or prove) that the nth derivative of a product AB of two functions A(x) and B(x) is given by *Leibniz's rule*:

$$(AB)^{(n)} = \sum_{j=0}^{n} \binom{n}{j} A^{(j)} B^{(n-j)}.$$

By applying this result to the product $(1+x^2)H(x)$, deduce that

$$(1+k^2)H^{(n)}(k) + 2nkH^{(n-1)}(k) + n(n-1)H^{(n-2)}(k) = 0.$$

Because the coefficients in this recurrence relation depend on n, we cannot solve it using the technique of Ex. 30 on p. 55.

(iii) Deduce from the previous part that the recurrence relation for the c_j's is

$$(1+k^2)c_n + 2kc_{n-1} + c_{n-2} = 0,$$

which *does* have constant coefficients.

(iv) Solve this recurrence relation, and hence recover the result (2.17) on p. 85.

10 Reconsider the series (2.18) on p. 85.

(i) Show that we recover the correct series (missing the odd powers of x) when the centre k of the series is at the origin.

(ii) Find a value of k such that the series is missing all the powers X^n, where $n = 2, 5, 8, 11, 14, \ldots$. Check your answer using a computer.

11 Show that each of the following series has the unit circle as its circle of convergence, then investigate the convergence *on* the unit circle. You can guess the correct answers by "drawing the series" in the manner of [2.18] on p. 89.

$$\text{(i)} \sum_{n=0}^{\infty} z^n, \quad \text{(ii)} \sum_{n=1}^{\infty} \frac{z^n}{n}, \quad \text{(iii)} \sum_{n=1}^{\infty} \frac{z^n}{n^2}.$$

[By virtue of (2.29), note that the second series is $-\text{Log}\,(1-z)$.]

12 Consider the geometric series $P(z) = \sum_{j=0}^{\infty} z^j$, which converges to $1/(1-z)$ inside the unit disc. The approximating polynomials in this case are $P_m(z) = \sum_{j=0}^{m} z^j$.

(i) Show that the error $E_m(z) \equiv |P(z) - P_m(z)|$ is given by

$$E_m(z) = \frac{|z|^{m+1}}{|1-z|}.$$

(ii) If z is any fixed point in the disc of convergence, what happens to the error as m tends to infinity?

(iii) If we fix m, what happens to the error as z approaches the boundary point $z = 1$?

(iv) Suppose we want to approximate this series in the disc $|z| \leqslant 0.9$, and further suppose that the maximum error we will tolerate is $\epsilon = 0.01$. Find the lowest degree polynomial $P_m(z)$ that approximates $P(z)$ with the desired accuracy throughout the disc.

13 We have seen that if we set $P_n(z) = z^n$, then the representation of a complex function $f(z)$ as an infinite series $\sum_{n=0}^{\infty} c_n P_n(z)$ (i.e., a power series) is unique. This is *not* true, however, if $P_n(z)$ is just any old set of polynomials. The following example is taken from Boas and Boas (2010). Defining

$$P_0(z) = -1, \qquad \text{and} \qquad P_n(z) = \frac{z^{n-1}}{(n-1)!} - \frac{z^n}{n!} \qquad (n = 1, 2, 3, \ldots),$$

show that

$$-2P_0 - P_1 + P_3 + 2P_4 + 3P_5 + \cdots = e^z = P_1 + 2P_2 + 3P_3 + 4P_4 + \cdots.$$

14 Consider two power series, $P(z) = \sum_{j=0}^{\infty} p_j z^j$ and $Q(z) = \sum_{j=0}^{\infty} q_j z^j$, which have approximating polynomials $P_n(z) = \sum_{j=0}^{n} p_j z^j$ and $Q_m(z) = \sum_{j=0}^{m} q_j z^j$. If the radii of convergence of $P(z)$ and $Q(z)$ are R_1 and R_2 then both series are uniformly convergent in the disc $|z| \leqslant r$, where $r < \min\{R_1, R_2\}$. Thus if ϵ is the maximum error we will tolerate in this disc, we can find a sufficiently large n such that

$$P_n(z) = P(z) + \mathcal{E}_1(z) \quad \text{and} \quad Q_n(z) = Q(z) + \mathcal{E}_2(z),$$

where the (complex) errors $\mathcal{E}_{1,2}(z)$ both have lengths less than ϵ. Use this to show that by taking a sufficiently high value of n we can approximate $[P(z) + Q(z)]$ and $P(z)\,Q(z)$ with arbitrarily high precision using $[P_n(z) + Q_n(z)]$ and $P_n(z)\,Q_n(z)$, respectively.

15 Give an example of a pair of origin-centred power series, say $P(z)$ and $Q(z)$, such that the disc of convergence for the product $P(z)Q(z)$ is *larger* than either of the two discs of convergence for $P(z)$ and $Q(z)$. [*Hint:* think in terms of rational functions, such as $[z^2/(5-z)^3]$, which are known to be expressible as power series.]

16 Our aim is to give a *combinatorial* explanation of the Binomial Theorem (2.14) for all negative integer values of n. The simple yet crucial first step is to write $n = -m$ and to change z to $-z$. Check that the desired result (2.14) now takes the form $(1-z)^{-m} = \sum_{r=0}^{\infty} c_r\, z^r$, where c_r is the binomial coefficient

$$c_r = \binom{m+r-1}{r}. \tag{2.42}$$

[Note that this says that the coefficients c_r are obtained by reading Pascal's triangle *diagonally*, instead of horizontally.] To begin to understand this, consider the special case $m = 3$. Using the geometric series for $(1-z)^{-1}$, we may express $(1-z)^{-3}$ as

$$[1+z+z^2+z^3+\cdots] \bullet [1+z+z^2+z^3+\cdots] \bullet [1+z+z^2+z^3+\cdots],$$

where $\bullet$ simply denotes multiplication. Suppose we want the coefficient c_9 of z^9. One way to get z^9 is to take z^3 from the first bracket, z^4 from the second, and z^2 from the third.

(i) Write this way of obtaining z^9 as the sequence $zzz \bullet zzzz \bullet zz$ of 9 z's and 2 $\bullet$'s, where the latter keep track of which power of z came from which bracket. [I got this nice idea from my friend Paul Zeitz.] Explain why c_9 is the number of distinguishable rearrangements of this sequence of 11 symbols. Be sure to address the meaning of sequences in which a $\bullet$ comes first, last, or is adjacent to the other $\bullet$.

(ii) Deduce that $c_9 = \binom{11}{9}$, in agreement with (2.42).

(iii) Generalize this argument and thereby deduce (2.42).

17 Here is an inductive approach to the result of the previous exercise.

(i) Write down the first few rows of Pascal's triangle and circle the numbers $\binom{5}{3}, \binom{4}{2}, \binom{3}{1}, \binom{2}{0}$. Check that the sum of these numbers is $\binom{6}{3}$. Explain this.

(ii) Generalize your argument to show that

$$\binom{n}{r} = \binom{n-1}{r} + \binom{n-2}{r-1} + \binom{n-3}{r-2} + \cdots + 1.$$

(iii) Assume that $(1-z)^{-M} = \sum_{r=0}^{\infty} \binom{M+r-1}{r} z^r$ holds for some positive integer M. Now multiply this series by the geometric series for $(1-z)^{-1}$ to find $(1-z)^{-(M+1)}$. Deduce that the binomial series is valid for all negative integer powers.

18 The basic idea of the following argument is due to Euler. Initially, let n be any real (possibly irrational) number, and define

$$B(z,n) \equiv \sum_{r=0}^{\infty} \binom{n}{r} z^r \quad \text{where} \quad \binom{n}{r} \equiv \frac{n(n-1)(n-2)\ldots(n-r+1)}{r!},$$

and $\binom{n}{0} \equiv 1$. We know from elementary algebra that if n is a positive integer then $B(z,n) = (1+z)^n$. To establish the Binomial Theorem (2.14) for rational powers, we must show that if p and q are integers then $B(z, \frac{p}{q})$ is the principal branch of $(1+z)^{\frac{p}{q}}$.

(i) With a fixed value of n, use the ratio test to show that $B(z, n)$ converges in the unit disc, $|z| < 1$.

(ii) By multiplying the two power series, deduce that

$$B(z, n)\, B(z, m) = \sum_{r=0}^{\infty} C_r(n, m)\, z^r \quad \text{where} \quad C_r(n, m) = \sum_{j=0}^{r} \binom{n}{j}\binom{m}{r-j}.$$

(iii) If m and n are positive integers, then show that

$$B(z, n)\, B(z, m) = B(z, n+m), \tag{2.43}$$

and deduce that $C_r(n, m) = \binom{n+m}{r}$. But $C_r(n, m)$ and $\binom{n+m}{r}$ are simply *polynomials* in n and m, and so the fact that they agree at infinitely many values of m and n [positive integers] implies that *they must be equal for all real values of* m *and* n. Thus the key formula (2.43) is valid for all real values of m and n.

(iv) By substituting $n = -m$ in (2.43), deduce the Binomial Theorem for negative integer values of n.

(v) Use (2.43) to show that if q is an integer then $[B(z, \frac{1}{q})]^q = (1+z)$. Deduce that $B(z, \frac{1}{q})$ is the principal branch of $(1+z)^{\frac{1}{q}}$.

(vi) Finally, show that if p and q are integers, then $B(z, \frac{p}{q})$ is indeed the principal branch of $(1+z)^{\frac{p}{q}}$.

19 Show that the ratio test cannot be used to find the radius of convergence of the power series (2.18) on p. 85. Use the root test to confirm that $R = \sqrt{1+k^2}$.

20 Show that if m and n are integers, then $\int_0^{2\pi} \cos m\theta \cos n\theta \, d\theta$ vanishes unless $m = n$, in which case it equals π. Likewise, establish a similar result for $\int_0^{2\pi} \sin m\theta \sin n\theta \, d\theta$. Use these facts to verify (2.19), at least formally.

21 Do the following problems by first substituting $z = r\, e^{i\theta}$ into the power series for e^z, then equating real and imaginary parts.

(i) Show that the Fourier series for $[\cos(\sin\theta)]\, e^{\cos\theta}$ is $\sum_{n=0}^{\infty} \frac{\cos n\theta}{n!}$, and write down the Fourier series for $[\sin(\sin\theta)]\, e^{\cos\theta}$.

(ii) Deduce that $\int_0^{2\pi} e^{\cos\theta}\, [\cos(\sin\theta)] \cos m\theta \, d\theta = (\pi/m!)$, where m is a positive integer.

(iii) By writing $x = (r/\sqrt{2})$, find the power series for $f(x) = e^x \sin x$.

(iv) Check the first few terms of the series for $f(x)$ by multiplying the series for e^x and $\sin x$.

(v) Calculate the nth derivative $f^{(n)}(0)$ using (1.14) on p. 25 of Chapter 1. By using these derivatives in Taylor's Theorem, verify your answer to part (iii).

22 Reconsider the formula,

$$e^{z} = \lim_{n\to\infty} P_n(z), \qquad \text{where} \qquad P_n(z) = \left(1 + \frac{z}{n}\right)^n.$$

(i) Check that $P_n(z)$ is the composition of a translation by n, followed by a contraction by $(1/n)$, followed by the power mapping $z \mapsto z^n$.

(ii) Referring to figure [2.4] on p. 64, use the previous part to sketch the images under $P_n(z)$ of circular arcs centred at $-n$, and of rays emanating from $-n$.

(iii) Let S be an origin-centred square (say of unit side) in the z-plane. With a large value of n, sketch just those portions of the arcs and rays (considered in the previous part) that lie within S.

(iv) Use the previous two parts to quantitatively explain figure [2.19] on p. 90.

23 If you did not do so earlier, sketch the image of a vertical line $x = k$ under $z \mapsto w = \cos z$ by drawing the analogue of [2.26]. Deduce that the asymptotes of this hyperbola are $\arg w = \pm k$. Check this using the equation of the hyperbola.

24 Consider the multifunction $f(z) = \sqrt{z-1}\,\sqrt[3]{z-i}$.

(i) Where are the branch points and what are their orders?

(ii) Why is it *not* possible to define branches using a single branch cut of the type shown in [2.35b]?

(iii) How many values does $f(z)$ have at a typical point z? Find and then plot all the values of $f(0)$.

(iv) Choose one of the values of $f(0)$ which you have just plotted, and label it p. Sketch a loop L that starts and ends at the origin such that if $f(0)$ is initially chosen to be -1, then as z travels along L and returns to the origin, $f(z)$ travels along a path from -1 to p. Do the same for each of the other possible values of $f(0)$.

25 Describe the branch points of the function $f(z) = 1/\sqrt{1-z^4}$. What is the smallest number of branch cuts that may be used to obtain single-valued branches of $f(z)$? Sketch an example of such cuts. [*Remark*: This function is historically important, owing to the fact (Ex. 20, p. 242) that $\int f(x)\,dx$ represents the arc length of the lemniscate. This integral (the *lemniscatic integral*) cannot be evaluated in terms of elementary functions—it is an example of a new kind of function called an *elliptic integral*. See Stillwell (2010), for more background and detail.]

26 For each function $f(z)$ below, find and then plot all the branch points and singularities. Assuming that these functions may be expressed as power series centred at k [in fact they can be], use the result (2.27) on p. 107 to verify the stated value of the radius of convergence R.

(i) If $f(z) = 1/(e^{\pi z} - 1)$ and $k = (1 + 2i)$, then $R = 1$.

(ii) If $f(z)$ is a branch of $\sqrt[5]{z^4 - 1}$ and $k = 3i$, then $R = 2$.

(iii) If $f(z)$ is a branch of $\sqrt{z - i}/(z - 1)$ and $k = -1$, then $R = \sqrt{2}$.

27 Until Euler cleared up the whole mess, the complex logarithm was a source of tremendous confusion. For example, show that $\log(z)$ and $\log(-z)$ have no common values, then consider the following argument of John Bernoulli:

$$\begin{aligned}\log[(-z)^2] = \log[z^2] &\implies \log(-z) + \log(-z) = \log(z) + \log(z)\\ &\implies 2\log(-z) = 2\log(z)\\ &\implies \log(-z) = \log(z).\end{aligned}$$

What is wrong with this argument?!

28 What value does z^i take at $z = -1$ if we start with the principal value at $z = 1$ (i.e., $1^i = 1$), and then let z travel one and a half revolutions clockwise round the origin?

29 In this exercise you will see that the "multifunction" k^z is quite different in character from all the other multifunctions we have discussed. For integer values of n, define $l_n \equiv [\text{Log}\,(k) + 2n\pi i]$.

(i) Show that the "branches" of k^z are $e^{l_n z}$.

(ii) Suppose that z travels along an arbitrary loop, beginning and ending at $z = p$. If we initially choose the value $e^{l_2 p}$ for k^z, then what value of k^z do we arrive at when z returns to p? Deduce that k^z has no branch points.

Since we cannot change one value of k^z into another by travelling round a loop, we should view its "branches" $\{\ldots, e^{l_{-1} z}, e^{l_0 z}, e^{l_1 z}, \ldots\}$ as an infinite set of *completely unrelated* single-valued functions.

30 Show that all the values of i^i are real! Are there any other points z such that z^i is real?

31 In the case of a real variable, the logarithmic power series was originally discovered [see next exercise] as follows. First check that $\ln(1+X)$ can be written as $\int_0^X [1/(1+x)]\,dx$, and then expand $[1/(1+x)]$ as a power series in x. Finally, integrate your series term by term. [Later in the book we will be able to generalize this argument to the complex plane.]

32 Here is another approach to the logarithmic power series. As before, let $L(z) = \text{Log}\,(1+z)$. Since $L(0) = 0$, the power series for $L(z)$ must be of the form $L(z) = az + bz^2 + cz^3 + dz^4 + \cdots$. Substitute this into the equation

$$1+z = e^L = 1 + L + \tfrac{1}{2!}L^2 + \tfrac{1}{3!}L^3 + \tfrac{1}{4!}L^4 + \cdots,$$

then find a, b, c, and d by equating powers of z. [Historically the logarithmic series came first—both Mercator and Newton discovered it using the method in the previous exercise—then Newton reversed the reasoning of the present exercise to obtain the series for e^x. See Stillwell (2010).]

33 (i) Use [2.26] to discuss the branch points of the multifunction $\cos^{-1}(z)$.

(ii) Rewrite the equation $w = \cos z$ as a quadratic in e^{iz}. By solving this equation, deduce that $\cos^{-1}(z) = -i\log[z + \sqrt{z^2-1}]$. [Why do we not need to bother to write $\pm$ in front of the square root?]

(iii) Show that as z travels along a loop that goes once round either 1 or -1 (but not both), the value of $[z + \sqrt{z^2-1}]$ changes to $1/[z+\sqrt{z^2-1}]$.

(iv) Use the previous part to show that the formula in part (ii) is in accord with the discussion in part (i).

34 Write down the origin-centred power series for $(1 - \cos z)$. Use the Binomial Theorem to write down the power series (centred at $Z = 0$) for the principal branch of $\sqrt{1-Z}$, then substitute $Z = (1-\cos z)$. Hence show that if we choose the branch of $\sqrt{\cos z}$ that maps 0 to 1, then

$$\sqrt{\cos z} = 1 - \frac{z^2}{4} - \frac{z^4}{96} - \frac{19z^6}{5760} - \cdots.$$

Verify this using a computer. Where does this series converge?

35 What value does $(z/\sin z)$ approach as z approaches the origin? Use the series for $\sin z$ to find the first few terms of the origin-centred power series for $(z/\sin z)$. Check your answer using a computer. Where does this series converge?

36 By considering $\text{Log}\,(1+ix)$, where x is a real number lying between ± 1, deduce that

$$\tan^{-1}(x) = x - \frac{x^3}{3} + \frac{x^5}{5} - \frac{x^7}{7} + \frac{x^9}{9} - \frac{x^{11}}{11} + \cdots.$$

In what range does this value of $\tan^{-1}(x)$ lie? Give another derivation of the series using the idea in Ex. 31.

37 (i) Show geometrically that as $z = e^{i\theta}$ goes round and round the unit circle (with ever increasing θ), $\text{Im}\,[\text{Log}\,(1+z)] = (\Theta/2)$, where Θ is the principal value of θ, i.e., $-\pi < \Theta \leqslant \pi$.

(ii) Consider the periodic "saw tooth" function $F(\theta)$ whose graph is shown below. By substituting $z = e^{i\theta}$ in the logarithmic series (2.29), use the previous part to deduce the following Fourier series:

$$F(\theta) = \sin\theta - \frac{\sin 2\theta}{2} + \frac{\sin 3\theta}{3} - \frac{\sin 4\theta}{4} + \cdots.$$

(iii) Check this Fourier series by directly evaluating the integrals (2.19).

(iv) Use a computer to draw graphs of the partial sums of the Fourier series. As you increase the number of terms, observe the magical convergence of this sum of smooth waves to the jagged graph below. If only Fourier could have seen this on the screen, not just in his mind's eye!

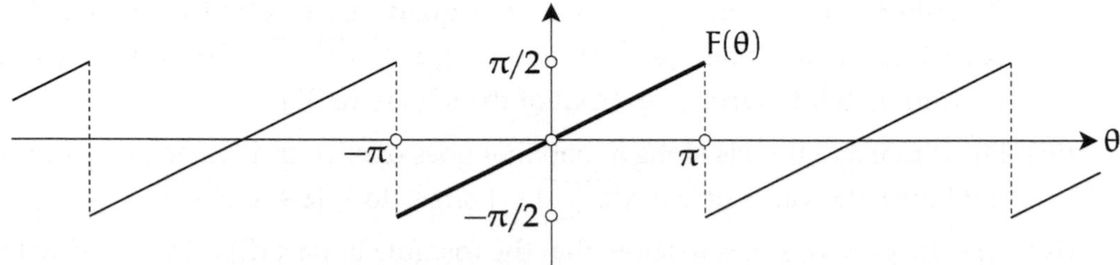

38 As in the previous exercise, let $\Theta = \text{Arg}\,(z)$.

(i) Use (2.29) to show that

$$\frac{1}{2}\,\text{Log}\left[\frac{1+z}{1-z}\right] = z + \frac{z^3}{3} + \frac{z^5}{5} + \frac{z^7}{7} + \cdots.$$

(ii) Show geometrically that as $z = e^{i\theta}$ goes round and round the unit circle,

$$\text{Im}\left\{\frac{1}{2}\,\text{Log}\left[\frac{1+z}{1-z}\right]\right\} = (\text{sign of } \Theta)\left[\frac{\pi}{4}\right].$$

(iii) Consider the periodic "square wave" function $G(\theta)$ whose graph is shown below. Use the previous two parts to deduce that its Fourier series is

$$G(\theta) = \sin\theta + \frac{\sin 3\theta}{3} + \frac{\sin 5\theta}{5} + \frac{\sin 7\theta}{7} + \cdots.$$

Finally, repeat parts (iii) and (iv) of the previous exercise.

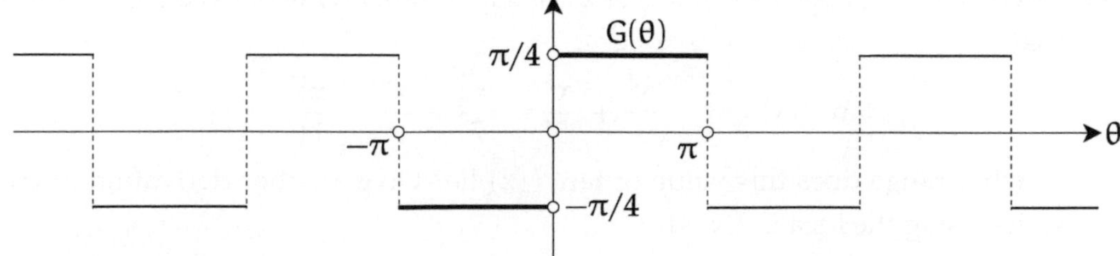

39 Show that (2.32) is still true even if the (positive) masses of the particles are not all equal.

40 Here is another simple way of deriving (2.34). If the vertices of the origin-centred regular n-gon are rotated by ϕ, then their centroid Z rotates with them to $e^{i\phi}Z$. By choosing $\phi = (2\pi/n)$, deduce that $Z = 0$.

41 To establish (2.36), let $z_0, z_1, z_2, \ldots, z_{n-1}$ be the vertices (labelled counterclockwise) of the regular n-gon, and let C be the circumscribing circle. Also, let $w_j = z_j^m$ be the image of vertex z_j under the mapping $z \mapsto w = z^m$. Think of z as a particle that starts at z_0 and orbits counterclockwise round C, so that the image particle $w = z^m$ starts at w_0 and orbits round another circle *with* m *times the angular speed of* z.

(i) Show that each time z travels from one vertex to the next, w executes (m/n) of a revolution. Thus as z travels from z_0 to z_k, w executes $k(m/n)$ revolutions as it travels from w_0 to w_k.

(ii) Let w_k be the first point in the sequence w_1, w_2, etc., such that $w_k = w_0$. Deduce that if (M/N) is (m/n) reduced to lowest terms, then $k = N$. Note that $N = (n$ divided by the highest common factor of m and $n)$.

(iii) Explain why $w_{N+1} = w_1, w_{N+2} = w_2$, etc.

(iv) Show that $w_0, w_1, \ldots, w_{N-1}$ are distinct.

(v) Show that $w_0, w_1, \ldots, w_{N-1}$ are the vertices of a regular N-gon.

42 Consider the mapping $z \mapsto w = P_n(z)$, where $P_n(z)$ is a general polynomial of degree $n \geqslant 2$. Let S_q be the set of points in the z-plane that are mapped to a particular point q in the w-plane. Show that *the centroid of* S_q *is independent of the choice of* q, and is therefore a property of the polynomial itself. [*Hint:* This is another way of looking at a familiar fact about the sum of the roots of a polynomial.]

43 Use Gauss's Mean Value Theorem [p. 78] to find the average of $\cos z$ over the circle $|z| = r$. Deduce (and check with a computer) that for all real values of r,

$$\int_0^{2\pi} \cos[r\cos\theta]\,\cosh[r\sin\theta]\,d\theta = 2\pi.$$

CHAPTER 3

Möbius Transformations and Inversion

3.1 Introduction

3.1.1 Definition and Significance of Möbius Transformations

A *Möbius transformation*[1] is a mapping of the form

$$M(z) = \frac{az + b}{cz + d}, \tag{3.1}$$

where a, b, c, d are complex constants. These mappings have many beautiful properties, and they find very varied application throughout complex analysis. Despite their apparent simplicity, Möbius transformations lie at the heart of several exciting areas of modern mathematical research. This is due in large part to their intimate and somewhat miraculous connection with the non-Euclidean geometries alluded to in Chapter 1. [This connection is the subject of Chapter 6.] Moreover, these transformations are also intimately connected[2] with Einstein's Theory of Relativity! This connection has been exploited with remarkable success by Sir Roger Penrose; see Penrose and Rindler (1984).

Two centuries have passed since August Ferdinand Möbius first studied the transformations that now bear his name, but it is fair to say that the rich vein of knowledge which he thereby exposed is still far from being exhausted. For this reason, we shall investigate Möbius transformations in considerably greater depth than is customary.

[1] Also known as a "linear", "bilinear", "linear-fractional", or "homographic" transformation.
[2] According to Coxeter (1967), this connection was first recognized by H. Liebmann in 1905, the very year that Einstein discovered Special Relativity!

Visual Complex Analysis. 25th Anniversary Edition. Tristan Needham, Oxford University Press.
 DOI: 10.1093/oso/9780192868916.003.0003

3.1.2 The Connection with Einstein's Theory of Relativity*

Clearly it would be neither appropriate nor feasible for us to explore this connection in detail, but let us at least briefly indicate how Möbius transformations are related to Einstein's Theory of Relativity.

In that theory, the time T and the 3-dimensional Cartesian coordinates (X, Y, Z) of an event are combined into a single *4-vector* (T, X, Y, Z) in 4-dimensional *spacetime*. Of course the spatial components of this vector have no absolute significance: rotating the coordinate axes yields different coordinates $(\widetilde{X}, \widetilde{Y}, \widetilde{Z})$ for one and the same point in space. But if two people choose different axes, they will nevertheless agree on the value of $\widetilde{X}^2 + \widetilde{Y}^2 + \widetilde{Z}^2 = X^2 + Y^2 + Z^2$, for this represents the square of the distance to the point.

In contrast to this, we are accustomed to thinking that the *time* component T *does* have an absolute significance. However, Einstein's theory—confirmed by innumerable experiments—tells us that this is wrong. *If two (momentarily coincident) observers are in relative motion, they will disagree about the times at which events occur.* Furthermore, they will no longer agree about the value of $(X^2 + Y^2 + Z^2)$—this is the famous *Lorentz contraction*. Is there *any* aspect of spacetime that has absolute significance and on which two observers in relative motion must agree? Yes: making a convenient choice of units in which the speed of light is equal to 1, Einstein discovered that both observers will agree on the value of

$$\widetilde{T}^2 - (\widetilde{X}^2 + \widetilde{Y}^2 + \widetilde{Z}^2) = T^2 - (X^2 + Y^2 + Z^2).$$

A *Lorentz transformation* $\mathcal{L}$ is a linear transformation of spacetime (a 4×4 matrix) that maps one observer's description (T, X, Y, Z) of an event to another observer's description $(\widetilde{T}, \widetilde{X}, \widetilde{Y}, \widetilde{Z})$ of the same event. Put differently, $\mathcal{L}$ is a linear transformation that preserves the quantity $T^2 - (X^2 + Y^2 + Z^2)$, upon which both observers must agree.

Now imagine that the spacetime coordinate origin emits a flash of light—an origin-centred sphere whose radius increases at the speed of light. It turns out that any given $\mathcal{L}$ is completely determined by its effect on the coordinates of the light rays that make up this flash. Here is the next crucial idea: in Ex. 8 we explain how *we may set up a one-to-one correspondence between these light rays and complex numbers.* Thus each Lorentz transformation of spacetime induces a definite mapping of the complex plane. What kinds of complex mappings do we obtain in this way? The miraculous answer turns out to be this:

> *The complex mappings that correspond to the Lorentz transformations are the Möbius transformations! Conversely, every Möbius transformation of $\mathbb{C}$ yields a unique Lorentz transformation of spacetime.* (3.2)

Even among professional physicists, this "miracle" is not as well known as it should be.

The connection exhibited in (3.2) is deep and powerful. Just for starters, it means that any result we establish concerning Möbius transformations will immediately yield a corresponding result in Einstein's Theory of Relativity. Furthermore, these Möbius transformation proofs turn out to be considerably more elegant than direct spacetime proofs.

To really understand the above claims, we strongly recommend that after reading this chapter you consult Penrose and Rindler (1984, Ch. 1), as well as Shaw (2006, Ch. 23) and Needham (2021, §6.4).

3.1.3 Decomposition into Simple Transformations

As a first step towards making sense of (3.1), let us decompose $M(z)$ [exercise] into the following sequence of transformations:

$$\left.\begin{array}{ll}\text{(i)} & z \mapsto z+\frac{d}{c}, \text{ which is a translation;}\\ \text{(ii)} & z \mapsto (1/z);\\ \text{(iii)} & z \mapsto -\frac{(ad-bc)}{c^2}z, \text{ which is an expansion and a rotation;}\\ \text{(iv)} & z \mapsto z+\frac{a}{c}, \text{ which is another translation.}\end{array}\right\} \quad (3.3)$$

Note that if $(ad-bc)=0$ then $M(z)$ is an uninteresting constant mapping, sending every point z to the same image point (a/c); in this exceptional case $M(z)$ is called *singular*. In discussing Möbius transformations we shall therefore always assume that $M(z)$ is *non-singular*, meaning that $(ad-bc) \neq 0$.

Of the four transformations above, only the second one has not yet been investigated. This mapping $z \mapsto (1/z)$ holds the key to understanding Möbius transformations; we shall call it *complex inversion*. The next section examines its many remarkable and powerful properties.

3.2 Inversion

3.2.1 Preliminary Definitions and Facts

The image of $z = r\,e^{i\theta}$ under complex inversion is $1/(r\,e^{i\theta}) = (1/r)\,e^{-i\theta}$: the new length is the reciprocal of the original, and the new angle is the negative of the original. See [3.1a]. Note how a point outside the unit circle C is mapped to a point inside C, and vice versa.

Figure [3.1a] also illustrates a particularly fruitful way of decomposing complex inversion into a two-stage process:

(i) Send $z = r\,e^{i\theta}$ to the point that is in the same direction as z but that has reciprocal length, namely the point $(1/r)\,e^{i\theta} = (1/\overline{z})$.

(ii) Apply complex conjugation (i.e., reflection in the real axis), which sends $(1/\overline{z})$ to $\overline{(1/\overline{z})} = (1/z)$.

Check for yourself that the order in which we apply these mappings is immaterial.

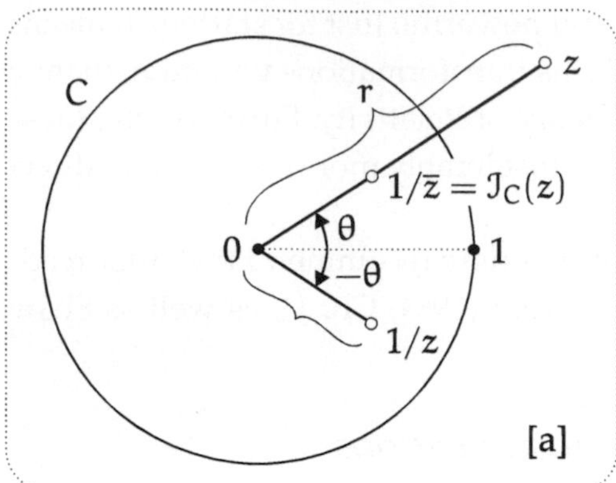

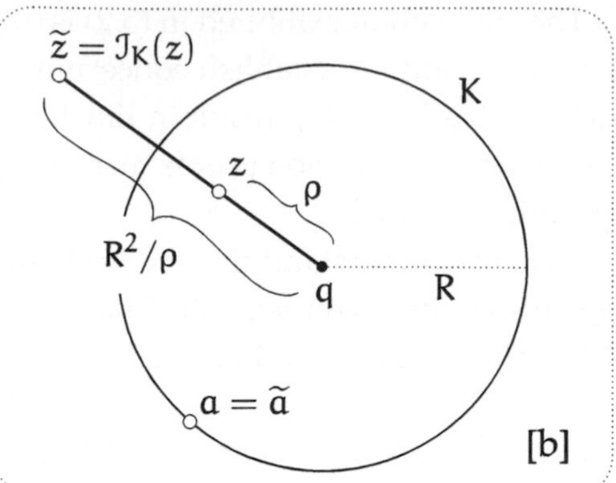

[3.1] [a] Geometric inversion in the unit circle C: $z \mapsto \mathcal{I}_C(z) \equiv (1/\bar{z})$ *leaves the points of* C *fixed, and swaps the interior and exterior. When composed with conjugation, it yields* complex inversion, $z \mapsto (1/z)$. **[b]** *Definition of geometric inversion* $z \mapsto \mathcal{I}_K(z)$ *in a general circle* K.

While stage (ii) is geometrically trivial, we shall see that the mapping in stage (i) is filled with surprises; it is called[3] *geometric inversion*, or simply *inversion*. Clearly, the unit circle C plays a special role for this mapping: the inversion interchanges the interior and exterior of C, while each point *on* C remains fixed (i.e., is mapped to itself). For this reason we write the mapping as $z \mapsto \mathcal{I}_C(z) = (1/\bar{z})$, and we call $\mathcal{I}_C$ (a little more precisely than before) "inversion in C".

This added precision in terminology is important because, as illustrated in [3.1b], there is a natural way of generalizing $\mathcal{I}_C$ to inversion in an *arbitrary* circle K (say with centre q and radius R). Clearly, this "inversion in K", written $z \mapsto \tilde{z} = \mathcal{I}_K(z)$, should be such that the interior and exterior of K are interchanged, while each point on K remains fixed. If ρ is the distance from q to z, then *we define* $\tilde{z} = \mathcal{I}_K(z)$ *to be the point in the same direction from* q *as* z, *and at distance* (R^2/ρ) *from* q. [Check for yourself that this definition does indeed perform as advertised.]

As usual, we invite you to use a computer to verify empirically the many results we shall derive concerning inversion. However, in the case of this particular mapping, you can also construct (fairly easily) a *mechanical* instrument that will carry out the mapping for you; see Ex. 2.

Although we shall not need it for a while, it is easy enough to obtain a *formula* for $\mathcal{I}_K(z)$. Because the connecting complex numbers from q to z and to $\tilde{z}$ both have the

[3] In older works it is often called "transformation by reciprocal radii".

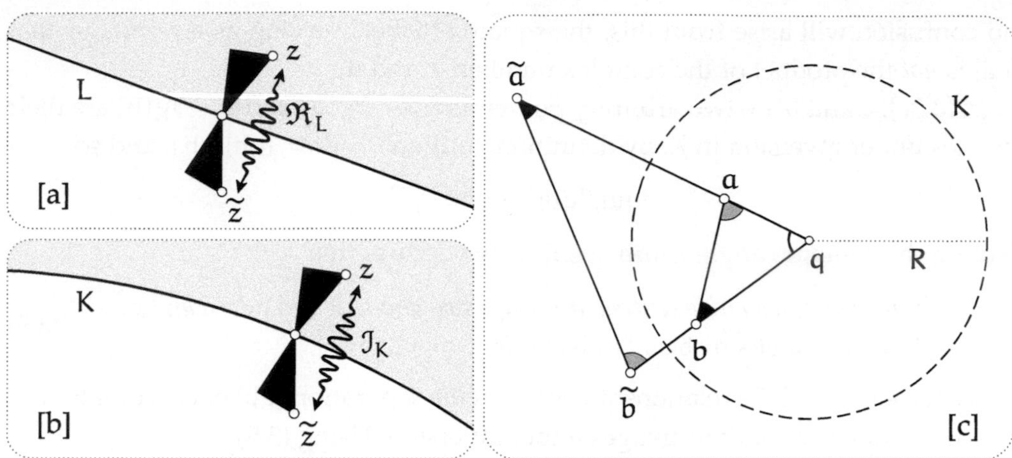

[3.2] [a] *Ordinary reflection in a line.* **[b]** *If the radius of* K *is large, so that it starts to look straight, then geometric inversion* $z \mapsto \widetilde{z} = \mathfrak{I}_K(z)$ *looks very much like ordinary reflection. More precisely, as* z *approaches* K, *then* $\mathfrak{I}_K(z)$ **is ultimately its reflection in the tangent to K. [c]** *Under inversion in* K, *the triangle* aqb is similar *to the triangle* $\widetilde{b}q\widetilde{a}$.

same direction, and their lengths are ρ and (R^2/ρ), it follows that $(\widetilde{z}-q)\overline{(z-q)} = R^2$. Solving for $\widetilde{z}$,

$$\mathfrak{I}_K(z) = \frac{R^2}{\overline{z} - \overline{q}} + q = \frac{q\,\overline{z} + (R^2 - |q|^2)}{\overline{z} - \overline{q}}. \tag{3.4}$$

For example, if we put $q = 0$ and $R = 1$, then we recover $\mathfrak{I}_C(z) = (1/\overline{z})$.

There is a very interesting similarity (which will deepen as we go on) between inversion $\mathfrak{I}_K(z)$ in a circle K and reflection $\mathfrak{R}_L(z)$ in a line L. See [3.2a] and [3.2b]. First, L divides the plane into two pieces, or "components", which are interchanged by $\mathfrak{R}_L(z)$; second, each point on the boundary between the components remains fixed; third, $\mathfrak{R}_L(z)$ is *involutory* or *self-inverse*, meaning that $\mathfrak{R}_L \circ \mathfrak{R}_L$ is the identity mapping, leaving every point fixed. To put this last property differently, consider a point z and its reflection $\widetilde{z} = \mathfrak{R}_L(z)$ in L. Such a pair are said to be "mirror images", or to be "symmetric with respect to L". The involutory property says that the reflection causes such a pair of points to *swap places*.

Check for yourself that $\mathfrak{I}_K(z)$ shares all three of these properties. Furthermore, the black triangle in [3.2b] illustrates the fact that if K is large then the effect of $\mathfrak{I}_K$ on a small shape close to K looks very much like ordinary reflection. [We will explain this later, but you might like to check this empirically using a computer.] For these reasons, and others still to come, $\mathfrak{I}_K(z)$ is often also called *reflection in a circle*, and the pair of points z and $\widetilde{z} = \mathfrak{I}_K(z)$ are said to be *symmetric with respect to* K.

We end this subsection with two simple properties of inversion, the first of which will serve as the springboard for the investigations that follow. Let us use the symbol [cd] to stand for the distance $|c - d|$ between two points c and d. We hope that no confusion will arise from this, the square brackets serving as a reminder that [cd] is *not* the product of the complex numbers c and d.

In [3.2c], a and b are two arbitrary points, and $\widetilde{a} = \mathcal{I}_K(a)$ and $\widetilde{b} = \mathcal{I}_K(b)$ are their images under inversion in K. By definition, $[qa][q\widetilde{a}] = R^2 = [qb][q\widetilde{b}]$, and so

$$[qa]/[qb] = [q\widetilde{b}]/[q\widetilde{a}].$$

Noting the common angle $\angle aqb = \angle \widetilde{a}q\widetilde{b}$, we deduce that

If inversion in a circle centred at q *maps two points* a *and* b *to* $\widetilde{a}$ *and* $\widetilde{b}$, *then the triangles* aqb *and* $\widetilde{b}q\widetilde{a}$ *are similar.* (3.5)

Lastly, let us find the relationship between the separation [ab] of two points, and the separation $[\widetilde{a}\widetilde{b}]$ of their images under inversion. Using (3.5),

$$[\widetilde{a}\widetilde{b}]/[ab] = [q\widetilde{b}]/[qa] = R^2/[qa][qb],$$

and so the separation of the image points is given by

$$[\widetilde{a}\widetilde{b}] = \left(\frac{R^2}{[qa][qb]}\right)[ab]. \tag{3.6}$$

3.2.2 Preservation of Circles

Let us examine the effect of $\mathcal{I}_K$ on lines and then on circles. If a line L passes through the centre q of K, then clearly $\mathcal{I}_K$ maps L to itself, which we may write as $\mathcal{I}_K(L) = L$. Of course we don't mean that each point of L remains fixed, for $\mathcal{I}_K$ interchanges the portions of L interior and exterior to K; the only *points* of L that remain fixed are the two places where it intersects K.

Matters become much more interesting when we consider a general line L that does not pass through q. Figure [3.3] provides the surprising answer:

If a line L *does not pass through the centre* q *of* K, *then inversion in* K *maps* L *to a circle that passes through* q. (3.7)

Here b is an arbitrary point on L, while a is the intersection of L with the perpendicular line through q. By virtue of (3.5), $\angle q\widetilde{b}\widetilde{a} = \angle qab = (\pi/2)$, so $\widetilde{b}$ lies on the circle having the line-segment $q\widetilde{a}$ as diameter. Done. Notice, incidentally, that the tangent at q of the image circle is parallel to L.

Note that (3.7) makes no mention of the radius R of K. You may therefore be concerned that in [3.3] we have chosen R so that K does not intersect L; what happens if K *does* intersect L? Check for yourself that, while the picture looks somewhat different in this case, the geometric argument above continues to apply without any modification.

We now give a less direct, but more instructive way of understanding why (3.7) does not depend on the size of K. We will show that if the result holds for one

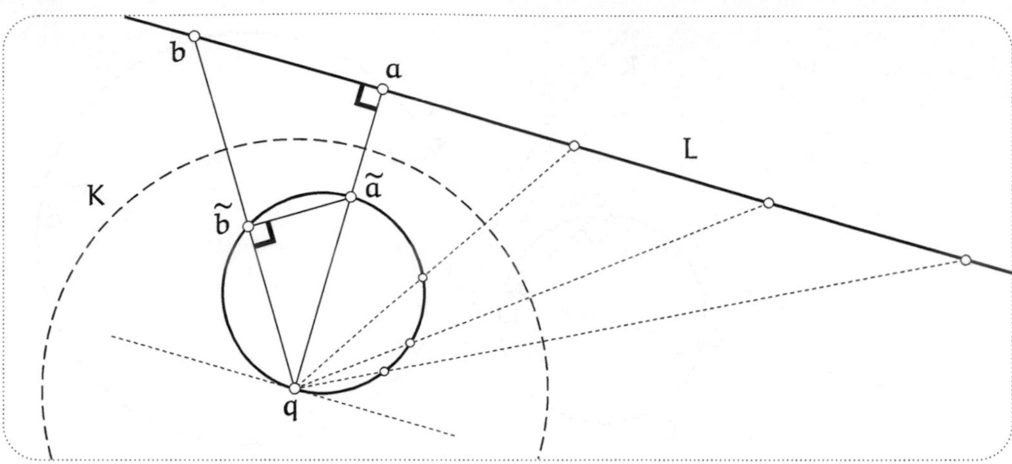

[3.3] The geometric inverse of a line is a circle through the centre of inversion. *By virtue of [3.2c], the illustrated right triangles are similar, so we see that the geometric inverse of a line* L *is a circle that passes through the centre of the inversion, and that its tangent there is parallel to* L.

circle K_1 (radius R_1) centred at q, then it will hold for any other circle K_2 (radius R_2) centred at q.

Let z be an arbitrary point, and let $\widetilde{z}_1 = \mathcal{I}_{K_1}(z)$ and $\widetilde{z}_2 = \mathcal{I}_{K_2}(z)$. Obviously $\widetilde{z}_1$ and $\widetilde{z}_2$ are both in the same direction from q as z, and you can easily check that *the ratio of their distances from* q *is independent of the location of* z:

$$[q\,\widetilde{z}_2]/[q\,\widetilde{z}_1] = (R_2/R_1)^2 \equiv k, \text{ say.}$$

Thus

$$\mathcal{I}_{K_2} = \mathcal{D}_q^k \circ \mathcal{I}_{K_1}, \tag{3.8}$$

where the "central dilation" $\mathcal{D}_q^k$ [see p. 45] is an expansion (centred at q) of the plane by a factor of k. It follows [exercise] that if (3.7) holds for K_1 then it also holds for K_2.

Look again at [3.3]. Since $\mathcal{I}_K$ is involutory, it simply *swaps* the line and circle, and so *the image of any circle through* q *is a line not passing through* q. But what happens to a general circle C that does not pass through q? Initially, suppose that C does not contain q in its interior. Figure [3.4] provides the beautiful answer:

If a circle C *does not pass through the centre* q *of* K, *then inversion in* K *maps* C *to another circle not passing through* q. (3.9)

This fundamental result is often described by saying that inversion "preserves circles".

It follows from (3.8) that if (3.9) is true for one choice of K, then it will be true for any choice of K. We may therefore conveniently choose K so that C lies inside

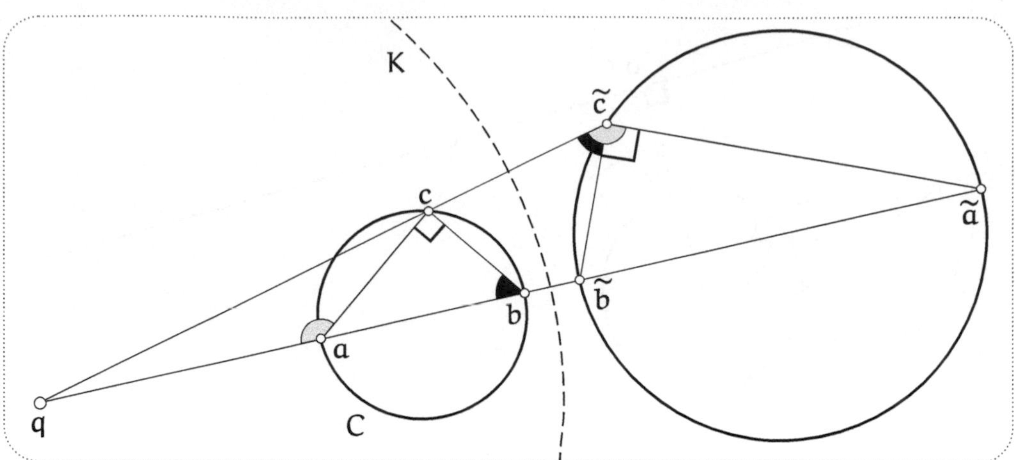

[3.4] Geometric inversion maps circles to circles! *This beautiful and important fact is readily confirmed by first observing that [3.2c] implies that the two grey angles are equal to each other, and, likewise, the two black angles are equal to each other. But this means that the right angle at* c*, being the difference of the grey and black angles, implies the illustrated right angle at* $\widetilde{c}$*. Done!*

it, as illustrated. Here a and b are the ends of a diameter of C, and they therefore subtend a right angle at a general point c on C. To understand (3.9), first use (3.5) to check that both the shaded angles are equal, and that both the black angles are equal. Next look at the triangle abc, and observe that the external shaded angle at a is the sum of the two illustrated internal angles: the right angle at c and the black angle at b. It follows that $\angle \widetilde{a}\widetilde{c}\widetilde{b} = (\pi/2)$, and hence that $\widetilde{a}$ and $\widetilde{b}$ are the ends of a diameter of a circle through $\widetilde{c}$. Thus we have demonstrated (3.9) in the case where C does not contain q. We leave it to you to check that the same line of reasoning establishes the result in the case where C does contain q.

The result (3.7) is in fact a special limiting case of (3.9). Figure [3.5] shows a line L, the point p on L closest to the centre q of the inversion, and a circle C tangent to L at p. As its radius tends to infinity, C tends to L, and the image circle $\widetilde{C} = \mathcal{I}_K(C)$ tends to a circle through q.

Later we will be able to give a much cleaner way of seeing that (3.7) and (3.9) are two aspects of a single result.

3.2.3 Constructing Inverse Points Using Orthogonal Circles

Consider [3.6a]. The circle C cuts the circle of inversion K at right angles at a and b. In other words, the tangent T to C at a (for example) passes through q. Under inversion in K, a and b remain fixed, and T is mapped into itself. Thus the image of C must be a circle that again passes through a and b and that is again orthogonal to K. But clearly there is only *one* circle with these properties, namely C itself. Thus,

$$\textit{Under inversion in } K, \textit{ every circle orthogonal to } K \textit{ is mapped to itself.} \tag{3.10}$$

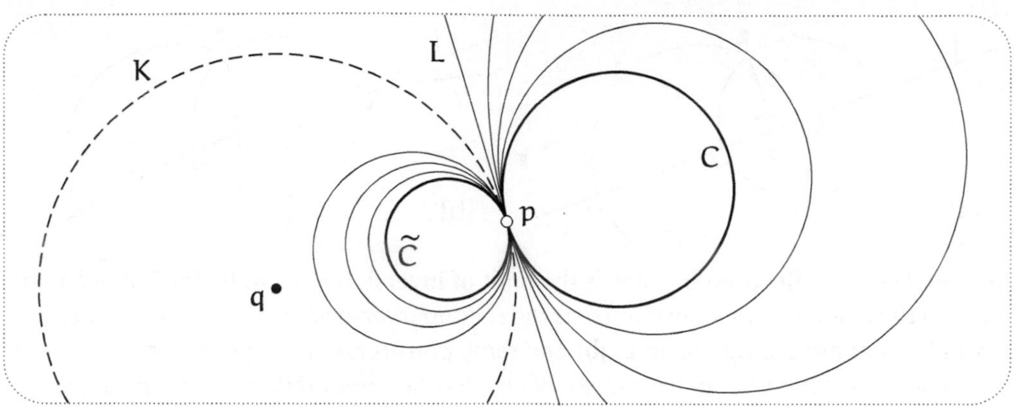

[3.5] Inversion of a line is the limit of inversion of a circle. *Here we see that [3.3] is in fact merely a special limiting case of the general result shown in [3.4]. As the radius of* C *tends to infinity, so that* C *becomes the line* L, *the image circle* $\widetilde{C} = \mathcal{I}_K(C)$ *becomes a circle through* q.

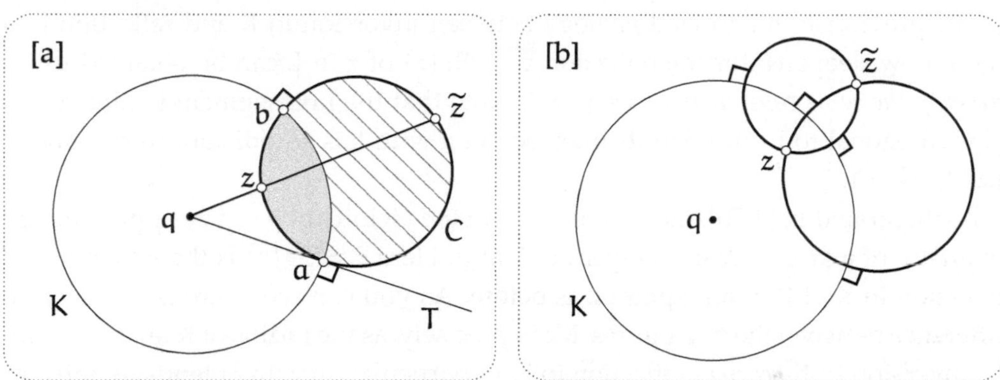

[3.6] Constructing the geometric inverse using orthogonal circles. [a] *If* C *is orthogonal to* K, *then it is mapped to itself, because the image must be another circle that again passes through the fixed points* a *and* b, *and the tangent* T *at* a *is mapped into itself. Thus it can only be* C *itself.* **[b]** *It follows that the geometric inverse* $\widetilde{z}$ *of a point* z *may be constructed as the second intersection point of any two circles through* z *that are orthogonal to* K.

Figure [3.6a] illustrates two immediate consequences of this result. First, the disc bounded by C is also mapped to itself, the shaded and hatched regions into which K divides it being swapped by the inversion. Second, a line from q through a point z on C intersects C for the second time at the inverse point $\widetilde{z}$.

Another consequence (the key result of this subsection) is the geometric construction shown in [3.6b], the verification of which is left to you.

> *The inverse* $\widetilde{z}$ *of* z *in* K *is the second intersection point of any two circles that pass through* z *and are orthogonal to* K.

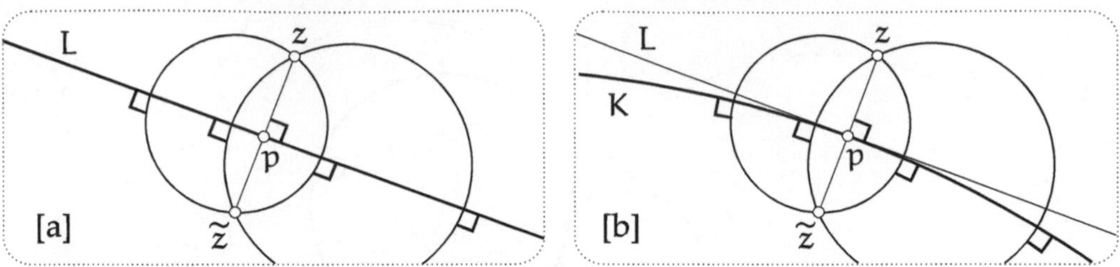

[3.7] Inversion (reflection) in a line is the limit of inversion in a circle. [a] *The reflection $\widetilde{z}$ of z in the line L can be constructed as the second intersection point of any two circles through z that are orthogonal to L.* **[b]** *The same construction for geometric inversion in the circle K. As the radius of K tends to infinity, this becomes reflection in the tangent L.*

Note that the construction of $\widetilde{z}$ in [3.6a] is the special limiting case in which the radius of one of the circles tends to infinity, and so becomes a line through q. For other, less important, geometric constructions of inversion, see Ex. 1.

The previously mentioned analogy between inversion in K and reflection in a line L now deepens, for the reflection $\widetilde{z} = \mathfrak{R}_L(z)$ of z in L can be obtained using *precisely the same construction*; see [3.7a]. Note that the line-segment joining z and $\widetilde{z}$ is orthogonal to L, and that its intersection p with L is equidistant from z and $\widetilde{z}$: $[pz]/[p\,\widetilde{z}] = 1$.

As illustrated in [3.7b], the segment of L in the vicinity of p can be approximated by an arc of a large circle K tangent to L at p. Here $\widetilde{z} = \mathfrak{I}_K(z)$ is the image under inversion in K of the same point z as before. As you can see, there is virtually no difference between the two figures. More precisely, as the radius of K tends to infinity, inversion in K *becomes* reflection in L. In particular, $[pz]/[p\,\widetilde{z}]$ tends to unity, or equivalently, $[p\,\widetilde{z}]$ is "ultimately equal" to [pz]. We can now understand what was happening in figure [3.2b].

We can also check this result algebraically. First, though, observe that from the geometric point of view it is sufficient to demonstrate the result for a single choice of the line L and a single point p on it. Let us therefore choose L to be the real axis, and let p be the origin. The circle K of radius R centred at $q = iR$ is therefore tangent to L at p. Using (3.4), we obtain [exercise]

$$\mathfrak{I}_K(z) = \frac{\overline{z}}{1 - (i\overline{z}/R)}.$$

Thus as R tends to infinity we find that $\mathfrak{I}_K(z)$ is ultimately equal to $\mathfrak{R}_L(z) = \overline{z}$, as was to be shown.

Here is another way of looking at the result. Instead of making K larger and larger, let z move closer and closer to an arbitrary point p on a circle K of fixed size. As z approaches p from any direction, $\mathfrak{I}_K(z)$ is ultimately equal to $\mathfrak{R}_T(z)$, where T is the tangent to K at p.

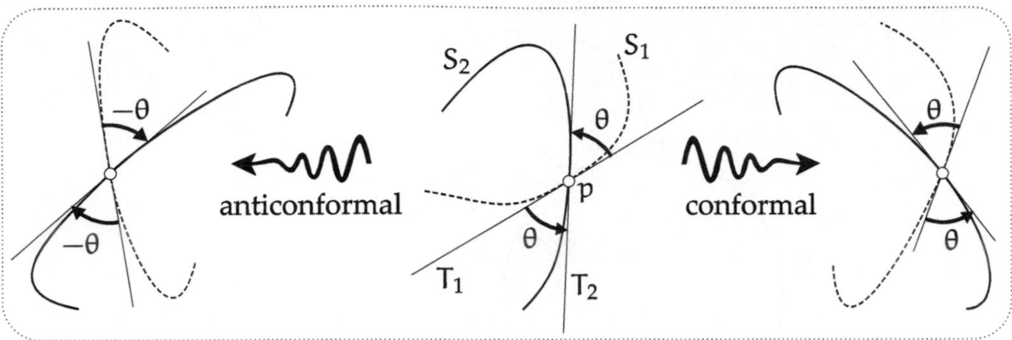

[3.8] Conformal and anticonformal mappings *preserve and reverse angles, respectively.*

Again, we can also get this algebraically using the above equation. If R is *fixed* and $|z| < R$, then [exercise]

$$\mathfrak{I}_K(z) = \bar{z} + \frac{i\bar{z}^2}{R} - \frac{\bar{z}^3}{R^2} + \cdots.$$

Thus as z approaches $p = 0$, $\mathfrak{I}_K(z)$ is ultimately equal to $\mathfrak{R}_L(z) = \bar{z}$, which is reflection in the tangent to K at p.

3.2.4 Preservation of Angles

Let us begin by discussing what is meant by "preservation of angles". In the centre of [3.8] are two curves S_1 and S_2 intersecting at a point p. Provided these curves are sufficiently smooth at p, then, as illustrated, we may draw their tangent lines T_1 and T_2 at p. We now define the "angle between S_1 and S_2" at p to be the acute angle θ *from* T_1 *to* T_2. Thus this angle θ has a *sign* attached to it: the angle between S_2 and S_1 is *minus* the illustrated angle between S_1 and S_2. If we now apply a sufficiently smooth transformation to the curves, then the image curves will again possess tangents at the image of p, and so there will be a well-defined angle between these image curves.

If the angle between the image curves is the same as the angle between the original curves through p, then we say that the transformation has "preserved" the angle at p. It is perfectly possible that the transformation preserves the angle between one pair of curves through p, but not *every* pair through p. However, if the transformation *does* preserve the angle between *every* pair of curves through p, then we say that it is *conformal* at p. We stress that this means that both the magnitude *and the sign* of the angles are preserved; see the right of [3.8]. If every angle at p is instead mapped to an angle of equal magnitude but opposite sign, then we say that the mapping is *anticonformal* at p; see the left of [3.8]. If the mapping is conformal at every point in the region where it is defined, then we call it a *conformal mapping*; if it is instead anticonformal at every point, then we call it an *anticonformal mapping*. Finally, if a mapping is known to preserve the *magnitude* of angles, but we

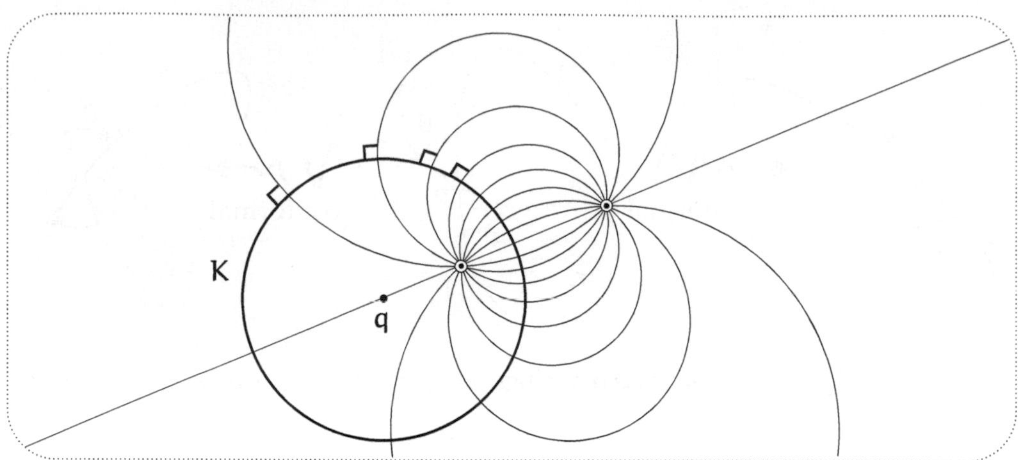

[3.9] Through any given point z not on K, there is precisely one circle orthogonal to K that passes through z in any given direction.

are unable to say whether or not it preserves their sense, then we call it an *isogonal mapping*.

It is easy enough to think of concrete mappings that are either conformal or anticonformal. For example, a translation $z \mapsto (z+c)$ is conformal, as is a rotation and expansion of the plane given by $z \mapsto az$. On the other hand, $z \mapsto \bar{z}$ is anticonformal, as is any reflection in a line. The analogy between such a reflection and inversion in a circle now gets even deeper, for

Inversion in a circle is an anticonformal mapping.

To see this, first look at [3.9]. This illustrates the fact that given any point z not on K, there is precisely one circle orthogonal to K that passes through z in any given direction. [Given the point and the direction, can you think how to *construct* this circle?]

As in [3.8], suppose that two curves S_1 and S_2 intersect at p, and that their tangents there are T_1 and T_2, the angle between them being θ. To find out what happens to this angle under inversion in K, let us replace S_1 and S_2 with the unique circles orthogonal to K that pass through p in the same directions as directions S_1 and S_2, i.e., circles whose tangents at p are T_1 and T_2. See [3.10a]. Since inversion in K maps each of these circles to themselves, the new angle at $\widetilde{p} = \mathcal{I}_K(p)$ is $-\theta$. Done.

Figure [3.10b] illustrates the effect of $z \mapsto (1/z)$ on angles. Since this mapping is equivalent to reflection (i.e., inversion) in the unit circle followed by reflection in the real axis (both of which are anticonformal), we see that their composition reverses the angle twice, restoring it to its original value:

Complex inversion, $z \mapsto (1/z)$, is conformal.

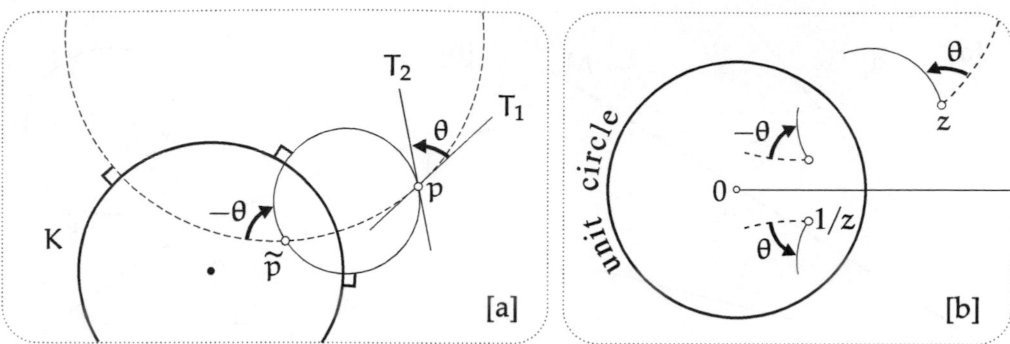

[3.10] Geometric inversion is anticonformal, so complex inversion is conformal. **[a]** *When two circles intersect, the magnitudes of both angles of intersection are the same, by symmetry. But, taking account of the sense of the angles, it follows that geometric inversion is anticonformal.* **[b]** *Since complex conjugation is also anticonformal, its composition with geometric inversion reverses angles twice. Therefore, complex inversion,* $z \mapsto (1/z)$, *is conformal.*

By the same reasoning, it follows more generally that

> *The composition of an even number of reflections (in lines or circles) is a conformal mapping, while the composition of an odd number of such reflections is an anticonformal mapping.*

3.2.5 *Preservation of Symmetry*

Consider [3.11a], which shows two points a and b that are symmetric with respect to a line L. If reflection in a line M maps a to $\widetilde{a}$, b to $\widetilde{b}$, and L to $\widetilde{L}$, then clearly the image points $\widetilde{a}$ and $\widetilde{b}$ are again symmetric with respect to the image line $\widetilde{L}$. In brief, reflection in lines "preserves symmetry" with respect to lines.

We now show that reflection in circles also preserves symmetry with respect to circles:

> *If* a *and* b *are symmetric with respect to a circle* K*, then their images* $\widetilde{a}$ *and* $\widetilde{b}$ *under inversion in any circle* J *are again symmetric with respect to the image* $\widetilde{K}$ *of* K.

To understand this, first note that, since inversion is anticonformal, (3.10) is just a special case of the following more general result:

> *Inversion maps any pair of orthogonal circles to another pair of orthogonal circles.*

Of course if one of the circles passes through the centre of inversion then its image will be a line. However, if we think of lines as merely being circles of infinite radius then the result is true without qualification.

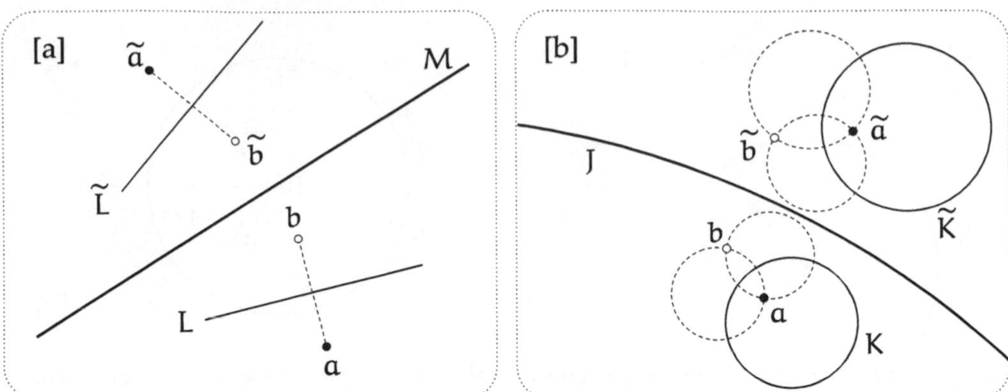

[3.11] Geometric inversion preserves symmetry. **[a]** *If* a *and* b *are symmetric with respect to a line* L, *then their reflections* $\widetilde{a}$ *and* $\widetilde{b}$ *in a line* M *are* still *symmetric with respect to the reflection* $\widetilde{L}$ *of* L*: we say that symmetry is* preserved. **[b]** *Since geometric inversion preserves circles and orthogonality, we see that if* a *and* b *are symmetric with respect to a circle* K, *then their "reflections"* $\widetilde{a}$ *and* $\widetilde{b}$ *in a circle* J *are* still *symmetric with respect to the reflection* $\widetilde{K}$ *of* K*: here, too, symmetry is preserved!*

The preservation of symmetry result is now easily understood. See [3.11b]. Since the two dashed circles through a and b are orthogonal to K, their images under inversion in J are likewise orthogonal to $\widetilde{K}$, and they therefore intersect in a pair of points that are symmetric with respect to $\widetilde{K}$.

3.2.6 Inversion in a Sphere

Inversion $\mathcal{I}_S$ of three-dimensional space in a sphere S (radius R and centre q) is defined in the obvious way: if p is a point in space at distance ρ from q, then $\mathcal{I}_S(p)$ is the point in the same direction from q as p, and at distance (R^2/ρ) from q. We should explain that this is not generalization for its own sake; soon we will see how this three-dimensional inversion sheds new light on two-dimensional inversion in $\mathbb{C}$.

Without any additional work, we may immediately generalize most of the above results on inversion in circles to results on inversion in spheres. For example, reconsider [3.3]. If we rotate this figure (in space) about the line through q and a, then we obtain [3.12], in which the circle of inversion K has swept out a sphere of inversion S, and the line has swept out a plane Π. Thus we have the following result:

> *Under inversion in a sphere centred at* q, *a plane* Π *that does not contain* q *is mapped to a sphere that contains* q *and whose tangent plane there is parallel to* Π. *Conversely, a sphere containing* q *is mapped to a plane that is parallel to the tangent plane of that sphere at* q. (3.11)

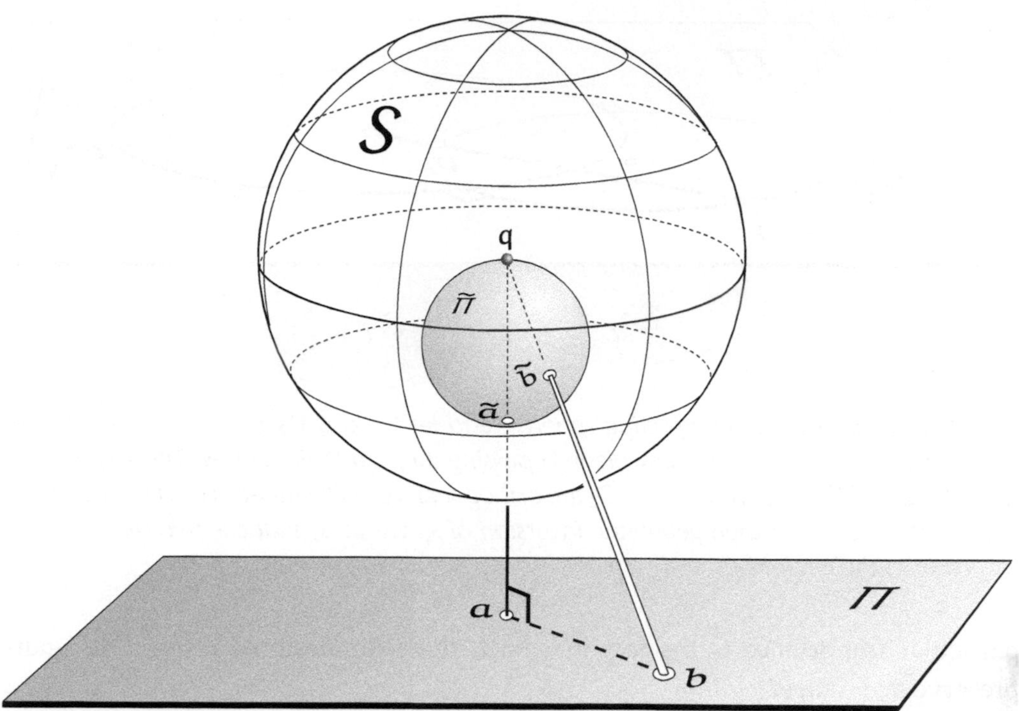

[3.12] Inversion in a sphere sends a plane to a sphere through the centre of inversion. *Rotating [3.3] (in space) about the line through* q *and* a*, we see that inversion in the sphere* S *maps the plane* Π *to a sphere that contains* q *and whose tangent plane there is parallel to* Π.

By the same token, if we rotate figure [3.4] about the line through q and a, then we find that

> *Under inversion in a sphere, the image of a sphere that does not contain the centre of inversion is another sphere that does not contain the centre of inversion.*

This result immediately tells us what will happen to a circle in space under inversion in a sphere, for such a circle may be thought of as the intersection of two spheres. Thus we easily deduce [exercise] the following result:

> *Under inversion in a sphere, the image of a circle* C *that does not pass through the centre* q *of inversion is another circle that does not pass through* q*. If* C *does pass through* q *then the image is a line parallel to the tangent of* C *at* q. (3.12)

The close connection between inversion in a circle and reflection in a line also persists: reflection in a plane is a limiting case of inversion in a sphere. For this reason, inversion in a sphere is also called "reflection in a sphere". Of

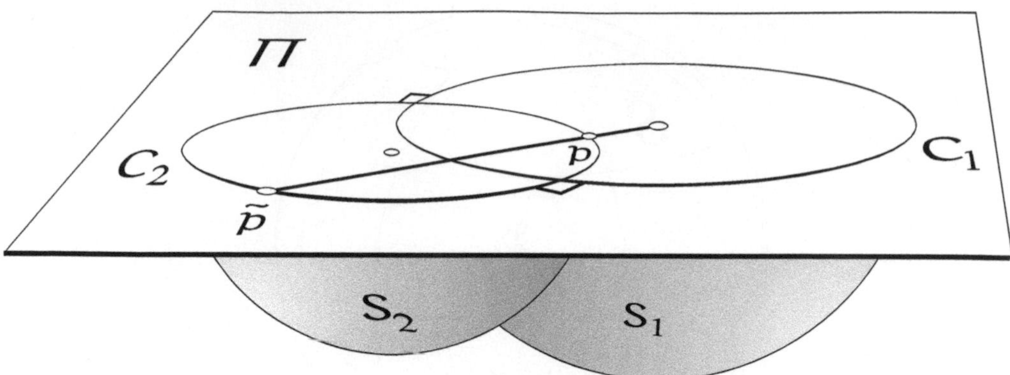

[3.13] *Let S_1 and S_2 be intersecting spheres, and let C_1 and C_2 be the great circles in which these spheres intersect any plane Π passing through their centres. Then S_1 and S_2 are orthogonal if and only if C_1 and C_2 are orthogonal. Also illustrated is the fact that if we restrict attention to Π then geometric inversion of space in S_1 induces two-dimensional inversion in the circle C_1.*

particular importance is the fact that such three-dimensional reflections again preserve symmetry:

> *Let K be a plane or sphere, and let a and b be symmetric points with respect to K. Under a three-dimensional reflection in any plane or sphere, the images of a and b are again symmetric with respect to the image of K.* (3.13)

We now describe the steps leading to this result; they are closely analogous to the steps leading to the two-dimensional preservation of symmetry result.

If we rotate figure [3.6a] about the line joining the centres of K and C, we deduce that

> *Under inversion in a sphere K, every sphere orthogonal to K is mapped to itself.* (3.14)

When we say that spheres are "orthogonal" we mean that their tangent planes are orthogonal at each point of their circle of intersection. However, in order to be able to easily draw on previous results, let us rephrase this three-dimensional description in two-dimensional terms:

> *Let S_1 and S_2 be intersecting spheres, and let C_1 and C_2 be the great circles in which these spheres intersect any plane Π passing through their centres. Then S_1 and S_2 are orthogonal if and only if C_1 and C_2 are orthogonal.*

See [3.13]. This figure is also intended to help you see that if we restrict attention to Π then the three-dimensional inversion in S_1 is identical to the two-dimensional inversion in C_1. This way of viewing inversion in spheres allows us to quickly generalize earlier results.

For example, referring back to [3.6b], we find—make sure you see this—that if p lies in Π then $\widetilde{p} = \mathcal{I}_{S_1}(p)$ may be constructed as the second intersection point of any two circles like C_2 that (i) lie in Π, (ii) are orthogonal to C_1, and (iii) pass through p.

Next, suppose that S_1 and S_2 in [3.13] are subjected to inversion in a third sphere K. Choose Π to be the unique plane passing through the centres of S_1, S_2, K, and let C be the great circle in which K intersects Π. Since $\mathcal{I}_C$ maps C_1 and C_2 to orthogonal circles, we deduce [exercise] that (3.14) is a special case of the following result:

Orthogonal spheres invert to orthogonal spheres. (3.15)

Here we are considering a plane to be a limiting case of a sphere.

Putting these facts together, you should now be able to see the truth of (3.13).

3.3 Three Illustrative Applications of Inversion

3.3.1 A Problem on Touching Circles

For our first problem, consider [3.14], in which we imagine that we are given two circles A and B that touch at q. As illustrated, we now construct the circle C_0 that touches A and B and whose centre lies on the horizontal line L through the centres of A and B. Finally, we construct the chain of circles C_1, C_2, etc., such that C_{n+1} touches C_n, A, and B.

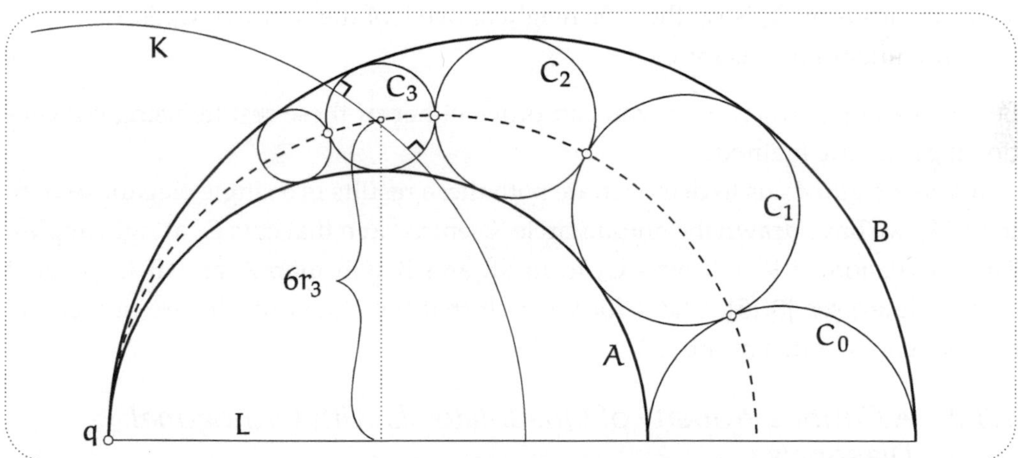

[3.14] *Given any two circles A and B that touch at q, as shown, construct touching circles, C_i, as shown. Then, very surprisingly, (1) the points of contact of the chain C_i all lie on a circle that also touches A and B at q, and (2) the height of the centre of C_n above L is 2n times the radius of C_n, as illustrated for $n = 3$.*

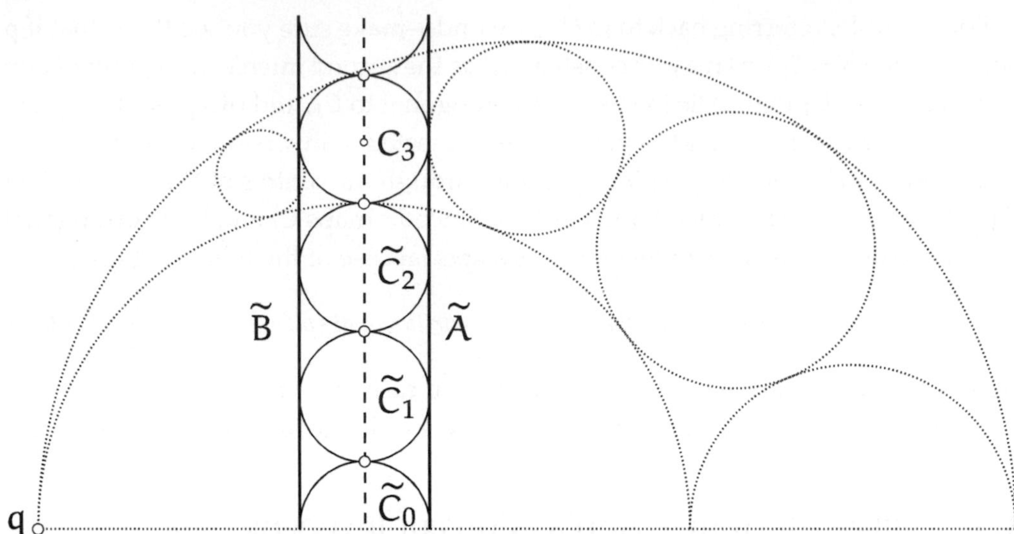

[3.15] *Referring to the previous figure, consider the illustrated (unique) circle K centred at q that cuts C_3 at right angles. Geometric inversion in K maps C_3 to itself, and it maps A and B to parallel vertical lines. The truth of* both *of the mysterious, bulleted facts below suddenly becomes obvious!*

The figure illustrates two remarkable facts about this chain of circles:

- The points of contact of the chain C_0, C_1, C_2, etc., all lie on a circle [dashed] touching A and B at q.
- If the radius of C_n is r_n, then the height above L of the centre of C_n is $2nr_n$. The figure illustrates this for C_3.

Before reading further, see if you can prove either of these results using conventional geometric methods.

Inversion allows us to demonstrate both these results in a single elegant swoop. In [3.14], we have drawn the unique circle K centred at q that cuts C_3 at right angles. Thus inversion in K will map C_3 to itself, and it will map A and B to parallel vertical lines; see [3.15]. Check for yourself that the stated results are immediate consequences of this figure.

3.3.2 A Curious Property of Quadrilaterals with Orthogonal Diagonals

Figure [3.16] shows a shaded quadrilateral whose diagonals intersect orthogonally at q. If we now reflect q in each of the edges of the quadrilateral, then we obtain four new points. Very surprisingly, *these four points lie on a circle*[4]. As with the previous problem, see if you can prove this by ordinary means.

[4] I am grateful to my friend Paul Zeitz for challenging me with this problem, which appeared in the USA Mathematical Olympiad.

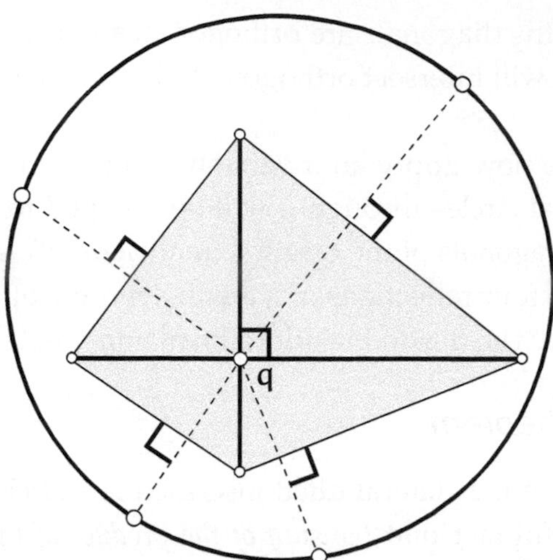

[3.16] A curious property of quadrilaterals with orthogonal diagonals. *Consider a quadrilateral whose diagonals intersect* orthogonally at q. *If we reflect* q *in each of the edges of the quadrilateral, then we obtain four new points. Very surprisingly, these four points lie on a* circle!

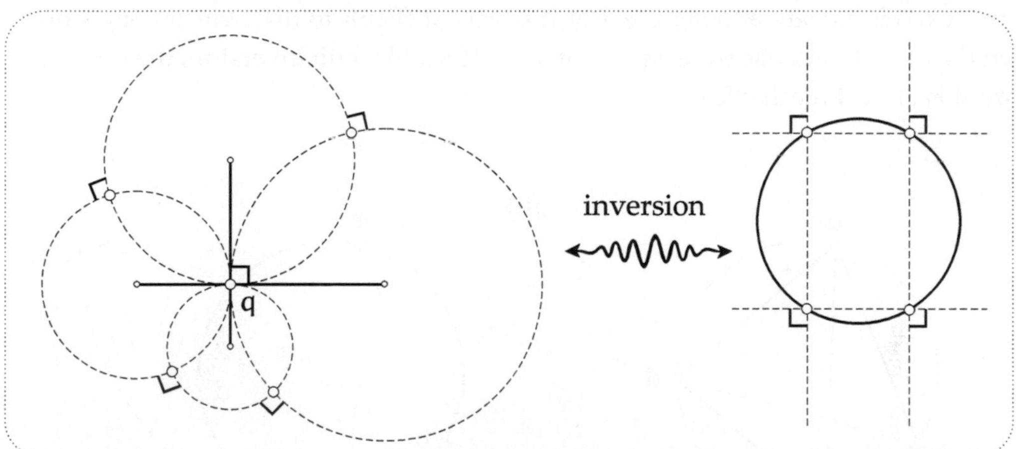

[3.17] Geometric explanation of the previous figure. *First, on the left, we use the construction in [3.7a] to represent the reflection of* q *in an edge as the second intersection point of any two circles through* q *whose centres lie on that edge. Here we have chosen the centres of these circles to be the* vertices *of the quadrilateral. Shown on the right is the result of performing an inversion in any circle centred at* q. *The orthogonal circles become orthogonal lines, intersecting in a rectangle, so these four points lie on a circle. But this means that the* original *four points were* also *on a circle. Done!*

To demonstrate the result using inversion, we first use the construction in [3.7a] to represent the reflection of q in an edge as the second intersection point of any two circles through q whose centres lie on that edge. More precisely, let us choose the centres of these circles to be the *vertices* of the quadrilateral; see the LHS of [3.17].

Note that, because the diagonals are orthogonal, a pair of these circles centred at the ends of an edge will intersect orthogonally both at q and at the reflection of q in that edge.

It follows that if we now apply an inversion in any circle centred at q, then a pair of such orthogonal circles through q will be mapped to a pair of orthogonal lines (parallel to the diagonals of the original quadrilateral); see the RHS of [3.17]. Thus the images of the four reflections of q are the vertices of a rectangle, and they therefore lie on a circle. The desired result follows immediately. Why?

3.3.3 *Ptolemy's Theorem*

Figure [3.18a] shows a quadrilateral abcd inscribed in a circle. Ptolemy (c. CE 125) discovered the beautiful fact that *the sum of the product of the opposite sides is the product of the diagonals*. In symbols,

$$[ad]\,[bc] + [ab]\,[cd] = [ac]\,[bd].$$

We note that for Ptolemy this was not merely interesting, it was a crucial tool for doing *astronomy!* See Ex. 9. His original proof (which is reproduced in most geometry texts) is elegant and simple, but it is very difficult to discover on one's own. On the other hand, once one has become comfortable with inversion, the following proof is almost mechanical.

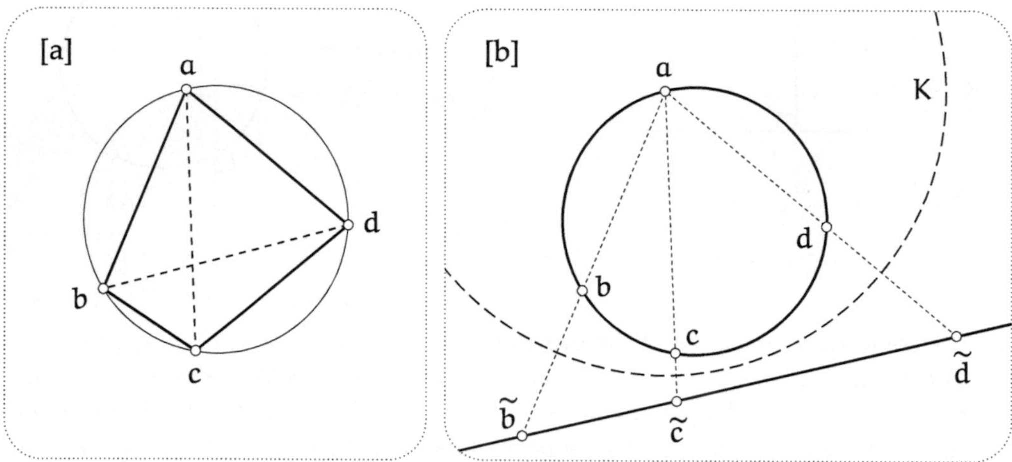

[3.18] [a] *A cyclic quadrilateral is subject to* **Ptolemy's Theorem***:* $[ad]\,[bc] + [ab]\,[cd] = [ac]\,[bd]$. **[b]** *Here is the result of inverting figure [a] in any circle* K *centred at* a*. Clearly,* $[\widetilde{b}\,\widetilde{c}] + [\widetilde{c}\,\widetilde{d}] = [\widetilde{b}\,\widetilde{d}]$. *But (3.6) dictates how the separations of these inverted points are related to the separations of the original points, and Ptolemy's Theorem then follows immediately!*

Inverting figure [3.18a] in a circle K centred at one of the vertices (say a), we obtain [3.18b], in which

$$[\widetilde{b}\,\widetilde{c}] + [\widetilde{c}\,\widetilde{d}] = [\widetilde{b}\,\widetilde{d}].$$

Recalling that (3.6) tells us how the separation of two inverted points is related to the separation of the original points, we deduce that

$$\frac{[bc]}{[ab][ac]} + \frac{[cd]}{[ac][ad]} = \frac{[bd]}{[ab][ad]}.$$

Multiplying both sides by $([ab]\,[ac]\,[ad])$, we deduce Ptolemy's Theorem.

3.4 The Riemann Sphere

3.4.1 *The Point at Infinity*

In discussing inversion we saw that results about lines could always be understood as special limiting cases of results about circles, simply by letting the radius tend to infinity. This limiting process is nevertheless tiresome and clumsy; how much better it would be if lines could literally be described as circles of infinite radius.

Here is another, related inconvenience. Inversion in the unit circle is a one-to-one mapping of the plane to itself that swaps pairs of points. The same is true of the mapping $z \mapsto (1/z)$. However, there are exceptions: no image point is presently associated with $z = 0$, nor is 0 to be found among the image points.

To resolve both these difficulties, note that as z moves further and further away from the origin, $(1/z)$ moves closer and closer to 0. Thus as z travels to ever greater distances (in any direction), it is as though it were approaching a single *point at infinity*, written ∞, whose image is 0. Thus, by definition, this point ∞ satisfies the following equations:

$$\frac{1}{\infty} = 0, \qquad \frac{1}{0} = \infty.$$

The addition of this single point at infinity turns the complex plane into the so-called *extended complex plane*. Thus we may now say, without qualification, that $z \mapsto (1/z)$ is a one-to-one mapping of the extended plane to itself.

If a curve passes through $z = 0$ then (by definition) the image curve under $z \mapsto (1/z)$ will be a curve through the point at infinity. Conversely, if the image curve passes through 0 then the original curve passed through the point ∞. Since $z \mapsto (1/z)$ swaps a circle through 0 with a line, we may now say that a line is just a circle that happens to pass through the point at infinity, and (without further qualification) inversion in a "circle" sends "circles" to "circles".

This is all very tidy, but it leaves one feeling none the wiser. We are accustomed to using the symbol ∞ only in conjunction with a limiting process, not as a thing in its own right; how are we to grasp its new meaning as a definite point that is infinitely far away?

3.4.2 Stereographic Projection

Riemann's profoundly beautiful answer to this question was to interpret complex numbers as points on a *sphere* Σ, instead of as points in a plane. Throughout the following discussion, imagine the complex plane positioned *horizontally* in space. In order to be definite about which way up the plane is, suppose that when we look down on $\mathbb{C}$ from above, a *positive* (i.e., counterclockwise) rotation of $(\pi/2)$ carries 1 to i. Now let Σ be the sphere centred at the origin of $\mathbb{C}$, and let it have unit radius so its "equator" coincides with the unit circle[5].

We now seek to set up a correspondence between points on Σ and points in $\mathbb{C}$. If we think of Σ as the surface of the Earth, then this is the ancient problem of how to draw a geographical map. In an atlas you will find many different ways of drawing maps, the reason for the variety being that no single map can faithfully represent every aspect of a curved[6] surface on a flat piece of paper. Although distortions of some kind are inevitably introduced, different maps can "preserve" or "faithfully represent" some (but not all) features of the curved surface. For example, a map can preserve angles at the expense of distorting areas.

Ptolemy (c. CE 125) was the first to construct such a map, which he used to plot the positions of heavenly bodies on the "celestial sphere". His method is called *stereographic projection*, and we will soon see how perfectly it is adapted to our needs. Figure [3.19] illustrates its definition. From the north pole N of the sphere Σ, draw the line through the point p in $\mathbb{C}$; the *stereographic image* of p on Σ is the point $\widehat{p}$ where this line intersects Σ. Since this gives us a one-to-one correspondence between points in $\mathbb{C}$ and points on Σ, let us also say that p is the stereographic image of $\widehat{p}$. No confusion should arise from this, the context making it clear whether we are mapping $\mathbb{C}$ to Σ, or vice versa.

Note the following immediate facts: (i) the interior of the unit circle is mapped to the southern hemisphere of Σ, and in particular 0 is mapped to the south pole, S; (ii) each point on the unit circle is mapped to itself, now viewed as lying on the equator of Σ; (iii) the exterior of the unit circle is mapped to the northern hemisphere of Σ, *except* that N is not the image of any finite point in the plane.

However, it is clear that as p moves further and further away from the origin (in any direction), $\widehat{p}$ moves closer and closer to N. This strongly suggests that N *is the stereographic image of the point at infinity*. Thus stereographic projection establishes a one-to-one correspondence between every point of the *extended* complex plane and every point of Σ. Instead of merely speaking of a "correspondence" between

[5] Some works instead define Σ to be tangent to the complex plane at its south pole.

[6] This concept of "curvature" will be defined more precisely in Chapter 6.

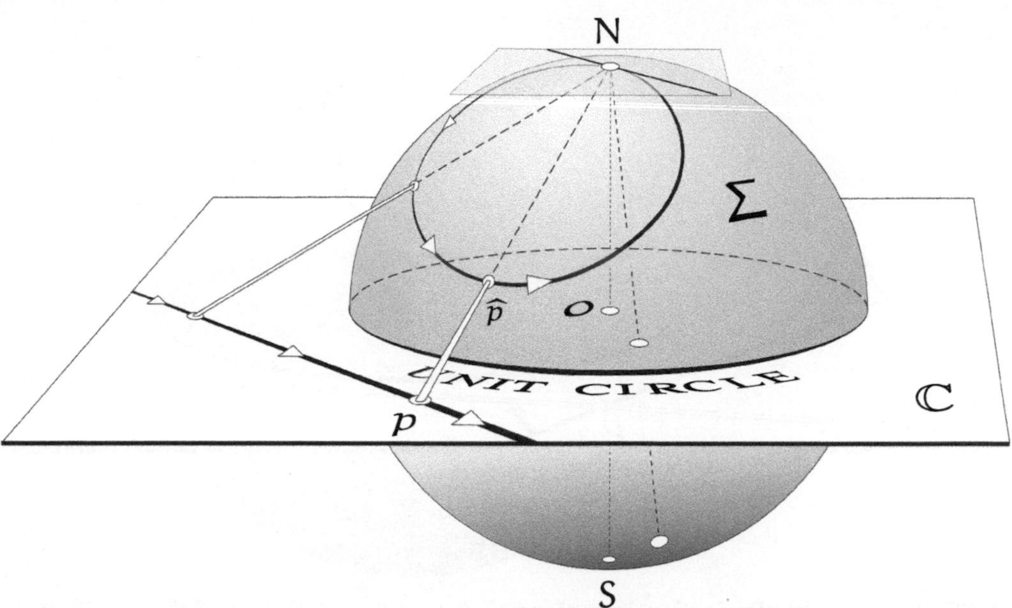

[3.19] Stereographic Projection *maps the surface of the (clear glass) sphere to its equatorial plane. Standing at the north pole,* N, *fire a laser beam at a point* $\widehat{p}$ *on the surface of the sphere: it goes on to hit the plane at* p, *called the* stereographic projection *(or* image*) of* $\widehat{p}$. *Conversely,* $\widehat{p}$ *is symmetrically called the* stereographic projection *(or image) of* p. *If* p *moves along a line in the plane, the laser beam sweeps out a plane, so the beam cuts the sphere in a* circle *that passes through* N, *and its tangent there is parallel to the line in the plane. This is because the equatorial plane and the tangent plane to the sphere at* N *are parallel—both are horizontal—so the plane swept out by the laser beam intersects these parallel planes in parallel lines.*

complex numbers and points of Σ, we can imagine that the points of Σ *are* the complex numbers. For example, $S = 0$ and $N = \infty$. Once stereographic projection has been used to label each point of Σ with a complex number, Σ is called the *Riemann sphere*.

We have already discussed the fact that a line in ℂ may be viewed as a circle passing through the point at infinity. The Riemann sphere now transforms this abstract idea into a literal fact:

> *The stereographic image of a line in the plane is a circle on* Σ *passing through* $N = \infty$. (3.16)

To see this, observe that as p moves along the line shown in [3.19], the line connecting N to p sweeps out a plane through N. Thus $\widehat{p}$ moves along the intersection of this plane with Σ, which is a circle passing through N. Done. In addition, note that *the tangent to this circle at* N *is parallel to the original line.* This is explained in the caption.

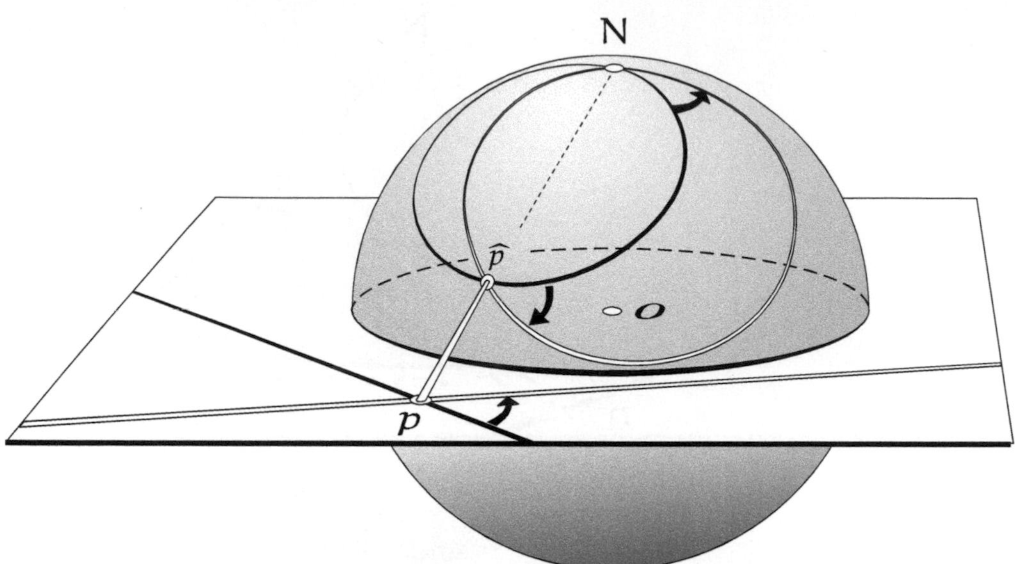

[3.20] Stereographic Projection is conformal. *If two lines intersect in the plane, their stereographic projections on the sphere are two circles that intersect at the* same angle *at* N, *for the tangents to these circles at* N *are parallel to the lines in the plane. But, by symmetry, the angle between these circles at* $\widehat{p}$ *is the same as their angle at* N, *which we have just seen is the same as the angle between the lines in the plane. Done!*

From this last fact it follows that *stereographic projection preserves angles.* Consider [3.20], which shows two lines intersecting at p, together with their circular, stereographic projections. By symmetry, the magnitude of the angle of intersection between the circles is the same at their two intersection points, $\widehat{p}$ and N. Since their tangents at N are parallel to the original lines in the plane, it follows that the illustrated angles at p and $\widehat{p}$ are of equal *magnitude*. But before we can say that stereographic projection is "conformal", we must assign a *sense* to the angle on the sphere.

According to our convention, the illustrated angle at p (from the black curve to the white one) is *positive*, i.e., it is counterclockwise when viewed from above the plane. From the perspective from which we have drawn [3.20], the angle at $\widehat{p}$ is negative, i.e., clockwise. However, if we were looking at this angle from *inside* the sphere then it would be positive. Thus

> *If we define the sense of an angle on Σ by its appearance to an observer inside Σ, then stereographic projection is conformal.* (3.17)

[HISTORICAL NOTE: It is quite remarkable that although stereographic projection had been well known since Ptolemy first put it to practical use around 125 CE, its beautiful and fundamentally important *conformality* was not discovered

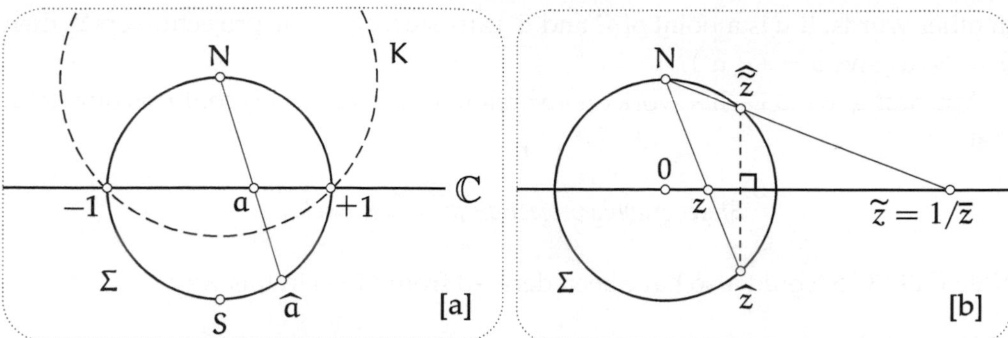

[3.21] [a] Two-dimensional stereographic projection of the Riemann sphere Σ to the plane ℂ is merely the restriction to Σ of the action of *three*-dimensional geometric inversion in the sphere K *of radius $\sqrt{2}$ (here shown in cross section) centred at* N *that cuts through* Σ *along its unit-circle equator in* ℂ. **[b]** *Inversion of* ℂ *in the unit circle induces a reflection of the Riemann sphere in its equatorial plane,* ℂ. *If we compose this with conjugation, which reflects* Σ *in the vertical plane through the real axis, then we obtain complex inversion $z \mapsto (1/z)$. But reflection across two perpendicular planes is equivalent to a rotation of space about their line of intersection through angle π. Thus,* **the mapping $z \mapsto (1/z)$ in ℂ induces a rotation of the Riemann sphere about the real axis through an angle of π.**

for another 1500 years! This was first done around 1590 by Thomas Harriot.[7] In Chapter 6 we shall meet Harriot again, for in 1603 he made another profoundly important discovery about the geometry of the sphere, this time relating the angles in a spherical triangle to the triangle's area—(6.9) on page 317.]

Clearly, any origin-centred circle in the plane is mapped to a horizontal circle on Σ, but what happens to a general circle? The startling answer is that *it too is mapped to a circle on the Riemann sphere!* This is quite difficult to see if we stick to our original definition of stereographic projection, but it suddenly becomes obvious if we change our point of view. Look again at [3.12], and observe how closely it resembles the definition of stereographic projection.

To make the connection precise, let K be the sphere centred at the north pole N of Σ that intersects Σ along its equator (the unit circle of ℂ). Figure [3.21a] shows a vertical cross section (through N and the real axis), of K, Σ, and ℂ. The full three-dimensional picture is obtained by rotating this figure about the line through N and S. We now see that

> *If* K *is the sphere of radius $\sqrt{2}$ centred at* N, *then stereographic projection is the restriction to* ℂ *or* Σ *of inversion in* K.

[7] See Stillwell (2010, §16.2). For a short sketch of Harriot's life, see Stillwell (2010, §17.7). The *first* full biography of this very remarkable, unsung hero of mathematics and science—born within a mile of Stephen Hawking's place of birth (and mine!)—has at last been published: Arianrhod (2019).

In other words, if a is a point of $\mathbb{C}$ and $\widehat{a}$ is its stereographic projection on Σ, then $\widehat{a} = \mathcal{J}_K(a)$ and $a = \mathcal{J}_K(\widehat{a})$.

Appealing to our earlier work on inversion in spheres, (3.12) confirms our claim that

Stereographic projection preserves circles.

Note that (3.16) could also have been derived from (3.12) in this way.

3.4.3 Transferring Complex Functions to the Sphere

Stereographic projection enables us to transfer the action of any complex function to the Riemann sphere. Given a complex mapping $z \mapsto w = f(z)$ of $\mathbb{C}$ to itself, we obtain a corresponding mapping $\widehat{z} \mapsto \widehat{w}$ of Σ to itself, where $\widehat{z}$ and $\widehat{w}$ are the stereographic images of z and w. We shall say that $z \mapsto w$ *induces* the mapping $\widehat{z} \mapsto \widehat{w}$ of Σ.

For example, consider what happens if we transfer $f(z) = \overline{z}$ to Σ. Clearly [exercise],

Complex conjugation in $\mathbb{C}$ induces a reflection of the Riemann sphere in the vertical plane passing through the real axis.

For our next example, consider $z \mapsto \widetilde{z} = (1/\overline{z})$, which is inversion in the unit circle C. Figure [3.21b] shows a vertical cross section of Σ taken through N and the point z in $\mathbb{C}$. This figure also illustrates the very surprising result of transferring this inversion to Σ:

Inversion of $\mathbb{C}$ in the unit circle induces a reflection of the Riemann sphere in its equatorial plane, $\mathbb{C}$. (3.18)

Here is an elegant way of seeing this. First note that not only are the pair of points z and $\widetilde{z}$ symmetric (in the two-dimensional sense) with respect to C, but they are also symmetric (in the three-dimensional sense) with respect to the sphere Σ. Now apply the three-dimensional preservation of symmetry result (3.13). Since z and $\widetilde{z}$ are symmetric with respect to Σ, their stereographic images $\widehat{z} = \mathcal{J}_K(z)$ and $\widehat{\widetilde{z}} = \mathcal{J}_K(\widetilde{z})$ will be symmetric with respect to $\mathcal{J}_K(\Sigma)$. But $\mathcal{J}_K(\Sigma) = \mathbb{C}$. Done! A more elementary (but less illuminating) derivation may be found in Ex. 6.

By combining the above results, we can now find the effect of complex inversion on the Riemann sphere. In $\mathbb{C}$, we know that $z \mapsto (1/z)$ is equivalent to inversion in the unit circle, followed by complex conjugation. The induced mapping on Σ is therefore the composition of *two reflections in perpendicular planes through the real axis*—one horizontal, the other vertical. However, it is not hard to see (perhaps with the aid of an orange) that the net effect of successively reflecting Σ in *any* two perpendicular planes through the real axis is a *rotation* of Σ about the real axis through angle π. Thus we have shown that

The mapping $z \mapsto (1/z)$ in $\mathbb{C}$ induces a rotation of the Riemann sphere about the real axis through an angle of π. (3.19)

Recall that the point ∞ was originally defined by the property that it be swapped with 0 under complex inversion, $z \mapsto (1/z)$. The result (3.19) vividly illustrates the correctness of identifying N with the point at infinity, for the point 0 in $\mathbb{C}$ corresponds to the south pole S of Σ, and the rotation of π about the real axis does indeed swap S with N.

3.4.4 Behaviour of Functions at Infinity

Suppose two curves in $\mathbb{C}$ extend to arbitrarily large distances from the origin. Abstractly, one would say that they meet at the point at infinity. On Σ this becomes a literal intersection at N, and if each of the curves arrives at N in a well defined direction, then one can even assign an "intersection angle at ∞". For example, [3.20] illustrates that if two lines in $\mathbb{C}$ intersect at a finite point and contain an angle α there, then they intersect for a second time at ∞ and they contain an angle $-\alpha$ at that point.

Transferring a complex function to the Riemann sphere enables one to examine its behaviour "at infinity" exactly as one would at any other point. In particular, one can look to see if the function preserves the angle between any two curves passing through ∞. For example, the result (3.19) shows that complex inversion does preserve such angles at N, and it is therefore said to be "conformal at infinity". By the same token, this rotation of Σ will also preserve the angle between two curves that pass through the singularity $z = 0$ of $z \mapsto (1/z)$, so complex inversion is conformal there too. In brief, *complex inversion is conformal throughout the extended complex plane.*

In this chapter we have found it convenient to depict $z \mapsto w$ as a mapping of $\mathbb{C}$ to *itself*, and in the above example we have likewise interpreted the induced mapping $\widehat{z} \mapsto \widehat{w}$ as sending points on the sphere to other points on the *same* sphere. However, it is often better to revert to the convention of the previous chapter, whereby the mapping sends points in the z-plane to image points residing in a second copy of $\mathbb{C}$, the w-plane. In the same spirit, the induced mapping $\widehat{z} \mapsto \widehat{w}$ may be viewed as mapping points in one sphere (the z-sphere) to points in a second sphere (the w-sphere). We illustrate this with an example.

Consider $z \mapsto w = z^n$, where n is a positive integer. The top half of [3.22] illustrates the effect of the mapping (in the case $n = 2$) on a grid of small "squares" abutting the unit circle and two rays containing an angle θ. Very mysteriously, the images of these "squares" in the w-plane are *again* almost square. In the next chapter we will show that this is just one consequence of a more basic mystery, namely, that $z \mapsto w = z^n$ is *conformal*. Indeed, we will show that if a mapping is conformal, then any infinitesimal shape is mapped to a *similar* infinitesimal shape.

Since stereographic projection is known to be conformal, we would therefore anticipate that when we transfer the grid from the z-plane to the z-sphere, the result will again be a grid of "squares". That this does indeed happen can be seen at the bottom left of [3.22]; the bottom right of [3.22] illustrates the same phenomenon as we pass from the image grid in the w-plane to the image grid on the w-sphere. Quite generally, any conformal mapping of $\mathbb{C}$ will induce a conformal mapping of Σ that will (as one consequence) map a grid of infinitesimal squares to another grid of infinitesimal squares.

Figure [3.22] not only manifests the conformality of $z \mapsto w = z^2$, but it also illustrates that there exist points at which this conformality *breaks down*. Clearly, the angle θ at the origin is *doubled*; more generally, $z \mapsto w = z^n$ multiplies angles at 0 by n. Quite generally, if the conformality of an otherwise conformal mapping breaks down at a particular point p, then p is called a *critical point* of the mapping. Thus we may say that 0 is a critical point of $z \mapsto w = z^n$.

If we restrict ourselves to $\mathbb{C}$ then this is the *only* critical point of this mapping. However, if we look at the induced mapping of Σ, then the figure makes it clear that in the *extended* complex plane there is a second critical point at infinity: angles there are multiplied by n, just as they were at 0. Thus, more precisely than before, the claim is that $z \mapsto w = z^n$ is a conformal mapping whose only critical points are 0 and ∞.

Next, we discuss how the behaviour of a complex mapping at infinity may be investigated *algebraically*. Complex inversion rotates Σ so that a neighbourhood of $N = \infty$ becomes a neighbourhood of $S = 0$. Thus to examine behaviour near infinity we may first apply complex inversion and then examine the neighbourhood of the origin. Algebraically, this means that to study $f(z)$ at infinity we should study $F(z) \equiv f(1/z)$ at the origin. For example, $f(z)$ is conformal at infinity if and only if $F(z)$ is conformal at the origin.

For example, if $f(z) = (z+1)^3/(z^5 - z)$, then $F(z) = z^2(1+z)^3/(1-z^4)$, which has a double root at 0. Thus instead of merely saying that $f(z)$ "dies away to zero like $(1/z^2)$ as z tends to infinity", we can now say (more precisely) that $f(z)$ has a double root at $z = \infty$.

This process can also be used to extend the concept of a branch point of a multifunction to the point at infinity. For example, if $f(z) = \log(z)$ then $F(z) = -\log(z)$. Thus $f(z)$ not only has a logarithmic branch point at $z = 0$, it also has one at $z = \infty$.

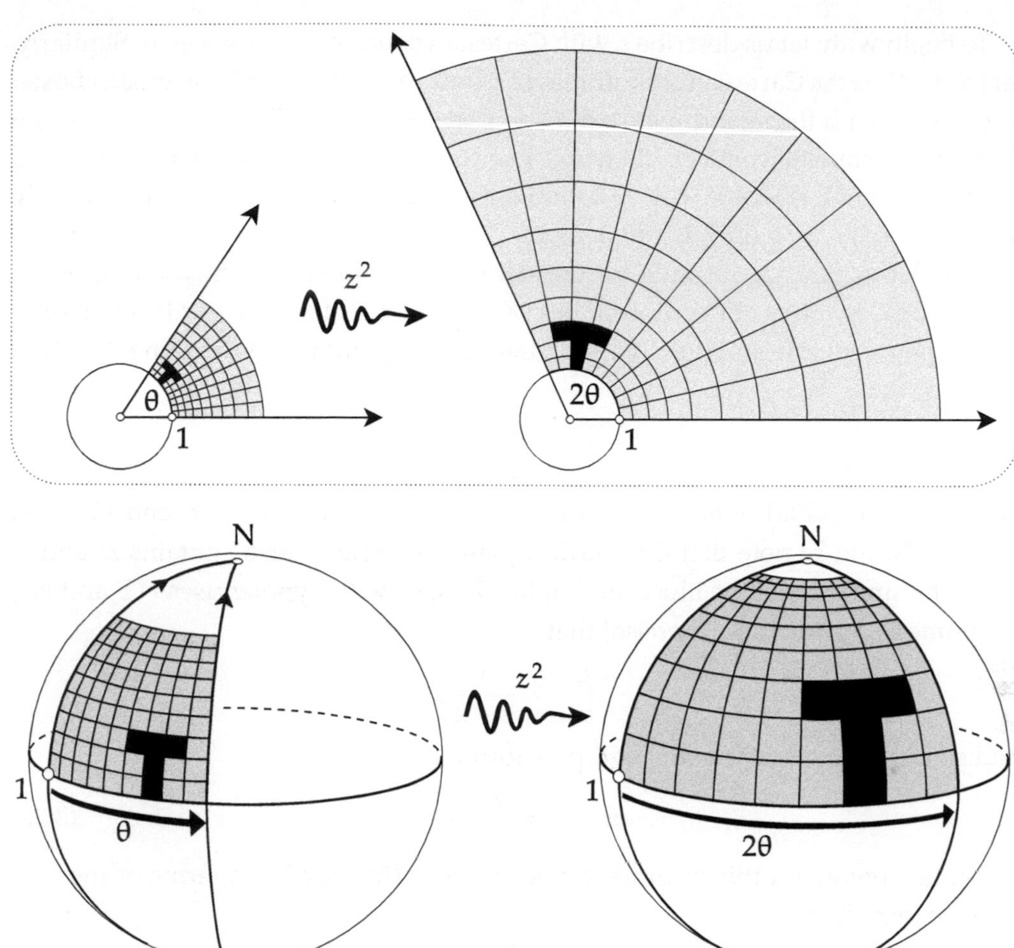

[3.22] The conformality of $z \mapsto z^n$. *In the next chapter we will begin to address the remarkable and mysterious fact that $z \mapsto z^n$ is conformal, here illustrated in the case $n = 2$ [TOP]. We will also see that conformality suffices to explain the illustrated fact that a grid of small squares is ultimately mapped to a grid of squares, in the limit that they vanish. Note, however, that it is* not conformal at 0: the angles between the rays are doubled. *We see this too at the south pole of the Riemann sphere,* and at the north pole, *corresponding to ∞. Since stereographic projection has been proven to be conformal, it follows that the conformality of $z \mapsto z^2$ is transferred to the induced mapping of the Riemann sphere [BOTTOM], and therefore it, too, ultimately maps squares to squares.*

*3.4.5 Stereographic Formulae**

In this subsection we derive explicit formulae connecting the coordinates of a point z in $\mathbb{C}$ and its stereographic projection $\widehat{z}$ on Σ. These formulae will prove useful in investigating non-Euclidean geometry, but if you don't plan to study Chapter 6 then you should feel free to skip this subsection.

To begin with, let us describe z with Cartesian coordinates: $z = x + iy$. Similarly, let (X, Y, Z) be the Cartesian coordinates of $\widehat{z}$ on Σ; here the X- and Y-axes are chosen to coincide with the x- and y-axes of $\mathbb{C}$, so that the positive Z-axis passes through N. To make yourself comfortable with these coordinates, check the following facts: the equation of Σ is $X^2+Y^2+Z^2 = 1$, the coordinates of N are $(0, 0, 1)$, and similarly $S = (0, 0, -1)$, $1 = (1, 0, 0)$, $i = (0, 1, 0)$, etc.

Now let us find the formula for the stereographic projection $z = x + iy$ of the point $\widehat{z}$ on Σ in terms of the coordinates (X, Y, Z) of $\widehat{z}$. Let $z' = X + iY$ be the foot of the perpendicular from $\widehat{z}$ to $\mathbb{C}$. Clearly, the desired point z is in the same direction as z', so

$$z = \frac{|z|}{|z'|}\, z'.$$

Now look at [3.23a], which shows the vertical cross section of Σ and $\mathbb{C}$ taken through N and $\widehat{z}$; note that this vertical plane necessarily also contains z' and z. From the similarity of the illustrated right triangles with hypotenuses $N\widehat{z}$ and Nz, we immediately deduce [exercise] that

$$\frac{|z|}{|z'|} = \frac{1}{1-Z},$$

and so we obtain our first stereographic formula:

$$x + iy = \frac{X + iY}{1 - Z}. \tag{3.20}$$

Let us now invert this formula to find the coordinates of $\widehat{z}$ in terms of those of z. Since [exercise]

$$|z|^2 = \frac{1+Z}{1-Z},$$

we obtain [exercise]

$$X + iY = \frac{2z}{1+|z|^2} = \frac{2x + i2y}{1 + x^2 + y^2}, \quad \text{and} \quad Z = \frac{|z|^2 - 1}{|z|^2 + 1}. \tag{3.21}$$

Although it is often useful to describe the points of Σ with the three coordinates (X, Y, Z), this is certainly unnatural, for the sphere is intrinsically *two* dimensional. If we instead describe $\widehat{z}$ with the more natural (two-dimensional) spherical polar coordinates (ϕ, θ) then we obtain a particularly neat stereographic formula.

First recall[8] that θ measures angle around the Z-axis, with $\theta = 0$ being assigned to the vertical half-plane through the positive X-axis: thus for a point z in $\mathbb{C}$, the angle θ is simply the usual angle from the positive real axis to z. The definition of ϕ is illustrated in [3.23b]—it is the angle subtended at the centre of Σ by the points N and $\widehat{z}$: for example, the equator corresponds to $\phi = (\pi/2)$. By convention, $0 \leqslant \phi \leqslant \pi$.

[8] This is the American convention; in my native England the roles of θ and ϕ are the reverse of those stated here.

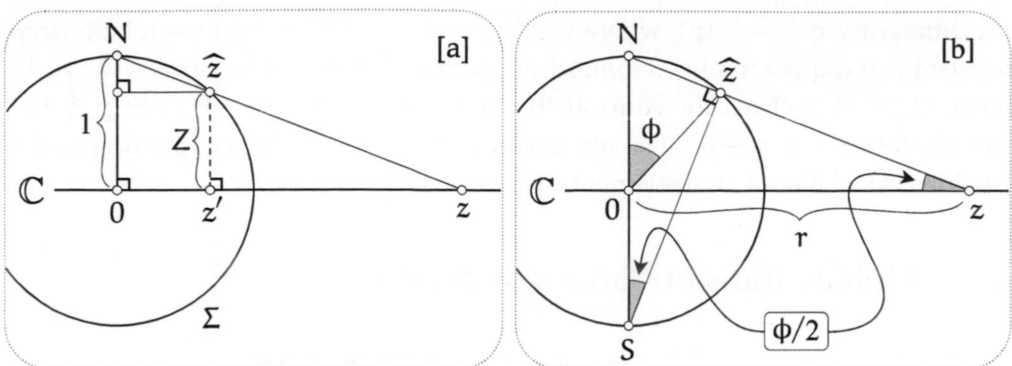

[3.23] Two stereographic formulae. [a] *From the similarity of the illustrated triangles, we immediately obtain a Cartesian formula for* $z = x + iy = \frac{X+iY}{1-Z}$ *in terms of the Cartesian coordinates* (X, Y, Z) *of the point* $\widehat{z}$ *on the sphere. This can be inverted to find* $\widehat{z}$ *in terms of* z*; see (3.21).* **[b]** *If we instead use geographical coordinates* (θ, ϕ) *to describe* $\widehat{z}$ *on the sphere, the figure shows that we obtain a particularly elegant and useful polar formula:* $z = \cot(\phi/2)\, e^{i\theta}$.

If z is the stereographic projection of the point $\widehat{z}$ having coordinates (ϕ, θ), then clearly $z = r\, e^{i\theta}$, and so it only remains to find r as a function of ϕ. From [3.23b] it is clear [exercise] that the triangles $N\,\widehat{z}\,S$ and $N0\,z$ are similar, and because the angle $\angle NS\,\widehat{z} = (\phi/2)$, it follows [exercise] that $r = \cot(\phi/2)$. Thus our new stereographic formula is

$$z = \cot(\phi/2)\, e^{i\theta}. \tag{3.22}$$

We will now illustrate this formula with two applications. In Ex. 8 we also show how this formula may be used to establish a beautiful alternative interpretation of stereographic projection, due to Sir Roger Penrose.

As our first application, let us rederive the result (3.19). As above, let $\widehat{z}$ be a general point of Σ having coordinates (ϕ, θ), and let $\widetilde{\widehat{z}}$ be the point to which it is carried when we rotate Σ by π about the real axis. Check for yourself (perhaps with the aid of an orange) that the coordinates of $\widetilde{\widehat{z}}$ are $(\pi - \phi, -\theta)$. Thus if $\widetilde{z}$ is the stereographic image of $\widetilde{\widehat{z}}$, then

$$\widetilde{z} = \cot\left[\frac{\pi}{2} - \frac{\phi}{2}\right] e^{-i\theta} = \frac{1}{\cot(\phi/2)}\, e^{-i\theta} = \frac{1}{z},$$

as was to be shown.

For our second application, recall that if two points on a sphere are diametrically opposite each other (such as the north and south poles) then they are said to be *antipodal*. Let us show that

> *If* $\widehat{p}$ *and* $\widehat{q}$ *are antipodal points of* Σ, *then their stereographic projections* p *and* q *are related by the following formula:* (3.23)
>
> $$q = -(1/\overline{p}).$$

Put differently, $q = -\mathfrak{I}_C(p)$, where C is the unit circle. Note that the relationship between p and q is actually symmetrical (as clearly it should be): $p = -(1/\overline{q})$. To verify (3.23), first check for yourself that if $\widehat{p}$ has coordinates (ϕ, θ) then $\widehat{q}$ has coordinates $(\pi - \phi, \pi + \theta)$. The remainder of the proof is almost identical to the previous calculation. For an elementary geometric proof, see Ex. 6.

3.5 Möbius Transformations: Basic Results

3.5.1 *Preservation of Circles, Angles, and Symmetry*

From (3.3) we know that a general Möbius transformation $M(z) = \frac{az+b}{cz+d}$ can be decomposed into the following sequence of more elementary transformations: a translation, complex inversion, a rotation, an expansion, and a second translation. Since each of these transformations preserves circles, angles, and symmetry, we immediately deduce the following fundamental results:

- *Möbius transformations map circles to circles.*[9]
- *Möbius transformations are conformal.*
- *If two points are symmetric with respect to a circle, then their images under a Möbius transformation are symmetric with respect to the image circle. This is called the "Symmetry Principle".*

We know that a circle C will map to a circle—of course lines are now included as "circles"—but what will happen to the *disc* bounded by C? First we give a useful way of thinking about this disc. Imagine yourself walking round C moving *counterclockwise*; your motion gives C what is a called a positive *sense* or *orientation*. Of the two regions into which this positively oriented circle divides the plane, the disc may now be identified as the one lying to your *left*.

Now consider the effect of the four transformations in (3.3) on the disc and on the positively oriented circle bounding it. Translations, rotations, and expansions all preserve the orientation of C and map the interior of C to the interior of the image $\widetilde{C}$ of C. However, the effect of complex inversion on C depends on whether or not C contains the origin. If C does not contain the origin, then $\widetilde{C}$ has the same orientation as C, and the interior of C is mapped to the interior of $\widetilde{C}$. This is easily understood by looking at [3.24].

If C does contain the origin then $\widetilde{C}$ has the opposite orientation and the interior of C is mapped to the exterior of $\widetilde{C}$. If C passes through the origin then its interior is mapped to the half-plane lying to the left of the oriented line $\widetilde{C}$. See [3.25].

[9] Remarkably, Carathéodory (1937) proved that this property actually *characterizes* the Möbius transformations: they alone have this property!

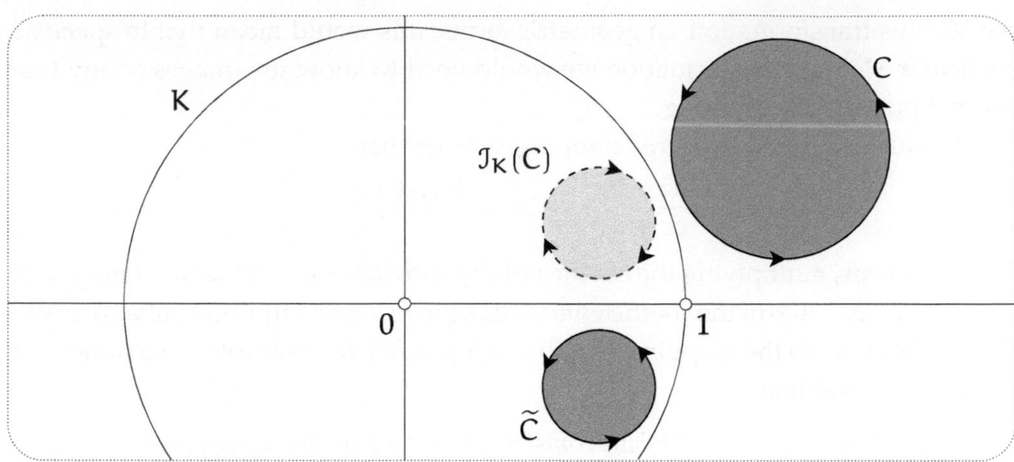

[3.24] *A Möbius transformation maps an oriented circle* C *to an oriented circle* $\widetilde{C}$ *in such a way that the region to the left of* C *is mapped to the region to the left of* $\widetilde{C}$.

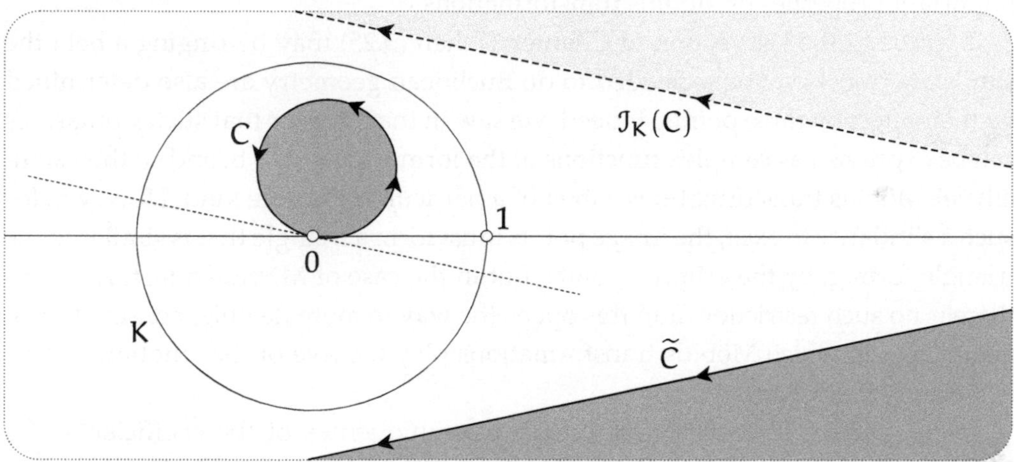

[3.25] *A limiting form of the previous figure has the image disc expand to become a half-plane, but it is still true that the region to the left of* C *is mapped to the region to the left of* $\widetilde{C}$. *In the next chapter we will see that this a more universal phenomenon, having little to do with the specific geometry of Möbius transformations.*

To summarize,

A Möbius transformation maps an oriented circle C *to an oriented circle* $\widetilde{C}$ *in such a way that the region to the left of* C *is mapped to the region to the left of* $\widetilde{C}$. (3.24)

3.5.2 Non-Uniqueness of the Coefficients

To specify a particular Möbius transformation $M(z) = \frac{az+b}{cz+d}$ it seems that we need to specify the *four* complex numbers a, b, c, and d, which we call the *coefficients* of the Möbius transformation. In geometric terms, this would mean that to specify a particular Möbius transformation we would need to know the images of any four distinct points. This is wrong.

If k is an arbitrary (non-zero) complex number then

$$\frac{az+b}{cz+d} = M(z) = \frac{kaz+kb}{kcz+kd}.$$

In other words, multiplying the coefficients by k yields one and the same mapping, and so only the *ratios* of the coefficients matter. Since *three* complex numbers are sufficient to pin down the mapping—(a/b), (b/c), (c/d), for example—we conjecture (and later prove) that

> *There exists a unique Möbius transformation sending any three points to any other three points.* (3.25)

In the course of gradually establishing this one result we shall be led to further important properties of Möbius transformations.

If you read the last section of Chapter 1, then (3.25) may be ringing a bell: the similarity transformations needed to do Euclidean geometry are also determined by their effect on three points. Indeed, we saw in that chapter that such similarities can be expressed as complex functions of the form $f(z) = az + b$, and so they actually *are* Möbius transformations, albeit of a particularly simple kind. However, for such a similarity to exist, the image points must form a triangle that is similar to the triangle formed by the original points. But in the case of Möbius transformations there is no such restriction, and this opens the way to more flexible, non-Euclidean geometries in which Möbius transformations play the role of the "motions". This is the subject of Chapter 6.

Let us make a further remark on the non-uniqueness of the coefficients of a Möbius transformation. Recall from the beginning of this chapter that the interesting Möbius transformations are the non-singular ones, for which $(ad - bc) \neq 0$. For if $(ad - bc) = 0$ then $M(z) = \frac{az+b}{cz+d}$ crushes the entire plane down to the single point (a/c). If M is non-singular, then we may multiply its coefficients by $k = \pm 1/\sqrt{ad - bc}$, in which case the *new* coefficients satisfy

$$(ad - bc) = 1;$$

the Möbius transformation is then said to be *normalized*. When investigating the properties of a *general* Möbius transformation, it turns out to be very convenient to work with this normalized form. However, when doing calculations with *specific* Möbius transformations, it is usually best *not* to normalize them.

3.5.3 The Group Property

In addition to preserving circles, angles, and symmetry, the mapping

$$z \mapsto w = M(z) = \frac{az+b}{cz+d} \qquad (ad - bc) \neq 0$$

is also *one-to-one* and *onto*. This means that if we are given any point w in the w-plane, there is one (and only one) point z in the z-plane that is mapped to w. We can show this by explicitly finding the inverse transformation $w \mapsto z = M^{-1}(w)$. Solving the above equation for z in terms of w, we find [exercise] that M^{-1} is *also* a Möbius transformation:

$$M^{-1}(z) = \frac{dz - b}{-cz + a}. \tag{3.26}$$

Note that if M is normalized, then this formula for M^{-1} is *automatically* normalized as well.

If we look at the induced mapping on the Riemann sphere, then we find that a Möbius transformation actually establishes a one-to-one correspondence between points of the complete z-sphere and points of the complete w-sphere, including their points at infinity. Indeed you may easily convince yourself that

$$M(\infty) = (a/c) \quad \text{and} \quad M(-d/c) = \infty.$$

Using (3.26), you may check for yourself that $M^{-1}(a/c) = \infty$ and $M^{-1}(\infty) = -(d/c)$.

Next, consider the composition $M \equiv (M_2 \circ M_1)$ of two Möbius transformations,

$$M_2(z) = \frac{a_2 z + b_2}{c_2 z + d_2} \quad \text{and} \quad M_1(z) = \frac{a_1 z + b_1}{c_1 z + d_1}.$$

A simple calculation [exercise] shows that M is *also* a Möbius transformation:

$$M(z) = (M_2 \circ M_1)(z) = \frac{(a_2 a_1 + b_2 c_1)z + (a_2 b_1 + b_2 d_1)}{(c_2 a_1 + d_2 c_1)z + (c_2 b_1 + d_2 d_1)}. \tag{3.27}$$

It is clear geometrically that if M_1 and M_2 are non-singular, then so is M. This is certainly not obvious algebraically, but later in this section we shall introduce a new algebraic approach that does make it obvious.

If you have studied "groups", or if you read the final section of Chapter 1, then you will realize that we have now established the following: *The set of non-singular Möbius transformations forms a group under composition.* For, (i) the identity mapping $\mathcal{E}(z) = z$ belongs to the set; (ii) the composition of two members of the set yields a third member of the set; (iii) every member of the set possesses an inverse that also lies in the set.

3.5.4 Fixed Points

As another step towards establishing (3.25), let us show that *if* a Möbius transformation exists mapping three given points to three other given points, then it is *unique*. To this end, we now introduce the extremely important concept of the *fixed points* of a Möbius transformation. Quite generally, p is called a fixed point of a mapping f if $f(p) = p$, in which case one may also say that p is "mapped to itself", or that it "remains fixed". Note that under the identity mapping, $z \mapsto \mathcal{E}(z) = z$, *every* point is a fixed point.

By definition, then, the fixed points of a general Möbius transformation $M(z)$ are the solutions of

$$z = M(z) = \frac{az+b}{cz+d}.$$

Since this is merely a quadratic in disguise, we deduce that

With the exception of the identity mapping, a Möbius transformation has at most two fixed points.

From the above result it follows that if a Möbius transformation is known to have more than two fixed points, then it must be the identity. This enables us to establish the uniqueness part of (3.25). Suppose that M and N are two Möbius transformations that both map the three given points (say q, r, s) to the three given image points. Since $(N^{-1} \circ M)$ is a Möbius transformation that has q, r, and s as fixed points, we deduce that it must be the identity mapping, and so $N = M$. Done.

We now describe the fixed points explicitly. If $M(z)$ is *normalized*, then the two fixed points ξ_+, ξ_- are given by [exercise]

$$\xi_{\pm} = \frac{(a-d) \pm \sqrt{(a+d)^2 - 4}}{2c}. \tag{3.28}$$

In the exceptional case where $(a+d) = \pm 2$, the two fixed points $\xi_{\pm}$ coalesce into the single fixed point $\xi = (a-d)/2c$. In this case the Möbius transformation is called *parabolic*.

3.5.5 Fixed Points at Infinity

Let us now briefly outline how the fixed point can be used to classify the Möbius transformations into just four achetypes. The full mathematical details are worked out later, in Section 3.7. Recall our earlier observation (3.2), that, miraculously, the Möbius transformations exactly correspond to the symmetries (called *Lorentz transformations*) of Minkowski and Einstein's spacetime. Thus this classification of the Möbius transformations is important for relativity theory, too! We cannot explore this in detail here, but see Needham (2021, §6.4) for the technical details of the spacetime interpretation via Lorentz transformations

Provided $c \neq 0$ then the fixed points both lie in the finite plane; we now discuss the fact that if $c = 0$ then at least one fixed point is at infinity. If $c = 0$ then the Möbius transformation takes the form $M(z) = Az+B$, which represents, as we have mentioned, the most general "direct" (i.e., conformal) similarity transformation of the plane. If we write $A = \rho\, e^{i\alpha}$ then this may be viewed as the composition of an

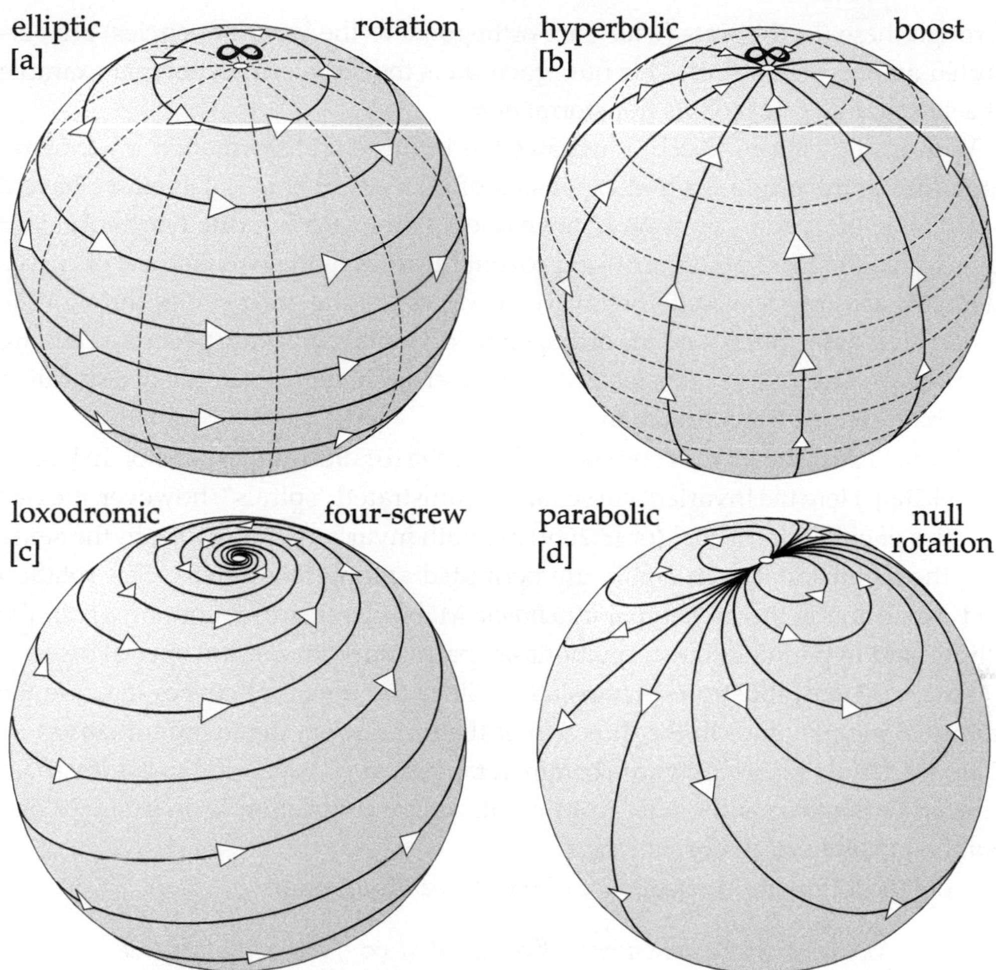

[3.26] Classification of Möbius and Lorentz transformations. *Each of the four types of transformation has two names, depending on whether it is viewed as acting on* $\mathbb{C}$ *[name on the left] or on spacetime [name on the right]. The mathematical classification depends on analysing the fixed points, and this is done in detail in Section 3.7. For the technical details of the spacetime interpretation via Lorentz transformations, see Needham (2021, §6.4).*

origin-centred rotation of α, an origin-centred expansion by ρ, and finally a translation of B. Let us visualize each of these three transformations on the Riemann sphere.

With $\alpha > 0$, figure [3.26a] illustrates that the rotation $z \mapsto e^{i\alpha}z$ in $\mathbb{C}$ induces an equal rotation of Σ about the vertical axis through its centre. Horizontal circles on Σ rotate (in the direction of the arrows) into themselves and are therefore called *invariant curves* of the transformation. This figure makes it vividly clear that the fixed points of such a rotation are 0 and ∞. Note also that the (great) circles through these fixed points (which are orthogonal to the invariant circles) are permuted among themselves. This pure rotation is the simplest, archetypal example of a so-called *elliptic* Möbius transformation.

With $\rho > 1$, figure [3.26b] illustrates the induced transformation on Σ corresponding to the origin-centred expansion of $\mathbb{C}$, $z \mapsto \rho z$. If $\rho < 1$ then we have a contraction of $\mathbb{C}$, and points on Σ move due South instead of due North. Again it is clear that the fixed points are 0 and ∞, but the roles of the two families of curves in [3.26a] are now reversed: the invariant curves are the great circles through the fixed points at the poles, and the orthogonal horizontal circles are permuted among themselves. This pure expansion is the simplest, archetypal example of a so-called *hyperbolic* Möbius transformation.

Figure [3.26c] shows the combined effect of the rotation and expansion in [3.26a] and [3.26b]. Here the invariant curves are the illustrated "spirals"; however, the two families of circles in [3.26a] (or [3.26b]) are both invariant as a whole, in the sense that the members of each family are permuted among themselves. This rotation and expansion is the archetypal *loxodromic* Möbius transformation, of which the elliptic and hyperbolic transformations are particularly important special cases.

Finally, [3.26d] illustrates a translation. Since the invariant curves in $\mathbb{C}$ are the family of parallel lines in the direction of the translation, the invariant curves on Σ are the family of circles whose common tangent at ∞ is parallel to the invariant lines in $\mathbb{C}$. Since ∞ is the only fixed point, a pure translation is an example of a parabolic Möbius transformation.

Note the following consequence of the above discussion:

> *A Möbius transformation has a fixed point at* ∞ *if and only if it is a similarity,* $M(z) = (az + b)$*. Furthermore,* ∞ *is the sole fixed point if and only if* $M(z)$ *is a translation,* $M(z) = (z + b)$*.* (3.29)

Later we will use this to show that each Möbius transformation is equivalent, in a certain sense, to one (and only one) of the four types shown in [3.26].

3.5.6 The Cross-Ratio

Returning to (3.25), we have already established that if we can find a Möbius transformation M that maps three given points q, r, s to three other given points $\widetilde{q}, \widetilde{r}, \widetilde{s}$, then M is unique. It thus remains to show that such an M always exists.

To see this, first let us arbitrarily choose three points q', r', s', *once and for all.* Next, suppose we can write down a Möbius transformation mapping three arbitrary points q, r, s to these particular three points, q', r', s'; let $M_{qrs}(z)$ denote this Möbius transformation. In exactly the same way we could also write down $M_{\widetilde{q}\widetilde{r}\widetilde{s}}(z)$. By virtue of the group property, it is now easy to see that

$$M = M_{\widetilde{q}\widetilde{r}\widetilde{s}}^{-1} \circ M_{qrs}$$

is a Möbius transformation mapping q, r, s to q′, r′, s′ and thence to $\widetilde{q}, \widetilde{r}, \widetilde{s}$, as was desired.

Now the real trick is to choose q′, r′, s′ in such a way as to make it easy to write down $M_{qrs}(z)$. We don't like to pull rabbits out of hats, but try $q' = 0$, $r' = 1$, and $s' = \infty$. Along with this special choice comes a special, standard notation: *the unique Möbius transformation mapping three given points* q, r, s *to* 0, 1, ∞ *(respectively) is written* [z, q, r, s].

In order to map q to $q' = 0$ and s to $s' = \infty$, the numerator and denominator of [z, q, r, s] must be proportional to $(z-q)$ and $(z-s)$, respectively. Thus $[z, q, r, s] = k\left(\frac{z-q}{z-s}\right)$, where k is a constant. Finally, since $k\left(\frac{r-q}{r-s}\right) = [r, q, r, s] \equiv 1$, we deduce that

$$[z, q, r, s] = \frac{(z-q)(r-s)}{(z-s)(r-q)}.$$

This is not quite so rabbit-like as it appears. Two hundred years prior to Möbius' investigations, Girard Desargues had discovered the importance of the expression [z, q, r, s] within the subject of projective geometry, where it was christened the *cross-ratio* of z, q, r, s (in this order[10]). Its significance in that context is briefly explained in Ex. 14, but the reader is urged to consult Stillwell (2010) for greater detail and background. We can now restate (3.25) in a more explicit form:

> *The unique Möbius transformation* $z \mapsto w = M(z)$ *sending three points* q, r, s *to any other three points* $\widetilde{q}, \widetilde{r}, \widetilde{s}$ *is given by*
>
> $$\frac{(w-\widetilde{q})(\widetilde{r}-\widetilde{s})}{(w-\widetilde{s})(\widetilde{r}-\widetilde{q})} = [w, \widetilde{q}, \widetilde{r}, \widetilde{s}] = [z, q, r, s] = \frac{(z-q)(r-s)}{(z-s)(r-q)}. \tag{3.30}$$

Although we have not done so, in any concrete case one could easily go on to solve this equation for w, thereby obtaining an explicit formula for $w = M(z)$.

The result (3.30) may be rephrased in various helpful ways. For example, if a Möbius transformation maps four points p, q, r, s to $\widetilde{p}, \widetilde{q}, \widetilde{r}, \widetilde{s}$ (respectively) then the cross-ratio is invariant: $[\widetilde{p}, \widetilde{q}, \widetilde{r}, \widetilde{s}] = [p, q, r, s]$. Conversely, p, q, r, s can be mapped to $\widetilde{p}, \widetilde{q}, \widetilde{r}, \widetilde{s}$ by a Möbius transformation if their cross-ratios are equal.

Recalling (3.24), we also obtain the following:

> *Let* C *be the unique circle through the points* q, r, s *in the z-plane, oriented so that these points succeed one another in the stated order. Likewise, let* $\widetilde{C}$ *be the unique oriented circle through* $\widetilde{q}, \widetilde{r}, \widetilde{s}$ *in the w-plane. Then the Möbius transformation given by (3.30) maps* C *to* $\widetilde{C}$*, and it maps the region lying to the left of* C *to the region lying to the left of* $\widetilde{C}$. (3.31)

[10] Different orders yield different values; see Ex. 16. Unfortunately, there is no firm convention as to which of these values is "the" cross-ratio. For example, our definition agrees with Carathéodory (1964), Penrose and Rindler (1984), and Jones and Singerman (1987), but it is different from the equally common definition of Ahlfors (1979).

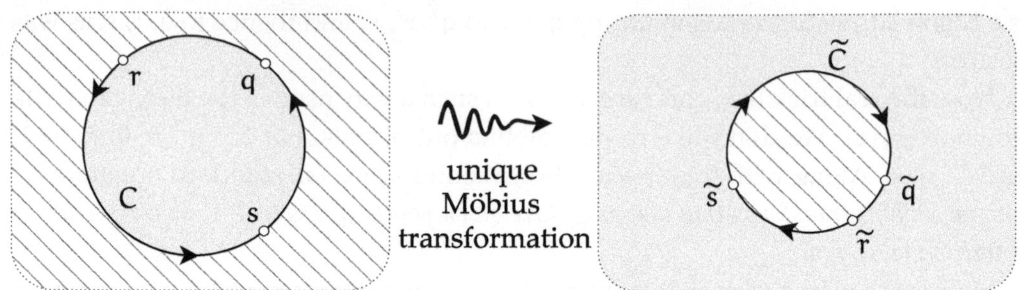

[3.27] *There exists a unique Möbius transformation that maps three ordered points on the circle* C *to three other ordered points on the circle* $\widetilde{C}$*, and the region to the left of* C *is mapped to the region to the left of* $\widetilde{C}$.

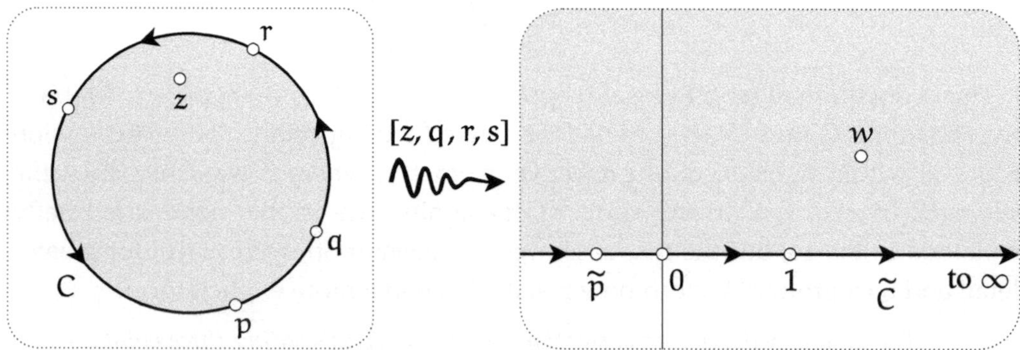

[3.28] Geometric interpretation of the cross-ratio. *The cross-ratio* $w = [z, q, r, s]$ *can be vividly pictured as the image of* z *under the unique Möbius transformation that maps the oriented circle* C *through* q, r, s *to the real axis in such a way that these three points map to* 0, 1, ∞. *Also illustrated is the fact that a point* p *lies on the circle* C *through* q, r, s *if and only if* $\operatorname{Im}[p, q, r, s] = 0$.

This is illustrated in [3.27].

This in turn gives us a more vivid picture of the cross-ratio: $w = [z, q, r, s]$ is the image of z under the unique Möbius transformation that maps the oriented circle C through q, r, s to the real axis in such a way that these three points map to 0, 1, ∞. If q, r, s induce a positive orientation on C then the interior of C is mapped to the upper half-plane; if they induce a negative orientation, then the image is the lower half-plane. This is illustrated in [3.28], from which we immediately deduce a neat equation for the circle C:

A point p *lies on the circle* C *through* q, r, s *if and only if*

$$\operatorname{Im}[p, q, r, s] = 0. \tag{3.32}$$

Furthermore, if q, r, s *induce a positive orientation on* C *(as in [3.28]), then* p *lies inside* C *if and only if* $\operatorname{Im}[p, q, r, s] > 0$. *If the orientation of* C *is negative, then the inequality is reversed.*

For a more elementary proof of (3.32), see Ex. 15.

3.6 Möbius Transformations as Matrices*

3.6.1 *Empirical Evidence of a Link with Linear Algebra*

As you were reading about the group property of Möbius transformations, you may well have experienced *déjà vu,* for the results we obtained were remarkably reminiscent of the behaviour of matrices in *linear algebra*. Before explaining the *reason* for this connection between Möbius transformations and linear algebra, let us be more explicit about the empirical evidence for believing that such a connection exists.

We begin by associating with every Möbius transformation $M(z)$ a corresponding 2×2 matrix $[M]$:

$$M(z) = \frac{az+b}{cz+d} \qquad \longleftrightarrow \qquad [M] = \begin{bmatrix} a & b \\ c & d \end{bmatrix}.$$

Since the coefficients of the Möbius transformation are not unique, neither is the corresponding matrix: if k is any non-zero constant, then the matrix $k[M]$ corresponds to the same Möbius transformation as $[M]$. However, if $[M]$ is normalized by imposing $(ad - bc) = 1$, then there are just two possible matrices associated with a given Möbius transformation: if one is called $[M]$, the other is $-[M]$; in other words, the matrix is determined "uniquely up to sign". This apparently trivial fact turns out to have deep significance in both mathematics and physics; see Penrose and Rindler (1984, Ch. 1) and Penrose (2005).

At this point there exists a strong possibility of confusion, so we issue the following *WARNING:* In linear algebra we are—or should be!—accustomed to thinking of a real 2×2 matrix as representing a linear transformation of $\mathbb{R}^2$. For example, $\left(\begin{smallmatrix} 0 & -1 \\ 1 & 0 \end{smallmatrix}\right)$ represents a rotation of the plane through $(\pi/2)$. That is, when we apply it to a vector $\left(\begin{smallmatrix} x \\ y \end{smallmatrix}\right)$ in $\mathbb{R}^2$, we obtain

$$\begin{pmatrix} 0 & -1 \\ 1 & 0 \end{pmatrix}\begin{pmatrix} x \\ y \end{pmatrix} = \begin{pmatrix} -y \\ x \end{pmatrix} = \left\{\begin{pmatrix} x \\ y \end{pmatrix} \text{ rotated by } (\pi/2)\right\}.$$

In stark contrast, the matrix $\left[\begin{smallmatrix} a & b \\ c & d \end{smallmatrix}\right]$ corresponding to a Möbius transformation generally has *complex* numbers as its entries, and so it *cannot* be interpreted as a linear transformation of $\mathbb{R}^2$. Even if the entries *are* real, it must not be thought of in this way. For example, the matrix $\begin{pmatrix} 0 & -1 \\ 1 & 0 \end{pmatrix}$ corresponds to the Möbius transformation $M(z) = -(1/z)$, which is certainly not a linear transformation of $\mathbb{C}$. To avoid confusion, we will adopt the following notational convention: *We use (ROUND) brackets for a real matrix corresponding to a linear transformation of* $\mathbb{R}^2$ *or of* $\mathbb{C}$*, and we use [SQUARE] brackets for a (generally) complex matrix corresponding to a Möbius transformation of* $\mathbb{C}$.

Despite this warning, we have the following striking parallels between the behaviour of Möbius transformations and the matrices that represent them:

- The identity Möbius transformation $\mathcal{E}(z) = z$ corresponds to the familiar identity matrix, $[\mathcal{E}] = \begin{bmatrix} 1 & 0 \\ 0 & 1 \end{bmatrix}$.
- The Möbius transformation $M(z)$ with matrix $[M] = \begin{bmatrix} a & b \\ c & d \end{bmatrix}$ possesses an inverse if and only if the *matrix* possesses an inverse. For recall that $[M]$ is non-singular if and only if its determinant $\det[M] = (ad - bc)$ is non-zero.
- If we look at (3.26), we see that the matrix of the inverse Möbius transformation $M^{-1}(z)$ is the same as the inverse matrix $[M]^{-1}$. To put this succinctly,

$$[M^{-1}] = [M]^{-1}.$$

- In linear algebra we compose two linear transformations by multiplying their matrices; indeed, this is the origin of the multiplication rule. If we multiply the matrices $[M_2]$ and $[M_1]$ corresponding to the two Möbius transformations $M_2(z)$ and $M_1(z)$, then we obtain

$$\begin{bmatrix} a_2 & b_2 \\ c_2 & d_2 \end{bmatrix} \begin{bmatrix} a_1 & b_1 \\ c_1 & d_1 \end{bmatrix} = \begin{bmatrix} a_2a_1 + b_2c_1 & a_2b_1 + b_2d_1 \\ c_2a_1 + d_2c_1 & c_2b_1 + d_2d_1 \end{bmatrix}.$$

But look at (3.27)! This is simply the matrix of the composite Möbius transformation $(M_2 \circ M_1)(z)$. Thus *multiplication of Möbius matrices corresponds to composition of Möbius transformations*:

$$[M_2]\,[M_1] = [M_2 \circ M_1].$$

3.6.2 *The Explanation: Homogeneous Coordinates*

Clearly this cannot all be coincidence, but what is really going on here?! The answer is simple, yet subtle. To see it we must first describe the complex plane with a completely new kind of coordinate system. Instead of expressing $z = x + iy$ in terms of two real numbers, we write it as the *ratio of two complex numbers*, $\mathfrak{z}_1$ and $\mathfrak{z}_2$:

$$z = \frac{\mathfrak{z}_1}{\mathfrak{z}_2}.$$

The ordered pair of complex numbers $[\mathfrak{z}_1, \mathfrak{z}_2]$ are called *homogeneous coordinates* of z. In order that this ratio be well defined we demand that $[\mathfrak{z}_1, \mathfrak{z}_2] \neq [0, 0]$. To each ordered pair $[\mathfrak{z}_1 \text{ arbitrary}, \mathfrak{z}_2 \neq 0]$ there corresponds precisely one point $z = (\mathfrak{z}_1/\mathfrak{z}_2)$, but to each point z there corresponds an infinite set of homogeneous coordinates, $[k\mathfrak{z}_1, k\mathfrak{z}_2] = k[\mathfrak{z}_1, \mathfrak{z}_2]$, where k is an arbitrary non-zero complex number.

What about a pair of the form $[\mathfrak{z}_1, 0]$? By holding $\mathfrak{z}_1$ fixed as $\mathfrak{z}_2$ tends to 0, it is clear that $[\mathfrak{z}_1, 0]$ must be identified with the point at infinity. Thus the totality of pairs $[\mathfrak{z}_1, \mathfrak{z}_2]$ provide coordinates for the *extended* complex plane. The introduction of homogeneous coordinates thereby accomplishes for algebra what the Riemann sphere accomplishes for geometry—it does away with the exceptional role of ∞.

Just as we use the symbol $\mathbb{R}^2$ to denote the set of pairs (x, y) of *real* numbers, so we use the symbol $\mathbb{C}^2$ to denote the set of pairs $[\mathfrak{z}_1, \mathfrak{z}_2]$ of *complex* numbers. To highlight the distinction between $\mathbb{R}^2$ and $\mathbb{C}^2$, we use conventional round brackets when writing down an element (x, y) of $\mathbb{R}^2$, but we use square brackets for an element $[\mathfrak{z}_1, \mathfrak{z}_2]$ of $\mathbb{C}^2$.

Just as a linear transformation of $\mathbb{R}^2$ is represented by a real 2×2 matrix, so a linear transformation of $\mathbb{C}^2$ is represented by a complex 2×2 matrix:

$$\begin{bmatrix} \mathfrak{z}_1 \\ \mathfrak{z}_2 \end{bmatrix} \longmapsto \begin{bmatrix} \mathfrak{w}_1 \\ \mathfrak{w}_2 \end{bmatrix} = \begin{bmatrix} a & b \\ c & d \end{bmatrix} \begin{bmatrix} \mathfrak{z}_1 \\ \mathfrak{z}_2 \end{bmatrix} = \begin{bmatrix} a\,\mathfrak{z}_1 + b\,\mathfrak{z}_2 \\ c\,\mathfrak{z}_1 + d\,\mathfrak{z}_2 \end{bmatrix}.$$

But if $[\mathfrak{z}_1, \mathfrak{z}_2]$ and $[\mathfrak{w}_1, \mathfrak{w}_2]$ are thought of as the homogeneous coordinates in $\mathbb{C}^2$ of the point $z = (\mathfrak{z}_1/\mathfrak{z}_2)$ in $\mathbb{C}$ and its image point $w = (\mathfrak{w}_1/\mathfrak{w}_2)$, then the above linear transformation of $\mathbb{C}^2$ induces the following (non-linear) transformation of $\mathbb{C}$:

$$z = \frac{\mathfrak{z}_1}{\mathfrak{z}_2} \quad \longmapsto \quad w = \frac{\mathfrak{w}_1}{\mathfrak{w}_2} = \frac{a\,\mathfrak{z}_1 + b\,\mathfrak{z}_2}{c\,\mathfrak{z}_1 + d\,\mathfrak{z}_2} = \frac{a\,(\mathfrak{z}_1/\mathfrak{z}_2) + b}{c\,(\mathfrak{z}_1/\mathfrak{z}_2) + d} = \frac{az + b}{cz + d}.$$

This is none other than the most general Möbius transformation!

We have thus explained why Möbius transformations in $\mathbb{C}$ behave so much like linear transformations—they *are* linear transformations, only they act on the homogeneous coordinates in $\mathbb{C}^2$, rather than directly on the points of $\mathbb{C}$ itself.

As with the cross-ratio, homogeneous coordinates first arose in projective geometry, and for this reason they are often also called *projective* coordinates. See Stillwell (2010) for greater detail on the history of the idea. We cannot move on without mentioning that in recent times these homogeneous coordinates have provided the key to great conceptual advances (and powerful new computational techniques) in Einstein's Theory of Relativity. This pioneering body of work on *2-spinors* is due to

Sir Roger Penrose. See Penrose and Rindler (1984, Ch. 1), Penrose (2005), and Shaw (2006, Ch. 23).

3.6.3 *Eigenvectors and Eigenvalues**

The above representation of Möbius transformations as matrices provides an elegant and practical method of doing concrete calculations. More significantly, however, it also means that in developing the theory of Möbius transformations we suddenly have access to a whole range of new ideas and techniques taken from linear algebra.

We begin with something very simple. We previously remarked that while it is geometrically obvious that the composition of two non-singular Möbius transformations is again non-singular, it is far from obvious algebraically. Our new point of view rectifies this, for recall the following elementary property of determinants:

$$\det\{[M_2]\,[M_1]\} = \det[M_2]\,\det[M_1].$$

Thus if $\det[M_2] \neq 0$ and $\det[M_1] \neq 0$, then $\det\{[M_2]\,[M_1]\} \neq 0$, as was to be shown. This also sheds further light on the virtue of working with normalized Möbius transformations. For if $\det[M_2] = 1$ and $\det[M_1] = 1$, then $\det\{[M_2]\,[M_1]\} = 1$. Thus the set of normalized 2×2 matrices form a *group*—a "subgroup" of the full group of non-singular matrices.

For our second example, consider the *eigenvectors* of a linear transformation $[M] = \begin{bmatrix} a & b \\ c & d \end{bmatrix}$ of $\mathbb{C}^2$. By definition, an eigenvector is a vector $\mathfrak{z} = \begin{bmatrix} \mathfrak{z}_1 \\ \mathfrak{z}_2 \end{bmatrix}$ whose "direction" is unaltered by the transformation, in the sense that its image is simply a multiple $\lambda\mathfrak{z}$ of the original; this multiple λ is called the *eigenvalue* of the eigenvector. In other words, an eigenvector satisfies the equation

$$\begin{bmatrix} a & b \\ c & d \end{bmatrix} \begin{bmatrix} \mathfrak{z}_1 \\ \mathfrak{z}_2 \end{bmatrix} = \lambda \begin{bmatrix} \mathfrak{z}_1 \\ \mathfrak{z}_2 \end{bmatrix}.$$

In terms of the corresponding Möbius transformation in $\mathbb{C}$, this means that $z = (\mathfrak{z}_1/\mathfrak{z}_2)$ is mapped to $M(z) = (\lambda\mathfrak{z}_1/\lambda\mathfrak{z}_2) = z$, and so

$$z = (\mathfrak{z}_1/\mathfrak{z}_2) \textit{ is a fixed point of } M(z) \textit{ if and only if } \mathfrak{z} = \begin{bmatrix} \mathfrak{z}_1 \\ \mathfrak{z}_2 \end{bmatrix} \textit{ is an eigenvector of } [M]. \tag{3.33}$$

Note that one immediate benefit of this approach is that there is no longer any real distinction between a finite fixed point and a fixed point at ∞, for the latter merely corresponds to an eigenvector of the form $\begin{bmatrix} \mathfrak{z}_1 \\ 0 \end{bmatrix}$. For example, consider how elegantly we may rederive the fact that ∞ is a fixed point if and only if $M(z)$ is a similarity transformation. If ∞ is a fixed point then

$$\lambda \begin{bmatrix} \mathfrak{z}_1 \\ 0 \end{bmatrix} = \begin{bmatrix} a & b \\ c & d \end{bmatrix} \begin{bmatrix} \mathfrak{z}_1 \\ 0 \end{bmatrix} = \begin{bmatrix} a\,\mathfrak{z}_1 \\ c\,\mathfrak{z}_1 \end{bmatrix}.$$

Thus $c = 0$, $\lambda = a$, and $M(z) = (a/d)z + (b/d)$.

Recall that if the matrix $[M]$ represents the Möbius transformation $M(z)$, then so does the matrix $k[M]$ obtained by multiplying the entries by k. The fact that eigenvectors carry geometric information about $M(z)$ shows up in the fact that they are independent of the choice of k. Indeed, if $\mathfrak{z}$ is an eigenvector of $[M]$ (with eigenvalue λ) then it is also an eigenvector of $k[M]$, but with eigenvalue $k\lambda$:

$$\{k[M]\}\,\mathfrak{z} = k\lambda\,\mathfrak{z}.$$

Since the eigenvalue *does* depend on the arbitrary choice of k, it appears that its value can have no bearing on the geometric nature of the mapping $M(z)$. Very surprisingly, however, if $[M]$ is *normalized* then the exact opposite is true! In the next section we will show that *the eigenvalues of the normalized matrix* $[M]$ *completely determine the geometric nature of the corresponding Möbius transformation* $M(z)$. In anticipation of this result, let us investigate the eigenvalues further.

Recall the fact that the eigenvalues of $[M]$ are the solutions of the so-called *characteristic equation*, $\det\{[M]-\lambda[\mathcal{E}]\} = 0$, where $[\mathcal{E}]$ is the identity matrix $\begin{bmatrix} 1 & 0 \\ 0 & 1 \end{bmatrix}$. Using the fact that $[M]$ is normalized, we find [exercise] that the characteristic equation is

$$\lambda^2 - (a+d)\lambda + 1 = 0,$$

which (for later use) may be written as

$$\lambda + \frac{1}{\lambda} = a + d. \tag{3.34}$$

The first thing we notice about this equation is that there are typically two eigenvalues, λ_1 and λ_2, and they are determined solely by the value of $(a + d)$. By inspecting the coefficients of the quadratic we immediately deduce that

$$\lambda_1\lambda_2 = 1 \quad \text{and} \quad \lambda_1 + \lambda_2 = (a+d). \tag{3.35}$$

Thus if we know λ_1, then $\lambda_2 = (1/\lambda_1)$. We emphasize this point because it is not obvious when we simply write down the formula for the eigenvalues:

$$\lambda_1,\ \lambda_2 = \tfrac{1}{2}\left\{(a+d) \pm \sqrt{(a+d)^2 - 4}\right\}.$$

Aficionados of linear algebra will recognize (3.35) as a special case of the following general result on the eigenvalues $\lambda_1, \lambda_2, \ldots, \lambda_n$ of any $n \times n$ matrix N:

$$\lambda_1 \lambda_2 \ldots \lambda_n = \det N \qquad \text{and} \qquad \lambda_1 + \lambda_2 + \cdots + \lambda_n = \operatorname{tr} N,$$

where $\operatorname{tr} N \equiv$ (the sum of the diagonal elements of N) is called the *trace* of N. For future use, recall the following nice property of the trace function: *If* N *and* P *are both* $n \times n$ *matrices, then*

$$\operatorname{tr}\{NP\} = \operatorname{tr}\{PN\}. \tag{3.36}$$

In the case of 2×2 matrices (which is all that we shall ever need) this is easily verified by a direct calculation [exercise].

*3.6.4 Rotations of the Sphere as Möbius Transformations**

This subsection is optional because its main result is only needed in Chapter 6. Furthermore, in that chapter we shall treat the same result in a much better and simpler way; the only purpose of this subsection is to further illustrate the connections that exist between Möbius transformations and linear algebra.

Let us investigate what it might mean to say that two vectors $\mathfrak{p}$ and $\mathfrak{q}$ in $\mathbb{C}^2$ are "orthogonal". Two vectors $\mathbf{p}$ and $\mathbf{q}$ in $\mathbb{R}^2$ are orthogonal if and only if their dot product vanishes:

$$\mathbf{p}\cdot\mathbf{q} = \begin{pmatrix} p_1 \\ p_2 \end{pmatrix} \cdot \begin{pmatrix} q_1 \\ q_2 \end{pmatrix} = p_1 q_1 + p_2 q_2 = 0.$$

Thus it would seem natural to say that $\mathfrak{p}$ and $\mathfrak{q}$ are "orthogonal" if $\mathfrak{p}\cdot\mathfrak{q} = 0$. This will not do. In particular, whereas we would like the dot product of any nonzero vector with itself to be a positive real number, we find that $\begin{bmatrix} 1 \\ i \end{bmatrix} \cdot \begin{bmatrix} 1 \\ i \end{bmatrix} = 0$, for example. As it stands, the dot product is not suitable for use in $\mathbb{C}^2$.

The standard solution to this difficulty is to generalize the dot product $\mathfrak{p}\cdot\mathfrak{q}$ to the so-called *inner product*, $\langle \mathfrak{p}, \mathfrak{q} \rangle \equiv \overline{\mathfrak{p}}\cdot\mathfrak{q}$:

$$\langle \mathfrak{p}, \mathfrak{q} \rangle = \left\langle \begin{bmatrix} p_1 \\ p_2 \end{bmatrix}, \begin{bmatrix} q_1 \\ q_2 \end{bmatrix} \right\rangle = \overline{p_1}\, q_1 + \overline{p_2}\, q_2.$$

We cannot go into all the reasons why this is the "right" generalization, but observe that it shares the following desirable properties of the dot product:

$$\langle \mathfrak{p}, \mathfrak{p} \rangle \geqslant 0 \quad \text{and} \quad \langle \mathfrak{p}, \mathfrak{p} \rangle = 0 \text{ if and only if } p_1 = 0 = p_2;$$
$$\langle \mathfrak{p} + \mathfrak{q}, \mathfrak{r} \rangle = \langle \mathfrak{p}, \mathfrak{r} \rangle + \langle \mathfrak{q}, \mathfrak{r} \rangle \quad \text{and} \quad \langle \mathfrak{r}, \mathfrak{p} + \mathfrak{q} \rangle = \langle \mathfrak{r}, \mathfrak{p} \rangle + \langle \mathfrak{r}, \mathfrak{q} \rangle.$$

Note, however, that it is not commutative: $\langle \mathfrak{q}, \mathfrak{p} \rangle = \overline{\langle \mathfrak{p}, \mathfrak{q} \rangle}$.

We now agree that $\mathfrak{p}$ and $\mathfrak{q}$ are "orthogonal" if and only if

$$\langle \mathfrak{p}, \mathfrak{q} \rangle = \overline{\mathfrak{p}_1}\, \mathfrak{q}_1 + \overline{\mathfrak{p}_2}\, \mathfrak{q}_2 = 0.$$

What does this "orthogonality" mean in terms of the points $p = (\mathfrak{p}_1/\mathfrak{p}_2)$ and $q = (\mathfrak{q}_1/\mathfrak{q}_2)$ whose homogeneous coordinate vectors are $\mathfrak{p}$ and $\mathfrak{q}$? The answer is surprising. As you may easily check, the above equation says that $q = -(1/\overline{p})$, and so from (3.23) we deduce that

> *Two vectors in $\mathbb{C}^2$ are orthogonal if and only if they are the homogeneous coordinates of antipodal points on the Riemann sphere.*

Suppose we could find a linear transformation $[R]$ of $\mathbb{C}^2$ that were analogous to a rotation—what transformation of the Riemann sphere Σ would be induced by the corresponding Möbius transformation $R(z)$? By "analogous to a rotation", we mean that $[R]$ preserves the inner product:

$$\langle [R]\mathfrak{p}, [R]\mathfrak{q} \rangle = \langle \mathfrak{p}, \mathfrak{q} \rangle . \tag{3.37}$$

In particular, $[R]$ maps every pair of orthogonal vectors to another such pair, and $R(z)$ therefore maps every pair of antipodal points on Σ to another such pair. We shall not attempt a real proof, but since the transformation of Σ is also known to be continuous and conformal[11], it can only be a *rotation* of Σ.

The desired invariance of the inner product (3.37) may be neatly rephrased using an operation called the *conjugate transpose*, denoted by a superscript $*$. This operation takes the complex conjugate of each element in a matrix and then interchanges the rows and columns:

$$\mathfrak{p}^* = \begin{bmatrix} \mathfrak{p}_1 \\ \mathfrak{p}_2 \end{bmatrix}^* = [\overline{\mathfrak{p}_1}, \overline{\mathfrak{p}_2}] \quad \text{and} \quad [R]^* = \begin{bmatrix} a & b \\ c & d \end{bmatrix}^* = \begin{bmatrix} \overline{a} & \overline{c} \\ \overline{b} & \overline{d} \end{bmatrix}.$$

Since the inner product can now be expressed in terms of ordinary matrix multiplication as $\langle \mathfrak{p}, \mathfrak{q} \rangle = \mathfrak{p}^* \mathfrak{q}$, and since [exercise] $\{[R]\mathfrak{p}\}^* = \mathfrak{p}^*[R]^*$, we find that (3.37) takes the form

$$\mathfrak{p}^* \{[R]^*[R]\}\, \mathfrak{q} = \mathfrak{p}^* \mathfrak{q}.$$

Clearly this is satisfied if

$$[R]^*[R] = [\mathcal{E}], \tag{3.38}$$

and in linear algebra it is shown that this is also a *necessary* condition.

Matrices satisfying equation (3.38) are extremely important in both mathematics and physics—they are called *unitary matrices*. In the present case of normalized 2×2

[11] If it were not continuous then it could, for example, exchange points on two antipodal patches of Σ while leaving the remainder fixed. If it were continuous but *anti*conformal, then it could map each point to its antipodal point, or to its reflection in a plane through the centre of Σ.

matrices, we can easily find the most general unitary matrix [R] by re-expressing (3.38) as $[R]^* = [R]^{-1}$:

$$\begin{bmatrix} \overline{a} & \overline{c} \\ \overline{b} & \overline{d} \end{bmatrix} = \begin{bmatrix} d & -b \\ -c & a \end{bmatrix} \implies [R] = \begin{bmatrix} a & b \\ -\overline{b} & \overline{a} \end{bmatrix}.$$

Although we have left some unsatisfactory gaps in the above reasoning, we have nevertheless arrived at an important truth: *The most general rotation of the Riemann sphere can be expressed as a Möbius transformation of the form*

$$R(z) = \frac{az + b}{-\overline{b}z + \overline{a}}. \tag{3.39}$$

This was first discovered by Gauss, around 1819.

3.7 Visualization and Classification*

3.7.1 The Main Idea

Although the decomposition (3.3) of a general Möbius transformation $M(z)$ has proved valuable in obtaining results, it makes $M(z)$ appear much more complicated than it is. In this section we will reveal this hidden simplicity by examining the fixed points in greater detail; this will enable us to visualize Möbius transformations in a particularly vivid way. In the process we will clarify our earlier remark that Möbius transformations can be classified into four types, each $M(z)$ being "equivalent" to one (and only one) of the four types of transformation illustrated in [3.26]. The lovely idea behind this classification scheme is due to Felix Klein.

To begin with, suppose that $M(z)$ has two distinct fixed points, ξ_+ and ξ_-. Now look at the LHS of [3.29], and in particular at the family $\mathcal{C}_1$ of circles [shown dashed] passing through the fixed points. If we think of $M(z)$ as a mapping $z \mapsto w = M(z)$ of this figure to itself, then *each member of* $\mathcal{C}_1$ *is mapped to another member of* $\mathcal{C}_1$. Why?

Still with reference to the LHS of [3.29], suppose that p [not shown] is an arbitrary point on the line through ξ_+ and ξ_-, but lying outside the segment connecting the fixed points. If K is the circle of radius $\sqrt{[p\xi_+][p\xi_-]}$ centred at p, then ξ_+ and ξ_- are symmetric with respect to K. Thus K cuts each member of $\mathcal{C}_1$ at right angles (cf. [3.9]). By varying the position of p we thus obtain a family $\mathcal{C}_2$ of circles [shown solid] such that *ξ_+ and ξ_- are symmetric with respect to each member of $\mathcal{C}_2$, and each member of $\mathcal{C}_2$ is orthogonal to each member of $\mathcal{C}_1$.*

Now we come to the main idea: to the LHS of [3.29] we *apply a Möbius transformation $F(z)$ that sends one fixed point (say ξ_+) to 0, and the other fixed point (ξ_-)* to ∞. The

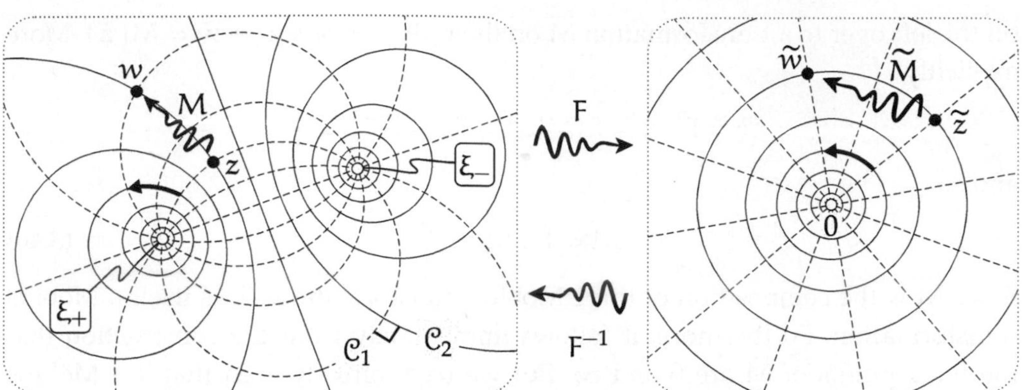

[3.29] The geometric idea behind the classification of the Möbius transformations. *Given a Möbius transformation $M(z)$ with two fixed points, ξ_+ and ξ_-, two families of circles naturally arise: the dashed family $\mathcal{C}_1$ passing through the fixed points, and the orthogonal family $\mathcal{C}_2$, such that ξ_+ and ξ_- are symmetric with respect to each member of $\mathcal{C}_2$. Note that $M(z)$ must map members of $\mathcal{C}_1$ amongst themselves, and likewise members of $\mathcal{C}_2$ are mapped amongst themselves. The key idea is to now apply a Möbius transformation $F(z) = \frac{z-\xi_+}{z-\xi_-}$ that sends one fixed point to the origin (south pole of the Riemann sphere) and the other to ∞ (the north pole). Thus, on the right, the circles $\mathcal{C}_1$ become rays, and the circles $\mathcal{C}_2$ become origin-centred circles. Then the induced mapping $\widetilde{M} = F \circ M \circ F^{-1}$ clearly has fixed points 0 and ∞, and must therefore take the form $\widetilde{M}(\widetilde{z}) = m\,\widetilde{z}$, where $m = \rho\, e^{i\alpha}$ is called the* multiplier; *in the illustrated case, $\rho = 1$ and $\alpha = (\pi/3)$. This multiplier m completely characterizes the* original *Möbius transformation, M, and therefore serves to classify it. In this manner, we obtain the first three archetypes shown in [3.26].*

RHS of [3.29] shows the image of the LHS under such a Möbius transformation, the simplest example of which is

$$F(z) = \frac{z-\xi_+}{z-\xi_-}.$$

[Note that we have not bothered to write this in normalized form.] Since F is a Möbius transformation, it must map the members of $\mathcal{C}_1$ to the circles passing through 0 and ∞, i.e., to lines through the origin [shown dashed]. Furthermore, since F is conformal, two such lines must contain the same angle at 0 as the corresponding $\mathcal{C}_1$ circles do at ξ_+. We have tried to make this easy to see in our picture by drawing $\mathcal{C}_1$ circles passing through ξ_+ in evenly spaced directions, each one making an angle of $(\pi/6)$ with the next.

As an aside, observe that we now have a second, simpler explanation of the existence of the family $\mathcal{C}_2$ of circles orthogonal to $\mathcal{C}_1$. Since the illustrated set of origin-centred circles are orthogonal to lines through 0, their images under F^{-1} must be circles orthogonal to each member of $\mathcal{C}_1$.

Next, let $\widetilde{z} = F(z)$ and $\widetilde{w} = F(w)$ be the images under F of z and $w = M(z)$. We may now think of F as carrying the original Möbius transformation $z \mapsto w = M(z)$

on the left over to a transformation $\widetilde{M}$ on the right, namely $\widetilde{z} \mapsto \widetilde{w} = \widetilde{M}(\widetilde{z})$. More explicitly,

$$\widetilde{w} = F(w) = F(M[z]) = F\left(M\left[F^{-1}(\widetilde{z})\right]\right),$$

and so

$$\widetilde{M} = F \circ M \circ F^{-1}. \tag{3.40}$$

Since $\widetilde{M}$ is the composition of three Möbius transformations, it is itself a Möbius transformation. Furthermore, it follows immediately from the construction that the fixed points of $\widetilde{M}$ are 0 and ∞. But we have already seen that if a Möbius transformation leaves these points fixed, it can only be of the form

$$\widetilde{M}(\widetilde{z}) = m\widetilde{z},$$

where $m = \rho\, e^{i\alpha}$ is simply a complex number. Geometrically, $\widetilde{M}$ is just a rotation by α combined with an expansion by ρ.

This complex number m not only constitutes a complete description of the mapping $\widetilde{M}$ but, as we will see shortly, it also completely characterizes the geometric nature of the original Möbius transformation M. The number m is called the *multiplier* of $M(z)$.

3.7.2 Elliptic, Hyperbolic, and Loxodromic Transformations

Before reading on, refresh your memory of the classification (shown in [3.26a,b,c]) of Möbius transformations of the form $\widetilde{M}(\widetilde{z}) = m\,\widetilde{z}$.

We call $M(z)$ an *elliptic Möbius transformation* if $\widetilde{M}$ is elliptic, meaning that the latter is a pure rotation corresponding to $m = e^{i\alpha}$. Since $\widetilde{M}$ is a rotation if and only if it maps each origin-centred circle to itself, $M(z)$ is elliptic if and only if it maps each $\mathcal{C}_2$ circle to itself. With $\alpha = (\pi/3)$, the RHS of [3.29] illustrates the effect of $\widetilde{M}$ on the point $\widetilde{z}$. On the LHS you can see the corresponding, unambiguous effect of M: it moves z along its $\mathcal{C}_2$ circle till it lies on the $\mathcal{C}_1$ circle making angle $(\pi/3)$ with the original $\mathcal{C}_1$ through z.

Figure [3.30][12] is intended to give a more vivid impression of this same elliptic transformation. Each shaded "rectangle" is mapped by $M(z)$ to the next one in the direction of the arrows—some of these regions have been filled with black to emphasize this. This figure may be viewed as typical, with one exception. Because we have chosen $\alpha = (\pi/3)$, six successive applications of M yield the identity, and one therefore says that M has *period* 6. More generally, if $\alpha = (m/n)2\pi$, where (m/n) is a fraction reduced to lowest terms, then M has period n. Of course this is not typical. In general $(\alpha/2\pi)$ will be irrational, and no matter how many times we apply M we will never obtain the identity.

[12] Shading inspired by Ford (1929).

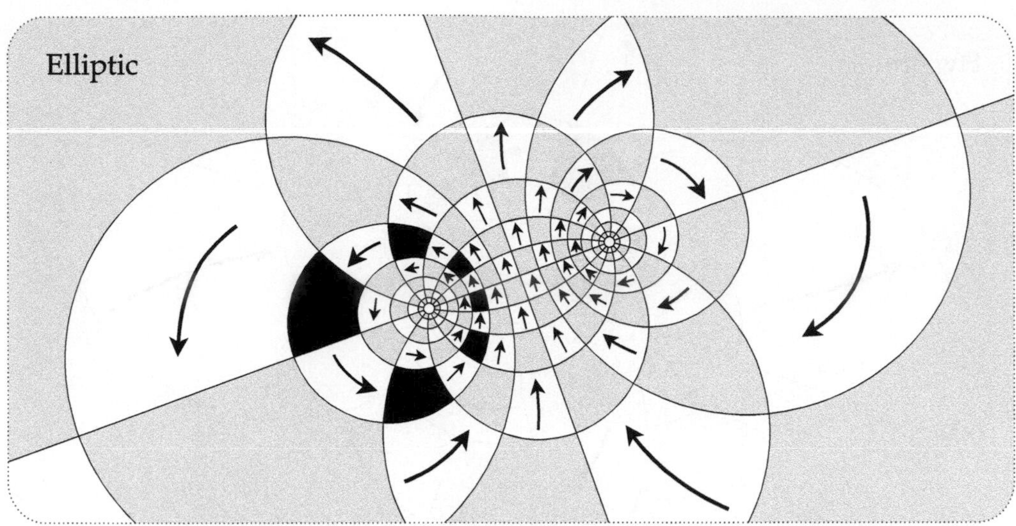

[3.30] An Elliptic Möbius transformation has multiplier $\mathfrak{m} = e^{i\alpha}$, *so that $\widetilde{M}$ is a pure rotation. See [3.26a]. Each shaded "rectangle" is mapped by $M(z)$ to the next one in the direction of the arrows—some of these regions have been filled with black to emphasize this. This figure may be viewed as typical, with one exception. Because we have chosen $\alpha = (\pi/3)$, six successive applications of M yield the identity, and one therefore says that M has* period 6.

We call $M(z)$ a *hyperbolic Möbius transformation* if $\widetilde{M}$ is hyperbolic, meaning that the latter is a pure expansion corresponding to $\mathfrak{m} = \rho \neq 1$. Since $\widetilde{M}$ is an expansion if and only if it maps each line through the origin to itself, $M(z)$ is hyperbolic if and only if it maps each $\mathcal{C}_1$ circle to itself. Figure [3.31] illustrates such a transformation with $\rho > 1$. Note that if we repeatedly apply this mapping then any shape (such as the small black square near ξ_+) is repelled away from ξ_+, eventually being sucked into ξ_-. In this case ξ_+ is called the *repulsive fixed point* and ξ_- is called the *attractive fixed point*; if $\mathfrak{m} = \rho < 1$ then the roles of ξ_+ and ξ_- are reversed.

Finally, if $\mathfrak{m} = \rho\, e^{i\alpha}$ has a general value, and $\widetilde{M}$ is the composition of both a rotation and an expansion, then M is called a *loxodromic Möbius transformation*. In this case neither the $\mathcal{C}_1$ circles, nor the $\mathcal{C}_2$ circles are invariant. The curves that *are* invariant are illustrated in [3.32], which also shows the effect of successive applications of M to a small square near ξ_+. In studying this figure, you may find it helpful to note that

> *The loxodromic Möbius transformation with fixed points $\xi_\pm$ and multiplier $\mathfrak{m} = \rho\, e^{i\alpha}$ is the composition (in either order) of (i) the elliptic Möbius transformation with multiplier $\mathfrak{m} = e^{i\alpha}$ and fixed points $\xi_\pm$; (ii) the hyperbolic Möbius transformation with multiplier $\mathfrak{m} = \rho$ and fixed points $\xi_\pm$.* (3.41)

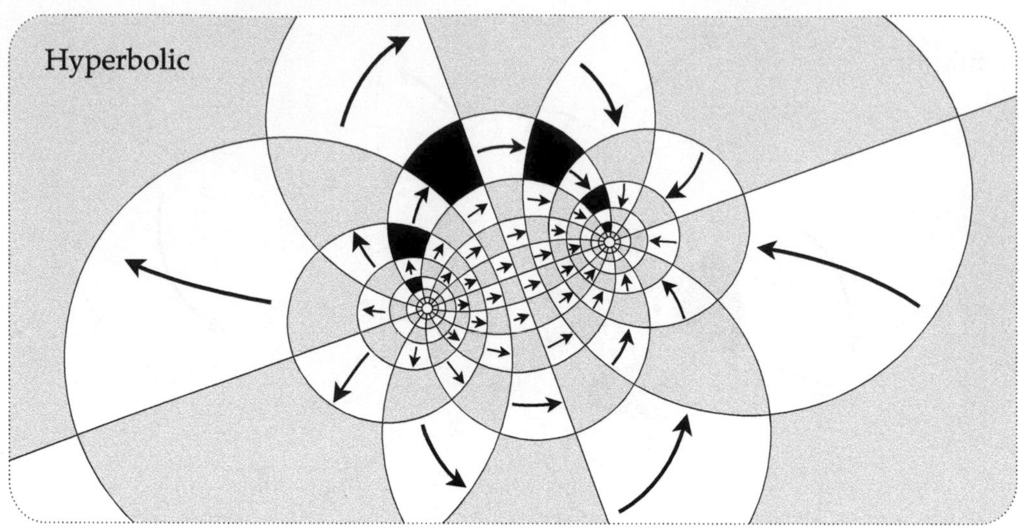

[3.31] A Hyperbolic Möbius transformation has multiplier $\mathfrak{m} = \rho \neq 1$, *so that* $\widetilde{M}$ *is a pure expansion. See [3.26b]. Iterating the mapping pushes everything away from the* repulsive fixed point *and sucks everything towards the* attractive fixed point.

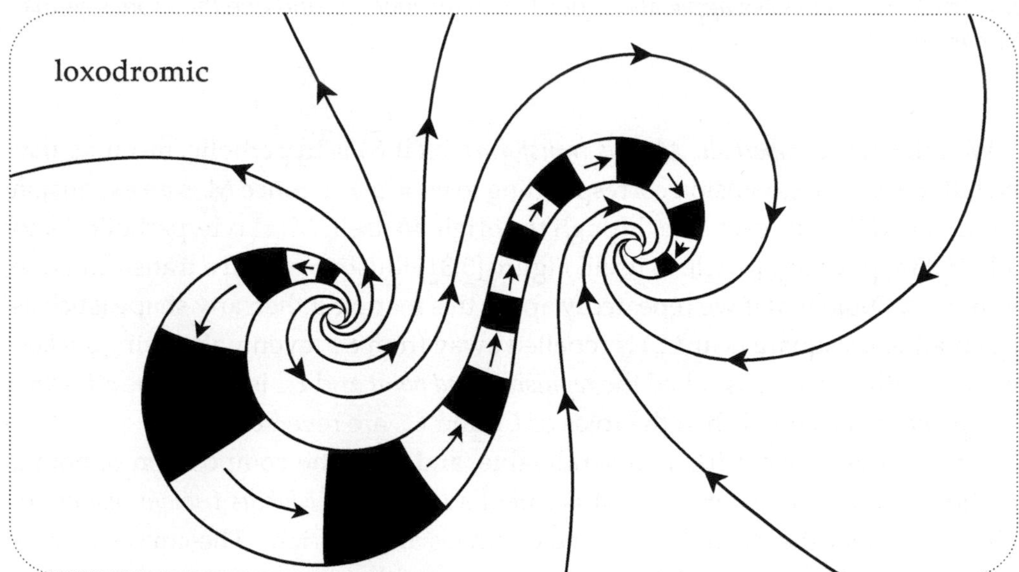

[3.32] A Loxodromoic Möbius transformation has general multiplier $\mathfrak{m} = \rho e^{i\alpha}$, *with* $\rho \neq 1$ *and* $\alpha \neq 0$*, so that* $\widetilde{M}$ *expands by* ρ *and rotates by* α*. See [3.26c]. Iterating the mapping pushes everything away from the* repulsive fixed point *and sucks everything towards the* attractive fixed point*, moving along the illustrated invariant curves.*

Just as in the case of a hyperbolic transformation, note that one fixed point is repulsive while the other is attractive. In this figure we have taken $\alpha > 0$ and $\rho > 1$; how would it look if α were negative, or if ρ were less than one?

3.7.3 Local Geometric Interpretation of the Multiplier

In [3.29] we *arbitrarily* elected to send ξ_+ to 0, rather than ξ_-. In this sense our definition of $\mathfrak{m}$ is clearly ambiguous. How would the new value of $\mathfrak{m}$ be related to the old one if we were to instead send ξ_- to 0?

Note that (3.40) may be expressed as $(F \circ M) = (\widetilde{M} \circ F)$. Writing $w = M(z)$, and recalling the definition of F, we therefore have

$$\frac{w - \xi_+}{w - \xi_-} = \mathfrak{m}\left(\frac{z - \xi_+}{z - \xi_-}\right). \tag{3.42}$$

[This formula is often called the *normal form* of the Möbius transformation.] Interchanging ξ_+ and ξ_- in this formula is equivalent to sending ξ_- to 0 and ξ_+ to ∞, in which case we obtain

$$\frac{w - \xi_+}{w - \xi_-} = \frac{1}{\mathfrak{m}}\left(\frac{z - \xi_+}{z - \xi_-}\right).$$

Thus the multiplier has changed from $\mathfrak{m}$ to $(1/\mathfrak{m})$, and both of these values can lay equal claim to being called "the" multiplier. Let us therefore refine our language and call the number $\mathfrak{m}$ occurring in (3.42) the multiplier *associated with* ξ_+; we will sometimes write it as $\mathfrak{m}_+$ to emphasize this. In these terms, we have just shown that the *multipliers associated with the two fixed points are the reciprocals of one another.* Let us try to understand this more geometrically.

Reconsider [3.29], in which the multiplier associated with ξ_+ is $\mathfrak{m} = e^{i(\pi/3)}$. We now seek to interpret $\mathfrak{m}$ directly in terms of [3.30], without the assistance of the RHS of [3.29]. The closer we are to ξ_+, the more closely do the members of $\mathcal{C}_2$ resemble tiny concentric circles centred at ξ_+. This is easy to understand: (A) as we examine smaller and smaller neighbourhoods of ξ_+, the $\mathcal{C}_1$ circles look more and more like their tangent lines at ξ_+; (B) by definition, each $\mathcal{C}_2$ cuts every $\mathcal{C}_1$ circle orthogonally.

From these remarks, it is now clear that the *local* effect of M (in an infinitesimal neighbourhood of ξ_+) is a rotation centred at ξ_+ through angle $(\pi/3)$—*this* is the meaning of the multiplier $\mathfrak{m}_+ = e^{i(\pi/3)}$ associated with ξ_+. Of course exactly the same reasoning applies to the infinitesimal neighbourhood of ξ_-, but we see from [3.30] that the positive rotation at ξ_+ forces an *equal and opposite* rotation at ξ_-. Thus the local effect of M in the neighbourhood of ξ_- is a rotation of $-(\pi/3)$, and the associated multiplier $\mathfrak{m}_-$ is $e^{-i(\pi/3)} = (1/\mathfrak{m}_+)$, as was to be explained.

If we look at [3.31], then we can see the same phenomenon at work in the case of a hyperbolic transformation. In this figure the multiplier associated with ξ_+ is $\mathfrak{m} = \rho > 1$, and this can now be interpreted as saying that the local effect of M in an infinitesimal neighbourhood of ξ_+ is an expansion centred at that point—we will verify in a moment that the "local expansion factor" is precisely ρ. It is also clear from the figure that the local effect of M in an infinitesimal neighbourhood of ξ_- is a *contraction*, so that the multiplier associated with that point is real and less than one. However, it is not so clear that this number is precisely $(1/\rho)$, as we know it must be. This too can be demonstrated geometrically, but let us instead content ourselves with showing how our original algebraic argument may be reinterpreted geometrically in terms of the "local effect" of M in the vicinity of each of the fixed points.

Let us write $Z = (z - \xi_+)$ and $W = (w - \xi_+)$ for the complex numbers emanating from ξ_+ connecting that point to z and to its image $w = M(z)$. We have claimed (and partially verified) that if Z is *infinitesimal* then the effect of M is to rotate Z by α and to expand it by ρ: in other words, $W = \mathfrak{m}\, Z$. To verify this, note that (3.42) can be rewritten as

$$\frac{W}{Z} = \mathfrak{m}\left(\frac{w - \xi_-}{z - \xi_-}\right).$$

As Z tends to zero, both z and w tend to ξ_+, and so the fraction on the right is ultimately equal to $\mathfrak{m}$. Thus W is ultimately equal to $\mathfrak{m}\, Z$, as was to be shown.

After you have read the next chapter, you will be able to look back at what we have just done and recognize it as an example of *differentiating* a complex function.

3.7.4 Parabolic Transformations

We now possess an excellent understanding of Möbius transformations with two fixed points, so all that remains is to treat the case where M has only one fixed point ξ, in which case M is called a *parabolic Möbius transformation*.

Consider the LHS of [3.33], but ignore the arrows for the time being. Here we have drawn two families of circles: the solid ones all pass through the fixed point ξ in one direction, and the dashed ones all pass through ξ in the perpendicular direction. Note that since the two types of circles are orthogonal at ξ, they are also

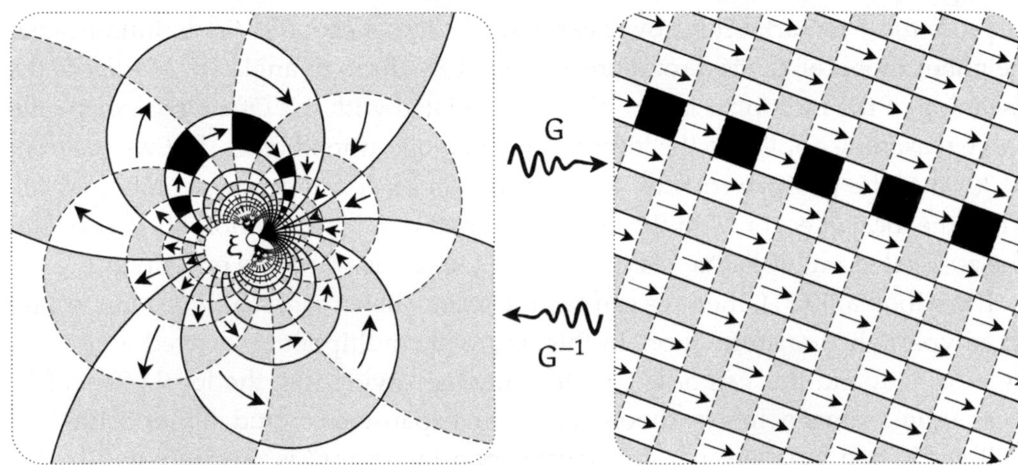

[3.33] A Parabolic Möbius transformation has a single fixed point, and $\widetilde{M}(\tilde{z}) = \tilde{z} + T$ *is a translation, with its single fixed point at ∞, as illustrated in [3.26d].*

(by symmetry) orthogonal at their second intersection point. The RHS illustrates what happens when we send ξ to ∞ by means of the Möbius transformation

$$G(z) = \frac{1}{z - \xi}.$$

Clearly [exercise], the two orthogonal families of circles become two orthogonal families of parallel lines. Conversely, if we apply G^{-1} to any two orthogonal families of lines on the right, then on the left we get two orthogonal families of circles through ξ.

As before, let $\widetilde{z} = G(z)$ and $\widetilde{w} = G(w)$ be the images on the RHS of z and $w = M(z)$. Thus the Möbius transformation $z \mapsto w = M(z)$ on the LHS induces another Möbius transformation $\widetilde{z} \mapsto \widetilde{w} = \widetilde{M}(\widetilde{z})$ on the RHS, where

$$\widetilde{M} = G \circ M \circ G^{-1}.$$

Since ∞ is the sole fixed point of $\widetilde{M}$, we deduce that $\widetilde{M}$ can only be a *translation*:

$$\widetilde{M}(\widetilde{z}) = \widetilde{z} + T.$$

Now suppose that the arrows on the RHS of [3.33] represent the direction of the translation T. As illustrated, we now draw a grid aligned with T, each shaded square being carried into the next by $\widetilde{M}$. On the LHS of [3.33] we thus obtain a vivid picture of the action of the original parabolic Möbius transformation M: each solid circle is carried into itself; each dashed circle is carried into another dashed circle; and each shaded region is carried into the next in the direction of the arrows.

If $M(z) = \frac{az+b}{cz+d}$ is *normalized*, then we know from (3.28) that it is parabolic if and only if $(a+d) = \pm 2$, in which case $\xi = (a-d)/2c$. Now let us determine the corresponding translation T in terms of the coefficients. Since $(G \circ M) = (\widetilde{M} \circ G)$, the so-called normal form of M is given by

$$\frac{1}{w - \xi} = \frac{1}{z - \xi} + T.$$

Since M maps $z = \infty$ to $w = (a/c)$, we deduce that

$$T = \frac{1}{(a/c) - \xi} = \pm c,$$

where the "$\pm$" is the arbitrarily chosen sign of $(a+d)$.

3.7.5 Computing the Multiplier*

We have seen how the multiplier m determines the character of a Möbius transformation, and we now show how we can determine the character of m directly from the coefficients of $M(z) = \frac{az+b}{cz+d}$.

Suppose we have already calculated the fixed points $\xi_{\pm}$ using (3.28), for example. Since M maps $z = \infty$ to $w = (a/c)$, we deduce from the normal form

(3.42) that the multiplier associated with ξ_+ is

$$\mathfrak{m} = \frac{a - c\,\xi_+}{a - c\,\xi_-}. \tag{3.43}$$

For example, consider complex inversion, $z \mapsto (1/z)$. The fixed points are the solutions of $z = (1/z)$, namely, $\xi_\pm = \pm 1$. Thus the multiplier associated with $\xi_+ = 1$ is $\mathfrak{m} = -1 = e^{i\pi}$, which happens to be the same as the multiplier $(1/\mathfrak{m})$ associated with $\xi_- = -1$. Thus complex inversion is elliptic, and an infinitesimal neighbourhood of either fixed point is simply rotated about that point through angle π. Try using a computer to check this prediction.

If desired, we can obtain a completely explicit formula for $\mathfrak{m}$ by substituting (3.28) into (3.43). If we only want to know the character of the Möbius transformation, then we can proceed as follows.

It turns out—we will prove it in a moment—that $\mathfrak{m}$ is related to the coefficients of the *normalized* Möbius transformation by the equation,

$$\sqrt{\mathfrak{m}} + \frac{1}{\sqrt{\mathfrak{m}}} = a + d. \tag{3.44}$$

Note that the symmetry of this equation implies that if $\mathfrak{m}$ is a solution, then so is $(1/\mathfrak{m})$; this is just as it should be. Without bothering to solve (3.44) for $\mathfrak{m}$, we now obtain [exercise] the following algebraic classification: *The normalized Möbius transformation* $M(z) = \frac{az+b}{cz+d}$ *is*

$$\left.\begin{array}{ll} \textit{elliptic}, & \text{iff } (a+d) \text{ is real and } |a+d| < 2; \\ \textit{parabolic}, & \text{iff } (a+d) = \pm 2; \\ \textit{hyperbolic}, & \text{iff } (a+d) \text{ is real and } |a+d| > 2; \\ \textit{loxodromic}, & \text{iff } (a+d) \text{ is complex.} \end{array}\right\} \tag{3.45}$$

Hint: you can get a better feel for this by sketching the graph of $y = x + (1/x)$.

In order to derive (3.44) elegantly, let us use matrices. Rewriting (3.40),

$$[\widetilde{M}] = [F]\,[M]\,[F]^{-1} \quad\Longrightarrow\quad \det[\widetilde{M}] = \det\left\{[F][F]^{-1}\right\}\det[M] = \det[M].$$

Thus, regardless of whether or not $[F]$ is normalized, $[M]$ is normalized if and only if $[\widetilde{M}]$ is normalized. Since $\widetilde{M}(z) = \mathfrak{m}\,z$, its normalized matrix is [exercise] $[\widetilde{M}] = \begin{bmatrix} \sqrt{\mathfrak{m}} & 0 \\ 0 & 1/\sqrt{\mathfrak{m}} \end{bmatrix}$. Recalling (3.36), we deduce that

$$\sqrt{\mathfrak{m}} + \frac{1}{\sqrt{\mathfrak{m}}} = \operatorname{tr}\left\{[F]\,[M]\,[F]^{-1}\right\} = \operatorname{tr}\left\{[F]^{-1}[F][M]\right\} = \operatorname{tr}[M] = a + d,$$

as was to be shown.

3.7.6 *Eigenvalue Interpretation of the Multiplier**

If $[M]$ is a linear transformation of $\mathbb{C}^2$, then we saw in (3.33) that its eigenvectors are the homogeneous coordinates of the fixed points of the corresponding

Möbius transformation $M(z)$. We also claimed that if $[M]$ is normalized then the eigenvalues completely determine the character of $M(z)$. We can now be more precise:

If a fixed point of $M(z)$ *is represented as an eigenvector (with eigenvalue* λ*) of the normalized matrix* $[M]$*, then the multiplier* m *associated with the fixed point is given by* $m = 1/\lambda^2$. (3.46)

Before proving this result, we illustrate it with the example of complex inversion, $z \mapsto (1/z)$. We already know that the fixed points are ± 1, that the associated multipliers are both given by $m = -1$, and we easily find [exercise] that the normalized matrix is $\begin{bmatrix} 0 & i \\ i & 0 \end{bmatrix}$. If we choose the homogeneous coordinate vector of a finite point z to be $\begin{bmatrix} z \\ 1 \end{bmatrix}$, then the eigenvectors corresponding to the fixed points $z = \pm 1$ are $\begin{bmatrix} \pm 1 \\ 1 \end{bmatrix}$. Since

$$\begin{bmatrix} 0 & i \\ i & 0 \end{bmatrix}\begin{bmatrix} 1 \\ 1 \end{bmatrix} = i\begin{bmatrix} 1 \\ 1 \end{bmatrix} \quad \text{and} \quad \begin{bmatrix} 0 & i \\ i & 0 \end{bmatrix}\begin{bmatrix} -1 \\ 1 \end{bmatrix} = -i\begin{bmatrix} -1 \\ 1 \end{bmatrix},$$

we see that the eigenvalues are given by $\lambda = \pm i$, in agreement with (3.46).

Returning to the general case, comparison of (3.34) and (3.44) reveals that $\sqrt{m}$ and λ satisfy the same quadratic, so we immediately deduce most of (3.46): the two reciprocal values of m are equal to the two reciprocal values of λ^2. However, this does not tell us which value of λ^2 yields which value of m, nor is this line of attack very illuminating. Here, then, is a more transparent approach.

We begin by recalling a standard result of linear algebra, which is valid for $n \times n$ matrices:

If $\mathbf{e}$ *is an eigenvector of* $[A]$ *with eigenvalue* λ, *then* $\widetilde{\mathbf{e}} \equiv [B]\mathbf{e}$ *is an eigenvector of* $[\widetilde{A}] \equiv [B][A][B]^{-1}$, *and its eigenvalue is also* λ.

This is verified easily:

$$[\widetilde{A}]\,\widetilde{\mathbf{e}} = \left\{[B][A][B]^{-1}\right\}[B]\mathbf{e} = [B][A]\mathbf{e} = [B]\lambda\mathbf{e} = \lambda\,\widetilde{\mathbf{e}}.$$

Let us return to [3.29], in which the fixed point ξ_+ of M (with associated multiplier m_+) was mapped to the fixed point 0 of $\widetilde{M} = (F \circ M \circ F^{-1})$ by means of $z \mapsto \widetilde{z} = F(z) = \frac{z-\xi_+}{z-\xi_-}$. In terms of linear transformations of $\mathbb{C}^2$, the eigenvector $\begin{bmatrix} \xi_+ \\ 1 \end{bmatrix}$ of $[M]$ is being mapped by $[F]$ to the eigenvector $\begin{bmatrix} 0 \\ 1 \end{bmatrix}$ of

$$[\widetilde{M}] = [F]\,[M]\,[F]^{-1}.$$

The linear algebra result now tells us that if λ_+ denotes the eigenvalue of $\begin{bmatrix} \xi_+ \\ 1 \end{bmatrix}$, then

$$[\widetilde{M}] \begin{bmatrix} 0 \\ 1 \end{bmatrix} = \lambda_+ \begin{bmatrix} 0 \\ 1 \end{bmatrix}.$$

This is true irrespective of whether or not any of the matrices in the above equation are normalized.

Now suppose that [M] is normalized, as demanded in (3.46). Irrespective of whether or not [F] is normalized, we have already noted that [M] is normalized if and only if $[\widetilde{M}]$ is normalized. Since the normalized matrix of $\widetilde{M}(\widetilde{z}) = m_+ \widetilde{z}$ is given by $[\widetilde{M}] = \begin{bmatrix} \sqrt{m_+} & 0 \\ 0 & 1/\sqrt{m_+} \end{bmatrix}$, we deduce that

$$\lambda_+ \begin{bmatrix} 0 \\ 1 \end{bmatrix} = \begin{bmatrix} \sqrt{m_+} & 0 \\ 0 & 1/\sqrt{m_+} \end{bmatrix} \begin{bmatrix} 0 \\ 1 \end{bmatrix} = \frac{1}{\sqrt{m_+}} \begin{bmatrix} 0 \\ 1 \end{bmatrix}.$$

Thus $m_+ = 1/\lambda_+^2$, as was to be shown.

3.8 Decomposition into 2 or 4 Reflections*

3.8.1 *Introduction*

Recall from (3.4) that the formula for inversion or "reflection" in a circle K has the form

$$\mathcal{I}_K(z) = \frac{A\,\overline{z} + B}{C\,\overline{z} + D}.$$

It follows easily that the composition of any two reflections (in circles or lines) is a Möbius transformation. Since the composition of two Möbius transformations is another Möbius transformation, it follows more generally that *the composition of an even number of reflections is a Möbius transformation.*

Conversely, in this section we will use the Symmetry Principle [see p. 168] to show that

> *Every non-loxodromic Möbius transformation can be expressed as the composition of two reflections, and every loxodromic Möbius transformation can be expressed as the composition of four reflections.*

In the following, it would be helpful (but not essential) for you to have read the final section of Chapter 1.

3.8.2 *Elliptic Case*

Consider [3.34], which depicts the same elliptic transformation shown in [3.29] and [3.30]. Recall that the LHS shows a Möbius transformation M such that after

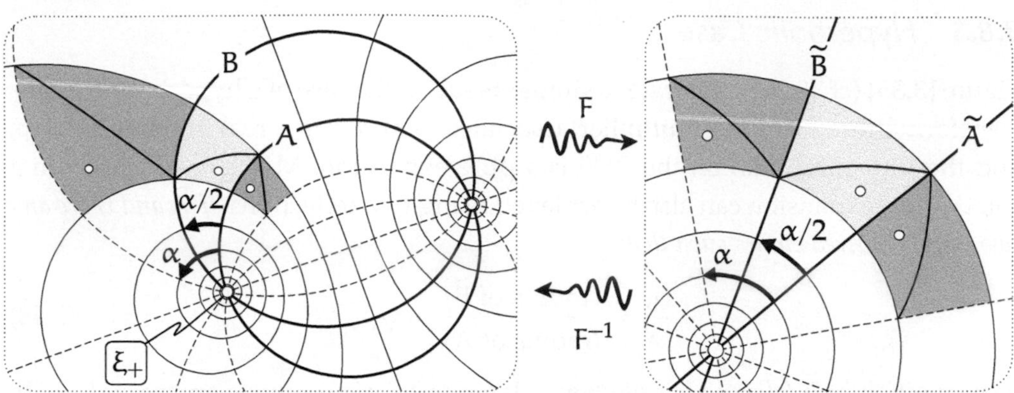

[3.34] Decomposition of an Elliptic Möbius transformation into two reflections. *If* M *is an elliptic Möbius transformation, and the multiplier associated with one of the fixed points* ξ_+ *is* $\mathfrak{m} = e^{i\alpha}$, *then* $\mathrm{M} = \mathfrak{I}_{\mathrm{B}} \circ \mathfrak{I}_{\mathrm{A}}$, *where* A *and* B *are any two circles through the fixed points such that the angle from* A *to* B *at* ξ_+ *is* $(\alpha/2)$.

sending ξ_+ and ξ_- to 0 and ∞ by means of $F(z) = (z-\xi_+)/(z-\xi_-)$, the new transformation on the RHS is a pure rotation $\widetilde{\mathrm{M}}(\widetilde{z}) = e^{i\alpha}\,\widetilde{z}$. In the illustrated example, $\alpha = (\pi/3)$ and the dark "rectangle" abutting the line $\widetilde{\mathrm{A}}$ is carried into the dark "rectangle" abutting the line $\widetilde{\mathrm{B}}$.

As we discussed in Chapter 1 [see p. 42], this origin-centred rotation of α is equivalent to successively reflecting in any two lines containing angle $(\alpha/2)$ at 0, such as the illustrated lines $\widetilde{\mathrm{A}}$ and $\widetilde{\mathrm{B}}$. In symbols,

$$\widetilde{\mathrm{M}} = \mathfrak{R}_{\widetilde{\mathrm{B}}} \circ \mathfrak{R}_{\widetilde{\mathrm{A}}}.$$

In particular, $\mathfrak{R}_{\widetilde{\mathrm{A}}}$ maps the dark "rectangle" abutting the line $\widetilde{\mathrm{A}}$ to the light "rectangle", then $\mathfrak{R}_{\widetilde{\mathrm{B}}}$ maps this to the dark "rectangle" abutting the line $\widetilde{\mathrm{B}}$. The figure tries to make this clear by also showing the successive images of both a point and a diagonal circular arc of the original dark "rectangle".

Now think what this means on the LHS of [3.34]. The Symmetry Principle tells us that if two points are symmetric with respect to the line $\widetilde{\mathrm{A}}$ then their images under the Möbius transformation F^{-1} are symmetric with respect to the circle $\mathrm{A} = F^{-1}(\widetilde{\mathrm{A}})$ through the fixed points. [Recall that in [3.29] the family of such circles was called $\mathcal{C}_1$.] Thus reflection in $\widetilde{\mathrm{A}}$ on the RHS becomes reflection (i.e., inversion) in A on the LHS. Of course the same goes for the second reflection in $\widetilde{\mathrm{B}}$. Thus we have shown the following:

> *If* M *is an elliptic Möbius transformation, and the multiplier associated with one of the fixed points* ξ_+ *is* $\mathfrak{m} = e^{i\alpha}$, *then* $\mathrm{M} = \mathfrak{I}_{\mathrm{B}} \circ \mathfrak{I}_{\mathrm{A}}$ *where* A *and* B *are any two circles through the fixed points such that the angle from* A *to* B *at* ξ_+ *is* $(\alpha/2)$. (3.47)

3.8.3 Hyperbolic Case

Figure [3.35] (cf. [3.31]) illustrates a similar result in the case of a hyperbolic Möbius transformation. Here the multiplier associated with ξ_+ is a real number $m = \rho$, and the transformation on the RHS is a pure expansion, $\widetilde{M}(\widetilde{z}) = \rho\,\widetilde{z}$. As with a rotation, an expansion can also be achieved using two reflections: *if* $\widetilde{A}$ *and* $\widetilde{B}$ *are any two origin-centred circles such that*

$$\frac{r_B}{r_A} = \frac{(\text{radius of } \widetilde{B})}{(\text{radius of } \widetilde{A})} = \sqrt{\rho}, \tag{3.48}$$

then reflection in $\widetilde{A}$ *followed by reflection in* $\widetilde{B}$ *yields an origin-centred expansion by* ρ. In symbols, this result—which is really the same as (3.8)—says that

$$\widetilde{M} = \mathcal{I}_{\widetilde{B}} \circ \mathcal{I}_{\widetilde{A}}.$$

As in [3.34], the RHS of [3.35] illustrates the successive effect of these two reflections on a dark rectangle abutting $\widetilde{A}$. Just as before, the Symmetry Principle applied to F^{-1} tells us that the original Möbius transformation on the LHS can be expressed as

$$M = \mathcal{I}_B \circ \mathcal{I}_A.$$

Recall from [3.29] that A and B belong to the family $\mathcal{C}_2$ of circles orthogonal to the family $\mathcal{C}_1$ of circles through the fixed points. At the time, we pointed out an equivalent property of $\mathcal{C}_2$, namely, that the fixed points $\xi_\pm$ are symmetric with respect to each member of $\mathcal{C}_2$; this enables us to explain how it is that $(\mathcal{I}_B \circ \mathcal{I}_A)$ leaves ξ_+ and ξ_- fixed. In the case of [3.34], this was obvious because each reflection separately left those points fixed; in the present case, however, $\mathcal{I}_A$ *swaps* the points, then $\mathcal{I}_B$ swaps them back again, the net effect being to leave them fixed.

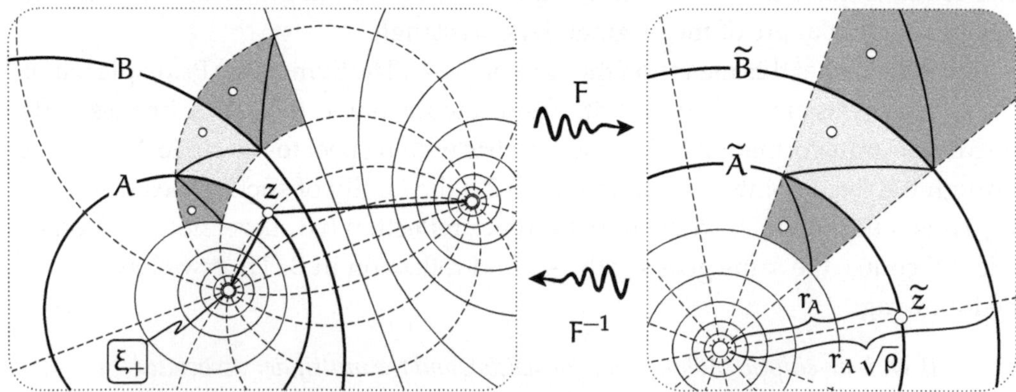

[3.35] Decomposition of a Hyperbolic Möbius transformation into two reflections. *If* M *is a hyperbolic Möbius transformation, and the multiplier associated with one of the fixed points* ξ_+ *is* $m = \rho$*, then* $M = \mathcal{I}_B \circ \mathcal{I}_A$*, where* A *and* B *are any two circles of Apollonius with limit points* $\xi_\pm$ *such that* $(r_B/r_A) = \sqrt{\rho}$.

In the case of an elliptic transformation, (3.47) describes how to pick out a pair of $\mathcal{C}_1$ circles corresponding to any given angle α. In the present case of a hyperbolic transformation, how are we to pick out a pair of $\mathcal{C}_2$ circles corresponding to any given value of ρ? The answer depends on a third characterizing property of the $\mathcal{C}_2$ circles: they are the *circles of Apollonius* with *limit points* $\xi_{\pm}$.

This terminology reflects Apollonius' remarkable discovery (c. 250 BCE) that if a point z moves in such a way that the ratio of the distances of z from two fixed points $\xi_{\pm}$ remains constant, then z moves on a *circle*. Figure [3.35] makes this easy to understand. As z travels round A, $\widetilde{z} = F(z)$ travels round the origin-centred circle $\widetilde{A}$ of radius r_A. But this constant r_A is none other than the ratio of the distances of z from two fixed points $\xi_{\pm}$:

$$r_A = |\widetilde{z}| = |F(z)| = \frac{|z - \xi_+|}{|z - \xi_-|}.$$

Note that this also explains the "limit point" terminology: as the ratio r_A tends to 0, the corresponding Apollonian circle A shrinks down towards the limit point ξ_+; as r_A tends to infinity, A shrinks down towards the other limit point ξ_-. Another bonus of our discussion is a result that is frequently not mentioned in geometry texts: *the limit points defining a family of Apollonian circles are symmetric with respect to each of these circles.*

Since the quantities r_A and r_B occurring in (3.48) are now expressible purely in terms of the geometry of the LHS of [3.35], we have solved the problem of picking an appropriate pair of $\mathcal{C}_2$ circles:

> *If* M *is a hyperbolic Möbius transformation, and the multiplier associated with one of the fixed points* ξ_+ *is* $m = \rho$, *then* $M = \mathfrak{I}_B \circ \mathfrak{I}_A$, *where* A *and* B *are any two circles of Apollonius with limit points* $\xi_{\pm}$ *such that* $(r_B/r_A) = \sqrt{\rho}$.

3.8.4 Parabolic Case

Figure [3.36] is a modified copy of [3.33], and it illustrates how the same idea applies to a parabolic transformation. Recall that after we have sent the solitary fixed point ξ to ∞ by means of the Möbius transformation $z \mapsto \widetilde{z} = G(z) = 1/(z-\xi)$, the new transformation on the RHS is a translation, $\widetilde{M}(\widetilde{z}) = \widetilde{z} + T$.

As we discussed on p. 42, this translation can be expressed as $\widetilde{M} = \mathfrak{R}_{\widetilde{B}} \circ \mathfrak{R}_{\widetilde{A}}$, where $\widetilde{A}$ and $\widetilde{B}$ are any two parallel lines such that the perpendicular connecting complex number from $\widetilde{A}$ to $\widetilde{B}$ is $(T/2)$. Applying the Symmetry Principle to the Möbius transformation G^{-1}, we deduce that (on the LHS)

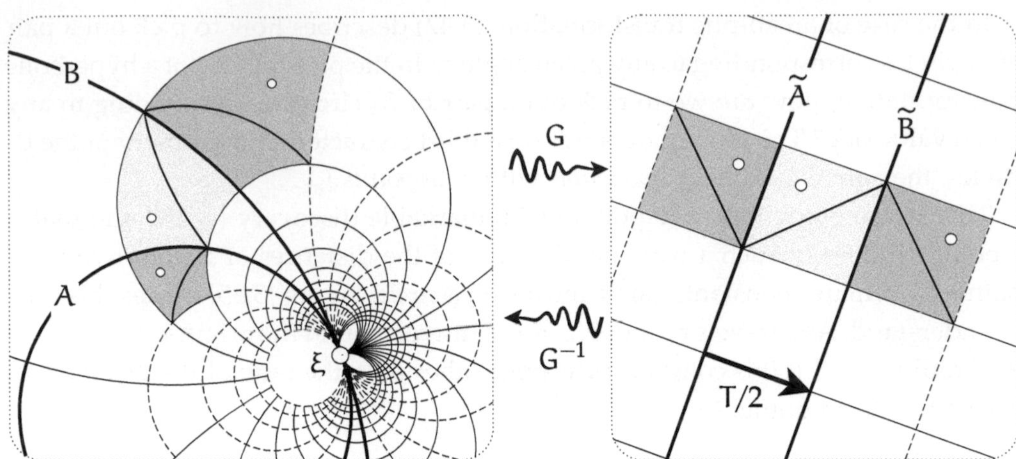

[3.36] Decomposition of a Parabolic Möbius transformation into two reflections. *A parabolic Möbius transformation* M *with fixed point* ξ *can be expressed as* $M = \mathfrak{I}_B \circ \mathfrak{I}_A$, *where* A *and* B *are circles that touch each other at* ξ.

> *A parabolic Möbius transformation* M *with fixed point* ξ *can be expressed as* $M = \mathfrak{I}_B \circ \mathfrak{I}_A$, *where* A *and* B *are circles that touch each other at* ξ.

3.8.5 Summary

Lest the details obscure the simplicity of what we have discovered, we summarize our results as follows:

> *A non-loxodromic Möbius transformation* M *can always be decomposed into two reflections in circles* A *and* B *that are orthogonal to the invariant circles of* M. *Furthermore,* M *is elliptic, parabolic, or hyperbolic according as* A *and* B *intersect, touch, or do not intersect.* (3.49)

Recalling (3.41), we also deduce that *a loxodromic Möbius transformation* M *can always be decomposed into four reflections in circles:*

$$M = \{\mathfrak{I}_{B'} \circ \mathfrak{I}_{A'}\} \circ \{\mathfrak{I}_B \circ \mathfrak{I}_A\} = \{\mathfrak{I}_B \circ \mathfrak{I}_A\} \circ \{\mathfrak{I}_{B'} \circ \mathfrak{I}_{A'}\},$$

where A *and* B *both pass through the fixed points, and where* A′ *and* B′ *are both orthogonal to* A *and* B.

We should stress that these results concern the *least* number of reflections into which a Möbius transformation can be decomposed. Thus if a particular Möbius transformation is expressible as the composition of four reflections, this does *not* necessarily imply that it is loxodromic—one might be able to reduce the number of reflections from four to two. For example, if A and B are lines containing angle $(\pi/12)$ at 0, and A′ and B′ are lines containing angle $(\pi/6)$ at 0, then the

Möbius transformation $(\mathfrak{R}_{B'} \circ \mathfrak{R}_{A'} \circ \mathfrak{R}_B \circ \mathfrak{R}_A)$ represents a rotation of $(\pi/2)$, which can be reduced to two reflections in lines containing angle $(\pi/4)$. As a more extreme example of this idea of redundant reflections, check for yourself that (3.3) represents a decomposition of a general Möbius transformation into *ten* reflections!

3.9 Automorphisms of the Unit Disc*

3.9.1 Counting Degrees of Freedom

An *automorphism* of a region R of the complex plane is a one-to-one, conformal mapping of R to itself. If R is a disc (or a half-plane) then clearly we can map it to itself with a Möbius transformation M, and since M is one-to-one and conformal, it is (by definition) an automorphism. In this subsection we will find all possible Möbius automorphisms of the unit disc. These Möbius transformations are important for at least two reasons: (i) in Chapter 6 we will see that they play a central role in non-Euclidean geometry; (ii) in Chapter 7 we will see that they are the *only* automorphisms of the disc!

In the following, let C denote the unit circle, let D denote the unit disc (including C), and let $M(z)$ denote a Möbius transformation of D to itself. Before we try to find a formula for the most general M, let us see "how many" such Möbius transformations there are. In other words, how many real numbers (*parameters*) are required to specify a particular M?

To illustrate how such counting may be done, let us first show that the set of *all* Möbius transformations forms a "six parameter family". Once we have chosen three points in $\mathbb{C}$, there is a unique Möbius transformation that maps them to three arbitrary image points, and each of these 3 image points $w = u + iv$ requires 2 real numbers (u and v) for its specification. If we think of the three original points as having fixed locations, and the three image points as freely movable, then the total number of parameters needed to specify a particular Möbius transformation is thus $3 \times 2 = 6$. Another suggestive way of describing this fact is to say that the most general Möbius transformation has six *degrees of freedom*.

Returning to the original problem, it is clear that we will lose some of these six degrees of freedom when we impose the condition that $M(z)$ map D to itself. In fact we lose half of them:

$$\textit{Möbius automorphisms of } \mathrm{D} \textit{ have three degrees of freedom.} \tag{3.50}$$

Figure [3.37a] gives one way of seeing this. Here q, r, s may be viewed as having fixed locations on C, while $\widetilde{q}, \widetilde{r}, \widetilde{s}$ are thought of as freely movable. Provided (as illustrated) that $\widetilde{q}, \widetilde{r}, \widetilde{s}$ induce the same orientation of C as q, r, s, we know from

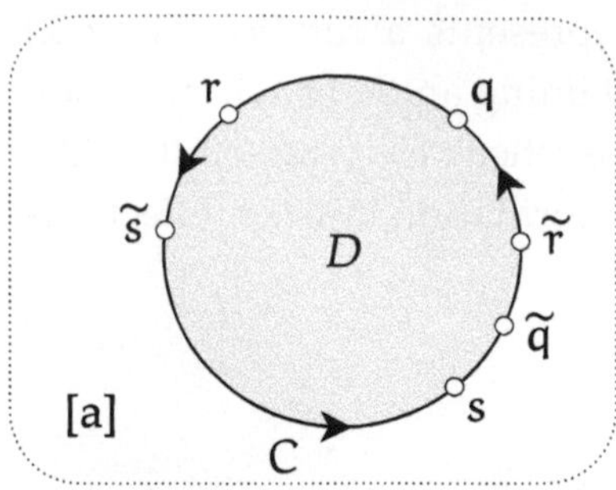

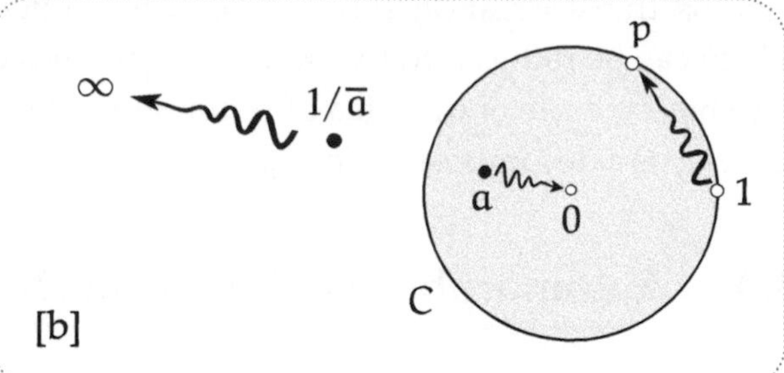

[3.37] [a] Möbius automorphisms of D have three real degrees of freedom. *Think of* q, r, s *as fixed on* C, *but* $\widetilde{q}$, $\widetilde{r}$, $\widetilde{s}$ *as freely movable, their locations specified by* three angles. *Then we know from (3.31) that the* unique *Möbius automorphism of* D *mapping* q, r, s *to* $\widetilde{q}$, $\widetilde{r}$, $\widetilde{s}$, *is given by* $z \mapsto \widetilde{z} = M(z)$, *where* $[\widetilde{z}, \widetilde{q}, \widetilde{r}, \widetilde{s}] = [z, q, r, s]$.
[b] Finding the most general Möbius automorphism via the Symmetry Principle. *If a Möbius automorphism of* D *sends* a *to* 0, *then the Symmetry Principle tells us that it sends* $(1/\overline{a})$ *to* ∞. *It then follows easily [see text] that the most general Möbius automorphism of* D *is* $M_a^{\phi}(z) = e^{i\phi}\left[\frac{z-a}{\overline{a}z-1}\right]$, *which does indeed have three real degrees of freedom.*

(3.31) that the unique Möbius automorphism of D mapping q, r, s to $\widetilde{q}, \widetilde{r}, \widetilde{s}$, is given by $z \mapsto \widetilde{z} = M(z)$, where

$$[\widetilde{z}, \widetilde{q}, \widetilde{r}, \widetilde{s}] = [z, q, r, s].$$

Since *three* real numbers are needed to specify $\widetilde{q}, \widetilde{r}, \widetilde{s}$—their angles, for example—this establishes (3.50).

3.9.2 Finding the Formula via the Symmetry Principle

According to (3.50), the specification of a particular M requires three bits of information. However, we are not obliged to give this information in the form of three points on C—any data that are equivalent to three real numbers will do equally well. A particularly useful alternative of this kind is shown in [3.37b]. We specify which point a inside D is to be mapped to the origin, and we also specify which point p on C is to be the image of the point 1 (or of some other definite point on C). Choosing a uses up two degrees of freedom; choosing p uses up the third and last degree of freedom.

Before pursuing this, we note another consequence of (3.50): we cannot generally find a Möbius automorphism that simultaneously sends the interior point a to 0 *and* sends another interior point to some other interior point. These requirements amount to *four* conditions on M, while (3.50) tells us that only *three* such conditions can be accommodated. It is very much as if we were seeking to draw a circle through four arbitrary points—it can't be done! However, suppose in this analogy that we are lucky, and that the four points just happen to be concyclic, then the circle that passes through them is *unique*. By the same token,

If two Möbius automorphisms M and N map two interior points to the same image points, then $M = N$. (3.51)

Returning to [3.37b], note that since C is mapped to itself by M, the Symmetry Principle tells us that if a pair of points are symmetric with respect to C, then so are their images. Now we apply this to the symmetric pair of points, a and $(1/\overline{a})$ shown in [3.37b]. Since a is mapped to 0, $(1/\overline{a})$ must be mapped to the reflection of 0 in C, namely, ∞. Thus M must have the form

$$M(z) = k\left(\frac{z-a}{\overline{a}z-1}\right),$$

where k is a constant. Finally, we require that $p = M(1)$ be a point on C, so

$$1 = |p| = |k|\frac{|1-a|}{|\overline{a}-1|} = |k| \qquad \Longrightarrow \qquad k = e^{i\phi}.$$

Thus the choice of p is equivalent to the choice of ϕ. Using the angle ϕ and the point a to label the transformation, we have discovered that the most general Möbius automorphism of D is

$$M_a^{\phi}(z) = e^{i\phi}\left(\frac{z-a}{\overline{a}z-1}\right). \tag{3.52}$$

Note that $M_0^{\phi}(z) = -e^{i\phi}z = e^{i(\pi+\phi)}z$ simply rotates D about its centre 0 through angle $(\pi+\phi)$. The general Möbius automorphism M_a^{ϕ} may be interpreted as M_a^0 followed by a rotation of ϕ, and from this point of view the really interesting part of the transformation is M_a^0, which we will now abbreviate to M_a. This is the same M_a whose properties you were asked to investigate algebraically in Chapter 2, Ex. 3.

3.9.3 *Interpreting the Simplest Formula Geometrically**

To find the geometric meaning of

$$M_a(z) = \frac{z-a}{\overline{a}z-1}, \tag{3.53}$$

we could simply apply our whole arsenal of classification techniques. We ask that you try this yourself in Ex. 26.

Here we will instead attempt to make sense of M_a "with our bare hands", as it were. This is probably more illuminating, and it certainly provides better geometric sport! Begin by noting that M_a has the property that it *swaps* a and 0: not only is $M_a(a) = 0$, but also $M_a(0) = a$. According to (3.51), this is the *only* Möbius automorphism with this property, so if we can geometrically construct a Möbius automorphism that swaps a and 0, then it must *be* M_a.

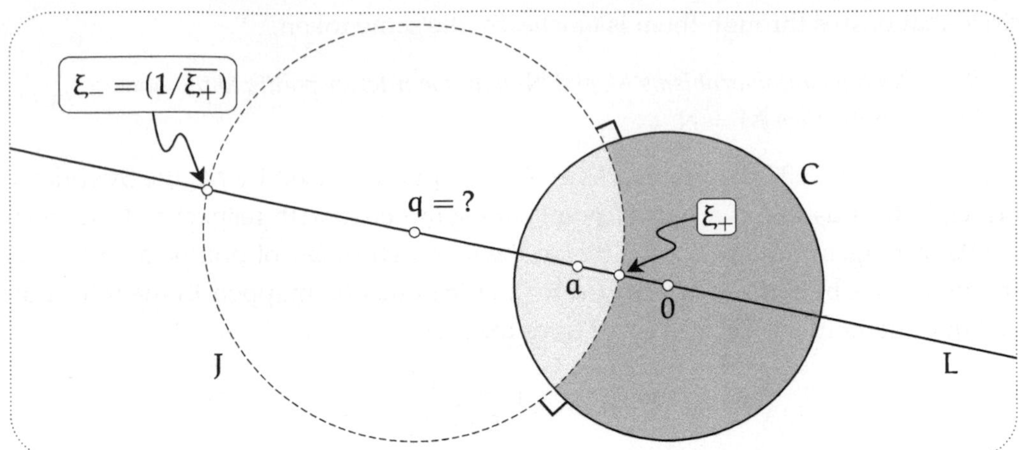

[3.38] Geometric meaning of the most general Möbius automorphism. *The most general Möbius automorphism $M_a^{\phi}(z)$ (above) is the composition of a rotation with the fundamental mapping $M_a(z) = \frac{z-a}{\overline{a}z-1}$. To discover the geometric* meaning *of $M_a(z)$, we first observe that M_a* swaps *a and 0. The Symmetry Principle then guides us [see text] to the answer: $M_a = \mathfrak{R}_L \circ \mathfrak{I}_J$, the order of the reflections being immaterial. The fact that M_a swaps a and 0 can now be recognized as a special case of the fact that M_a is* involutory*: $(M_a \circ M_a) = \mathcal{E}$, and* every *pair of points $\{z, M_a(z)\}$ is swapped by M_a.*

As was explained earlier in [3.6] on page 145, the reflection $\mathfrak{I}_J$ in any circle J orthogonal to C will map D to itself, the two regions into which D is divided by J being swapped. See [3.38]. At this point the obvious thing to do is to find the circle J such that $\mathfrak{I}_J$ swaps a and 0. Clearly the centre q of J must lie on the line L through a and 0, but where?

We can answer this question with the same symmetry argument that we used earlier. Since a and $(1/\overline{a})$ are symmetric with respect to C, their images under $\mathfrak{I}_J$ are symmetric with respect to $\mathfrak{I}_J(C) = C$. Because we want $\mathfrak{I}_J(a) = 0$, we deduce that $\mathfrak{I}_J(1/\overline{a}) = \infty$. But the point that is mapped to infinity by $\mathfrak{I}_J$ is the centre of J, so $q = (1/\overline{a})$.

Of course $\mathfrak{I}_J$ is an anticonformal mapping; to obtain a conformal Möbius automorphism we must compose it with another reflection. However, we have already successfully swapped a and 0, so this second reflection must leave these points fixed. The obvious (and only) choice is thus reflection in L. Here, then, is our geometric interpretation of M_a:

$$M_a = \mathfrak{R}_L \circ \mathfrak{I}_J.$$

Incidentally, observe [exercise] that the order of these reflections doesn't matter: we may also write $M_a = \mathfrak{I}_J \circ \mathfrak{R}_L$.

Clearly the fixed points $\xi_\pm$ are the intersection points of J and L, and so they are symmetric with respect to C. Since the reflections occur in orthogonal circles

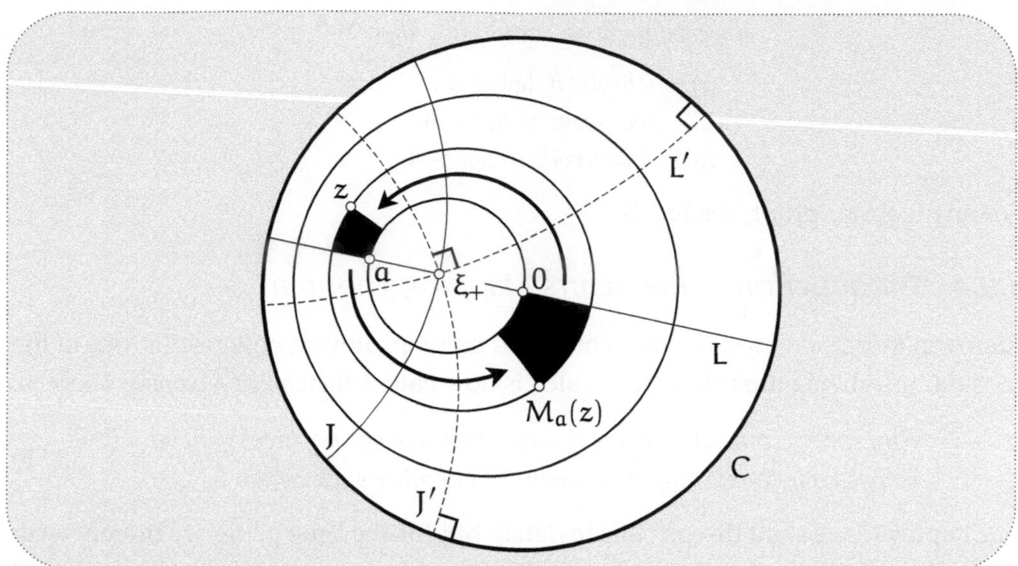

[3.39] The *true* meaning of M_a awaits in Chapter 6: it is a half-turn of the hyperbolic plane about ξ_+. *We can also express* M_a *as* $(\mathcal{I}_{L'} \circ \mathcal{I}_{J'})$*, where* J' *and* L' *are any two orthogonal circles through* ξ_+ *that are orthogonal to* C*. Here we illustrate some of the invariant circles, together with the effect of* M_a *on a "square". The deeper meaning of all this will become clear once we have encountered* hyperbolic geometry *in Chapter 6. We will then recognize that for inhabitants of the hyperbolic plane,* M_a is nothing more than an ordinary rotation of the plane about ξ_+ through angle π*! See Section 6.3.11.*

through these points, M_a is elliptic, and the multipliers associated with $\xi_\pm$ are both given by $m = e^{i\pi} = -1$. Thus the effect of M_a on an infinitesimal neighbourhood of the interior fixed point ξ_+ is a rotation of π. The fact that M_a swaps a and 0 can now be recognized as a special case of the fact that M_a is *involutory*: $(M_a \circ M_a) = \mathcal{E}$, and *every* pair of points z, $M_a(z)$ is swapped by M_a. Finally, note that we can also express M_a as $(\mathcal{I}_{L'} \circ \mathcal{I}_{J'})$, where J' and L' are any two orthogonal circles through ξ_+ that are orthogonal to C. All this is illustrated in [3.39], which also shows some of the invariant circles, together with the effect of M_a on a "square".

The deeper meaning of all this will become clear once we have encountered Hyperbolic Geometry in Chapter 6. We will then recognize that for inhabitants of the hyperbolic plane, M_a *is nothing more than an ordinary rotation of the plane about* ξ_+ *through angle* π! See Section 6.3.11.

While we will return to the geometry of the general Möbius automorphisms M_a^{ϕ} in Chapter 6, we remark here that they can only be elliptic, parabolic, or hyperbolic. This is because (by construction) they leave C invariant, while a loxodromic Möbius transformation has no invariant circles. To be more precise, in Chapter 6 we will use the above interpretation of M_a to show geometrically that

If we define $\Phi \equiv 2\cos^{-1}|a|$, *then* M_a^ϕ is

$$\begin{aligned} &\text{(i)} \quad \textit{elliptic if } |\phi| < \Phi, \\ &\text{(ii)} \quad \textit{parabolic if } |\phi| = \Phi, \\ &\text{(iii)} \quad \textit{hyperbolic if } |\phi| > \Phi. \end{aligned} \tag{3.54}$$

For an algebraic proof, see Ex. 27.

3.9.4 Introduction to Riemann's Mapping Theorem

Riemann's doctoral thesis of 1851 contained many profound new results, one of the most famous being the following, which is now called *Riemann's Mapping Theorem:*

> *Any simply connected region* R *(other than the entire plane) may be mapped one-to-one and conformally to any other such region* S. (3.55)

In Chapter 12 we shall discuss this in detail, but for the time being we merely wish to point out some connections between Riemann's result and what we have learnt concerning automorphisms of the disc.

First note that to establish (3.55) in general, it is sufficient to establish it in the special case that S is the unit disc D. For if F_R is a one-to-one conformal mapping from R to D, and F_S is likewise a one-to-one conformal mapping of S to D, then $F_S^{-1} \circ F_R$ is a one-to-one conformal mapping of R to S, as required.

If M is an arbitrary automorphism of D, then $M \circ F_R$ is clearly another one-to-one conformal mapping from R to D. In fact *every* such mapping must be of this form. For if $\widetilde{F}_R$ were any other such mapping, then $\widetilde{F}_R \circ F_R^{-1}$ would be some automorphism M of D, in which case $\widetilde{F}_R = M \circ F_R$.

Thus the number of one-to-one conformal mappings from R to S is equal to the number from R to D, which in turn is equal to the number of automorphisms of D. As we have already said, in Chapter 7 we will show that these automorphisms are the Möbius transformations M_a^ϕ, which form a 3-parameter family. Thus (3.55) in fact implies that *the one-to-one conformal mappings from* R *to* S *form a three-parameter family.*

3.10 Exercises

1 In each of the figures below, show that p and $\widetilde{p}$ are symmetric with respect to the circle. The dashed lines are not strictly part of the constructions, rather they are intended to be helpful or suggestive.

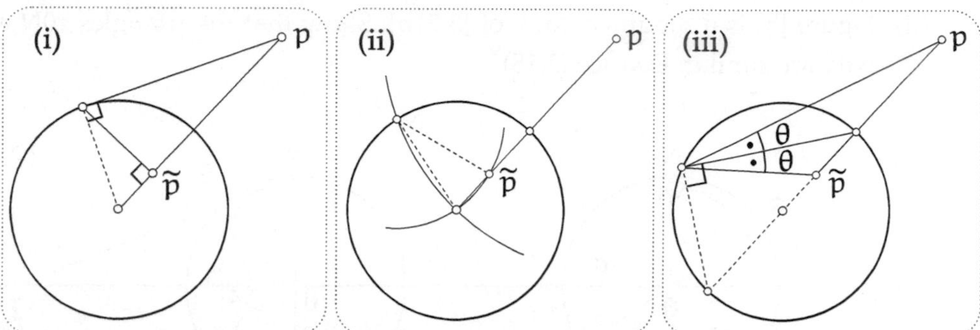

2 In 1864 a French officer named Peaucellier caused a sensation by discovering a simple mechanism (*Peaucellier's linkage*) for transforming linear motion (say of a piston) into circular motion (say of a wheel). The figure below shows six rods hinged at the white dots, and anchored at o. Two of the rods have length l, and the other four have length r. With the assistance of the dashed circle, show that $\widetilde{p} = \mathcal{I}_K(p)$, where K is the circle of radius $\sqrt{l^2 - r^2}$ centred at o. Construct this mechanism—perhaps using strips of fairly stiff cardboard for rods, and drawing pins for hinges—and use it to verify properties of inversion. In particular, try moving p along a line.

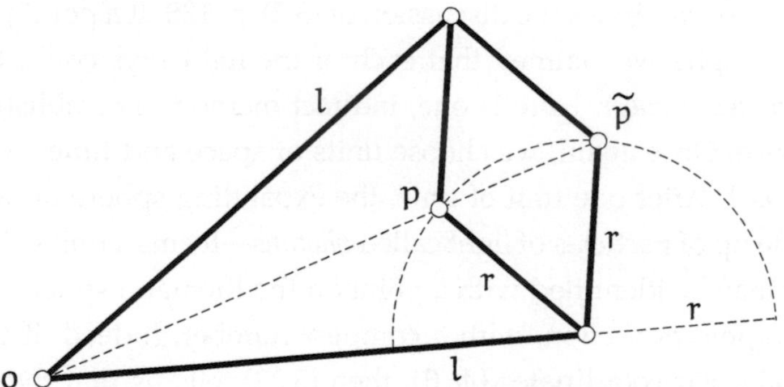

3 Let S be a sphere, and let p be a point not on S. Explain why $\mathcal{I}_S(p)$ may be constructed as the second intersection point of any three spheres that pass through p and are orthogonal to S. Explain the preservation of three-dimensional symmetry in terms of this construction.

4 Deduce (3.23), p. 167 directly from (3.18), p. 162.

5 Consider the following two-stage mapping: first stereographically project $\mathbb{C}$ onto the Riemann sphere Σ in the usual way; now stereographically project Σ

back to $\mathbb{C}$, *but from the south pole* instead of the north pole. The net effect of this is some complex mapping $z \mapsto f(z)$ of $\mathbb{C}$ to itself. What is f?

6 Both figures below show vertical cross sections of the Riemann sphere.

(i) In figure [a], show that the triangles p0N and N0q are similar. Deduce (3.23).

(ii) Figure [b] is a modified copy of [3.21b]. Show that the triangles z0N and N0$\tilde{z}$ are similar. Deduce (3.18).

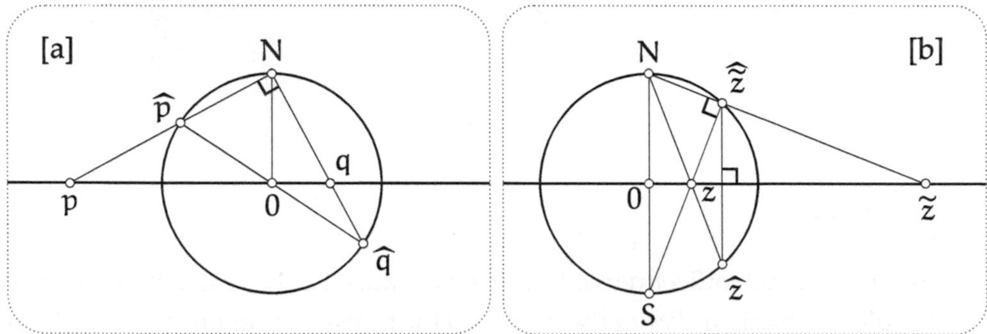

7 (i) Use a computer to draw the images in $\mathbb{C}$ of several origin-centred circles under the exponential mapping, $z \mapsto e^z$. Explain the obvious symmetry of these image curves with respect to the real axis.

(ii) Now use the computer to draw these same image curves on the Riemann sphere, instead of in $\mathbb{C}$. Note the surprising new symmetry!

(iii) Use (3.19) to explain this extra symmetry.

8 This exercise continues the discussion of (3.2), p. 138. If a point p in space emits a flash of light, we claimed that each of the light rays could be represented by a *complex number.* Here is one, indirect method of establishing this correspondence. Once again, we choose units of space and time so that the speed of light is 1. After one unit of time, the expanding sphere of light emitted by p—made up of particles of light called *photons*—forms a unit sphere. Thus each photon may be identified with a point on the Riemann sphere, and hence, via stereographic projection, with a complex number. Indeed, if the photon has spherical polar coordinates (ϕ, θ), then (3.22) tells us that the corresponding complex number is $z = \cot(\phi/2)\, e^{i\theta}$.

Sir Roger Penrose (see Penrose and Rindler (1984, p. 13)) discovered the following remarkable method of passing from a light ray to the associated complex number *directly*, without the assistance of the Riemann sphere. Imagine that p is one unit vertically above the origin of the (horizontal) complex plane. At the instant that p emits its flash, let $\mathbb{C}$ begin to travel straight up (in the direction $\phi = 0$) at the speed of light ($= 1$) towards p. Decompose the velocity of the photon F emitted by p in the direction (ϕ, θ) into components perpendicular and parallel to $\mathbb{C}$. Hence find the time at which F hits $\mathbb{C}$. Deduce that F hits $\mathbb{C}$ at the point $z = \cot(\phi/2)\, e^{i\theta}$. Amazingly, we see that *Penrose's construction is equivalent to stereographic projection!*

9 In order to analyse astronomical data, Ptolemy required accurate trigonometric tables, which he constructed using the addition formulae for sine and cosine. The figures below explain how he discovered these key addition formulae. Both the circles have unit radius.

(i) In figure [a], show that $A = 2\sin\theta$ and $B = 2\cos\theta$.

(ii) In figure [b], apply Ptolemy's Theorem to the illustrated quadrilateral, and deduce that $\sin(\theta+\phi) = \sin\theta\cos\phi + \sin\phi\cos\theta$.

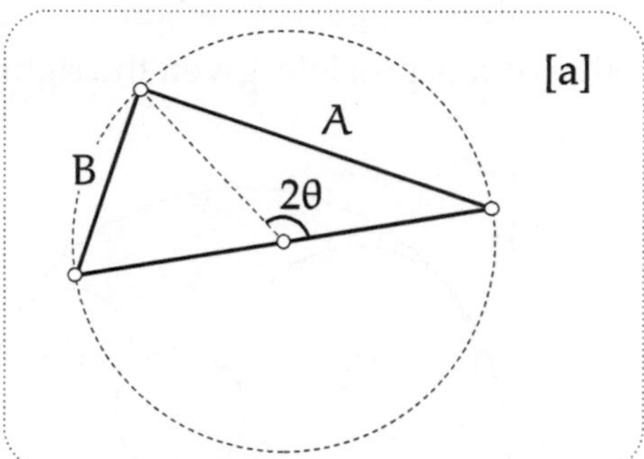

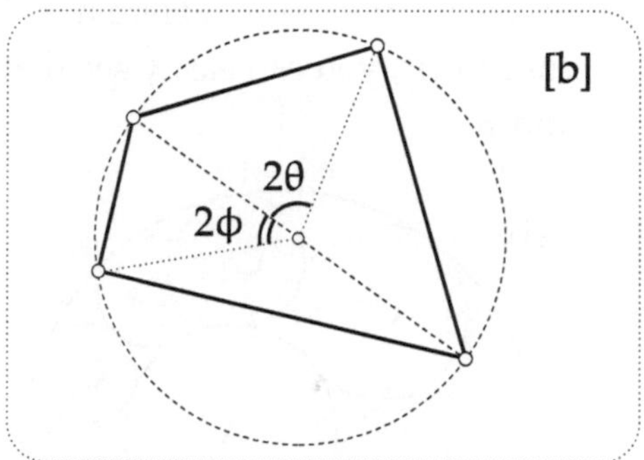

10 The aim of this question is to understand the following result:

> *Any two non-intersecting, non-concentric circles can be mapped to concentric circles by means of a suitable Möbius transformation.*

(i) If A and B are the two circles in question, show that there exists a pair of points $\xi_{\pm}$ that are symmetric with respect to both A and B.

(ii) Deduce that if $F(z) = (z - \xi_+)/(z - \xi_-)$, then F(A) and F(B) are concentric circles, as was desired.

11 This exercise yields a more intuitive proof of the result of the previous exercise. Using different colours for each, draw two non-intersecting, non-concentric circles, A and B, then draw the line L through their centres. Label as p and q the intersection points of B with L.

(i) Using corresponding colours, draw a fresh picture showing the images $\widetilde{A}$, $\widetilde{B}, \widetilde{L}, \widetilde{q}$ of A, B, L, q *under inversion in any circle centred at* p. To get you started, note that $\widetilde{L} = L$.

(ii) Now add to your figure by drawing the circle K, centred at $\widetilde{q}$, that cuts $\widetilde{A}$ at right angles, and let g and h be the intersection points of K and L.

(iii) Now draw a new picture showing the images K', L', h' of K, L, h *under inversion in any circle centred at* g.

(iv) By appealing to the anticonformal nature of inversion, deduce that $\widetilde{A}'$, $\widetilde{B}'$ are *concentric* circles centred at h'.

Since the composition of two inversions is a Möbius transformation, you have proved the result of the previous exercise.

12 Figure (i) below shows two non-intersecting, non-concentric circles A and B, together with a chain of circles $C_1, C_2, \ldots$ that touch one another successively, and that all touch A and B. As you would expect, the chain fails to "close up": C_8 *overlaps* C_1 instead of touching it. Figure (ii) shows that this failure to close is not inevitable. Given a different pair A, B, it is possible to obtain a closed chain where C_n touches C_1. Here $n = 5$, but by considering the case where A, B are concentric, you can easily see that any value of n is possible, given the right A and B.

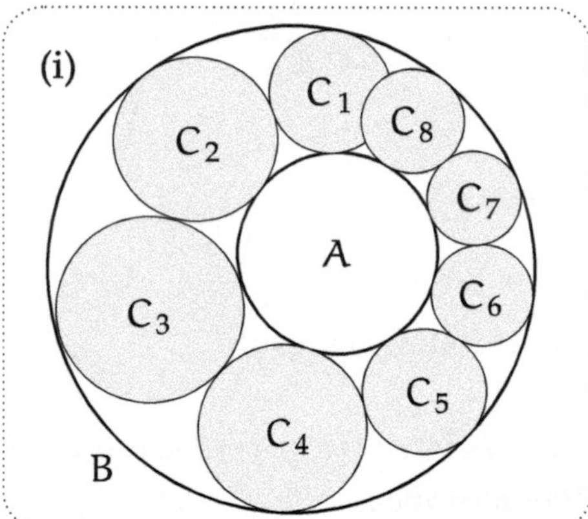

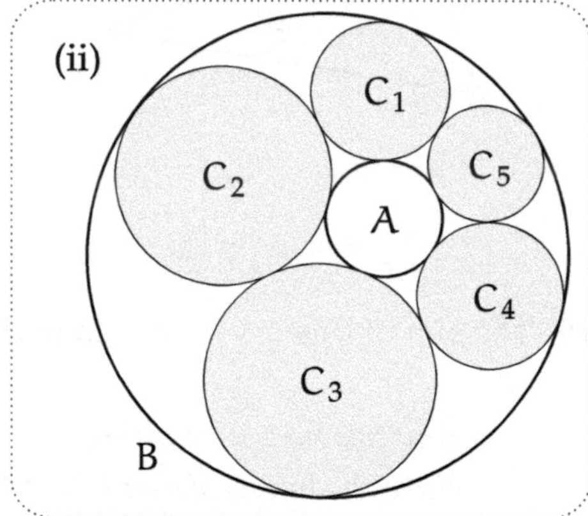

Steiner discovered, very surprisingly, that if the chain closes for one choice of C_1, then it closes for *every* choice of C_1, and the resulting chain always contains the same number of touching circles. Explain this using the result of Ex. 10.

13 (i) Let P be a sphere resting on the flat surface Q of a table. Let $S_1, S_2, \ldots$ be a string of spheres touching one another successively and all the same size as P. If each S-sphere touches both P and Q, show that S_6 touches S_1, so that we have a closed "necklace" of six spheres around P.

(ii) Let A, B, C be three spheres (not necessarily of equal size) all touching one another. As in the previous part, let $S_1, S_2, \ldots$ be a string of spheres (now of unequal size) touching one another successively, and all touching A, B, C. Astonishingly (cf. previous exercise), S_6 will always touch S_1, forming a closed "necklace" of six spheres interlocked with A, B, C. Prove this by first applying an inversion centred at the point of contact of A and B, then appealing to part (i).

The chain of six spheres in part (ii) is called *Soddy's Hexlet*, after the amateur mathematician Frederick Soddy—fellow graduate of Merton College, Oxford!—who discovered it (*without inversion!*). For further information on Soddy's Hexlet, see Ogilvy (1990). Soddy's full time job was chemistry—in 1921 he won the Nobel Prize for his discovery of isotopes!

14 The figure below shows four collinear points a, b, c, d, together with the (necessarily coplanar) light rays from those points to an observer. Imagine that the collinear points lie in the complex plane, and that the observer is above the plane looking down. Show that the cross-ratio $[a, b, c, d]$ can be expressed purely in terms of the directions of these light rays; more precisely, show that

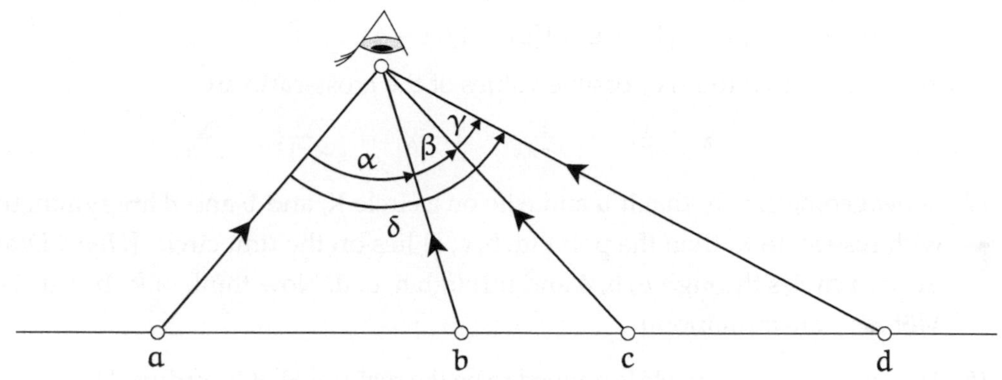

$$[a, b, c, d] = -\frac{\sin\alpha \sin\gamma}{\sin\beta \sin\delta}.$$

Suppose the observer now does a *perspective drawing* on a glass "canvas plane" C (arbitrarily positioned between himself and $\mathbb{C}$). That is, for each point p in $\mathbb{C}$ he draws a point p' where the light ray from p to his eye hits C. Use the above result to show that although angles and distances are both distorted in his drawing, *cross-ratios of collinear points are preserved:* $[a', b', c', d'] = [a, b, c, d]$.

15 Show that in both of the figures below, $\text{Arg}\,[z, q, r, s] = \theta + \phi$. Hence deduce (3.32), p. 156.

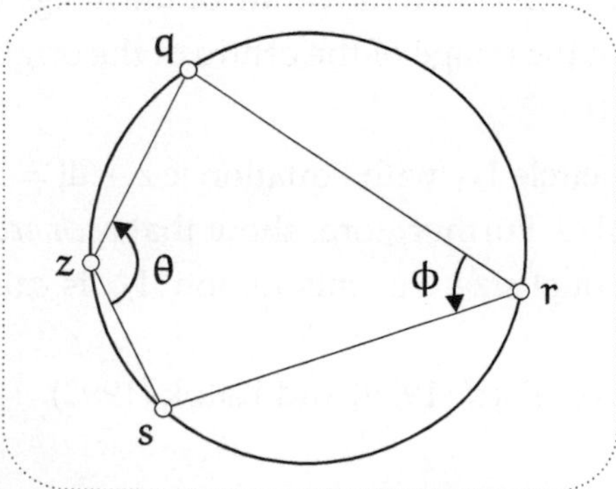

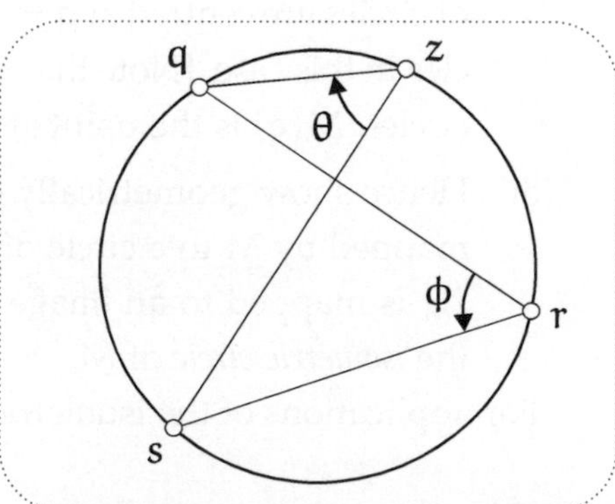

16 As in figure [3.28], think of the cross-ratio $[z, q, r, s]$ as a Möbius transformation.

(i) Explain geometrically why permuting q, r, s, in $[z, q, r, s]$ yields six different Möbius transformations.

(ii) If $I(z)$ is the Möbius transformation that leaves 1 fixed and that swaps 0 with ∞, explain geometrically why $I \circ [z, q, r, s] = [z, s, r, q]$.

(iii) If $J(z)$ is the Möbius transformation that sends $0, 1, \infty$ to $1, \infty, 0$, respectively, explain geometrically why $J \circ [z, q, r, s] = [z, s, q, r]$.

(iv) Employing the abbreviation $\chi \equiv [z, q, r, s]$, explain why the six Möbius transformations in part (i) can be expressed as

$$\chi, \quad I \circ \chi, \quad J \circ \chi, \quad I \circ J \circ \chi, \quad J \circ I \circ \chi, \quad I \circ J \circ I \circ \chi.$$

(v) Show that $I(z) = (1/z)$ and $J(z) = 1/(1 - z)$.

(vi) Deduce that the six possible values of the cross-ratio are

$$\chi, \quad \frac{1}{\chi}, \quad \frac{1}{1-\chi}, \quad 1-\chi, \quad \frac{\chi}{\chi-1}, \quad \frac{\chi-1}{\chi}.$$

17 Show geometrically that if a and c lie on a circle K, and b and d are symmetric with respect to K, then the point $[a, b, c, d]$ lies on the unit circle. [*Hints:* Draw the two circles through a, b, d and through b, c, d. Now think of $[z, b, c, d]$ as a Möbius transformation.]

18 The *curvature* κ of a circle is defined to be the reciprocal of its radius. Let $M(z) = \frac{az+b}{cz+d}$ be normalized. Use (3.3) to show geometrically that M maps the real line to a circle of curvature

$$\kappa = \left| 2c^2 \operatorname{Im}\left(\frac{d}{c}\right) \right|.$$

19 Let $M(z) = \frac{az+b}{cz+d}$ be normalized.

(i) Using (3.3), draw diagrams to illustrate the successive effects of these transformations on a family of *concentric* circles. Note that the image circles are generally *not* concentric.

(ii) Deduce that the image circles *are* concentric if and only if the original family of circles are centred at $q = -(d/c)$. Write down the centre of the image circles in this case. [Note that this is not the image of the centre of the original circles: $M(q)$ is the point at infinity!]

(iii) Hence show geometrically that the circle I_M with equation $|cz + d| = 1$ is mapped by M to a circle of equal size. Furthermore, show that *each arc* of I_M is mapped to an image arc of equal size. For this reason, I_M is called the *isometric circle* of M.

For applications of the isometric circle, see Ford (1929) and Katok (1992).

20 (i) Show that every Möbius transformation of the form

$$M(z) = \frac{pz + q}{\overline{q}\,z + \overline{p}} \qquad \text{where } |p| > |q|$$

can be rewritten in the form

$$M(z) = e^{i\theta}\left(\frac{z - a}{\overline{a}\,z - 1}\right), \qquad \text{where } |a| < 1.$$

[Notice that the converse is also true. In other words, the two sets of functions are the same.]

(ii) Use the matrix representation of the first equation to show that this set of Möbius transformations forms a group under the operation of composition.

(iii) Use the disc-automorphism interpretation of these transformations to give a geometric explanation of the fact that they form a group.

21 (i) Use the matrix representation to show algebraically that the set of Möbius transformations

$$R(z) = \frac{az + b}{-\overline{b}z + \overline{a}} \qquad \text{with} \qquad |a|^2 + |b|^2 = 1$$

forms a group under the operation of composition.

(ii) Using the interpretation of these functions given on page 184, explain part (i) geometrically.

22 Let H be the rectangular hyperbola with Cartesian equation $x^2 - y^2 = 1$. Show that $z \mapsto w = z^2$ maps H to the line $\mathrm{Re}(w) = 1$. What is the image of this line under complex inversion, $w \mapsto (1/w)$? Referring back to figure [2.9], p. 69, deduce that complex inversion maps H to a *lemniscate!*
[*Hint:* Think of complex inversion as $z \mapsto \sqrt{(1/z^2)}$.]

23 From the simple fact that $z \mapsto (1/z)$ is involutory, deduce that it is elliptic, with multiplier -1.

24 (i) Use the Symmetry Principle to show that the most general Möbius transformation of the upper half-plane to the unit disc has the form

$$M(z) = e^{i\theta}\left(\frac{z - a}{z - \overline{a}}\right),$$

where $\mathrm{Im}\, a > 0$.

(ii) The most general Möbius transformation back *from* the unit disc *to* the upper half-plane will therefore be the inverse of $M(z)$. Let's call this inverse $N(z)$. Use the matrix form of M to show that

$$N(z) = M^{-1}(z) = \frac{\overline{a}\,z - a\,e^{i\theta}}{z - e^{i\theta}}.$$

(iii) Explain why the Symmetry Principle implies that $N(1/\overline{z}) = \overline{N(z)}$.

(iv) Show by direct calculation that the formula for N in part (ii) does indeed satisfy the equation in part (iii).

25 Let $M(z)$ be the general Möbius automorphism of the upper half-plane.

(i) Observing that M maps the real axis into itself, use (3.30) to show that the coefficients of M are *real*.

(ii) By considering $\text{Im}[M(i)]$, deduce that the only restriction on these real coefficients is that they have positive determinant: $(ad - bc) > 0$.

(iii) Explain (both algebraically and geometrically) why these Möbius transformations form a group under composition.

(iv) How many degrees of freedom does M have? Why does this make sense?

26 Reconsider (3.53), p. 201.

(i) Use (3.45), p. 192 to show that M_a is elliptic.

(ii) Use (3.44), p. 192 to show that both multipliers are given by $m = -1$.

(iii) Calculate the matrix product $[M_a]\,[M_a]$, and thereby verify that M_a is involutory.

(iv) Use (3.28), p. 172 to calculate the fixed points of M_a.

(v) Show that the result of the previous part is in accord with figure [3.38].

27 Use (3.45), p. 192 to verify (3.54), p. 204.

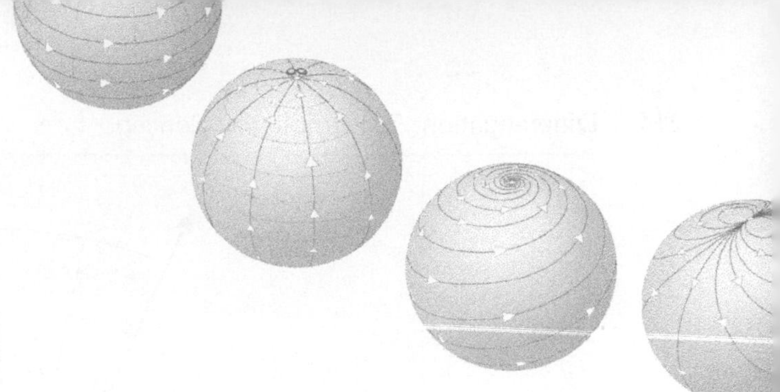

CHAPTER 4

Differentiation: The Amplitwist Concept

4.1 Introduction

Having studied functions of complex numbers, we now turn to the *calculus* of such functions.

To know the graph of an ordinary real function is to know the function completely, and so to understand curves is to understand real functions. The key insight of differential calculus is that if we take a common or garden curve, place it under a microscope and examine it using lenses of greater and greater magnifying power, each little piece looks like a *straight line*. When produced, these infinitesimal pieces of straight line are the tangents to the curve, and their directions describe the local behaviour of the curve. Thinking of the curve as the graph of $f(x)$, these directions are in turn described by the derivative, $f'(x)$.

Despite the fact that we cannot draw the graph of a complex function, in this chapter we shall see how it is still possible to describe the local behaviour of a complex mapping by means of a complex analogue of the ordinary derivative—the "amplitwist".

4.2 A Puzzling Phenomenon

Throughout Chapter 2 we witnessed a very strange phenomenon. Whenever we generalized a familiar real function to a corresponding complex function, *the mapping sent infinitesimal squares to infinitesimal squares*. At present this is a purely empirical observation based on using a computer to draw pictures of the mappings. In this chapter we begin to explore the theoretical underpinnings of the phenomenon.

Let's go back and take a closer look at a simple mapping like $z \mapsto w = z^2$. As we already know, this maps the origin-centred circle $|z| = r$ into the circle $|w| = r^2$, and it maps the ray $\arg(z) = \theta$ into the ray $\arg(w) = 2\theta$. An obvious consequence of this

Visual Complex Analysis. 25th Anniversary Edition. Tristan Needham, Oxford University Press.
 DOI: 10.1093/oso/9780192868916.003.0004

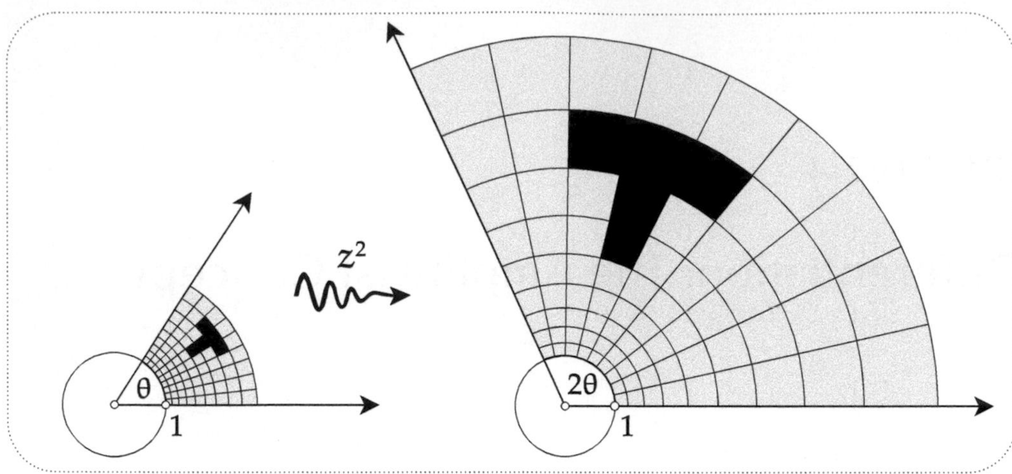

[4.1] The tip of the iceberg: small squares map to small squares—but *why?* *Clearly the mapping $z \mapsto z^2$ preserves the orthogonality of origin-centred rays and circles. So the small, ultimately vanishing "squares" on the left must be mapped to small, ultimately vanishing* rectangles *on the right. Very surprisingly, though, squares are ultimately mapped to* squares! *As we shall see, this is a consequence of the fact that the mapping is* conformal.

is that the right angle of intersection between such circles and rays in the z-plane is preserved by the mapping, which is to say that their images in the w-plane also meet at right angles. As illustrated in [4.1], a grid of infinitesimal squares formed from such circles and rays must therefore be mapped to an image grid composed of infinitesimal rectangles. However, this does not explain why these image rectangles must again be *squares*.

As we will explain shortly, the fact that infinitesimal squares are preserved is just one consequence of the fact that $z \mapsto w = z^2$ is *conformal* everywhere except at the two critical points $z = 0$ and $z = \infty$, where angles are doubled. In particular, any pair of orthogonal curves is mapped to another pair of orthogonal curves. In order to give another example of this, we first dismember our mapping into its real and imaginary parts. Writing $z = x + iy$ and $w = u + iv$, we obtain

$$u + iv = w = z^2 = (x + iy)^2 = (x^2 - y^2) + i\,2xy.$$

Thus the new coordinates are given in terms of the old ones by

$$\begin{aligned} u &= x^2 - y^2, \\ v &= 2xy. \end{aligned} \tag{4.1}$$

We now forget (temporarily!) that we are in $\mathbb{C}$, and think of (4.1) as simply representing a mapping of $\mathbb{R}^2$ to $\mathbb{R}^2$. If we let our point (x, y) slide along any of the rectangular hyperbolas with equation $2xy = \text{const.}$, then we see from (4.1) that its

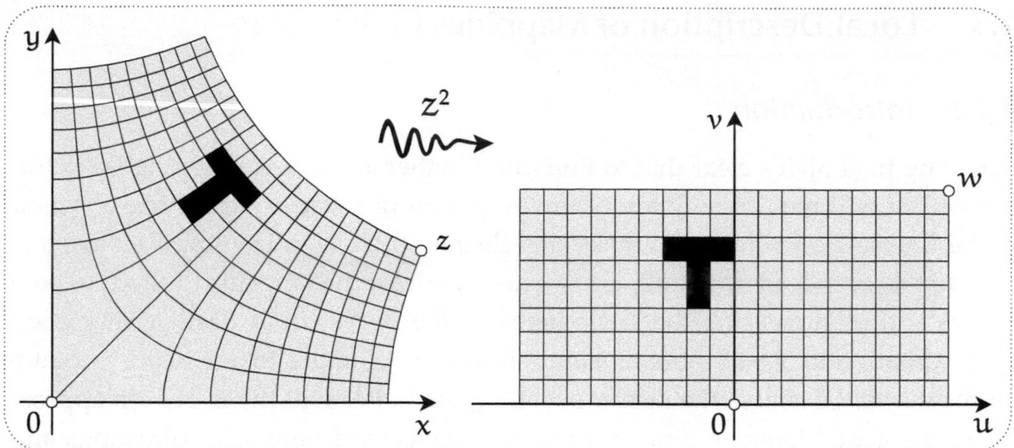

[4.2] Again, small squares map to small squares—but *why?* *The pre-images of the horizontal and vertical lines on the right are the two families of hyperbolas on the left, which a computation confirms are orthogonal to each other, as illustrated. Thus the squares on the right must have pre-images that are rectangles on the left. But, once again, surprisingly, the pre-images of the squares are actually* squares!

image (u, v) will move on a horizontal line $v = \text{const}$. Likewise, the preimages of the vertical lines $u = \text{const}$. will be another family of rectangular hyperbolas with equations $(x^2 - y^2) = \text{const}$. Since their images are orthogonal, the claimed conformality of $z \mapsto z^2$ implies that these two kinds of hyperbolas should themselves be orthogonal.

Figure [4.2] makes it clear that they are indeed orthogonal. We may verify this mathematically by recalling that two curves are orthogonal at a point of intersection if the product of their slopes at that point is equal to -1. Implicitly differentiating the equations of the hyperbolas, we find that

$$x^2 - y^2 = \text{const.} \quad \Rightarrow \quad x - yy' = 0 \quad \Rightarrow \quad y' = +(x/y),$$
$$2xy = \text{const.} \quad \Rightarrow \quad y + xy' = 0 \quad \Rightarrow \quad y' = -(y/x).$$

Thus the product of the slopes of the two kinds of hyperbola at a point of intersection is -1, as was to be shown.

Clearly we could carry on in this way, analysing the effect of the mapping on one pair of curves after another, but what is really needed is a general argument showing that if two curves meet at some arbitrary angle ϕ, then their images under (4.1) will *also* meet at angle ϕ. To obtain such an argument, we shall continue to pretend that we are living in the less rich structure of $\mathbb{R}^2$ (rather than our own home $\mathbb{C}$) and investigate the local properties of a general mapping of the plane to itself.

4.3 Local Description of Mappings in the Plane

4.3.1 Introduction

Referring to [4.3], it's clear that to find out whether any given mapping is conformal or not will require only a local investigation of what is happening *very near* to the intersection point q. To make this clearer still, recognize that if we wish to measure ϕ, or indeed even define it, we need to draw the tangents [dotted] to both curves and then measure the angle between them. We could draw a very good approximation to one of these tangents simply by joining q to any nearby point p on the curve. Of course the nearer p is to q, the better will the chord qp approximate the actual tangent. Since we are only concerned here with directions and angles (rather than positions) we may dispense with the tangent itself, and instead use the infinitesimal vector $\overrightarrow{qp}$ that points along it. Likewise, after we have performed the mapping, we are not interested in the positions of the image points Q and P themselves; rather, we want the infinitesimal connecting vector $\overrightarrow{QP}$ that describes the direction of the new tangent at Q. We will call this infinitesimal vector $\overrightarrow{QP}$ the *image* of the vector $\overrightarrow{qp}$. However natural this terminology may seem, note that this really is a new sense of the word "image".

Let us now summarize our strategy. Given formulae such as (4.1), which describe the mapping of the points to their image points, we wish to discover the induced mapping of infinitesimal vectors emanating from a point q to their image vectors emanating from the image point Q. In principle, we could then apply the latter mapping to $\overrightarrow{qp}$ and to $\overrightarrow{qs}$, yielding their images $\overrightarrow{QP}$ and $\overrightarrow{QS}$, and hence the angle of intersection of the image curves through Q.

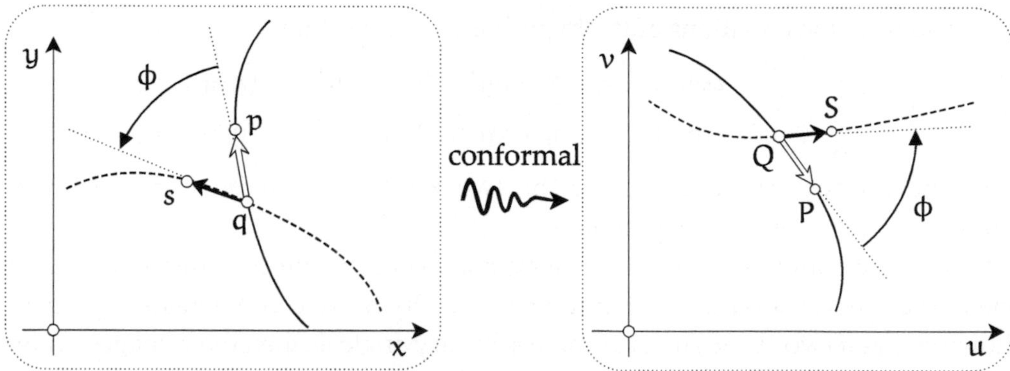

[4.3] Conformal mappings. *The angle ϕ between the curves passing through q on the left is defined to be the angle between their tangents there. But as p and s merge with q, this angle $\phi \asymp \angle pqs$. Let the images of p, q, s be P, Q, S, respectively. Then to prove conformality, we need to show that $\angle PQS \asymp \phi \asymp \angle pqs$. To do so, we need to show that the transformation that carries the initial connecting vectors on the left to their "images" on the right (ultimately) does not alter the angle between them. This (linear) transformation is called the Jacobian, and its coordinate description in the case of a general mapping is derived via the next figure.*

4.3.2 The Jacobian Matrix

Consider [4.4]. As discussed, the direction of the illustrated curve through q is being described with an infinitesimal vector $\begin{pmatrix} dx \\ dy \end{pmatrix}$; the infinitesimal image vector $\begin{pmatrix} du \\ dv \end{pmatrix}$ gives the direction of the image curve through Q. We can determine the component du of $\begin{pmatrix} du \\ dv \end{pmatrix}$ as follows:

$$
\begin{aligned}
du &= \text{total change in u due to moving along} \begin{pmatrix} dx \\ dy \end{pmatrix} \\
&= \text{(change in u produced by moving dx in the x-direction)} \\
&\qquad + \\
&\quad \text{(change in u produced by moving dy in the y-direction)} \\
&= \text{(rate of change of u with x)} \cdot \text{(change dx in x)} \\
&\qquad + \\
&\quad \text{(rate of change of u with y)} \cdot \text{(change dy in y)} \\
&= (\partial_x u)\, dx + (\partial_y u)\, dy,
\end{aligned}
$$

where $\partial_x = \partial/\partial x$ etc. Likewise, we find that the vertical component is given by the formula

$$dv = (\partial_x v)\, dx + (\partial_y v)\, dy.$$

Since these expressions are linear in dx and dy, it follows (assuming that not all the partial derivatives vanish) that the infinitesimal vectors are carried to their images by a *linear transformation*. The general significance of this will be discussed later, but for the moment it means that the local effect of our mapping is completely described by a matrix J called the *Jacobian*. Thus,

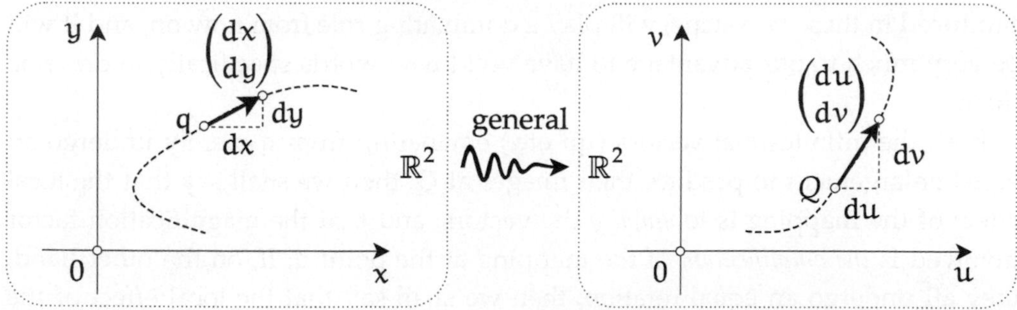

[4.4] The local linear (Jacobian) transformation induced by a general mapping. *An infinitesimal vector along a curve is carried to its image along the image curve by a* linear transformation *called the Jacobian,* J. *As the text explains,* $du = (\partial_x u)\, dx + (\partial_y u)\, dy$, *and similarly for* dv, *leading to the matrix representation of* J *as (4.2).*

$$\begin{pmatrix} dx \\ dy \end{pmatrix} \longmapsto \begin{pmatrix} du \\ dv \end{pmatrix} = J \begin{pmatrix} dx \\ dy \end{pmatrix},$$

where the Jacobian matrix is

$$J = \begin{pmatrix} \partial_x u & \partial_y u \\ \partial_x v & \partial_y v \end{pmatrix}. \tag{4.2}$$

We are now in a position to return to the specific mapping $z \mapsto z^2$, or more precisely to the mapping of $\mathbb{R}^2$ that we extracted from it. If we evaluate (4.2) for the mapping (4.1), we find that

$$J = \begin{pmatrix} 2x & -2y \\ 2y & 2x \end{pmatrix}.$$

The geometric effect of this matrix is perhaps more clearly seen if we switch to polar coordinates. At the point $z = r\,e^{i\theta}$—or rather $(r\cos\theta,\ r\sin\theta)$, since for the moment we are still in $\mathbb{R}^2$—we have

$$J = 2r \begin{pmatrix} \cos\theta & -\sin\theta \\ \sin\theta & \cos\theta \end{pmatrix}.$$

The effect of the $2r$ is merely to expand all the vectors by this factor. This clearly does not affect the angle between any two of them. The remaining matrix is probably familiar to you as producing a rotation of θ, and hence it too does not alter the angle between vectors. Since both stages of the transformation preserve angles, we have in fact verified the previous claim: the net transformation *is* conformal.

4.3.3 The Amplitwist Concept

We have just seen that the local effect of $z \mapsto z^2$ on infinitesimal vectors is to expand them and to rotate them. Transformations of this type (i.e., whose local effect is produced in these two steps) will play a dominating role from now on, and it will be very much to our advantage to have vivid new words specifically to describe them.

If all the infinitesimal vectors ($\overrightarrow{qp}$ etc.) emanating from q merely undergo an equal enlargement to produce their images at Q, then we shall say that the local effect of the mapping is to *amplify* the vectors, and that the magnification factor involved is *the amplification* of the mapping at the point q. If, on the other hand, they all undergo an equal rotation, then we shall say that the local effect of the mapping is to *twist* the vectors, and that the angle of rotation involved is *the twist* of the mapping at the point q. More generally, the kind of mapping that will concern us will locally both amplify *and* twist infinitesimal vectors—we say that such a transformation is locally *an amplitwist*. Thus "an amplitwist" is synonymous with "a (direct) similarity", *except* that the former refers to the transformation of *infinitesimal* vectors, whereas "a similarity" has no such connotation.

[We remind the reader of the discussion in both Prefaces: "infinitesimal" is being used here in a definite, technical sense—small and ultimately vanishing, the relationship to other infinitesimals being expressed via Newtonian "ultimate equalities".]

We can illustrate the new terminology with reference to the concrete case we have just analysed: The mapping $z \mapsto z^2$ is locally an amplitwist with amplification $2r$ and twist θ. See [4.5]. Quite generally, this figure makes it clear that *if a mapping is locally an amplitwist then it is automatically conformal*—the angle ϕ between the infinitesimal complex numbers is preserved.

Returning to [4.1] and [4.2], we now understand why infinitesimal squares were mapped to infinitesimal squares. Indeed, an infinitesimal region of *arbitrary* shape located at z will be "amplitwisted" (amplified and twisted) to a similar shape at z^2. Note that here we are extending our terminology still further: henceforth we will freely employ the verb "to amplitwist", meaning to amplify and to twist an infinitesimal geometric object.

All we really have at the moment is one simple mapping that turned out to be locally an amplitwist. In order to appreciate how truly fundamental this amplitwist concept is, we must return to $\mathbb{C}$ and begin from scratch to develop the idea of complex differentiation.

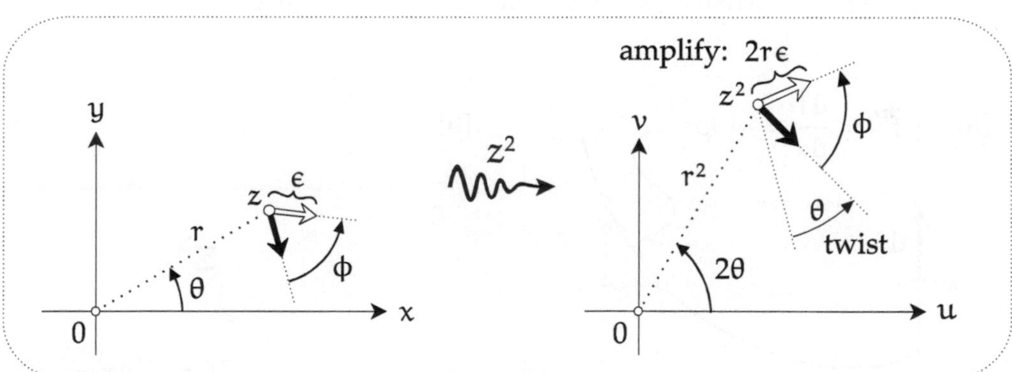

[4.5] Amplitwist terminology. *The local effect of* $z \mapsto z^2$ *on infinitesimal complex numbers emanating from* $z = r\,e^{i\theta}$ *is to* amplitwist *them. Specifically, here we must* amplify *them by* $2r$*—this magnification factor is called* the amplification*—and we must* twist *them by* θ*—this rotation angle is called* the twist. *The net transformation is called* the amplitwist: *it is equivalent to multiplication by* (amplification) $e^{i(\text{twist})} = 2r\,e^{i\theta}$, *in this example. If a complex mapping is locally an amplitwist, as this one is, then it is also* conformal, *as illustrated. We reserve this terminology for* infinitesimal *complex numbers, i.e., for small complex numbers in their moment of vanishing.*

4.4 The Complex Derivative as Amplitwist

4.4.1 The Real Derivative Re-examined

In the ordinary real calculus we have a potent means of visualizing the derivative f' of a function f from $\mathbb{R}$ to $\mathbb{R}$, namely, as the slope of the graph $y = f(x)$. See [4.6a]. Unfortunately, due to our lack of four-dimensional imagination, we can't draw the graph of a complex function, and hence we cannot generalize this particular conception of the derivative in any obvious way.

As a first step towards a successful generalization, we simply split the axes apart, so that [4.6a] becomes [4.6b]. Note that we have drawn both copies of $\mathbb{R}$ in a horizontal position, in anticipation of their being viewed as merely the real axes of two complex planes.

Next, continuing in the spirit of the previous section, we observe that $|f'(x)|$ describes how much the initial infinitesimal vector at x must be expanded to obtain its image at $f(x)$. More algebraically, $f'(x)$ is that real number by which we must multiply the initial vector to obtain its image:

$$f'(x) \cdot \rightarrow \; = \; \longrightarrow. \tag{4.3}$$

If $f'(x) > 0$ (as in [4.6b]) then the image of positive dx is a positive df, but if $f'(x) < 0$ then the infinitesimal image vector df is negative and points to the left, as illustrated in [4.7]. In this case, df can be obtained by first expanding dx by $|f'(x)|$, then *rotating it by* π. If we think of $f'(x)$ as a point on the real axis of $\mathbb{C}$, then $\arg[f'(x)] = 0$ when $f'(x) > 0$, and $\arg[f'(x)] = \pi$ when $f'(x) < 0$. Thus,

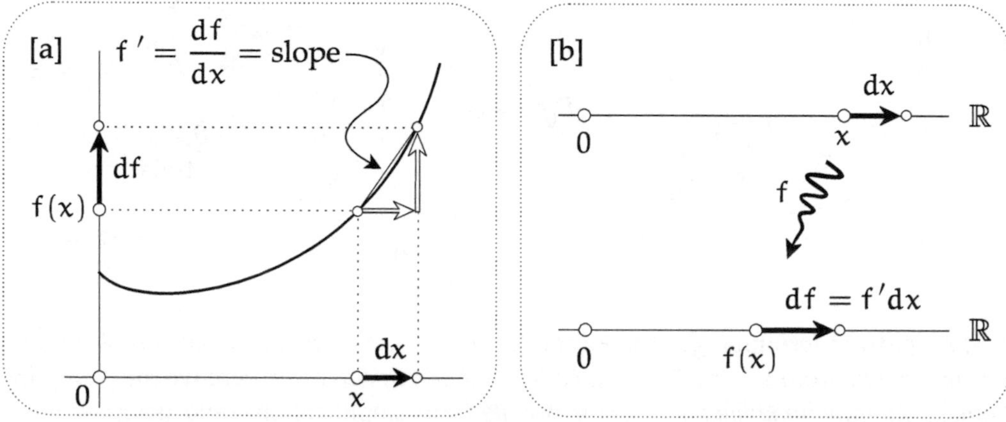

[4.6] The real derivative re-examined. [a] *The traditional picture of the derivative* f' *as the slope of the tangent to the graph* $y = f(x)$. **[b]** *Let us instead break apart the two copies of the real axis, as we are forced to break apart the two copies of* $\mathbb{C}$ *in the case of a complex function,* $z \mapsto f(z)$. *Then an infinitesimal movement* dx *is mapped to* $df = f'dx$*: in other words, the local transformation is an amplification by* f'.

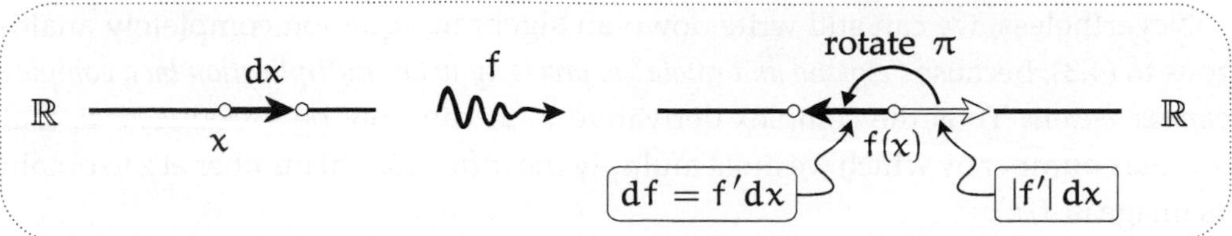

[4.7] *If we think of* $f'(x)$ *as a point on the real axis of* $\mathbb{C}$, *then* $\arg[f'(x)] = 0$ *when* $f'(x) > 0$, *and* $\arg[f'(x)] = \pi$ *when* $f'(x) < 0$. *Thus, regardless of whether* $f'(x)$ *is positive or negative, we see that the local effect of* f *on an infinitesimal vector* dx *at* x *is to expand it by* $|f'(x)|$ *and to rotate it by* $\arg[f'(x)]$.

regardless of whether $f'(x)$ is positive or negative, we see that the local effect of f on an infinitesimal vector dx at x is to expand it by $|f'(x)|$ and to rotate it by $\arg[f'(x)]$.

With all this fresh in our minds, we now attempt to generalize the notion of "derivative" to mappings of $\mathbb{C}$.

4.4.2 The Complex Derivative

Consider the effect of a complex mapping $f(z)$ on an infinitesimal complex number emanating from z. Its image (i.e., the connecting complex number between the two image points) will be an infinitesimal complex number emanating from $f(z)$. The generalization of [4.6b] or [4.7] is now [4.8]. On the right, we have drawn this image complex number in black, and we have also drawn a copy [white] at $f(z)$ of the original arrow at z. To transform the white arrow into the black image arrow now requires not only an expansion, but also a rotation. In figure [4.8] it looks as though we must expand the white arrow by 2, and rotate it by $(3\pi/4)$. Contrast this with the case of a real function, where the required rotation angle could only be 0 or π; in the case of a complex function we need rotations through arbitrary angles.

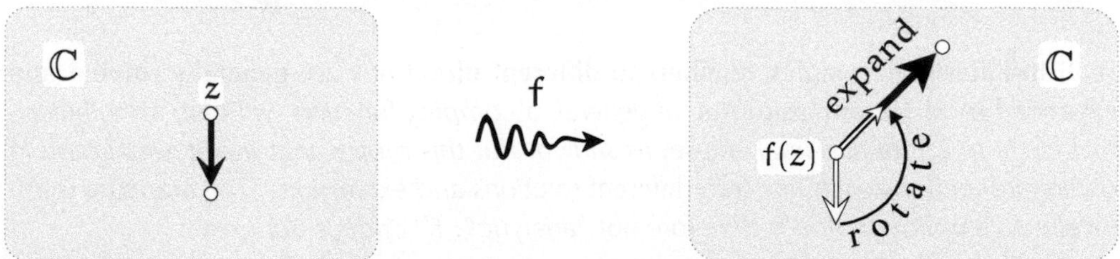

[4.8] The local effect of a general complex mapping. *The effect of a complex function on an infinitesimal arrow is to rotate it and expand it, which is equivalent to multiplication by a complex number* = (expansion factor)$e^{i(\text{rotation angle})} = 2\,e^{i(3\pi/4)}$, *in the illustrated example.*

Nevertheless, we can still write down an algebraic equation completely analogous to (4.3), because *"expand and rotate" is precisely what multiplication by a complex number means.* Thus the complex derivative $f'(z)$ can now be introduced as that complex number by which we must multiply the infinitesimal number at z to obtain its image at $f(z)$:

$$f'(z) \cdot \downarrow = \nearrow . \tag{4.4}$$

In order to produce the correct effect, *the length of* $f'(z)$ *must be the magnification factor, and the argument of* $f'(z)$ *must be the angle of rotation.* For example, at the particular point shown in [4.8] we would have $f'(z) = 2\,e^{i(3\pi/4)}$. In fact, in the spirit of Chapter 1, we need not even distinguish between the local transformation and the complex number that represents it.

To find $f'(z)$ we have looked at the image of a specific arrow at z, but (unlike the case of $\mathbb{R}$) there are now infinitely many possible directions for such arrows. What if we had looked at an arrow in a different direction from the illustrated one?

We are immediately in trouble, because a typical mapping[1] will do what you see in [4.9]. Clearly the magnification factor differs for the various arrows, and likewise each arrow needs to be rotated a different amount to obtain its image. While we could still use a complex number in (4.4) to describe the transformation of the arrows, it would have to be a *different* number for each arrow. There would therefore be no single complex number we could assign to this point as being *the* derivative of f at z. We have arrived at an apparently gloomy impasse: a typical mapping of $\mathbb{C}$ simply cannot be differentiated.

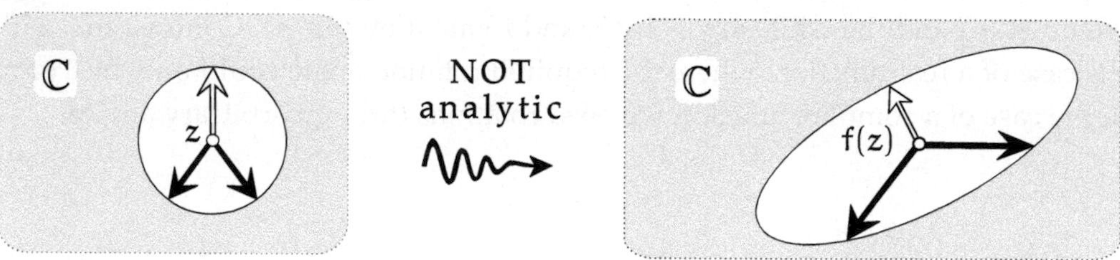

[4.9] Infinitesimal complex numbers in different directions are generally rotated and expanded by different amounts. *In general, a complex function will map an infinitesimal circle to an infinitesimal ellipse, as shown. But this means that infinitesimal radii in different directions each undergo* different *rotations and expansions. The mapping is* not *locally an amplitwist, and is therefore* not *"analytic":* $f'(z)$ *does not exist.*

[1] Shortly we will justify certain details of [4.9], such as the fact that an infinitesimal circle is mapped to an infinitesimal ellipse.

4.4.3 *Analytic Functions*

We get around the above obstacle in Zen-like fashion—by ignoring it! That is, from now on we concentrate almost exclusively on those very special mappings that *can* be differentiated. Such functions are called *analytic* (or *holomorphic*). From the previous discussion it follows that

> *Analytic mappings are precisely those whose local effect is an amplitwist: all the infinitesimal complex numbers emanating from a single point are amplified and twisted the same amount.*

In contrast to [4.9], the effect of an analytic mapping can be seen in [4.10]. For such a mapping the derivative exists, and simply *is* the amplitwist, or, if you prefer, the complex number representing the amplitwist.

At this point you might quite reasonably fear that however interesting such mappings might be, they would be too exotic to include any familiar or useful functions. However, a ray of hope is held out by the humble-looking mapping $z \mapsto z^2$, for we have already established that it is locally an amplitwist, as illustrated in [4.5], and so it now gains admittance into the select set of analytic functions. In fact, quite amazingly, we will discover in the next chapter that virtually *every* function we have met in this book is analytic! Of course we have already seen plenty of empirical evidence of this in our many pictures showing small "squares" being mapped to small "squares".

It should perhaps be stressed that all our recent pictures have been concerned with local properties, and hence with *infinitesimal* arrows and figures. For example, it's clear from [4.10] that any analytic mapping will send infinitesimal circles to other infinitesimal circles; however, this does *not* mean that such mappings typically send circles to circles. Figure [4.11] (which contains [4.10] at its centre) illustrates the fact that if we start with an infinitesimal circle and then expand it, its image will generally distort out of all semblance of circularity. Of course, an important exception to this is provided by the Möbius transformations, for these

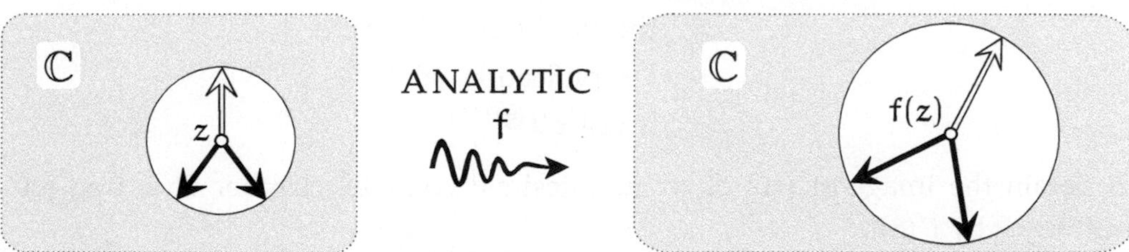

[4.10] Analytic = locally an amplitwist, *expanding and rotating all infinitesimal complex numbers* equally. *By definition, an* analytic *mapping is locally an amplitwist, and the derivative* $f'(z)$ *simply* is *the amplitwist.*

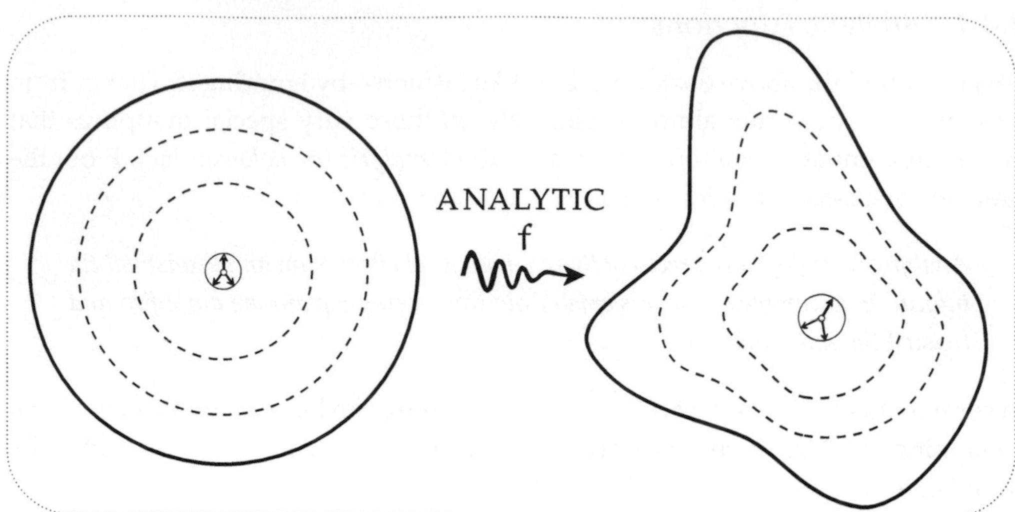

[4.11] Analytic mappings only send infinitesimal circles to circles. *While an analytic mapping must send* infinitesimal *circles to circles, it will generally deform an expanding circle into a shape that becomes less and less circular. The* only *analytic mappings that preserve circles of all sizes are the Möbius transformations.*

precisely do preserve circles of all sizes. In fact it can be shown that the Möbius transformations are the *only*[2] ones with this property.

4.4.4 A Brief Summary

The principal kinds of mapping we wish to study in this book are the analytic (complex-differentiable) ones. Although these will turn out to include almost all the useful functions, they are nevertheless very special. Their effect on an infinitesimal disc centred at z is, after translation to $f(z)$, simply to amplify and twist it. The *amplification* is the expansion factor, and the *twist* is the angle of rotation. The local effect of f is then completely encoded in the single complex number $f'(z)$, the derivative of f, or (as we will often prefer to call it) the *amplitwist* of f:

$$\begin{aligned} f'(z) &= \text{the derivative of f at } z \\ &= \text{the amplitwist of f at } z \\ &= (\text{amplification})\ e^{i(\text{twist})} \\ &= |f'(z)|\ e^{i \arg[f'(z)]}. \end{aligned}$$

To obtain the image at $f(z)$ of an infinitesimal complex number at z, you just multiply it by $f'(z)$.

Two last points. We have introduced the word "amplitwist" (in addition to "derivative") because it is suggestive, and because it will make later reasoning

[2] This was proved by Carathéodory (1937).

much easier to explain. However, the student meeting this subject for the first time should be made aware of the fact that in all other books *only* the word derivative is used. Also, note that the two words are synonymous only to the extent (cf. Chapter 1) that a complex number can be identified with the similarity transformation it produces when each point is multiplied by it. Thus, for example, "to differentiate" will *not* mean the same as "to amplitwist": the former refers to the act of finding the derivative of a function, while the latter refers to the act of "amplifying and twisting" an infinitesimal geometric figure.

4.5 Some Simple Examples

In the following examples we have superimposed the image copy of $\mathbb{C}$ on the original one.

$$\underline{z \mapsto z + c.}$$

This represents a translation of the points by c. As we see in [4.12a], the length of complex numbers emanating from z is preserved, and hence the amplification is unity. Equally clearly, since no rotation is induced, the twist is zero. Hence

$$(z+c)' = \text{amplitwist of } (z+c) = 1\, e^{i0} = 1.$$

Notice how this is in complete accord with the familiar rule of real calculus, namely, that $\frac{d}{dx}(x+c) = 1$.

$$\underline{z \mapsto Az.}$$

If $A = a\, e^{i\alpha}$, then this represents the combination of an origin-centred expansion by a, and a rotation by α. It is clear in [4.12b] that any arrow at z (in particular an infinitesimal one) will suffer precisely the same amplification and twist as do the points of the plane themselves. Hence

$$(Az)' = \text{amplitwist of } (Az) = A.$$

While the meaning is richer, this is once again formally identical to the familiar result $\frac{d}{dx}(Ax) = A$.

$$\underline{z \mapsto z^2.}$$

Our earlier investigation revealed that at the point $z = r\, e^{i\theta}$ this mapping is locally an amplitwist with amplification $2r$ and twist θ, as illustrated in [4.5]. Hence,

$$(z^2)' = \text{amplitwist of } (z^2) = (\text{amplification})\; e^{i(\text{twist})} = 2r\, e^{i\theta} = 2z\,.$$

Once again, note that this result is formally identical to the formula $(x^2)' = 2x$ of ordinary calculus. In the next chapter we shall obtain a directly complex and geometrical demonstration of this fact.

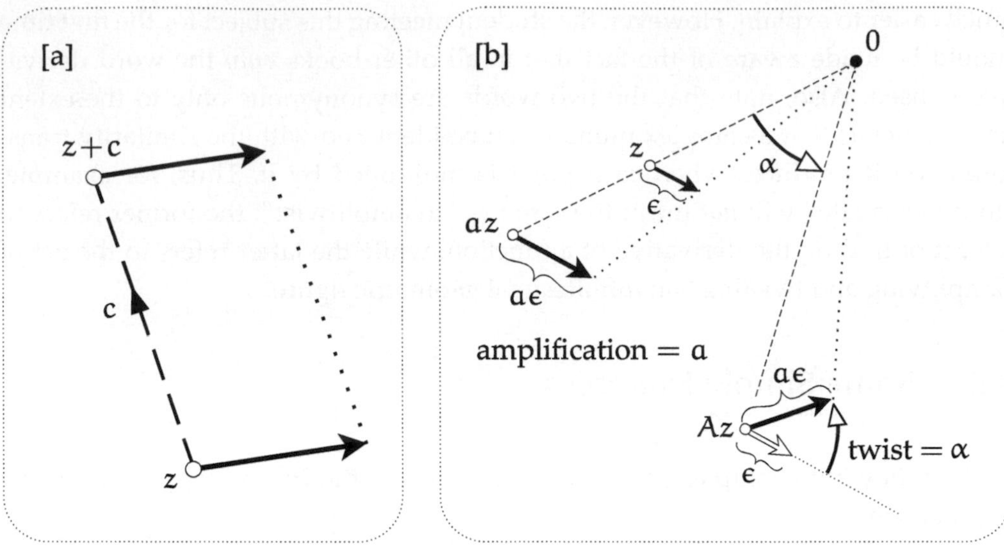

[4.12] The complex amplitwist formula agrees with the real derivative. [a] *The translation* $z \mapsto z + c$ *has* amplification 1 *and* twist 0, *so its amplitwist is* 1, *in accordance with the real formula,* $(x+c)' = 1$. **[b]** *The mapping* $z \mapsto Az = a\,e^{i\alpha}z$ *has amplification* a *and twist* α, *so its amplitwist is* $a\,e^{i\alpha} = A$, *in accordance with the real formula,* $(Ax)' = A$.

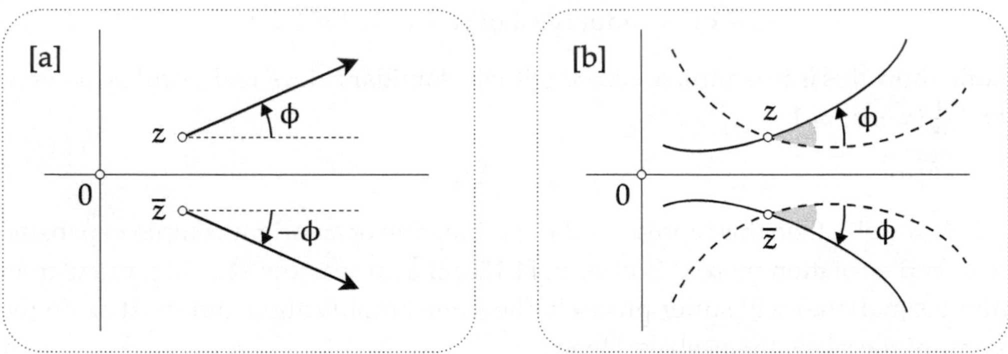

[4.13] Conjugation is anticonformal and therefore not analytic. *The mapping* $z \mapsto \overline{z}$ *is anticonformal, so it is not locally an amplitwist, and the derivative does not exist.*

$$\underline{z \mapsto \overline{z}.}$$

Since this mapping is anticonformal, it clearly cannot be analytic, for we have already observed that if a mapping is locally an amplitwist, then it is automatically conformal. Figure [4.13a] pinpoints the trouble. From the picture we see that the image at $\overline{z}$ of any complex number emanating from z has the same length as the original, and hence the amplification is unity. The problem lies in the fact that an arrow at angle ϕ must be rotated by -2ϕ to obtain its image arrow at angle $-\phi$. Thus different arrows must be rotated different amounts (which is not a twist) and hence there is no amplitwist.

4.6 Conformal = Analytic

4.6.1 Introduction

In [4.5] we saw clearly that any mapping that is locally an amplitwist is also automatically conformal. In terms of complex differentiation, we can now rephrase this by saying that all analytic functions are conformal. The question then naturally arises as to whether the converse might also be true. Is every conformal mapping analytic, or, in other words, is the local effect of every conformal mapping nothing more complicated than an amplitwist? If this were the case then the two concepts would be equivalent and we would have a new way of recognizing, and perhaps reasoning about, analytic functions. A tempting prospect!

To dismiss this as a possibility would only require the discovery of a single function that is conformal and yet whose local effect is not an amplitwist. The example of complex conjugation, illustrated in [4.13b], shows how important it is that we take into account the fact that the mapping preserves not only the magnitude of angles, but also their *sense*. For $z \mapsto \overline{z}$ is not analytic, but it is also not a counterexample to the conjecture, because it is *anti*conformal.

We have seen that although conjugation does possess an amplification, it fails to be analytic because it doesn't have a twist. Let us now consider instead a function that does possess a twist, but which again fails to be analytic, this time by virtue of not having an amplification. The effect of such a mapping at a particular point is illustrated in [4.14]. The three curves on the LHS intersect at equal angles of $(\pi/3)$, and on the RHS their images do too. But the picture clearly shows that we are not dealing with an amplitwist. Imagine that the infinitesimal tangent complex numbers to the curves are first twisted, but then rather than being amplified, as they would be by an analytic function, they are expanded by *different* factors. Despite

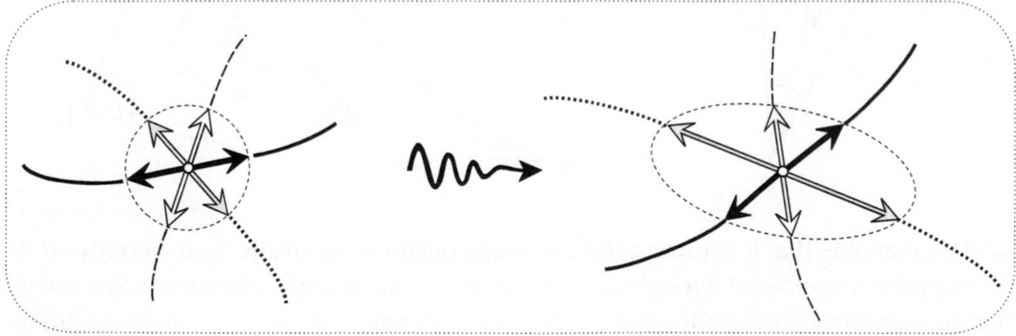

[4.14] A twist without an amplitwist. *At a single point it is possible for a mapping to possess a twist, and therefore to be conformal at that point, and yet* not *possess an amplification, and therefore* not *be an amplitwist.*

this, however, the initial twist ensures that the angle between two curves is preserved both in magnitude and sense: the mapping is genuinely conformal at this point.

4.6.2 Conformality Throughout a Region

If we only insist on conformality at isolated points then such counterexamples do indeed exist (we've drawn one!), but if we require the mapping to be conformal *throughout a region* then this nonanalytic behaviour cannot occur.

Imagine that we have a region throughout which the mapping is (i) conformal, and (ii) sufficiently non-pathological that an infinitesimal line-segment is mapped to another infinitesimal line-segment. In fact, re-examination of [4.3] reveals that (ii) must be presupposed in order for (i) even to make sense. For if the infinitesimal straight piece of curve from q to p did not map to another of the same kind at Q, then we could not even speak of an angle of intersection at Q, let alone of its possible equality with ϕ.

Now look at [4.15]. In our conformal region we have drawn a large (i.e., not infinitesimal) triangle abc, along with its image ABC. Notice that while the straight edges of abc are completely distorted to produce the curvilinear edges of ABC, the angles of this "triangular" image are identical with those of the original. Now imagine shrinking abc down towards an arbitrary point in the region. As we do so, the sides of its shrinking image will increasingly resemble straight lines [by virtue

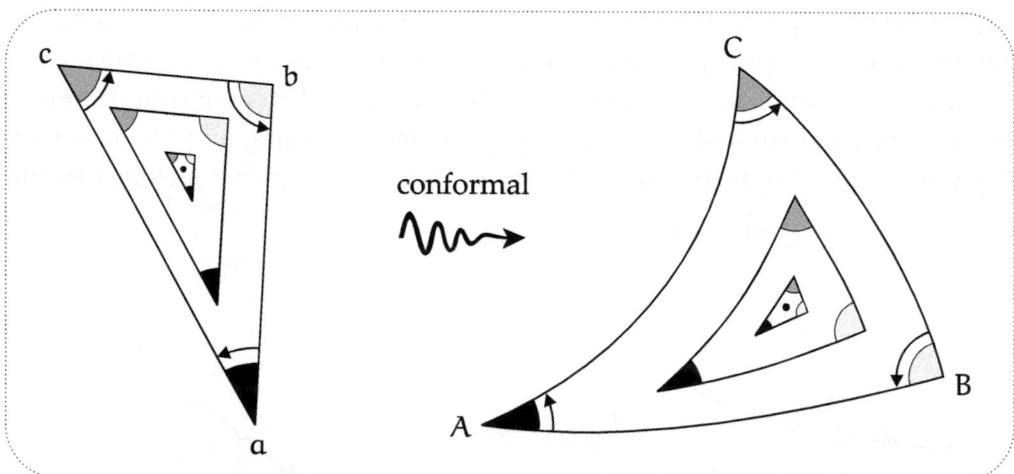

[4.15] A mapping that is conformal throughout a region must locally be an amplitwist. *If the mapping is conformal through the region in which the triangle* abc *shrinks to a point, then its curvilinear triangular image* ABC *must ultimately become a* similar *infinitesimal triangle, which can therefore be obtained by amplitwisting the original infinitesimal triangle.*

of (ii)], and all the while the angles will remain the same as the original's. Thus, any infinitesimal triangle in this region is mapped to another infinitesimal *similar* triangle. Since the image triangle merely has a different size and alignment on the page, it is indeed obtained by amplitwisting the original.

We have thus established the sought-after equivalence of conformal and analytic mappings:

> *A mapping is locally an amplitwist at a point* p *if it is conformal throughout an infinitesimal neighbourhood of* p.

For this reason, the conventional definition of f being "analytic" at p is that f′ exist at p and at all points in an infinitesimal neighbourhood of p.

From this result we can immediately deduce, for example, that complex inversion $z \mapsto (1/z)$ is analytic, for we have already demonstrated geometrically that it is conformal. By the same token, it follows more generally that all Möbius transformations are analytic.

For no extra charge, we can obtain a further equivalence simply by concentrating on distances rather than angles. What we have just seen is that a mapping cannot possess a twist throughout a region without also having an amplification. In order to investigate the converse, suppose that a mapping is only known to possess an amplification throughout a region. Re-examine [4.15] from this point of view. Unlike the previous case, there is no longer any *a priori* reason for the image ABC to betray any features common to the original. However, as we carry out the same shrinking process as before, the local existence of amplifications begins to reveal itself.

As the triangle becomes very small, we may consider two of its sides, for example ab and ac, to be infinitesimal arrows emanating from a vertex. While we may not yet know anything of angles, we do know that these arrows both undergo the same amplification to produce their images AB and AC. But if we now apply this reasoning at one of the other vertices, we immediately find that in order to be consistent, *all three* sides must undergo the same[3] amplification. Once again we have been able to deduce that the image triangle is similar to the original.

However, this time all we know is that the *magnitude* of the angles in the infinitesimal image triangle are the same as those in the original. If the sense of the angles also agree, then the image is obtained by amplitwisting the original, just as before. But if the angles are reversed, then we must flip the original triangle over as well

[3] We only mean "same" in the sense that the variations in amplification are of the same infinitesimal order as the dimensions of abc. If the amplifications were *precisely* the same, then extending our argument to a whole network of closely spaced vertices, we would conclude that the amplification was constant throughout the region.

as amplitwisting it. This "flip" may be accomplished by reflecting in any line; in particular, we may employ reflection in the real axis, $z \mapsto \overline{z}$. Thus if $f(z)$ is a mapping that is known to possess an amplification throughout an infinitesimal neighbourhood of a point p, then either $f(z)$ is analytic at p, or else $\overline{f(z)}$ is analytic at p.

It is interesting to note that the use of triangles in the above arguments was not incidental, but instead crucial. Rectangles, for example, would simply not have sufficed. Take the first argument. Certainly conformality still guarantees us that an infinitesimal rectangle maps to another infinitesimal rectangle. However, this image rectangle could in principle have very different proportions from the original, and hence not be obtainable via an amplitwist. Try the second argument for yourself and see how it too fails.

For a computational approach to the above results, see Ahlfors (1979, p. 73).

4.6.3 *Conformality and the Riemann Sphere*

In the previous chapter we addressed a twin question: "How are we to visualize the effect of a mapping on infinitely remote parts of the complex plane, or the effect of a mapping that hurls finite points into the infinite distance?" Our answer was to replace both complex planes (original and image) with Riemann spheres. We could then visualize the mapping as taking place between the two spheres, rather than between the two planes. To a large extent the success of this merely depended upon the fact that we had gathered up the infinite reaches of the plane to a single point on the sphere. It did not depend on the precise manner in which we chose to do this. Why then the insistence on accomplishing this with stereographic projection, rather than in some other way? Several reasons emerged in the previous chapter, but the present discussion shows that another compelling reason is that stereographic projection is *conformal*.

Only now can we fully appreciate this point, for we have seen that analytic functions are the conformal mappings of the plane. As illustrated in [4.16], the conformality of stereographic projection now enables us to translate this directly into a statement about Riemann spheres:

> *A mapping between spheres represents an analytic function if and only if it is conformal.*

We have drawn the spheres separate from the planes to reinforce the idea that we are entitled to let the plane fade from our minds, and to adopt instead the sphere as a logically independent base of operations. Indeed, at this stage we could consider complex analysis to be nothing more than the study of conformal maps between spheres. But in works on Riemann surfaces it is shown that in order to embrace the global aspects of many-one functions and their inverses, one must extend this conception to conformal mappings between more general surfaces, such as doughnuts.

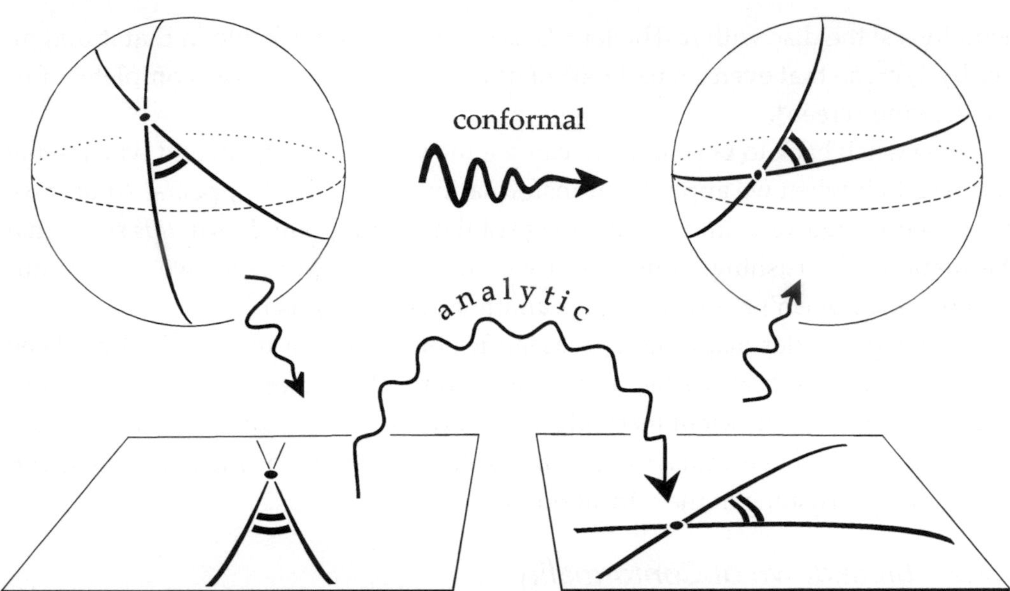

[4.16] Analytic mappings of $\mathbb{C}$ induce conformal mappings of the Riemann sphere, *for the simple reason that stereographic projection is conformal.*

4.7 Critical Points

4.7.1 Degrees of Crushing

We return to the mapping z^2 and note that at $z = 0$, $(z^2)' = 2z = 0$. A place such as this, where the derivative vanishes, is called a *critical point*. Recall that in the previous chapter we defined the term "critical point" differently, as a point at which the conformality of an otherwise conformal mapping breaks down. These two definitions are not at odds with one another. If the derivative $f'(z)$ of an analytic mapping f is not zero at $z = p$, then we know that f is conformal at p, so conformality can only break down at points where $f'(z) = 0$. Although it is not obvious, later we will be able to prove the converse fact that if $f'(p) = 0$ then f *cannot* be conformal at p. Thus the two definitions are equivalent.

In terms of the amplitwist concept a critical point could equally well be defined as a point of zero amplification. This suggests that the effect of an analytic mapping on an infinitesimal disc centred at a critical point is to "crush it down to a single image point". The statement in quotes is not to be taken literally, rather it is to be understood in the following sense.

Imagine that the disc (radius ϵ) is so tiny that it must be placed under a microscope in order to be seen. Suppose that we have available a whole family of lenses of increasing power with which to view it: $L_0, L_1, L_2, L_3, \ldots$. For example, L_0 has magnification $1/\epsilon^0 = 1$, so that it's really no better than using the naked eye. On the other hand L_1 has magnification $1/\epsilon^1$, and it is thus so powerful that we can actually see the disc with it. The lens L_2 is even more remarkable in that it magnifies by $1/\epsilon^2$, so that even a small part of our microscopic disc now completely fills the viewing screen[4].

Let's switch back to L_1 so that we can see the whole disc again, and watch what happens to it when we apply the transformation $z \mapsto z^2$. It disappears! At best we might see a single dot sitting at the image of the critical point. It is in *this* sense that the mapping is crushing. However, if we now attach L_2 instead, we can see our mistake: the dot isn't a dot, in fact it's another disc of radius ϵ^2.

For this particular mapping, L_2 was sufficient to see that the disc had not been completely crushed. However, at a critical point of another mapping, even this might not provide sufficient magnification, and we would require a stronger lens, say L_m, to reveal that the image of the disc isn't just a point. The integer m measures the degree of crushing at the critical point.

4.7.2 Breakdown of Conformality

In addition to being locally crushing, we have stated (but not yet proved) that the conformality of an analytic function breaks down at its critical points. We can see this in our example. When the z^2 mapping acts on a pair of rays through the critical point $z = 0$, it fails to preserve the angle between them; in fact it doubles it. Thus, just at the critical point, the conformality of z^2 breaks down. This is a general property. In fact we will show later that the behaviour of a mapping very near to a critical point is essentially given by z^m, $m \geqslant 2$. Rather than being conserved, angles at the critical point $z = 0$ are consequently multiplied by m. We quantify the degree[5] of this strange behaviour by saying that $z = 0$ is a critical point of *order* $(m - 1)$. Notice that this m is the same one as in the previous paragraph: in order to see the image we have to use the L_m lens.

Despite the fact that conformality breaks down at critical points, we shall continue to make such bald statements as, "z^2 is conformal". The tacit assumption is that critical points are being excluded. Indeed we were making this assumption throughout the previous section, for we only concerned ourselves there with typical points. Later we will see that critical points are, in a mathematically precise sense, "few and far between", and this is our excuse for the scant attention we are presently paying them. Nevertheless, we may safely skirt around this issue

[4] In terms of this analogy we could say that most of the diagrams in this chapter, indeed in the rest of the book, show views of the image complex plane taken through L_1. For example [4.10] depicts a tiny circle being amplitwisted to produce another circle. However, if we viewed part of this image 'circle' with L_2 instead of L_1, then deviations from circularity would become visible. Of course the smaller we make the preimage circle, the smaller these deviations will be.

[5] The reason we define it to be $(m - 1)$, rather than m, is that this properly reflects the multiplicity of the root of the derivative.

only so long as we focus on the effect of the function on separate chunks of its domain. When one studies Riemann surfaces, one tries to fit all this partial information into a global picture of the mapping, and in achieving this the critical points will play a crucial role. They do so by virtue of yet another aspect of the peculiar behaviour of a mapping in the vicinity of such points, and it is to this feature that we now turn.

In the previous chapter we discussed the possibility of critical points being located at infinity. In particular, we considered $z \mapsto z^m$. On the Riemann sphere we drew two straight lines passing through the origin, and we thereby saw that angles at both $z = 0$ and $z = \infty$ were multiplied by m. We therefore conclude that ∞ is a critical point of z^m of order $(m - 1)$, just like the origin. Actually, except for $m = 2$, we don't yet know if z^m is conformal *anywhere!* However, in the next chapter we will see that it is conformal everywhere except at the two critical points we have just discussed.

4.7.3 Branch Points

First consider the case of a real function $R(x)$ from $\mathbb{R}$ to $\mathbb{R}$. In solving problems on maxima and minima, we learn from an early age the importance of finding the places where $R'(x) = 0$. Figure [4.17] shows an ordinary graph of $y = R(x)$, emphasizing a different aspect of the behaviour of R near to a "critical point" c where $R'(c) = 0$. Above a typical point t, for which $R'(t) \neq 0$, the graph is either going up or going down, so the function is locally one-to-one. However, near c it is clearly *two*-to-one.

An analogous significance holds for complex mappings. Typically $f'(z) \neq 0$, and so an infinitesimal neighbourhood of z is amplitwisted to an infinitesimal image neighbourhood of $w = f(z)$, and the two neighbourhoods are clearly in one-to-one correspondence. However, if $f'(z_0) = 0$ then (according to our earlier claim) near to z_0 the function behaves like z^m, with $m \geqslant 2$. Thus, if a point is in a close orbit around z_0, its image will orbit w_0 at m-times the angular speed, and corresponding to each point near w_0 there will be m preimages near z_0. Thus w_0 is a branch point of order $(m - 1)$. We conclude that *a critical point of a given order maps to a branch point of the same order.*

We began this idea by using an analogy with real functions, but we should also note an important difference. A real function $R(x)$ is necessarily one-to-one when $R'(x) \neq 0$, but (unlike the complex case) it *need not* be many-to-one when $R'(x) = 0$. The graph of x^3, for example, is flat at the origin and yet it is still one-to-one in an infinitesimal neighbourhood of that point. In contrast to this, the complex mapping $z \mapsto z^3$ is three-to-one near the origin, due to the existence of complex cube roots.

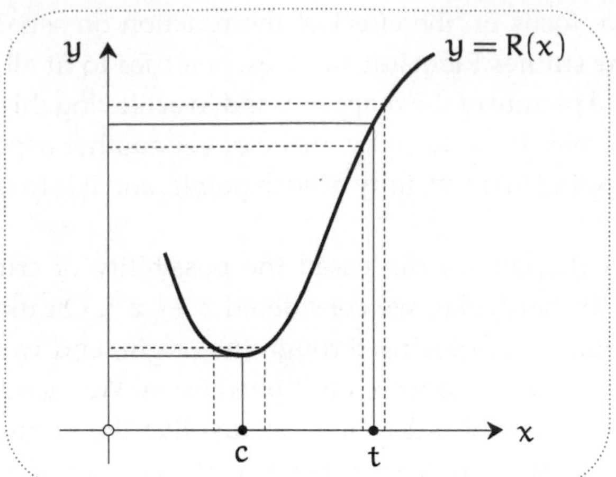

[4.17] Breakdown of one-to-oneness near critical points. *When* $R'(x) \neq 0$ *(as at* $x = t$*) the graph is either going up or down, so the function is one-to-one in the immediate vicinity of the point. But near a "critical point"* $x = c$*, defined by* $R'(c) = 0$*, we see that* R need not *be one-to-one. The same turns out to be true for complex analytic mappings, only more so: if* $f'(c) = 0$*, then* f *is* necessarily not *one-to-one in the immediate vicinity of* $z = c$.

4.8 The Cauchy–Riemann Equations

4.8.1 Introduction

To end this chapter we will try to gain a better perspective on where the analytic functions lie within the hierarchy of mappings of the plane. A benefit of this will be the discovery of another way (the third!) of characterizing analytic functions, this time in terms of their real and imaginary parts.

The first thing to do is realize that the "general" mappings $(x, y) \mapsto (u, v)$ that we considered earlier were not really as general as they could have been. Picture part of the plane as being a rolled out piece of pastry on a table. A general mapping corresponds to "doing something" to the pastry, thereby moving its points to new locations (the images) on the table. For example, we might cut the pastry in half and move the two pieces away from each other. This is much more general than anything we contemplated earlier, for it does not even possess the rudimentary quality of *continuity*. That is, if two points are on either side of the cut, then no matter how close we move them together, their images will remain far apart.

Even if we do insist on continuity, the resulting mappings are still more general than those we have considered. For example, imagine pressing down the rolling-pin somewhere in the middle of the pastry, and, in a single roll, stretching the far side to twice its former size. This certainly is continuous, for bringing two points together always brings their images together. The problem now lies in the fact that if two infinitesimal, diametrically opposed arrows emanate from a point beneath the starting position of the pin, then they each undergo a quite different transformation. Thus, in an obvious sense (not a subtle complex-differentiation sense) the mapping isn't differentiable at this point. Nevertheless, provided we stay away from this line, the mapping is differentiable in the real sense, and hence subject to our earlier analysis using the Jacobian matrix.

Another interesting kind of mapping arises from the commonplace operation of *folding* the pastry. Suppose we fold it like a letter being placed in an envelope [two creases]. Three different points will end up above a single point of the table, and the mapping is thus three-to-one. However, at the creases themselves the mapping is only one-to-one, and furthermore, differentiability also breaks down there. Nevertheless, provided that we only look at the fold-free portions of the pastry, such many-one functions are still subject to our previous analysis.

Suppose we play with the pastry in an ordinary way, rolling it (not necessarily evenly) now in this direction, now in another, then turning it, folding it, rolling it again, and so forth; then, provided we suitably restrict the domain, we can still apply our old analysis. While such a mapping is indeed very general, we hope that this discussion has revealed that (being continuous and differentiable in the real sense) it is, in fact, already quite high up the evolutionary ladder. It will therefore not come as such a surprise to learn that the local geometric effect of such a mapping is remarkably simple, though naturally not *as* simple as an amplitwist.

4.8.2 The Geometry of Linear Transformations

We pick up our earlier investigation where we left it. The local effect of the mapping is to perform the linear transformation encoded in the Jacobian matrix (4.2). If we can first understand the effect of a *uniform* linear transformation—corresponding to a constant matrix—then we shall be finished. For we need only then remember that our analysis is only applicable *locally*, the actual linear transformation varying as it does from one place to the next.

Consider the effect of a uniform linear transformation on a circle C. Since the Cartesian equation of C is quadratic, the linear change of coordinates induced by the transformation will lead to another quadratic equation for the image curve. The image curve E is thus a conic section, and since the finite points of C are not sent to infinity, this conic must be an *ellipse*. See [4.18], and compare this also with [4.9], where the local consequence of this result was illustrated for a *non*-uniform transformation acting on an infinitesimal circle.

We have just used an algebraic statement of linearity. The fundamental *geometric* fact is that it makes no difference if we add two vectors and then map the result, or if we map the vectors first and *then* add them. Convince yourself of these two simple consequences:

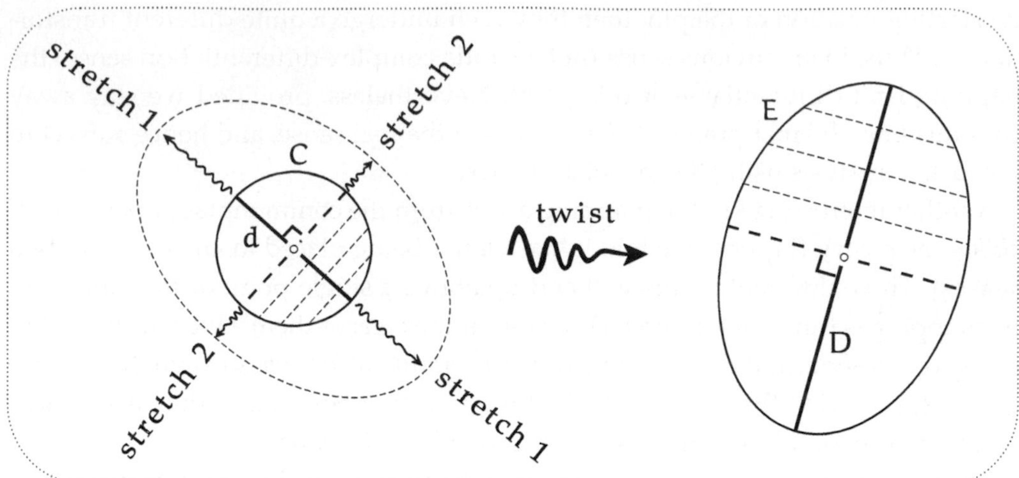

[4.18] Singular Value Decomposition. *A linear transformation maps the circle* C *to the ellipse* E. *Let* d *be the diameter of* C *that is mapped to the major axis* D. *Consider the chords of* C *[dashed] that are perpendicular to* d. *Since these are all bisected by* d, *their images must be a family of parallel chords of* E *such that* D *is their common bisector. They must therefore be the family perpendicular to* D. *We thereby deduce the* Singular Value Decomposition*:* A linear transformation is a stretch in the direction of d, another stretch perpendicular to it, and finally a twist.

- *Parallel lines map to parallel lines.*
- *The midpoint of a line-segment maps to the midpoint of the image line-segment.*

We now apply these facts to E.

Since all the diameters of C are bisected by the centre of C, it follows that the image chords of E must all pass through a common point of bisection. Thus the centre of C is mapped to the centre of E. Drawn in the same heavy line as its image is the particular diameter d of the circle that is mapped to the major axis D of the ellipse. Now consider the chords of C [dashed] that are perpendicular to d. Since these are all bisected by d, their images must be a family of parallel chords of E such that D is their common bisector. They must therefore be the family perpendicular to D. All this is summarized in [4.18].

Thus we have obtained the following result:

> ***Local Singular Value Decomposition.*** *The local linear transformation induced by a (non-linear) complex mapping is a stretch in the direction of* d, *another stretch perpendicular to it, and finally a twist.* (4.5)

This result also makes sense at the level of counting degrees of freedom. Just as the matrix has four independent entries, so the specification of our transformation also requires four bits of information: the direction of d, the stretch factor in this direction, the perpendicular stretch factor, and the twist.

[REMARKS: This *Singular Value Decomposition* (SVD for short) is of enormous importance in Linear Algebra and its real-world applications. It was discovered by Eugenio Beltrami,[6] whom we shall meet again in Chapter 6, because he also made pivotal discoveries in Hyperbolic Geometry. In the context of the standard SVD theorem, the two orthogonal expansion factors are called the *singular values* of the transformation. In VDGF[7] we elaborate on these ideas, and use them to provide a *simple geometrical interpretation of the transpose of a matrix*, which we have not found in any standard Linear Algebra textbook. This in turn can be used to provide geometrical explanations of other standard results in Linear Algebra.]

The ultimate specialization to analytic functions now simply requires that the two stretch factors be put equal. This apparently reduces the number of degrees of freedom from four to three. However, since we are now producing an equal expansion in all directions, the direction chosen for d becomes irrelevant, and we are left with only *two* genuine degrees of freedom: the amplification and the twist.

Note that we now have the following:

> *An orientation preserving mapping is conformal if and only if it sends infinitesimal circles to infinitesimal circles.*

If a mapping preserves circles in general, then, in particular, it must send infinitesimal circles to infinitesimal circles, and hence it must be conformal[8]. Bypassing the detailed investigation of the previous chapter, we now see that the conformality/analyticity of Möbius transformations follows from the mere fact that they preserve circles.

4.8.3 The Cauchy–Riemann Equations

We obtain another characterization of analytic functions if we now ask how we may recognize a Jacobian matrix for which both expansion factors are equal. This is most easily answered by considering what kind of matrix corresponds to multiplication by a complex number, for we already know that this produces the desired type of linear transformation. Multiplying $z = (x + iy)$ by $(a + ib)$, we get

$$(x + iy) \mapsto (a + ib)\,(x + iy) = (ax - by) + i\,(bx + ay).$$

This corresponds to the multiplication of a vector in $\mathbb{R}^2$ by the matrix

$$\begin{pmatrix} a & -b \\ b & a \end{pmatrix}. \tag{4.6}$$

[6] See Stewart (1993).
[7] See Needham (2021, §15.4).
[8] For a different proof of this fact, see Sommerville (1958, p. 237).

Compare this with the Jacobian matrix (4.2),

$$J = \begin{pmatrix} \partial_x u & \partial_y u \\ \partial_x v & \partial_y v \end{pmatrix}.$$

In order for the effect of J to reduce to an amplitwist, it must have the same form as (4.6), and thus

$$\begin{aligned} \partial_x u &= +\partial_y v, \\ \partial_x v &= -\partial_y u. \end{aligned} \tag{4.7}$$

These are the celebrated *Cauchy–Riemann equations*. They provide us with a third way of recognizing an analytic function. However, as with the underlying amplitwist concept, these equations must be satisfied throughout an infinitesimal neighbourhood of a point in order that the mapping be analytic there [see Ex. 12].

Since $(a + ib)$ is playing the role of the amplitwist, comparison of (4.6) and (4.2) now yields two formulae for the derivative:

$$f' = \partial_x u + i\,\partial_x v = \partial_x f, \tag{4.8}$$

and

$$f' = \partial_y v - i\,\partial_y u = -i\,\partial_y f. \tag{4.9}$$

By way of example, consider $z \mapsto z^3$. Multiplying this out we obtain a rather haphazard looking mess:

$$u + iv = (x^3 - 3xy^2) + i\,(3x^2y - y^3).$$

However, differentiating the real and imaginary parts, we obtain

$$\begin{aligned} \partial_x u &= 3x^2 - 3y^2 &= +\partial_y v, \\ \partial_x v &= 6xy &= -\partial_y u, \end{aligned}$$

and so the Cauchy–Riemann equations are satisfied. Thus, far from being haphazard, the special forms of u and v have ensured that the mapping is analytic. Using (4.8) we can calculate the amplitwist:

$$(z^3)' = 3(x^2 - y^2) + i\,6xy = 3z^2,$$

just as in ordinary calculus. Check that (4.9) gives the same answer.

In the next chapter we will sever our umbilical cord to $\mathbb{R}^2$ and discover how the above results can be better understood by directly appealing to the geometry of the complex plane.

4.9 Exercises

1 Use the Cauchy–Riemann equations to verify that $z \mapsto \bar{z}$ is not analytic.

2 The mapping $z \mapsto z^3$ acts on an infinitesimal shape and the image is examined. It is found that the shape has been rotated by π, and its linear dimensions expanded by 12. Where was the shape originally located? [There are two possibilities.]

3 Consider $z \mapsto \Omega(z) = \bar{z}^2/z$. By writing z in polar form, find out the geometric effect of Ω. Using two colours, draw two very small arrows of equal length emanating from a typical point z: one parallel to z; the other perpendicular to z. Draw their images emanating from $\Omega(z)$. Deduce that Ω fails to produce an amplitwist. [Your picture should show this in *two* ways.]

4 The picture shows the shaded interior of a curve being mapped by an analytic function to the exterior of the image curve. If z travels round the curve counterclockwise, then which way does its image w travel round the image curve? [*Hint*: Draw some infinitesimal arrows emanating from z, including one in the direction of motion.]

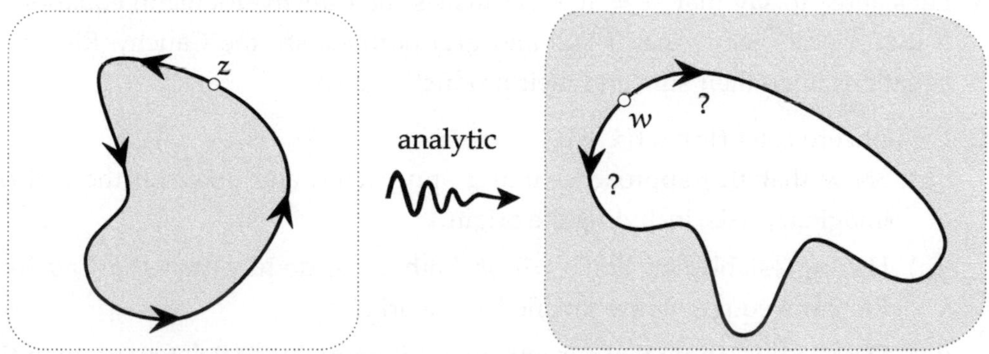

5 Consider $f(x + iy) = (x^2 + y^2) + i(y/x)$. Find and sketch the curves that are mapped by f into (a) horizontal lines, and (b) vertical lines. Notice from your answers that f appears to be conformal. Show that it is *not* in two ways: (i) by explicitly finding some curves whose angle of intersection isn't preserved; and (ii) by using the Cauchy–Riemann equations.

6 Continuing from the previous exercise, show that *no* choice of v can make $f(x + iy) = (x^2 + y^2) + iv$ analytic.

7 (i) If $g(z) = 3 + 2i$ then explain geometrically why $g'(z) \equiv 0$.

(ii) Show that if the amplification of an analytic function is identically zero (i.e., $f'(z) \equiv 0$) on some *connected* region, then the function is constant there.

(iii) Give a simple counterexample to show that this conclusion does not follow if the region is instead made up of disconnected components.

8 Use pictures to explain why if $f(z)$ is analytic on some *connected* region, each of the following conditions forces it to reduce to a constant.

(i) $\mathrm{Re} f(z) = 0$

(ii) $|f(z)| = \text{const.}$

(iii) Not only is $f(z)$ analytic, but $\overline{f(z)}$ is too.

9 Use the Cauchy–Riemann equations to give rigorous computational proofs of the results of the previous two exercises.

10 Instead of writing a mapping in terms of its real and imaginary parts (i.e. $f = u+iv$), it is sometimes more convenient to write it in terms of length and angle:

$$f(z) = R\,e^{i\Psi},$$

where R and Ψ are functions of z. Show that the equations that characterize an analytic f are now

$$\partial_x R = R\,\partial_y \Psi \quad \text{and} \quad \partial_y R = -R\,\partial_x \Psi.$$

11 Let's agree to say that "$f = u + iv$ satisfies the Cauchy–Riemann equations" if u and v do. Show that if $f(z)$ and $g(z)$ both satisfy the Cauchy–Riemann equations, then their sum and their product do also.

12 For nonzero z, let $f(z) = f(x+iy) = xy/\overline{z}$.

(i) Show that $f(z)$ approaches 0 as z approaches any point on the real or imaginary axis, including the origin.

(ii) Having established that $f=0$ on both axes, deduce that the Cauchy–Riemann equations are satisfied at the origin.

(iii) Despite this, show that f is not even differentiable at 0, let alone analytic there! To do so, find the image of an infinitesimal arrow emanating from 0 and pointing in the direction $e^{i\phi}$. Deduce that while f does have a twist at 0, it fails to have an amplification there.

13 Verify that $z \mapsto e^z$ satisfies the Cauchy–Riemann equations, and find $(e^z)'$.

14 By sketching the image of an infinitesimal rectangle under an analytic mapping, deduce that the local magnification factor for area is the square of the amplification. Rederive this fact by looking at the determinant of the Jacobian matrix.

15 Let us define S to be the square region given by

$$a - b \leqslant \mathrm{Re}(z) \leqslant a + b \quad \text{and} \quad -b \leqslant \mathrm{Im}(z) \leqslant b.$$

(i) Sketch a typical S for which $b < a$. Now sketch its image $\widetilde{S}$ under the mapping $z \mapsto e^z$.

(ii) Deduce the area of $\widetilde{S}$ from your sketch, and write down the ratio

$$\Lambda \equiv \left(\frac{\text{area of } \widetilde{S}}{\text{area of } S} \right).$$

(iii) Using the results of the previous two exercises, what limit should Λ approach as b shrinks to nothing?

(iv) Find $\lim_{b \to 0} \Lambda$ from your expression in part (ii), and check that it agrees with your geometric answer in part (iii).

16 Consider the complex inversion mapping $I(z) = (1/z)$. Since I is conformal, its local effect must be an amplitwist. By considering the image of an arc of an origin-centred circle, deduce that $|(1/z)'| = 1/|z|^2$.

17 Consider the complex inversion mapping $I(z) = (1/z)$.

(i) If $z = x + iy$ and $I = u + iv$, express u and v in terms of x and y.

(ii) Show that the Cauchy–Riemann equations are satisfied everywhere except the origin, so that I is analytic except at this point.

(iii) Find the Jacobian matrix, and by expressing it in terms of polar coordinates, find the local geometric effect of I.

(iv) Use (4.8) to show that the amplitwist is $-(1/z^2)$, just as in ordinary calculus, and in accord with the previous exercise. Use this to confirm the result of part (iii).

18 Recall Ex. 19, p. 210, where you showed that a general Möbius transformation

$$M(z) = \frac{az + b}{cz + d},$$

maps concentric circles to concentric circles if and only if the original family (call it $\mathcal{F}$) is centred at $q = -(d/c)$. Let $\rho = |z - q|$ be the distance from q to z, so that the members of $\mathcal{F}$ are $\rho = \text{const}$.

(i) By considering orthogonal connecting vectors from one member of $\mathcal{F}$ to an infinitesimally larger member of $\mathcal{F}$, deduce that the amplification of M is constant on each circle of $\mathcal{F}$. Deduce that $|M'|$ must be a function of ρ alone.

(ii) By considering the image of an infinitesimal shape that starts far from q and then travels to a point very close to q, deduce that at some point in the journey the image and preimage are *congruent*.

(iii) Combine the above results to deduce that there is a special member I_M of $\mathcal{F}$ such that infinitesimal shapes on I_M are mapped to congruent image shapes on the image circle $M(I_M)$. Recall that I_M is called the *isometric circle* of M.

(iv) Use the previous part to explain why $M(I_M)$ has the same radius as I_M.

(v) Explain why $I_{M^{-1}} = M(I_M)$.

(vi) Suppose that M is normalized. Using the idea in Ex. 16, show that the amplification of M is

$$|M'(z)| = \frac{1}{|c|^2\,\rho^2}.$$

19 Consider the mapping $f(z) = z^4$, illustrated below. On the left is a particle p travelling upwards along a segment of the line $x = 1$, while on the right is the image path traced by $f(p)$.

(i) Copy this diagram, and by considering the length and angle of p as it continues its upward journey, sketch the continuation of the image path.

(ii) Show that $A = i\,\sec^4(\pi/8)$.

(iii) Find and mark on your picture the two positions (call them b_1 and b_2) of p that map to the self-intersection point B of the image path.

(iv) Assuming the result $f'(z) = 4z^3$, find the *twist* at b_1 and also at b_2.

(v) Using the previous part, show that (as indicated at B) the image path cuts itself at *right angles*.

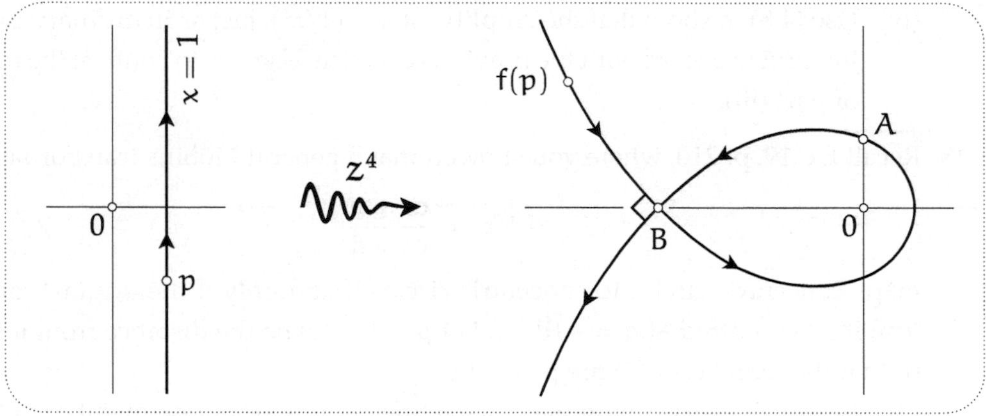

20 The figure below is a copy of [2.9], p. 69.

(i) Show geometrically that if z moves distance ds along the lemniscate, increasing θ by dθ, then $w = z^2$ moves a distance 4 dθ along the circle.

(ii) Using the fact that $(z^2)' = 2z$, deduce geometrically that $ds = 2\,d\theta/r$.

(iii) Using the fact that $r^2 = 2\cos 2\theta$, show by calculation that

$$r\,dr = 2\sqrt{1-(r^4/4)}\;d\theta.$$

(iv) Let s represent the length of the segment of the lemniscate connecting the origin to the point z. Deduce from the previous two parts that

$$s = \int_0^r \frac{dr}{\sqrt{1 - (r^4/4)}},$$

hence the name, *lemniscatic integral*.

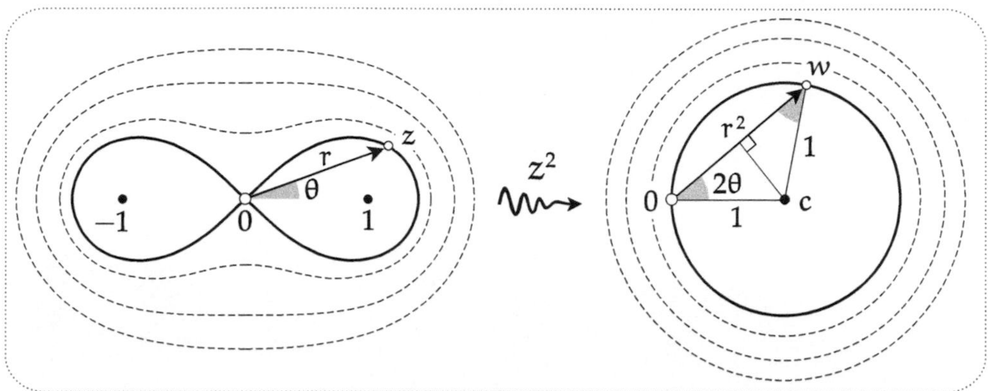

21 (i) **Three-Dimensional SVD:** By extending the argument given in the text, show that in *three*-dimensional space the effect of a linear transformation is to stretch space in three mutually perpendicular directions (generally by three different factors), then to rotate it.

(ii) Deduce that a mapping of three-dimensional space to itself is locally a three-dimensional amplitwist if and only if it maps infinitesimal spheres to infinitesimal spheres.

(iii) Deduce that inversion in a sphere preserves the magnitude of the angle contained by two intersecting curves in space.

(iv) Deduce that stereographic projection is conformal.

Remark: In stark contrast to the bountiful conformal mappings of the plane, Liouville and Maxwell independently discovered that the *only* angle-preserving transformation of space is an inversion, or perhaps the composition of several inversions.

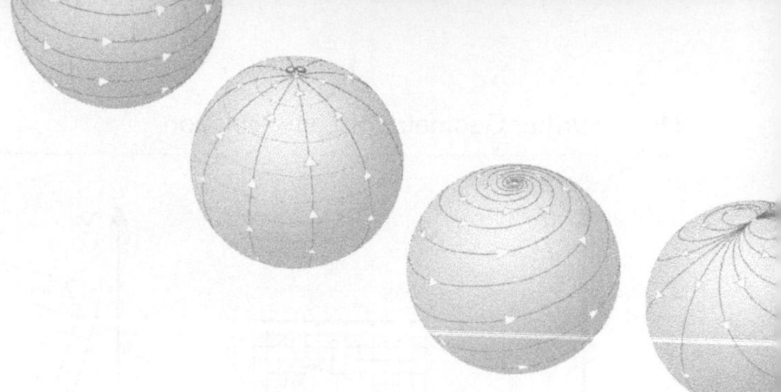

CHAPTER 5

Further Geometry of Differentiation

5.1 Cauchy–Riemann Revealed

5.1.1 Introduction

In the previous chapter we began to investigate the remarkable nature of analytic functions in $\mathbb{C}$ by studying mappings in the less structured realm of $\mathbb{R}^2$. In particular, the Jacobian provided us with a painless way of deriving the Cauchy–Riemann characterization of analytic functions, and also of computing their amplitwists. However, this approach was rather indirect.

In this chapter we will instead study differentiation directly in the complex plane, primarily through the use of *ultimate equality*—the powerful geometrical tool that Newton unleashed in his great *Principia* of 1687; see both the original and the new Preface. However, in the original edition these arguments were expressed mainly in the *language* of infinitesimals, and we have left this mode of expression largely intact. But in the *captions*, all of which are new to this edition, we have felt free to employ the $\asymp$-notation.

Our first application of this approach will be the rederivation of the Cauchy–Riemann (henceforth "CR") equations, and the discovery of new forms that they can take on.

5.1.2 The Cartesian Form

Consider a very fine mesh of squares aligned with the real and imaginary axes. See the top left of [5.1]. Under an analytic mapping each infinitesimal square will be amplitwisted to produce an image that is also square. We will show that the CR equations are nothing more than a symbolic restatement of this geometric fact.

Visual Complex Analysis. 25th Anniversary Edition. Tristan Needham, Oxford University Press.
 DOI: 10.1093/oso/9780192868916.003.0005

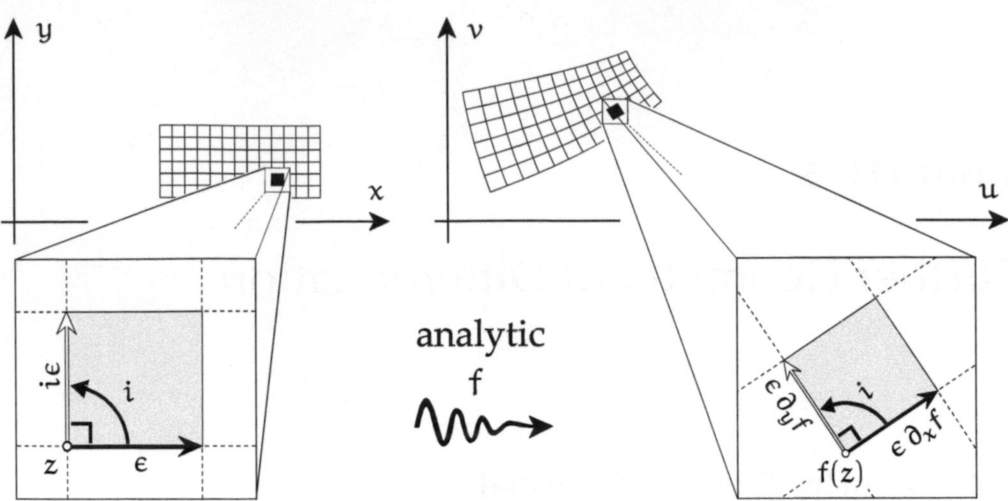

[5.1] Geometric derivation of the Cartesian Cauchy–Riemann equations. *The image of the real edge ϵ of the small (ultimately vanishing) square on the left is ultimately equal to $\epsilon\, \partial_x f$ on the right. Likewise, the image of the imaginary edge $i\epsilon$ of the square on the left is ultimately equal to $\epsilon\, \partial_y f$ on the right. But if f is analytic, then the image of the square on the left must ultimately be another square on the right. So one edge of the image is ultimately equal to the other edge rotated by $(\pi/2)$, that is, multiplied by i. Thus, $\epsilon\, \partial_y f \asymp i\epsilon\, \partial_x f \implies \partial_y f = i\, \partial_x f$, which is the compact Cartesian form of the celebrated Cauchy–Riemann equations.*

Zoom in on an individual square and its image, as depicted in the bottom half of [5.1]. Suppose, as drawn, that the initial square has side ϵ. If we start at z and then move a distance ϵ in the x-direction, the image will move along a complex number given by

$$\text{(change in x)}\cdot\text{(rate of change with x of the image f)} = \epsilon\, \partial_x f.$$

Similarly, if the point moves along the vertical edge by going ϵ in the y-direction, then its image will move along $\epsilon\, \partial_y f$. Now since these two image vectors span a square they must be related by a simple rotation of $\pi/2$, that is by multiplication with i. After cancelling ϵ, we thus obtain

$$i\partial_x f = \partial_y f, \tag{5.1}$$

et voilà! That this is indeed a compact form of the CR equations may be seen by inserting $f = u + iv$:

$$i\partial_x(u + iv) = \partial_y(u + iv),$$

and then equating real and imaginary parts to yield

$$\partial_x u = \partial_y v \quad \text{and} \quad \partial_x v = -\partial_y u, \tag{5.2}$$

just as before.

To obtain the amplitwist itself, we recall that each infinitesimal arrow is taken to its image by multiplication with f'. Now, since we know what the images are for the two sides of the square, we easily deduce

$$\begin{aligned} \epsilon \longmapsto \epsilon f' &= \epsilon\, \partial_x f \\ \Rightarrow f' &= \partial_x f \end{aligned}$$

and

$$\begin{aligned} i\epsilon \longmapsto i\epsilon f' &= \epsilon\, \partial_y f \\ \Rightarrow f' &= -i\, \partial_y f. \end{aligned}$$

These are the same formulae (4.8) and (4.9) that we previously obtained, but *now* we understand their geometrical *meaning*.

5.1.3 The Polar Form

Equation (5.2) is the most common way of writing CR, but it isn't the only way. It took this form because we chose to describe both complex planes in terms of their real and imaginary parts, that is with Cartesian coordinates. Thus we could briefly describe (5.2) as being the Cart.–Cart. form. In Ex. 10, p. 240 we retained Cartesian coordinates for the first plane but employed polar coordinates in the image plane; this led to another form (Cart.–Polar) of CR. As the next example of our geometric method we will derive the Polar-Cart. form of the equations.

In order to do this, we begin with an infinitesimal square adapted to polar coordinates. See [5.2]. If we start at z and increase r by dr, then we obtain $e^{i\theta}\, dr$ as the

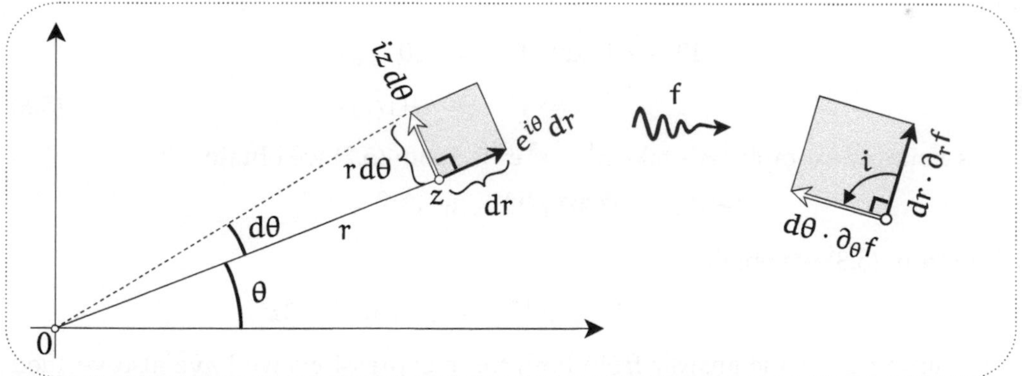

[5.2] Geometric derivation of the polar Cauchy–Riemann equations. *If z moves radially by dr then its image ultimately moves $dr \cdot \partial_r f$. If the angle of z increases by $d\theta$, ultimately causing z to move $r\, d\theta$ orthogonally, then its image ultimately moves $d\theta \cdot \partial_\theta f$, as illustrated. If the initial shape is ultimately a square (so that $dr \asymp r\, d\theta$) then an analytic f will ultimately map it to another* square, *so $d\theta \cdot \partial_\theta f \asymp i dr \cdot \partial_r f \asymp i r d\theta \cdot \partial_r f$.* **Thus the Cauchy–Riemann equations take the form $\partial_\theta f = i r\, \partial_r f$.**

radial edge. If, on the other hand, we increase θ by $d\theta$, then the point will move in the perpendicular direction given by $ie^{i\theta}$. As $d\theta$ tends to zero, this edge is ultimately equal to an infinitesimal arc of circle of length $r\,d\theta$; the complex number describing it will therefore be $i\,e^{i\theta}\,r\,d\theta = i\,z\,d\theta$. It's also clear from our picture that

$$\text{initially square} \quad \Longleftrightarrow \quad dr = r\,d\theta. \tag{5.3}$$

Now look at the image. Just as before, if we increase r by dr then the image will move along $dr \cdot \partial_r f$; likewise, changing θ by $d\theta$ will move the image along $d\theta \cdot \partial_\theta f$. If the mapping is analytic then these again span a square, and so the latter must be i times the former:

$$d\theta \cdot \partial_\theta f = i dr \cdot \partial_r f.$$

Substituting (5.3) into this, and cancelling $d\theta$, we obtain

$$\partial_\theta f = i r\,\partial_r f \tag{5.4}$$

as the new compact form of CR. By inserting $f = u + iv$, the reader may verify that (5.4) is equivalent to the following pair of Polar–Cart. equations:

$$\partial_\theta v = +r\,\partial_r u \tag{5.5}$$

$$\partial_\theta u = -r\,\partial_r v. \tag{5.6}$$

By examining the amplitwist that carries each arrow to its image we can also obtain two expressions for the derivative:

$$\begin{aligned} e^{i\theta}\,dr \longmapsto e^{i\theta}\,dr \cdot f' &= dr \cdot \partial_r f \\ \Longrightarrow f' &= e^{-i\theta}\,\partial_r f \end{aligned} \tag{5.7}$$

and

$$\begin{aligned} i\,z\,d\theta \longmapsto i\,z\,d\theta \cdot f' &= d\theta \cdot \partial_\theta f \\ \Longrightarrow f' &= -(i/z)\,\partial_\theta f. \end{aligned} \tag{5.8}$$

As a simple example let's take $z^3 = r^3\,e^{3i\theta}$. From (5.7) we obtain

$$(z^3)' = e^{-i\theta}\,3r^2\,e^{3i\theta} = 3r^2\,e^{2i\theta} = 3z^2,$$

while from (5.8) we obtain

$$(z^3)' = -(i/z)\,r^3\,3ie^{3i\theta} = -(i/z)\,3i\,z^3 = 3z^2.$$

In obtaining the same answer from both these expressions we have also verified that z^3 actually was analytic in the first place.

Of the four possible ways of writing CR, only one now remains to be found, namely the Polar-Polar form. We leave it to the reader to verify that if we write $f = R\,e^{i\Psi}$ (cf. Ex. 10, p. 240) then CR takes the form

$$\partial_\theta R = -rR\,\partial_r \Psi \quad \text{and} \quad R\,\partial_\theta \Psi = r\,\partial_r R.$$

5.2 An Intimation of Rigidity

A recurring theme in complex analysis is the "rigidity" of analytic functions. By this we mean that their highly structured nature (everywhere locally an amplitwist) enables us to pin down their precise behaviour from very limited information. For example, even if we are only told the effect of an analytic function on a small region, then its definition can be extended beyond these confines in a unique way—like a crystal grown from a seed. In fact, given even the meagre knowledge of how an analytic mapping affects a closed curve (just the points *on* the curve, mind you), we can predict precisely what happens to each point *inside*! See [5.3].

Later we will justify these wild claims, and in Chapter 9 we will even find an explicit formula (called *Cauchy's Formula*) for w in terms of A, B, C, etc. For the moment, though, we will obtain our first glimpse of this rigidity by considering a different kind of partial information.

Consider [5.4]. Origin-centred circles are being mapped to vertical lines, and the larger the circle, the further to the right is the image, but with no restriction on how the lines are spaced. How much information do you think we can gather about an *analytic* mapping possessing this property? Try meditating on this before reading further.

Well, we know that f is conformal and that its local effect is just an amplitwist. Consider the rays emanating from the origin. Since these cut through all the circles at right angles, their images must cut through the vertical lines at right angles, and

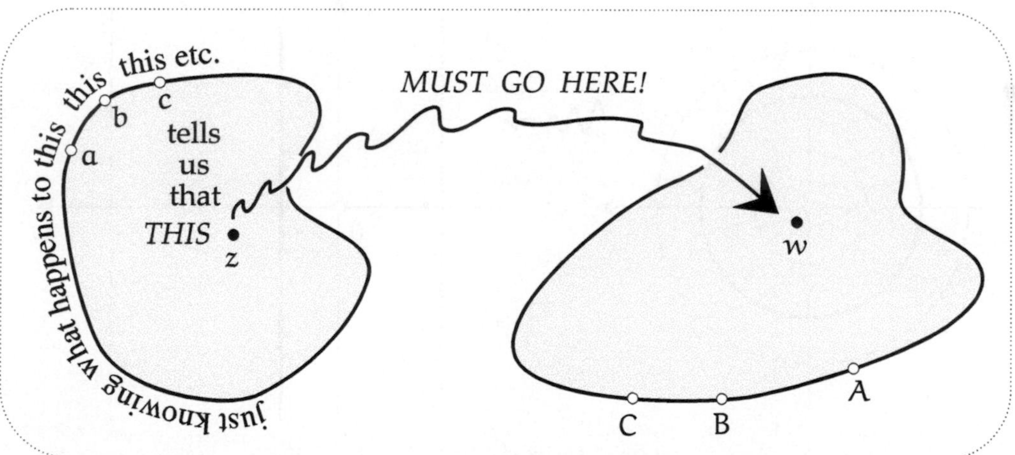

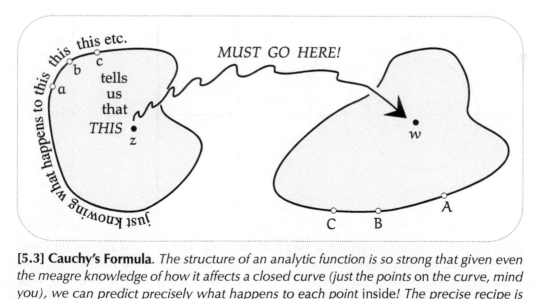

[5.3] Cauchy's Formula. *The structure of an analytic function is so strong that given even the meagre knowledge of how it affects a closed curve (just the points on the curve, mind you), we can predict precisely what happens to each point* inside! *The precise recipe is given by* Cauchy's Formula, *(9.1), page 485.*

they are thus horizontal lines. In fact, if we swing the ray around counter-clockwise, we can even tell whether its image line will move up or down. Look at [5.5], which depicts the fate of a infinitesimal square bounded by two circles and two rays. We know that the infinitesimal radial arrow connecting the two circles must map to a connecting arrow between the lines going from *left to right*. But since the square is to be amplitwisted, its image must be positioned as shown. Thus we find that a positive rotation of the ray will translate the image line *upwards*.

We have made some good progress, but that we cannot yet have fully captured the consequences of analyticity can be seen from Ex. 5, p. 239. Despite not being analytic, the mapping $(x + iy) \mapsto (x^2 + y^2) + i(y/x)$ was there shown to possess all the above desiderata. Indeed it would be easy to write down an infinitude of nonanalytic functions that would be consistent with the known facts. Very remarkably, and in stark contrast to this, if we restrict ourselves to *analytic* mappings, then we will prove that *there is only one!* To show this we must turn to the CR equations.

In [5.4] we are mapping natural polar objects to natural Cartesian objects, so it's clear that we should employ the Polar–Cart. form, namely (5.5) and (5.6). In order to put them to use, we must first translate [5.4] into 'Equationspeak'. We could describe the figure by saying that rotating the point only moves the image up and down, not side to side; in other words, varying θ produces no change in u: $\partial_\theta u = 0$. It follows from (5.6) that $\partial_r v = 0$. This says that moving the point radially outwards does not affect the height of the image, and thus that rays are mapped to horizontal

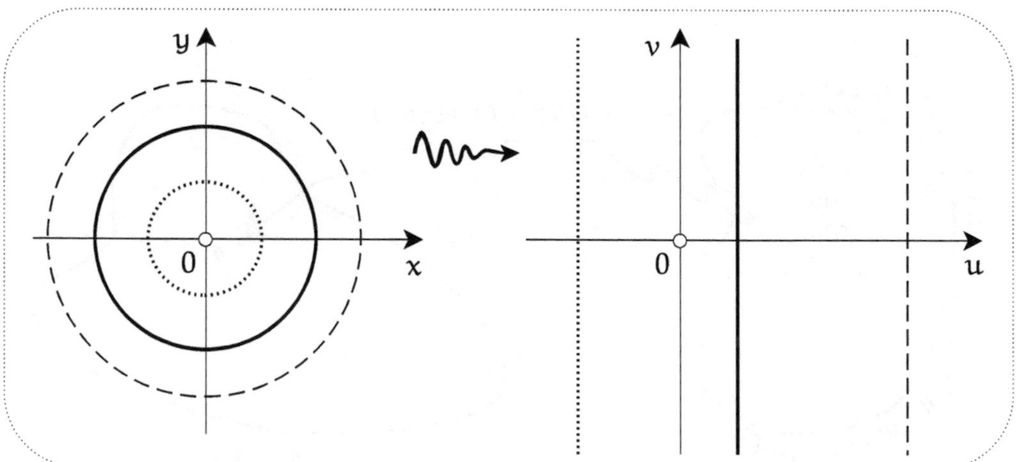

[5.4] There is only *one* analytic mapping that does this. *Suppose that a mapping sends origin-centred circles to vertical lines, and the larger the circle, the further to the right is the image line, but with no restriction on how the lines are spaced. It is easy to write down an infinitude of different non-analytic mappings that do this, but, very remarkably, if we restrict ourselves to* analytic *mappings, then we can prove that* there is only one!

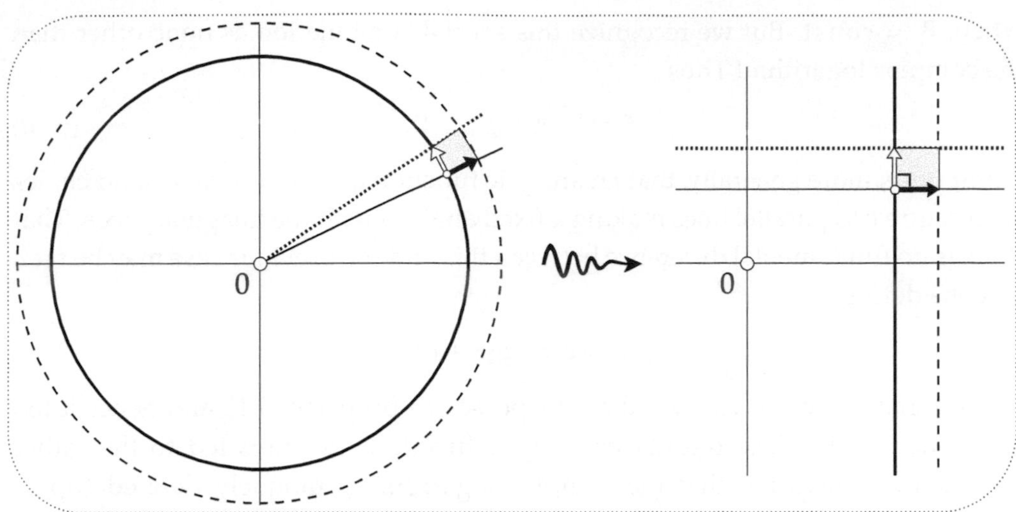

[5.5] Geometric proof of the claim made in the previous figure. *Let us now restrict ourselves to* analytic *mappings possessing the property in the previous figure. Since rays are orthogonal to origin-centred circles, their images under a conformal mapping must be orthogonal to vertical lines*—so the images of rays are horizontal lines. *Furthermore, we know that the infinitesimal* outward *radial arrow connecting the two illustrated circles must map to a connecting arrow between vertical lines going from* left to right. *Since the square is* amplitwisted *to its image, we deduce that* as the ray rotates counterclockwise, its horizontal image line moves upwards.

lines. This is old news to us seasoned geometers, but fortunately we have another equation left:

$$\partial_\theta V(\theta) = r\,\partial_r U(r). \tag{5.9}$$

Here we have written $v = V(\theta)$ to stress that it is known to depend only on θ; similarly for $u = U(r)$.

Now (5.9) looks like an impossible equation, for the LHS quite explicitly depends only on θ, while the RHS is equally emphatic about only depending on r. The only way out of this is for both of these real quantities to equal a constant, say A. Dispensing with the superfluous partial derivatives, we thereby obtain

$$r\frac{dU}{dr} = A \quad \text{and} \quad \frac{dV}{d\theta} = A.$$

Integrating these equations we find that

$$U = A\ln r + \text{const.} \quad \text{and} \quad V = A\theta + \text{const.},$$

and hence

$$U + iV = A\,(\ln r + i\theta) + B,$$

where $B = \text{const}$. But we recognize this special combination as none other than the complex logarithm! Thus

$$f(z) = A \log z + B. \tag{5.10}$$

Suppose, more generally, that an analytic function $g(z)$ is known to send circles with centre c to parallel lines making a fixed angle ϕ with the imaginary axis. That there is no fundamental difference between this and the previous case may be seen by considering

$$z \mapsto e^{-i\phi}\, g(z + c);$$

for you may convince yourself that this possesses property [5.4], and hence it too must equal (5.10). The rigidity of analytic functions has thus led to the rather remarkable conclusion that the complex logarithm is uniquely defined (up to constants) as *the* conformal mapping sending concentric circles to parallel lines.

5.3 Visual Differentiation of log(z)

A fringe benefit of the previous section was the discovery that $\log(z)$ actually is analytic. Since this multifunction finds its simplest representation in Polar–Cart. form, namely

$$\log z = \ln r + i(\theta + 2m\pi),$$

we can easily find its derivative using (5.7) or (5.8). For purposes of illustration, we will now use them both:

$$(\log z)' = e^{-i\theta}\, \partial_r \log z = e^{-i\theta}(1/r) = 1/z,$$

and

$$(\log z)' = -(i/z)\, \partial_\theta \log z = -(i/z)\, i = 1/z. \tag{5.11}$$

You notice, of course, how this is formally identical to the case of the ordinary, real logarithm.

You may be wondering how our previous discussion of the branches of this multifunction affects all this. For example, it's interesting how m (which labels the different branches) does not appear in the result (5.11). The basic philosophy of this book is that while it often takes more imagination and effort to find a picture than to do a calculation, the picture will always reward you by bringing you nearer to the Truth. In this spirit, we now find a visual explanation of (5.11) that will also make it clear that the answer does not depend on m.

Equations (5.7) and (5.8) were derived by examining the infinitesimal geometry of a general analytic mapping. Why not then apply this idea to the geometry of a *specific* mapping, and thereby evaluate its amplitwist directly?

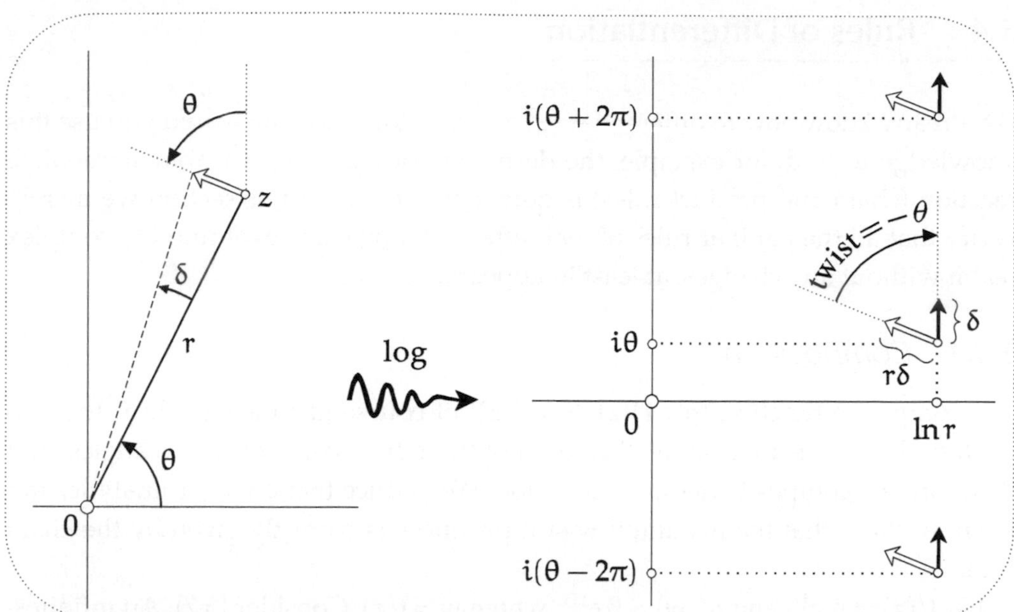

[5.6] Geometric demonstration that $[\log(z)]' = (1/z)$. *Rotating z by the small, ultimately vanishing angle δ moves it along the white complex number, which ultimately has length* $r\delta$. *All of the infinitely many images under* $\log(z)$ *move upwards together, along* $i\delta$. *Thus, for all branches, the* amplification $= (1/r)$, *the* twist $= -\theta$, *and therefore the* amplitwist $= (1/r)e^{-i\theta} = (1/z)$.

Consider [5.6], which shows a typical point z and a few of its infinitely many images under log. In order to find the amplitwist we need only find the image of a single arrow emanating from z. The easiest one to find is shown in [5.6], namely an arrow perpendicular to z. Notice how if z makes an angle θ with the horizontal, then the perpendicular vector will make an angle θ with the vertical. Also, if it subtends an infinitesimal angle δ at the origin, then—because it is like a small arc of a circle—its length will be $r\delta$. Now look at the images of z. Since we have purely rotated z, its images will all move vertically up through a distance equal to the angle of rotation δ. To make it easier to see what amplitwist carries the arrow at z into its image, we have drawn copies of the original arrow at each image point. It is now evident from the picture that

$$\begin{aligned} \text{amplification} &= 1/r \\ \text{twist} &= -\theta \\ \Longrightarrow \text{amplitwist} &= (1/r)\, e^{-i\theta} = 1/z. \end{aligned}$$

Although all the image vectors emanate from different points in the different branches, they are all identical as vectors, and so it is clear that the amplitwist does not depend on which branch we look at.

5.4 Rules of Differentiation

We already know how to differentiate z^2 and also $\log z$, so how would you use this knowledge to find, for example, the derivative of $\log(z^2 \log z)$? Your immediate reaction (chain and product rules) is quite correct, and in this section we merely verify that all the familiar rules of real differentiation carry over into the complex realm without any changes, at least in appearance.

5.4.1 Composition

The composite function $(g \circ f)(z) = g[f(z)]$ of course just means "do f, then do g". If both f and g are analytic then each of these two steps conserves angles, and therefore the composite mapping does too. We deduce that $g[f(z)]$ is analytic, and we now show that the net amplitwist it produces is correctly given by the chain rule.

Let $f'(z) = A\,e^{i\alpha}$ and $g'(w) = B\,e^{i\beta}$, where $w = f(z)$. Consider [5.7]. An infinitesimal arrow at z is amplitwisted by f to produce an image at w; then this, in its turn, is amplitwisted by g to produce the final image at $g(w)$. It is clear from the picture that

$$\begin{aligned} \text{net amplification} &= AB \\ \text{net twist} &= \alpha + \beta \\ \Longrightarrow \text{net amplitwist} &= AB\,e^{i(\alpha+\beta)}, \end{aligned}$$

and thus we obtain the familiar chain rule:

$$\{g[f(z)]\}' = g'(w) \cdot f'(z). \tag{5.12}$$

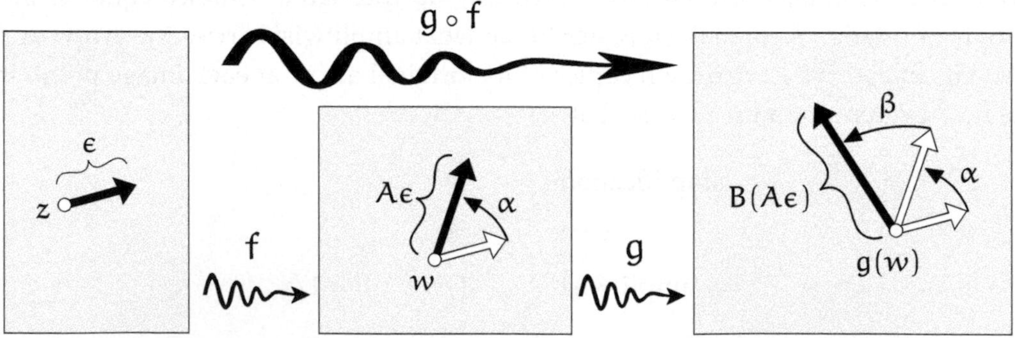

[5.7] Geometry of the Chain Rule. *The local effect of the composition of two analytic mappings is an amplitwist that is the composition of their separate amplitwists, so we obtain the familiar chain rule.*

As an example of this we may put $g(z) = kz$. In the last chapter we showed that $g'(z) = k$, and so we now conclude from (5.12) that

$$[k\,f(z)]' = k\,f'(z).$$

5.4.2 Inverse Functions

Provided we are not at a critical point (where the derivative vanishes), an infinitesimal disc at z will be amplitwisted to produce an image disc at $w = f(z)$, and these two discs will be in one-to-one correspondence. See [5.8]. An analytic function thus always possesses a local inverse in this sense, and we wish to know its derivative.

Clearly, the amplitwist that returns the image disc to its original state has reciprocal amplification, and opposite twist:

$$\begin{aligned} \text{amplification of } f^{-1} \text{ at } w &= 1/(\text{amplification of f at } z) = 1/|f'(z)| \\ \text{twist of } f^{-1} \text{ at } w &= -(\text{twist of f at } z) = \arg[1/f'(z)] \\ \Longrightarrow [f^{-1}(w)]' &= 1/f'(z). \end{aligned} \tag{5.13}$$

By way of example, consider $w = f(z) = \log z$, for which $z = f^{-1}(w) = e^w$. From (5.13) we find that

$$(e^w)' = 1/(\log z)' = z = e^w, \tag{5.14}$$

in agreement with your calculation in Ex. 13, p. 240. Later we will give a visual derivation of (5.14).

Both (5.12) and (5.13) could have been derived even more quickly if we had directly employed the algebraic idea of the image arrow being f' times the original one. We chose instead to keep the geometry to the fore, and reserved the algebra of multiplication for the final encoding of the results as (5.12) and (5.13). However, to derive the next two rules by pure geometry would be cumbersome, so we will use a little algebra.

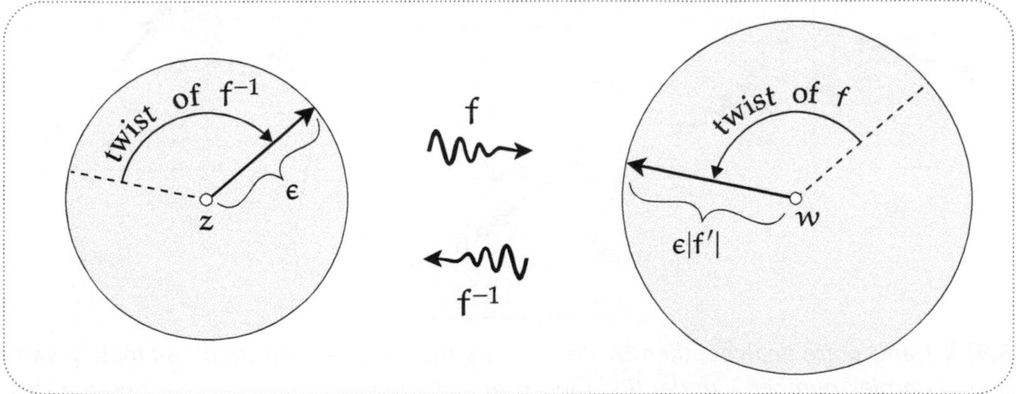

[5.8] The amplitwist of the inverse of an analytic mapping *has the opposite twist and the reciprocal amplification, and so we obtain the familiar rule* $[f^{-1}(w)]' = 1/f'(z)$.

5.4.3 Addition and Multiplication

On the far left of [5.9] we see an infinitesimal arrow ξ connecting z to a neighbouring point. The images of these two points under f and (separately) g, are shown in the middle of the figure. Lastly, we either add or multiply these points to obtain the two points on the far right. By examining the image vector connecting these final points we can deduce the amplitwists of $(f+g)$ and fg, respectively. In [5.9] we see that

$$A - a \asymp \xi f' \quad \text{and} \quad B - b \asymp \xi g',$$

so that

$$(A+B)-(a+b) \asymp \underbrace{\xi(f'+g'),}_{\text{ultimate image of } \xi}$$

and hence we obtain the addition rule:

$$(f+g)' = f' + g'. \tag{5.15}$$

Likewise, we find that

$$AB - ab \asymp \underbrace{\xi(f'b + ag'),}_{\text{ultimate image of } \xi}$$

and thereby obtain the product rule:

$$(fg)' = f'g + fg'. \tag{5.16}$$

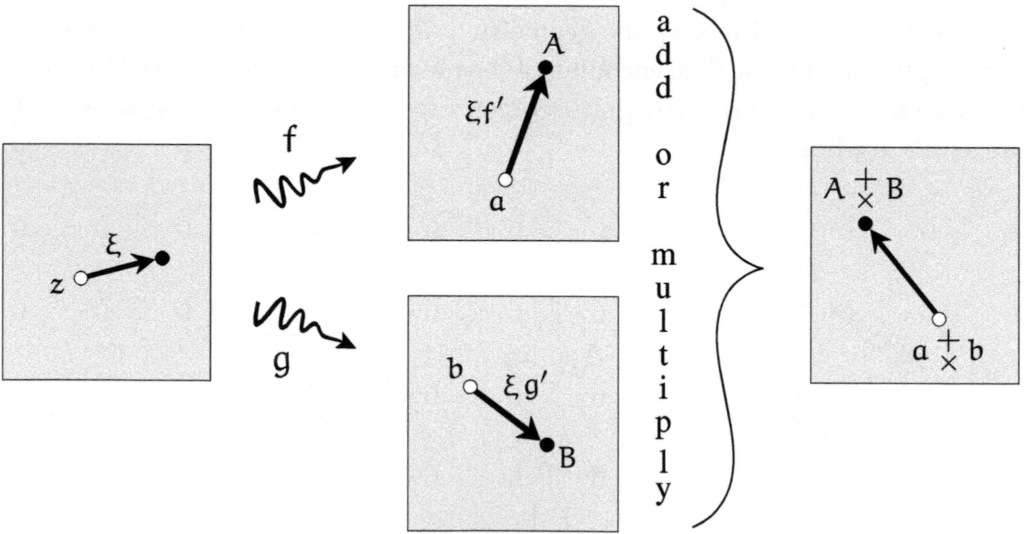

[5.9] *If f and g are analytic, then by considering the image of the small, ultimately vanishing complex number ξ under first their sum and then their product, we deduce that $(f+g)' = f' + g'$, and $(fg)' = f'g + fg'$, just as in elementary real calculus.*

5.5 Polynomials, Power Series, and Rational Functions

5.5.1 Polynomials

We can look at the rules of the previous section from a slightly different point of view. Take rule (5.16), for example. In a way, what is on the RHS is less important than the fact that there *is* a RHS. By this we mean that we have here a recipe for creating new analytic functions: 'given two such functions, form their product'. Likewise, each of our other rules can be thought of as a means of producing new analytic functions from old. The analytic functions are indeed the aristocrats of the complex plane, but provided they only mate with their own kind, and only in ways sanctioned by the rules (which allow many forms of incest!), their offspring will also be aristocrats. For example, suppose we start with only the mapping $z \mapsto z$, which is known to be analytic. Our rules now quickly generate $z^2, z^3, \ldots$, and thence any polynomial.

Consider a typical polynomial of degree n:

$$S_n(z) = a_0 + a_1 z + a_2 z^2 + \cdots + a_n z^n.$$

We have just seen that this is analytic, and thus it maps an infinitesimal disc at p to another at $S_n(p)$. Furthermore, the amplitwist that transforms the former into the latter is, according to (5.15),

$$S_n'(z) = (a_0)' + (a_1 z)' + (a_2 z^2)' + \cdots + (a_n z^n)'.$$

We already know how to differentiate the first four terms, and in the next section we will confirm that in general $(z^m)' = m z^{m-1}$, as you no doubt anticipated. Thus

$$S_n'(z) = a_1 + 2a_2 z + 3a_3 z^2 + \cdots + n a_n z^{n-1}. \tag{5.17}$$

5.5.2 Power Series

This discussion of polynomials naturally leads to the investigation of power series. In Chapter 2 we discussed how a convergent power series[1]

$$S(z) = a_0 + a_1 z + a_2 z^2 + a_3 z^3 + \cdots \tag{5.18}$$

could be approximated by a polynomial S_n. We explained how the effect of S within its circle of convergence could be mimicked by S_n, with arbitrarily high accuracy, simply by taking a sufficiently high value of n.

Of course the question we now face is whether power series are analytic, and if they are, how are we to calculate their derivatives? We will see that the answers to these questions are "yes" and "(5.17)".

[1] For simplicity's sake we shall use a power series centred at the origin. However, as we pointed out in Chapter 2, this does not involve any loss of generality.

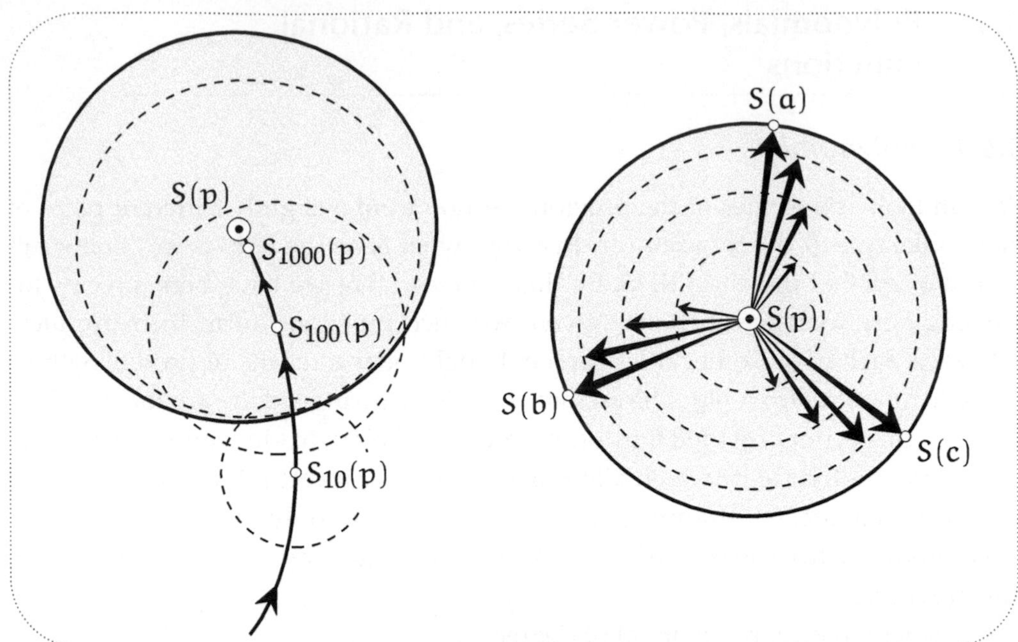

[5.10] Power series are analytic. *On the left, we see the images of a small, ultimately vanishing disc* D *centred at* p *under the higher and higher degree polynomial approximations* $S_n(z)$ *of a power series,* $S(z)$. *On the right, we see these same image discs, now translated to have their common centre at* $S(p)$. *All these polynomial approximations are analytic, so three equally spaced points on the rim of* D *will be amplitwisted to their images on the rims of the image discs, and so it becomes clear that* $S(z)$ *is indeed analytic, and that its amplitwist is the limit of the amplitwists of* S_n *as* n *goes to infinity. [For ease of visualization, here the amplification and twist are both depicted as monotonically increasing with* n*, but this will not be the case in general.]*

Consider an infinitesimal disc D with centre p. If p is inside the circle of convergence of S, then so is a sufficiently small D. The series (5.18) therefore converges at all points of D, and thus S maps the disc to some infinitesimal unknown shape $S(D)$ covering $S(p)$. Now look at the left of [5.10]. This shows a magnified view of the successive images of D [itself not shown in the figure] under S_{10}, S_{100}, S_{1000}, etc. Since each of these polynomials is known to be analytic, each image is a disc. However, it is also known that these images will coincide, ever more perfectly, with $S(D)$. Thus S sends infinitesimal discs to other discs, and it is therefore analytic.

We have tried to make this plainer still on the right of [5.10]. Since we are now only interested in amplitwists, the actual image points are unimportant—we only care about the connecting arrows between them. To make it easier to watch what is happening to these arrows we have translated the discs—which doesn't affect the vectors—so that their centres all coincide at $S(p)$. By way of illustration, we now consider the fate, as n increases, of three equally spaced vectors from p to three equally spaced points (a, b, c) on the rim of D. Each of the analytic mappings S_n

amplitwists these vectors to three equally spaced image vectors. The figure shows the gradual evolution[2] of these images towards their final state (given by S) as we successively apply S_{10}, S_{100}, etc. The amplitwist that carries the arrows of D into these images therefore undergoes a corresponding evolution towards a final value. The amplitwist S' that carries the original vectors of D to their ultimate images is thus mimicked with arbitrarily high accuracy by S'_n, as n increases. Therefore

$$S'(z) = a_1 + 2a_2\,z + 3a_3\,z^2 + 4a_4\,z^3 + \cdots . \tag{5.19}$$

We have reached a very important conclusion. Any power series is analytic within its disc of convergence, and its derivative is obtained simply by differentiating the series term by term. Since the result of this process (5.19) is yet another convergent power series with the same radius of convergence, there is nothing to stop us differentiating again. Continuing in this manner, we discover that a power series is infinitely differentiable within its disc of convergence. The reason this is so important is that we will be able to show later that *every* analytic function can be represented locally as a power series, and thus *analytic functions are infinitely differentiable.*

This result is in sharp contrast to the case of real functions. For example, the mileage displayed on the dash of your car is a differentiable function of the time displayed on the clock. In fact the derivative is itself displayed on the speedometer. However, in the instant that you hit the brakes, the second derivative (acceleration) does not exist. More generally, consider the real function that vanishes for negative x, and that equals x^m for non-negative x. This is differentiable $(m-1)$ times everywhere, but not m times at the origin. Our complex aristocrats will be shown to be quite incapable of stooping to this sort of behaviour.

5.5.3 Rational Functions

Earlier we established that the product rule applies to complex analytic functions, but we neglected to check the quotient rule. We invite you to verify this now, using the same kind of reasoning that led to (5.16). If you get stuck, there is a hint in Ex. 9. In any event, the important point is that the quotient of two analytic functions is also analytic except at the points where it has singularities. In particular, if we apply this result to polynomials then we can conclude that the rational functions are analytic.

The fact that the quotient of two analytic functions is again analytic can be looked at in a rather more geometric way. Let $I(z) = (1/z)$ be the complex inversion mapping. As we discussed at such length in Chapter 3, $I(z)$ is conformal, and hence it is analytic. It follows that if $g(z)$ is analytic, then so is $[1/g(z)]$, because this is the

[2] For ease of visualization, we have taken both the amplification and the twist to be steadily increasing with n. In general they could exhibit damped oscillations as they settled down to their final values.

composition $(I \circ g)$ of two analytic functions. Finally, if $f(z)$ is analytic, the product rule tells us that $f(z) \cdot [1/g(z)] = [f(z)/g(z)]$ is too.

5.6 Visual Differentiation of the Power Function

We saw in the last section that z^2, z^3, z^4, ... were all analytic. Composing with complex inversion, it follows that z^{-2}, z^{-3}, z^{-4}, ... are too. Since the inverse functions (in the sense of [5.8]) are branches of the multifunctions $z^{\pm 1/2}$, $z^{\pm 1/3}$, ... discussed in Chapter 2, it follows that these too are analytic. Composing z^p with $z^{1/q}$ (p, q integers), it follows that any rational power is analytic. Furthermore, since the geometric effect of any real power can be reproduced with arbitrary accuracy by rational powers, it follows that these real powers are also analytic.

The calculation of the derivative of a real power z^a is similar to the example z^3 given on p. 248. We find that

$$(z^a)' = a\, z^{a-1}, \tag{5.20}$$

just as in ordinary calculus. In fact the real formula $(x^a)' = a\, x^{a-1}$ can be thought of as the specialization of (5.20) that results when both z and the infinitesimal arrow emanating from it are taken to be on the real axis (cf. [4.7], p. 221).

Just as in the case of the complex logarithm, we do not rest at the result (5.20) of a calculation, but rather we stalk the thing to its geometric lair. Since the amplitwist is the same for all arrows, we need only find the image of a single arrow in the direction of our choice. As a first (ill-fated) attempt, consider [5.11], in which we have chosen an arrow parallel to z. To facilitate comparison, we have drawn a copy of the initial arrow at the image point. You can see from the picture that

$$\text{twist} = (a-1)\theta \quad \text{BUT} \quad \text{amplification} = ???????$$

We are thus half-thwarted, for we cannot see how long the image arrow is. In fact to figure this out would require precisely the same calculation (general Binomial Theorem) as is needed in the real case. Oh well, "If at first you don't succeed, ..."

"Try, try an arrow *perpendicular* to z!" From [5.12], we see that this arrow originally makes an angle θ with the vertical, and so after magnifying the angle of z by a, it will make an angle $a\theta$ with the vertical. Once again we see that the twist $= (a-1)\theta$. However, this time we *can* see the amplification, simply by recognizing that each arrow is an infinitesimal arc of a circle. The angle subtended by the arc has been magnified by a, while the radius of the circle has been magnified by r^{a-1}. The net amplification of the arc is therefore $a\, r^{a-1}$. Thus

$$\begin{aligned} \text{amplification} &= a r^{a-1} \\ \text{twist} &= (a-1)\theta \\ \Longrightarrow \text{amplitwist} &= a r^{a-1}\, e^{i(a-1)\theta} = a\, z^{a-1}. \end{aligned}$$

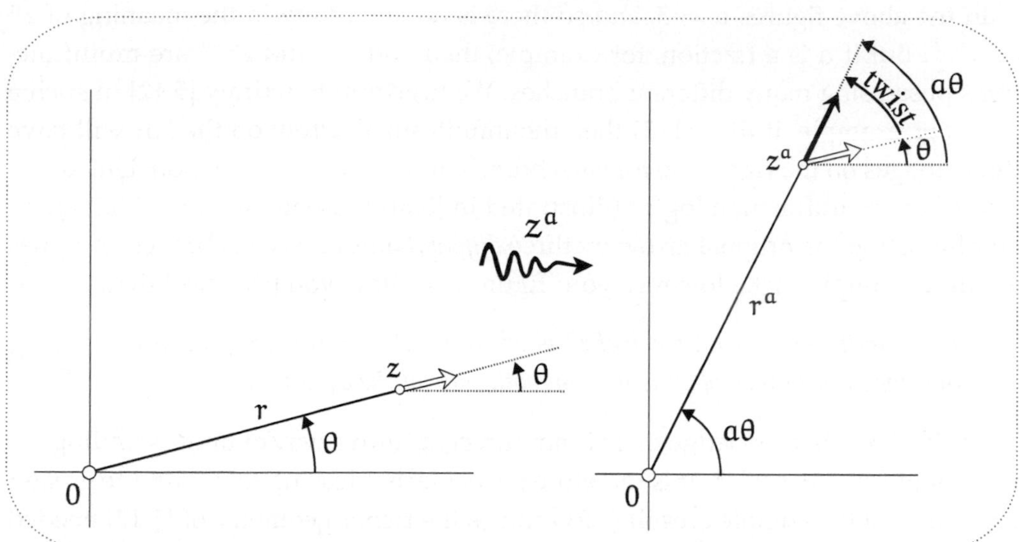

[5.11] First (ill-fated) attempt to geometrically evaluate the amplitwist of $z \mapsto z^a$. *An infinitesimal radial complex number emanating from* $z = r\,e^{i\theta}$ *is amplitwisted to a radial complex number emanating from* $z = r^a\,e^{ia\theta}$*, and therefore the* twist $= (a-1)\theta$. *However, the amplification* cannot *be geometrically determined from this figure.*

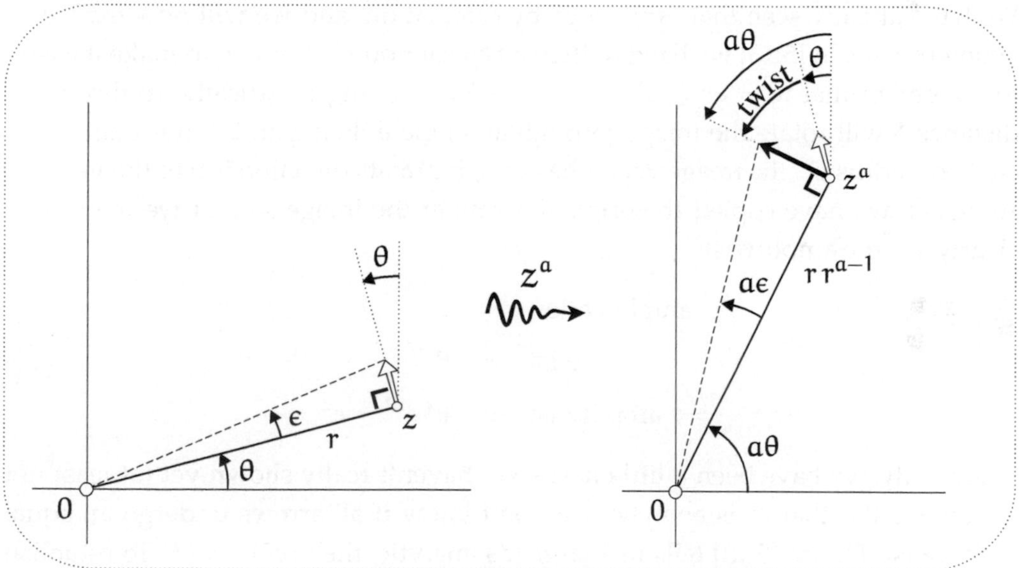

[5.12] Second (successful!) attempt to geometrically evaluate the amplitwist of $z \mapsto z^a$. *If we instead consider an infinitesimal complex number* orthogonal *to* $z = r\,e^{i\theta}$*, then it is amplitwisted to an infinitesimal complex number* orthogonal *to* $z = r^a\,e^{ia\theta}$*. Therefore, once again, we see that the* twist $= (a-1)\theta$. *However, this time the amplification* can *be geometrically determined: the initial complex number has length* $\asymp r\epsilon$*, while its image has length* $\asymp r^a(a\epsilon)$*, so the* amplification $= a\,r^{a-1}$*. Therefore, the* amplitwist $= (z^a)' = a\,r^{a-1}e^{i(a-1)\theta} = a\,z^{a-1}$.

In the above figures $a = 3$, and so there is no ambiguity in the meaning of z^a or z^{a-1}. But if a is a fraction, for example, then both z^a and z^{a-1} are multifunctions possessing many different branches. We urge you to redraw [5.12] in such a case. For example, if $a = (1/3)$ then the infinitesimal arrow on the left will have three images on the right, one for each branch of the cube root function. Unlike the case of the multifunction $\log(z)$ (illustrated in [5.6]) these images are obtained by amplitwisting the original arrow by three *different* amounts: each branch of z^a has a different amplitwist. However, your figure will show you [exercise] that

The amplitwist of each branch of z^a is given by $(z^a)' = a\,z^a/z$, provided that the same branch of z^a is used on both sides of the equation. (5.21)

To the best of our knowledge there is no[3] direct, intuitive way of understanding the real result $(x^a)' = a\,x^{a-1}$. It is therefore particularly pleasing that with the greater generality of the complex result (5.21) comes the richer geometry of [5.12] needed to see its truth.

5.7 Visual Differentiation of exp(z)

We have already seen that $(e^z)' = e^z$ by calculation, and we will now explain it geometrically. In [5.13] we have written a typical point $z = x + i\theta$ to make it easier to remember that $w = e^z = e^x e^{i\theta}$ has angle θ. Moving z vertically up through a distance δ will rotate the image through an angle δ. Being an infinitesimal arc of circle of radius e^x, the image vector has length $e^x\delta$; its direction is θ to the vertical. As usual, we have copied the original arrow at the image so that we may more clearly see the amplitwist:

$$\begin{aligned} \text{amplification} &= e^x \\ \text{twist} &= \theta \\ \Longrightarrow \text{amplitwist} &= e^x e^{i\theta} = e^z. \end{aligned}$$

Actually, we have been a little hasty. We haven't really shown yet (at least not geometrically) that e^z is analytic: we don't know if all arrows undergo an equal amplitwist. Figure [5.13] tells us that *if* it's analytic, then $(e^z)' = e^z$. To establish analyticity we need only see that one other arrow is affected in the same way.

In [5.14] we move z an infinitesimal distance δ in the x-direction, thereby moving the image radially outwards. Now, from ordinary calculus, the amplification produced by e^x along the real axis is e^x (cf. [4.6], p. 220), so the length of this image

[3] In special cases there are ways. For example, consider a cube of side x. It is easy to visualize that if we increase the separation of one of the three pairs of faces by δ, we add a layer of volume $x^2\delta$. The result $(x^3)' = 3x^2$ follows.

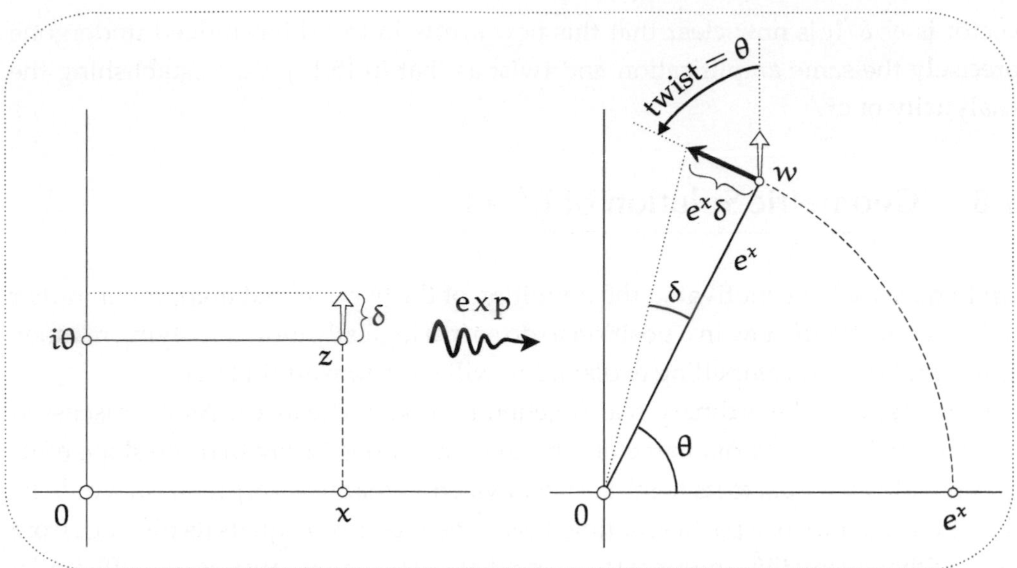

[5.13] Geometrical evaluation the amplitwist of $z \mapsto e^z$. *An infinitesimal vertical complex number* $i\delta$ *emanating from* $z = x + i\theta$ *is amplitwisted to a complex number orthogonal to* $w = e^z = e^x e^{i\theta}$ *whose length is* $\asymp e^x \delta$. *Therefore the* amplification $= e^x$ *and the* twist $= \theta$, *so the* amplitwist $= (e^z)' = e^x e^{i\theta} = e^z$.

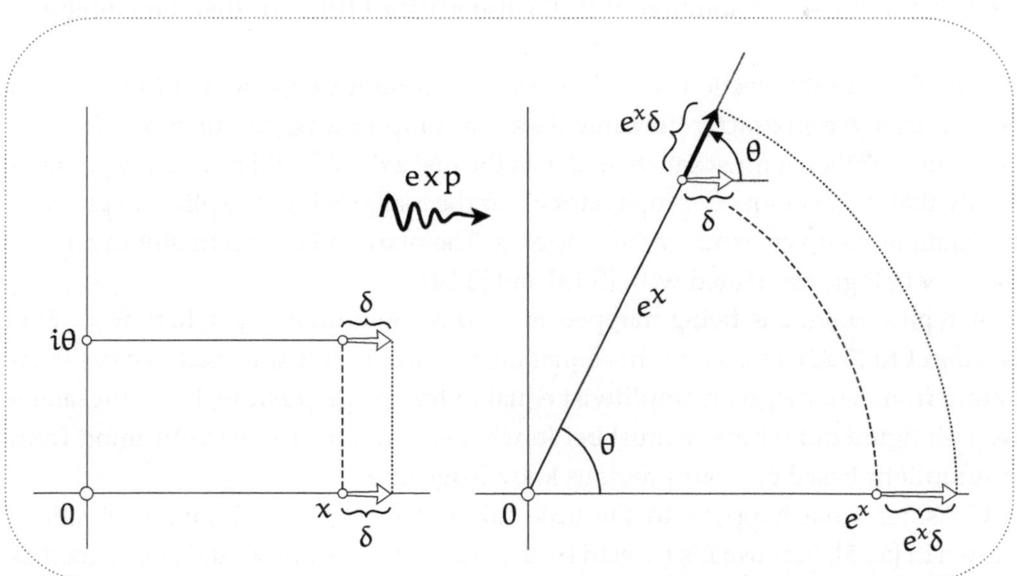

[5.14] Verification that $z \mapsto e^z$ is locally an amplitwist. *If we instead look at the image of an infinitesimal horizontal complex number emanating from z, we see that it, too, undergoes the same amplitwist as did the vertical one in the previous figure, confirming that* all *infinitesimal complex numbers undergo the same amplitwist.*

vector is $e^x\,\delta$. It is now clear that this new arrow in [5.14] has indeed undergone precisely the same amplification and twist as that in [5.13], thus establishing the analyticity of e^z.

5.8 Geometric Solution of E′ = E

Up to now we have motivated the definition of the exponential mapping in rather *ad hoc* ways. We are now in a position to do so in a logically more satisfying manner, although *the* most compelling explanation will have to wait till later.

Consider first the ordinary real function that we write as e^x. As we discussed in Chapter 2, one way of characterizing this function is to say that the slope of its graph is always equal to its height. An equivalent dynamic interpretation would be that if the distance of a particle at time t is e^t, then its speed equals its distance from us. In either event, this amounts to saying that the function satisfies the differential equation

$$E' = E. \tag{5.22}$$

Of course this doesn't quite pin it down since $k\,e^x$ also obeys (5.22); however, if we insist that the real solution of (5.22) also satisfy $E(0) = 1$, then no ambiguity remains.

The object of this section is to show that the complex exponential function can be characterized in exactly the same way. If a complex-analytic function $E(z)$ is to generalize e^x then it must satisfy (5.22) on the real axis. We will now show geometrically that (5.22) uniquely propagates e^x off the real axis into the plane to produce the familiar complex exponential mapping. The plan will be essentially to reverse the flow of logic associated with [5.13] and [5.14].

A typical point z is being mapped to an unknown image w, where $w = E(z)$ is subject to (5.22). Decoding this equation, we find that it says that vectors emanating from z undergo an amplitwist equal to the image point w. From this alone we will figure out where w must be! In what follows, try to free your mind from assumptions based on your previous knowledge of e^z.

Consider what happens to the little (ultimately vanishing) square of side ϵ shown in [5.15]. Because it's twisted by the angle of w, its horizontal edge becomes parallel to w, while its vertical edge becomes orthogonal to w. Thus horizontal movement of z results in radial movement of the image, while vertical movement results in rotation of the image. The question that now remains is exactly how swift these radial and rotational motions are. Having used the twist, we now turn to the amplification.

If z moves at unit speed in the x-direction, then since the amplification is r, the image moves radially with speed equal to its distance from the origin. But this

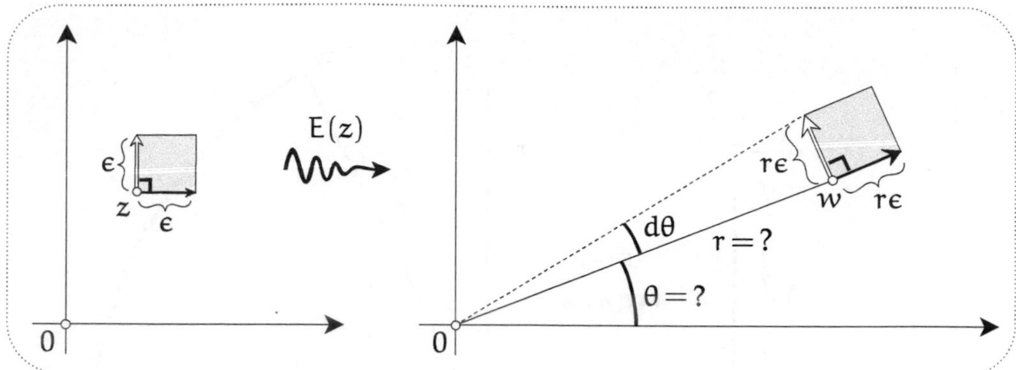

[5.15] Geometric solution of $E' = E$. *The small, ultimately vanishing square is amplitwisted by $E'(z) = E(z) = w$, so the image of the horizontal edge is radial, and the sides are amplified by $r = |w|$. Thus $d\theta \asymp \epsilon$: vertical movement of z produces a numerically equal rotation of w. If z moves at unit speed in the x-direction, then since the amplification is r, the image moves radially with speed equal to its distance from the origin. But this is just the familiar property of the ordinary exponential function. Thus E maps horizontal lines exponentially onto rays.*

is just the familiar property of the ordinary exponential function. Thus E maps horizontal lines exponentially onto rays. If we now insist that $E(0) = 1$, then the real axis maps to the real axis, and we thereby recover the ordinary exponential function. We also know that translating a horizontal line upwards will rotate its image ray counter-clockwise, but we don't yet know how fast. In [5.15] $d\theta$ is the infinitesimal rotation produced by moving z through a distance ϵ along the vertical edge of the square. But since the amplification is r, we know that the image of this edge has length $r\epsilon$, and consequently $d\theta = \epsilon$. In other words,

An infinitesimal vertical translation produces a numerically equal rotation. (5.23)

We can now completely describe the mapping produced by $E(z)$. Imagine watching the image as we move from the origin to a typical point $z = x + i\theta$ in a two-legged journey: first along the real axis to x, then straight up to z. See [5.16]. As we move to x, the image moves along the real axis from 1 to e^x. Repeated application of (5.23) then tells us that moving up a distance θ will rotate the image through an angle θ. For example, we find that $E(z)$ wraps the imaginary axis around the unit circle in such a way that

$$E(i\theta) = \cos\theta + i\sin\theta.$$

This is our old friend, the celebrated Euler Formula. It also follows directly from this geometry that the mapping has the property

$$E(a + b) = E(a) \cdot E(b).$$

It is now entirely logical to define "e^z" to be $E(z)$, and our work is done.

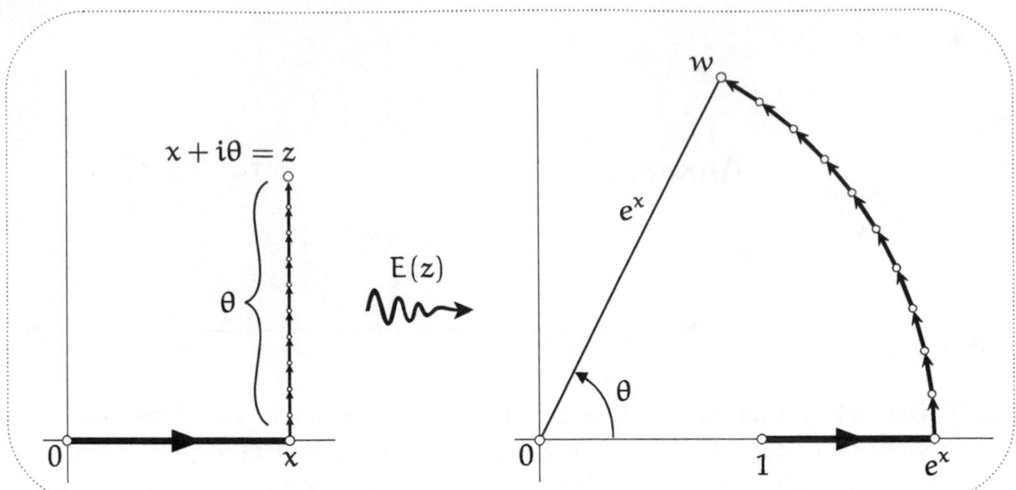

[5.16] Geometry of the exponential. *If we insist that* $E(0) = 1$, *then combining the two facts we deduced in the previous figure, we see that we have arrived at the familiar geometry of* $z \mapsto e^z$.

As we indicated at the start of this section, there is in fact an even more compelling explanation than the above. We have just used a very natural differential equation to propagate e^x off the real axis; however, it will turn out that even this equation is superfluous. The rigidity of analytic functions is so great that merely knowing the values of e^x on the real axis uniquely determines its "analytic continuation" into the complex realm.

5.9 An Application of Higher Derivatives: Curvature*

5.9.1 Introduction

Earlier we alluded to the remarkable fact that analytic functions are infinitely differentiable. In other words, if f is analytic then f'' exists. In this section we seek to shed geometric light on the meaning and existence of this second derivative f''. We shall do so by answering the following question:

> *If an analytic mapping* f *acts on a curve* K *of known curvature* κ *at* p, *then what is the curvature* $\widetilde{\kappa}$ *of the image curve* $\widetilde{K}$ *at* $f(p)$?

In the next section we shall see that the solution to this problem provides a novel insight into (of all things!) the elliptical orbits of the planets round the sun at one focus.

At the risk of ruining the suspense, here is the answer to our question:

$$\widetilde{\kappa} = \frac{1}{|f'(p)|}\left(\text{Im}\left[\frac{f''(p)\,\widehat{\xi}}{f'(p)}\right] + \kappa\right), \tag{5.24}$$

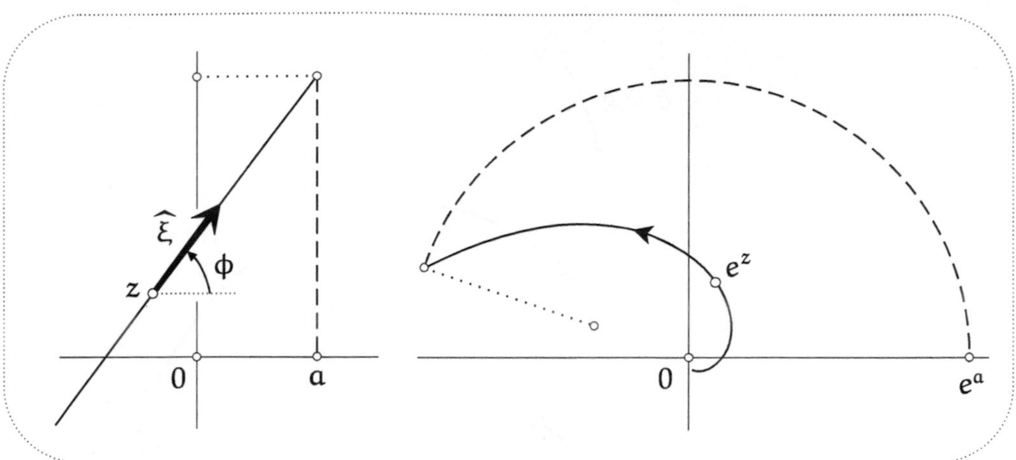

[5.17] Transformation of curvature under $z \mapsto e^z$. *Applying the general formula (5.24) for the transformation of curvature, the image under $z \mapsto e^z$ of the straight line at angle ϕ is given by $\widetilde{\kappa} = e^{-x} \sin\phi$. Check for yourself that this is in accord with the empirical evidence in this figure.*

where $\widehat{\xi}$ denotes the unit complex number tangent to the original curve at p. Before explaining this result, let us simply test it on an example.

On the left of [5.17] we have drawn three line-segments, and on the right their images under $f(z) = e^z$. The segments are distinguished by the value of the angle ϕ that each makes with the horizontal: $\phi = 0$ for the dotted one; $\phi = (\pi/2)$ for the dashed one; and the solid one represents a general value of ϕ. Now look at the curvature of their images: $\widetilde{\kappa} = 0$ for the dotted one; $\widetilde{\kappa} = e^{-a}$ for the dashed one; and on the solid image, $\widetilde{\kappa}$ starts out large and then dies away as we spiral out from the origin.

In order to compare these empirical observations with our formula, write the unit tangent as $\widehat{\xi} = e^{i\phi}$ and note that if $f(z) = e^z$ then $f'' = f' = e^z$. With $z = x + iy$, formula (5.24) therefore reduces to

$$\widetilde{\kappa} = e^{-x}(\sin\phi + \kappa).$$

Using the fact that $\kappa = 0$ for our line-segments, and that ϕ is constant on each, you may now easily check the accord between this formula and figure [5.17].

5.9.2 Analytic Transformation of Curvature

We now turn to the explanation of (5.24). The presence of an imaginary part in this rather daunting formula would seem to bode ill for a purely geometric attack. Surprisingly, this isn't the case. Consider [5.18].

On the left is the curve K, with curvature κ at p. Note that we have arbitrarily assigned a sense to K so as to give κ a definite sign. At the top of the figure is the

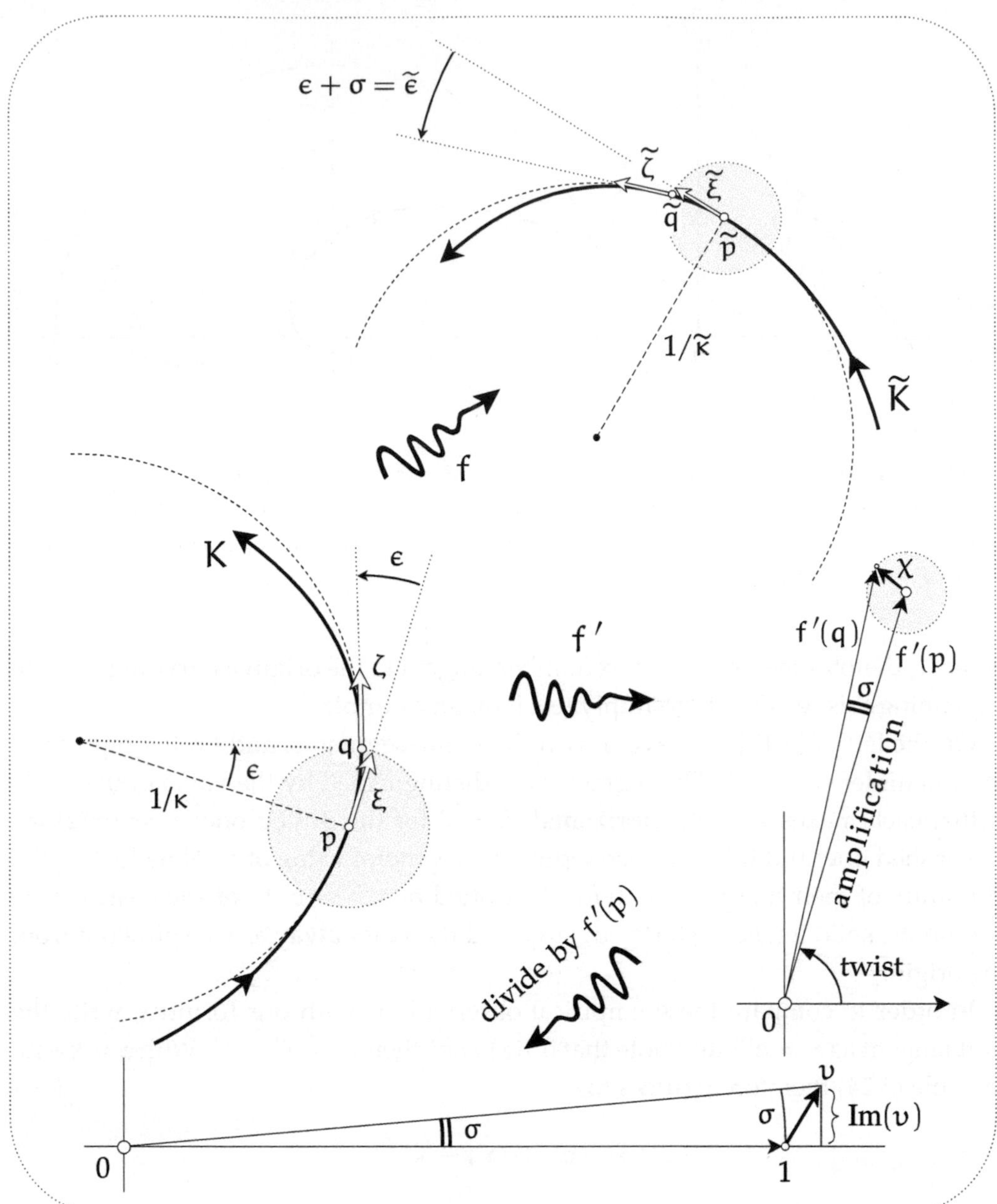

[5.18] Geometric derivation of the transformation law for curvature. *The analytic mapping* f *transforms the curve* K *into the curve* $\widetilde{K}$. *Curvature is the rate of rotation of the tangent, so* $\kappa \asymp (\epsilon/|\xi|)$ *at* p *is transformed into* $\widetilde{\kappa} \asymp (\widetilde{\epsilon}/|\widetilde{\xi}|) \asymp (\widetilde{\epsilon}/|f'(p)\xi|)$ *at* $\widetilde{p} = f(p)$. *If the twists at* p *and at* q *were the same, then the turning angle* ε *would not change, but in fact (as we see on the right) the amplitwists at* p *and* q *differ by* $\chi \asymp f''(p)\xi$, *so there is an* extra *twist* $\sigma \asymp$ (angle subtended by χ). *To geometrically determine this extra twist* σ, *divide the figure on the right by* f′(p) *to obtain the figure at the bottom. We now see that* $\sigma \asymp \mathrm{Im}(\nu) = \mathrm{Im}\left[\frac{\chi}{f'(p)}\right] \asymp \mathrm{Im}\left[\frac{f''(p)\xi}{f'(p)}\right]$, *thereby completing the proof of the curvature transformation law, (5.24).*

image curve $\widetilde{K}$ under the mapping f; note that its sense is determined by that of K. It is the curvature of $\widetilde{K}$ at $\widetilde{p} = f(p)$ that (5.24) purports to describe.

As illustrated, ξ is a small (ultimately infinitesimal) complex number tangent to K at p. With centre at p we have drawn a circle through the tip of ξ cutting K in q. At q we have drawn another small (ultimately infinitesimal) tangent complex number ζ, and we have marked the angle ϵ of rotation from ξ to ζ. Recall that the curvature κ at p is, by definition, the rate of rotation of the tangent with respect to distance along K. Since for infinitesimal ξ the arc pq equals $|\xi|$, the curvature at p is therefore

$$\kappa = \frac{\epsilon}{|\xi|}. \tag{5.25}$$

Likewise, at the image points $\widetilde{p}$ and $\widetilde{q}$ on the image curve $\widetilde{K}$ we have drawn the image complex numbers $\widetilde{\xi}$ and $\widetilde{\zeta}$, the rotation from $\widetilde{\xi}$ to $\widetilde{\zeta}$ being $\widetilde{\epsilon}$. Thus the image curvature is

$$\widetilde{\kappa} = \frac{\widetilde{\epsilon}}{|\widetilde{\xi}|}. \tag{5.26}$$

Our problem therefore reduces to finding $\widetilde{\epsilon}$ and $|\widetilde{\xi}|$.

Since $|\widetilde{\xi}|$ is the length of the amplitwisted image of ξ,

$$|\widetilde{\xi}| = (\text{amplification}) \cdot |\xi| = |f'(p)| \cdot |\xi|. \tag{5.27}$$

The more interesting and difficult part of the problem is to find $\widetilde{\epsilon}$.

If ξ and ζ both underwent precisely the same twist, then the turning angle $\widetilde{\epsilon}$ for the images would equal the original turning angle ϵ. However, the twist at q will differ very slightly, say by σ, from that at p. Thus

$$\widetilde{\epsilon} = \epsilon + (\text{extra twist}) = \epsilon + \sigma. \tag{5.28}$$

This is how f'' enters the picture, for it describes how the amplitwist varies.

The function f' is a perfectly respectable mapping in its own right, and it may be drawn like any other. The right-hand side of [5.18] is precisely such a picture. Each point z is mapped to the complex number that amplitwists infinitesimal complex numbers emanating from z. In particular, we have drawn the images $f'(p)$ and $f'(q)$ of p and q. The statement about infinite differentiability can now be recast in a more blatantly astonishing form: *if* f *is locally an amplitwist, then* f' *automatically is too*. We have indicated this in the picture by showing the disc at p being mapped by f' to *another disc* at $f'(p)$. This startling fact will now yield to us the value of σ.

The amplitwist that carries the disc at p to the disc at $f'(p)$ is $f''(p)$. In particular, ξ is amplitwisted to

$$\chi = f''(p)\,\xi.$$

But looking at the triangle on the right, the sides of which are the known quantities $f'(p)$ and χ, we see that the angle at the origin is precisely the extra twist σ that we seek.

It is easier to obtain an expression for this angle if we first rotate the triangle to the real axis. This rotation is achieved quite naturally (see the bottom figure) by dividing by $f'(p)$; the sides of the triangle now become 1 and $\nu = [\chi/f'(p)]$. Because σ equals the (ultimately) vertical arc through 1, the figure tells us that

$$\sigma = \text{arc} = \text{Im}(\nu) = \text{Im}\left[\frac{\chi}{f'(p)}\right] = \text{Im}\left[\frac{f''(p)\,\xi}{f'(p)}\right].$$

Thus, from (5.26), (5.27), and (5.28), and taking evaluation at p as understood, we obtain

$$\widetilde{\kappa} = \frac{\left(\text{Im}\left[\frac{f''\,\xi}{f'}\right] + \epsilon\right)}{|f'|\,|\xi|}.$$

Finally, using (5.25) and noting that $\widehat{\xi} = (\xi/|\xi|)$ is the unit tangent at p, we do indeed obtain formula (5.24).

Finally, we note that, alternatively, a very short (but unilluminating) computational proof can be obtained [exercise] by differentiation of the equation,

$$\text{Twist} = \text{Im}\log f'.$$

5.9.3 Complex Curvature

Let us take a closer look at formula (5.24), which may be written

$$\widetilde{\kappa} = \text{Im}\left[\frac{f''\,\widehat{\xi}}{f'\,|f'|}\right] + \frac{\kappa}{|f'|}.$$

The presence of the second term can be understood as follows. If the plane were to undergo a uniform expansion by factor R then a circle of radius $(1/\kappa)$ would become a circle of radius (R/κ), that is of curvature (κ/R). But a small piece of a general curve resembles an arc of its *circle of curvature*[4], and the principal local effect of f (apart from a curvature-preserving twist) is an expansion by factor $|f'|$.

In addition to this phenomenon, the first term says that the mapping will introduce curvature even when none is originally present: the curvature k of the image of a straight line (as a function of its direction) is

$$k(\widehat{\xi}) = \text{Im}\left[\frac{f''\,\widehat{\xi}}{f'\,|f'|}\right].$$

Now consider the fate of all the curves that pass through p in the direction $\widehat{\xi}$. The general formula says that f will not only scale their curvatures by $(1/|f'|)$ (as previously explained), but it will also increase their curvatures by the *fixed* amount $k(\widehat{\xi})$. In this sense, the first term corresponds to an intrinsic property of the mapping f.

[4] The circle that touches the curve at the point in question, and whose curvature $\kappa = (1/\text{radius})$ agrees with that of the curve at that point.

However, $k(\widehat{\xi})$ is not really intrinsic to f since it retains a vestige of the original curves, namely, their direction $\widehat{\xi}$. It would appear that the most natural intrinsic quantity that can be abstracted from $k(\widehat{\xi})$ is

$$\mathcal{K} \equiv \frac{i\,\overline{f''}}{\overline{f'}\,|f'|}. \tag{5.29}$$

We propose to call this complex function $\mathcal{K}$ (which does not appear to have been investigated previously) the *complex curvature*[5] of f.

To see that the complex curvature is indeed a natural quantity, picture $\mathcal{K}(p)$ as a vector emanating from p. We will show that

The projection of $\mathcal{K}(p)$ onto a line through p is the curvature of the image of that line at $f(p)$. (5.30)

See [5.19], in which $\mathcal{K}$ has also been drawn at two additional points. Note how the increasing length of the projection of $\mathcal{K}$ onto the line corresponds to increasing curvature along the image.

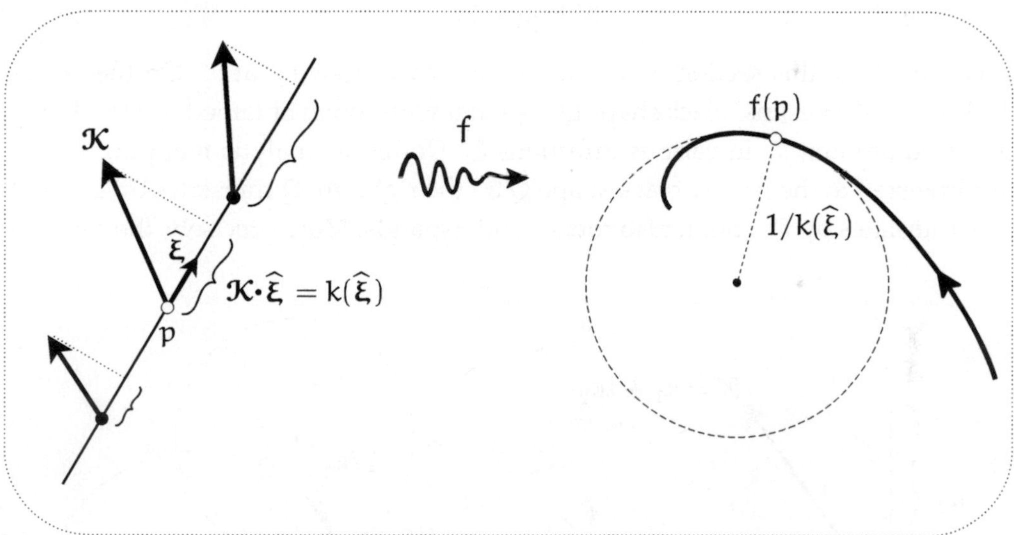

[5.19] Geometric interpretation of the complex curvature. *The projection of the complex curvature $\mathcal{K}(p)$ onto a line through p is the curvature of the image of that line at $f(p)$. Here, the projection grows as we move along the line, so the image curve bends more tightly.*

[5] NOTE added in this *25th Anniversary Edition*: I am delighted that my discovery of $\mathcal{K}$ and its mathematical properties has since found several applications (particularly in physics) that go beyond the two applications that I originally discovered and published here in VCA, namely, to dual central force fields [this chapter] and to the geometry of harmonic functions [Chapter 12]. For example, see Vitelli and Nelson (2006) and Mughal and Weaire (2009).

To prove (5.30), recall how the scalar product in $\mathbb{R}^2$ can be expressed in terms of complex multiplication:

$$\mathcal{K}\cdot\widehat{\xi} = \text{Re}\left[\overline{\mathcal{K}}\,\widehat{\xi}\right] = \text{Im}\left[i\overline{\mathcal{K}}\,\widehat{\xi}\right] = \text{Im}\left[\frac{f''\,\widehat{\xi}}{f'\,|f'|}\right] = k(\,\widehat{\xi}\,),$$

as was to be shown. This result yields a neater and more intelligible form of (5.24):

$$\widetilde{\kappa} = \mathcal{K}\cdot\widehat{\xi} + \frac{\kappa}{|f'|}. \tag{5.31}$$

To see how $\mathcal{K}(p)$ may be determined geometrically, imagine a short, directed line-segment S rotating about p. The image f(S) rotates with equal speed about f(p), and its curvature oscillates sinusoidally: it reaches its maximum value $|\mathcal{K}(p)|$ when S points in the direction of $\mathcal{K}(p)$, while it vanishes when S is perpendicular to $\mathcal{K}(p)$.

In fact, to reconstruct $\mathcal{K}(p)$ it is sufficient to know the image curvatures κ_1 and κ_2 for just two positions S_1 and S_2 of the line-segment. Figure [5.20] illustrates this in the particularly simple case that S_1 and S_2 are horizontal and vertical, respectively. We then have

$$\mathcal{K} = \kappa_1 + i\kappa_2.$$

We conclude this section with a different way of looking at $\mathcal{K}$. On the left of [5.21] is an infinitesimal black shape Q, together with copies obtained by translating Q a fixed amount $|\xi|$ in various directions ξ. Under an analytic mapping f, Q is amplitwisted to the similar black shape $\widetilde{Q}$ on the right. As Q translates by ξ, $\widetilde{Q}$ not only translates by $f'\xi$, but it also rotates and expands. More precisely, the rotation

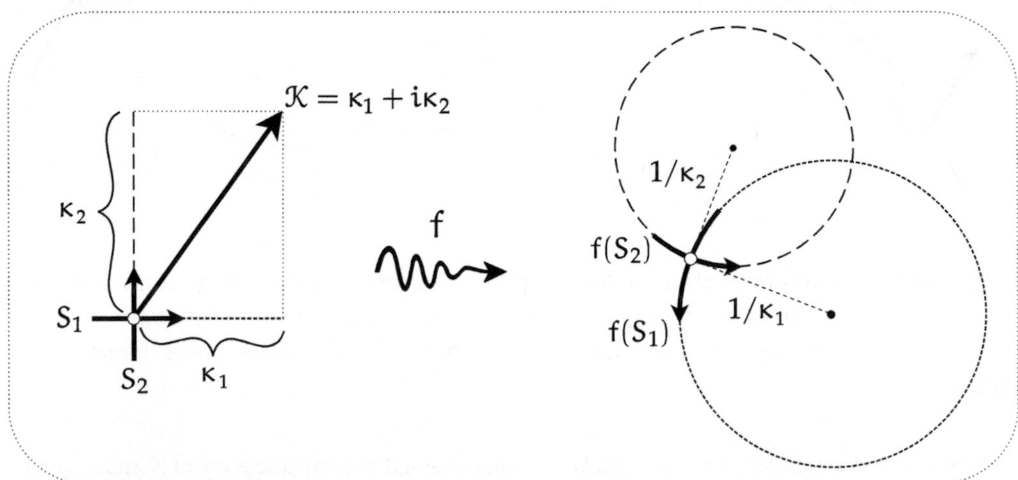

[5.20] Geometric interpretation of the real and imaginary parts of the complex curvature. *The real and imaginary parts of the complex curvature* $\mathcal{K}(p)$ *are the curvatures of the images of the real and imaginary line segments through* p.

of $\widetilde{Q}$ is just the angle σ on the RHS of [5.18]. This rotation is clearly greatest when χ is perpendicular to $f'(p)$, pointing counterclockwise along the circle $|f'| = \text{const}$. This occurs when ξ is in the direction of $\mathcal{K}$, for then

$$\chi \propto f'' \mathcal{K} \propto i/\overline{f'} \propto if'.$$

If we turn the direction of motion of Q by $-(\pi/2)$, then χ also turns by $-(\pi/2)$ to point radially outwards along the ray $\arg f' = \text{const.}$, thereby producing the greatest increase in $|f'|$.

We now understand [5.21] in greater detail:

> *Let Q be an infinitesimal shape, and let $\widetilde{Q}$ be its image under an analytic mapping f. Then $\widetilde{Q}$ rotates most rapidly, and its size remains constant, when Q moves in the direction of $\mathcal{K}$. On the other hand, $\widetilde{Q}$ expands most rapidly, and does not rotate, when Q moves in the orthogonal direction $-i\mathcal{K}$.* (5.32)

In still greater detail, as Q begins to translate in an arbitrary direction $\widehat{\xi}$, let $\mathcal{R}$ denote the rate of rotation of $\widetilde{Q}$ with respect to the distance *it* moves. Then

$$\mathcal{R} = \boldsymbol{\mathcal{K}} \boldsymbol{\cdot} \widehat{\xi}.$$

This achieves its maximum value $\mathcal{R}_{\text{max}} = |\mathcal{K}|$ when Q moves in the direction of $\mathcal{K}$. Similarly, consider the expansion of $\widetilde{Q}$. Let $\mathcal{E}$ denote the rate of increase of the size[6]

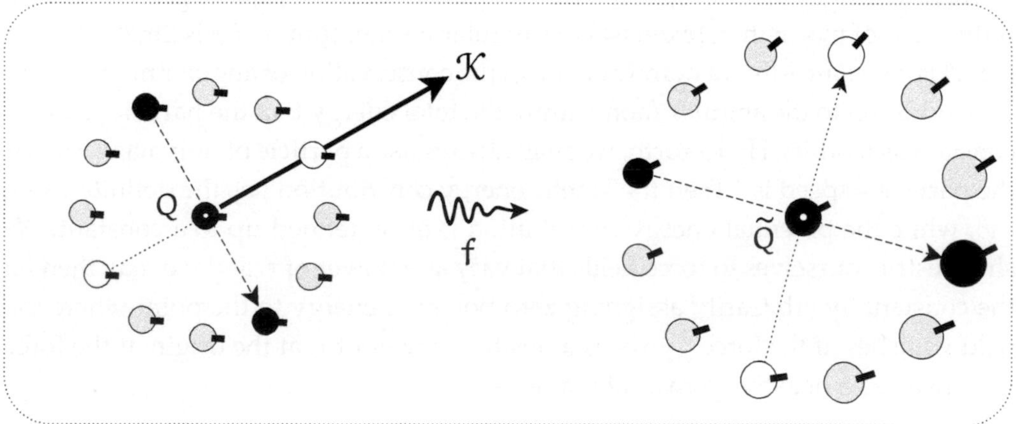

[5.21] Another geometric interpretation of the complex curvature. *Let Q be an infinitesimal shape, and let $\widetilde{Q}$ be its image under an analytic mapping f. Then $\widetilde{Q}$ rotates most rapidly, and its size remains constant, when Q moves in the direction of the complex curvature $\mathcal{K}$. On the other hand, $\widetilde{Q}$ expands most rapidly, and does not rotate, when Q moves in the orthogonal direction $-i\mathcal{K}$.*

[6] Here we mean the linear dimensions of $\widetilde{Q}$. For example, if $\widetilde{Q}$ were a disc then we could take its "size" to be its radius.

of $\widetilde{Q}$ (again with respect to the distance *it* moves) as a fraction of $\widetilde{Q}$'s initial size. Then [exercise]

$$\mathcal{E} = \widehat{\xi} \times \mathcal{K}.$$

This achieves its maximum value $\mathcal{E}_{max} = |\mathcal{K}|$ when Q moves in the direction $-i\mathcal{K}$. These two results may be viewed as two facets of the single complex equation

$$\overline{\widehat{\xi}}\, \mathcal{K} = \mathcal{R} + i\mathcal{E}.$$

In Chapter 12, having developed the physical concepts of "flux" and "circulation", we shall return to the complex curvature and see that it has other elegant properties and applications.

5.10 Celestial Mechanics*

5.10.1 Central Force Fields

If a particle p, moving through space, is constantly being pulled towards (or pushed away from) a fixed point o with a force that depends only on its distance r from o then we say that it is in a *central force field* and that o is the *centre of force*. No matter how the force varies with r, it is not hard to show [exercise] that the orbit of p will always lie in a plane through o.

Another feature of motion in *any* central force field is that the radius op sweeps out area at a constant rate $\mathcal{A}$, called the *areal speed*. A proof of this is given in Ex. 24. If the mass of p is m then [exercise] the angular momentum h of p is $2m\mathcal{A}$. The fact that $\mathcal{A}$ is constant is thus a manifestation of the conservation of angular momentum.

In addition to the angular momentum, the total energy E of the particle remains constant as it orbits. Henceforth, we shall always use a particle of unit mass. Thus if the particle's speed is v then the kinetic energy contribution has the definite value $\frac{1}{2}v^2$, while the potential energy contribution is only defined up to a constant. We shall restrict ourselves to force fields that vary as a power of r, and we may then fix the constant by arbitrarily assigning zero potential energy to the point where the field vanishes: if the force grows as a positive power of r, at the origin; if the force dies away as a negative power of r, at infinity.

5.10.2 Two Kinds of Elliptical Orbit

Consider the attractive *linear* force field in which, by definition, the force towards o is proportional to r. This linear force law is extremely important in physics, for if almost any physical system is slightly disturbed from equilibrium then the restoring force is precisely of this kind. Here is a simple example of what we mean; it will enable you to experimentally investigate motion in a linear force field. You are encouraged to do the following, not merely to imagine it.

Take a small weight W and suspend it just above a point o of a horizontal table using several feet of thread, perhaps attached to the ceiling. If you pull W to the side by just an inch or two then, because the thread is long, W barely rises above the table's surface and we may idealize this to a movement *on* the table. Furthermore, although the forces acting on W in this displaced position are actually gravity and the tension in the thread, the net effect [exercise] is as though o were magically pulling W towards it with a force proportional to r, as was required. To avoid the possibility of confusion later, we stress that gravity is playing absolutely no essential role here; it is merely providing one particularly convenient way of simulating a linear force field.

Now pull W a little bit away from o and give it a gentle flick in a random direction. You see that the orbit of W is a closed curve traversed again and again—a beautifully symmetrical oval shape centred at o. But exactly what is this oval?

It is an ellipse! To demonstrate this, take the tabletop to be $\mathbb{C}$ with o as its origin. Once again take W to have unit mass, and let its location at time t be $z(t)$. For simplicity's sake, let the force directed towards the origin *equal* the distance $|z|$. The differential equation governing the motion of W will therefore be $\ddot{z} = -z$, the two basic solutions of which are $z = e^{\pm it}$. These represent counter-rotating motions of unit speed around the unit circle. [Try launching W so as to produce these solutions.] The general solution is then obtained as a linear combination of these motions:

$$z = p\, e^{it} + q\, e^{-it}, \tag{5.33}$$

where p and q may, without any real loss of generality, be taken as real and satisfying $p > q$.

As is illustrated in [5.22], the addition of such counter-rotating circular motions results in elliptical motion with the attracting point at the centre. This becomes clear if we rewrite (5.33) as

$$z = a \cos t + ib \sin t,$$

where $a = p + q$ and $b = p - q$. Each of these numbers has a double significance: a is both the semimajor axis and the point of launch; b is both the semiminor axis and the speed of launch. Note that the foci are at $\pm\sqrt{a^2 - b^2} = \pm 2\sqrt{pq}$.

Finally, for future use, let us calculate the constant energy E of a particle orbiting in this field. The potential energy is the work needed to pull the particle away from the origin out to a distance of r, namely, [exercise] $(r^2/2)$. Thus

$$E = \tfrac{1}{2}(v^2 + r^2).$$

As the particle orbits round the ellipse in [5.22], we see that this expression always equals $\frac{1}{2}(a^2 + b^2)$.

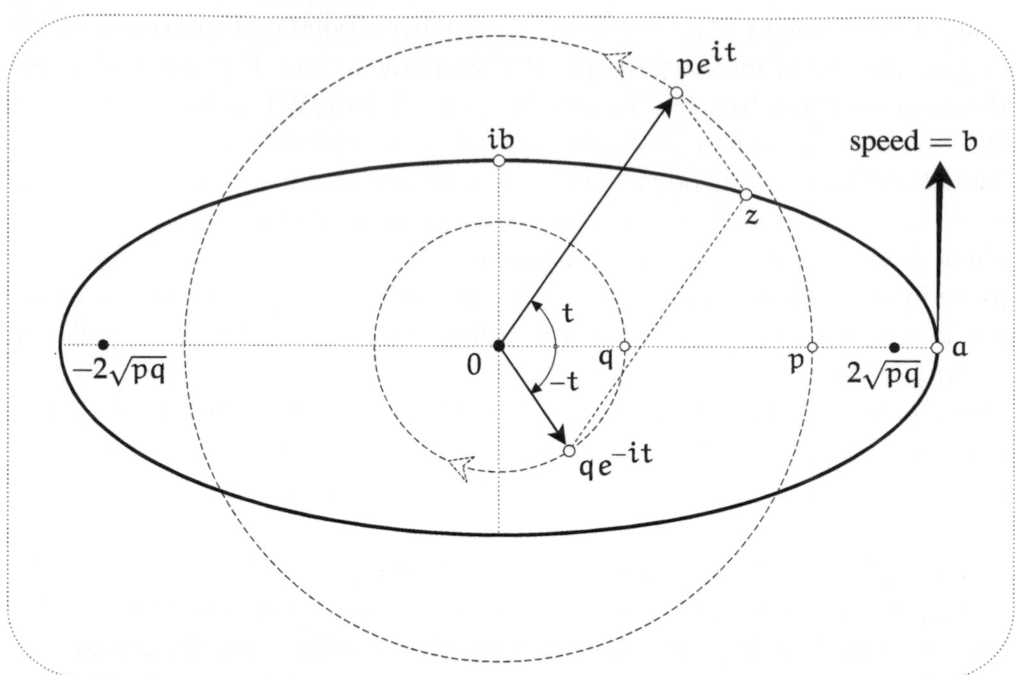

[5.22] Addition of counter-rotating circular motions results in elliptical motion.

We now turn to a second, more famous example of elliptical motion in a central force field: the orbits of the planets around the sun. There are two fundamental differences between this phenomenon and the one above. First, instead of the force of attraction increasing linearly with distance, here the force of gravity dies away as the square of the distance from the sun. Second, instead of the centre of attraction being at the centre of the elliptical orbit, here the sun is at one of the foci.

The ancient Greeks discovered that the ellipse has beautiful *mathematical* properties; two thousand years later Newton revealed that it has equally beautiful *physical* significance. He discovered that if, and *only*[7] if, the force field is linear or inverse-square, then elliptical orbits result. In the *Principia* Newton explicitly drew attention to this coincidence, calling it "very remarkable". As the Nobel physicist Subrahmanyan Chandrasekhar (1995, p. 287) observed, "nowhere else in the *Principia* has Newton allowed himself a similar expression of surprise."

We are left with something of a mystery. There appears to be some special connection between the linear and inverse-square force fields, but what could it possibly be? Newton himself was able to find a connection, and we shall use complex analysis to find another. For more on both these connections, see Arnol'd (1990), Needham (1993), and Chandrasekhar (1995).

[7] Newton assumed that the force varies as a power of the distance, but it has since been discovered that the result is still true if we drop this requirement.

5.10.3 Changing the First into the Second

The geometry of complex numbers was not yet understood in the time of Newton; had it been, he would surely have discovered the following surprising fact. If we apply the mapping $z \mapsto z^2$ to an origin-centred ellipse, then the image is not some strange ugly shape, as one might expect, but rather another *perfect ellipse*; furthermore, this ellipse automatically has one focus at the *origin*! See [5.23]. Before exploring the implications, let us verify this fact: squaring (5.33),

$$z \mapsto z^2 = (p\, e^{it} + q\, e^{-it})^2 = p^2\, e^{i2t} + q^2\, e^{-i2t} + 2pq.$$

The first two terms correspond to an origin-centred ellipse with foci at $\pm 2pq$; the last term therefore translates the left-hand focus to the origin.

Expressed in dynamical terms, this geometric result states that while leaving the attracting point fixed at the origin, $z \mapsto z^2$ transforms an orbit of the linear field into an orbit of the inverse square field. However, we are only in a position to state the result in this way because we already know what the orbits in the two fields look like. *Is there instead some a priori reason why $z \mapsto z^2$ should map orbits of the linear field to orbits of the gravitational field?* If there were such a reason then [5.23] could be viewed as a novel derivation, or explanation, of the elliptical motion of planets about the sun as focus.

That there is indeed such a reason was discovered in the first decade of the 20th century. Several people deserve credit for this beautiful result which, at the time of this writing, is still not widely known. Apparently Bohlin (1911) was the first to publish it, not knowing that Kasner (1913) had already discovered a more general result in 1909. Finally, knowing only of Bohlin's work, Arnol'd (1990) rediscovered Kasner's general theorem.

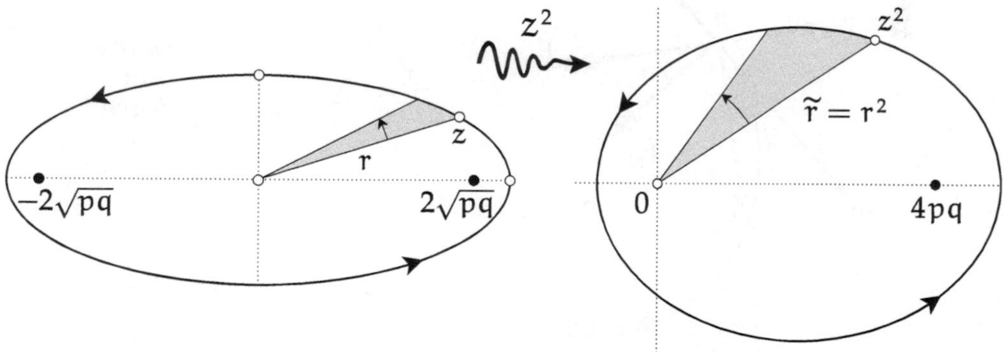

[5.23] Transformation of the linear force field into the inverse-square force field. *Very remarkably, squaring an origin-centred ellipse results in another ellipse. Furthermore, one of the foci of the image ellipse is located at the origin! As the text explains, this provides a novel explanation of the elliptical orbits of the planets around the Sun at one focus.*

Before embarking on the details of the explanation, here (following Needham (1993)) is our plan of attack. In the absence of force a particle will move in a straight line; *bending* is therefore the manifestation of force, and this can be quantified in terms of the curvature of the orbit. Since the mapping $z \mapsto z^2$ is analytic, we may use the results of the previous section to find the relationship between the curvature of an orbit and the curvature of the image orbit produced by the mapping. This will enable us to find the relationship between the forces that hold the preimage and image in their respective orbits.

5.10.4 *The Geometry of Force*

Given an orbit and a centre of force, our aim is to find a purely geometric formula for the magnitude F of the force $\boldsymbol{F}$ that holds the particle in that orbit. Consider figure [5.24]. As illustrated, it is conceptually very helpful to decompose $\boldsymbol{F}$ into components $\boldsymbol{F}_{\mathrm{T}}$ and $\boldsymbol{F}_{\mathrm{N}}$ that are tangential and normal to the orbit, respectively. The effect of the component $\boldsymbol{F}_{\mathrm{T}}$ is to change the speed v of p without altering its course. The effect of the component $\boldsymbol{F}_{\mathrm{N}}$ is to bend the orbit of p without altering its speed.

From elementary mechanics we know that if a particle of unit mass moves at constant speed v round a circle of radius ρ then the force directed towards the centre is (v^2/ρ). Thus if the curvature of the orbit is κ (as illustrated) then $\mathsf{F}_{\mathrm{N}} = \kappa v^2$. If we call the acute angle between the radius and the normal γ, then it follows that the

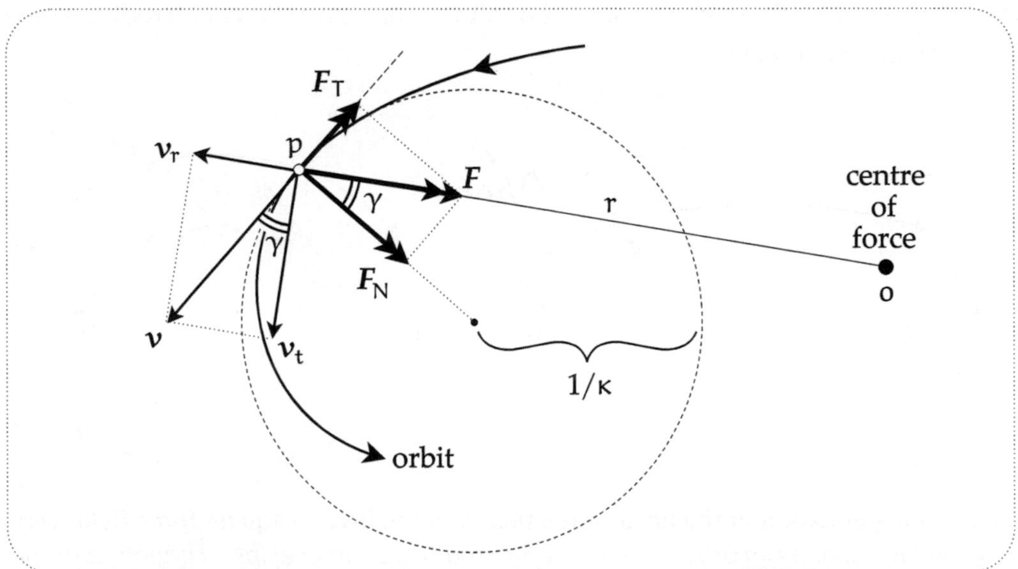

[5.24] Geometric determination of the central force law from the orbit. *If* h *is the (constant) angular momentum of the orbiting particle, the magnitude of the central force field that holds the particle in its orbit of curvature* κ *is given by* $\mathsf{F} = \mathsf{h}^2 \left(\frac{\kappa \sec^3 \gamma}{r^2}\right)$.

total force acting on p is

$$F = F_N \sec\gamma = \kappa v^2 \sec\gamma.$$

In order to fully reduce this formula to geometric terms, we need to express v in geometric terms. This is made possible by the constancy of the angular momentum $h = 2\mathcal{A}$. If we decompose the velocity v into radial and transverse components v_r and v_t, then clearly only the latter generates area. More precisely, $h = 2\mathcal{A} = r\,v_t = r\,v\cos\gamma$, so

$$v = h\left(\frac{\sec\gamma}{r}\right). \tag{5.34}$$

Substituting for v in the previous result, we obtain the desired geometric formula for the force:

$$F = h^2\left(\frac{\kappa\sec^3\gamma}{r^2}\right). \tag{5.35}$$

This result is essentially due to Newton (1687, Prop. VII). Observe that the concept of time has almost disappeared in this formula, the only vestige being the constant h that specifies how fast the orbit is traversed.

5.10.5 An Explanation

As z describes an arbitrary orbit, (5.35) tells us the force F needed to hold it in that orbit. Now apply the mapping $z \mapsto z^2$, and let a tilde denote a quantity associated with the image, e.g., $\widetilde{r} = r^2$. The force $\widetilde{F}$ needed to hold the image in its orbit is

$$\widetilde{F} = \widetilde{h}^2\left(\frac{\widetilde{\kappa}\sec^3\widetilde{\gamma}}{\widetilde{r}^2}\right),$$

and we now seek to relate this to the original force F.

First, to find $\widetilde{\kappa}$, simply put $f(z) = z^2$ into (5.24) and thereby obtain [exercise]

$$\widetilde{\kappa} = \frac{1}{2}\left[\frac{\cos\gamma}{r^2} + \frac{\kappa}{r}\right].$$

Next, observe that since the ray from 0 to z maps to the ray from 0 to z^2, the conformality of the mapping implies that $\widetilde{\gamma} = \gamma$.

Putting these facts into the formula for $\widetilde{F}$, and substituting for the original speed and force from (5.34) and (5.35), we get

$$\widetilde{F} = \left(\frac{\widetilde{h}}{h}\right)^2 \frac{\left[\frac{1}{2}v^2 + \frac{1}{2}rF\right]}{\widetilde{r}^2}. \tag{5.36}$$

Even if F is a simple power law, generally this $\widetilde{F}$ will not be. However, if and *only* if the original force field is *linear*[8], the numerator in the above expression magically

[8] However, as we shall see in a moment, it could be the *repulsive* linear field $F = -r$ instead of the attractive one $F = +r$.

becomes the constant total energy E of the particle in the original field:

$$\widetilde{F} = \left(\frac{\widetilde{h}}{h}\right)^2 \frac{E}{\widetilde{r}^2}! \tag{5.37}$$

The image therefore moves in a field that is *inverse-square*, as was to be shown.

Here is a fact which may have been bothering you already. The only gravitational orbits we have managed to explain in this way are the ellipses; where are the hyperbolic orbits which we know are also possible in a gravitational field? In fact the geometry of $z \mapsto z^2$ *does* explain these, the resolution being that gravitational orbits arise not only as the images of orbits in a linear field that is attractive, but also of orbits in a linear field that is *repulsive*, $F = -r$. The orbits in this field are hyperbolae with centre (i.e., intersection of asymptotes) at the origin, and $z \mapsto z^2$ maps these to hyperbolae with one focus at the origin.

The dynamical explanation is almost unchanged: the constant total energy of the particle in the original repulsive linear field is now given by $E = \frac{1}{2}(v^2 - r^2)$, so inserting $F = -r$ in (5.36) once again yields (5.37). See Needham (1993) for more on this, as well as the general result we are about to state, which may be proved in exactly the same way as the special case above.

5.10.6 The Kasner–Arnol'd Theorem

The power laws $F \propto r$ and $\widetilde{F} \propto \widetilde{r}^{-2}$ are examples of what Arnol'd calls *dual* force laws, and both he and Kasner discovered that they constitute just one example of duality. Here is the general result:

> *Associated with each power law $F \propto r^A$ there is precisely one power law $\widetilde{F} \propto \widetilde{r}^{\widetilde{A}}$ that is dual in the sense that orbits of the former are mapped to orbits of the latter by $z \mapsto z^m$, and the relationships between the forces and the mapping are:*

$$(A+3)(\widetilde{A}+3) = 4 \quad \text{and} \quad m = \frac{(A+3)}{2}.$$

To their result we add the following point of clarification on the role of energy:

> *In general, positive energy orbits in either the attractive or repulsive field $F \propto r^A$ map to attractive orbits in the dual field, while negative energy orbits map to repulsive ones. However, if $-3 < A < -1$ (e.g., gravity) then these roles are reversed. In all cases, zero energy orbits map to force-free rectilinear orbits.*

5.11 Analytic Continuation*

5.11.1 *Introduction*

Throughout this book we have stressed how functions may be viewed as geometric entities that need not be expressed (nor even be expressible) in terms of formulae. As an illustration of the limitations of formulae, consider

$$G(z) = 1 + z + z^2 + z^3 + \cdots .$$

This power series converges inside the unit circle $|z| = 1$, and consequently it is analytic there. Figure [5.25] shows a grid of little squares inside this circle being amplitwisted by $z \mapsto w = G(z)$ to another such grid lying to the right of the vertical line $\text{Re}\,(w) = \frac{1}{2}$, which itself is the image of the circle. Now this circle is certainly a barrier to the power series formula for G, since it clearly diverges at 1: geometrically, the image of the circle extends to ∞. However, the circle is *not* a barrier to the geometric entity that the formula is attempting to describe.

Consider a somewhat different-looking power series centred at -1:

$$H(z) = \frac{1}{2}\left[1 + \left(\frac{z+1}{2}\right) + \left(\frac{z+1}{2}\right)^2 + \cdots\right].$$

This series is analytic inside a larger circle of convergence $|z+1| = 2$. Despite the apparent difference, $H(z)$ maps the previously considered solid grid inside $|z| = 1$ to precisely the same grid on the right of $x = (1/2)$ as G did: $H = G$ inside $|z| = 1$. But now the grid may be extended to the dotted one lying *outside* $|z| = 1$, and H amplitwists it to the dotted grid lying to the left of $x = (1/2)$. We say that H is an *analytic continuation* of G to the larger disc. An obvious question is whether H is the *only* analytic continuation of G to this region. As we hope [5.25] makes palpable, the rigidity imposed by being locally an amplitwist does indeed force the mapping to grow in a unique way.

The objective of this final section will be to make this rigidity clearer, and also to describe one method (due to Schwarz) of explicitly finding the mapping in regions beyond its original definition. Before doing this, however, we will complete our discussion of [5.25].

The figure makes it plain that H is no more the end of the line than G was: it too can be continued. But if we cling to power series then the scope of our description of the mapping that underlies both G and H will be strictly limited. This is because such series only converge inside discs, and if we try to expand any disc then it will eventually hit the singularity at $z = 1$ and then be unable to go round it. Thus any power series will necessarily miss out at least half of the potential domain of the mapping. On the other hand, as you may have already noticed, the Möbius transformation $1/(1-z)$ is analytic everywhere except $z = 1$, and it agrees with both

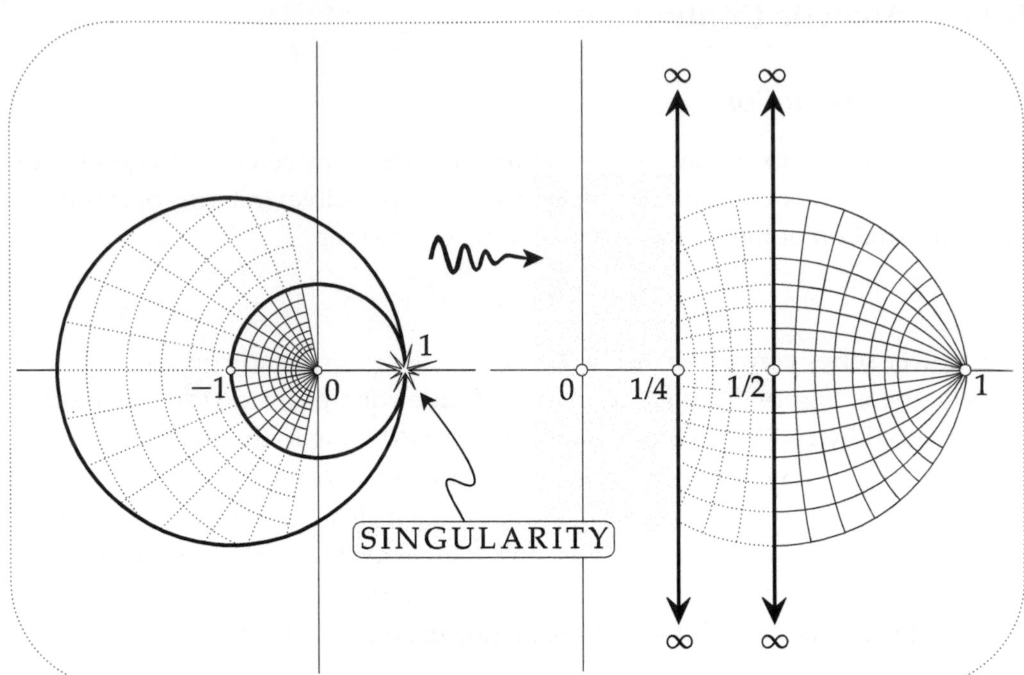

[5.25] Analytic Continuation. *The origin-centred power series* $G(z) = 1+z+z^2+z^3+\cdots$ *only converges inside the unit disc. The image* $w = G(z)$ *of the grid of small "squares" on the left (inside the unit disc) is shown [solid] on the right, in the region* $\text{Re}(w) > \frac{1}{2}$. *But if we extend the grid [dotted] on the left* beyond *the unit circle, then there is a unique* analytic continuation *of the image grid [dotted] on the right into the region* $\frac{1}{4} < \text{Re}(w) < \frac{1}{2}$, *expressible as a power series that converges inside the larger circle* $|z+1| = 2$. *If we abandon power series altogether, then the grid can in fact be extended uniquely to the entire plane, the complete analytic continuation being given by* $z \mapsto 1/(1-z)$, *with the singularity* $z = 1$ *being mapped to the point at* ∞.

G and H within their circles of convergence; it thus constitutes the complete analytic continuation of the mapping. [We encourage you to use this fact to check the details of the figure.] The simplicity of this example is perhaps misleading. Usually one cannot hope to capture the entire geometric mapping within a single closed expression such as $1/(1-z)$.

When one stares at a figure like [5.25] one starts to sense the rigid growth of the mapping due to the analytic requirement that an expanding mesh of tiny squares must map to another such mesh. It also becomes clear how the mapping itself is oblivious to the different formulae with which we try to describe it. Indeed we have seen that the two circles—such formidable and impenetrable barriers to the power series—have only a slight significance for the mapping itself: both hit $z = 1$, so both images extend to ∞.

5.11.2 Rigidity

The essential character of analytic rigidity is captured in the following result:

> *If even an arbitrarily small segment of curve is crushed to a point by an analytic mapping, then its* ***entire domain*** *will be collapsed down to that point.*

The theory of integration to be developed in the following chapters will provide a convincing explanation of this fact. For the present, though, we can obtain a good measure of insight into its truth by extending our previous discussion of critical points (page 231). This may give the illusion of dispensing with integration theory, but as we pointed out at the time, that discussion *also* had to draw on future results. We now recap the relevant facts concerning critical points.

The amplification vanishes at a critical point p, leading to the impression that an infinitesimal disc centred there is crushed down to a point. However, this is merely a 'trick of the light' due to low magnification of the image plane. If the order of p is $(m-1)$, so that the mapping locally resembles z^m, then an infinitesimal disc at p of radius ϵ will be mapped [m-fold] onto a vastly smaller disc of radius ϵ^m. In terms of the microscope analogy this means that we must use the L_m lens to see that the image isn't a perfect point. The greater the order, the greater the degree of crushing at p, and the greater the power of the first lens that will reveal the nonpointlike image.

Now observe that, calculationally speaking, the role of the increasingly high-powered lenses that fail to resolve the image is taken over by the increasingly high-order derivatives that vanish at p:

$$f(z) \sim z^m \iff \begin{cases} L_1, \ldots, L_{m-1} \text{ show nothing, but image visible with } L_m \\ f'(p) = 0, f''(p) = 0, \ldots, f^{(m-1)}(p) = 0, \text{ but } f^{(m)}(p) \neq 0 \end{cases}$$

In short, the higher the derivative that vanishes at p, the greater the degree of crushing at p.

We now apply this insight to the given situation. Let s be the (possibly) tiny segment that is crushed by $f(z)$. The amplification of f at a point of s may be read off by looking in any direction. By choosing to look *along* s we find that the amplification vanishes at each point of s. The entire segment is therefore made up of critical points for which $f' = 0$. Now think of f' as an analytic mapping in its own right, just as we did in [5.18]. We have just seen that this mapping automatically possesses the same property as f did: it crushes s to a point. We conclude that *its* derivative must also vanish on s. Clearly there is no end to this; *all* the derivatives of f must vanish, and, correspondingly, infinitesimal discs centred on s must be *totally* crushed.

This means that there is at least a sheathlike region surrounding s which is completely crushed by f. But if we take a new curve lying in this region, the whole line of thought may be repeated to deduce that f must crush a still larger region. The collapse of the function therefore proceeds outwards (at the speed of thought!) to the entire domain.

5.11.3 Uniqueness

Suppose that $A(z)$ and $B(z)$ are both analytic functions defined on a region that happens to be the same size and shape as California. Suppose, further, that A and B both happen to have the same effect on a tiny piece of curve, say a fallen eyelash lying in a San Francisco street. This tiny measure of agreement instantly forces them into *total* agreement, even hundreds of miles away in Los Angeles! For $(A - B)$ is analytic throughout California, and since it crushes the eyelash to 0, it must do the same to the entire state.

We can express this slightly differently. If we arbitrarily specify the image points of a small piece of curve s, then in general there will not exist an analytic function that sends s to this image. However, the previous paragraph assures us that if we can find such a function on a domain including s, then it is *unique*.

This is the "compelling reason" we referred to earlier in connection with the uniqueness of the generalization of e^x to complex values. For if an analytic generalization $E(z)$ exists, then we see that it will be uniquely determined by the values of e^x on even a small piece of the real axis. Of course knowing this does not help in the least to find out what $E(z)$ actually *is*. The value of our previous derivations of explicit expressions for $E(z)$ therefore remains undiminished. On the other hand, the new knowledge is not without practical implications. Consider these three very different-looking expressions:

$$\lim_{n\to\infty}\left(1+\frac{z}{n}\right)^n, \quad e^x(\cos y + i\sin y), \quad 1+z+z^2/2!+z^3/3!+\cdots.$$

They are all analytic, and they all agree with e^x when z is real. Thus, without further calculation, we know they must all be equal to each other, for they can only be different ways of expressing the unique analytic continuation of e^x.

New and important aspects of uniqueness emerge when we consider domains that merely overlap, rather than coincide. Let $g(z)$ and $h(z)$ be analytic functions defined on the sets P and Q shown in [5.26a]. If they agree on even a small segment s in $P \cap Q$ then they will agree throughout $P \cap Q$. If we imagine that we initially only know about g on P, then we may think of h as describing the same geometric mapping as g but with the domain P extended to encompass Q. We are encouraged in this view by the fact that g uniquely determines this analytic continuation. For suppose h^* were another continuation of g into Q. On s we would then have $h^* = g = h$, but this forces $h^* = h$ throughout their common domain Q.

The functions $G(z)$ and $H(z)$ of the introduction furnish a concrete example of the above, where P happens to lie wholly within Q. The function $1/(1-z)$ then constitutes the analytic continuation of H to the rest of the plane.

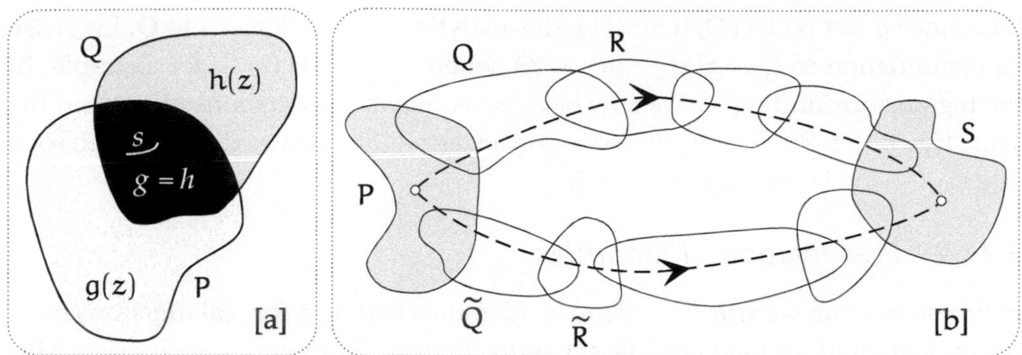

[5.26] Analytic continuation along alternative routes. [a] *If* g(z) *and* h(z) *are analytic functions defined in overlapping regions* P *and* Q, *and they agree on even a small segment* s *in this overlap, then they must agree* throughout $P \cap Q$, *and we are entitled to think of* h(z) *as the* unique *analytic continuation of* g(z) *from* P *into* Q. **[b]** *Although the analytic continuations along* $PQR\cdots S$ *and along* $P\widetilde{Q}\widetilde{R}\cdots S$ *are both uniquely determined, they need* not *agree on* S*: in this case we have a multifunction.*

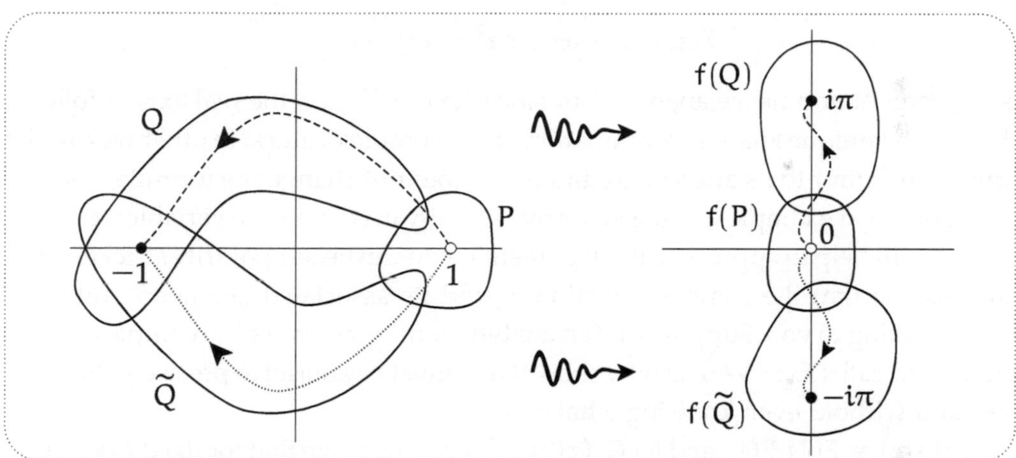

[5.27] Analytic continuation of log z. *In* P *we can define a single-valued branch of* log z *given by* $f(z) = \ln r + i\theta$, *where* $-\pi < \theta \leqslant \pi$. *The figure shows how* 1 *maps to* 0, *and how* P *maps to the surrounding region* f(P). *The definition of* log z *then has unique analytic continuations into* Q *and into* $\widetilde{Q}$, *but they* disagree at −1, *for example.*

Just as g was continued from P to Q, so we may continue the process along a whole chain of overlapping sets $P, Q, R, \ldots, S$, as in [5.26b]. We thereby obtain a unique analytic continuation of g to S. But what if we chose an alternative route such as $P, \widetilde{Q}, \widetilde{R}, \ldots, S$? Once again the continuation of g to S is unique, *but this need not agree with the first continuation.* The idea of analytic continuation has thus led very naturally to the idea of multifunctions.

Consider [5.27]. In P we can define a single-valued branch of log z given by $f(z) = \ln r + i\theta$, where $-\pi < \theta \leqslant \pi$. The figure shows how 1 maps to 0, and how P maps to the surrounding region f(P). If we define $g(z) = \ln r + i\Theta$ (with $-\frac{\pi}{2} < \Theta \leqslant \frac{3\pi}{2}$)

then since $g = f$ on $P \cap Q$, it must be the analytic continuation of f to Q. Likewise its continuation to $\widetilde{Q}$ is $\tilde{g}(z) = \ln r + i\widetilde{\Theta}$, where $-\frac{3\pi}{2} < \widetilde{\Theta} \leqslant \frac{\pi}{2}$, for example. In the region surrounding -1 we now have two unique continuations of one and the same function f. But despite this common ancestry, they clearly disagree with each other: $g(-1) = i\pi$, while $\tilde{g}(-1) = -i\pi$.

5.11.4 Preservation of Identities

In this subsection we will show that any identities that hold for real functions must continue to hold for their analytic generalizations to $\mathbb{C}$ (assuming such exist). This is easiest to explain through examples.

First we consider an important example dealing with power series. Suppose that the real function $f(x)$ can be represented by a convergent power series

$$f(x) = a + bx + c\,x^2 + d\,x^3 + \cdots.$$

We therefore know that the complex series

$$F(z) = a + bz + c\,z^2 + d\,z^3 + \cdots.$$

is convergent and hence analytic. But since $F(x) = f(x)$ on the real axis, it follows that F is the unique analytic continuation of f to complex values. In other words, the transition from f to its analytic continuation does not change the formula (series).

For our next example we consider a real identity involving *two* variables: $e^x \cdot e^y = e^{x+y}$. It will help to appreciate the argument if you can be temporarily stricken with amnesia, so that the complex function e^z and its associated geometry suddenly mean nothing to you. Suppose that an analytic continuation of e^x to complex values exists, and call it $E(z)$. We can now show that E must be subject to precisely the same law, and without even knowing what E is!

Let $F_\zeta(z) \equiv E(\zeta)\cdot E(z)$, and let $G_\zeta(z) \equiv E(\zeta{+}z)$. First note that for fixed ζ both $F(z)$ and $G(z)$ are analytic functions of z. Now suppose that ζ is real, so that $E(\zeta) = e^\zeta$. If z now moves on a segment of the real axis then it follows from the real identity that $F(z) = G(z)$; but from our recent results we know this implies that they are equal everywhere. If we hold z fixed instead, then analogous reasoning yields $F_\zeta = G_\zeta$, and we conclude that

$$E(\zeta) \cdot E(z) = E(\zeta + z)$$

for complex values of both ζ and z. It should be clear that this reasoning extends to any identity, even one involving more than two variables.

5.11.5 Analytic Continuation via Reflections

Quite distinct from questions of existence and uniqueness is the problem of actually *finding* an analytic continuation. The above ideas and results are mute on this

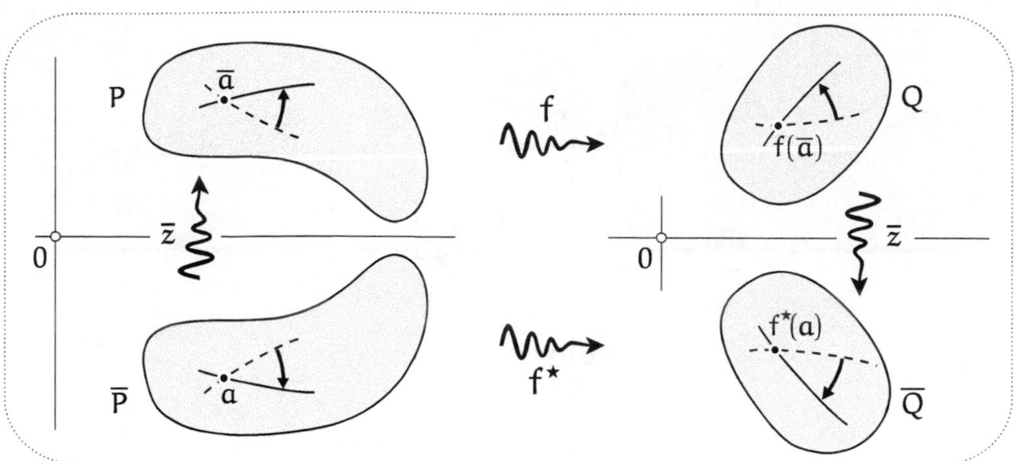

[5.28] Extending an analytic mapping via reflection. *If the analytic function* f *maps* P *to* Q*, then we can use it to construct an analytic mapping from* $\overline{P}$ *to* $\overline{Q}$*, given by* $f^{\star}(z) \equiv \overline{f(\overline{z})}$*. As we see, the two reflections reverse angles twice, resulting in a conformal mapping.*

issue, although it could reasonably be claimed that the persistence of identities is a practical help. We next explain a *Symmetry Principle* (due to Schwarz) which enables one to find a continuation easily and explicitly, albeit under rather special circumstances.

We first describe how it is possible to use two reflections to construct a new analytic function from an old one. Suppose an analytic function f is defined on a region P, the image of which is Q (see [5.28]). Let $\overline{P}$ and $\overline{Q}$ be the reflections of these regions across the real axis. We can now use f to construct an analytic mapping from $\overline{P}$ to $\overline{Q}$, namely

$$f^{\star}(z) \equiv \overline{f(\overline{z})}.$$

The figure explains why $f^{\star}$ is conformal, and hence analytic. All three stages, $a \mapsto \overline{a} \mapsto f(\overline{a}) \mapsto \overline{f(\overline{a})}$, preserve the magnitude of an angle at a; the first reflection reverses the sense, then f preserves the reversed sense, and finally the second reflection undoes the damage, restoring the angle to pristine condition at $f^{\star}(a)$.

In general this mapping $f^{\star}$ will not be a continuation of f in any sense, rather it is an entirely new mapping. This should become clear if you imagine moving P downwards until some of it crosses the real axis. P and $\overline{P}$ now overlap so that $P \cap \overline{P}$ constitutes a common domain for f and $f^{\star}$, but we hope you can see that there is no reason for them to agree with each other. This is clearer still if we take an example:

$$f = (\text{rotation of } \phi) \quad \overset{\text{exercise}}{\Longrightarrow} \quad f^{\star} = (\text{rotation of } -\phi).$$

Although it is generally not a continuation of f, this new mapping $f^{\star}$ (together with its soon to be introduced generalization to circles) is very useful in its own right. In

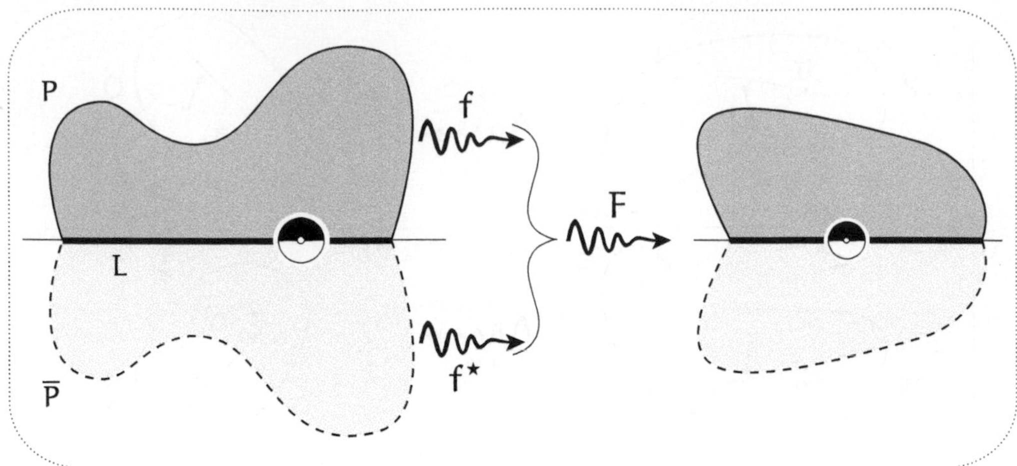

[5.29] Analytic continuation via reflection. *Suppose that the analytic function* f *defined in* P *is the complex extension of a real function defined on* L*. Then the function* $f^{\star}(z) \equiv \overline{f(\overline{z})}$ *introduced in the previous figure is the analytic continuation of* f *into* $\overline{P}$*. Generalizing, if* f *maps a line-segment* L *(not necessarily real) to another line-segment* $\widehat{L}$*, then we can analytically continue from one side of* L *to the other by using the fact that* points symmetric in L map to points symmetric in $\widehat{L}$.

Chapter 12 we will show that it is intimately connected with the so-called *method of images* of electrostatics and fluid dynamics.

We now turn to the special circumstance under which $f^{\star}$ *is* the analytic continuation of f. Suppose that f is itself the complex generalization of a real function, and let P have a part L of its boundary along the real axis, as in [5.29]. Since f is real on L, the image set Q will also border on the real axis. Unlike the general situation previously considered, f and $f^{\star}$ will now automatically agree on their common domain $P \cap \overline{P} = L$, for if z is real then

$$f^{\star}(z) = \overline{f(\overline{z})} = \overline{f(z)} = f(z).$$

We can now think of f and $f^{\star}$ as being two parts of a single analytic mapping F on $P \cup \overline{P}$. Indeed, by considering what happens to the two halves of an infinitesimal disc centred on L, it's clear that F is analytic there, for the image is another infinitesimal disc. [What happens if we are at a critical point?] Once again, notice how different this is from the case of real functions, for we could easily join two pieces of graph together with a kink at the join; their values would then agree, while their derivatives would not.

Of course if f is already defined in $\overline{P}$ (as well as P) then $f^{\star}$ must simply reproduce the mapping that's already there. For example, the formula for the complex generalization $\sin z$ is valid everywhere, so it should be subject to the symmetry $f^{\star}(z) = f(z)$. Indeed if we follow the three steps of $a \mapsto f^{\star}(a)$ then we do find that

$$a \mapsto \overline{a} \mapsto \frac{e^{i\overline{a}} - e^{-i\overline{a}}}{2i} \mapsto \frac{e^{-ia} - e^{ia}}{-2i} = \sin a\,.$$

We can rephrase our result in a more symmetric and slightly generalized [exercise] form. If f maps a line-segment L (not necessarily real) to another line-segment $\widehat{L}$, then we can analytically continue from one side of L to the other by using the fact that *points symmetric in* L *map to points symmetric in* $\widehat{L}$.

This sounds very reminiscent of the conservation of symmetry by Möbius transformations that we discovered in Chapter 3, and indeed by fusing these two symmetry principles we can obtain a significant generalization of our result. Suppose that instead of mapping a particular[9] line to a line, f sends a part C of a circle to a part $\widehat{C}$ of another circle. We can reduce this to the previous case by using two Möbius transformations to send $C \mapsto L$, and $\widehat{C} \mapsto \widehat{L}$. We deduce that *points symmetric in* C *map to points symmetric in* $\widehat{C}$.

As a mixed example, imagine that f maps part of the unit circle to part of the real axis. If f is only known inside the circle then the above result tells us [exercise] that there is an analytic continuation to the exterior given by

$$f^{\dagger}(z) \equiv \overline{f\left(\frac{1}{\overline{z}}\right)}.$$

The complete analytic function F is then defined to be f inside the circle, and $f^{\dagger}$ outside the circle. By construction, this function sends symmetric pairs of points to conjugate images: $F^{\dagger}(z) = F(z)$.

Using what is now known as *Schwarzian reflection*[10], in 1870 H. A. Schwarz was able to generalize his Reflection Principle beyond lines and circles to more general curves. We end this chapter with a description of Schwarz's simple, yet fascinating, idea. The key is to use an analytic function to fake conjugation.

We know that reflecting every point across the real axis ($z \mapsto \overline{z}$) is not an analytic function. However, given a sufficiently smooth[11] curve K, it is possible to find an analytic function $\mathcal{S}_K(z)$ that *selectively* sends just the points of K to their conjugates:

$$z \in K \quad \Longrightarrow \quad \mathcal{S}_K(z) = \overline{z}.$$

Davis and Pollak (1958) christened $\mathcal{S}_K$ the *Schwarz function* of K. We can now define the Schwarzian reflection of z across K to be $\widetilde{z} = \mathfrak{R}_K(z)$, where

$$\mathfrak{R}_K(z) \equiv \overline{\mathcal{S}_K(z)}.$$

To see why this is a good idea, consider [5.30]. First note that points on K are unaffected, in accord with the ordinary notion of reflection, e.g.,

$$\widetilde{q} = \overline{\mathcal{S}_K(q)} = \overline{(\overline{q})} = q.$$

[9] We stress "particular", because if a general line were sent to a line then the mapping could only be linear. Similarly, in the new case, if a general circle were sent to a circle then the mapping would have to be a Möbius transformation.

[10] Schwarz (1972a).

[11] The curve must in fact be "analytic". On this point, see Davis (1974), which also contains many interesting applications of the Schwarz function. A more advanced work is Shapiro (1992).

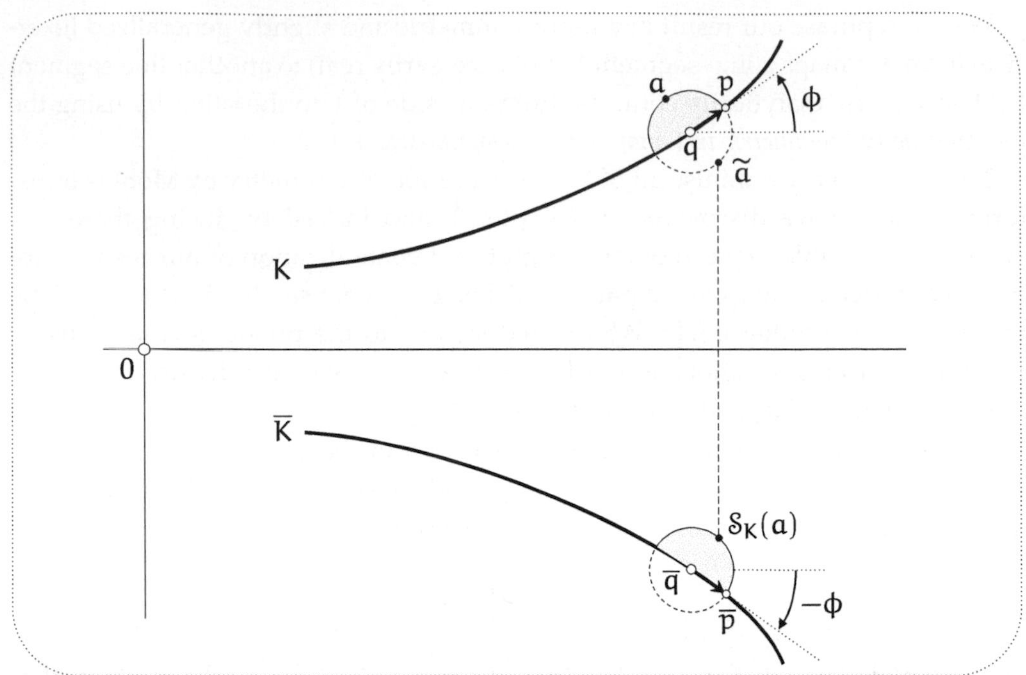

[5.30] Schwarzian Reflection. *The* Schwarz function $\mathcal{S}_K$ *of a curve* K *is defined to be the analytic mapping that mimics conjugation on* K*: if* z *lies on* K*, then* $\mathcal{S}_K(z) = \overline{z}$*. We can then define* Schwarzian reflection *across* K *to be* $z \mapsto \widetilde{z} = \mathfrak{R}_K(z) \equiv \overline{\mathcal{S}_K(z)}$*. As we see, very close to* K*, Schwarzian reflection becomes ordinary reflection. If* K *is a line or circle, then Schwarzian reflection is precisely reflection/inversion in* K*.*

Next, observe that since $\mathcal{S}_K$ is analytic, an infinitesimal disc centred at q is amplitwisted (*not* reflected) to a disc centred at $\overline{q}$. Furthermore, by noting how $\overrightarrow{qp}$ is mapped to $\overrightarrow{\overline{qp}}$, it follows that on K

$$\text{amplification} = 1 \quad \text{and} \quad \text{twist} = -2\phi \quad \Longrightarrow \quad \mathcal{S}'_K = e^{-i2\phi},$$

where ϕ is the angle that the tangent to K makes with the horizontal. It is now clear from the symmetry of the figure that if the point a is on the infinitesimal circle, then $\widetilde{a}$ is indeed its reflection across the tangent of K. Thus, at least very close to K, $z \mapsto \widetilde{z}$ is a reasonable generalization of the reflection concept. Furthermore, reflecting in K twice yields the identity mapping, as it should. For since $\mathfrak{R}_K$ is anticonformal, $\mathfrak{R}_K \circ \mathfrak{R}_K$ is conformal, i.e., analytic. But since this function maps each point of K to itself, and since an analytic function is determined by its values on a curve, $\mathfrak{R}_K \circ \mathfrak{R}_K$ must be the identity mapping.

We leave it to the exercises for you to show that if K is a line or a circle then $\widetilde{z}$ is just the ordinary reflection, even if z is far from K. For example, the unit circle C may be written as $\overline{z}z = |z|^2 = 1$, so that on C we have $\overline{z} = (1/z)$. Thus its Schwarz function is $\mathcal{S}_C(z) = (1/z)$, and so $\mathfrak{R}_C(z) = (1/\overline{z})$, which is just inversion in C.

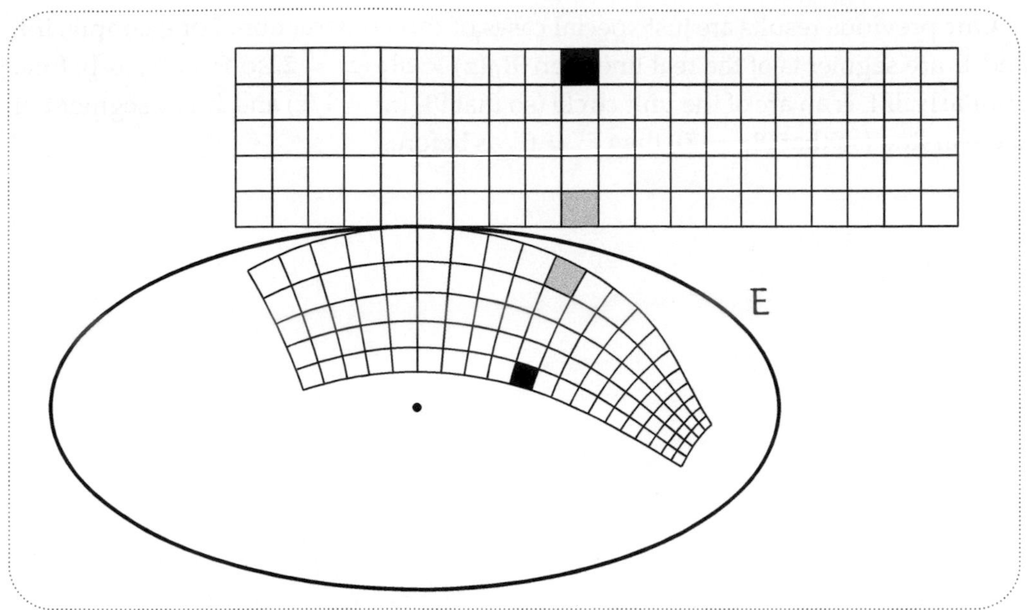

[5.31] Schwarzian reflection in an ellipse.

Let us give a less trivial example, namely, reflection in the ellipse E with equation $(x/a)^2 + (y/b)^2 = 1$. Writing $x = \frac{1}{2}(z+\bar{z})$ and $y = \frac{1}{2i}(z-\bar{z})$, then solving for $\bar{z}$ in terms of z, we find [exercise] that

$$\mathcal{S}_E(z) = \frac{1}{a^2-b^2}\left[(a^2+b^2)\,z - 2ab\,\sqrt{z^2+b^2-a^2}\,\right].$$

With $a = 2$ and $b = 1$, for example, Schwarzian reflection is given by

$$\mathfrak{R}_E(z) = \tfrac{1}{3}\left[5\,\bar{z} - 4\sqrt{\bar{z}^2-3}\,\right],$$

which is illustrated in [5.31]. We encourage you to verify this figure with your computer, as well as to examine the effect of $\mathfrak{R}_E$ on other shapes.

With the proper concept of reflection in hand, we may now generalize the above method of analytic continuation across lines and circles to more general curves. Let f be an analytic mapping defined in a region P bordering on a curve L that is smooth enough to possess a Schwarz function, and let $\widehat{L} = f(L)$ be the image of L under f. Much as in figures [5.28] and [5.29], we may now analytically continue f across L by demanding that points that are symmetric in L map to points that are symmetric in $\widehat{L}$. Thus [draw a picture!] the continuation $f^\ddagger$ of f to $\widetilde{P} = \mathfrak{R}_L(P)$ is

$$f^\ddagger = \mathfrak{R}_{\widehat{L}} \circ f \circ \mathfrak{R}_L.$$

By the same argument as in [5.28], this is indeed analytic in $\widetilde{P}$, for it is the composition of one conformal mapping with two anticonformal mappings. Also, $f^\ddagger = f$ on L. The complete analytic function F given by f in P and $f^\ddagger$ in $\widetilde{P}$ is then subject to the symmetry $F^\ddagger = F$. This is *Schwarz's Symmetry Principle.*

Our previous results are just special cases of this construction. For example, if L and $\widehat{L}$ are segments of the real line then $\mathfrak{R}_L(z) = \mathfrak{R}_{\widehat{L}}(z) = \overline{z}$, so $f^{\ddagger} = f^{\star}$, as before. Similarly, if L is an arc of the unit circle (so that $\mathfrak{R}_L(z) = 1/\overline{z}$) and $\widehat{L}$ is a segment of the real line (so that $\mathfrak{R}_{\widehat{L}} = \overline{z}$), then $f^{\ddagger} = f^{\dagger}$, as before.

5.12 Exercises

1 Show that if $f = u + iv$ is analytic then $(\nabla u)\cdot(\nabla v) = 0$, where ∇ is the "gradient operator" of vector calculus. Explain this geometrically.

2 Show that the real and imaginary parts of an analytic function are *harmonic*, i.e., they both automatically satisfy *Laplace's equation*:

$$\Delta\Phi = 0\,,$$

where Δ (which is often instead written ∇^2) is defined by $\Delta \equiv \partial_x^2 + \partial_y^2$ and is called the *Laplacian*. [In Chapter 12 we will see that this equation represents a crucial link between analytic functions and physics.]

3 Use the previous exercise (*not* calculation) to show that each of the following is "harmonic".

(i) $e^x \cos y$.

(ii) $e^{(x^2-y^2)} \cos 2xy$.

(iii) $\ln|f(z)|$, where $f(z)$ is analytic.

4 What is the most general function $u = a\,x^2 + b\,xy + c\,y^2$ that is the real part of an analytic function? Construct this analytic function, and express it in terms of z.

5 Which of the following are analytic?

(i) $e^{-y}(\cos x + i \sin x)$.

(ii) $\cos x - i \sin y$.

(iii) $r^3 + i3\theta$.

(iv) $\left[r\, e^{r\cos\theta}\right] e^{i(\theta + r\cos\theta)}$.

6 Solve the Polar CR equations given that $\partial_\theta v \equiv 0$. Express your answer in terms of a familiar function, and interpret everything you have done geometrically.

7 Use the Cartesian CR equations to show that the *only* analytic mapping that sends parallel lines to parallel lines is the linear mapping. [*Hint*: Begin with the case of horizontal lines being mapped to horizontal lines. How does this translate into 'Equationspeak'? Now solve CR.]

8 Calculate, then draw on a picture, a possible location for $\log(1+i)$. Draw a small shape at $1+i$. Use the amplitwist of $\log(z)$ to draw its image. Verify this using your computer.

9 Derive the quotient rule in an analogous way to the product rule (see page 256). [*Hint*: Multiply top and bottom of (A/B) by $(b - \xi g')$.]

10 Consider the polynomial $P(z) = (z - a_1)(z - a_2)\ldots(z - a_n)$.

(i) Show that the critical points of $P(z)$ are the solutions of

$$\frac{1}{z-a_1} + \frac{1}{z-a_2} + \cdots + \frac{1}{z-a_n} = 0\,.$$

(ii) Let K be a circle with centre p. By considering the conjugate of the equation in (i), deduce that p is a critical point if and only if it is the centre of mass of the inverted points $\mathcal{I}_K(a_j)$.

(iii) Show that the equation in (i) is equivalent to

$$\frac{z-a_1}{|z-a_1|^2} + \frac{z-a_2}{|z-a_2|^2} + \cdots + \frac{z-a_n}{|z-a_n|^2} = 0,$$

and by interpreting the LHS as a (positively) weighted sum of the vectors from z to the roots of $P(z)$, deduce *Lucas' Theorem*: *The critical points of a polynomial in* $\mathbb{C}$ *must all lie within the convex hull of its zeros.* This is a complex generalization of Rolle's Theorem in ordinary calculus. [*Hint*: Use the fact that (2.32) on page 116 is still valid even if the masses are not equal.]

11 Use $(e^z)' = e^z$ to show that the derivatives of all the trig functions are given by the familiar rules of real analysis.

12 Provided it is properly interpreted, show that $(z^\mu)' = \mu z^{\mu-1}$ is still true even if μ is complex.

13 (i) If a is an arbitrary constant, show that the series

$$f(z) = 1 + az + \frac{a(a-1)}{2!}z^2 + \frac{a(a-1)(a-2)}{3!}z^3 + \cdots$$

converges inside the unit circle.

(ii) Show that $(1+z)f' = af$.

(iii) Deduce that $[(1+z)^{-a}f]' = 0$.

(iv) Conclude that $f(z) = (1+z)^a$.

14 As we pointed out in Chapter 3, stereographic projection has a very practical use in drawing a conformal map of the world. Once we have this map we can go on to generate further conformal maps, simply by applying different analytic functions to it. One particularly useful one was discovered (using other means) by Gerhard Mercator in 1569. We can describe it (though he could not have) as the result of applying $\log(z)$ to the stereographic map.

(i) Look up both a stereographic map and a Mercator map in an atlas, and make sure you can relate the changes in shape you see to your understanding of the complex logarithm.

(ii) Imagine plotting a straight-line course on a Mercator map and then actually travelling it on the high seas. Show that as you sail, the reading of your compass never changes.

15 (i) By noting that the unit tangent (in the counterclockwise direction) to an origin-centred circle can be written as $\widehat{\xi} = i(z/|z|)$, show that formula (5.24) for the curvature of the image of such a circle can be written as

$$\widetilde{\kappa} = \frac{1 + \text{Re}\left[\frac{zf''}{f'}\right]}{|zf'|}.$$

(ii) What should this formula yield if $f(z) = \log z$? Check that it does.

(iii) What should this formula yield if $f(z) = z^m$? Check that it does. What is the significance of the negative value of $\widetilde{\kappa}$ when m is negative? [*Hint*: Which way does the velocity complex number of the image rotate as z travels counterclockwise round the original circle?]

16 As illustrated below, a region is called *convex* if all of it is visible from an arbitrary vantage point inside. Let an analytic mapping f act on an origin-centred circle C to produce a simple image curve f(C), the interior of which is convex.

(i) From the formula of Ex. 15, deduce that if f maps the interior of C to the *interior* of f(C), then the following inequality holds at all points z of C:

$$\text{Re}\left[\frac{zf''}{f'}\right] \geqslant -1.$$

(ii) What is the analogous inequality when f maps the interior of C to the *exterior* of f(C). [*Hint*: Ex. 4, p. 239.]

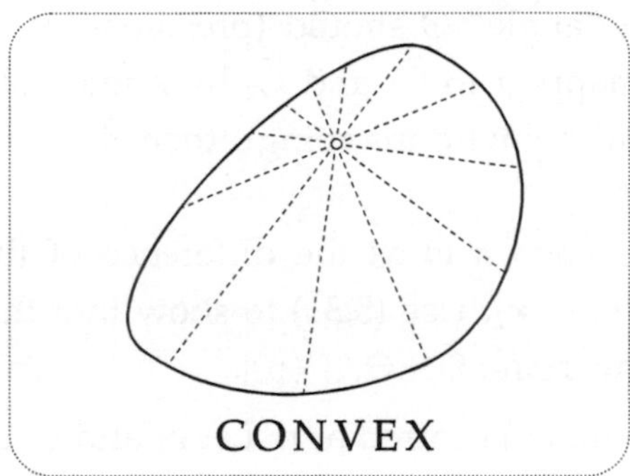

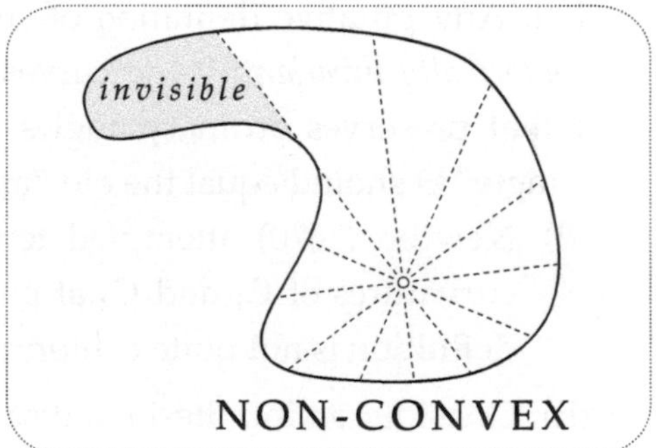

17 Let S be a directed line-segment through a point $p = x + iy$.

(i) Let $f(z) = e^z$. Without calculation, decide which direction of S yields an image f(S) having vanishing curvature at f(p).

(ii) The complex curvature $\mathcal{K}$ must therefore point in one of the two orthogonal directions. Which? By considering the image of S when it points in this direction, deduce the value of $|\mathcal{K}|$, and thereby conclude that $\mathcal{K}(p) = ie^{-x}$.

(iii) Use (5.29) to verify this formula.

(iv) Repeat as much as possible of the above analysis in the cases $f(z) = \log(z)$ and $f(z) = z^m$, where m is a positive integer. [In neither of these cases will you be able to see the exact value of $|\mathcal{K}(p)|$.]

(v) According to the geometric reasoning in Ex. 18, p. 241, the amplification of a Möbius transformation $M(z) = \frac{az+b}{cz+d}$ is constant on each circle centred at $-(d/c)$. Thus the complex curvature of M should be tangent to these concentric circles. Verify this by calculating $\mathcal{K}$.

(vi) Use a computer to verify figure [5.21] for all four mappings above.

18 Let two curves C_1 and C_2 emerge from a point p in the same direction. Two examples are illustrated below.

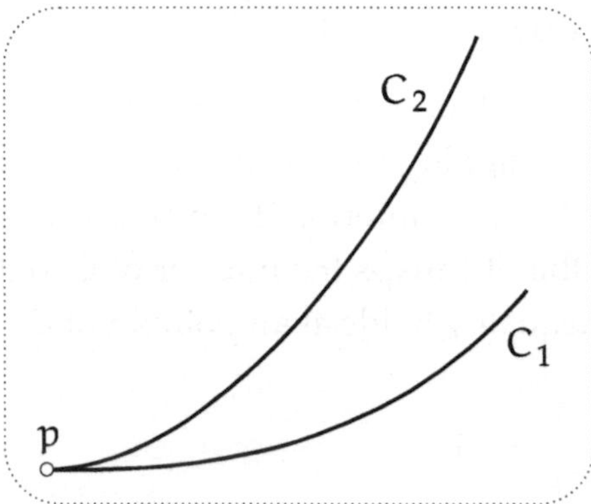

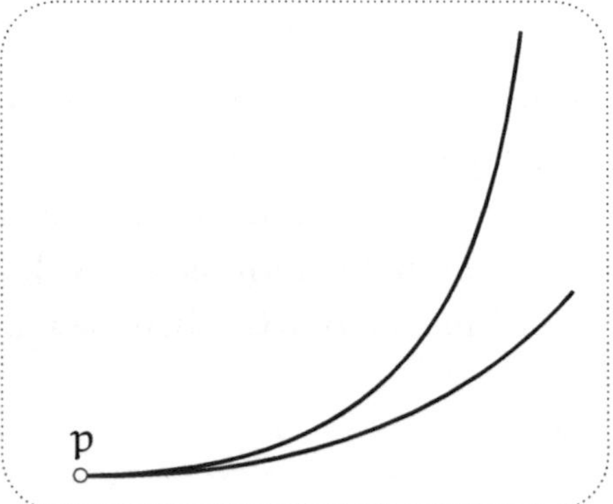

Although in both cases the angle at p is zero, there is a great temptation to say that the curves on the right meet at a smaller "angle" than those on the left. Any putative definition of such an "angle" Θ should (presumably) be *conformally invariant*: if the curves are mapped to $\widetilde{C}_1$ and $\widetilde{C}_2$ by a mapping f that preserves ordinary angles (i.e., an analytic mapping), then the new "angle" $\widetilde{\Theta}$ should equal the old "angle" Θ.

(i) Newton (1670) attempted to define such a Θ as the difference of the curvatures of C_1 and C_2 at p: $\Theta \equiv \kappa_1 - \kappa_2$. Use (5.31) to show that this definition is not quite conformally invariant: $\widetilde{\Theta} = \Theta/|f'(p)|$.

(ii) Consider an infinitesimal disc D (radius ϵ) centred at p. Let c_1 and c_2 be the centres of curvature of C_1 and C_2, and let $\mathcal{D}$ be the difference between the angular sizes of D as seen from c_1 and c_2. Show that $\mathcal{D} = \epsilon\Theta$. If a conformal mapping f is applied to C_1, C_2, *and* D, deduce that $\widetilde{\mathcal{D}} = \mathcal{D}$. [Of course this is not what we were after: $\mathcal{D}$ is (a) infinitesimal, and (b) not defined by the curves alone. The discovery of a true conformal invariant had to await Kasner (1912). See Ex. 10, p. 649.]

19 In more advanced work on Möbius transformations (e.g., Nehari (1952) and Beardon (1984)), an important role is played by the so-called *Schwarzian derivative* $\{f(z), z\}$ of an analytic function $f(z)$ with respect to z:

$$\{f(z), z\} \equiv \left(\frac{f''}{f'}\right)' - \frac{1}{2}\left(\frac{f''}{f'}\right)^2.$$

(i) Show that the Schwarzian derivative may also be written as

$$\{f(z), z\} = \frac{f'''}{f'} - \frac{3}{2}\left(\frac{f''}{f'}\right)^2.$$

(ii) Show that $\{az + b, z\} = 0 = \{(1/z), z\}$.

(iii) Let f and g be analytic functions, and write $w = f(z)$. Show that the Schwarzian derivative of the composite function $g[f(z)] = g[w]$ is given by the following "chain rule":

$$\{g(w), z\} = [f'(z)]^2 \{g(w), w\} + \{f(z), z\}.$$

(iv) Use the previous two parts to show that all Möbius transformations have vanishing Schwarzian derivative. [*Hint*: Recall that the mappings in part (ii) generate (via composition) the set of all Möbius transformations.] *Remark:* Ex. 19, p. 481 shows that the converse is also true: If $\{f(z), z\} = 0$ then $f =$ Möbius. Thus Möbius transformations are completely characterized by their vanishing Schwarzian derivative.

(v) Use the previous two parts to show that the Schwarzian derivative is "invariant under Möbius transformations", in the following sense: if M is a Möbius transformation, and f is analytic, then

$$\{M[f(z)], z\} = \{f(z), z\}.$$

20 Think of the real axis as representing time t, and let a moving particle $w = f(t)$ (where $f(z)$ is analytic) trace an orbit curve C. The velocity is then $v = \dot{w} = f'(t)$.

(i) Use (5.24) to show that the curvature of C is

$$\widetilde{\kappa} = \frac{\operatorname{Im}(\dot{v}/v)}{|v|}.$$

(ii) Argue that this result does not in fact depend on C being produced by an analytic mapping, but is instead true of any motion for which the velocity v and acceleration $\dot{v}$ are well-defined.

(iii) Show that the formula may be rewritten as

$$\widetilde{\kappa} = \frac{\operatorname{Im}(\overline{v}\,\dot{v})}{|v|^3}.$$

(iv) Deduce that it may also be written vectorially as

$$\widetilde{\kappa} = \frac{|\mathbf{v} \times \dot{\mathbf{v}}|}{|\mathbf{v}|^3}.$$

By considering $\mathbb{C}$ to be the "osculating plane"—see Needham (2021)—of a curve in 3-dimensional space, we see that this formula holds in that case also.

21 In 3-dimensional space, let (X, Y, Z) be the coordinates of a moving particle. If $X = a \cos \omega t$, $Y = a \sin \omega t$, $Z = bt$, then the path traced by the particle is a helix.

(i) Give interpretations for the numbers a, ω, and b.

(ii) If a and ω remain fixed, what does the helix look like in the two limiting cases of b becoming very small or very large? What if a and b remain fixed while ω becomes very small or very large?

(iii) What limiting values would you anticipate for the curvature of the helix for each of the limiting cases considered in (ii)?

(iv) Use Ex. 20(iv) to show that the curvature of the helix is

$$\widetilde{\kappa} = \frac{a\,\omega^2}{b^2 + a^2\,\omega^2},$$

and use this to confirm your hunches in (iii).

22 Continuing from Ex. 20, take $f(z)$ to be a general Möbius transformation:

$$w = f(t) = \frac{at + b}{ct + d},$$

where $\Delta \equiv (ad - bc) \neq 0$. Show that the curvature of this path is

$$\widetilde{\kappa} = \left| \left(\frac{2c^2}{\Delta} \right) \operatorname{Im} \left(\frac{d}{c} \right) \right|,$$

in agreement with Ex. 18, p. 210. The fact that this is constant provides a new proof that the image is a circle, for only circles have constant curvature.

23 As another continuation of Ex. 20, let us see how the Schwarzian derivative $\{f(z), z\}$ of Ex. 19 arises rather naturally in the context of curvature.

(i) Show that

$$\frac{d\widetilde{\kappa}}{dt} = \frac{1}{|f'|} \operatorname{Im}\{f(z), z\}.$$

[This formula was discovered by G. Pick. For elegant applications, see Beardon (1987). For another connection between curvature and Schwarzian derivatives, see Ex. 28(iii).]

(ii) Use Ex. 19, part (iv) to deduce that if $f(z)$ is a Möbius transformation then $\frac{d}{dt}\widetilde{\kappa} = 0$. Why is this result geometrically obvious?

24 Let the position at time t of a moving particle in $\mathbb{C}$ be $z(t) = r(t)\, e^{i\theta(t)}$.

(i) Show that the acceleration of the particle is

$$\ddot{z} = \left[\ddot{r} - r\,\dot{\theta}^2\right] e^{i\theta} + [2\dot{r}\dot{\theta} + r\,\ddot{\theta}]\, ie^{i\theta}.$$

(ii) What are the radial and transverse components of the acceleration?

(iii) If the particle is moving in a central force field, with the centre of force at the origin, deduce that the areal speed $\mathcal{A} = (r^2\dot{\theta}/2)$ is constant. For a beautiful geometric proof of this fact, see Newton (1687, Prop. 1), a comic-strip rendition of which may be found in Needham (2021, §11.7.3).

25 Sometimes the circle of convergence of a power series is so densely packed with singularities that it becomes a genuine barrier for the geometric mapping, beyond which it cannot be continued. This is called a *natural boundary*. An example of this is furnished by

$$f(z) = z + z^2 + z^4 + z^8 + z^{16} + \cdots ,$$

which converges inside the unit circle. Show that every point of $|z| = 1$ is either a singularity itself, or else has singularities arbitrarily near to it. [*Hint*: What is $f(1)$? Now note that $f(z) = z + f(z^2)$, and deduce that f is singular when $z^2 = 1$. Continuing in this manner, show that the 2^n-th roots of unity are all singular.]

26 Unlike inversion in a circle, show that Schwarzian reflection in an ellipse E (see figure [5.31]) does not interchange the interior and the exterior. Indeed, how does $\mathfrak{R}_E(z)$ behave for large values of $|z|$?

27 (i) If L is a line passing through the real point X, and making an angle σ with the horizontal, then show that its Schwarz function is

$$\mathcal{S}_L(z) = z\,e^{-i2\sigma} + X(1 - e^{-i2\sigma})\,.$$

(ii) If C is a circle with centre p and radius r, show that its Schwarz function is

$$\mathcal{S}_C(z) = \overline{p} + \frac{r^2}{z - p}\,.$$

(iii) Verify the claim that in both these cases $z \mapsto \mathfrak{R}(z)$ is the ordinary reflection, even if z is far from the curve.

28 Let a be a point on a (directed) curve K having Schwarz function $\mathcal{S}(z)$.

(i) Show that the curvature of K at a is

$$\kappa \equiv \dot{\phi} = \frac{i}{2} \cdot \frac{\mathcal{S}''(a)}{[\mathcal{S}'(a)]^{3/2}},$$

where ϕ is the angle in [5.30], and the dot denotes differentiation with respect to distance l along K (in the given sense). Deduce that

$$|\kappa| = |\mathcal{S}''/2|.$$

[*Hints*: Since $\mathcal{S}$ is analytic, so is $\mathcal{S}'$. Thus to calculate $\mathcal{S}'' = d\mathcal{S}'/dz$ we need only find the change $d\mathcal{S}'$ in $\mathcal{S}'$ produced by an infinitesimal movement dz of z, *taken in any one direction of our choosing*. At a let us choose dz along K, so that $dz = e^{i\phi}\,dl$. The corresponding change in $\mathcal{S}'$ is then determined solely by the shape of K, for the values of $\mathcal{S}'$ on K are given by $\mathcal{S}' = e^{-2i\phi}$.]

(ii) Deduce that the centre of curvature of K at a is $\{a + 2[\mathcal{S}'(a)/\mathcal{S}''(a)]\}$.

(iii) Show that the rate of change of the curvature of K is given by the "Schwarzian derivative" [Ex. 19] of the Schwarz function:

$$\dot{\kappa} = \frac{i}{2\mathcal{S}'}\{\mathcal{S}(z), z\}.$$

29 Check the result of Ex. 28 (i) by applying it to the results of Ex. 27.

30 Let a be a point on a curve K having Schwarz function $\mathcal{S}(z)$. By the still unproven result on the infinite differentiability of analytic functions, $\mathcal{S}(z)$ may be expanded into a Taylor series in the vicinity of a:

$$\mathcal{S}(z) = \mathcal{S}(a) + \mathcal{S}'(a)(z-a) + \frac{1}{2!}\mathcal{S}''(a)(z-a)^2 + \frac{1}{3!}\mathcal{S}'''(a)(z-a)^3 + \cdots.$$

(i) Show that the Schwarz function of the tangent line to K at a is given by the first two terms of the series above. This reconfirms something we saw in [5.30]: very close to a, reflection in the tangent is a good approximation to Schwarzian reflection.

(ii) It is natural to suspect that a better approximation to $\mathfrak{R}_K(z)$ would be inversion in the circle of curvature (call it C) of K at a. Let's verify this. Use Ex. 28(ii) and Ex. 27(ii) to find $\mathcal{S}_C$, and show that it may be written as

$$\mathcal{S}_C(z) = \overline{a} + \frac{2\overline{\mathcal{S}'}}{\overline{\mathcal{S}''}} + \frac{2(\mathcal{S}')^2}{\mathcal{S}''}\left[1 - \frac{\mathcal{S}''}{2\mathcal{S}'}(z-a)\right]^{-1},$$

where it is understood that the derivatives are all evaluated at a. Show that the first three terms in the binomial expansion of $\mathcal{S}_C$ agree with those of $\mathcal{S}$, but that they generally differ thereafter. [*Hint*: You will need the fact that $\overline{(\mathcal{S}'/\mathcal{S}'')} = -(\mathcal{S}')^2/\mathcal{S}''$ on K. Prove this.]

(iii) If the curvature κ of K were constant then K would be identical to its circle of curvature. The fact that $\mathcal{S}$ and $\mathcal{S}_C$ disagree beyond the third term thus reflects the fact that κ does change. One is thus led to guess that the faster κ changes, the greater the discrepancy between $\mathfrak{R}_K$ and inversion in C. Continuing from the last part, use Ex. 28(iii) to verify this hunch in the following precise form:

$$\mathcal{S}_C(z) - \mathcal{S}(z) \approx (i/3)[\mathcal{S}']^2\,\dot{\kappa}\,(z-a)^3.$$

31 Let C and D be intersecting circles. Let us say that "D is symmetric in C" if reflection (inversion) in C maps D into itself. We know this occurs if and only if D is orthogonal to C, so

$$\text{D is symmetric in C} \iff \text{C is symmetric in D.}$$

Briefly, we may simply say that "C and D are symmetric". Let's see what happens if we generalize C and D to intersecting arcs possessing Schwarz functions, and generalize inversion to Schwarzian reflection.

(i) Explain why the statement "D is symmetric in C" is the same as "if the point d lies on D then $\mathfrak{R}_D[\mathfrak{R}_C(d)] = \mathfrak{R}_C(d)$". Must the arcs be orthogonal?

(ii) If D is symmetric in C, deduce that the mappings $(\mathfrak{R}_D \circ \mathfrak{R}_C)$ and $(\mathfrak{R}_C \circ \mathfrak{R}_D)$ are equal at points of D.

(iii) Using the fact that these two mappings are analytic (why?), deduce that C must also be symmetric in D. Thus, as with circles, we may simply say that C and D are symmetric.

CHAPTER 6

Non-Euclidean Geometry*

6.1 Introduction

6.1.1 *The Parallel Axiom*

We have previously alluded to the remarkable discovery (made in the 19th century) that there exist geometries other than Euclid's. In this optional chapter we begin to explore the beautiful connections that exist between these so-called non-Euclidean geometries and the complex numbers. Since this Introduction summarizes many of the key ideas and results, you may wish to read it even if you cannot afford the time to read the entire chapter.

One way to approach Euclidean geometry is to begin with definitions of such things as "points" and "lines", together with a few assumptions (*axioms*) concerning their properties. From there one goes on, using nothing but logic, to deduce further properties of these objects that are necessary consequences of the initial axioms. This is the path followed in Euclid's famous book, *The Elements*, which was published around 300 BCE.

Of course Euclidean geometry did not suddenly spring into existence as a fully formed logical system of axioms and theorems. It was instead developed gradually as an idealized description of physical measurements performed on physically constructed lines, triangles, circles, etc. Though the ancients did not think of it in this way, Euclidean geometry is thus not simply mathematics, it is a physical theory of space—a *fantastically accurate* theory of space.

Euclidean geometry is not, however, a perfect theory: modern experiments have revealed extremely small discrepancies between the predictions of Euclidean geometry and the measured geometric properties of figures constructed in physical space. These departures from Euclidean geometry are now known to be governed, in a precise mathematical way, by the distribution of matter and energy in space. This is the essence of a revolutionary theory of gravity (*General Relativity*) discovered by Einstein in 1915.

Visual Complex Analysis. 25th Anniversary Edition. Tristan Needham, Oxford University Press.
 DOI: 10.1093/oso/9780192868916.003.0006

It turns out that the larger the figures examined, the larger the deviations from the predictions of Euclidean geometry. However, it's important to realize just how small these deviations typically are for figures of reasonable size. For example, suppose we measure the circumference of a circle having a radius of one meter. Even if our measuring device were capable of detecting a discrepancy the size of a single atom of matter, no deviation from Euclidean geometry would be found! Little wonder, then, that for two thousand years mathematicians were seduced into believing that Euclidean geometry was the *only logically possible* geometry.

It is a marvellous tribute to the power of human mathematical thought that non-Euclidean geometry was discovered a full century *before* Einstein found that it was needed to describe gravity. To locate the seeds of this mathematical discovery, let us return to ancient Greece.

Euclid began with just five axioms, the first four of which never aroused controversy. The first axiom, for example, merely states that there exists a unique line passing through any two given points. However, the status of the fifth axiom (the so-called *parallel axiom*) was less clear, and it became the subject of investigations that ultimately led to the discovery of non-Euclidean geometry:

> ***Parallel Axiom.*** *Through any point* p *not on the line* L *there exists precisely one line* L′ *that does not meet* L. (6.1)

Figure [6.1a] illustrates the parallel axiom, and it also explains why this axiom cannot be experimentally tested, at least as stated. As the line M rotates towards L′, the intersection point q moves further and further away along L. Our geometric intuition is based on figures drawn in a finite portion of the plane, but to verify that L′ never meets L, we need an infinite plane. We can certainly try to *imagine* what an infinite plane would be like, but we have no first hand experience to back up our hunches.

These are very modern doubts we are expressing. Historically, mathematicians fervently believed in (6.1), so much so that they thought it must be a logically necessary property of straight lines. But in that case they ought to be able to prove it outright, instead of merely assuming it as Euclid had done.

Many attempts were made to deduce (6.1) from the first four axioms, one of the most penetrating being that of Girolamo Saccheri in 1733.[1] His idea was to show that if (6.1) were not true, then a contradiction would necessarily arise. He divided the denial of (6.1) into two alternatives:

> ***Spherical Axiom.*** *There is no line through* p *that does not meet* L. (6.2)

or

> ***Hyperbolic Axiom.*** *There are at least two lines through* p *that do not meet* L. (6.3)

[1] See Stillwell (2010).

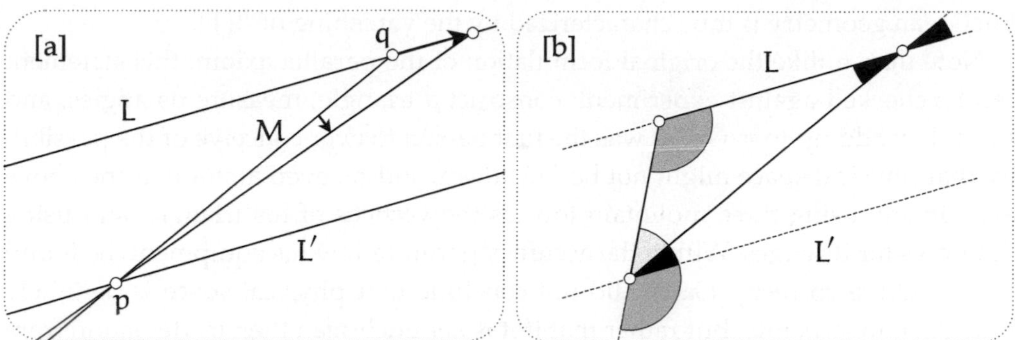

[6.1] Two forms of the Parallel Axiom. **[a]** *The Parallel Axiom seems beyond empirical verification: As the line* M *rotates towards* L′*, the intersection point* q *moves further and further away along* L*, so to be certain that the supposedly unique parallel line* L′ never *intersects* L*, we need to go all the way to* infinity! **[b]** *However, an alternative formulation of the Parallel Axiom* is *subject to empirical verification or refutation: The angles in every triangle must sum to* π*, so that* $\mathcal{E} = 0$.

Our naming of (6.2) will become clear shortly, but the use of "hyperbolic" in connection with (6.3) is more obscure, though standard.

In the case of (6.2), Saccheri was indeed able to obtain a contradiction, provided "lines" are assumed to have infinite length. If we drop this requirement, then we obtain a *non-Euclidean geometry* called *spherical geometry*. This is the subject of the following section.

In the case of (6.3), Saccheri and later mathematicians were able to derive very strange conclusions, but they were not able to find a contradiction. As we now know, this is because (6.3) yields another viable non-Euclidean geometry, called *hyperbolic geometry*. Of the two non-Euclidean geometries obtained from (6.2) and (6.3), hyperbolic geometry is by far the more intriguing and important: it is an essential tool in many areas of contemporary research. Furthermore, there is even a sense (to be discussed later) in which hyperbolic geometry subsumes both Euclidean and spherical geometry.

6.1.2 Some Facts from Non-Euclidean Geometry

Let's take our first look at how these new geometries differ from Euclid's. A very familiar theorem of Euclidean geometry states that in any triangle T,

$$(\text{Angle sum of T}) = \pi.$$

As indicated in [6.1b], this result is actually equivalent to the parallel axiom. It follows that in non-Euclidean geometry the angle sum of a triangle differs from π. To measure this difference, we introduce the so-called *angular excess* $\mathcal{E}$:

$$\mathcal{E}(\text{T}) \equiv (\text{Angle sum of T}) - \pi.$$

Euclidean geometry is thus characterized by the vanishing of $\mathcal{E}(T)$.

Note that, unlike the original formulation of the parallel axiom, this statement can be checked against experiment: construct a triangle, measure its angles, and see if they add up to π. Gauss was the first person to ever conceive of the possibility that physical space might not be Euclidean, and he even attempted the above experiment, using three mountain tops as the vertices of his triangle, and using light rays for its edges. Within the accuracy permitted by his equipment, he found $\mathcal{E} = 0$. Quite correctly, Gauss did not conclude that physical space is definitely Euclidean in structure, but rather that if it is *not* Euclidean then its deviation from Euclidean geometry is extremely small.

Let us return from physics to mathematics. Using pure logic to work out the consequences of (6.2) and (6.3), both Gauss and Johann Heinrich Lambert independently discovered that the two non-Euclidean geometries departed from Euclid's in opposite ways:

- In spherical geometry the angle sum is greater than π: $\mathcal{E} > 0$.
- In hyperbolic geometry the angle sum is less than π: $\mathcal{E} < 0$.

Furthermore, they discovered the striking fact that $\mathcal{E}(T)$ is completely determined by the *size* of the triangle. More precisely, $\mathcal{E}(T)$ is simply proportional to the *area* $\mathcal{A}(T)$ of the triangle T:

$$\mathcal{E}(T) = \mathcal{K}\mathcal{A}(T), \tag{6.4}$$

where $\mathcal{K}$ is a constant that is positive in spherical geometry, and negative in hyperbolic geometry.

Several interesting points can be made in connection with this result:

- Although there are no qualitative differences between them, there are nevertheless *infinitely many* different spherical geometries, depending on the value of the positive constant $\mathcal{K}$. Likewise, each negative value of $\mathcal{K}$ yields a different hyperbolic geometry.
- Since the angle sum of a triangle cannot be negative, $\mathcal{E} \geqslant -\pi$. Thus in hyperbolic geometry ($\mathcal{K} < 0$) no triangle can have an area greater than $|(\pi/\mathcal{K})|$.
- In non-Euclidean geometry, similar triangles do not exist! This is because (6.4) tells us that two triangles of different size cannot have the same angles.
- Closely related to the previous point, in non-Euclidean geometry there exists an *absolute unit of length*. For example, in spherical geometry we could define it to be the side of the equilateral triangle having angle sum 1.01π. Similarly, in hyperbolic geometry we could define it to be the side of the equilateral triangle having angle sum 0.99π.

- A somewhat more natural way of defining the absolute unit of length is in terms of the constant $\mathcal{K}$. Since the radian measure of angle is defined as a ratio of lengths, $\mathcal{E}$ is a pure number. On the other hand, the area $\mathcal{A}$ has units of $(\text{length})^2$. It follows that $\mathcal{K}$ has units of $1/(\text{length})^2$ and so it can be written as follows in terms of a length R: $\mathcal{K} = +(1/R^2)$ in spherical geometry; $\mathcal{K} = -(1/R^2)$ in hyperbolic geometry. Later we will see that this length R can be given a very intuitive interpretation P.
- The smaller the triangle, the harder it is to distinguish it from a Euclidean triangle: only when the linear dimensions are a significant fraction of R will the difference become obvious. This is why Gauss chose the biggest triangle he could in his experiment. Einstein's theory explains why Gauss's triangle was nevertheless much too small: the weak gravitational field in the space surrounding the earth corresponds to a microscopic value of $\mathcal{K}$ and hence to an enormous value of R. It would have been a different story if Gauss had been able to perform his experiment in the vicinity of a small black hole!

6.1.3 Geometry on a Curved Surface

We began this book by discussing how the complex numbers met with enormous initial resistance, and how they were finally accepted only after they were given a *concrete interpretation*, via the complex plane. The story of non-Euclidean geometry is remarkably similar.

Gauss never published his revolutionary ideas on non-Euclidean geometry, and the two men who are usually credited for their independent discovery of hyperbolic geometry are János Bolyai (1832) and Nikolai Lobachevsky (1829). Indeed, hyperbolic geometry is frequently also called Lobachevskian geometry, perhaps because Lobachevsky's investigations went somewhat deeper than Bolyai's. However, in the decades that followed their discoveries, Bolyai's work was completely ignored, and Lobachevsky's met only with vicious attacks.

The decisive figure in the acceptance of non-Euclidean geometry was Eugenio Beltrami. In 1868 he discovered that hyperbolic geometry could be given a *concrete interpretation*, via "differential geometry". For our purposes, differential geometry is the study of curved surfaces by means of ideas from calculus. What Beltrami discovered was that there exists a surface (the so-called *pseudosphere* shown in [6.2]) such that figures drawn on it automatically obey the rules of hyperbolic geometry[2]. Psychologically, Beltrami's pseudosphere was to hyperbolic geometry as the complex plane had been to the theory of complex numbers.

[2] This oversimplification does not do justice to Beltrami's accomplishments. Later in this chapter we shall see what Beltrami *really* did!

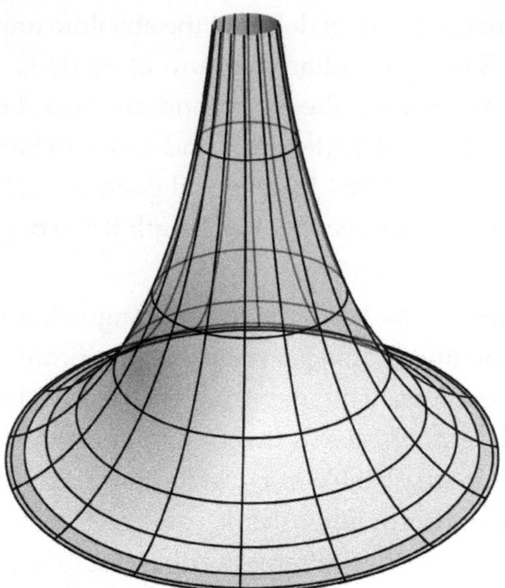

[6.2] The Pseudosphere. *In 1868 Eugenio Beltrami finally removed the mystery and scepticism that had overshadowed the strange, abstract laws of hyperbolic geometry ever since their discovery, 40 years earlier. He did so by recognizing that the hyperbolic laws simply* are *the intrinsic laws of geometry that hold true within a surface called the* pseudosphere *(shown here) that has constant negative Gaussian curvature. The pseudosphere extends upward indefinitely, but the base circle is a boundary beyond which it cannot be extended.*

To explain what we mean by this, let us first discuss how we may "do geometry" on a more general surface, such as the surface of the strange looking vegetable[3] shown in [6.3]. The idea of doing geometry on such a surface is essentially due to Gauss and (in greater generality) to Riemann.

The first thing we must do is to replace the concept of a straight line with that of a *geodesic*. Just as a line-segment in a flat plane may be defined as the shortest route between two points, so a geodesic segment connecting two points on a curved surface may be defined (provisionally) as the shortest connecting route *within the surface*. For example, if you were an ant living on the surface in [6.3], and you wanted to travel from a to b as quickly as possible, then you would follow the illustrated geodesic segment. The figure also shows the geodesic segment connecting another pair of points, c and d.

Here is a simple way you can actually construct such geodesic segments: take a thread and stretch it tightly over the surface to connect the points a and b. Provided

[3] European readers may think this an imaginary vegetable, but Americans can buy it in the supermarket.

that the thread can slide around on the surface easily, the tension in the thread ensures that the resulting path is as short as possible. Note that in the case of cd, we must imagine that the thread runs over the inside of the surface. In order to deal with all possible pairs of points in a uniform way, it is therefore best to imagine the surface as made up of two thinly separated layers, with the thread trapped between them.

In fact there exists a wonderfully simple and *practical* (yet surprisingly little known) method of constructing geodesics on *any* part of a surface. Cut a long, straight, *narrow* strip of masking tape (aka painter's tape) and start to lay one end down on the surface in the direction you want your geodesic to go. Holding the free end up away from the surface with one hand, use your other hand to gently run your finger along the tape, allowing it to choose its own path as it sticks itself to the surface as it goes. Behold! You have created a *geodesic*, and this construction works equally well on ab and cd! In the case of ab, you can easily check the validity of the construction by stretching a string over the surface: it will coincide with path of your sticky tape! For the explanation of *why* this works (and many applications) see Needham (2021).

It is now obvious how we should define distance in this geometry: the distance between a and b is the length of the geodesic segment connecting them. Figure [6.3] shows how we can then define, for example, a circle of radius r and centre p as the locus of points at distance r from p. To construct this circle we may take a piece of thread of length r, hold one end fixed at p, then (keeping the thread taut) drag the other end round on the surface.

Given three points on the surface, we may join them with geodesics to form a triangle; [6.3] shows two such triangles, Δ_1 and Δ_2. Now look at the angles in Δ_1. Clearly $\mathcal{E}(\Delta_1) > 0$, like a triangle in spherical geometry, while $\mathcal{E}(\Delta_2) < 0$, like a triangle in hyperbolic geometry.

6.1.4 *Intrinsic versus Extrinsic Geometry*

Clearly it is the curvature of the surface that causes $\mathcal{E}(\Delta_1)$ and $\mathcal{E}(\Delta_2)$ to differ from their Euclidean value $\mathcal{E} = 0$. However, it cannot be the precise shape of the surface in space that is involved here. To see this, imagine that from the vegetable in [6.3] we were to cut out a patch of the skin containing Δ_1. Suppose that this patch is made of fairly stiff material that does not stretch if we try to bend it a little. [As it happens, the skin of this vegetable is actually like this!] We can now gently bend the patch into infinitely many slightly different shapes: its so-called *extrinsic geometry* has been changed by our stretch-free bending. For example, the curves in space making up the edges of Δ_1 are no longer the same shape as before.

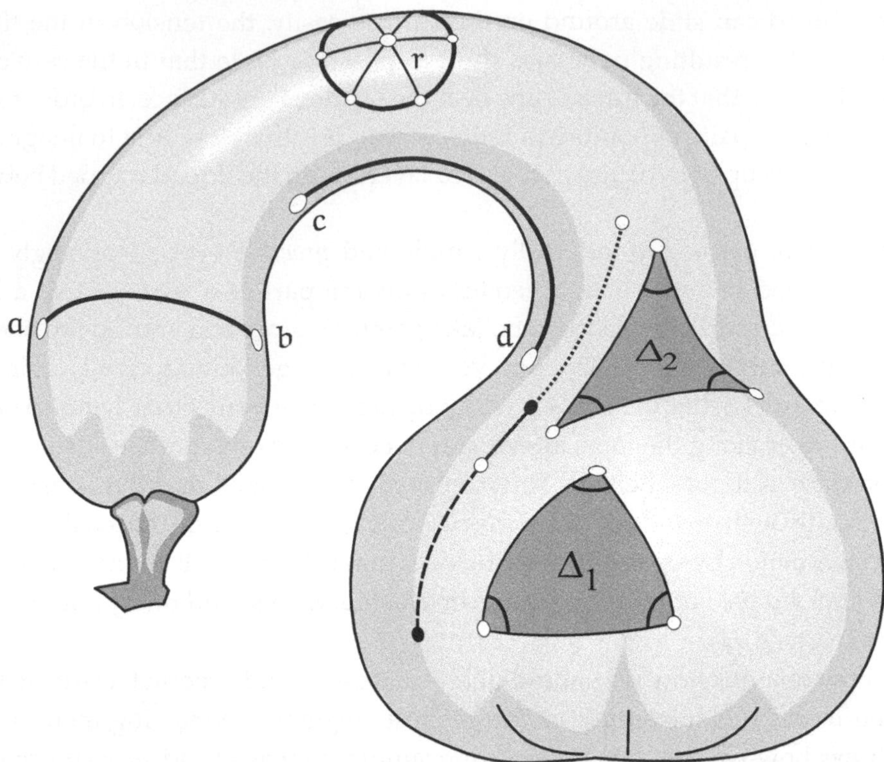

[6.3] The intrinsic geometry of a surface *is the geometry experienced by small, 2-dimensional creatures living within the surface. For them, a "straight line" is the shortest route between two points, such as* ab *or* cd. *A "circle" of radius* r *is the set of points that lie at a fixed distance* r *from its centre. "Triangles" are no longer subject to the Parallel Axiom, so now their angles need* not *add up to* π: $\mathcal{E}(\Delta_1) > 0$, *and* $\mathcal{E}(\Delta_2) < 0$.

On the other hand, if you were an intelligent ant living on this patch, no geometric experiment you could perform within the surface would reveal that any change had taken place whatsoever. We say that the *intrinsic geometry* has not changed. For example, the curves into which the edges of Δ_1 have been deformed are *still* the shortest routes on the surface. Correspondingly, the value of $\mathcal{E}$ is unaffected by stretch-free bending: $\mathcal{E}$ is governed by intrinsic (not extrinsic) curvature.

To highlight this fact, consider [6.4]. On the left is a flat piece of paper on which we have drawn a triangle T with angles $(\pi/2)$, $(\pi/6)$, $(\pi/3)$. Of course $\mathcal{E}(T) = 0$. Clearly we can bend such a flat piece of paper into either of the two (extrinsically) curved surfaces on the right[4]. However, *intrinsically* these surfaces have undergone no change at all—they are both as flat as a pancake! The illustrated triangles on these surfaces (into which T is carried by our stretch-free bending of the paper) are

[4] Of course the conical example on the far right cannot be obtained by bending a rectangle.

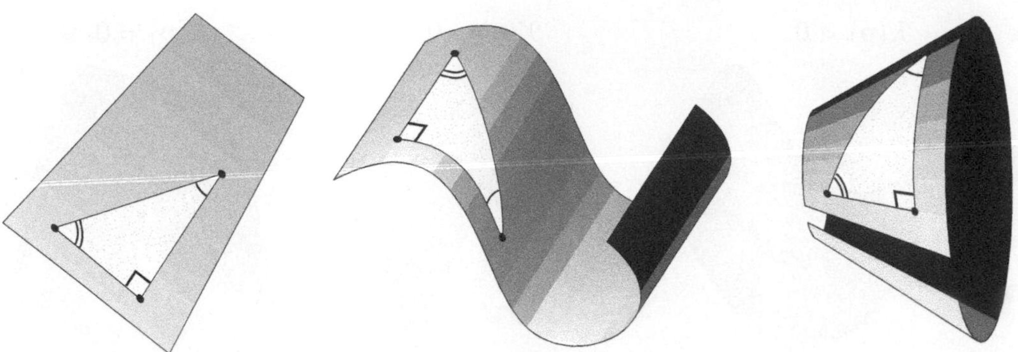

[6.4] A bent piece of paper remains intrinsically flat. *A flat piece of paper can be bent into all kinds of extrinsically curved shapes in space, but its internal,* intrinsic *geometry remains totally flat, and triangles constructed within the surface continue to obey all the normal laws of Euclidean geometry, such as Pythagoras's Theorem, and* $\mathcal{E} = 0$.

identical to the ones that intelligent ants would construct using geodesics, and in both cases $\mathcal{E} = 0$: geometry on these surfaces is Euclidean.

6.1.5 Gaussian Curvature

In 1827 Gauss published a beautiful and revolutionary analysis[5] of the intrinsic and extrinsic geometry of surfaces, in which he revealed that remarkable connections exist between the two. Here we will simply state some of his most important conclusions, in their most general form. For explanations of these general results we refer you to works on differential geometry; see the recommendations at the end of this chapter. However, only special cases of the general results are needed to understand non-Euclidean geometry, and these will be separately verified in the course of this chapter.

For a surface such as [6.3], it is clear that some parts are more curved than others. Furthermore, the *kind* of bending also varies from place to place. To quantify the amount (and type) of bending of the surface at a point p, Gauss introduced a quantity $\mathcal{K}(\mathrm{p})$. This function $\mathcal{K}(\mathrm{p})$, whose precise definition will be given in a moment, is called the *Gaussian curvature*[6]. The greater the magnitude of $\mathcal{K}(\mathrm{p})$, the more curved the surface is at p. The sign of $\mathcal{K}(\mathrm{p})$ tells us qualitatively what the surface is like in the immediate neighbourhood of p. See [6.5]. If $\mathcal{K}(\mathrm{p}) < 0$ then the neighbourhood of p resembles a saddle: it bends upwards in some directions, and downwards in others. If $\mathcal{K}(\mathrm{p}) > 0$ then it bends the same way in all directions, like a piece of a sphere.

[5] Gauss (1827).

[6] Other names are *intrinsic curvature, total curvature,* or just plain *curvature.*

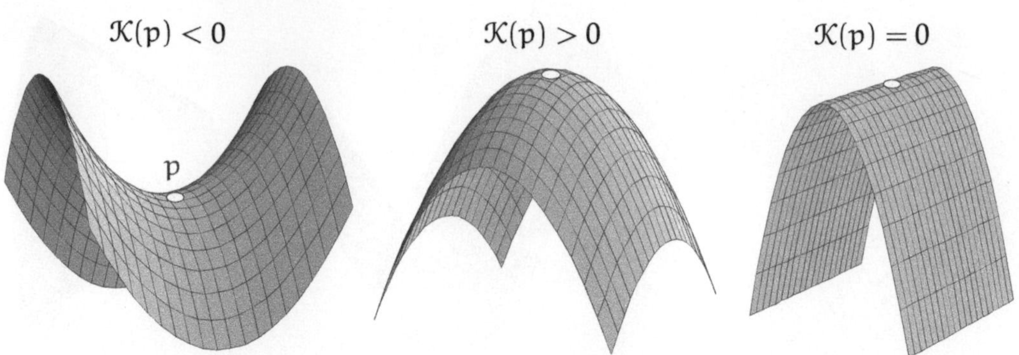

[6.5] The sign of the Gaussian curvature $\mathcal{K}(p)$ tells us the shape of the surface at p.

As we will now start to explain, it is no accident that we have used the same symbol to represent Gaussian curvature as we earlier used for the constant occurring in (6.4)—they are the same thing!

Gauss originally defined $\mathcal{K}(p)$ as follows. Let Π be a plane containing the normal vector $\mathbf{n}$ to the surface at p, and let κ be the (signed) curvature at p of the curve in which Π intersects the surface. The sign of κ depends on whether the centre of curvature is in the direction $\mathbf{n}$ or $-\mathbf{n}$. The so-called *principal curvatures* are the minimum κ_{min} and the maximum κ_{max} values of κ as Π rotates about $\mathbf{n}$. [Incidentally, Euler had previously made the important discovery that these principal curvatures occur in two *perpendicular* directions.] Gauss defined $\mathcal{K}$ as the product of the principal curvatures:

$$\mathcal{K} \equiv \kappa_{min}\,\kappa_{max}.$$

Note that this definition is in terms of the precise shape of the surface in space (extrinsic geometry). However, Gauss (1827) went on to make the astonishing discovery that $\mathcal{K}(p)$ actually measures the *intrinsic* curvature of the surface, that is, $\mathcal{K}$ *is invariant under bending!* Gauss was justifiably proud of this result, calling it the *Theorema Egregium* (Latin for "remarkable theorem"). As an example of the result, you may visually convince yourself that $\mathcal{K} = 0$ everywhere on each of the intrinsically flat surfaces in [6.4].

The intrinsic significance of $\mathcal{K}$ is exhibited in the following fundamental result: *If Δ is an infinitesimal triangle of area $d\mathcal{A}$ located at the point p, then*

$$\mathcal{E}(\Delta) = \mathcal{K}(p)\, d\mathcal{A}. \tag{6.5}$$

Since $\mathcal{E}$ and $d\mathcal{A}$ are defined by the intrinsic geometry, so is $\mathcal{K} = (\mathcal{E}/d\mathcal{A})$. Once again, we refer you to works on differential geometry for a proof of (6.5).

It follows from (6.5) [see Ex. 1] that the angular excess of a non-infinitesimal triangle T is obtained by adding up (i.e., integrating) the Gaussian curvature over the interior of T:

$$\mathcal{E}(T) = \iint_T \mathcal{K}(p)\, d\mathcal{A}. \tag{6.6}$$

As Beltrami recognized, and as we now explain, this lovely result of differential geometry brings us very close to a concrete interpretation of the non-Euclidean geometries.

6.1.6 Surfaces of Constant Curvature

Consider a surface such that $\mathcal{K}(p)$ has the same value $\mathcal{K}$ at every point p; we call this a *surface of constant curvature*. For example, a plane is a surface of constant curvature $\mathcal{K} = 0$, as are the other surfaces in [6.4]; a sphere is an example (not the only one) of a surface of constant positive curvature; and the pseudosphere in [6.2] is an example (not the only one) of a surface of constant negative curvature.

In the case of a surface of constant curvature (and only in this case) we find that (6.6) takes the form,

$$\mathcal{E}(T) = \mathcal{K} \iint_T d\mathcal{A} = \mathcal{K}\,\mathcal{A}(T).$$

But this is identical to the fundamental formula (6.4) of non-Euclidean geometry! Thus, as Beltrami realized,

> *Euclidean, spherical, and hyperbolic geometry can all be interpreted concretely as the intrinsic geometry of surfaces of constant vanishing, positive, or negative curvature.*

Figure [6.6] illustrates this using the simplest surfaces of each type. To obtain an added bonus, recall that we previously associated an absolute unit of length R with a non-Euclidean geometry by writing $\mathcal{K} = \pm(1/R^2)$. The bonus is that this length R now takes on vivid meaning: in spherical geometry R is simply the radius of the sphere, while in hyperbolic geometry it is the radius of the circular base of the pseudosphere (called the *radius* of the pseudosphere). These two interpretations will be justified later.

The requirement of constant curvature can be understood more intuitively by reconsidering the discussion at the end of Chapter 1. There we saw that a central idea in Euclidean geometry is that of *a group of motions* of the plane: one-to-one mappings that preserve the distance between all pairs of points. For example, two figures are congruent if and only if there exists a motion that carries the first into coincidence with the second. In order that this basic concept of equality be available in non-Euclidean geometry, we require that our surface admits an analogous group of motions. If we take one of the triangles on the surface in [6.3], it's clear that we cannot slide it to a new location and still have it fit the surface snugly, because the way in which the surface is curved at the new location is different: *variation in the curvature is the obstruction to motion.*

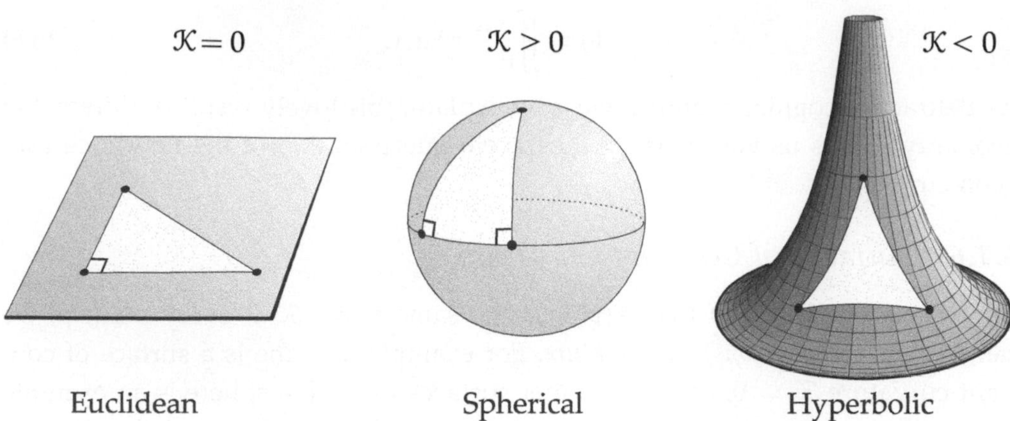

[6.6] Euclidean, Spherical, and Hyperbolic geometry are the intrinsic geometries of surfaces of constant Gaussian curvature.

This intuitive explanation can be clarified by appealing to (6.5). First, though, we wish to eliminate a possible confusion. The triangle on the flat plane in [6.6] can clearly be slid about and rotated freely, but what about the triangles on the (extrinsically) curved surfaces in [6.4]? After all, these surfaces are intrinsically flat, and so Beltrami would have us believe they are therefore just as good as the plane for doing Euclidean geometry. If we imagine these triangles as *completely* rigid then it's clear that if we try to move them to another location on the surface, they will no longer fit snugly against the surface. But if the triangle is instead cut out of a piece of ordinary (bendable but unstretchable) paper, then it *can* be slid about and rotated freely, always fitting perfectly against the surface. *This* is the kind of motion we are concerned with.

In order to clarify the connection between constant curvature and the existence of motions, consider an infinitesimal (bendable but unstretchable) triangle located at p. If its angular excess is $\mathcal{E}$ and its area is $d\mathcal{A}$, then (6.5) tells us that the Gaussian curvature of the surface at p is given by $\mathcal{K}(p) = (\mathcal{E}/d\mathcal{A})$. Now suppose that there exists a motion that carries this triangle to an arbitrary point q on the surface. We may have to bend the triangle to make it fit against the surface at q, but since we are not allowed to stretch it, the values of $\mathcal{E}$ and $d\mathcal{A}$ do not alter. Thus $\mathcal{K}(q) = (\mathcal{E}/d\mathcal{A}) = \mathcal{K}(p)$, and the surface has constant curvature.

Finally, let us return to the specific models of spherical and hyperbolic geometry shown in [6.6]. Clearly the triangle on the sphere can be slid about and rotated freely. In fact here, as on the plane, no bending is needed at all, because the sphere not only has constant intrinsic curvature, it also has constant extrinsic curvature.

What about hyperbolic geometry on the pseudosphere? It is certainly much less obvious, but the fact [to be proved later] that the pseudosphere has constant curvature guarantees that a bendable but unstretchable triangle *can* be slid about and rotated freely, always fitting perfectly snugly against the surface. Exercise 15 shows how you can build your own pseudosphere; once built, you can verify this surprising claim experimentally.

6.1.7 The Connection with Möbius Transformations

As we established in Chapter 1, if the Euclidean plane is identified with $\mathbb{C}$ then its motions (and similarities) are represented by the particularly simple Möbius transformations of the form $M(z) = az + b$. One of the principal miracles we wish to explain in this chapter is that the motions of spherical and hyperbolic geometry are *also* Möbius transformations!

The most general (direct) motion of the sphere is a rotation about its centre. Stereographic projection onto $\mathbb{C}$ yields a conformal map of the sphere, and the rotations of the sphere thus become complex functions acting on this map. As we showed algebraically in Ex. 21 on page 211 of Chapter 3, they are the Möbius transformations of the form

$$M(z) = \frac{az + b}{-\overline{b}z + \overline{a}}.$$

This was first discovered by Gauss, around 1819. In the next section we will rederive this result in a more illuminating way, and we will also explore the connection with Hamilton's "quaternions", which we first met on page 49.

Following the same pattern, it is also possible to construct conformal maps (in $\mathbb{C}$) of the pseudosphere, thereby transforming its motions into complex functions. One of the most convenient of these conformal maps is constructed in the unit disc. The motions of hyperbolic geometry then turn out to be the Möbius automorphisms of this circular map, which we first derived in (3.52) and then further investigated in Ex. 20 on page 210:

$$M(z) = \frac{az + b}{\overline{b}z + \overline{a}}.$$

This beautiful discovery was made by Henri Poincaré in 1882; see Poincaré (1985).

It seems magical enough that the motions of all three of the two-dimensional geometries are represented by special kinds of Möbius transformations, but there's more! In Chapter 3 [see (3.2) on page 138] we saw that the *general* Möbius transformation

$$M(z) = \frac{az + b}{cz + d}$$

has deep significance for physics: it corresponds to the most general Lorentz transformation of spacetime. Might it also have significance in non-Euclidean geometry? As we will explain at the end of this chapter, in 1883 Poincaré made the startling discovery that it represents the most general (direct) motion of *three*-dimensional hyperbolic space!

6.2 Spherical Geometry

6.2.1 *The Angular Excess of a Spherical Triangle*

The geodesics on the sphere are the *great circles*, that is the intersections of the sphere with planes through its centre. Thus if you were an ant living on the sphere, these great circles are what you would call "lines".

Figure [6.7a] illustrates a general triangle Δ on a sphere of radius R obtained by joining three points using such "lines". Without appealing to (6.6), which is a deep result in differential geometry, let us show directly that the angular excess $\mathcal{E}(\Delta)$ obeys the law (6.4), and that the constant $\mathcal{K}$ is indeed the Gaussian curvature, $\mathcal{K} = (1/R^2)$. The elegant argument that follows is usually attributed to Euler, who rediscovered it, but it was in fact first discovered by Thomas Harriot in 1603.[7] Yes, this is the *same* Thomas Harriot who discovered the fundamentally important conformality of stereographic projection, (3.17)!

Prolonging the sides of Δ divides the surface of the sphere into eight triangles, the four triangles labelled $\Delta, \Delta_\alpha, \Delta_\beta, \Delta_\gamma$ each being paired with a congruent antipodal triangle. This is clearer in [6.7b]. Since the area of the sphere is $4\pi R^2$, we deduce that

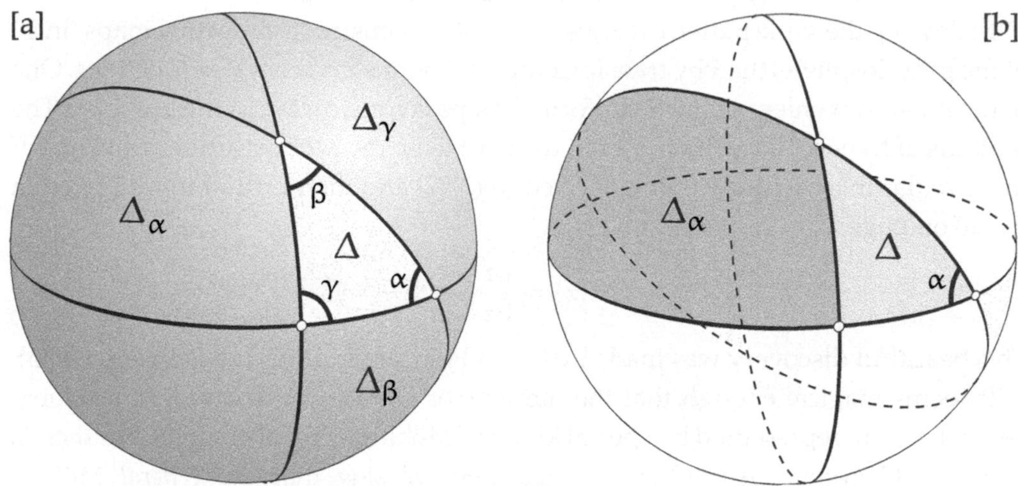

[6.7] Harriot's 1603 proof that $\mathcal{E}(\Delta) = (1/R^2)\,\mathcal{A}(\Delta)$. *Extending the sides of Δ that contain the angle α, we obtain the shaded wedge on the right, which occupies $(\alpha/2\pi)$ of the complete sphere, so $\mathcal{A}(\Delta) + \mathcal{A}(\Delta_\alpha) = 2\alpha R^2$; likewise for β and γ. Since each of the four triangles on the left has an antipodal partner of equal area—as can be seen more clearly on the right—it follows that together they occupy half the sphere: $\mathcal{A}(\Delta)+\mathcal{A}(\Delta_\alpha)+\mathcal{A}(\Delta_\beta) + \mathcal{A}(\Delta_\gamma) = 2\pi R^2$, and the result follows immediately.*

[7] See Stillwell (2010, §17.6).

$$\mathcal{A}(\Delta) + \mathcal{A}(\Delta_\alpha) + \mathcal{A}(\Delta_\beta) + \mathcal{A}(\Delta_\gamma) = 2\pi R^2. \tag{6.7}$$

On the other hand, it is clear in [6.7b] that Δ and Δ_α together form a wedge whose area is $(\alpha/2\pi)$ times the area of the sphere:

$$\mathcal{A}(\Delta) + \mathcal{A}(\Delta_\alpha) = 2\alpha R^2.$$

Similarly,

$$\begin{aligned} \mathcal{A}(\Delta) + \mathcal{A}(\Delta_\beta) &= 2\beta R^2, \\ \mathcal{A}(\Delta) + \mathcal{A}(\Delta_\gamma) &= 2\gamma R^2. \end{aligned}$$

Adding these last three equations, we find that

$$3\mathcal{A}(\Delta) + \mathcal{A}(\Delta_\alpha) + \mathcal{A}(\Delta_\beta) + \mathcal{A}(\Delta_\gamma) = 2(\alpha + \beta + \gamma)R^2. \tag{6.8}$$

Finally, subtracting (6.7) from (6.8), we get

$$\mathcal{A}(\Delta) = (\alpha + \beta + \gamma - \pi)R^2.$$

In other words,

$$\mathcal{E}(\Delta) = \mathcal{K}\,\mathcal{A}(\Delta), \quad \text{where} \quad \mathcal{K} = (1/R^2), \tag{6.9}$$

as was to be shown.

6.2.2 *Motions of the Sphere: Spatial Rotations and Reflections*

In order to understand the motions (i.e., one-to-one, distance-preserving mappings) of the sphere, we must first clarify the idea of "distance". If two points a and b are not antipodal then there exists a unique line (great circle) L passing through them, and a and b divide L into two arcs of unequal length. The "distance" between the points can now be defined as the length of the shorter arc. But if the points *are* antipodal then every line through a automatically passes through b, and the distance between the points is defined to be the length πR of any of the semicircular arcs connecting them.

We can now generalize the Euclidean arguments given in the final section of Chapter 1. There we saw that a motion of the plane is uniquely determined by the images a', b', c' of any three points a, b, c not on a line: the image of P is the unique point P' whose distances from a', b', c' equal the distances of P from a, b, c. We leave it to you to check that this result (and the reason for it) is still true on the sphere.

On the sphere, as on the plane, we may consistently attribute a sense to angles—by convention an angle is positive if it is counterclockwise when viewed from *outside* the sphere. As happened in the plane, this leads to a division of spherical motions into two types: *direct* (i.e., conformal) motions, and *opposite* (i.e., anticonformal) motions.

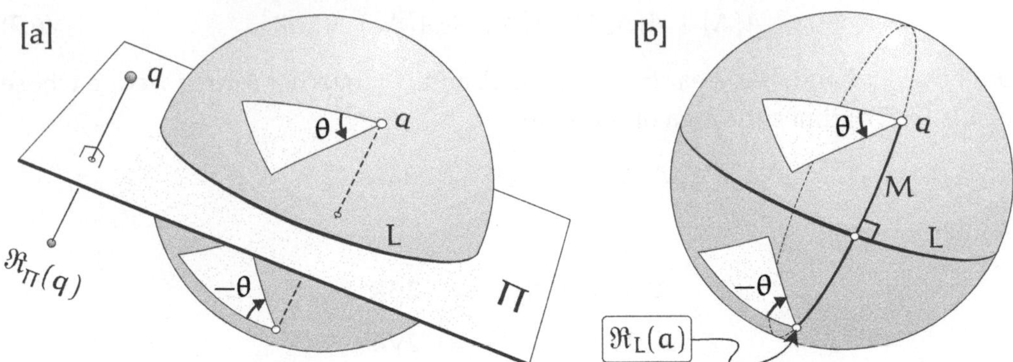

[6.8] Reflection of the sphere in a line. [a] *If Π is the plane through the line (great circle) L, then reflection $\mathfrak{R}_\Pi$ of space across Π induces the anticonformal reflection $\mathfrak{R}_L$ of the sphere across L.* **[b]** *This reflection $\mathfrak{R}_L(a)$ can instead be expressed in intrinsic terms that make sense to the inhabitants of the sphere: travel some distance d along the orthogonal "line" M from a until you arrive at L, then travel an equal distance d again.*

As in the plane, the simplest opposite motion of the sphere is *reflection* $\mathfrak{R}_L$ in a line L. This may be thought of as the transformation induced on the sphere by reflection $\mathfrak{R}_\Pi$ of *space* in the plane Π containing L. See [6.8a], which illustrates how the positive angle θ in the illustrated spherical triangle is reversed by $\mathfrak{R}_L$.

If you were an intelligent ant living on the sphere, the above construction of $\mathfrak{R}_L$ as the restriction of $\mathfrak{R}_\Pi$ to the sphere would be meaningless to you. However, it is not hard to re-express $\mathfrak{R}_L$ in intrinsically spherical terms. See [6.8b]. To reflect a in L, first draw the unique line M through a that cuts L at right angles[8]. If d is the distance we must crawl along M from a to reach L, then $\mathfrak{R}_L(a)$ is the point we reach after crawling a further distance of d. Of course M actually intersects L in two antipodal points, but we will arrive at the same $\mathfrak{R}_L(a)$ irrespective of which of these two points is used in the construction.

We now turn to direct motions. The obvious example of a direct motion is a rotation of the sphere about an axis V passing through its centre. Less obvious is the fact (to be proved shortly) that these rotations are the *only* direct motions. To avoid ambiguity in the description of such rotations, we introduce the following standard convention. First note that specifying the axis V is equivalent to specifying either of its antipodal intersection points (say p and q) with the sphere. Now pick one of these, say p. Suppose that the effect of the rotation on a small line-segment issuing from p is a positive rotation of θ—recall that this means *counterclockwise as seen from outside the sphere*. In this case the motion can be unambiguously described as a "positive rotation of θ about p"; see [6.9b]. We will write this rotation as $\mathcal{R}_p^\theta$. Check for yourself that $\mathcal{R}_p^\theta = \mathcal{R}_q^{-\theta}$.

[8] If L is thought of as the equator, then when a is one of the poles there are infinitely many M's—pick any one you like.

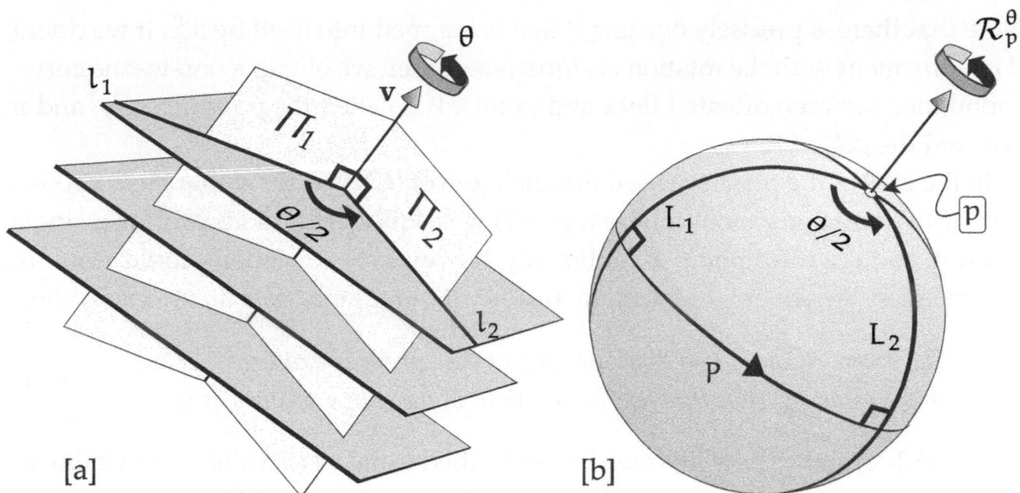

[6.9] Rotation of the sphere via two reflections. [a] *If the planes* Π_1 *and* Π_2 *meet at angle* $(\theta/2)$ *then* $(\mathfrak{R}_{\Pi_2} \circ \mathfrak{R}_{\Pi_1})$ *is a rotation of space through angle* θ *about their line of intersection.* **[b]** *The induced transformation of the sphere is therefore a rotation* $\mathcal{R}_p^\theta$ *by* θ *about the intersection point* p *of the lines on the sphere:* $\mathfrak{R}_{L_2} \circ \mathfrak{R}_{L_1} = \mathcal{R}_p^\theta$.

In Chapter 1 we saw that every direct motion of the plane was the composition of two reflections: a rotation if the lines intersected; a translation if the lines were parallel. We will now see that a similar phenomenon occurs on the sphere, but because *every* pair of lines intersect, the composition of two reflections is *always* a rotation—the sphere has no motions analogous to translations.

Figure [6.9a] illustrates the composition $(\mathfrak{R}_{\Pi_2} \circ \mathfrak{R}_{\Pi_1})$ of two reflections of space. Here the planes Π_1 and Π_2 intersect in a line with direction vector **v**, and the angle from Π_1 to Π_2 is $(\theta/2)$. Restricting attention to any one of the shaded planes orthogonal to **v**, we see that the transformation induced by $(\mathfrak{R}_{\Pi_2} \circ \mathfrak{R}_{\Pi_1})$ is $(\mathfrak{R}_{l_2} \circ \mathfrak{R}_{l_1})$, where l_1 and l_2 are the lines in which Π_1 and Π_2 intersect the plane. Since $(\mathfrak{R}_{l_2} \circ \mathfrak{R}_{l_1})$ is a rotation of the plane through θ about the intersection point of l_1 and l_2, it is now clear that $(\mathfrak{R}_{\Pi_2} \circ \mathfrak{R}_{\Pi_1})$ is a rotation of space through angle θ about the axis **v**.

Figure [6.9b] translates this idea into spherical terms. If Π_1 and Π_2 pass through the centre of the sphere, and the lines (great circles) in which they intersect the sphere are L_1 and L_2, then

$$\mathfrak{R}_{L_2} \circ \mathfrak{R}_{L_1} = \mathcal{R}_p^\theta.$$

In other words,

> *A rotation* $\mathcal{R}_p^\theta$ *of the sphere about a point* p *through angle* θ *may be expressed as the composition of reflections in any two spherical lines that pass through* p *and contain the angle* $(\theta/2)$. (6.10)

Note that there is precisely one line P that is mapped into itself by $\mathcal{R}_p^\theta$. If we orient P in agreement with the rotation (as illustrated) then we obtain a one-to-one correspondence between oriented lines and points: P is called the *polar line* of p, and p is called the *pole* of P.

In the case of the plane we used the analogue of (6.10) to show that the composition of two rotations about different points was equivalent (in general) to a single rotation about a third point; exceptionally, however, two rotations could result in a translation. As you might guess, in the case of the sphere there are no exceptions:

The composition of any two rotations of the sphere is equivalent to a single rotation. Thus the set of all rotations of the sphere forms a group. (6.11)

Figure [6.10a] shows how this may be established using exactly the same argument that was used in the plane. In order to find the net effect of $(\mathcal{R}_q^\phi \circ \mathcal{R}_p^\theta)$, draw the lines L, M, N in the illustrated manner. Then

$$\mathcal{R}_q^\phi \circ \mathcal{R}_p^\theta = (\mathfrak{R}_N \circ \mathfrak{R}_M) \circ (\mathfrak{R}_M \circ \mathfrak{R}_L) = \mathfrak{R}_N \circ \mathfrak{R}_L = \mathcal{R}_r^\psi.$$

This beautiful geometric method of composing spatial rotations was discovered by Olinde Rodrigues in 1840.

Note that in the plane the total amount of rotation produced by rotations of θ and ϕ is simply the sum $(\theta + \phi)$, but on the sphere we have a more complicated rule. If $\mathcal{A}$ is the area of the white spherical triangle, and $\mathcal{K} = (1/R^2)$ is the Gaussian curvature of the sphere, then the formula for the angular excess implies that

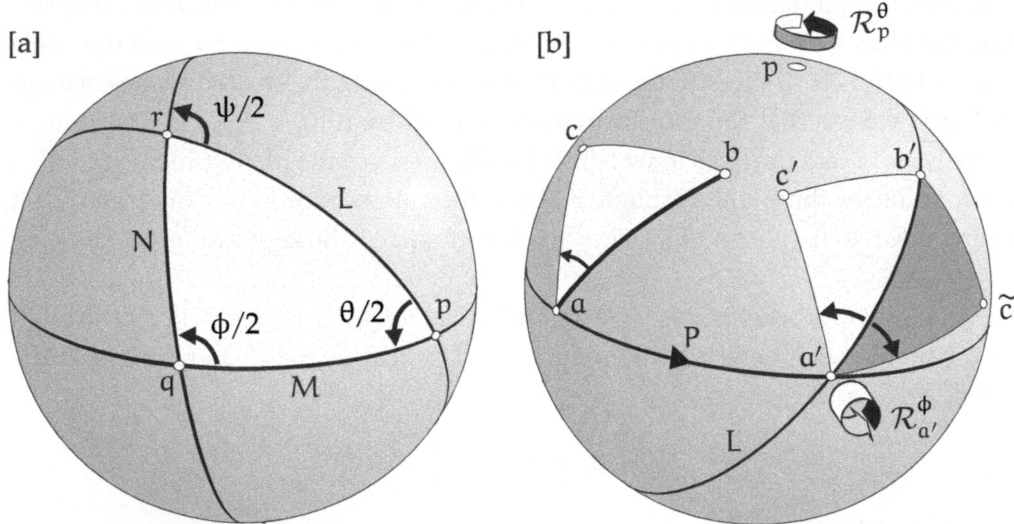

[6.10] [a] Composition of rotations: $\mathcal{R}_q^\phi \circ \mathcal{R}_p^\theta = (\mathfrak{R}_N \circ \mathfrak{R}_M) \circ (\mathfrak{R}_M \circ \mathfrak{R}_L) = \mathfrak{R}_N \circ \mathfrak{R}_L = \mathcal{R}_r^\psi$. **[b]** *There is exactly one direct motion $\mathcal{M}$ (and exactly one opposite motion $\widetilde{\mathcal{M}}$) that maps a given line-segment ab to another line-segment a′b′ of equal length. Furthermore, $\widetilde{\mathcal{M}} = (\mathfrak{R}_L \circ \mathcal{M})$, where L is the line through a′ and b′.*

$$\psi = \theta + \phi - 2\mathcal{K}\mathcal{A}.$$

We may now complete the classification of the motions of the sphere. As we have remarked, there is precisely one motion of the sphere that carries a given spherical triangle abc to a given congruent image triangle. Figure [6.10b] helps to refine this result. Using the same logic as was used in the plane, we see [exercise] that

There is exactly one direct motion $\mathcal{M}$ (and exactly one opposite motion $\widetilde{\mathcal{M}}$) that maps a given line-segment ab to another line-segment a′b′ of equal length. Furthermore, $\widetilde{\mathcal{M}} = (\mathfrak{R}_L \circ \mathcal{M})$, where L is the line through a′ and b′. (6.12)

Figure [6.10b] also shows how we may construct $\mathcal{M}$. Draw the line P through a and a′, and let p be its pole. With the appropriate value of θ, it's clear that $\mathcal{R}_p^\theta$ will carry the segment ab along P to a segment of equal length emanating from a′; finally, an appropriate rotation $\mathcal{R}_{a'}^\phi$ about a′ will carry this segment into a′b′. Thus $\mathcal{M} = (\mathcal{R}_{a'}^\phi \circ \mathcal{R}_p^\theta)$, which is equivalent to a single rotation by virtue of (6.11). Combining this fact with (6.12), we deduce that

Every direct motion of the sphere is a rotation, and every opposite motion is the composition of a rotation and a reflection. (6.13)

As a simple test of this result (and your grasp of it) consider the *antipodal mapping* that sends every point on the sphere to its antipodal point. Clearly this is a motion, but how does it accord with the above result?

6.2.3 A Conformal Map of the Sphere

The sphere merely provides one particularly simple model of what we have called spherical geometry. As Minding (1839) discovered, *any* surface of constant[9] Gaussian curvature $\mathcal{K} = (1/R^2)$ has exactly the same intrinsic geometry as a sphere of radius R. To see that such surfaces exist, take a Ping-Pong ball and cut it in half: as you gently flex one of the hemispheres you obtain infinitely many surfaces whose intrinsic geometry is identical to the original sphere.

Figure [6.11] illustrates that even if we restrict attention just to surfaces of revolution, the sphere is not the only one of constant positive curvature. Though they hardly look like spheres, an intelligent ant living on either of these surfaces would never know that he wasn't living on a sphere. Well, that's almost true: eventually he might discover points at which the surface is not smooth, or else he might run into an edge. In 1899 H. Liebmann proved that if a surface of constant positive curvature does not suffer from these defects then it can *only* be a sphere.

[9] If the curvature is *not* constant, two surfaces can have equal curvature at corresponding points and yet have different intrinsic geometry.

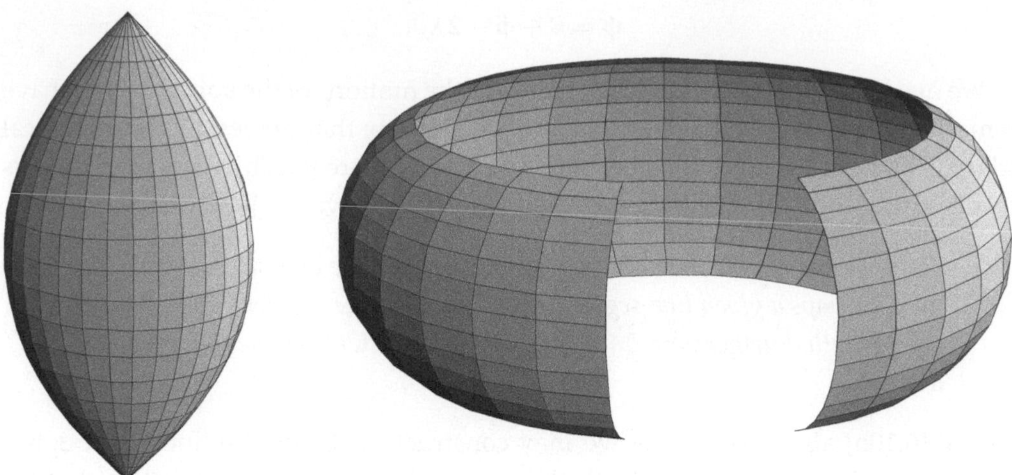

[6.11] Non-spherical surfaces of constant positive curvature. *There exist surfaces of constant positive curvature that are* not *a portion of the sphere, but Minding proved that their* intrinsic *geometry is identical to that of the sphere, at least so long as we stay away from the pointed tips on the left, or the edges on the right.*

The sphere also has the advantage of making it obvious that its intrinsic geometry admits a group of motions: in [6.11] it's certainly not clear that figures can be freely moved about and rotated on the surface without stretching them. Nevertheless, the above discussion shows that the actual shape of a surface in space is a distraction, and it would be better to have a more abstract model that captured the essence of all possible surfaces having the same intrinsic geometry.

By the "essence" we mean knowledge of the distance between any two points, for this and this alone determines the intrinsic geometry. In fact—and this is a fundamental insight of differential geometry—it is sufficient to have a rule for the *infinitesimal* distance between neighbouring points. Given this, we may determine the length of any curve as an infinite sum (i.e., integral) of the infinitesimal segments into which it may be divided. Consequently, we may also identify the "lines" of the geometry as shortest routes from one point to another, and we can likewise [exercise] determine angles.

This leads to the following strategy for capturing the essence of any curved surface S (not necessarily one of constant curvature). To avoid the distraction of the shape of the surface in space, we draw a map (in the sense of a geographical atlas) of S on a flat piece of paper. That is we set up a one-to-one correspondence between points $\widehat{z}$ on S and points z on the plane, which we will think of as the *complex plane*.

Now consider the distance $d\widehat{s}$ separating two neighbouring points $\widehat{z}$ and $\widehat{q}$ on S. In the map, these points will be represented by z and $q = z + dz$, separated by (Euclidean) distance $ds = |dz|$. Once we have a rule for calculating the actual separation $d\widehat{s}$ on S from the apparent separation ds in the map, then (in principle) we know everything there is to know about the intrinsic geometry of S.

The rule giving $d\widehat{s}$ in terms of ds is called the *metric*. In general $d\widehat{s}$ depends on the direction of dz as well as its length ds: writing $dz = e^{i\phi}\, ds$,

$$d\widehat{s} = \Lambda(z, \phi)\, ds. \tag{6.14}$$

According to this formula, $\Lambda(z, \phi)$ is the amount by which we must expand the apparent separation ds in the map—located at z, and in the direction ϕ—to obtain the true separation $d\widehat{s}$ on the surface S.

We will now carry out the above strategy for the sphere. It follows from (6.9) that it is impossible [exercise] to draw a map of the sphere that faithfully represents *every* aspect of its intrinsic geometry. How we choose to draw our map therefore depends on which features we wish to faithfully represent. For example, if we want lines (great circles) on the sphere to be represented by straight lines in the map, then we may employ the so-called *central projection*, in which points are projected from the centre of the sphere onto one of its tangent planes. This yields the so-called *projective map* or *projective model* of the sphere. See [6.12]. Here, the price that we pay for preserving the concept of lines is that angles are not faithfully represented: the angle at which two curves meet on the sphere is not (in general) the angle at which they meet on the map.

For most purposes it is much better to sacrifice straight lines in favour of preserving angles, thereby obtaining a *conformal map* of the surface. In terms of (6.14), a map is conformal if and only if the expansion factor Λ does not depend on the direction ϕ of the infinitesimal vector dz emanating from z:

$$d\widehat{s} = \Lambda(z)\, ds. \tag{6.15}$$

[Recall that we established this fact in Chapter 4.] The great advantage of such a map is that an infinitesimal shape on the surface is represented in the map by a *similar* shape that differs from the original only in size: the one on S is just Λ times bigger.

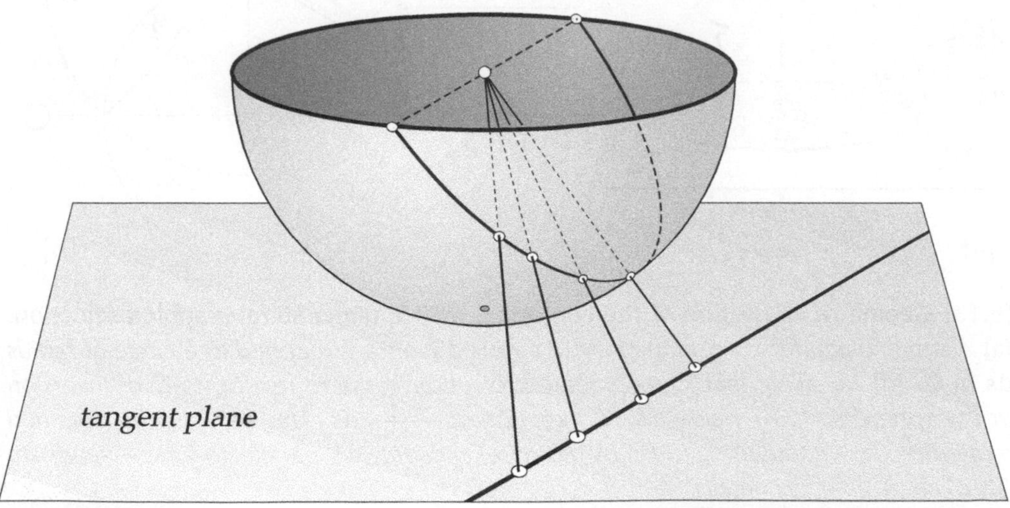

[6.12] Central projection *maps the lines of the sphere to the lines of the map.*

In the case of the sphere we already know of a simple method of constructing a conformal map, namely, via stereographic projection. For simplicity's sake, *henceforth we shall take the sphere to have unit radius* so that it may be identified with the Riemann sphere Σ of Chapter 3. Unlike [6.12], the "lines" of this conformal map do not appear as straight lines. In fact it's not too hard to see [exercise] that great circles on Σ are mapped to circles in $\mathbb{C}$ that intersect the unit circle at opposite points.

Formula (6.15) may be paraphrased as saying that a map is conformal if infinitesimal circles on S are represented in the map by infinitesimal *circles* (rather than ellipses). Of course stereographic projection satisfies this requirement since it preserves circles of all sizes. Figure [6.13a] illustrates this with an infinitesimal circle of radius $d\widehat{s}$ on Σ being mapped to an infinitesimal circle of radius ds in $\mathbb{C}$. To complete the stereographic map we must find its associated metric function Λ—that is the ratio of the two radii in [6.13a].

Consider the vertical cross section of [6.13a] shown in [6.13b], and recall that we showed in Chapter 3 [see p. 161] that stereographic projection is a special case of inversion:

> *If* K *is the sphere of radius* $\sqrt{2}$ *centred at* N, *then stereographic projection is the restriction to* $\mathbb{C}$ *or* Σ *of inversion in* K.

Next, consider (3.6) on p. 142, which describes the effect of inversion on the separation of two points. By taking the limit in which the two points coalesce, we may apply this result to [6.13b] to obtain [exercise]

$$d\widehat{s} = \frac{2}{[Nz]^2}\, ds.$$

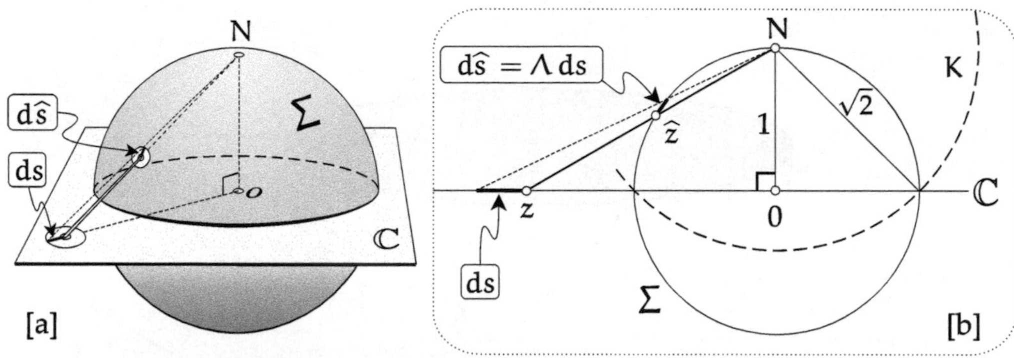

[6.13] Geometric derivation of the conformal metric under stereographic projection. **[a]** *A small, ultimately vanishing circle of radius* $d\widehat{s}$ *on* Σ *is mapped to a circle of radius* ds *in* $\mathbb{C}$. **[b]** *Recalling that stereographic projection is the restriction to* Σ *of inversion in the sphere* K, *(3.6) yields* $d\widehat{s} \asymp \frac{2}{[Nz]^2}\, ds \asymp \frac{2}{1+|z|^2}\, ds$. *This can also be obtained more directly, without using (3.6), by [exercise] choosing* $d\widehat{s}$ *parallel to* $\mathbb{C}$*; see Needham (2021, p. 47) for the details.*

This can also be obtained more directly, without using (3.6), by [exercise] choosing $d\widehat{s}$ parallel to $\mathbb{C}$. Finally, applying Pythagoras' Theorem to the triangle $Nz0$, we obtain

$$d\widehat{s} = \frac{2}{1+|z|^2}\,ds. \tag{6.16}$$

This flat conformal map with metric (6.16) is the desired abstract depiction of all possible surfaces of constant Gaussian curvature $\mathcal{K} = +1$.

6.2.4 Spatial Rotations as Möbius Transformations

Quite generally, suppose that S is a surface of constant Gaussian curvature (so that it possesses a group of motions) and suppose we have drawn a conformal map of S with metric (6.15). Any motion of S will induce a corresponding transformation of this map in $\mathbb{C}$. Since direct motions of the curved surface are conformal, the conformality of the map implies that the induced complex functions must also be conformal and hence analytic. Purely in terms of $\mathbb{C}$, we may therefore identify a function $f(z)$ as a motion if it is analytic and it "preserves the metric" (6.15). That is, suppose that the analytic function $z \mapsto \widetilde{z} = f(z)$ sends two infinitesimally separated points z and $(z+dz)$ to $\widetilde{z}$ and $(\widetilde{z}+d\widetilde{z})$. Then $f(z)$ is a motion if and only if the image separation $d\widetilde{s} = |d\widetilde{z}|$ is related to the original separation $ds = |dz|$ by

$$\Lambda(\widetilde{z})\,d\widetilde{s} = \Lambda(z)\,ds.$$

[Likewise, opposite motions of S correspond to the anticonformal mappings of $\mathbb{C}$ that satisfy this equation.] Since $d\widetilde{z} = f'(z)\,dz$, this is equivalent to demanding that f satisfy the following differential equation:

$$f'(z) = \frac{\Lambda(z)}{\Lambda[f(z)]}.$$

Returning to the particular case $S = \Sigma$, and to the particular conformal map obtained by stereographic projection, the direct motions of all possible surfaces of constant Gaussian curvature $\mathcal{K} = +1$ become the set of analytic complex functions that satisfy

$$f'(z) = \frac{1+|f(z)|^2}{1+|z|^2}. \tag{6.17}$$

In principle, we could find these complex functions without ever leaving $\mathbb{C}$. However, it is simpler and more illuminating to return to the motions of Σ, described by (6.10) and (6.13). When we apply stereographic projection to these motions, what complex functions are induced in $\mathbb{C}$?

The first step is clearly to find the complex function induced by a reflection $\mathfrak{R}_{\widehat{L}}$ of Σ in a line $\widehat{L}$. Consider [6.14a], which shows a new intrinsic method of constructing the reflection $\mathfrak{R}_{\widehat{L}}(\widehat{z})$ of a point $\widehat{z}$ on Σ, namely [exercise], as the second intersection

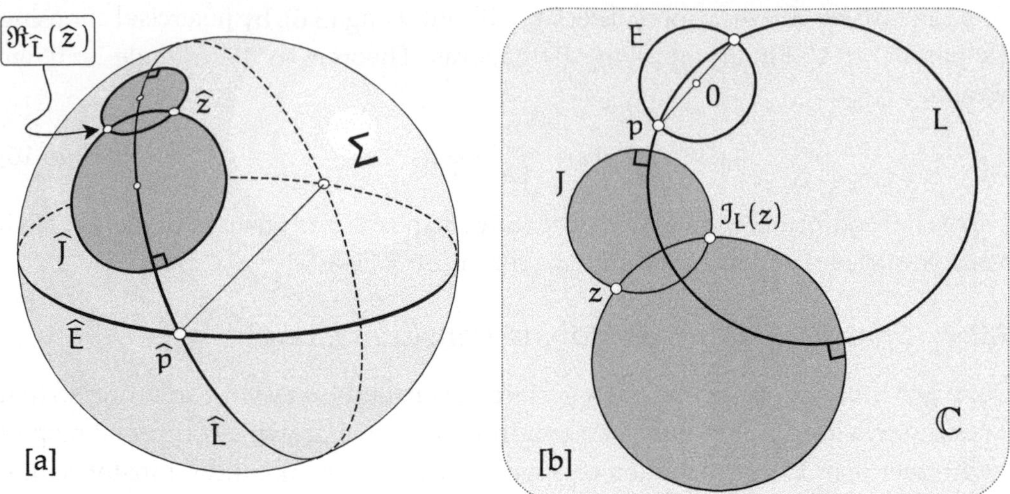

[6.14] Reflection of Σ in a line induces reflection (inversion) of ℂ in the stereographic image of that line. [a] *The reflection* $\mathfrak{R}_{\widehat{L}}(\widehat{z})$ *of a point* $\widehat{z}$ *on* Σ *across the line* $\widehat{L}$ *is the second intersection point of any two circles centred on* $\widehat{L}$ *that pass through* $\widehat{z}$. **[b]** *The same construction after stereographic projection to* $\mathbb{C}$, *revealing the standard orthogonal-circle construction of geometric inversion in* L.

point of any two circles centred on $\widehat{L}$ and passing through $\widehat{z}$. Note that these two circles are orthogonal to $\widehat{L}$. Figure [6.14b] shows what this construction looks like in the stereographic map. Since stereographic projection preserves circles and angles, the two circles orthogonal to $\widehat{L}$ and passing through $\widehat{z}$ are mapped to two circles orthogonal to L and passing through z. The second intersection point of these circles is thus the reflection $\mathfrak{I}_L(z)$ of z in L! To sum up,

Reflection of Σ in a line induces reflection (inversion) of ℂ in the stereographic image of that line. (6.18)

[For a different proof of (6.18), one that is perhaps even more natural than the one above, see Ex. 2.] As an important special case, note that if $\widehat{L}$ is the intersection of Σ with the vertical plane through the real axis, then reflection of Σ in $\widehat{L}$ induces complex conjugation, $z \mapsto \overline{z}$.

Now let's find the complex functions corresponding to rotations of Σ. Figure [6.15] illustrates a rotation $\mathcal{R}_{\widehat{a}}^{\psi}$ of Σ through angle ψ about the point $\widehat{a}$. Let $\widehat{b}$ be the antipodal point to $\widehat{a}$, so that $b = -(1/\overline{a})$ [see (3.23), p. 167]. These points $\widehat{a}$ and $\widehat{b}$ lie on the axis of the rotation and remain fixed; correspondingly, a and b will be the fixed points of the induced transformation of ℂ. Furthermore, it is clear geometrically that the effect of the induced transformation on an infinitesimal neighbourhood of a is a rotation about a [exercise], and, by virtue of our conventions, the rotation angle is *negative* ψ.

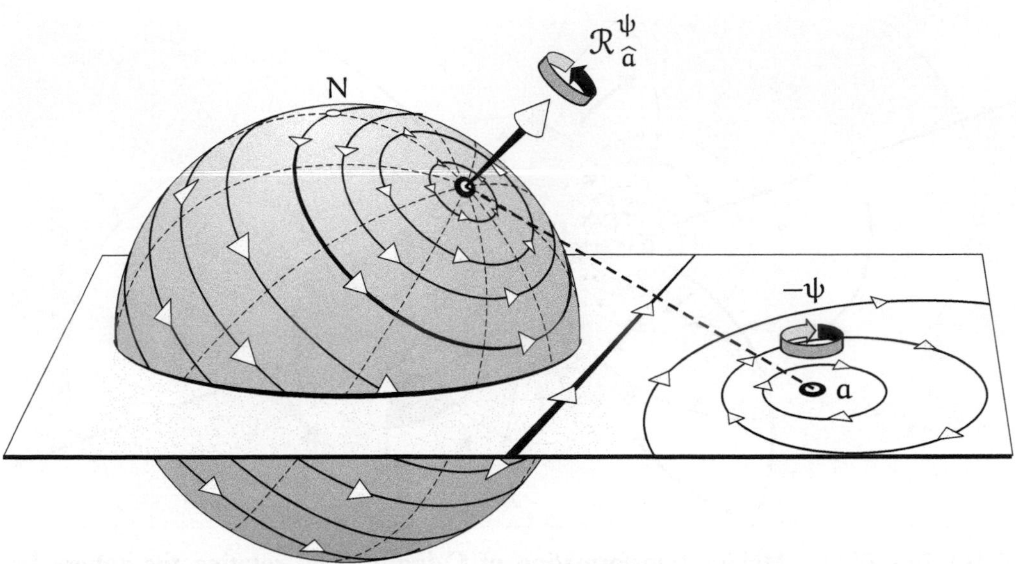

[6.15] A rotation of the sphere stereographically projects to an Elliptic Möbius transformation of $\mathbb{C}$. *According to (6.10), $\mathcal{R}^{\psi}_{\widehat{a}} = \mathfrak{R}_{\widehat{L}_2} \circ \mathfrak{R}_{\widehat{L}_1}$, where $\widehat{L}_1$ and $\widehat{L}_2$ are any two lines passing through $\widehat{a}$ such that the angle between them is $(\psi/2)$. Since stereographic projection preserves circles and angles, the images in $\mathbb{C}$ of these lines will be two circles passing through the fixed points a and b, and containing angle $(\psi/2)$ there. So (6.18) shows that the induced transformation of $\mathbb{C}$ is effected by reflecting across these two intersecting circles, yielding an Elliptic Möbius transformation with multiplier $m = e^{-i\psi}$.*

According to (6.10),

$$\mathcal{R}^{\psi}_{\widehat{a}} = \mathfrak{R}_{\widehat{L}_2} \circ \mathfrak{R}_{\widehat{L}_1},$$

where $\widehat{L}_1$ and $\widehat{L}_2$ are any two lines passing through $\widehat{a}$ (and hence also through $\widehat{b}$) such that the angle between them is $(\psi/2)$. Since stereographic projection preserves circles and angles, the images in $\mathbb{C}$ of these lines will be two circles L_1 and L_2 passing through the fixed points a and b, and containing angle $(\psi/2)$ there. It follows from (6.18) that the transformation R^{ψ}_{a} induced by the rotation $\mathcal{R}^{\psi}_{\widehat{a}}$ is

$$R^{\psi}_{a} = \mathfrak{I}_{L_2} \circ \mathfrak{I}_{L_1}.$$

Thus R^{ψ}_{a} is a Möbius transformation! See [6.16], which illustrates a rotation of $\psi = (\pi/3)$. Referring back to (3.47) on p. 195, and recalling that the "multiplier" describes the local effect of a Möbius transformation in the immediate neighbourhood of a fixed point, we have found that

> *A rotation $\mathcal{R}^{\psi}_{\widehat{a}}$ of Σ stereographically induces an elliptic Möbius transformation R^{ψ}_{a} of $\mathbb{C}$. The fixed points of R^{ψ}_{a} are a and $-(1/\overline{a})$, and the multiplier m associated with a is $m = e^{-i\psi}$.*

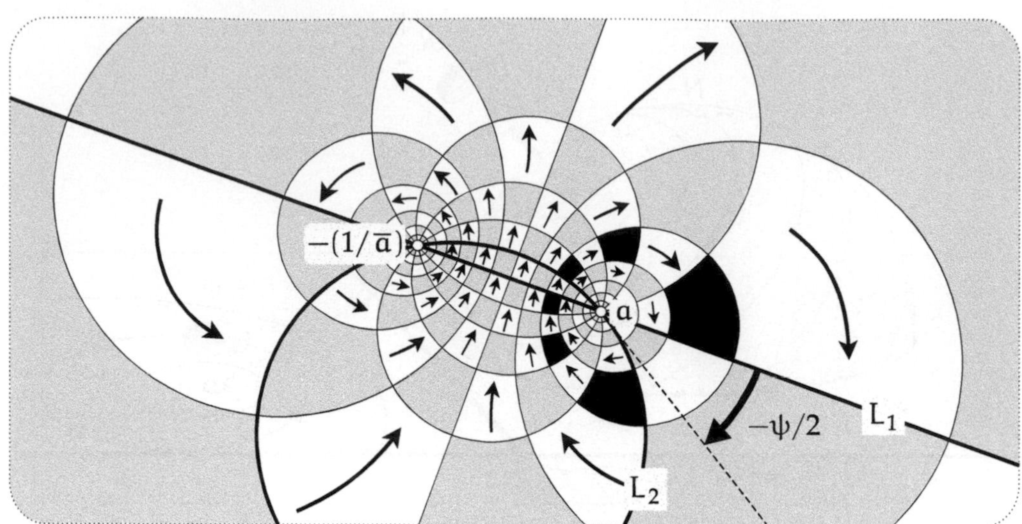

[6.16] The Elliptic Möbius transformation of $\mathbb{C}$ induced by rotating the sphere by $\psi = (\pi/3)$. *The fixed points of the rotation of the sphere are the antipodal points $\widehat{a}$ and $\widehat{b}$ on the axis of rotation, and these map to the fixed points of the Möbius transformation, a and $b = -(1/\overline{a})$. The Möbius transformation is effected by reflecting in any two circles through these fixed points making angle $-(\psi/2)$, such as L_1 and L_2.*

A straightforward matrix calculation [see Ex. 4] based on (3.42), p. 189, yields the following explicit formula for the matrix of R_a^{ψ}:

$$\left[R_a^{\psi}\right] = \begin{bmatrix} e^{i(\psi/2)}|a|^2 + e^{-i(\psi/2)} & 2i\,a\sin(\psi/2) \\ 2i\,\overline{a}\sin(\psi/2) & e^{-i(\psi/2)}|a|^2 + e^{i(\psi/2)} \end{bmatrix}. \tag{6.19}$$

Note that this is in agreement with (3.39), p. 184: rotations of Σ induce Möbius transformations of the form

$$R_a^{\psi}(z) = \frac{Az + B}{-\overline{B}z + \overline{A}}. \tag{6.20}$$

By virtue of (6.13), this formula represents the most general direct motion of Σ. We have already noted that $z \mapsto \overline{z}$ corresponds to a reflection of Σ, and it follows [exercise] that the most general opposite motion is represented by a function of the form

$$z \mapsto \left(\frac{A\overline{z} + B}{-\overline{B}\,\overline{z} + \overline{A}}\right).$$

Figure [6.10b] provided a very elegant geometric method of composing rotations of space. The above analysis now opens the way to an equally elegant method of *computing* the net rotation produced by $\left(\mathcal{R}_{\widehat{p}}^{\omega} \circ \mathcal{R}_{\widehat{a}}^{\psi}\right)$. All we need do is compose the corresponding Möbius transformations:

$$\left[R_p^{\omega} \circ R_a^{\psi}\right] = \left[R_p^{\omega}\right]\left[R_a^{\psi}\right].$$

An otherwise tricky problem has been reduced to multiplying 2×2 matrices!

In practice, rotations are frequently expressed in terms of a unit vector $\mathbf{v}$ pointing along the axis of rotation, with $\widehat{a}$ at its tip. However, (6.19) is currently expressed in terms of the stereographic image a of the point $\widehat{a}$. Let us therefore re-express $\left[\mathcal{R}_a^{\psi}\right]$ in terms of the components l, m, n of the unit vector

$$\mathbf{v} = l\,\mathbf{i} + m\,\mathbf{j} + n\,\mathbf{k}, \quad l^2 + m^2 + n^2 = 1.$$

Referring back to (3.20) on p. 166, we see that a and $\mathbf{v}$ are related as follows:

$$a = \frac{l + im}{1 - n} \quad \text{and} \quad |a|^2 = \frac{1 + n}{1 - n}.$$

Substituting these expressions into (6.19), and removing the common factor of $2/(1-n)$, we obtain [exercise]

$$\left[\mathcal{R}_{\mathbf{v}}^{\psi}\right] = \begin{bmatrix} \cos(\psi/2) + in\sin(\psi/2) & (-m + il)\sin(\psi/2) \\ (m + il)\sin(\psi/2) & \cos(\psi/2) - in\sin(\psi/2) \end{bmatrix}. \tag{6.21}$$

You may check for yourself that this matrix is "normalized": $\det\left[\mathcal{R}_{\mathbf{v}}^{\psi}\right] = 1$. This makes life that much easier, for when we multiply two such matrices the resulting matrix will be of precisely the same form. Thus, by comparing the result with (6.21), we may read off the net rotation.

For example, suppose we perform a rotation of $(\pi/2)$ about $\mathbf{i}$, followed by a rotation of $(\pi/2)$ about $\mathbf{j}$. The Möbius matrix of the net rotation will therefore be

$$\frac{1}{\sqrt{2}}\begin{bmatrix} 1 & -1 \\ 1 & 1 \end{bmatrix} \frac{1}{\sqrt{2}}\begin{bmatrix} 1 & i \\ i & 1 \end{bmatrix} = \frac{1}{2}\begin{bmatrix} 1-i & -1+i \\ 1+i & 1+i \end{bmatrix}. \tag{6.22}$$

Comparing this with (6.21), we see [exercise] that this is rotation of $\psi = (2\pi/3)$ about the axis $\mathbf{v} = \frac{1}{\sqrt{3}}(\mathbf{i} + \mathbf{j} - \mathbf{k})$.

6.2.5 *Spatial Rotations and Quaternions*

This is all rather elegant, but in fact the above method of composing rotations can be streamlined still further. To see how, let us resume the story of Hamilton's *quaternions*, which were introduced at the close of Chapter 1.

On the morning of Monday, 16 October 1843, Hamilton went for a walk with his wife. In the back of his mind was a problem with which he had wrestled fruitlessly for more than ten years—the search for a three-dimensional analogue of the complex numbers, one that would permit vectors in space to be multiplied and divided. As we indicated in Chapter 1, Hamilton was unable to solve this problem for the simple reason that no such analogue exists. However, as he passed Brougham Bridge, he suddenly realized that the prize which had eluded him in three-dimensional space was indeed attainable in *four*-dimensional space!

In the two-dimensional complex plane, we may think of 1 and i as unit basis "vectors" in terms of which a general complex number may be expressed as

$z = a\mathbf{1} + b\,\mathfrak{i}$. The algebra of $\mathbb{C}$ amounts to stipulating that multiplication distributes over addition, that 1 is the identity (i.e., $1z = z1 = z$), and that $\mathfrak{i}^2 = -1$.

In four-dimensional space, Hamilton introduced four basis vectors $\mathbf{1}, \mathbf{I}, \mathbf{J}, \mathbf{K}$ in terms of which a general vector $\mathbb{V}$ (which Hamilton called a quaternion) could be expressed as

$$\mathbb{V} = v\mathbf{1} + v_1\,\mathbf{I} + v_2\,\mathbf{J} + v_3\,\mathbf{K}, \tag{6.23}$$

where the coefficients are all real numbers. To define the product of two such quaternions, Hamilton took $\mathbf{1}$ to be the identity, and he took $\mathbf{I}, \mathbf{J}, \mathbf{K}$ to be three different square roots of $-\mathbf{1}$, each analogous to $\mathfrak{i}$:

$$\mathbf{I}^2 = \mathbf{J}^2 = \mathbf{K}^2 = -\mathbf{1}. \tag{6.24}$$

As in ordinary algebra, Hamilton insisted that multiplication distribute over addition, but in order to render division possible he was forced to make a leap that was revolutionary in its time: non-commutative multiplication. More precisely, Hamilton postulated that

$$\mathbf{IJ} = \mathbf{K} = -\mathbf{JI}, \quad \mathbf{JK} = \mathbf{I} = -\mathbf{KJ}, \quad \mathbf{KI} = \mathbf{J} = -\mathbf{IK}. \tag{6.25}$$

These relations probably look familiar: they are formally identical to the vector products of the basis vectors $\mathbf{i}, \mathbf{j}, \mathbf{k}$ in three-dimensional space. For example, $\mathbf{i}\times\mathbf{j} = \mathbf{k} = -\mathbf{j}\times\mathbf{i}$. We can use this analogy between $\mathbf{i}, \mathbf{j}, \mathbf{k}$ and $\mathbf{I}, \mathbf{J}, \mathbf{K}$ to express the product of two quaternions in a particularly simple way.

First, let's use the analogy to simplify the notation (6.23). As in ordinary algebra, we suppress the identity $\mathbf{1}$ in the first term and write $v\,\mathbf{1} = v$, which Hamilton called the *scalar part* of $\mathbb{V}$. Next we collect the remaining three terms into $\mathbf{V} \equiv v_1\,\mathbf{I} + v_2\,\mathbf{J} + v_3\,\mathbf{K}$, which Hamilton called the *vector part* of $\mathbb{V}$. Thus (6.23) becomes

$$\mathbb{V} = v + \mathbf{V}.$$

In the special case where the scalar part v vanishes, Hamilton called $\mathbb{V} = \mathbf{V}$ a *pure quaternion*. Historically, the concept of a pure quaternion was the forerunner of the idea of an ordinary vector in space. In fact the very word "vector" was coined by Hamilton in 1846 as a synonym for a "pure quaternion".

If we multiply $\mathbb{V}$ by another quaternion $\mathbb{W} = w + \mathbf{W}$, then (6.24) and (6.25) imply [exercise] that

$$\mathbb{V}\,\mathbb{W} = (vw - \mathbf{V}\cdot\mathbf{W}) + (v\,\mathbf{W} + w\,\mathbf{V} + \mathbf{V}\times\mathbf{W}). \tag{6.26}$$

In particular, if $\mathbb{V}$ and $\mathbb{W}$ are pure (i.e., $v = 0 = w$) then this reduces to

$$\mathbb{V}\,\mathbb{W} = -\mathbf{V}\cdot\mathbf{W} + \mathbf{V}\times\mathbf{W}. \tag{6.27}$$

Historically, this formula constituted the very first appearance in mathematics of the concepts of the dot and cross products. Thus, initially, these vectorial operations were viewed as merely two facets (the scalar and vector parts) of quaternion multiplication. However, it did not take physicists long to realize that the scalar product and the vector product were each important in their own right, independently of the quaternions from which they had both sprung.

Further results on quaternions will be derived in the exercises; here we wish only to explain the connection between quaternions and rotations of space. This connection hinges on the idea of a *binary rotation,* which means a rotation of space though an angle of π. The appropriateness of the word "binary" stems from the fact that if the same binary rotation is applied twice then the result is the identity.

According to (6.21), the Möbius transformation corresponding to the binary rotation about the axis $\mathbf{v} = l\,\mathbf{i} + m\,\mathbf{j} + n\,\mathbf{k}$ is

$$[\mathrm{R}_{\mathbf{v}}^{\pi}] = \begin{bmatrix} in & -m+il \\ m+il & -in \end{bmatrix}.$$

Now, forgetting about quaternions for a moment, let us redefine **1** to be the identity matrix, and **I**, **J**, **K** to be the binary rotation matrices about **i**, **j**, **k**, respectively. Thus

$$\mathbf{1} = \begin{bmatrix} 1 & 0 \\ 0 & 1 \end{bmatrix}, \quad \mathbf{I} = \begin{bmatrix} 0 & i \\ i & 0 \end{bmatrix}, \quad \mathbf{J} = \begin{bmatrix} 0 & -1 \\ 1 & 0 \end{bmatrix}, \quad \mathbf{K} = \begin{bmatrix} i & 0 \\ 0 & -i \end{bmatrix}.$$

As a simple check, note that the Möbius transformation corresponding to the Möbius matrix $\mathbf{K}$ is $K(z) = -z$. Make sure you can see why this is as it should be.

Now we can state the surprising connection with quaternions: *under matrix multiplication, these binary rotation matrices obey [exercise] exactly the same laws (6.24) and (6.25) as Hamilton's **I**, **J**, **K**.* It follows that quaternion multiplication is equivalent to multiplying the corresponding 2×2 matrices obtained by replacing Hamilton's **1**, **I**, **J**, **K** with the matrices above. Conversely, the general rotation matrix $\left[\mathrm{R}_{\mathbf{v}}^{\psi}\right]$ in (6.21) can be expressed [exercise] as the quaternion

$$\mathbb{R}_{\mathbf{v}}^{\psi} = \cos(\psi/2) + \mathbf{V}\sin(\psi/2), \tag{6.28}$$

where $\mathbf{V} = l\,\mathbf{I} + m\,\mathbf{J} + n\,\mathbf{K}$. This elegant formula is much easier to remember than (6.21)!

To compose two rotations of space, we need only multiply the corresponding quaternions. For example, the calculation (6.22)—in which a rotation of $(\pi/2)$ about $\mathbf{i}$ was followed by a rotation of $(\pi/2)$ about $\mathbf{j}$—now becomes

$$\tfrac{1}{\sqrt{2}}(1+\mathbf{J})\,\tfrac{1}{\sqrt{2}}(1+\mathbf{I}) = \tfrac{1}{2}(1+\mathbf{I}+\mathbf{J}-\mathbf{K}).$$

Once again, but more easily than before, we deduce that this is rotation of $\psi = (2\pi/3)$ about the axis $\mathbf{v} = \frac{1}{\sqrt{3}}(\mathbf{i}+\mathbf{j}-\mathbf{k})$.

Quaternions also yield a very compact formula for the effect of $\mathrm{R}_{\mathbf{v}}^{\psi}$ on the position vector $\mathbf{P} = X\,\mathbf{i} + Y\,\mathbf{j} + Z\,\mathbf{k}$ of a point in space. Suppose that $\mathrm{R}_{\mathbf{v}}^{\psi}$ rotates $\mathbf{P}$ to $\widetilde{\mathbf{P}}$. If we represent **P** by the pure quaternion $\mathbb{P} = X\mathbf{I} + Y\mathbf{J} + Z\mathbf{K}$, and likewise represent $\widetilde{\mathbf{P}}$ as $\widetilde{\mathbb{P}}$, then

$$\widetilde{\mathbb{P}} = \mathbb{R}_{\mathbf{v}}^{\psi}\, \mathbb{P}\, \mathbb{R}_{\mathbf{v}}^{-\psi}. \tag{6.29}$$

This result was first published by Arthur Cayley in 1845, though he later conceded priority to Hamilton. Not only is the result elegant, it is also practical. For example, Hoggar (1992) discusses how (6.29) can be used to smooth the motion of a rotating object in a computer animation, while Horn (1991) has used it in research connected with robotic vision![10]

Here we will give the most intuitive explanation of (6.29) that we have been able to think of; Exs. 7, 8 give two more. Begin by noting that any multiple of **P** is rotated to the same multiple of $\widetilde{\mathbf{P}}$. To establish (6.29) in general, it is therefore sufficient to establish it for the case where **P** and $\widetilde{\mathbf{P}}$ are unit vectors whose tips $\widehat{p}$ and $\widehat{\widetilde{p}}$ lie on the unit sphere. As before, let $\widehat{a}$ be the point at the tip of **v**.

Consider the following composition of three rotations: $\left(\mathcal{R}_{\widehat{a}}^{\psi} \circ \mathcal{R}_{\widehat{p}}^{\theta} \circ \mathcal{R}_{\widehat{a}}^{-\psi}\right)$. Certainly this is equivalent to a single rotation, and [6.17] helps us to see what it is. Let C be the invariant circle of $\mathcal{R}_{\widehat{a}}^{\psi}$ passing through $\widehat{p}$ and $\widehat{\widetilde{p}}$, and let w be an infinitesimal vector emanating from $\widehat{\widetilde{p}}$ and tangent to C. Note that any vector emanating from a point on C will be carried by $\mathcal{R}_{\widehat{a}}^{\pm\psi}$ into a vector making the same

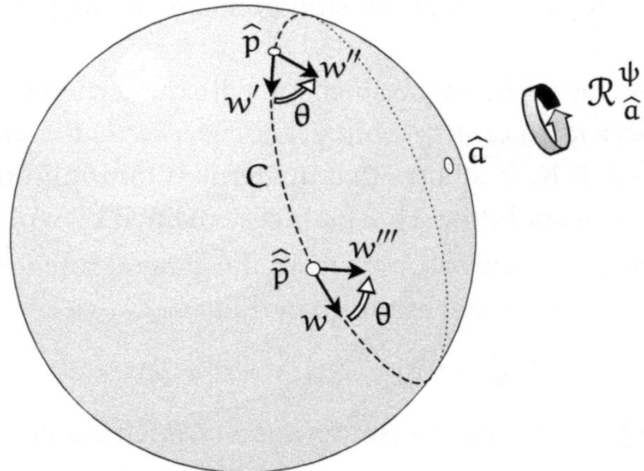

[6.17] Geometric derivation of the quaternion rotation formula. *A rotation of θ about $\widehat{\widetilde{p}}$, turning w into w''', can be expressed in terms of an equal rotation about $\widehat{p}$, because the effect of $\mathcal{R}_{\widehat{\widetilde{p}}}^{\theta} = \mathcal{R}_{\widehat{a}}^{\psi} \circ \mathcal{R}_{\widehat{p}}^{\theta} \circ \mathcal{R}_{\widehat{a}}^{-\psi}$ is $w \mapsto w' \mapsto w'' \mapsto w'''$. Expressed in terms of quaternions, $\mathbb{R}_{\widetilde{p}}^{\theta} = \mathbb{R}_{v}^{\psi}\, \mathbb{R}_{p}^{\theta}\, \mathbb{R}_{v}^{-\psi}$. Finally, putting $\theta = \pi$, the binary rotations $\mathbb{R}_{p}^{\pi}$ and $\mathbb{R}_{\widetilde{p}}^{\pi}$ become the pure quaternions $\mathbb{P}$ and $\widetilde{\mathbb{P}}$, thereby proving the quaternion rotation formula discovered by Hamilton and rediscovered by Cayley: $\widetilde{\mathbb{P}} = \mathbb{R}_{v}^{\psi}\, \mathbb{P}\, \mathbb{R}_{v}^{-\psi}$.*

[10] NOTE for the *25th Anniversary Edition*: There have no doubt been countless other important applications of this idea over the past 25 years, but I have not researched them for this new edition.

angle with C. This justifies the illustrated effect $w \mapsto w' \mapsto w'' \mapsto w'''$ of the three rotations. Thus the net effect $w \mapsto w'''$ is a rotation of θ about $\widehat{\widehat{p}}$:

$$\mathcal{R}^{\theta}_{\widehat{\widehat{p}}} = \mathcal{R}^{\psi}_{\widehat{a}} \circ \mathcal{R}^{\theta}_{\widehat{p}} \circ \mathcal{R}^{-\psi}_{\widehat{a}}.$$

This geometric fact may be expressed in terms of Möbius matrices, or equivalently in terms of quaternions:

$$\mathbb{R}^{\theta}_{\widetilde{\mathbf{p}}} = \mathbb{R}^{\psi}_{\mathbf{v}} \, \mathbb{R}^{\theta}_{\mathbf{p}} \, \mathbb{R}^{-\psi}_{\mathbf{v}}.$$

Finally, if we put $\theta = \pi$ then the binary rotations $\mathbb{R}^{\pi}_{\mathbf{p}}$ and $\mathbb{R}^{\pi}_{\widetilde{\mathbf{p}}}$ are simply the pure quaternions $\mathbb{P}$ and $\widetilde{\mathbb{P}}$, so we are done.

Further Reading. For more on the historical significance of (6.29), see Altmann (1989); for the details of how Hamilton was led to quaternions, see Waerden (1985); for discussion of the connections with modern mathematics and physics, see Penrose and Rindler (1984), Yaglom (1988), Stillwell (1992), and Penrose (2005).

6.3 Hyperbolic Geometry

6.3.1 The Tractrix and the Pseudosphere

Having studied the intrinsic geometry of surfaces of constant positive Gaussian curvature, we now turn to the intrinsic geometry of surfaces of constant *negative* curvature. Just as there are infinitely many surfaces with $\mathcal{K} > 0$, so there are infinitely many with $\mathcal{K} < 0$. Beltrami called such surfaces *pseudospherical*. According to the previously stated result of Minding, all pseudospherical surfaces having the same negative value of $\mathcal{K}$ possess the same intrinsic geometry. To begin to understand hyperbolic geometry, it is therefore sufficient to examine *any* pseudospherical surface. For our purposes, the simplest one is the pseudosphere, so let us explain how this surface may be constructed.

Try the following experiment. Take a small heavy object, such as a paperweight, and attach a length of string to it. Now place the object on a table and drag it by moving the free end of the string along the edge of the table. You will see that the object moves along a curve like that in [6.18a], where the Y-axis represents the edge of the table. This curve is called the *tractrix*, and the Y-axis (which the curve approaches asymptotically) is called the *axis*. The tractrix was first investigated by Newton, in 1676.

If the length of the string is R, then it follows that the tractrix has the following geometric property: *the segment of the tangent from the point of contact to the* Y-*axis has constant length* R. This was Newton's definition of the tractrix. As an interesting aside, it follows [exercise] that the tractrix can be constructed as shown in [6.18b], namely, as an orthogonal trajectory through the family of circles of radius R centred on the axis. This provides a good method of quickly sketching a fairly accurate tractrix.

Returning to [6.18a], let σ represent arc length along the tractrix, with $\sigma = 0$ corresponding to the starting position $X = R$ of the object we are dragging. Just as the object is about to pass through (X, Y), let dX denote the infinitesimal change in X that occurs while the object moves a distance $d\sigma$ along the tractrix. From the similarity of the illustrated triangles, we deduce that

$$\frac{dX}{d\sigma} = -\frac{X}{R} \quad \Longrightarrow \quad X = R\,e^{-\sigma/R}. \tag{6.30}$$

The *pseudosphere of radius* R may now be simultaneously defined and constructed as the surface obtained by rotating the tractrix about its axis. Remarkably, this surface was investigated as early as 1693 (by Christiaan Huygens), two centuries prior to its catalytic role in the acceptance of hyperbolic geometry.

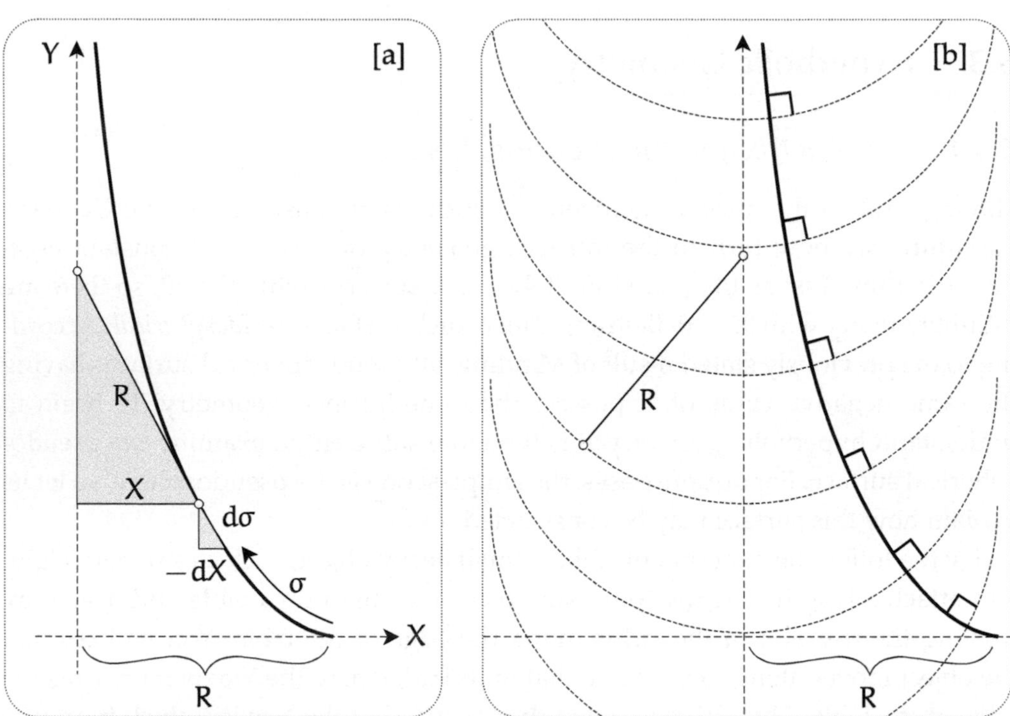

[6.18] Newton's *tractrix*. [a] *Tie a small paperweight to a piece of string of length* R. *On a table top, start with the string running along one edge, at right angles to the other edge, the* Y*-axis. If you move the free end of the string along the* Y*-axis edge of the table, the paperweight will be dragged along the illustrated curve, called the tractrix, which Newton first investigated in 1676. We see that* $\frac{dX}{d\sigma} = -\frac{X}{R}$ *and so* $X = R\,e^{-\sigma/R}$. **[b]** *The tractrix can also be constructed as an orthogonal trajectory through the family of circles of radius* R *centred on the* Y*-axis.*

6.3.2 *The Constant Negative Curvature of the Pseudosphere**

In this optional section we offer a purely geometric proof that the pseudosphere does indeed have constant negative Gaussian curvature. More precisely, we will use the *extrinsic* definition of $\mathcal{K}$ as the product of the principal curvatures to show that *the pseudosphere of radius* R *has constant negative curvature* $\mathcal{K} = -(1/R^2)$. Later we will give a purely intrinsic demonstration of this fact, so you won't miss too much if you skip the following argument.

Let r and $\widetilde{r}$ be the two principal radii of curvature of the pseudosphere of radius R. As with any surface of revolution, it follows by symmetry[11] that

$$\widetilde{r} = \textit{radius of curvature of the generating tractrix},$$

$$r = \textit{the segment of the normal from the surface to the axis},$$

as illustrated in [6.19a]. The problem of determining the Gaussian curvature

$$\mathcal{K} = -\frac{1}{r\,\widetilde{r}}$$

is thereby reduced to a problem in plane geometry, which is solved in [6.19b].

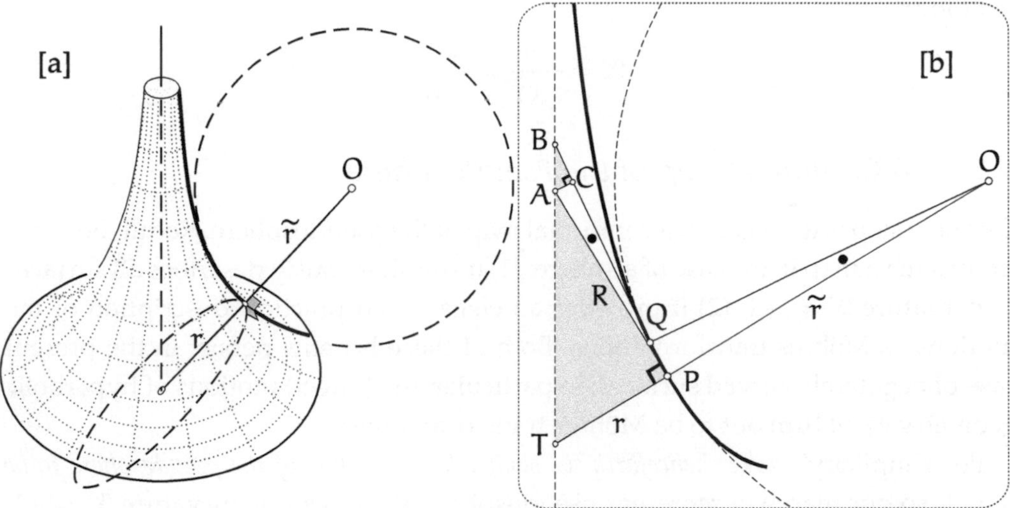

[6.19] Geometric proof that the pseudosphere has constant negative Gaussian curvature. [a] *The principal directions of maximum and minimum curvature will always occur in the orthogonal directions in which a surface has local mirror symmetry. [See Needham (2021).] Clearly, one direction in which the pseudosphere has mirror symmetry is straight up the pseudosphere, along a meridian tractrix generator. The other principal direction is the orthogonal sideways direction along a circle of latitude. Thus,* $\mathcal{K} = -1/(r\,\widetilde{r})$. **[b]** *From this figure we can deduce [see text] that the pseudosphere does indeed have constant negative Gaussian curvature,* $\mathcal{K} = -(1/R^2)$.

11 For a full discussion of all these ideas, see Needham (2021).

By definition, the tractrix in this figure has tangents of constant length R. At the neighbouring points P and Q, figure [6.19b] illustrates two such tangents, PA and QB, containing angle •. The corresponding normals PO and QO therefore contain the same angle •. Note that AC has been drawn perpendicular to QB.

Now let's watch what happens as Q coalesces with P, which itself remains fixed. In this limit, O is the centre of the circle of curvature, PQ is an arc of this circle, and AC is an arc of a circle of radius R centred at P. Thus,[12]

$$\widetilde{r} \asymp OP \quad \text{and} \quad \frac{PQ}{OP} \asymp \bullet \asymp \frac{AC}{R} \quad \Longrightarrow \quad \frac{AC}{PQ} \asymp \frac{R}{\widetilde{r}}.$$

Next we appeal to the defining property $PA = R = QB$ of the tractrix to deduce [exercise] that as Q coalesces with P,

$$BC \asymp PQ.$$

Finally, using the fact that as Q coalesces with P the triangle ABC is ultimately similar to the triangle TAP, we deduce that

$$\frac{r}{R} \asymp \frac{AC}{BC} \asymp \frac{AC}{PQ} \asymp \frac{R}{\widetilde{r}}.$$

Behold![13]

$$\mathcal{K} = -\frac{1}{r\widetilde{r}} = -\frac{1}{R^2}.$$

6.3.3 A Conformal Map of the Pseudosphere

Our next step is to construct a conformal map of the pseudosphere. Recall the benefits of such a map in the case of a sphere: (1) it simultaneously describes all surfaces of curvature $\mathcal{K} = +1$; (2) it provides an elegant and practical description of the motions as Möbius transformations. Both of these benefits persist in the present case of negatively curved surfaces; in particular, the (direct) motions of hyperbolic geometry *again* turn out to be Möbius transformations!

For simplicity's sake, *henceforth we shall take the radius of the pseudosphere to be* $R = 1$, so our map will represent pseudospherical surfaces of curvature $\mathcal{K} = -1$. [NOTE: This is the conventional choice in almost all texts on hyperbolic geometry.] As a first step towards a conformal map, [6.20a] introduces a fairly natural coordinate system (x, σ) on the pseudosphere.

The first coordinate x measures angle around the axis of the pseudosphere, say restricted to $0 \leqslant x < 2\pi$. The second coordinate σ measures arc length along each tractrix generator (as in [6.18a]). Thus the curves $x = \text{const.}$ are the tractrix

[12] Once again, "$\asymp$" denotes Newton's concept of *ultimate equality*; see the new Preface.

[13] One of my most treasured possessions is an email from Bill Thurston saying that he *liked* my Newtonian proof of this fundamentally important fact!

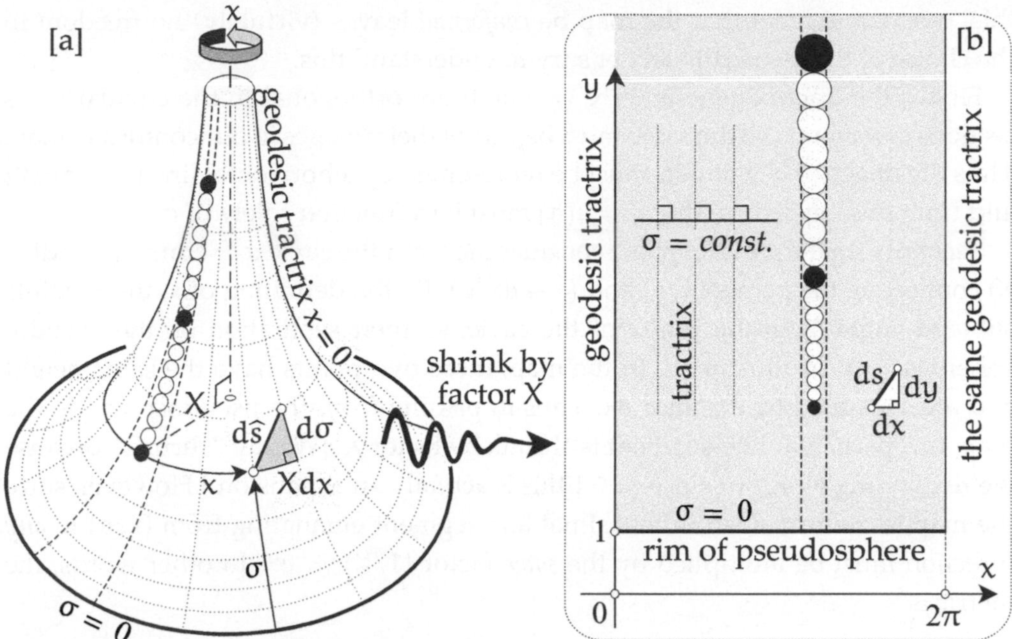

[6.20] Constructing a conformal map of the pseudosphere. [a] *Having specialized to the standard case,* $R = 1$, *we have* $X = e^{-\sigma}$. **[b]** *Let us decide to use the angle* x *around the axis of symmetry as the* x*-coordinate in this map. If we insist that the map be* conformal *then all small (ultimately vanishing) distances must be scaled* equally, *regardless of direction. Since the sideways movement* $X\,dx$ *on the left is divided by* X *in passing to the map on the right,* all *distances must be divided by* X, *so* $\frac{dy}{d\sigma} = \frac{1}{X} = e^{\sigma}$ *and therefore* $y = e^{\sigma} + k$. *Choosing* $k = 0$, *we find that the distance* $d\hat{s}$ *on the pseudosphere is related to the distance* ds *in the map by the conformal metric formula,* $d\hat{s} = ds/y = \sqrt{dx^2 + dy^2}/y$.

generators of the pseudosphere [note that these are clearly geodesics], and the curves $\sigma = \text{const.}$ are circular cross sections of the pseudosphere [note that these are clearly *not* geodesics]. Since the radius of such a circle is the same thing as the X-coordinate in [6.18a], it follows from (6.30) that

> *The radius* X *of the circle* $\sigma = \text{const.}$ *passing through the point* (x, σ) *is given by* $X = e^{-\sigma}$.

In our map, let us choose the angle x as our horizontal axis, so that the tractrix generators of the pseudosphere are represented by vertical lines. See [6.20b]. Thus a point on the pseudosphere with coordinates (x, σ) will be represented in the map by a point with Cartesian coordinates (x, y), which we will soon think of as the complex number $z = x + iy$.

If our map were not required to be special in any way, then we could simply choose $y = y(x, \sigma)$ to be an arbitrary function of x and σ. In stark contrast to this, our requirement that the map be *conformal* leaves (virtually) no freedom in the choice of the y-coordinate. Let's try to understand this.

Firstly, the tractrix generators $x = \text{const.}$ are orthogonal to the circular cross sections $\sigma = \text{const.}$, so the same must be true of their images in our conformal map. Thus the image of $\sigma = \text{const.}$ must be represented by a horizontal line $y = \text{const.}$, and from this we deduce that $y = y(\sigma)$ must be a function solely of σ.

Secondly, on the pseudosphere consider the arc of the circle $\sigma = \text{const.}$ (of radius X) connecting the points (x, σ) and $(x + dx, \sigma)$. By the definition of x, these points subtend angle dx at the centre of the circle, so their separation on the pseudosphere is $X\,dx$, as illustrated. In the map, these two points have the same height and are separated by distance dx. Thus in passing from the pseudosphere to the map, this particular line-segment is shrunk by factor X. [We say "shrunk" because we're dividing by X, but since $X \leqslant 1$ this is actually an expansion.] However, since the map is conformal, an infinitesimal line-segment emanating from (x, σ) in *any* direction must be multiplied by the *same* factor $(1/X) = e^{\sigma}$. In other words, the metric is

$$d\widehat{s} = X\,ds.$$

Thirdly, consider the uppermost black disc on the pseudosphere shown in [6.20a]. Think of this disc as infinitesimal, say of diameter ϵ. In the map, it will be represented by *another disc*, whose diameter (ϵ/X) may be interpreted more vividly as the angular width of the original disc as seen by an observer on the pseudosphere's axis. Now suppose we repeatedly translate the original disc towards the pseudosphere's rim, moving it a distance ϵ each time. Figure [6.20a] illustrates the resulting chain of touching, congruent discs. As the disc moves down the pseudosphere, it recedes from the axis, and its angular width as seen from the axis therefore diminishes. Thus the image disc in the map appears to gradually shrink as it moves downward, and the equal distances 8ϵ between the successive black discs certainly do not appear equal in the map.

Having developed a feel for how the map works, let's actually calculate the y-coordinate corresponding to the point (x, σ) on the pseudosphere. From the above observations (or directly from the requirement that the illustrated triangles be similar) we deduce that

$$\frac{dy}{d\sigma} = \frac{1}{X} = e^{\sigma} \quad \Longrightarrow \quad y = e^{\sigma} + \text{const.}$$

The standard choice of this constant is 0, so that

$$y = e^{\sigma} = (1/X).$$

Thus the entire pseudosphere is represented in the map by the shaded region lying above the line $y = 1$ (which itself represents the pseudosphere's rim), and the metric associated with the map is

$$d\widehat{s} = \frac{ds}{y} = \frac{\sqrt{dx^2 + dy^2}}{y}. \tag{6.31}$$

For future use, also note that an infinitesimal rectangle in the map with sides dx and dy represents a similar infinitesimal rectangle on the pseudosphere with sides (dx/y) and (dy/y). Thus the apparent area $dx\,dy$ in the map is related to the true area $d\mathcal{A}$ on the pseudosphere by

$$d\mathcal{A} = \frac{dx\,dy}{y^2}. \tag{6.32}$$

6.3.4 Beltrami's Hyperbolic Plane

In the Introduction we gave the impression that Beltrami had succeeded in interpreting abstract hyperbolic geometry as the intrinsic geometry of the pseudosphere. This is really not possible, and it is *not* what Beltrami claimed.

The abstract hyperbolic geometry discovered by Gauss, Bolyai, and Lobachevsky is understood to take place in a *hyperbolic plane* that is exactly like the Euclidean plane, *except* that lines within it obey the hyperbolic axiom (6.3):

> *Given a line* L *and a point* p *not on* L, *there are at least two lines through* p *that do not meet* L.

The constant negative curvature of the pseudosphere ensures that it faithfully represents all consequences of this axiom that deal only with a finite region of the hyperbolic plane. An example of such a consequence is the theorem that the angular excess of a triangle is a negative multiple of its area, and this does indeed hold on the pseudosphere.

However, the pseudosphere will not do as a model of the *entire* hyperbolic plane, because it departs from the Euclidean plane in two unacceptable ways:

- The pseudosphere is akin to a cylinder instead of a plane. For example, a closed loop in the plane can always be shrunk to a point, but a loop on the pseudosphere that wraps around the axis cannot be.
- In the hyperbolic plane, as in the Euclidean plane, a line-segment can be extended indefinitely in either direction. We have already remarked that the tractrix generators of the pseudosphere are clearly geodesic, and we would therefore like to interpret them as hyperbolic lines. But although such a tractrix extends indefinitely *up* the pseudosphere, in the other direction it terminates when it hits the rim.

Beltrami pointed out that the first of these problems can be resolved as follows. Imagine the pseudosphere covered by a thin stretchable sheet. To obtain the map in [6.20b], we cut this sheet along a tractrix generator and unwrap it onto the shaded region. Of course to make it lie flat and fit into this rectangular region, the sheet must be stretched—the metric (6.31) tells us *how much* stretching must be applied to each part. But now imagine the sheet as wrapping round and round the pseudosphere infinitely many times[14], like an endless roll of cling film[15]. By unwrapping this infinitely long sheet (stretching as we go) we can now cover the entire region above $y = 1$. According to this interpretation, a particle travelling along a horizontal line in the map would correspond to a particle travelling round and round a circle $\sigma = \text{const.}$ on the pseudosphere, executing one complete revolution for each movement of 2π along the line.

Now let us explain how the conformal map solves our second problem—the pseudosphere's edge. In terms of *extrinsic* geometry, this edge is an insurmountable obstacle: we cannot extend the pseudosphere smoothly beyond this edge while preserving its constant curvature. However, we only care about the pseudosphere's *intrinsic* geometry, and we have seen that if we measure distance using $d\widehat{s} = \frac{ds}{y}$, this is identical to the region $y > 1$ in [6.21].

Imagining yourself as a tiny two-dimensional being living in [6.21], walking down a line $x = \text{const.}$ is exactly the same thing as walking down a tractrix on the pseudosphere. Of course on the pseudosphere your walk is rudely interrupted at some point $\widehat{p}$ on the rim ($\sigma = 0$), corresponding to a point p on the line $y = 1$. But in the map this point p is just like any other, and there is absolutely nothing preventing you from continuing your walk all the way down to the point q on $y = 0$.

Why stop at q? The answer is that you will never even get that far, because q is infinitely far from p! Suppose that you are the illustrated small disc on the line $y = 2$, and that I am standing outside your hyperbolic world, watching as you walk at a steady pace towards $y = 0$. Of course you remain the same hyperbolic size as you walk, but to *me* you appear to shrink. This is made particularly vivid by the illustrated Euclidean interpretation [exercise] of your hyperbolic size $d\widehat{s} = \frac{ds}{y}$:

> *The hyperbolic diameter of an infinitesimal disc centred at* $(x + iy)$ *is the angle it subtends at the point* x *on the real axis.* (6.33)

Thus your apparent size must shrink so that you subtend a constant angle, and although all your hyperbolic strides are the same length, to me they look shorter and shorter, and you appear to be travelling more and more slowly.

For example, suppose you are walking at a steady speed of $\ln 2$. As illustrated, integration of (dy/y) shows [exercise] that you reach $y = 1$ after one unit of time, $y = (1/2)$ after two units of time, $y = (1/4)$ after three units of time, etc. Thus, viewed from outside your world, each successive unit of time only halves your

[14] Stillwell (1996) points out that this is probably the very first appearance in mathematics of what topologists now call a *universal cover*.

[15] For Americans, read "plastic wrap".

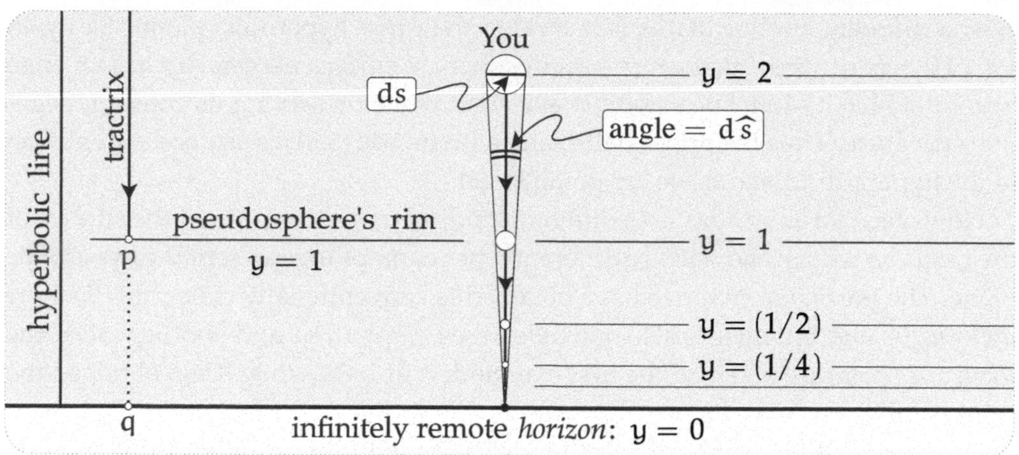

[6.21] The pseudosphere has an edge, but the hyperbolic plane is uniform and infinite. *Suppose "You" start at the point on the pseudosphere corresponding to* $y = 2$ *and walk down a tractrix generator at a steady rate of* $\ln 2$*. Then in the conformal* map *you reach* $y = 1$ *(the pseudosphere's rim—the edge of your world!) after just one unit of time. But if we imagine you living in the map, perceiving distances around you via the hyperbolic metric, then you can continue your downward journey, arriving at* $y = (1/2)$ *after two units of time,* $y = (1/4)$ *after three units of time, etc. Although you actually stay the same size, in the map you* appear *to shrink, for your true size is the angle* $d\hat{s}$ *that you subtend at the horizon. Thus, viewed from outside your world, you shrink with each successive unit of time, only halving your distance from the* horizon, $y = 0$*; therefore, you will never reach it. Your world is infinite and uniform: no place or direction seems any different from any other.*

distance from $y = 0$, and therefore you will never reach it. [An appropriate name for this phenomenon might be "Zeno's Revenge"!]

We now possess a concrete model of the *hyperbolic plane*, namely, the entire shaded half-plane $y > 0$ with metric $d\hat{s} = \frac{ds}{y}$. The points on the real axis are infinitely far from ordinary points and are not (strictly speaking) considered part of the hyperbolic plane. They are called *ideal points*, or *points at infinity*. The complete line $y = 0$ of points at infinity will be called the *horizon*[16].

Studying hyperbolic geometry by means of this map is like studying spherical geometry via a stereographic map, without ever having seen an actual sphere. This is not as bad as it sounds. After all, by constructing geographical maps through terrestrial measurements, man developed a good understanding of the surface of the Earth centuries before venturing into space and gazing down on its roundness!

Still, it would be nice to have the analogue of a globe instead of a mere atlas. The pseudosphere only models a portion of the hyperbolic plane, but might there

[16] For reasons that will be clear shortly, another name is the *circle at infinity*.

exist a different surface that is isometric to the *entire* hyperbolic plane? Sadly, in 1901 Hilbert proved[17] that every pseudospherical surface necessarily has an edge beyond which it cannot be smoothly extended while preserving its constant negative curvature. Thus the upper half-plane with metric (6.31) is as good a depiction of the hyperbolic plane as we are going to get.

However, just as an atlas uses different kinds of maps to represent the surface of the Earth, so we can and will use different types of maps to represent the hyperbolic plane. The particular map we have obtained is conventionally called the *Poincaré upper half-plane*, but there is also one called the *Poincaré disc*, and another called the *Klein disc*. Poincaré obtained the first two models in 1882, while Klein obtained the third in 1871.

We cannot let the names of these models pass without comment. Anyone with even a passing interest in the history of mathematics will know that ideas are frequently (usually?) named after the wrong person. In fact[18], *the three models above were all discovered by Beltrami!* As we shall see, Beltrami obtained these three models (in 1868, 14 years before Poincaré) in a beautifully unified way, from a fourth model consisting of a map drawn on a hemisphere. And in case you're wondering, yes, the hemisphere model is Beltrami's, too!

In this *25th Anniversary Edition* we shall dogedly attempt to set history straight, as we have previously attempted to do in VDGF, by giving both men *equal* credit, and renaming the conformal maps as the *Beltrami–Poincaré* half-plane and disc models.

6.3.5 Hyperbolic Lines and Reflections

Before we get going, let's indicate *where* we are going, focusing just on direct motions. In Euclidean geometry, every direct motion is the composition of reflections in two lines. We have seen that the same is true in spherical geometry, and we will soon show that it is again true in hyperbolic geometry. Since two Euclidean lines must intersect or be parallel, there are just two kinds of direct Euclidean motions: rotations and translations. The absence of parallel lines on the sphere implies that its direct motions can only be rotations. Conversely, the multitude of parallel lines in the hyperbolic plane yields a geometry that is *richer* than Euclid's, containing rotations, translations, and a *third* kind of motion that has no Euclidean counterpart.

To avoid confusion, let us use the prefix "h-" to distinguish hyperbolic concepts from their Euclidean descriptions in the map. For example, an "h-line" will mean a "hyperbolic line" (i.e., a geodesic), while a "line" will refer to an ordinary straight

[17] Hilbert (1965, Vol. 2, pp. 437–448), with English translation available in Hilbert (1902).
[18] See Milnor (1982), Beltrami (1868a), Beltrami (1868b), and Stillwell (2010).

line in the map. Let us also define $\mathcal{H}\{z_1, z_2\}$ to be the h-distance (measured using $d\hat{s} = \frac{ds}{y}$) between z_1 and z_2. For example, if dz is infinitesimal, then

$$\mathcal{H}\{z + dz, z\} = \frac{|dz|}{\text{Im } z}.$$

Finally, let us define an h-circle of h-radius ρ and h-centre c to be the locus of points z such that $\mathcal{H}\{z, c\} = \rho$.

Since tractrix generators of the pseudosphere are clearly geodesic, vertical lines in the map should also be geodesic, i.e., they should be examples of h-lines. Figure [6.22a] confirms this directly by showing that

The (unique) shortest route between two vertically separated points is the vertical line-segment L *connecting them.* (6.34)

To see this, compare L with any other route, such as M. Let ds_1 be an infinitesimal segment of L at height y, and let ds_2 be the corresponding element of M cut off by horizontal lines through the ends of ds_1. Since

$$d\hat{s}_1 = \frac{ds_1}{y} < \frac{ds_2}{y} = d\hat{s}_2,$$

the total hyperbolic length of L is less than M's. Done. From this we can deduce that

$$\mathcal{H}\{(x + iy_1), (x + iy_2)\} = |\ln(y_1/y_2)|. \tag{6.35}$$

Through a given point of the pseudosphere we obviously have geodesics in all directions, not just tractrix generators; what do these more general h-lines look like in the map? The answer is very beautiful and unexpected:

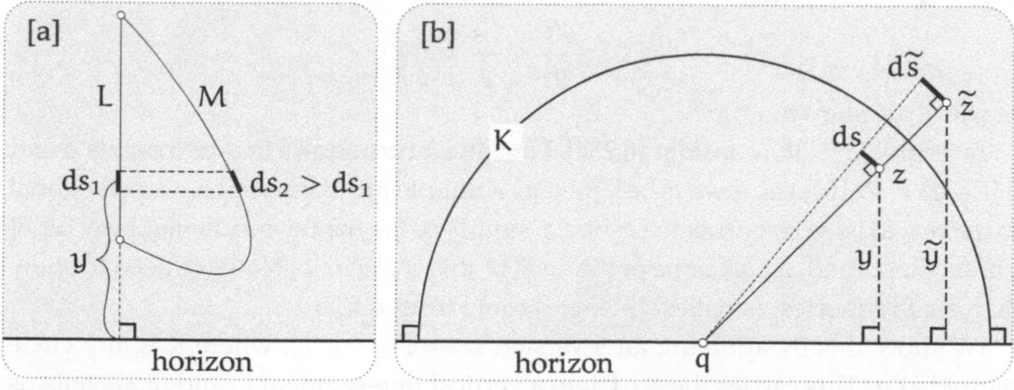

[6.22] [a] Vertical lines are h-lines (geodesics), *because* $d\hat{s}_1 = ds_1/y < ds_2/y = d\hat{s}_2$. **[b] Inversion in a semicircle orthogonal to the horizon preserves h-distance.** *Under* $z \mapsto \tilde{z} = \mathcal{I}_K(z)$, *we see that* $d\hat{\tilde{s}} = d\tilde{s}/\tilde{y} = ds/y = d\hat{s}$. *Since distance is preserved in this particular direction, it must be preserved in all directions.*

Every h-line is either a half-line orthogonal to the horizon, or else a semicircle orthogonal to the horizon. (6.36)

Before we prove this, it's important to realize that if you were an inhabitant of the hyperbolic plane, there would be no way for you to distinguish between the semicircular h-lines and the vertical h-lines: every line is exactly like every other, it's just our map that makes them look different. What about the fact that the semicircles have two ends on the horizon, whereas the vertical h-lines appear to only have one? The answer is that, in addition to the points on the real axis, there is one more point at infinity, and all the vertical h-lines meet there. According to (6.31), as we move upward along two neighbouring, vertical h-lines, the h-distance between them dies away like $(1/y)$, and they converge to a single point at infinity; this is particularly vivid on the pseudosphere. Finally, note that even in terms of the map, a vertical h-line may be viewed as just a special case of a semicircular h-line by allowing the radius to tend to infinity.

We will prove (6.36) by first establishing another equally beautiful fact, one that is fundamental to all that follows:

Inversion in a semicircle orthogonal to the horizon is an opposite motion of the hyperbolic plane. (6.37)

To see why this is true, consider the inversion $z \mapsto \tilde{z} = \mathcal{I}_K(z)$ illustrated in [6.22b]. We need to show that $\mathcal{I}_K(z)$ does not alter the h-length $d\widehat{s}$ of any infinitesimal line-segment ds emanating from z. However, because our model of the hyperbolic plane is *conformal*, we need only show that $\mathcal{I}_K(z)$ preserves the h-length of any *single* ds, in a direction of our choosing. Choosing ds orthogonal to the radius qz of K (as illustrated), the anticonformality of inversion implies that the image $d\widetilde{s}$ is also orthogonal to this radius. Thus, by virtue of the illustrated similar triangles, it follows [exercise] that

$$d\widehat{\widetilde{s}} = \frac{d\widetilde{s}}{\widetilde{y}} = \frac{ds}{y} = d\widehat{s}\,,$$

as was to be shown.

To establish (6.36), consider [6.23a]. First, the figure shows that two points a and b [$\mathrm{Re}(a) \neq \mathrm{Re}(b)$] can always be joined by a unique arc L of a semicircle orthogonal to the real axis: to construct the centre c, simply draw the perpendicular bisector of ab. As illustrated, let q be one of the ends of this semicircle. Now we need to show that L is the shortest (smallest h-length) route from a to b.

We show this by applying an inversion $z \mapsto \tilde{z} = \mathcal{I}_K(z)$, where K is any circle centred at q. This carries the arc L into a vertical line-segment $\widetilde{L}$, and (6.37) tells us that $\widetilde{L}$ and L have equal h-length. More generally, any route M from a to b has the same h-length as the route $\widetilde{M} = \mathcal{I}_K(M)$ from $\widetilde{a}$ to $\widetilde{b}$. Thus if L were not the shortest

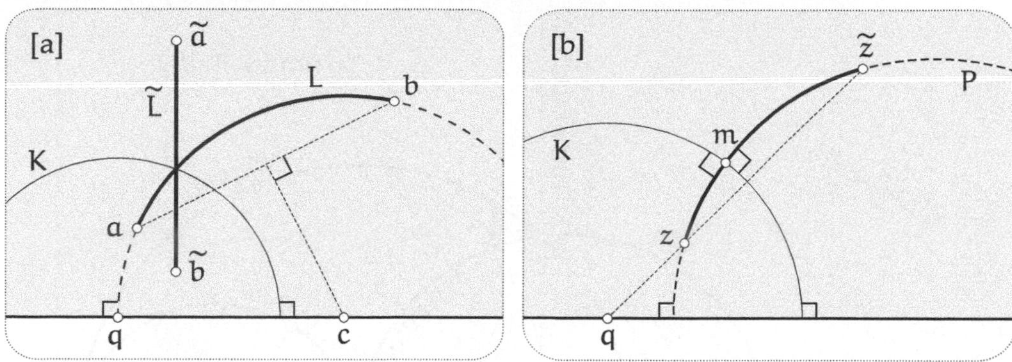

[6.23] [a] Geometric proof that semicircles orthogonal to the horizon are h-lines (geodesics). *Construct* L *as shown, then invert it in any circle* K *centred at* q, *obtaining* $\mathcal{I}_K(L) = \widetilde{L}$, *as shown. Since* $\widetilde{L}$ *has been proven to be the shortest route from* $\widetilde{a}$ *to* $\widetilde{b}$, *and since* $\mathcal{I}_K$ *preserves h-distance,* L *must be the shortest route from* a *to* b. **[b] If K is an h-line, then $\mathcal{I}_K = \mathfrak{R}_K$ is (literal) h-reflection in K.** *For if* $\widetilde{z} = \mathcal{I}_K(z)$ *then all circles through* z *and* $\widetilde{z}$ *cut* K *orthogonally, and so the illustrated h-line* P *does, too. The h-lengths of* zm *and* $m\widetilde{z}$ *are equal, for they are interchanged by the h-distance-preserving* $\mathcal{I}_K$, *so* $\widetilde{z} = \mathfrak{R}_K(z)$, *as claimed.*

route from a to b, then $\widetilde{L}$ would not be the shortest route from $\widetilde{a}$ to $\widetilde{b}$, in violation of (6.34). Done.

Incidentally, note that this construction also enables us (in principle) to calculate the h-distance between any two points in the hyperbolic plane:

$$\mathcal{H}\{a, b\} = \mathcal{H}\{\widetilde{a}, \widetilde{b}\} = \left|\ln\left(\frac{\operatorname{Im}\widetilde{a}}{\operatorname{Im}\widetilde{b}}\right)\right|,$$

by virtue of (6.35). Later we shall be able to derive a more explicit formula.

The fact that a semicircle orthogonal to the real axis is an h-line strongly suggests the following re-interpretation of (6.37): hyperbolic plane in the h-line

Inversion in a semicircle K *orthogonal to the horizon is a reflection* $\mathfrak{R}_K$ *of the hyperbolic plane in the h-line* K. (6.38)

In symbols, $\mathfrak{R}_K(z) = \mathcal{I}_K(z)$. Before proving this, let's be clear what we *mean* by reflection. Just as we would in Euclidean and spherical geometry, we begin the construction of $\mathfrak{R}_K(z)$ by drawing the h-line P that passes through z and cuts K perpendicularly, say at m. Then $\mathfrak{R}_K(z)$ is defined to be the point on P that is the same h-distance from m as z.

To prove (6.38), consider [6.23b], in which $\widetilde{z} = \mathcal{I}_K(z)$. First recall that every circle through z and $\widetilde{z}$ is automatically orthogonal to K. In particular, the unique h-line through z and $\widetilde{z}$ must be orthogonal to K, and hence it is the desired "P" of the previous paragraph. Finally, recall that $\mathcal{I}_K$ maps P into itself, swapping the segments zm and $\widetilde{z}$m. Thus, since $\mathcal{I}_K$ is a motion, these two h-line segments have equal h-length, as was to be shown.

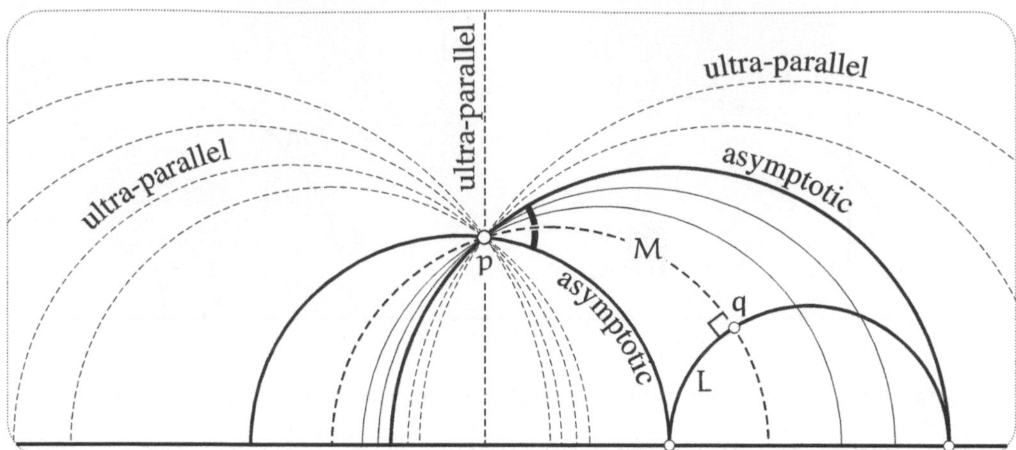

[6.24] The hyperbolic plane satisfies the Hyperbolic Axiom, *for we see that there are infinitely many h-lines [shown dashed] through* p *that fail to meet the h-line* L*: they are said to be* ultra-parallel *to* L*. The two h-lines that meet* L *at the horizon are called* asymptotic*. The h-distance of* p *from* L *is defined as in Euclidean geometry, as the length of the unique line segment* pq *that meets* L *at right angles.*

Conversely, if we are *given* any two points z and $\widetilde{z}$, then we may draw the perpendicular h-bisector K, and $\mathfrak{R}_K$ swaps z and $\widetilde{z}$. Also note that z and its reflection $\widetilde{z} = \mathfrak{R}_K(z)$ are the same h-distance from *every* point k on K, just as in Euclidean and spherical geometry. This is easily proved: since $\mathcal{I}_K$ is a motion, and $\widetilde{k} = \mathcal{I}_K(k) = k$, it follows that $\mathcal{H}\{z, k\} = \mathcal{H}\{\widetilde{z}, \widetilde{k}\} = \mathcal{H}\{\widetilde{z}, k\}$.

It is becoming clear that hyperbolic geometry has much in common with Euclidean geometry. However, now that we know what h-lines look like, [6.24] shows that hyperbolic geometry really is non-Euclidean: there are infinitely many h-lines through p [shown dashed] that do not meet the h-line L. Such h-lines are said to be *ultra-parallel* to L.

Separating the ultra-parallels from the h-lines that do intersect L, we see that there are precisely two h-lines that fail to meet L anywhere within the hyperbolic plane proper, but that do meet it on the horizon. These two h-lines are called *asymptotic*[19].

As in Euclidean geometry, the figure makes it clear that there is precisely one h-line M passing through p that cuts L at right angles (say at q). In fact [exercise] M may be constructed as the unique h-line through p and $\mathfrak{R}_L(p)$. The existence of M makes it possible to define the distance of a point p from a line L in the usual way, namely, as the h-length of the segment pq of M.

Since M and L are orthogonal, $\mathfrak{R}_M = \mathcal{I}_M$ maps L into itself, swapping the two ends on the horizon. It follows [exercise] that $\mathfrak{R}_M$ swaps the two asymptotic lines,

[19] Another commonly used name is *parallel*.

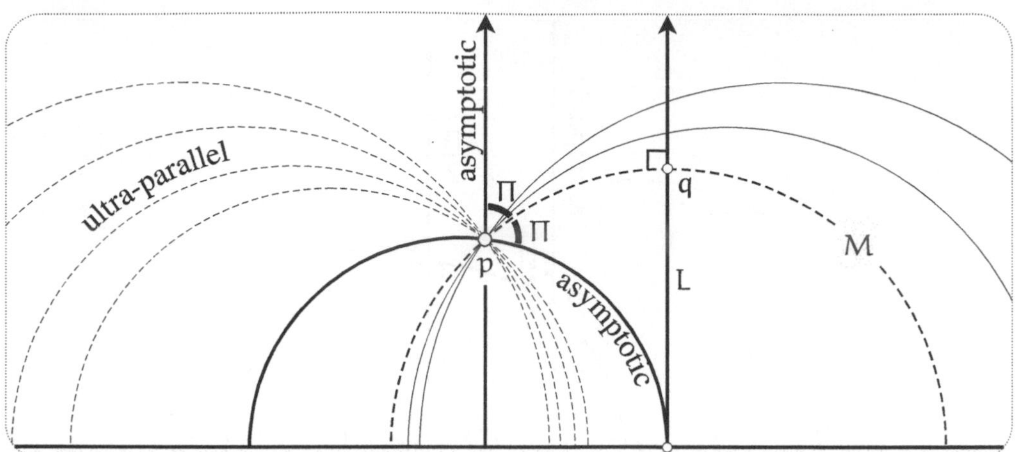

[6.25] *The same geometry as the previous figure, but in the case where* L *is a vertical half-line.*

and that M bisects the angle at p contained by the asymptotic lines. The angle between M and either asymptotic line is called the *angle of parallelism*, and is usually denoted Π. As one rotates the line M about p, its intersection point on L moves off towards infinity, and Π tells you how far you can rotate M before it starts missing L entirely.

Finally, [6.25] merely serves to illustrate the same concepts and terminology as [6.24], but in the case where the h-line L happens to be represented as a vertical half-line instead of a semicircle.

*6.3.6 The Bolyai–Lobachevsky Formula**

This brief, optional subsection nicely illustrates how the preceding ideas may be used to solve a significant, concrete problem: finding the angle of parallelism, Π.

In Euclidean geometry the analogue of the two asymptotic lines is the unique parallel line through p, and since this is perpendicular to M, the analogue of Π is a right angle. On the other hand, in hyperbolic geometry it is clear that Π is always acute, and that its value decreases as the distance $D \equiv \mathcal{H}\{p, q\}$ of p from L increases. More precisely, both Bolyai and Lobachevsky obtained this result:

$$\textit{Bolyai–Lobachevsky Formula}: \qquad \tan(\Pi/2) = e^{-D},$$

and from this they were able to derive many of their other results. We now give a simple geometric proof of this so-called *Bolyai–Lobachevsky Formula*. Greenberg (2008) has called this "one of the most remarkable formulas in the whole of mathematics", but for us it will be of only incidental interest.

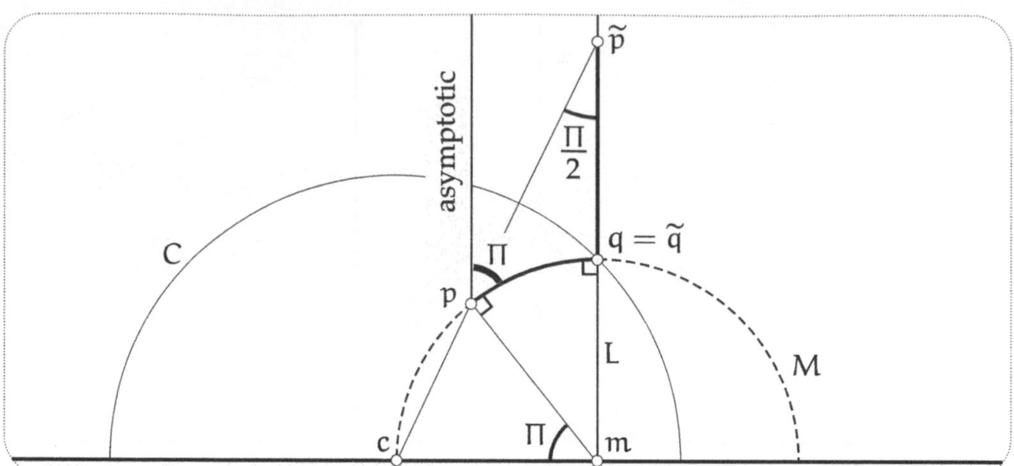

[6.26] Geometric proof of the Bolyai–Lobachevsky Formula. *To find the h-length* D *of the arc* pq*, apply* $z \mapsto \widetilde{z} = \mathfrak{R}_C(z)$. *This carries the arc* pq *into the illustrated vertical line-segment* $\widetilde{p}q$. *Applying (6.35),* $D = \left|\ln\left(\frac{[qm]}{[\widetilde{p}m]}\right)\right| = \left|\ln\left(\frac{[cm]}{[\widetilde{p}m]}\right)\right| = |\ln\tan(\Pi/2)| = -\ln\tan(\Pi/2)$, *the last equality following from the fact that* Π *is acute. Thus,* $\tan(\Pi/2) = e^{-D}$, *as was to be shown.*

First note that it is sufficient to establish the formula using [6.25], rather than [6.24]. This is because we may transform [6.24] into [6.25] by performing an inversion (i.e., a hyperbolic reflection) in any semicircle centred at one of the ends of L.

Figure [6.26] reproduces the essential elements of [6.25]. In order to find the h-length D of the arc pq, let us apply the h-reflection $z \mapsto \widetilde{z} = \mathfrak{R}_C(z)$, where C is the illustrated semicircle that is centred at the end c of M, and that passes through q. This carries the arc pq into the illustrated vertical line-segment $\widetilde{p}q$. By virtue of (6.35), it only remains to find the ratio of the y-coordinates of q and $\widetilde{p}$, i.e., the ratio of the Euclidean distances [qm] and $[\widetilde{p}m]$.

From the fact that the radius pm is orthogonal to the circle M it follows [exercise] that the angle pmc equals Π. It then follows [exercise] that the angle $c\widetilde{p}m$ equals (Π/2), as illustrated. Thus

$$D = \left|\ln\left(\frac{[qm]}{[\widetilde{p}m]}\right)\right| = \left|\ln\left(\frac{[cm]}{[\widetilde{p}m]}\right)\right| = |\ln\tan(\Pi/2)| = -\ln\tan(\Pi/2),$$

where the last equality follows from the fact that $\tan(\Pi/2) < 1$, because Π is acute. Thus $\tan(\Pi/2) = e^{-D}$, as was to be shown.

6.3.7 The Three Types of Direct Motion

As we have pointed out, the "Poincaré upper half-plane" was first discovered by Beltrami, and in this *25th Anniversary Edition* we are biting the bullet (as we already did in VDGF) calling it instead the *Beltrami–Poincaré upper half-plane*. What Poincaré *does* deserve sole credit for—enormous credit!—is the realization that hyperbolic geometry is intimately connected with complex analysis. The cornerstone of this connection is the fact that the (direct) motions of the hyperbolic plane are Möbius transformations. Let us outline how this comes about.

If L_1 and L_2 are two h-lines, then the composition

$$\mathcal{M} \equiv \mathfrak{R}_{L_2} \circ \mathfrak{R}_{L_1}$$

of h-reflection in these lines will be a direct motion of the hyperbolic plane. Since every h-reflection is represented in the map by inversion in a circle, we immediately deduce that any direct motion of the form $\mathcal{M}$ is represented by a (non-loxodromic) Möbius transformation $M(z)$. Furthermore, later we will show that *every* direct motion is of the form $\mathcal{M}$; indeed, we will even give an explicit geometric construction for decomposing an arbitrary direct motion into two h-reflections. Supposing this already done, we see that *every direct motion is represented as a (non-loxodromic) Möbius transformation.*

Conversely, suppose that $M(z)$ is an *arbitrary* Möbius transformation that maps the upper half-plane to itself. Then it follows that $M(z)$ must map the real axis (the horizon) into itself. But a loxodromic Möbius transformation cannot possess such an invariant line: its strangely shaped invariant curves were illustrated in [3.32] on p. 188. Thus $M(z)$ is non-loxodromic, and from (3.49), p. 198, we deduce that $M(z)$ is the composition of inversion in two circles orthogonal to the real axis. Thus *the most general Möbius transformation of the upper half-plane to itself represents a direct hyperbolic motion of the type $\mathcal{M}$ above.*

One way to discover the algebraic form of these Möbius transformations is to use the formula (3.4), p. 141: inversion in a circle K centred at the point q on the real axis, and of radius R, is given by

$$\mathcal{I}_K(z) = \frac{q\,\overline{z} + (R^2 - q^2)}{\overline{z} - q}.$$

Composing two such functions, we find [exercise] that a motion of type $\mathcal{M}$ corresponds to a Möbius transformation

$$M(z) = \frac{az + b}{cz + d}, \quad \text{where } a, b, c, d \text{ are real, and } (ad - bc) > 0. \tag{6.39}$$

Recall that in Ex. 25, p. 212, you showed that this is the form of the most general Möbius transformation of the upper half-plane to itself. Thus we have agreement with the conclusion of the previous paragraph.

So much for the overview—now let's look in detail at the direct motions $\mathcal{M}$. We know from [6.24] or [6.25] that there are just three possible configurations for the h-lines L_1 and L_2, and correspondingly $\mathcal{M} \equiv \mathfrak{R}_{L_2} \circ \mathfrak{R}_{L_2}$ is one of three fundamentally different types:

(i) If the h-lines *intersect*, then $\mathcal{M}$ is called a *hyperbolic rotation*.

(ii) If the h-lines are *asymptotic*, then $\mathcal{M}$ is a new kind of motion (peculiar to hyperbolic geometry) called a *limit rotation*.

(iii) If the h-lines are *ultra-parallel*, then $\mathcal{M}$ is called a *hyperbolic translation*.

We can now reap the rewards of all our hard work in Chapter 3, for these three types of motion are just the three types of non-loxodromic Möbius transformation: (i) h-rotations are the "elliptic" ones; (ii) limit rotations are the "parabolic" ones; and (iii) h-translations are the "hyperbolic"[20] ones. At this point, you might find it helpful to reread the discussion of these Möbius transformations at the end of Chapter 3.

We already understand these Möbius transformations, so it only remains to look at them afresh, through hyperbolic spectacles. That is, imagine that you belong to the race of *Beltrami–Poincarites*—tiny, intelligent, two-dimensional beings who inhabit the hyperbolic plane. To you and your fellow Beltrami–Poincarites, h-lines *really are* straight lines, the real axis *really is* infinitely far away, etc. What will you see if the above motions are applied to your world?

Let us begin with h-rotations. Figure [6.27] illustrates the elliptic Möbius transformation—let's call it $\mathcal{R}_a^\phi$—that arises in the case where the h-lines intersect at a, and the angle from L_1 to L_2 is $(\phi/2)$. [We have chosen to illustrate $\phi = (\pi/3)$.] Thus $\mathcal{R}_a^\phi$ has fixed points a and $\overline{a}$, and the multiplier associated with a is $m = e^{i\phi}$. As in

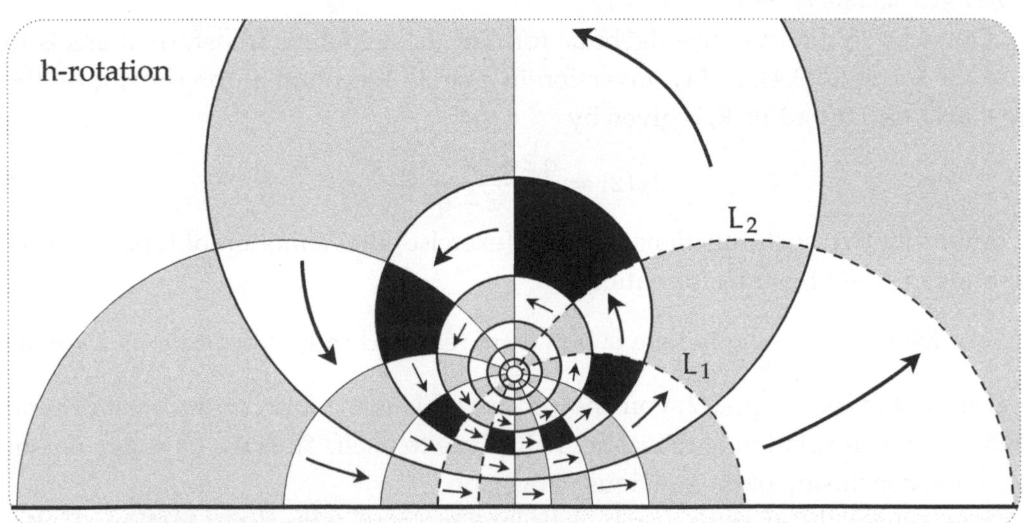

[6.27] A Hyperbolic Rotation, *results from reflecting across intersecting h-lines,* L_1 *and* L_2.

20 Try not to be confused by this unrelated use of the word "hyperbolic".

Chapter 3, each shaded "rectangle" is mapped by $\mathcal{R}_a^\phi$ to the next one in the direction of the arrows—some of these regions have been filled with black to emphasize this.

Consider how all this looks to you and your fellow Beltrami–Poincarites. For example, *you* see each black "rectangle" as being exactly the same shape and size as every other. To understand $\mathcal{R}_a^\phi$ better, we begin by noting that (in terms of the map) its effect on an infinitesimal neighbourhood of a is just a Euclidean rotation of ϕ about a. But since the map is conformal, this implies that a Poincarite standing at a will *also* see his immediate neighbourhood undergoing a rotation of ϕ.

More remarkably, however, the Poincarite at a will see the *entire* hyperbolic plane undergoing a perfect rotation of ϕ. Every h-line segment ap he constructs emanating from a is transformed by $z \mapsto \widetilde{z} = \mathcal{R}_a^\phi(z)$ into another h-line segment $a\widetilde{p}$ of equal length, making angle ϕ with the original. If the Poincarite gradually increases ϕ from 0 to 2π, then he sees $\widetilde{p}$ tracing out an *h-circle* centred at a, while in the map we see $\widetilde{p}$ miraculously tracing out a *Euclidean circle!* Thus the illustrated Euclidean circles orthogonal to the h-lines through a are all genuine hyperbolic circles, and a is their common h-centre. Let us record this remarkable result, adding a detail that is not too hard to prove [exercise]:

> *Every h-circle is represented in the map by a Euclidean circle, and its h-centre is the intersection of any two h-lines orthogonal to it. Algebraically, the h-circle with h-centre* $a = (x + iy)$ *and h-radius* ρ *is represented by the Euclidean circle with centre* $(x + iy\cosh\rho)$ *and radius* $y\sinh\rho$.

As a stepping stone to the limit rotations, [6.28] introduces a new type of curve in the hyperbolic plane. On a line L in Euclidean geometry, let p be a fixed point, let a be a moveable point, and let C be the circle centred at a that passes through p. If we let a recede to infinity along L, then the limiting form of C is a *line* (through p and perpendicular to L). Figure [6.28a] shows that it's a different story in the hyperbolic plane. As a recedes towards the infinitely remote point A on the real axis, the limiting form of C is a (Euclidean) circle that touches the real axis at A. This is neither an ordinary h-circle, nor an h-line: it is a new type of curve called a *horocycle*. Figure [6.28b] shows that horizontal (Euclidean) lines are also horocycles. Note that if K is any circle centred at A then the h-reflection $\mathfrak{R}_K = \mathfrak{I}_K$ transforms [6.28a] into [6.28b]. Thus the Beltrami–Poincarites cannot distinguish between these two types of horocycle.

Now consider [6.29], which illustrates the parabolic Möbius transformation that results from h-reflection in h-lines L_1 and L_2 that are asymptotic at A. Referring to [6.27] and [6.28], you can now understand why this is called a limit rotation: it may be viewed as the limit of the h-rotation $\mathcal{R}_a^\phi$ as a tends to the point A on the horizon. Note some of the interesting features of this picture: the invariant curves are horocycles touching at A; each such horocycle is orthogonal to every h-line that

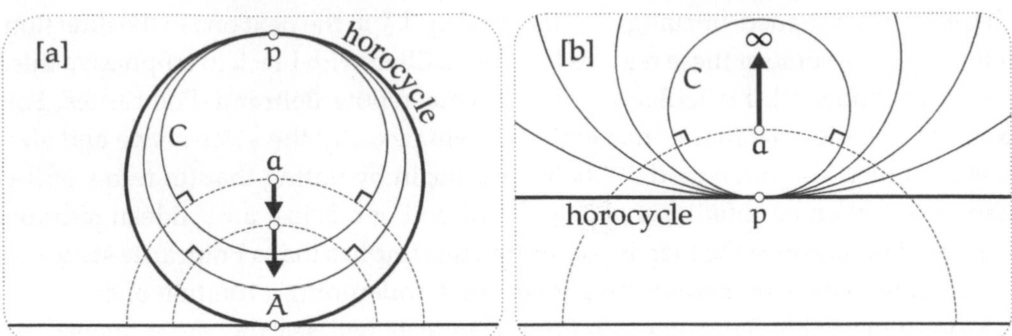

[6.28] Horocycles. [a] *Remarkably, an h-circle* C *appears in the map as an ordinary Euclidean circle, but its centre* $\mathfrak{a}$ *is the intersection of any two h-lines that meet* C *orthogonally. Keeping* $\mathfrak{p}$ *fixed, let us keep increasing the h-radius of* C *indefinitely. The centre* $\mathfrak{a}$ *gradually moves toward the point* A *on the horizon, and the circle ultimately becomes the illustrated* horocycle. **[b]** *If we instead let the centre* $\mathfrak{a}$ *move upward indefinitely, the limiting horocycle takes the form of a horizontal line. If* K *is any circle centred at* A *then the h-reflection* $\mathfrak{R}_K = \mathfrak{I}_K$ *transforms the first kind of horocycle in* **[a]** *into this new one, so inhabitants of the hyperbolic plane cannot tell them apart!*

ends at A; and any two such horocycles cut off the same h-length on every h-line that ends at A.

In terms of the map, the simplest limit rotation occurs when the asymptotic h-lines L_1 and L_2 are represented as vertical Euclidean half-lines, say separated by Euclidean distance $(\alpha/2)$. In this case, $\mathcal{M} = (\mathfrak{R}_{L_2} \circ \mathfrak{R}_{L_1})$ is represented in the map by the composition of two Euclidean reflections in parallel lines. Thus $\mathcal{M}$ is just a Euclidean translation $z \mapsto (z + \alpha)$ of the upper half-plane, and the invariant curves are horizontal lines, which are again horocycles, but now of the form shown in [6.28b]. Note that this Euclidean translation is *not* an h-translation. This is particularly clear if we visualize the effect of $\mathcal{M}$ on the pseudosphere, where it becomes a rotation through angle α about the pseudosphere's axis.

Figure [6.30] illustrates the third and final type of motion, the h-translation (hyperbolic Möbius transformation) resulting from h-reflection in two ultra-parallel h-lines. First note that there is precisely one h-line L that is orthogonal to both L_1 and L_2. Unlike a Euclidean translation, this h-line L is the *only* h-line that is mapped into itself; it is called the *axis* of the h-translation. Despite this difference, the name "h-translation" is appropriate, for every point on the h-line L is moved the same h-distance (say δ) along L. If we assume that the axis L has a direction assigned to it, then we may unambiguously denote this h-translation by $\mathcal{T}_L^\delta$.

In Euclidean geometry, the invariant curves of a translation are the parallel lines in the direction of the translation. However, [6.30] shows that the invariant curves of $\mathcal{T}_L^\delta$ are not h-lines, but rather arcs of Euclidean circles connecting the ends e_1 and

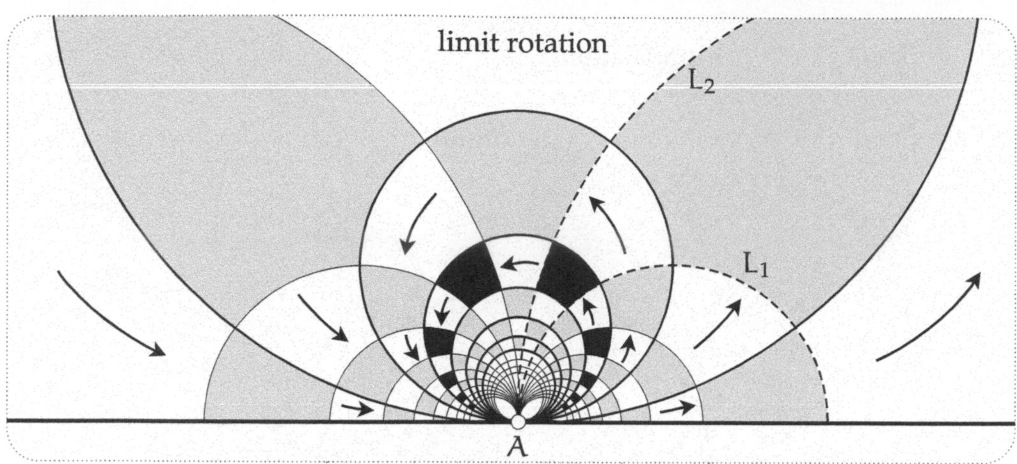

[6.29] A Hyperbolic Limit Rotation *results from reflecting across asymptotic h-lines,* L_1 *and* L_2*. This transformation has no analogue in Euclidean geometry, but it can be thought of as the limit of the h-rotation* $\mathcal{R}_a^{\phi}$ *as* a *tends to the point* A *on the horizon. Note that the invariant curves are horocycles.*

e_2 of L. These are called the *equidistant* curves of L, because every point on such a curve is the same h-distance from the h-line L. Make sure you can see this.

In terms of the map, the simplest h-translation occurs when the ultra-parallel h-lines L_1 and L_2 are represented by concentric Euclidean semicircles, say centred at the origin for convenience. In this case, the two h-reflections (i.e., inversions) yield a central dilation $z \mapsto kz$, where k is the real expansion factor. The axis of this h-translation is the vertical line through the origin (the y-axis), and the equidistant curves are all other (Euclidean) lines through the origin (cf. [6.20] and [6.21]). Note that this Euclidean expansion is a similarity transformation of the map, but it is *not* a similarity transformation of the hyperbolic plane—there are none!

Having completed our survey of these three types of direct motion, it's important to note that they not only look very different in terms of their effect on the map, but they also have unique fingerprints in terms of the intrinsic hyperbolic geometry. To put this another way, *Beltrami–Poincarites can tell these motions apart*. For example, of the three, only h-rotations have invariant h-circles, and only h-translations have an invariant h-line.

6.3.8 Decomposing an Arbitrary Direct Motion into Two Reflections

We will now show that the h-rotations, limit rotations, and h-translations are the *only* direct motions of the hyperbolic plane. That is, an *arbitrary* direct motion $\mathcal{M}$ can always be decomposed into two h-reflections: $\mathcal{M} \equiv (\mathfrak{R}_{L_2} \circ \mathfrak{R}_{L_1})$.

The first step is a familiar lemma: *an arbitrary hyperbolic motion* $\mathcal{M}$ *(not necessarily direct) is uniquely determined by its effect on any three non-collinear points*. As in

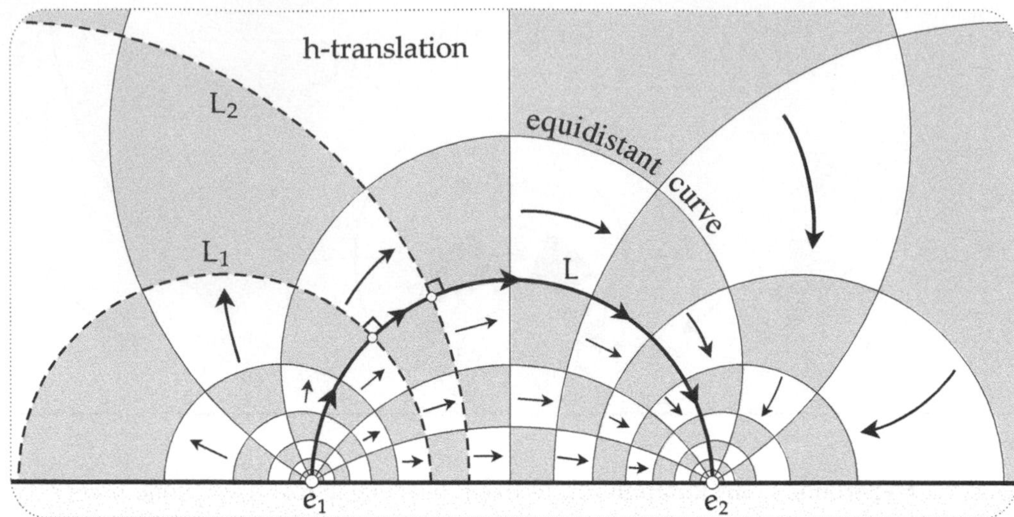

[6.30] A Hyperbolic Translation *results from reflecting across non-intersecting (ultra-parallel) h-lines,* L_1 *and* L_2*. Unlike Euclidean geometry, there is only* one *h-line* L *orthogonal to* L_1 *and* L_2*, and it is the* only *invariant h-line. The other invariant curves are arcs of Euclidean circles connecting the ends* e_1 *and* e_2*, and they are* not *h-lines. They are called* equidistant curves*, because every point on such a curve is the same h-distance from the h-line* L.

Euclidean geometry, this will be established if we can show that the location of a point p is uniquely determined by its h-distances from any three non-collinear points a, b, c. Consider [6.31a], in which we have supposed (for simplicity's sake only) that the h-line L through a and b is represented by a vertical line in the map. Through the point p, draw h-circles centred at a, b, and c. Since c does not lie on L (by assumption), we see that p is the *only* point at which the three circles intersect. Done.

Now suppose that an arbitrary motion carries two points a and b to the points a' and b' in [6.31b]. By the above result, the motion will be determined once we know the image of any third point p not on the line L through a and b. Drawing the illustrated h-circles with h-centres a' and b' and with h-radii $\mathcal{H}\{a, p\}$ and $\mathcal{H}\{b, p\}$, we see that the two intersection points p' and $\widetilde{p}$ are the only possible images for p. Furthermore, since the h-line L' through a' and b' is necessarily orthogonal to the h-circles centred at those points, we also see that p' and $\widetilde{p}$ are symmetric with respect to L', i.e., $\widetilde{p} = \mathcal{I}_{L'}(p') = \mathfrak{R}_{L'}(p')$. Thus we have shown that

> *There is exactly one direct motion* $\mathcal{M}$ *(and exactly one opposite motion* $\widetilde{\mathcal{M}}$*) that maps a given h-line segment* ab *to another h-line segment* $a'b'$ *of equal h-length. Furthermore,* $\widetilde{\mathcal{M}} = (\mathfrak{R}_{L'} \circ \mathcal{M})$*, where* L' *is the h-line through* a' *and* b'*.* (6.40)

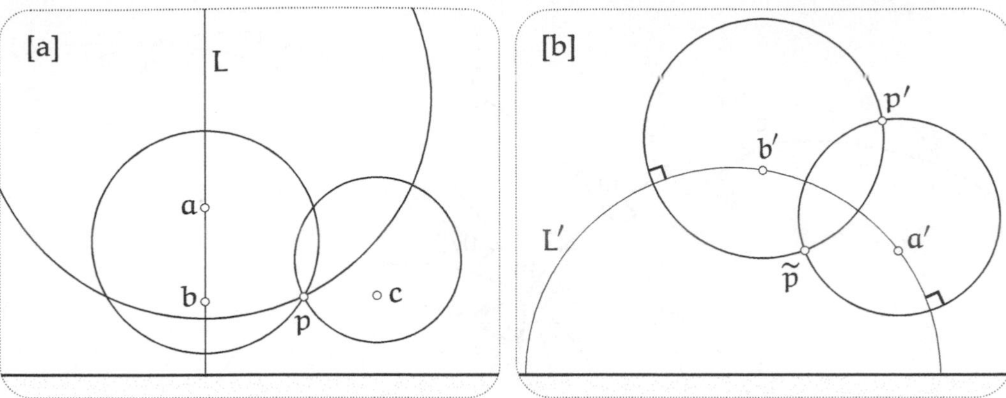

[6.31] An arbitrary hyperbolic motion $\mathcal{M}$ is uniquely determined by its effect on any three non-collinear points. **[a]** *The location of a point* p *is uniquely determined by its h-distances from any three non-collinear points* a, b, c. **[b]** *Let* $\mathcal{M}$ *be an arbitrary motion, and let* $a' = \mathcal{M}(a)$ *and* $b' = \mathcal{M}(b)$. *Then* $\mathcal{M}$ *will be determined once we know the image of any third point* p *not on the line* L *through* a *and* b. *Drawing the illustrated h-circles with h-centres* a' *and* b' *and with h-radii* $\mathcal{H}\{a,p\}$ *and* $\mathcal{H}\{b,p\}$, *we see that the two intersection points* p' *and* $\widetilde{p}$ *are the only possible images for* p, *and that these two possibilities are simply reflections of each other in* L'. *This completes the proof of (6.40).*

We shall now give an explicit geometric construction for decomposing an arbitrary direct motion $\mathcal{M}$ into two h-reflections. First note that (6.40) implies that $\mathcal{M}$ is determined by its effect on any two points, *no matter how close together they are*. Though it is not essential, the following construction is particularly clear if we choose the points to be infinitesimally separated.

Let us therefore take the two given points to be z and $(z + dz)$, and their given images under $\mathcal{M}$ to be $w = \mathcal{M}(z)$ and $(w + dw) = \mathcal{M}(z + dz)$. Figure [6.32] illustrates this idea. Our task is to find two h-reflections that will simultaneously carry z to w, and dz to dw. [Incidentally, since $\mathcal{M}$ must be conformal, it can be thought of as an analytic function, so we may write $dw = \mathcal{M}'(z)\,dz$.]

First, carry z to w using the h-translation $\mathcal{T}_L^{\delta}$, where $\delta = \mathcal{H}\{z, w\}$, and L is the unique h-line from z to w. Note that since $\mathcal{T}_L^{\delta}$ is conformal, it carries dz to an infinitesimal vector $d\widetilde{z}$ (of equal h-length) making the same angle with L as dz. Next, apply the h-rotation $\mathcal{R}_w^{\theta}$, where θ is the angle from $d\widetilde{z}$ to dw. This leaves w where it is, and it rotates $d\widetilde{z}$ to dw. Since the net transformation carries z to w, and dz to dw, it must be $\mathcal{M}$:

$$\mathcal{M} = \mathcal{R}_w^{\theta} \circ \mathcal{T}_L^{\delta}.$$

Implicitly, this formula decomposes $\mathcal{M}$ into four h-reflections, because $\mathcal{T}_L^{\delta}$ and $\mathcal{R}_w^{\theta}$ can both be decomposed into two h-reflections. However, [6.32] illustrates that we can always arrange for two of the four h-reflections to cancel. Defining m to

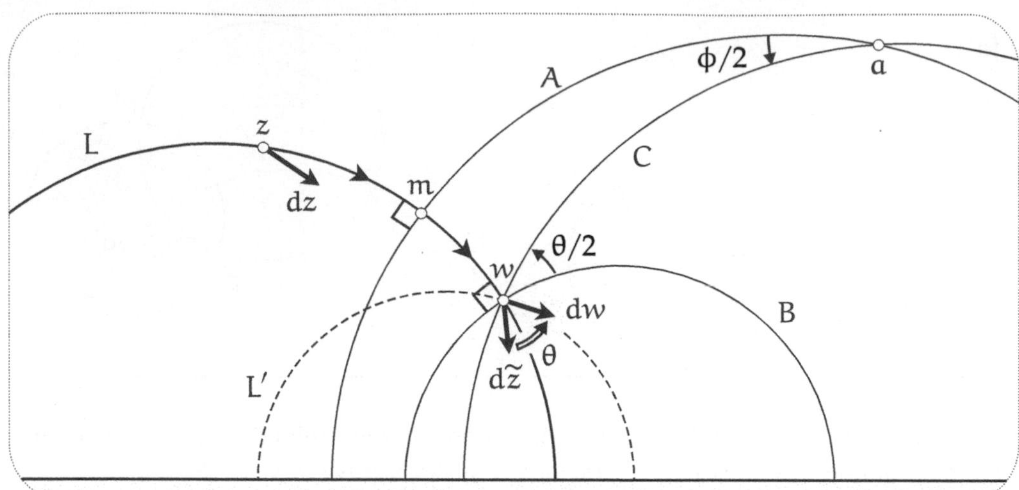

[6.32] Decomposing an arbitrary direct motion $\mathcal{M}$ into two h-reflections. *Imagine two points close together, z and z + dz, being mapped to their images w and w + dw. First carry z to w with the unique h-translation $\mathcal{T}_L^\delta$ (where $\delta = \mathcal{H}\{z, w\}$) along the h-line from z to w. This carries dz conformally to $d\tilde{z}$, still making the same angle with L. Now rotate $d\tilde{z}$ about w by θ to obtain dw. Thus, $\mathcal{M} = \mathcal{R}_w^\theta \circ \mathcal{T}_L^\delta$. This implicitly decomposes $\mathcal{M}$ into four reflections, but we can always arrange for the middle two to cancel. Here, $\mathcal{M} = (\mathfrak{R}_C \circ \mathfrak{R}_B) \circ (\mathfrak{R}_B \circ \mathfrak{R}_A) = \mathfrak{R}_C \circ \mathfrak{R}_A = \mathcal{R}_a^\phi$, but clearly this construction may just as easily yield an A and a C that are asymptotic or ultra-parallel, in which case $\mathcal{M}$ is a limit rotation or an h-translation.*

be the h-midpoint of the h-line segment zw, draw h-lines A and B orthogonal to L and passing through m and w, respectively. Then $\mathcal{T}_L^\delta = (\mathfrak{R}_B \circ \mathfrak{R}_A)$. If we now draw an h-line C through w making angle (θ/2) with B, then $\mathcal{R}_w^\theta = (\mathfrak{R}_C \circ \mathfrak{R}_B)$. Thus, as we set out to show, every direct motion can be decomposed into two h-reflections:

$$\mathcal{M} = (\mathfrak{R}_C \circ \mathfrak{R}_B) \circ (\mathfrak{R}_B \circ \mathfrak{R}_A) = \mathfrak{R}_C \circ \mathfrak{R}_A.$$

In the illustrated example, it so happens that the h-lines A and C intersect, and so the motion is an h-rotation: $\mathcal{M} = \mathcal{R}_a^\phi$, where a is the intersection of A and C, and (ϕ/2) is the angle between them. However, it is clear that this construction may just as easily yield an A and a C that are asymptotic or ultra-parallel, in which case $\mathcal{M}$ is a limit rotation or an h-translation.

Summarizing what we have shown, and recalling (6.39),

> *Every direct motion of the hyperbolic plane is the composition of two h-reflections, and is thus an h-rotation, a limit rotation, or an h-translation. In the Beltrami–Poincaré upper half-plane, all such motions are presented by Möbius transformations of the form*

$$M(z) = \frac{az+b}{cz+d}, \quad \textit{where } a, b, c, d \textit{ are real, and } (ad - bc) > 0.$$

Finally, returning to [6.32] and appealing to (6.40), the unique opposite motion $\widetilde{\mathcal{M}}$ carrying z to w and dz to dw is given by three h-reflections:

$$\widetilde{\mathcal{M}} = \mathfrak{R}_{L'} \circ \mathfrak{R}_C \circ \mathfrak{R}_A.$$

Here L' is the illustrated h-line passing through w and $(w + dw)$, i.e., passing through w in the direction dw. This decomposition does not, however, yield the simplest geometric interpretation of $\widetilde{\mathcal{M}}$; for that, and for the formula describing the general opposite motion, see Ex. 24.

6.3.9 *The Angular Excess of a Hyperbolic Triangle*

Joining three points in the hyperbolic plane with h-line segments yields (by definition) a *hyperbolic triangle*. Our objective will be to show that the angular excess $\mathcal{E}(T)$ of such a hyperbolic triangle T is given by

$$\mathcal{E}(T) = (-1)\mathcal{A}(T). \tag{6.41}$$

As we pointed out in the Introduction, this says (amongst other things) that the angles of Δ always add up to less than π, and that no matter how large we make Δ, its area can never exceed π. Referring to the differential geometry result (6.6), we also see that in establishing this formula we will have provided an intrinsic[21] proof of the fact that the hyperbolic plane is a surface of constant negative curvature $\mathcal{K} = -1$.

We have already remarked that Christiaan Huygens investigated the pseudosphere as early as 1693, and to get acquainted with hyperbolic area we will now confirm one of his surprising results: *the pseudosphere has finite area*. In the upper half-plane the pseudosphere is represented by the shaded region $\{0 \leqslant x < 2\pi, y \geqslant 1\}$ shown in [6.20], and (6.32) implies that this region of infinite Euclidean area does indeed have finite hyperbolic area:

$$\mathcal{A}(\text{pseudosphere}) = \iint d\mathcal{A} = \int_{x=0}^{2\pi}\int_{y=1}^{\infty} \frac{dx\,dy}{y^2} = \int_{x=0}^{2\pi} dx \int_{y=1}^{\infty} \frac{dy}{y^2} = 2\pi,$$

as Huygens discovered.

Figure [6.33a] illustrates a triangle on the pseudosphere. If the uppermost vertex moves up the pseudosphere indefinitely, then the angle at that vertex tends to zero, and the edges meeting at that vertex tend to asymptotic lines, namely, tractrix generators meeting at infinity. Such a limiting triangle, two of whose edges are asymptotic, is called an *asymptotic triangle*. In order to establish (6.41) for ordinary triangles, we first establish it for asymptotic triangles. Figure [6.33b] illustrates

[21] Recall that earlier we used the pseudosphere to give an *extrinsic* proof.

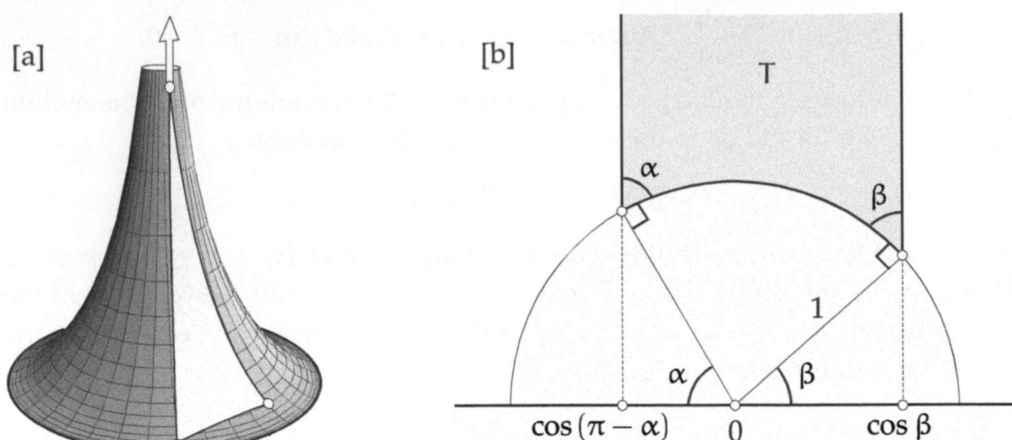

[6.33] [a] An asymptotic triangle *results when the top vertex moves up the pseudosphere indefinitely.* **[b] The angular excess of an asymptotic triangle.** *Suppose the finite edge of the asymptotic triangle* T *is an arc of the unit circle. A simple calculation [see text] then shows that* $\mathcal{A}(T) = \pi - \alpha - \beta$. *Taking the third angle of* T *to be zero, this indeed accords with* $\mathcal{E}(T) = (-1)\mathcal{A}(T)$.

such a triangle T in the upper half-plane, the asymptotic tractrix generators becoming vertical half-lines. By Huygens' result, T clearly has a finite area $\mathcal{A}(T)$, and because the asymptotic edges meet at angle zero, the result we wish to establish is $\mathcal{A}(T) = (\pi - \alpha - \beta)$.

To simplify the derivation of this result, [6.33b] supposes that the finite edge of T is an arc of the unit circle. This does not involve any loss of generality, because an arc of a circle of radius r centred at $x = X$ may be transformed into an arc of the unit circle by applying the limit rotation $z \mapsto (z - X)$, followed by the h-translation $z \mapsto (z/r)$. From [6.33b] we now deduce that

$$\mathcal{A}(T) = \int_{x=\cos(\pi-\alpha)}^{\cos\beta} \left[\int_{y=\sqrt{1-x^2}}^{\infty} \frac{dy}{y^2}\right] dx = \int_{x=\cos(\pi-\alpha)}^{\cos\beta} \frac{dx}{\sqrt{1-x^2}},$$

and writing $x = \cos\theta$ then yields the desired result:

$$\mathcal{A}(T) = \int_{\pi-\alpha}^{\beta} \frac{-\sin\theta\, d\theta}{\sin\theta} = \pi - \alpha - \beta.$$

On the left of [6.34] is a general triangle, say of area $\mathcal{A}$. By applying a suitable h-rotation about one of the vertices, we can bring one of the edges into a vertical position, as illustrated on the right of [6.34]. This makes it clear that the area $\mathcal{A}$ of the triangle may be viewed as the difference of the areas of two asymptotic triangles: one with angles α and $(\beta+\theta)$; the other with angles $(\pi-\gamma)$ and θ. Finally, applying the above result for asymptotic triangles, we deduce (6.41):

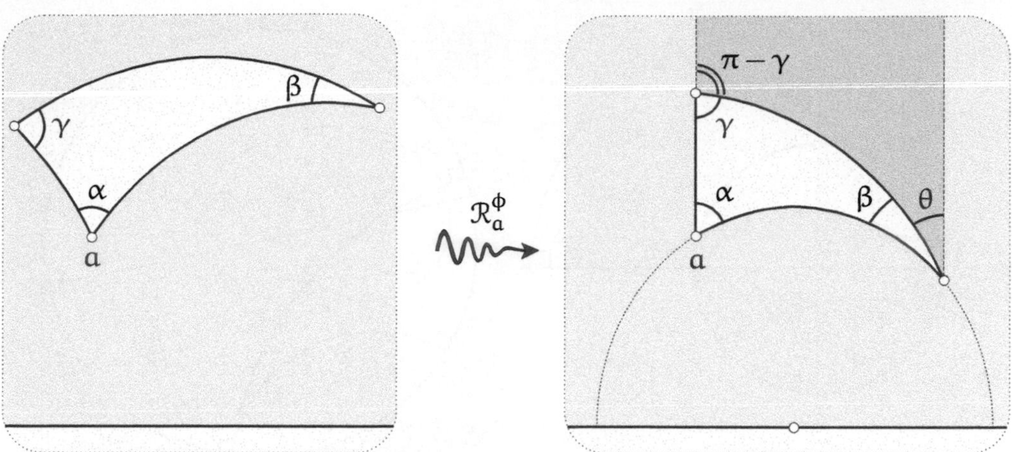

[6.34] The angular excess of a general hyperbolic triangle. [a] *A general hyperbolic triangle of area $\mathcal{A}$ can be rotated so that one edge becomes vertical.* **[b]** *But now $\mathcal{A}$ is clearly the difference of the areas of two asymptotic triangles: one with angles α and $(\beta+\theta)$; the other with angles $(\pi-\gamma)$ and θ. It follows immediately that $\mathcal{E}(T) = (-1)\mathcal{A}(T)$, thereby confirming that the hyperbolic plane has constant negative Gaussian curvature $\mathcal{K} = -1$.*

$$\begin{aligned}\mathcal{A} &= [\pi - \alpha - (\beta+\theta)] - [\pi - (\pi-\gamma) - \theta] \\ &= \pi - \alpha - \beta - \gamma \\ &= -\mathcal{E}.\end{aligned}$$

6.3.10 The Beltrami–Poincaré Disc

In addition to the upper half-plane model, Beltrami (1868b) constructed another extremely useful conformal map of the hyperbolic plane, this time inside the unit disc. Fourteen years later Poincaré rediscovered this map, which is now universally (and unfairly) known as the *Poincaré disc*. Following the example we set ourselves in VDGF, let us now attempt to set the record straight by given both men *equal* credit, calling it the *Beltrami–Poincaré disc*.

Figure [6.35a] illustrates the first step of the construction, which is to map the entire upper half-plane into the unit disc by means of the inversion

$$z \mapsto \widetilde{z} = \mathcal{I}_K(z),$$

where K is the illustrated circle centred at $-i$ and passing through ± 1. In order for this disc to represent the hyperbolic plane, its metric must be inherited from the upper half-plane. That is, we must define the h-separation $\mathcal{H}\{\widetilde{a}, \widetilde{b}\}$ of two points in the disc to be the h-separation $\mathcal{H}\{a, b\}$ of their preimages in the upper half-plane. Note that this implies [exercise] that the h-lines of the disc are precisely the images of h-lines in the upper half-plane.

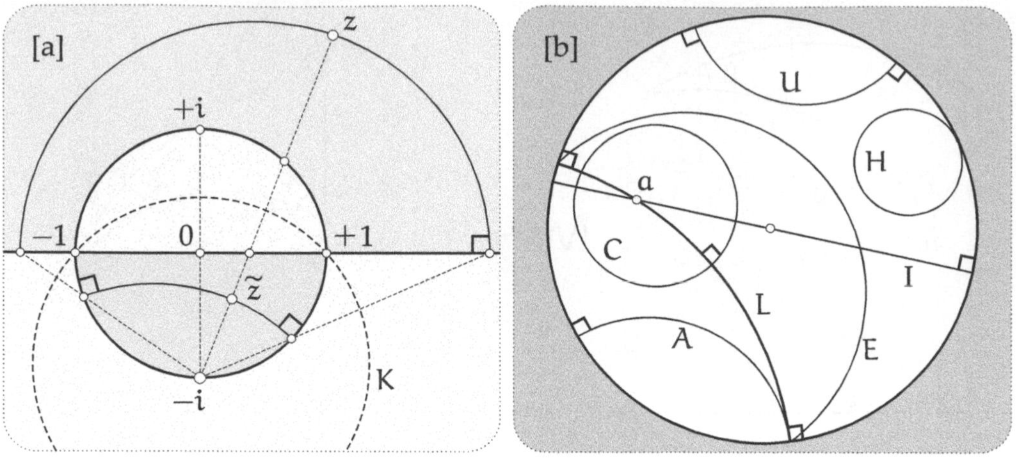

[6.35] Mapping the Beltrami–Poincaré Half-Plane to the Beltrami–Poincaré Disc. **[a]** *The anticonformal inversion* $z \mapsto \widetilde{z} = \mathcal{I}_K(z)$ *maps the upper half-plane to the interior of the unit disc. To retain the hyperbolic geometry, we define the h-separation* $\mathcal{H}\{\widetilde{a}, \widetilde{b}\}$ *of two points in the disc to be the h-separation* $\mathcal{H}\{a, b\}$ *of their preimages in the upper half-plane. H-lines now become arcs of circles meeting the unit circle at right angles, and the horizon is now the unit circle itself, also known as the* circle at infinity. **[b]** *Composing this inversion with conjugation, we recover a* conformal *mapping to the unit disc, which we call the* Beltrami–Poincaré disc model *of the hyperbolic plane, in which h-lines now take the form of arcs of circles orthogonal to the horizon, such as* L, A, U.

Before moving on, try staring at [6.35a] until the following details become clear: (i) ± 1 remain fixed and i is mapped to 0; (ii) the entire shaded part of the upper half-plane is mapped to the shaded bottom half of the unit disc; (iii) the remaining part of the upper half-plane (i.e., the top half of the unit disc) is mapped into itself; (iv) h-lines in the disc are the images of h-lines in the upper half-plane, and these are arcs of circles orthogonal to the unit circle; (v) the entire horizon of the hyperbolic plane is represented by the unit circle, with the common point at infinity of vertical h-lines in the upper half-plane being represented by $-i$.

At this point we have obtained a map of the hyperbolic plane within the unit disc. However, since $\mathcal{I}_K(z)$ is anticonformal, so is our map: angles in the upper half-plane are currently represented by equal but *opposite* angles in the disc. If we now apply $z \mapsto \overline{z}$, which reflects the disc across the real axis into itself, then angles are reversed a second time, and we obtain the conformal Beltrami–Poincaré disc.

The net transformation from the Beltrami–Poincaré upper half-plane to the Beltrami–Poincaré disc is thus the composition of $z \mapsto \mathcal{I}_K(z)$ and $z \mapsto \overline{z}$, and this is a Möbius transformation, say $D(z)$. Since $D(z)$ maps i to 0 and $-i$ to ∞, it is clear that $D(z)$ must be proportional to $(z - i)/(z + i)$. Finally, recalling that a Möbius transformation is uniquely determined by its effect on three points, and noting that ± 1 remain fixed, we deduce [exercise] that

$$D(z) = \frac{iz+1}{z+i}. \tag{6.42}$$

Alternatively, this may be derived by brute force [exercise] using the formula for inversion, (3.4), p. 141.

Since $D(z)$ preserves angles and circles, it is easy to transfer the basic types of curve in the hyperbolic plane from the Beltrami–Poincaré upper half-plane to the Beltrami–Poincaré disc. Figure [6.35b] illustrates that h-lines are represented by arcs of circles orthogonal to the unit circle (such as L, A, U), including diameters such as I. Incidentally, since the horizon is now represented by the unit circle, you can understand why the horizon is also called the *circle at infinity*.

The terminology for h-lines is the same as before: I intersects L, A is asymptotic to L, U is ultra-parallel to L, and a Euclidean circular arc E connecting the ends of L is an equidistant curve of L. It is also easy to see that a Euclidean circle C lying strictly inside the unit disc represents an h-circle, though its h-centre a does not generally coincide with its Euclidean centre. Finally, the horocycles in [6.28a] and [6.28b] are represented in the Beltrami–Poincaré disc by circles such as H that touch the unit circle.

Now let us find the metric in the Beltrami–Poincaré disc. Ex. 19 shows how this may be done by brute calculation, but the following geometric approach[22] is much more enlightening and powerful. First, [6.36a] recalls the earlier observation (6.33): if ds is the infinitesimal Euclidean length of a horizontal line-element emanating from z, then the angle between L and E is its hyperbolic length $d\hat{s} = [ds/\text{Im}(z)]$.

Note that in purely hyperbolic terms, L is an h-line orthogonal to ds, and E is an equidistant curve of L. If we apply an h-rotation $\mathcal{R}_z^{\phi}$ then L is carried into another h-line L′, and E is carried into an equidistant curve E′ of L′, and the angle between L′ and E′ is the same as before. Thus we have the following general construction:

Through one end of ds*, draw the h-line* l *orthogonal to* ds*, and through the other end of* ds *draw the equidistant curve* e*. Then the h-length* $d\hat{s}$ *of* ds *is the angle of intersection (on the horizon) of* l *and* e*.* (6.43)

Now the beauty of interpreting $d\hat{s}$ as an angle in this way is that the Möbius transformation D to the Beltrami–Poincaré disc is *conformal*, and so the above construction of $d\hat{s}$ is valid there too!

Figure [6.36b] illustrates an infinitesimal disc of Euclidean radius ds centred at $z = r\,e^{i\theta}$ in the Beltrami–Poincaré disc. Because the map is conformal, the h-length $d\hat{s}$ of ds is independent of the direction of ds, so we may simplify the construction (6.43) by choosing ds orthogonal to the diameter l through z. The equidistant curve e is then the illustrated arc of a Euclidean circle through the ends of l.

[22] We merely rediscovered this angular interpretation of hyperbolic distance, which we believe originates with Thurston (1997). However, our explanation (and our applications) of the idea differ somewhat from Thurston's.

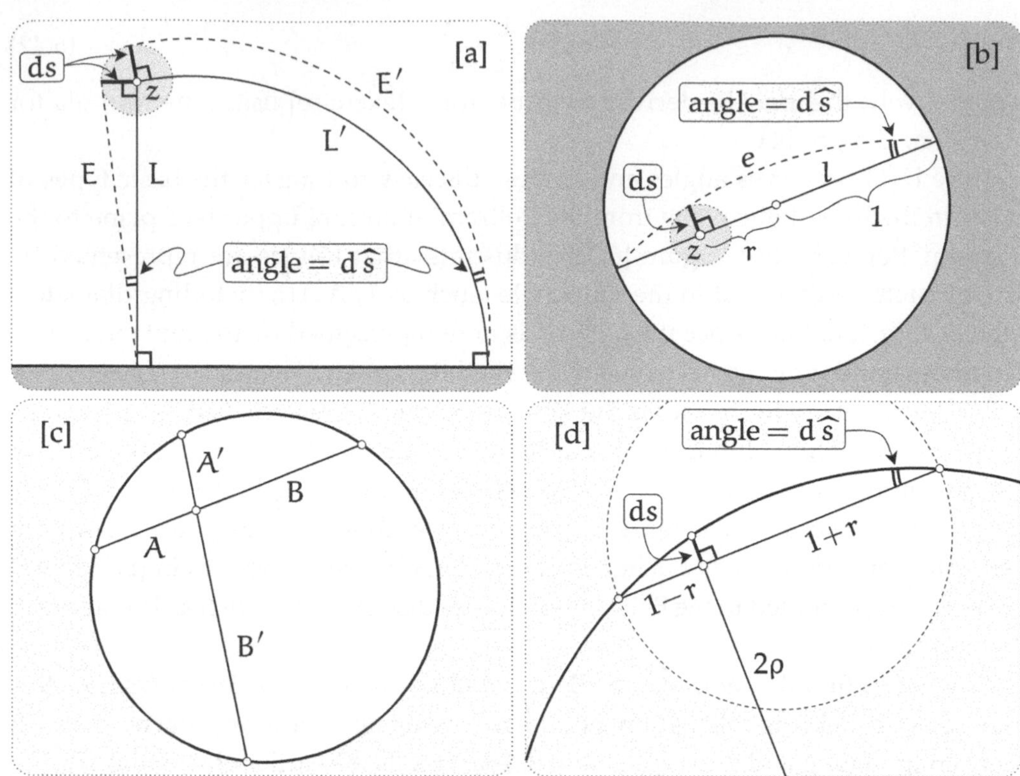

[6.36] Geometric derivation of the disc metric. [a] *If* ds *is the infinitesimal Euclidean length of a horizontal line-element emanating from* z*, then the angle between the orthogonal h-line* L *and the equidistant curve* E *is its hyperbolic length* $d\widehat{s} = [ds/\text{Im}(z)]$. *But performing a rotation about* z *we see that* $d\widehat{s}$ *may equally be measured as the angle between the h-line* L′ *and its equidistant curve* E′. **[b]** Since the mapping from the half-plane to the disc is conformal, this angular interpretation of hyperbolic distance applies here, too! *Choose* ds *orthogonal to the diameter* l *through* z. *The equidistant curve* e *is then the illustrated arc of a Euclidean circle of radius* ρ *through the ends of* l, *and* $\rho\, d\widehat{s} = 1$. **[c]** AB = A′B′. **[d]** *So,* $2\rho\, ds = (1-r)(1+r) = 1-|z|^2$, *and the hyperbolic disc metric is therefore* $d\widehat{s} = \frac{2}{1-|z|^2}\, ds$.

To turn this picture of $d\widehat{s}$ into a formula, begin by noting that if ρ is the radius of the circle containing the arc e, then [draw a picture!]

$$\rho\, d\widehat{s} = 1.$$

Next recall (or prove) the familiar property of circles illustrated in [6.36c], namely, that all chords passing through a fixed interior point are divided into two parts whose lengths have constant product: AB = A′B′. Applying this result to the copy of [6.36b] shown in [6.36d], we obtain [exercise]

$$2\rho\, ds = (1-r)(1+r) = 1-|z|^2.$$

Thus the metric of the Beltrami–Poincaré disc is

$$d\widehat{s} = \frac{2}{1-|z|^2}\, ds. \tag{6.44}$$

Note the remarkable similarity to the formula for the metric of the sphere under stereographic projection, (6.16)!

Since the Euclidean line-segment connecting 0 to z is also an h-line segment, we can now find the h-separation of these points by simple integration along the line-segment:

$$\mathcal{H}\{0, z\} = \int_0^{|z|} \frac{2\,dr}{1-r^2} = \int_0^{|z|} \left[\frac{1}{1+r} + \frac{1}{1-r}\right] dr,$$

and so

$$\mathcal{H}\{0, z\} = \ln\left(\frac{1+|z|}{1-|z|}\right). \tag{6.45}$$

As a simple check of this formula, note that as z moves toward the unit circle (the horizon), $\mathcal{H}\{0, z\}$ tends to infinity, as it should.

6.3.11 *Motions of the Beltrami–Poincaré Disc*

In the upper half-plane we found that every direct motion was the composition of two h-reflections, and every opposite motion was the composition of three h-reflections. Since the intrinsic geometry of the Beltrami–Poincaré disc is identical to the upper half-plane, this result must still be true, so it only remains to find out what h-reflection means in the disc. In the upper half-plane we saw that h-reflection in an h-line K meant geometric inversion in K, *and the same is true in the Beltrami–Poincaré disc!*

This is easy to understand. In the upper half-plane, q is the h-reflection of p in K means that p and q are symmetric (in the sense of inversion) in K. In order to make the Beltrami–Poincaré disc isometric to the upper half-plane, we insisted that the mapping $z \mapsto \widetilde{z} = D(z)$ preserve hyperbolic distance. In particular, $\widetilde{q}$ is the h-reflection of $\widetilde{p}$ in $\widetilde{K}$. But D(z) is a *Möbius transformation*, and so the Symmetry Principle [see p. 168] implies that $\widetilde{p}$ and $\widetilde{q}$ are symmetric in $\widetilde{K}$, as was to be shown.

Thus every direct motion $\mathcal{M}$ of the Beltrami–Poincaré disc has the form

$$\mathcal{M} = \mathfrak{R}_{L_2} \circ \mathfrak{R}_{L_1} = \mathcal{I}_{L_2} \circ \mathcal{I}_{L_1},$$

where L_1 and L_2 are h-lines, namely, arcs of circles orthogonal to the unit circle. As in the upper half-plane, every direct motion is therefore a non-loxodromic Möbius transformation. We already know that there are just three hyperbolically distinguishable types of direct motion, and the distinction between them in terms of L_1 and L_2 is the same as before: we get an h-rotation when they intersect, a limit rotation when they are asymptotic, and an h-translation when they are ultra-parallel.

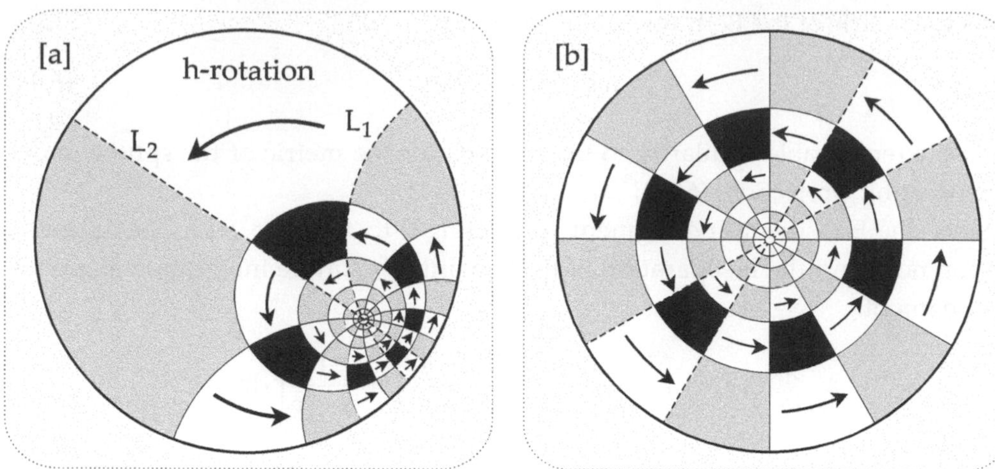

[6.37] A Hyperbolic Rotation *results from reflecting across intersecting h-lines,* L_1 *and* L_2*.* **[a]** *The typical case.* **[b]** *If the rotation is about the centre, then it reduces to an ordinary Euclidean rotation.*

We will discuss the formula for these Möbius transformations in a moment, but first let's draw pictures of them.

Figure [6.37a] shows a typical h-rotation; note the appearance of h-circles with a common h-centre. Figure [6.37b] illustrates the pleasant fact that if L_1 and L_2 intersect at the origin (in which case they are Euclidean diameters) then the resulting h-rotation manifests itself as a Euclidean rotation.

In this connection, we offer a word of warning. As Euclidean beings, we suffer from an almost overwhelming temptation to regard the centre of the Beltrami–Poincaré disc as being special in some way. One must therefore constantly remind oneself that to the Beltrami–Poincarites who inhabit the disc, every point is indistinguishable from every other point. In particular, the Beltrami–Poincarites do not see any difference between [6.37a] and [6.37b].

Figure [6.38a] illustrates a typical limit rotation generated by an L_1 and an L_2 that are asymptotic at a point A on the horizon. Once again note that the invariant curves are horocycles touching at A, and that these are orthogonal to the family of h-lines that are asymptotic at A.

Finally, [6.38b] illustrates a typical h-translation. Once again, note that there is precisely one invariant h-line [shown in bold], and that the invariant equidistant curves are arcs of circles through the ends of this axis.

From our work in the upper half-plane we know that the three types of motion pictured above are the *only* direct motions of the Beltrami–Poincaré disc, and we now turn to the formula that describes them. We know that every direct motion is a Möbius transformation that maps the unit disc into itself, and at the end of

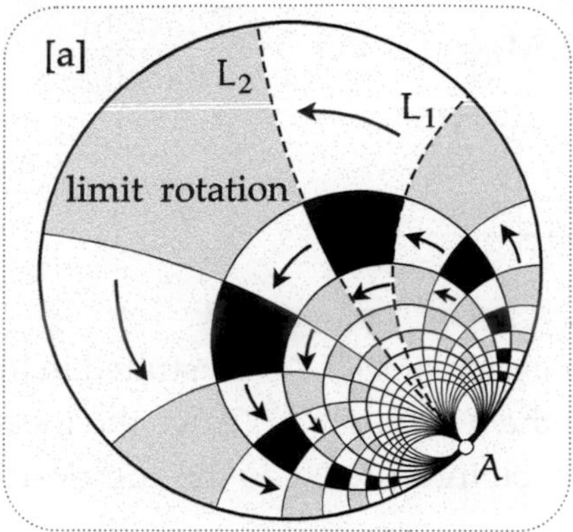

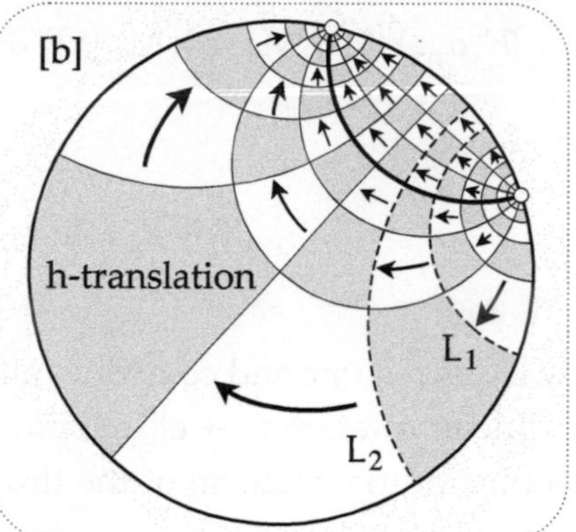

[6.38] [a] A Limit Rotation *results from reflecting across asymptotic h-lines,* L_1 *and* L_2. **[b] A Translation** *results from reflecting across ultra-parallel h-lines,* L_1 *and* L_2.

Chapter 3, with malice aforethought, we investigated these "Möbius automorphisms" of the unit disc. We found [see (3.52), p. 201] that the formula representing the most general one $M_a^\phi(z)$ is

$$M_a^\phi(z) = e^{i\phi} M_a(z), \quad \textit{where} \quad M_a(z) = \frac{z-a}{\overline{a}z-1}.$$

Thus M_a^ϕ is the composition of M_a and a rotation of ϕ about the origin.

Recall that M_a swaps a and 0: $M_a(a) = 0$ and $M_a(0) = a$. More generally, M_a swaps every pair of points z, $M_a(z)$: the transformation is *involutory*. This is explained by [6.39a], which recalls the result illustrated in [3.39], p. 203:

$$M_a = \mathcal{I}_B \circ \mathcal{I}_A,$$

where B is the diameter through a, and where A is the circle centred at $(1/\overline{a})$ that is orthogonal to the unit circle.

Hyperbolic geometry gives us a fresh perspective on this result: the intersection point m of A and B is the h-midpoint of 0 and a, and A itself is the perpendicular h-bisector of $0a$. Furthermore, the inversions in A and B are h-reflections. Thus M_a is the composition of two h-reflections in perpendicular h-lines through m, and so

> *The unique Möbius automorphism* M_a *that swaps* a *and* 0 *is the h-rotation* $\mathcal{R}_m^\pi$ *through angle* π *about the h-midpoint* m *of the h-line segment* $0a$.

An immediate benefit of this insight is that we can now easily find the formula for the h-separation of any two points, a and z. The h-rotation M_a brings a to the origin, and we already know the formula (6.45) for the h-distance of a point from there:

$$\mathcal{H}\{a, z\} = \mathcal{H}\{M_a(a), M_a(z)\} = \mathcal{H}\{0, M_a(z)\} = \ln\left(\frac{1 + |M_a(z)|}{1 - |M_a(z)|}\right),$$

and so

$$\mathcal{H}\{a, z\} = \ln\left(\frac{|\overline{a}\,z - 1| + |z - a|}{|\overline{a}\,z - 1| - |z - a|}\right). \tag{6.46}$$

Now let us resume and complete our discussion of M_a^ϕ. As illustrated in [6.37b], the Euclidean rotation $z \mapsto e^{i\phi}z$ represents the h-rotation $\mathcal{R}_0^\phi$. Thus the most general Möbius automorphism of the disc may be interpreted as the composition of two h-rotations:

$$M_a^\phi = \mathcal{R}_0^\phi \circ \mathcal{R}_m^\pi.$$

Figure [6.39b] shows how to compose these h-rotations, using the same idea as was used in both Euclidean and spherical geometry. The h-rotation $\mathcal{R}_0^\phi$ is the composition of h-reflections in any two h-lines through 0 (diameters) containing angle $(\phi/2)$. Thus, choosing the first h-line to be B, and calling the second h-line C, we deduce that

$$M_a^\phi = (\mathfrak{R}_C \circ \mathfrak{R}_B) \circ (\mathfrak{R}_B \circ \mathfrak{R}_A) = \mathfrak{R}_C \circ \mathfrak{R}_A.$$

Thus M_a^ϕ is an h-rotation, limit rotation, or h-translation according as A and C are intersecting, asymptotic, or ultra-parallel.

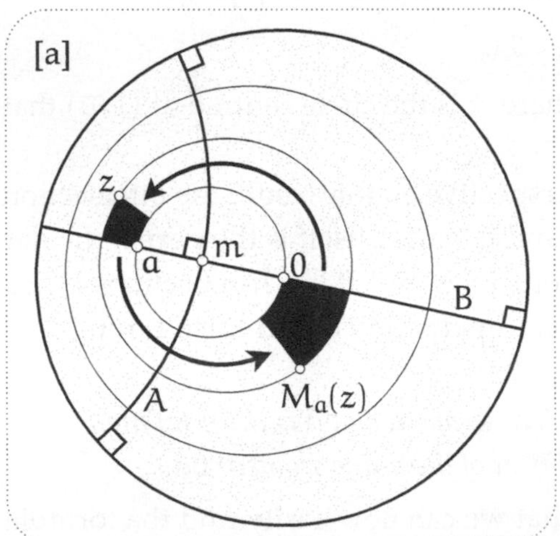

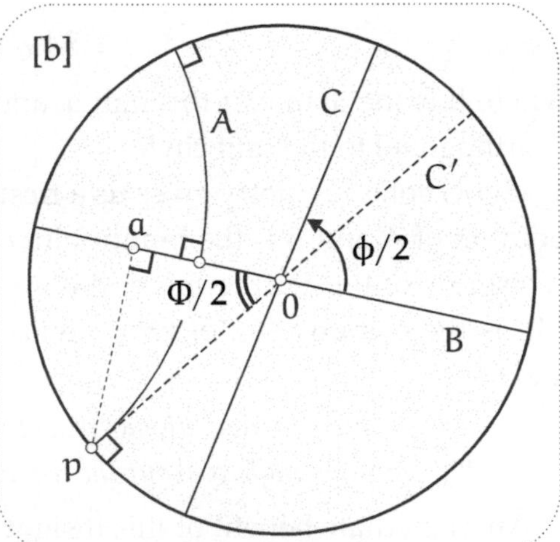

[6.39] Geometric interpretation and classification of the general Möbius automorphism, M_a^ϕ. **[a]** *The unique Möbius automorphism $M_a = \mathfrak{I}_B \circ \mathfrak{I}_A$ that swaps a and 0 is the h-rotation $\mathcal{R}_m^\pi$ through angle π about the h-midpoint m of the h-line segment $0a$.* **[b]** *The most general Möbius automorphism is $M_a^\phi = \mathcal{R}_0^\phi \circ \mathcal{R}_m^\pi = \mathfrak{I}_C \circ \mathfrak{I}_B \circ \mathfrak{I}_B \circ \mathfrak{I}_A = \mathfrak{I}_C \circ \mathfrak{I}_A$. Thus M_a^ϕ is an h-rotation, limit rotation, or h-translation according as A and C are intersecting, asymptotic, or ultra-parallel.*

Thinking of a as fixed and ϕ as variable, the critical value $\phi = \Phi$ separating the h-rotations from the h-translations occurs when C is in the position C′ [shown dashed] asymptotic to A at p. It is not hard to see [exercise] that the triangle $pa0$ is right angled, and so it follows that $\cos(\Phi/2) = |a|$, or

$$\Phi = 2\cos^{-1}|a|.$$

This explains the result (3.54), p. 204, which you proved algebraically in Ex. 27, p. 212.

To sum up,

> *The most general Möbius automorphism M_a^{ϕ} of the disc is a direct hyperbolic motion, and it is (i) an h-rotation if $\phi < \Phi$; (ii) a limit rotation if $\phi = \Phi$; and (iii) an h-translation if $\phi > \Phi$.*

Finally, recall from Ex. 20, p. 210, that the set of Möbius transformations of the form M_a^{ϕ} is identical to the set of the form

$$M(z) = \frac{Az+B}{\overline{B}z+\overline{A}}, \quad \text{where } |A| > |B|.$$

Comparing this with (6.20), we see that there are striking similarities not only between the conformal metrics of the sphere and the hyperbolic plane, but also between the Möbius transformations that represent their direct motions.

6.3.12 The Hemisphere Model and Hyperbolic Space

Figure [6.40a] illustrates how we may obtain two new models of the hyperbolic plane. Following Beltrami (1868b), let us stereographically project the Beltrami–Poincaré disc from the south pole S of the Riemann sphere onto the northern hemisphere. Defining the h-separation of two points to be the h-separation of their preimages in the disc, we have a new *conformal* map of the hyperbolic plane, called the *hemisphere model*. The h-lines of this model are the images of h-lines in the disc, and since stereographic projection preserves circles as well as angles, we deduce [exercise] that *h-lines are (semi-circular) vertical sections of the hemisphere*. What do equidistant curves and horocycles look like?

The hemisphere was Beltrami's primary model of the hyperbolic plane, and it was by applying the above stereographic projection *to* this hemisphere that he discovered the Beltrami–Poincaré disc. In fact by projecting his hemisphere in different ways, Beltrami obtained (in a unified way) almost all the models in current use.

For example, by projecting the hemisphere vertically down onto the complex plane (see [6.40a]) he obtained a new model of the hyperbolic plane inside the unit disc. This is traditionally called the "Klein model", but, again, we shall rename it the *Beltrami–Klein model*. Less controversially, it is also called the *projective model*. Since a small circle on the hemisphere is clearly projected to an ellipse in the disc, the

Beltrami–Klein model is *not* conformal. This is a serious disadvantage, but it is compensated for by the fact that the vertical sections of the hemisphere are projected to (Euclidean) straight lines: *h-lines in the Beltrami–Klein model are straight Euclidean chords of the unit circle.* Note the analogy with figure [6.12], in which geodesics on the sphere are represented by straight lines in the map; Ex. 14 reveals that this analogy is more than superficial.

Other properties of the Beltrami–Klein model will be explored in the exercises, but right now we have bigger fish to fry! Up to this point we have focused on developing the geometry of the hyperbolic *plane*, the negatively curved counterpart of the Euclidean plane. The geometry of this Euclidean plane may be thought of as being inherited from the geometry of three-dimensional Euclidean space. That is, if (X, Y, Z) are Cartesian coordinates in this space, then the Euclidean distance ds between two infinitesimally separated points is given by

$$ds = \sqrt{dX^2 + dY^2 + dZ^2},$$

and two-dimensional Euclidean geometry is obtained by restricting this formula to the points of an ordinary plane.

The question therefore arises whether there might exist a negatively curved (whatever that might mean) counterpart of three-dimensional Euclidean space, such that the geometry induced on each "plane" within this space would automatically be the geometry of the hyperbolic plane. We shall now show that this three-dimensional *hyperbolic space* does indeed exist.

To do so, let us find the metric of the hemisphere model. Because the stereographic projection of the Beltrami–Poincaré disc onto the hemisphere is conformal, it follows that $d\widehat{s}$ is once again given by the construction (6.43). Since $d\widehat{s}$ is independent of the direction of ds on the hemisphere, we may once again simplify the construction by choosing ds in an auspicious direction. In the Beltrami–Poincaré disc the best choice of ds was orthogonal to the diameter through the point of interest, and the best choice on the hemisphere is simply the stereographic projection of this configuration.

Thus in [6.40b] we have chosen the h-line l to be the vertical section of the hemisphere passing through the north pole and the point from which ds emanates. Thus l and e are both halves of great circles: the plane of l is vertical, the plane of e is inclined at angle $d\widehat{s}$ to the vertical, and the intersection of these planes is the illustrated diameter of the unit circle lying directly beneath l.

Now let the coordinates of the point at which we have drawn ds be (X, Y, Z), where the X and Y axes coincide with the real and imaginary axes of $\mathbb{C}$, so that Z measures the height of the point above $\mathbb{C}$. Since ds is orthogonal to l, and since the vertical plane of l is orthogonal to the hemisphere, we see that ds is *horizontal*. Thus the angle that ds subtends at the point $(X, Y, 0)$ directly beneath it is (ds/Z). But this angle is just the angle between the planes of l and e! Thus the metric of the hemisphere model is

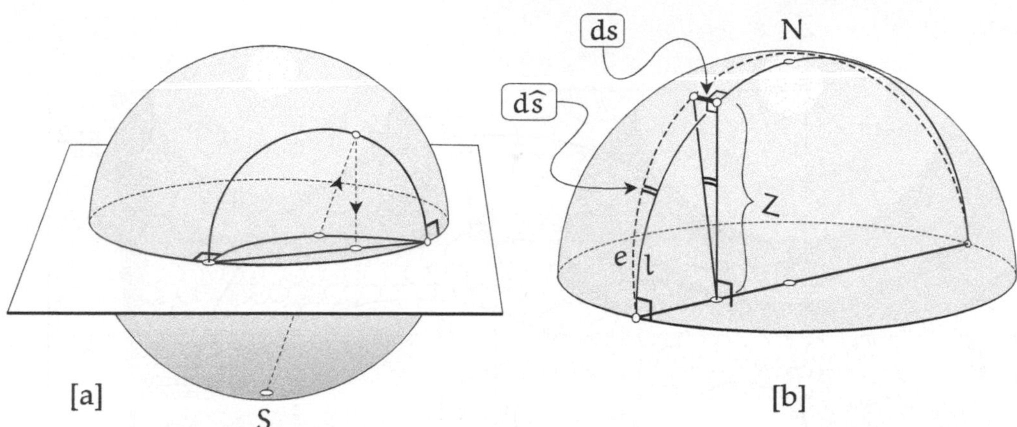

[6.40] [a] The Beltrami Hemisphere Model *is obtained by stereographically projecting the Beltrami–Poincaré disc model from the south pole, S, onto the northern hemisphere. Since stereographic projection preserves circles and angles,* h-lines are semi-circular vertical sections of the hemisphere. *Projecting vertically downwards yields the non-conformal* Beltrami–Klein model *(aka* projective model*) of the hyperbolic plane, in which h-lines now appear as straight Euclidean chords of the unit circle.* **[b]** *Conformality implies that the angular interpretation of hyperbolic distance given in [6.36] applies here, too. Therefore, the hyperbolic metric on the hemisphere is* $d\widehat{s} = (ds/Z)$.

$$d\widehat{s} = \frac{ds}{Z}. \tag{6.47}$$

This formula only describes the h-separation of points on the hemisphere, but there is nothing preventing us from using it to *define* the h-separation of *any* two infinitesimally separated points in the three-dimensional region $Z > 0$. This region lying above $\mathbb{C}$, with h-distance defined by (6.47), is called the *half-space model* of three-dimensional *hyperbolic space*, denoted $\mathbb{H}^3$. Without going into detail, it is clear from (6.47) that the points of $\mathbb{C}$ are infinitely h-distant from points that lie strictly above $\mathbb{C}$. Thus $\mathbb{C}$ represents the two-dimensional *horizon* or *sphere at infinity* of hyperbolic space.

At the moment it is a mere tautology that the geometry induced on the hemisphere by (6.47) is that of a hyperbolic plane. To begin to see that there is real meat on this idea, let us consider some simple motions of hyperbolic space. Clearly $d\widehat{s}$ is unaltered by a *translation parallel to* $\mathbb{C}$, so this is a motion. It is also clear that $d\widehat{s}$ is unaltered by a dilation $(X, Y, Z) \mapsto (kX, kY, kZ)$ centred at the origin. More generally, a *dilation centred at any point of* $\mathbb{C}$ will preserve $d\widehat{s}$, so this too is a motion.

By applying these two types of motion to the origin-centred unit hemisphere that we have been studying, we see that

In the half-space model, $\mathbb{H}^3$, *every hemisphere orthogonal to* $\mathbb{C}$ *is a hyperbolic plane,* $\mathbb{H}^2$. (6.48)

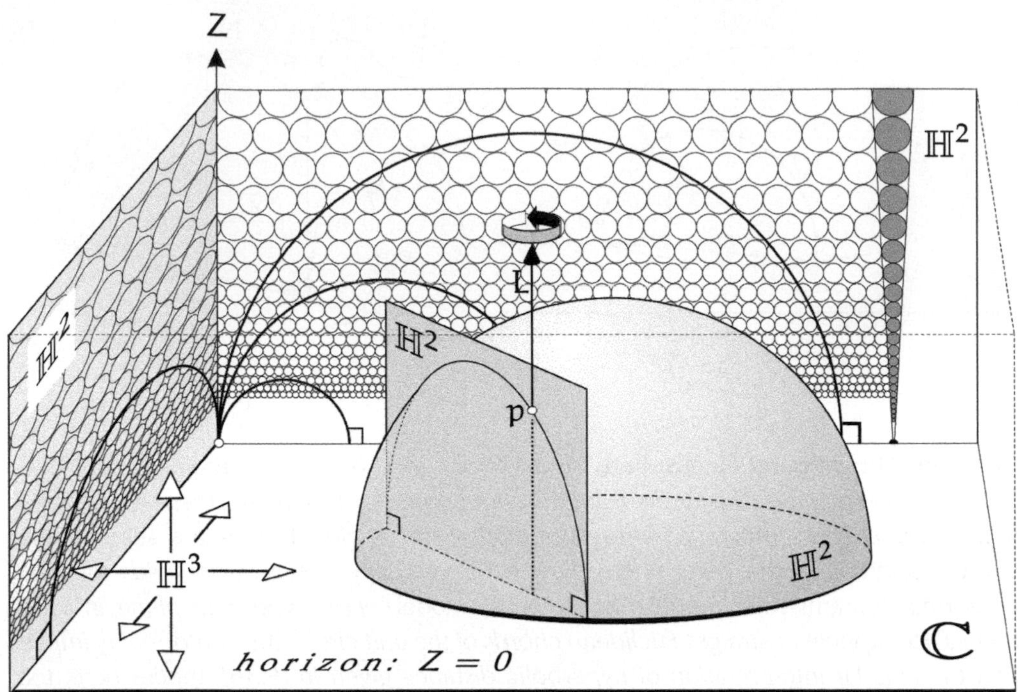

[6.41] Hyperbolic 3-space, $\mathbb{H}^3$. *Hyperbolic planes appear as vertical half-planes and as hemispheres orthogonal to $\mathbb{C}$. The intersection of two such $\mathbb{H}^2$ planes is a hyperbolic line: a vertical half-line, or a semicircle orthogonal to $\mathbb{C}$.*

In Euclidean geometry the intersection of two planes is a line, and this suggests that an h-line should be the intersection of two hyperbolic planes. Thus we anticipate that *every semicircle orthogonal to* $\mathbb{C}$ *is an h-line*, for every such semicircle is the intersection of two hemispheres orthogonal to $\mathbb{C}$. Note that this agrees with what we already know: the h-lines of the hemisphere model *are* semicircles orthogonal to $\mathbb{C}$. Figure [6.41] attempts to make this all much more vivid,[23] showing the vertical $\mathbb{H}^2$ walls filled with circles that are all the same hyperbolic size; compare this to [6.20], page 337.

Let us return to two-dimensional geometry for a moment. Figure [6.42] illustrates how Beltrami obtained the upper half-plane model from his hemisphere model. From a point q on the rim of the hemisphere, we stereographically project onto the tangent plane at the point antipodal to q—actually, any plane tangent to this one would do equally well. Since this preserves circles and angles, we see that a typical h-line of the hemisphere is mapped to a semicircle orthogonal to the bottom edge of the half-plane, while an h-line passing through q is mapped to a vertical line.

[23] This figure was not in the original edition of VCA. I drew it about 15 years later, for VDGF, where it appears as figure [6.6].

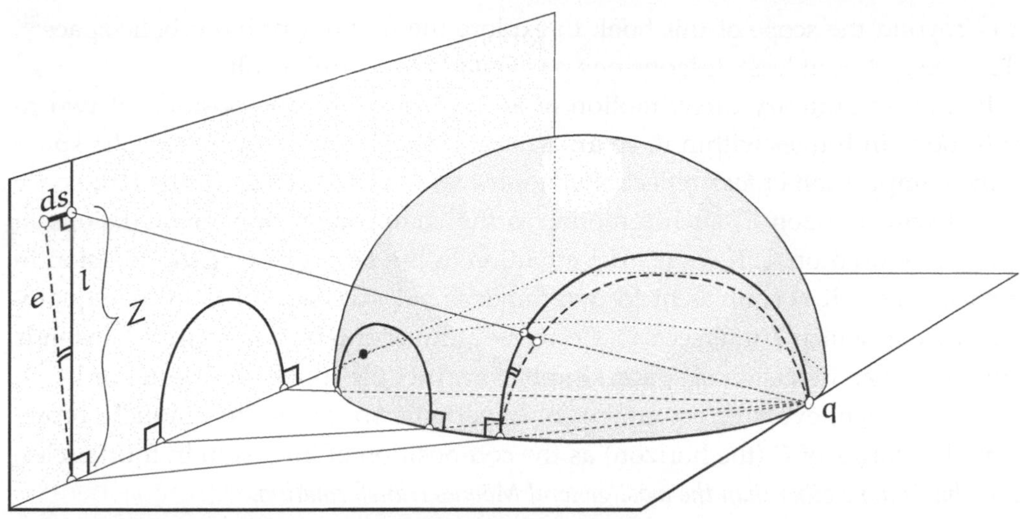

[6.42] How Beltrami discovered the upper half-plane model. *From a point* q *on the rim of the hemisphere, stereographically project onto the tangent plane at the point antipodal to* q*. Since this preserves circles and angles, we see that a typical h-line of the hemisphere is mapped to a semicircle orthogonal to the bottom edge of the half-plane, while an h-line passing through* q *is mapped to a vertical line. The angular interpretation of hyperbolic distance given in [6.36] applies here, too. We therefore recover the familiar hyperbolic metric in the upper half-plane:* $d\widehat{s} = (ds/Z)$.

Well, since these are the h-lines, it certainly *looks* like we have obtained the Beltrami–Poincaré upper half-plane, but to make sure, let's check that its metric is really given by (6.31). Since stereographic projection is conformal, we may yet again use the construction (6.43). Choosing l to be the image of an h-line through q, the figure immediately reveals that the metric is $d\widehat{s} = (ds/Z)$. Apart from a change of notation, this is indeed the same as (6.31).

We have thus returned to the half-plane that began our journey, but we have returned wiser than when we left. Looking at (6.47) we now recognize this half-plane orthogonal to $\mathbb{C}$ as a hyperbolic plane within hyperbolic space. This reveals the true role of the stereographic projection in [6.42].

We know that stereographic projection from q is just the restriction to the hemisphere of inversion $\mathcal{I}_K$ in a sphere K centred at q. Using the same argument (figure [6.22b]) as in the plane, we see that $\mathcal{I}_K$ preserves the metric (6.47), so it is a *motion* of hyperbolic space, carrying h-lines into h-lines and carrying h-planes into h-planes. Furthermore, (6.48) tells us that K is a hyperbolic plane in this hyperbolic space, and we therefore suspect that $\mathcal{I}_K$ is *reflection* in this h-plane. This can be confirmed [exercise] by generalizing the argument in [6.23b]. Thus we have the following generalization of (6.38):

> *Inversion in a hemisphere* K *orthogonal to the horizon is reflection* $\mathfrak{R}_K$ *of hyperbolic space in the h-plane* K.

It is beyond the scope of this book to explore the motions of hyperbolic space[24]. However, let us at least describe one particularly beautiful result.

Just as an arbitrary direct motion of an h-plane is the composition of two h-reflections in h-lines within it, so an arbitrary direct motion of hyperbolic space is the composition of *four* reflections in h-*planes* within it. Thus, in the half-space model with horizon $\mathbb{C}$, such a motion is the composition of four inversions in spheres centred on $\mathbb{C}$. If we restrict attention to the points of $\mathbb{C}$ then inversion in such a sphere K is equivalent to two-dimensional inversion of $\mathbb{C}$ in the equatorial circle in which K intersects $\mathbb{C}$. Conversely, inversion of $\mathbb{C}$ in a circle k extends uniquely to an inversion of space: simply construct the sphere with equator k.

Finally, then, every direct motion of hyperbolic space can be uniquely represented in terms of $\mathbb{C}$ (the horizon) as the composition of inversion in four circles, and *this is none other than the most general Möbius transformation,*

$$z \mapsto M(z) = \frac{az+b}{cz+d},$$

of the complex plane! Poincaré discovered this wonderful fact in 1883; see Poincaré (1985).

We have seen that the direct motions of the hyperbolic plane, the Euclidean plane and the sphere are subgroups of this group of general Möbius transformations. As we shall now see, this fact has a remarkable geometrical explanation.

Hilbert's result on surfaces of constant negative curvature shows that three-dimensional Euclidean geometry cannot accommodate a model of the hyperbolic plane. Amazingly, however, three-dimensional hyperbolic space *does* contain surfaces whose intrinsic geometry is *Euclidean!* In fact these surfaces are the *horospheres* that generalize the horocycles. Analogously to [6.28], horospheres are Euclidean spheres that touch $\mathbb{C}$, as well as planes $Z = \text{const.}$ that are parallel to $\mathbb{C}$.

Vertical planes orthogonal to $\mathbb{C}$ *look* flat in our model of hyperbolic space, but in reality they are intrinsically curved hyperbolic planes. However, a horosphere $Z = \text{const.}$ not only looks flat, it really *is* flat. For its metric, inherited from the metric (6.47) of the surrounding space, is just

$$d\widehat{s} = (\text{constant})\, ds,$$

and this is the metric of a Euclidean plane!

The motions of Euclidean plane geometry may now be viewed as the motions of hyperbolic space that map this intrinsically flat horosphere into itself. Clearly, these are the composition of reflections in vertical planes, i.e., h-reflections in h-planes orthogonal to the horosphere. In this manner, the direct motions of the Euclidean plane manifest themselves on the horizon $\mathbb{C}$ as a subgroup of the Möbius transformations.

[24] Thurston (1997) is an excellent source of information on these motions.

As for spherical geometry, we begin by defining an *h-sphere* as the set of points at constant h-distance from a given point (the h-centre). It is not hard to see that these h-spheres are represented in the half-space model by Euclidean spheres, though their h-centres do not coincide with their Euclidean centres.

Though it is not immediately obvious in this model, it can be shown [see Ex. 27] that the intrinsic geometry of such an h-sphere is the same as that of an ordinary sphere (of different radius) in Euclidean space. As with the horosphere, the motions of this h-sphere may be viewed as the motions of hyperbolic space that map the h-sphere into itself. Again, these are the composition of h-reflections in h-planes orthogonal to the h-sphere, and again we arrive at a subgroup of the Möbius transformations.

Clearly the motions of the hyperbolic plane may also be viewed in this way, and so we have arrived at a fitting high point on which to end this chapter:

> *Two-dimensional hyperbolic, Euclidean, and spherical geometry are all subsumed by three-dimensional hyperbolic geometry.*

Further Reading. For a short, masterful overview of Differential Geometry, see Penrose (1978). For a complete, elementary, and *geometrical* treatment of Differential Geometry (and Differential Forms), in the same spirit as this book, see Needham (2021). For more on its fundamental applications to physics, particularly General Relativity, see Misner, Thorne, and Wheeler (1973), Wald (1984), Ludvigsen (1999), Penrose (2005), Thorne and Blandford (2017), and Schutz (2022). For more on the use of Differential Forms in Differential Geometry, see do Carmo (1994) and O'Neill (2006), and for their applications to physics see Schutz (1980), Arnol'd (1989), and Dray (2015). For more on hyperbolic geometry itself, see Stillwell (1992), Stillwell (1996), Stillwell (2010), and Thurston (1997). Finally, for the connections between topology and differential geometry, see Hopf (1956).

6.4 Exercises

1 Draw a geodesic triangle Δ on the surface of a suitable fruit or vegetable. Now draw a geodesic segment from one of the vertices to an arbitrary point of the opposite side. This divides Δ into two geodesic triangles, say Δ_1 and Δ_2. Show that the angular excess function $\mathcal{E}$ is *additive*, i.e., $\mathcal{E}(\Delta) = \mathcal{E}(\Delta_1) + \mathcal{E}(\Delta_2)$. By continuing this process of subdivision, deduce that (6.5) implies (6.6).

2 Explain (6.18) by generalizing the argument that was used to obtain the special case (3.18), on p. 162. That is, think of reflection of the sphere in terms reflection of space in a plane Π, as in [6.8], p. 318. Also, think of stereographic projection as the restriction to the sphere of the three-dimensional inversion $\mathcal{I}_K$, where K is the sphere of radius $\sqrt{2}$ centred at the north pole of Σ (see [6.13b]). Now let a be a point on Σ, and consider the effect of $\mathcal{I}_K$ on a, Π, and $\mathfrak{R}_\Pi(a)$.

3 Let C be a circle in $\mathbb{C}$, and let $\widehat{C}$ be it stereographic image on Σ. If $\widehat{C}$ is a great circle, then (6.18) says that $\mathcal{I}_C$ stereographically induces reflection of Σ in $\widehat{C}$, but what transformation is induced if C is an *arbitrary* circle? Generalize the argument of figure [6.14] to show that $\mathcal{I}_C$ *becomes projection from the vertex* v *of the cone that touches* Σ *along* $\widehat{C}$. That is, if $w = \mathcal{I}_C(z)$ then $\widehat{w}$ is the second intersection point of Σ with the line in space that passes through the vertex v and the point $\widehat{z}$. Explain how (6.18) may be viewed as a limiting case of this more general result.

4 Use (3.42), p. 189 to show that if the Möbius transformation $M(z)$ has fixed points $\xi_\pm$, and the multiplier associated with ξ_+ is m, then

$$[M] = \begin{bmatrix} 1 & -\xi_+ \\ 1 & -\xi_- \end{bmatrix}^{-1} \begin{bmatrix} \sqrt{m} & 0 \\ 0 & 1/\sqrt{m} \end{bmatrix} \begin{bmatrix} 1 & -\xi_+ \\ 1 & -\xi_- \end{bmatrix}.$$

By putting $\xi_+ = a$, $m = e^{-i\psi}$, and $\xi_- = -(1/\overline{a})$, deduce (6.19), p. 328. [*Hint*: Remember that you are free to multiply a Möbius matrix by a constant.]

5 Show that the Möbius transformations (6.20) do indeed satisfy the differential equation (6.17).

6 (i) The *conjugate* $\overline{\mathbb{V}}$ of a quaternion $\mathbb{V} = v + \mathbf{V}$ is defined to be the conjugate transpose $\mathbb{V}^*$ of the corresponding matrix. Show that $\overline{\mathbb{V}} \equiv \mathbb{V}^* = v - \mathbf{V}$, and deduce that $\mathbb{V}$ is a *pure* quaternion (analogous to a purely imaginary complex number) if and only if $\overline{\mathbb{V}} = -\mathbb{V}$.

(ii) The *length* $|\mathbb{V}|$ of $\mathbb{V}$ is defined (by analogy with complex numbers) by $|\mathbb{V}|^2 = \mathbb{V}\,\overline{\mathbb{V}}$. Show that $|\mathbb{V}|^2 = v^2 + |\mathbf{V}|^2 = |\overline{\mathbb{V}}|^2$.

(iii) If $|\mathbb{V}| = 1$, then $\mathbb{V}$ is called a *unit quaternion*. Verify that $\mathbb{R}_{\mathbf{v}}^{\psi}$ [see (6.28)] is a unit quaternion, and that $\overline{\mathbb{R}_{\mathbf{v}}^{\psi}} = \left[\mathbb{R}_{\mathbf{v}}^{\psi}\right]^{*} = \mathbb{R}_{\mathbf{v}}^{-\psi}$.

(iv) Show that $\overline{\mathbb{V}\,\mathbb{W}} = \overline{\mathbb{W}}\;\overline{\mathbb{V}}$ and deduce that $|\mathbb{V}\,\mathbb{W}| = |\mathbb{V}|\,|\mathbb{W}|$. Thus, for example, the product of two unit quaternions is another unit quaternion.

(v) Show that $\mathbb{A}$ is a pure, unit quaternion if and only if $\mathbb{A}^2 = -1$.

(vi) Show that any quaternion $\mathbb{Q}$ can be expressed as $\mathbb{Q} = |\mathbb{Q}|\,\mathbb{R}_{\mathbf{v}}^{\psi}$ for some $\mathbf{v}$ and some ψ.

(vii) Suppose we generalize the transformation (6.29) to $\mathbb{P} \mapsto \widetilde{\mathbb{P}} = \mathbb{Q}\,\mathbb{P}\,\overline{\mathbb{Q}}$, where $\mathbb{Q}$ is an arbitrary quaternion. When interpreted in this way, deduce that $\mathbb{Q}$ represents a *dilative rotation of space*, and the product of two quaternions represents the composition of the corresponding dilative rotations. [This confirms the claim at the end of Chapter 1.]

7 [Do the previous exercise before this one.] The following proof of (6.29) is based on a paper of H. S. M. Coxeter (1946).

(i) Use (6.27) to show that the pure quaternions $\mathbb{P}$ and $\mathbb{A}$ are orthogonal if and only if $\mathbb{P}\,\mathbb{A} + \mathbb{A}\,\mathbb{P} = 0$.

(ii) If $\mathbb{A}$ has unit length, so that $\mathbb{A}^2 = -1$, deduce that the previous equation may be expressed as $\mathbb{P} = \mathbb{A}\,\mathbb{P}\,\mathbb{A}$.

(iii) Now keep the pure, unit quaternion $\mathbb{A}$ fixed, but let $\mathbb{P}$ represent an *arbitrary* pure quaternion. Let Π_{A} denote the plane with normal vector $\mathbf{A}$ that passes through the origin, so that its equation is $\mathbf{P}\cdot\mathbf{A} = 0$. Now consider the transformation

$$\mathbb{P} \mapsto \mathbb{P}' = \mathbb{A}\,\mathbb{P}\,\mathbb{A}. \tag{6.49}$$

Show that (a) $\mathbb{P}'$ is automatically pure, and $|\mathbb{P}'| = |\mathbb{P}|$, so that (6.49) represents a motion of space; (b) every point on Π_{A} remains fixed; (c) every vector orthogonal to Π_{A} is reversed. Deduce that *(6.49) represents reflection* $\mathfrak{R}_{\Pi_{\mathrm{A}}}$ *of space in the plane* Π_{A}.

(iv) Deduce that if the angle from Π_{A} to a second plane Π_{B} is $(\psi/2)$, and the unit vector along the intersection of the planes is $\mathbf{V}$, then the rotation $\mathcal{R}_{\mathbf{V}}^{\psi} = (\mathfrak{R}_{\Pi_{\mathrm{B}}} \circ \mathfrak{R}_{\Pi_{\mathrm{A}}})$ is given by

$$\mathbb{P} \mapsto \widetilde{\mathbb{P}} = (\mathbb{B}\,\mathbb{A})\,\mathbb{P}\,(\mathbb{A}\,\mathbb{B}) = (-\mathbb{B}\,\mathbb{A})\,\mathbb{P}\,(-\overline{\mathbb{B}\,\mathbb{A}}).$$

(v) Use (6.27) to show that $-\mathbb{B}\,\mathbb{A} = \cos(\psi/2) + \mathbf{V}\,\sin(\psi/2)$, thereby simultaneously proving (6.29) and (6.28).

8 Here is another proof of (6.29). As in the text, we shall assume that $\mathbf{P}$ is a unit vector with its tip at the point $\widehat{p}$ on the unit sphere. If we represent the stereographic images p and $\widetilde{p}$ of $\widehat{p}$ and $\widehat{\widetilde{p}}$ by their homogeneous coordinate vectors $\mathfrak{p}$ and $\widetilde{\mathfrak{p}}$ in $\mathbb{C}^2$, then we know that the rotation is represented as

$$\mathfrak{p} \mapsto \widetilde{\mathfrak{p}} = \mathbb{R}_{\mathbf{v}}^{\psi}\, \mathfrak{p},$$

where $\mathbb{R}_{\mathbf{v}}^{\psi}$ is being thought of as a 2×2 matrix.

(i) Show that in homogeneous coordinates, (3.21), p. 166, becomes

$$X + iY = \frac{2\,\mathfrak{p}_1\overline{\mathfrak{p}_2}}{|\mathfrak{p}_1|^2 + |\mathfrak{p}_2|^2} \quad \text{and} \quad Z = \frac{|\mathfrak{p}_1|^2 - |\mathfrak{p}_2|^2}{|\mathfrak{p}_1|^2 + |\mathfrak{p}_2|^2}.$$

(ii) To simplify this, recall that all multiples of $\mathfrak{p}$ describe the same point p in $\mathbb{C}$. We can therefore choose the "length" of $\mathfrak{p}$ to be $\sqrt{2}$:

$$\langle \mathfrak{p}\,, \mathfrak{p} \rangle \equiv |\mathfrak{p}_1|^2 + |\mathfrak{p}_2|^2 = 2.$$

With this choice, show that the above equations can be written as

$$\begin{bmatrix} 1+Z & X+iY \\ X-iY & 1-Z \end{bmatrix} = \begin{bmatrix} \mathfrak{p}_1\,\overline{\mathfrak{p}_1} & \mathfrak{p}_1\,\overline{\mathfrak{p}_2} \\ \mathfrak{p}_2\,\overline{\mathfrak{p}_1} & \mathfrak{p}_2\,\overline{\mathfrak{p}_2} \end{bmatrix} = \begin{bmatrix} \mathfrak{p}_1 \\ \mathfrak{p}_2 \end{bmatrix} [\overline{\mathfrak{p}_1} \;\; \overline{\mathfrak{p}_2}] = \mathfrak{p}\,\mathfrak{p}^*.$$

(iii) Verify that

$$\begin{bmatrix} 1+Z & X+iY \\ X-iY & 1-Z \end{bmatrix} = 1 - i\mathbb{P}.$$

(iv) Deduce that

$$1 - i\widetilde{\mathbb{P}} = \mathbb{R}_{\mathbf{v}}^{\psi}\,(1 - i\,\mathbb{P})\,\left[\mathbb{R}_{\mathbf{v}}^{\psi}\right]^* = 1 - i\,\mathbb{R}_{\mathbf{v}}^{\psi}\,\mathbb{P}\,\mathbb{R}_{\mathbf{v}}^{-\psi},$$

from which (6.29) follows immediately.

9 (i) Figure [6.40a] gave a two-step process for carrying a point z in the Beltrami–Poincaré disc to the corresponding point z' in the Klein model. Explain why the net mapping $z \mapsto z'$ of the disc to itself is the one shown in figure [a] below, where C is an arbitrary circle passing through z and orthogonal to the unit circle U.

(ii) Figure [b] is a vertical cross section of [6.40a] through z and z'. Deduce that

$$\frac{|z'|}{a} = \frac{1}{b} \quad \text{and} \quad \frac{a}{|z|} = \frac{2}{b}.$$

By multiplying these two equations, deduce that $z' = \frac{2z}{1+|z|^2}$. [Thus we have a geometric explanation of the result (3.21), p. 166.]

(iii) This formula can be derived directly from figure [a], without the assistance of the hemisphere. Redraw the figure with C chosen orthogonal to $0z$. Explain geometrically why the centre of C may be viewed as either $\mathcal{I}_U(z')$, or as the midpoint of z and $\mathcal{I}_U(z)$. Conclude that

$$1/\overline{z'} = \mathcal{I}_U(z') = \tfrac{1}{2}\,[z + \mathcal{I}_U(z)] = \tfrac{1}{2}\,[z + (1/\overline{z})]\,,$$

from which the result follows immediately.

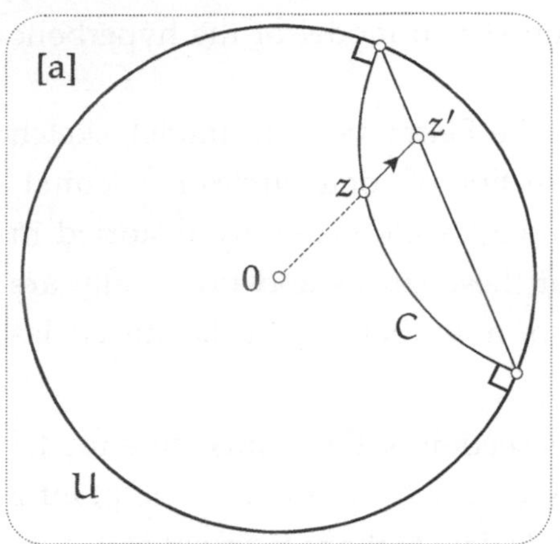

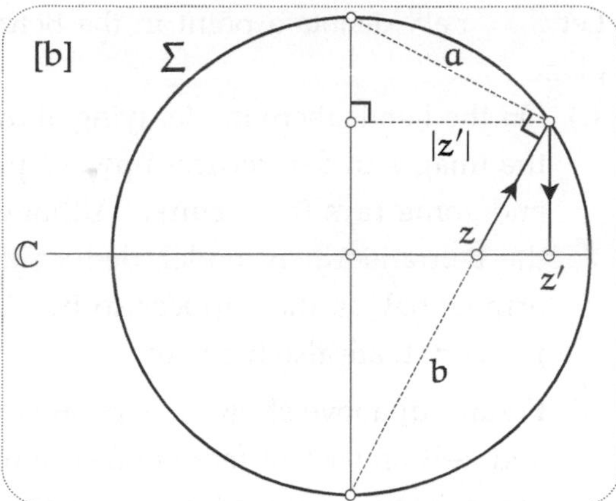

10 Think of the sphere as the surface of revolution generated by a semicircle. Construct a conformal map of the sphere by strict analogy with the construction of the map of the pseudosphere in [6.20]. Show that this is the Mercator map that you obtained in Ex. 14, p. 294.

11 (i) In the hyperbolic plane, show that the h-circumference of an h-circle of h-radius ρ is $2\pi \sinh \rho$. [*Hint*: Represent the h-circle as an origin-centred Euclidean circle in the Beltrami–Poincaré disc.]

(ii) Let the inhabitants of the sphere of radius R draw a circle of (intrinsic) radius ρ. Use elementary geometry to show that the circle's circumference is $2\pi R \sin(\rho/R)$. Show that if we take the radius of the sphere to be $R = i$, then this becomes the formula in part (i)! [Compare this with Ex. 14.]

12 Let L and M be two intersecting chords of the unit circle, and let l and m be the intersection points of the pairs of tangents drawn at the ends of these chords. See figure [a] below. In the Beltrami–Klein model, show that L and M represent orthogonal h-lines if and only if l lies on M (produced) and m lies on L (produced).

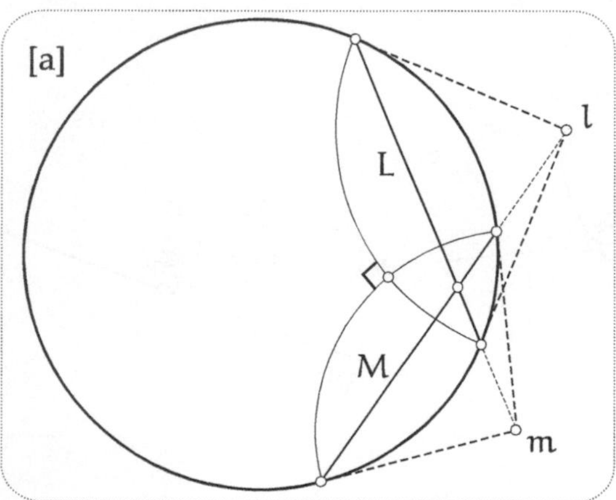

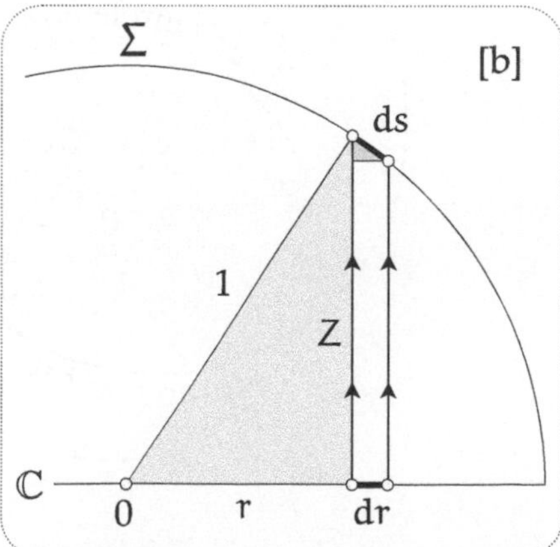

13 Let $z = r\,e^{i\theta}$ denote a point in the Beltrami–Klein model of the hyperbolic plane.

(i) On the hemisphere model lying above the Beltrami–Klein model, sketch the images under vertical upward projection of some circles $r = \text{const.}$ and some rays $\theta = \text{const.}$ Although angles are generally distorted in the Beltrami–Klein model, deduce that these circles and rays really are orthogonal, as they appear to be. Also, note that the Euclidean circles $r = \text{const.}$ are also h-circles.

(ii) Figure [b] above shows a vertical cross section of the hemisphere model and Beltrami–Klein model taken though a ray $\theta = \text{const.}$ If the point z moves outward along this ray by dr, let ds denote the movement of its vertical projection on the hemisphere. Explain why the two shaded triangles are similar, and deduce that $ds = (dr/Z)$.

(iii) Use the metric (6.47) of the hemisphere to conclude that the h-separation $d\widehat{s}_r$ of the points in the Beltrami–Klein model with polar coordinates (r, θ) and $(r + dr, \theta)$ is given by

$$d\widehat{s}_r = \frac{dr}{1 - r^2}.$$

[Remarkably, this means that the formula for $\mathcal{H}\{0, z\}$ differs from the formula (6.45) in the Beltrami–Poincaré disc by a mere factor of two!]

(iv) Use the same idea (of projecting onto the hemisphere) to show that the h-separation $d\widehat{s}_\theta$ of the points (r, θ) and $(r, \theta + d\theta)$ is given by

$$d\widehat{s}_\theta = \frac{r\,d\theta}{\sqrt{1 - r^2}}.$$

(v) Deduce that the h-separation $d\widehat{s}$ of the points (r, θ) and $(r + dr, \theta + d\theta)$ is given by the following formula of Beltrami (1868a):

$$d\widehat{s}^{\,2} = \frac{dr^2}{(1 - r^2)^2} + \frac{r^2\,d\theta^2}{1 - r^2}. \tag{6.50}$$

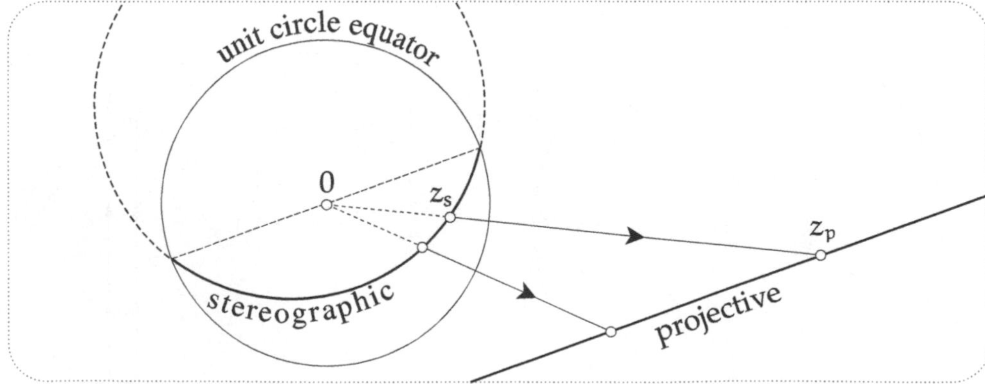

14 (i) The figure above superimposes the stereographic and projective [see p. 323] images of a great circle on the unit sphere. Let z_s and z_p be the stereographic and projective images of the point whose spherical-polar coordinates are (ϕ, θ). Referring to (3.22), p. 167, $z_s = \cot(\phi/2)\, e^{i\theta}$. Show that $z_p = [-\tan\phi]\, e^{i\theta}$, and deduce that

$$z_p = \frac{2z_s}{1 - |z_s|^2}.$$

Compare this with Ex. 9!

(ii) Sketch the curves on the hemisphere that are centrally projected to the circles $|z_p| = \texttt{const.}$ and to the rays $\arg z_p = \texttt{const}$. Although angles are generally distorted in the projective model, observe that these circles and rays really are orthogonal, as they appear to be.

(iii) *Now let the sphere have radius* R, and write $z_p = r\, e^{i\theta}$ for the projective image of the point (ϕ, θ). Thus $r = -R\tan\phi$. Show that if z_p moves a distance dr along the ray $\theta = \texttt{const.}$, then the corresponding point on the sphere moves a distance $d\widehat{s}_r$ given by

$$d\widehat{s}_r = \frac{dr}{1 + (r/R)^2}.$$

(iv) Likewise, show that the separation $d\widehat{s}_\theta$ of the points on the sphere corresponding to the points (r, θ) and $(r, \theta + d\theta)$ in the map is given by

$$d\widehat{s}_\theta = \frac{r\, d\theta}{\sqrt{1 + (r/R)^2}}.$$

(v) Deduce that the spherical separation $d\widehat{s}$ corresponding to the points (r, θ) and $(r + dr, \theta + d\theta)$ is given by

$$d\widehat{s}^{\,2} = \frac{dr^2}{(1 + (r/R)^2)^2} + \frac{r^2\, d\theta^2}{1 + (r/R)^2}.$$

(vi) Here is a crazy idea, essentially due to Lambert; see Penrose (2005, § 2.6). Perhaps we can get a surface of constant *negative* curvature $\mathcal{K} = -(1/R^2) = +1/(iR)^2$ by allowing the radius R of the sphere to take on the *imaginary* value iR. Verify that if we substitute $R = i$ into the above formula, then it becomes the Beltrami metric (6.50) of the hyperbolic plane! [To make true sense of this idea, one must turn to Einstein's relativity theory; see Thurston (1997) and especially Penrose (2005).]

15 Take a stack of ten sheets of paper and staple them together, placing staples along three of the edges. Use a pair of compasses to draw the largest circle that will fit comfortably inside the top sheet. Pierce through all ten sheets in the centre of the circle. With heavy scissors, cut along the circle to obtain ten identical discs, say of radius R. Repeat this whole process to double the number of discs to 20.

(i) Cut a narrow sector out of the first disc, and tape the edges together to form a shallow cone. Repeat this process with the remaining discs, steadily increasing the angle of the sector each time, so that the cones get sharper and taller. Ensure that by the end of the process you are making very narrow cones, using only a quarter disc or less.

(ii) Stack these cones in the order you made them. Explain how it is that *you have created a model of a portion of a pseudosphere of radius* R. Create weird new (extrinsically asymmetric) surfaces of constant negative curvature by holding the tip of your structure and moving it from side to side!

(iii) Use the same idea to create a disc-like piece of "hyperbolic paper", such as you would get if you could simply cut out a disc from your pseudosphere. Press it against the pseudosphere and verify that you can freely move it about and rotate it on the surface.

16 By holding a fairly short piece of string against the surface of the toy pseudosphere of the previous exercise, draw a segment of a typical geodesic. Extend this segment in both directions, one string-length at a time. Note the surprising way the geodesic only spirals a finite distance up the pseudosphere before spiralling down again.

(i) Use the upper half-plane to verify mathematically that the tractrix generators are the *only* geodesics that extend all the way up to the top.

(ii) Let L be a typical geodesic, and let α be the angle between L and the tractrix generator at the point where L hits the rim $\sigma = 0$. Show that the maximum distance σ_{max} that L travels up the pseudosphere is given by $\sigma_{max} = |\ln \sin \alpha|$.

17 Suppose we have a conformal map of a surface in the xy-plane, with the metric given by (6.15):

$$d\widehat{s} = \Lambda\, ds = \Lambda(x, y)\sqrt{dx^2 + dy^2}.$$

An elegant result from differential geometry [see Needham (2021, §4.4)] states that the Gaussian curvature at any point on the surface is given by

$$\mathcal{K} = -\frac{1}{\Lambda^2}\Delta(\ln \Lambda),$$

where $\Delta \equiv (\partial_x^2 + \partial_y^2)$ is the Laplacian. Try this out on the metric (6.16) of the stereographic map of the sphere, and on the metrics (6.31) and (6.44) of the half-plane and disc models of the hyperbolic plane.

18 Use the Beltrami–Poincaré disc to rederive the formula $\tan(\Pi/2) = e^{-D}$ for the angle of parallelism. [*Hint*: Let one of the h-lines be a diameter.]

19 To derive the metric (6.44), consider the mapping (6.42) $z \mapsto w = D(z)$ from the upper half-plane to the Beltrami–Poincaré disc. An infinitesimal vector dz emanating from z is amplitwisted to an infinitesimal vector $dw = D'(z)\,dz$ emanating from w, and (by definition) the h-length $d\widehat{s}$ of dw is the h-length of dz. Verify (6.44) by showing that

$$\frac{2\,|dw|}{1-|w|^2} = \frac{|dz|}{\operatorname{Im} z} = d\widehat{s}\,.$$

20 Consider the mapping $z \mapsto w = \mathcal{M}_a^{\phi}(z)$ of the Beltrami–Poincaré disc to itself. Use the calculational approach of the previous exercise to show that $z \mapsto w$ is a hyperbolic motion, i.e., it preserves the metric:

$$\frac{2\,|dw|}{1-|w|^2} = d\widehat{s} = \frac{2\,|dz|}{1-|z|^2}.$$

21 In the upper half-plane, the h-rotation $\mathcal{R}_i^{\phi}$ through angle ϕ about the point i is given by the following Möbius transformation:

$$\mathcal{R}_i^{\phi}(z) = \frac{c\,z+s}{-s\,z+c}\,, \quad \text{where } c = \cos(\phi/2) \text{ and } s = \sin(\phi/2).$$

Prove this in three ways:

(i) Show that $\mathcal{R}_i^{\phi}(i) = i$ and $\left\{\mathcal{R}_i^{\phi}\right\}'(i) = e^{i\phi}$. Why does this prove the result?

(ii) Use the formula for inversion [(3.4), p. 141] to calculate the composition $\mathcal{R}_i^{\phi} = (\mathfrak{R}_B \circ \mathfrak{R}_A)$, where A and B are h-lines through i, and the angle from A to B is $(\phi/2)$. [*Hint*: Take A to be the imaginary axis, and use a diagram to show that the semicircle B has centre $-(c/s)$ and radius $(1/s)$.]

(iii) Describe and explain the geometrical effect of applying $(D \circ \mathcal{R}_i^{\phi} \circ D^{-1})$ to the Beltrami–Poincaré disc, where D is the mapping (6.42) from the upper half-plane to the Beltrami–Poincaré disc. Deduce that

$$(D \circ \mathcal{R}_i^{\phi} \circ D^{-1})(z) = e^{i\phi}\,z.$$

Re-express this equation in terms of products of Möbius matrices, and solve for the matrix $[\mathcal{R}_i^{\phi}]$.

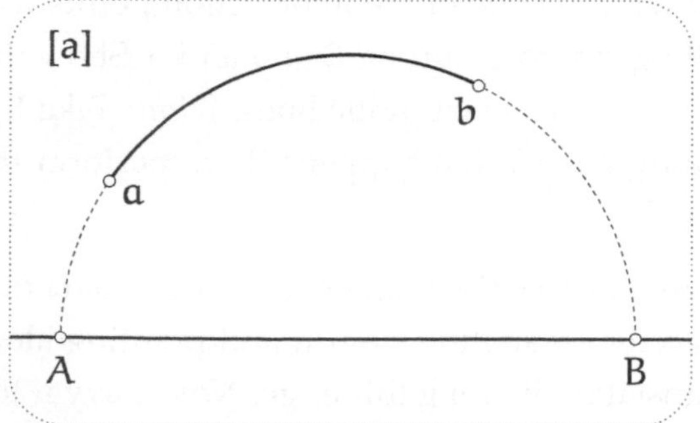

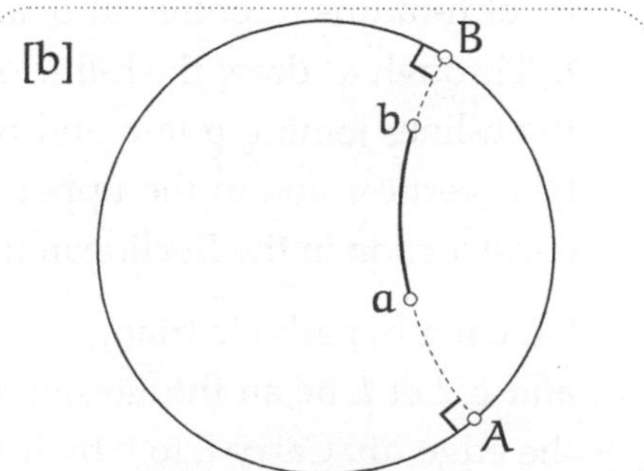

22 (i) Referring to figure [a] above, show that the h-separation of two points in the upper half-plane may be expressed in terms of a cross-ratio as

$$\mathcal{H}\{a, b\} = \ln[a, B, b, A].$$

[*Hint*: Apply an h-rotation centred at a to bring b into a position vertically above a.]

(ii) Show that the same formula applies to the Beltrami–Poincaré disc in figure [b].

23 (i) Let $\widehat{a}$ and $\widehat{z}$ be two point on Σ, and let a and z be their stereographic images in $\mathbb{C}$. If $S\{a, z\}$ is the distance (on the sphere) between $\widehat{a}$ and $\widehat{z}$, show that

$$S\{a, z\} = 2 \tan^{-1}\left|\frac{z - a}{\overline{a}\,z + 1}\right|.$$

[*Hint*: Use (6.20) to bring a to the origin.]

(ii) Show that the h-distance formula (6.46) in the Beltrami–Poincaré disc can be re-expressed as

$$\mathcal{H}\{a, z\} = 2 \tanh^{-1}\left|\frac{z - a}{\overline{a}\,z - 1}\right|.$$

24 (i) Use a simple sketch to show that $z \mapsto f(z) = -\overline{z}$ is an opposite motion of the Beltrami–Poincaré upper half-plane.

(ii) Let $\widetilde{\mathcal{M}}$ be an arbitrary opposite motion. By considering $(f \circ \widetilde{\mathcal{M}})$, show that

$$\widetilde{\mathcal{M}}(z) = \frac{a\,\overline{z} + b}{c\,\overline{z} + d}, \quad \text{where } a, b, c, d \text{ are real, and } (ad - bc) > 0.$$

(iii) Use this formula to show that $\widetilde{\mathcal{M}}$ has two fixed points on the horizon (the real axis). If L is the h-line whose ends are these two points, explain why L is invariant under $\widetilde{\mathcal{M}}$.

(iv) Deduce that $\widetilde{\mathcal{M}}$ is always a *glide reflection*: h-translation along L, followed by (or preceded by) h-reflection in L.

25 Given a point p not on the h-line L, draw an h-circle C of h-radius ρ centred at p. Draw the h-line orthogonal to L through p, cutting L in q. Draw an h-circle C' of h-radius ρ centred at q, and let q' be one of the intersection points with L. Through q' draw the h-line orthogonal to L, cutting C at a and b. Show that the h-lines joining p to a and b are the two asymptotic lines! [*Hint*: Take L to be a vertical line in the upper half-plane.] What happens if we perform this construction in the Euclidean plane?

26 Sketch a hyperbolic triangle Δ with vertices (in counterclockwise order) a, b, and c. Let ξ be an infinitesimal vector emanating from a and pointing along the edge ab. Carry ξ to b by h-translating it along this edge. Now carry it to c

along bc, and finally carry it home to a, along ca. In Euclidean geometry these three translations would simply cancel, and ξ would return home unaltered. Use your sketch to show that in hyperbolic geometry *the composition of these three h-translations is an h-rotation about vertex* a *through angle* $\mathcal{E}(\Delta)$.

[Suppose that Δ is instead a geodesic triangle on an *arbitrary* surface S of variable curvature. If an inhabitant of S wants to translate ξ along a "straight line" (a geodesic), all he has to do is keep its length constant, and keep the angle it makes with the line constant. This is called *parallel transport*, and it plays a central role in differential geometry; see Needham (2021, Act IV). The above argument still applies, and so when ξ is parallel transported round Δ, it returns home rotated through $\mathcal{E}(\Delta)$. By virtue of (6.6), this angle of rotation is the total amount of curvature inside Δ.]

27 Generalize the transformation from the upper half-plane to the Beltrami–Poincaré disc to obtain a model of hyperbolic space in the interior of the unit sphere. Describe the appearance of h-lines, h-planes, h-spheres, and horospheres in this model, and explain why an h-sphere is intrinsically the same as a Euclidean sphere of different radius.

CHAPTER 7

Winding Numbers and Topology

In this chapter we shall investigate a simple but immensely powerful concept—the number of times a loop winds around a point. In Chapter 2 we saw that this concept was needed to understand multifunctions, and in the next chapter we will see that it plays an equally crucial role in understanding complex integration. However, only the first two sections [up to (7.2)] of the present chapter are actually a prerequisite for that work; the rest may be read at any time. If you are in a rush to learn about integration, you may wish to skip the rest of the chapter and return to it later.

7.1 Winding Number

7.1.1 The Definition

As the name suggests, the *winding number* $\nu(L, 0)$ of a closed loop L about the origin 0 is simply the *net* number of revolutions of the direction of z as it traces out L once in its given sense. A nut on a bolt admirably illustrates the concept of "net rotation": spin the nut this way and that way for a while; the final distance of the nut from its starting point measures the net rotation it has undergone.

Figure [7.1] shows six loops and their corresponding winding numbers. You can verify these values by starting at a random point on each curve and tracing it out with your finger: starting with zero, add one after each positive (= counterclockwise) revolution of the vector connecting the origin to your finger, and subtract one after each negative (= clockwise) revolution. When you have returned to your starting point, the final count is the winding number of the loop.

It is often useful to consider the winding number of a loop about a point p other than the origin, and this is correspondingly written $\nu(L, p)$. Instead of counting the revolutions of z, we now count those of $(z - p)$. For example, the shaded region in [7.1] can be defined as all the positions of p for which $\nu(L, p) \neq 0$. Try shading this set for the other loops.

Visual Complex Analysis. 25th Anniversary Edition. Tristan Needham, Oxford University Press.
 DOI: 10.1093/oso/9780192868916.003.0007

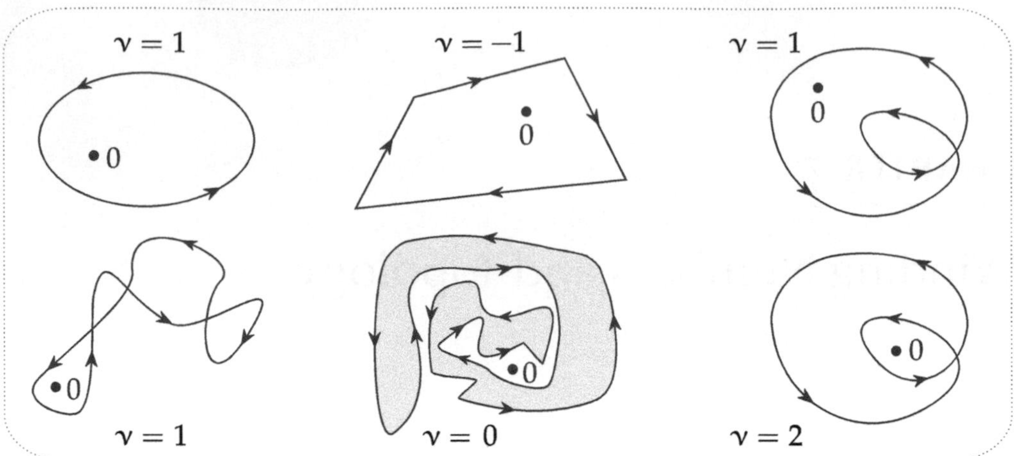

[7.1] The *winding number* ν *of a closed loop* L *counts the number of times* L *encircles a point.*

7.1.2 What Does "Inside" Mean?

A loop is called *simple* if it does not intersect itself; for example, circles, ellipses, and triangles are all simple. Although a simple loop can actually be very complicated [see Ex. 1] it seems clear, though it is hard to prove, that it will divide the plane into just two sets, its inside and its outside. However, in the case of a loop that is not simple, such as [7.2], it is no longer obvious which points are to be considered inside the loop, and which outside. The winding number concept allows us make the desired distinction clearly.

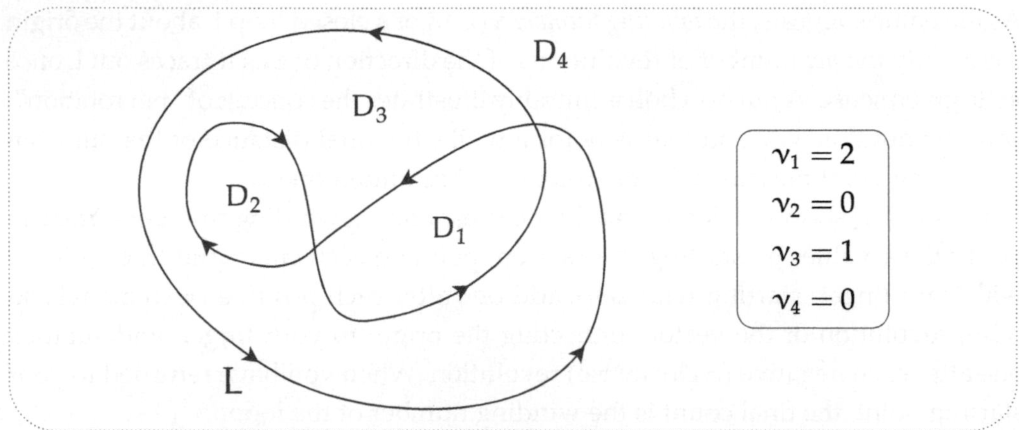

[7.2] Using the winding number to define the "inside" and the "outside" of a closed loop. *A non-simple loop* L *partitions the plane into multiple simply connected regions,* D_j*, and the winding number is constant in each one, taking the value* ν_j*. The "inside" of* L *is then defined to be the union* $D_1 \cup D_3$ *of the* D_j*'s with* $\nu_j \neq 0$*, and the "outside" is the union* $D_2 \cup D_4$ *of the* D_j*'s with* $\nu_j = 0$*.*

A typical loop such as L will partition the plane into a number of sets D_j (four in this case). If the point p wanders around within one of these sets then it seems plausible that the winding number $\nu(L, p)$ remains constant. Let's check this.

Concentrate on just a short segment of L. As z traverses it, the rotation of $(z - p)$ will depend continuously on p unless[1] p crosses L. In other words, if we move p a tiny bit then the rotation angle will likewise only change a tiny bit. Since the winding number of L is just the sum of the rotations due to all its segments, it follows that it too depends continuously on the location of p: a tiny movement of p to $\widetilde{p}$ can only produce a tiny change $[\nu(L, \widetilde{p}) - \nu(L, p)]$ in the winding number. But since this small difference is an *integer*, it must be exactly 0. Done.

Since L winds round each point of D_j the same number of times, it follows that we can attach a winding number ν_j to the set as a whole. Verify the values of ν_j given in the figure.

The "inside" can now be *defined* to consist of those D_j for which $\nu_j \neq 0$, while the remaining D_j constitute the "outside". Thus in [7.2] we find that $D_1 \cup D_3$ is the inside, while $D_2 \cup D_4$ is the outside.

The "correctness" of this definition will become apparent in the next chapter.

7.1.3 Finding Winding Numbers Quickly

In [7.2] we found the winding numbers directly from the definition: we strenuously followed the curve with our finger (or eye) and counted revolutions. For a really complicated loop this could literally become a headache. We now derive a much quicker and more elegant method of visually computing winding numbers.

If a point r moves around without crossing a loop K then $\nu(K, r)$ remains constant, but what happens when the point *does* cross K? Consider [7.3]. On the far left, close to the loop K, is the point r; the rest of K is off the picture, and the number of times it winds round r is $\nu(K, r)$. The time-lapse pictures in [7.3] show r moving towards the loop, which itself deforms so as to avoid being crossed, finally ending up at the point s.

Now since the moving point never crosses the loop, the winding number remains constant throughout the process. But on the far right, the new loop can be thought of as the union of the old loop K, together with the new circle L. Thus,

$$\begin{aligned} \nu(K, r) &= \nu(K, s) + \nu(L, s) = \nu(K, s) - 1 \\ \Rightarrow \nu(K, s) &= \nu(K, r) + 1 . \end{aligned}$$

[1] Consider the behaviour of the rotation due to a short segment of L as p crosses it.

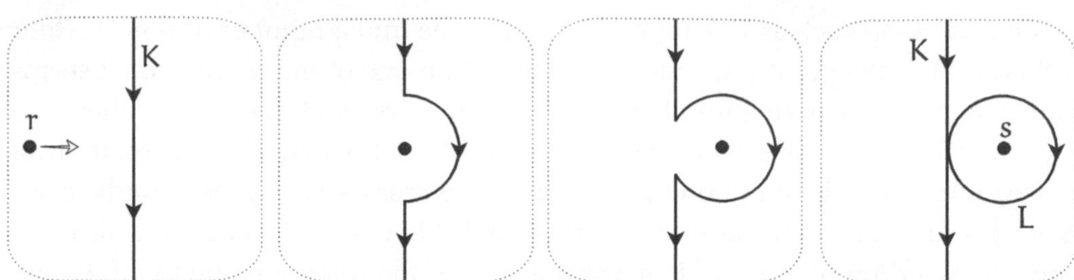

[7.3] Crossing Rule: If the loop is travelling from our left to our right as we cross it, $\boldsymbol{\nu} \rightsquigarrow \boldsymbol{\nu}+\mathbf{1}$. *For imagine that* K *deforms and moves out of our way, so that we never cross it, then the winding number does not change. But on the far right,* K *has evolved into* K *and a clockwise circle,* L*, so* $\nu(K,r) = \nu(K,s) + \nu(L,s) = \nu(K,s) - 1$*, and therefore* $\boldsymbol{\nu}(\mathbf{K},\mathbf{s}) = \boldsymbol{\nu}(\mathbf{K},\mathbf{r}) + \mathbf{1}$*, as claimed.*

Imagining ourselves at r, looking towards K as we approach it, we may express this result in the form of the following very useful *crossing rule*:

> *If* K *is moving from our left to our right* [our right to our left] *as we cross it, its winding number around us increases* [decreases] *by one.* (7.1)

Using this result, it is incredibly quick and easy to find the ν_j's for even the most complicated loop. Try it out on [7.2]. Starting your journey well outside L, where you know that the winding number is zero, move from region to region, using crossing rule (7.1) to add or subtract one at each crossing of L.

An immediate consequence of this idea is a connection between $n = \nu(K,p)$ and the number of intersection points of K with a ray emanating from p. Suppose that the ray is in general position in the sense that it doesn't pass through any self-intersection points of K, nor is it tangent to K. If a point q on this ray is sufficiently distant from p then clearly K cannot wind around it; thus as we move along the ray from p to q the winding number changes by n. But the winding number only changes when we cross K, and only one unit per crossing. The ray must therefore intersect K at least $|n|$ times. However, in addition to these $|n|$ necessary crossings there may be additional cancelling *pairs* of crossings. In general, then, the number of intersection points will be $|n|$, or $|n| + 2$, or $|n| + 4$, etc. Figure [7.4] illustrates these possibilities for a case in which $n = 2$, each intersection point being marked with $\oplus$ or $\ominus$ according as the winding number increases or decreases as it is crossed.

7.2 Hopf's Degree Theorem

7.2.1 *The Result*

We have discussed the fact that for a fixed loop and a continuously moving point, the winding number only changes when the point crosses the loop. But it is clear

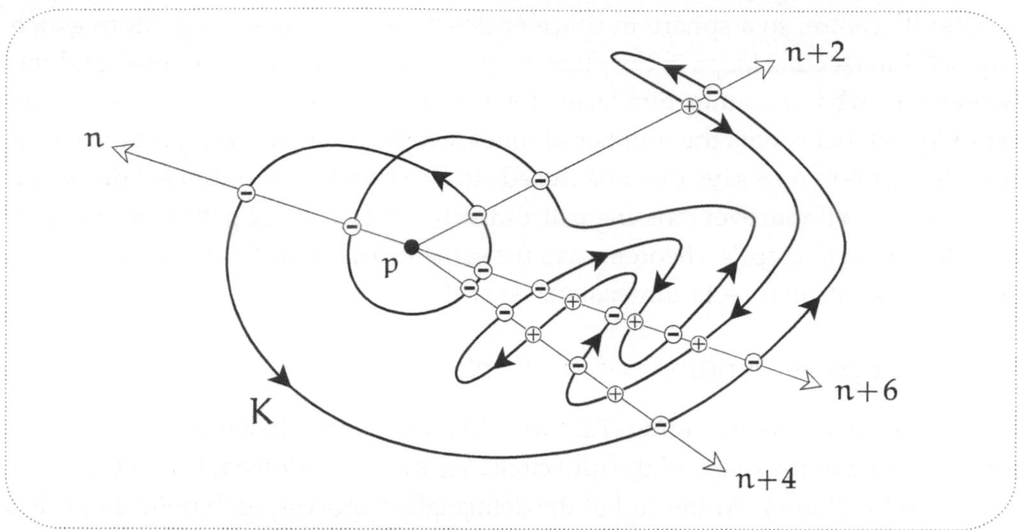

[7.4] A ray emanating from p intersects K at least $|\nu(K, p)|$ times. *For as we move along the ray towards infinity, the winding number must eventually drop to zero. But the initial value $n = \nu(K, p)$ only changes when we cross K, and only one unit at a time, so there must be at least $|n|$ such crossings. But, in addition to these $|n|$ necessary crossings, there may be additional cancelling* pairs *of crossings.*

that the same must be true of a fixed point and a continuously moving loop: the winding number of the evolving loop can only change if it crosses the point, and it changes by ± 1 according to the same crossing rule as before. Thus if a loop K can be continuously deformed into another loop L without ever crossing a point p, the winding numbers of K and L around p will be equal.

It is natural to ask if the converse is also true: if K and L wind round p the same number of times, is it always possible to deform K into L without ever crossing p? This is certainly a more subtle question, but by drawing examples you will be led to suspect that it is true. In this section we will confirm this hunch, so establishing that

> *A loop* K *may be continuously deformed into another loop* L, *without ever crossing the point* p, *if and only if* K *and* L *have the same winding number round* p. (7.2)

At the end of the next chapter, this will turn out to be the key to understanding one of the central results of complex analysis.

The result in (7.2) is the simplest example of a remarkable topological fact, called *Hopf's Degree Theorem*, that is valid in any number of dimensions. In the 2-dimensional complex plane, a point can be surrounded using a closed 1-dimensional curve—a loop. In 3-dimensional space, a point can be surrounded using a closed 2-dimensional surface. Just as a circle in the plane winds once around its centre, so a sphere in space encloses its centre just once. More generally, self-intersecting loops in the plane may enclose a point several times, and this is precisely what ν counts. Similarly, it is possible to define a more general concept (*degree*) that counts the number of times a surface surrounds a point in space. Hopf's Theorem now says that one closed surface may be continuously deformed into another, without ever crossing p, if and only if they enclose p the same number of times. Indeed, Hopf's Theorem says the same is true of n-dimensional surfaces enclosing points in $(n+1)$-dimensional space!

*7.2.2 Loops as Mappings of the Circle**

As a first step to understanding (7.2), we will look at loops in a new way. Let C be a rubber band in the shape of the unit circle. We may now deform C into the shape of any desired loop L. At the end of the deformation process, each point z of C has been brought to a definite image point w on L, and thus L may be thought of as the image of C under a continuous mapping $w = \mathcal{L}(z)$. See [7.5].

As θ varies from 0 to 2π, $z = e^{i\theta}$ moves round C once and w moves round L once, the length R and angle Φ of w varying continuously with θ. We may write

$$w = \mathcal{L}(e^{i\theta}) = R(\theta)\, e^{i\Phi(\theta)},$$

where $R(\theta)$ and $\Phi(\theta)$ are continuous functions. By rotating L (if necessary) we can ensure that $\mathcal{L}(e^{i0})$ is a positive real number, so that we may set $\Phi(0) = 0$. The net rotation of w after it has returned to its starting point is then given by $\Phi(2\pi) = 2\pi\nu$.

Clearly, the varying length of w is something of a red herring when it comes to understanding winding numbers, and we now remove this distraction by pulling each point w of L radially onto the point $\widehat{w} = w/|w|$ on the unit circle, so obtaining

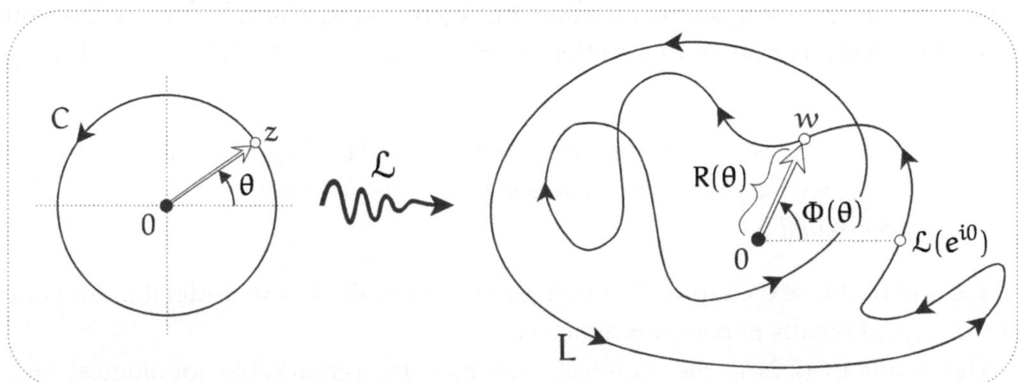

[7.5] A closed loop L may be viewed as the image of the unit circle C under a continuous mapping $\mathcal{L}$. *As θ goes from 0 to 2π, $z = e^{i\theta}$ traverses C once, and $w = \mathcal{L}(e^{i\theta}) = R(\theta)\, e^{i\Phi(\theta)}$ traverses L once. If $\Phi(0) = 0$ then the net rotation of w after it has returned to its starting point is given by $\Phi(2\pi) = 2\pi\nu$.*

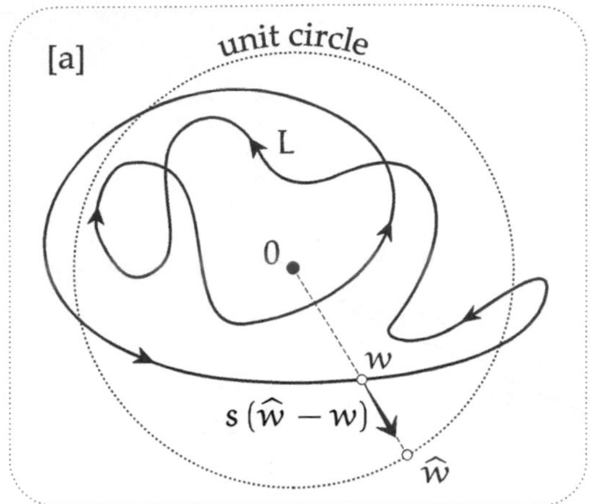

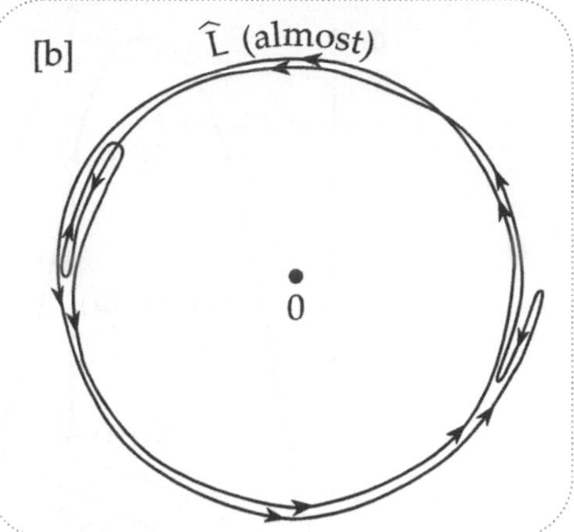

[7.6] [a] The loop L can be gradually pushed onto the unit circle by moving w along $(\widehat{w} - w)$. *As s increases from 0 to 1, $\mathcal{L}_s(z) = w + s\,(\widehat{w} - w)$ moves radially from w to $\widehat{w}$ on the unit circle.* **[b]** *With a value of s close to 1, the loop has almost evolved into its final form, $\widehat{\mathrm{L}}$, given by $\widehat{w} = \widehat{\mathcal{L}}\,(e^{i\theta}) = e^{i\Phi(\theta)}$.*

a standardized representation $\widehat{\mathrm{L}}$. We can even give an explicit prescription for gradually deforming L into $\widehat{\mathrm{L}}$. See [7.6a]. Since $(\widehat{w} - w)$ is the complex number from a point on L to its destination on $\widehat{\mathrm{L}}$, the point that is a fraction s of the way there is

$$\mathcal{L}_s(z) = w + s\,(\widehat{w} - w)\,. \tag{7.3}$$

As s varies from 0 to 1, $\mathcal{L}_s(\mathrm{C})$ gradually (and reversibly) changes from L into $\widehat{\mathrm{L}}$. Figure [7.6b] shows $\mathcal{L}_s(\mathrm{C})$ for a value of s close to 1. Finally, note the obvious fact that as we gradually pull L radially onto $\widehat{\mathrm{L}}$, the origin is not crossed.

With lengths disposed of in this way, we are now dealing with a mapping $\widehat{\mathcal{L}}$ from the unit circle C to the standardized loop $\widehat{\mathrm{L}}$ on the unit circle, where

$$\widehat{w} = \widehat{\mathcal{L}}\,(e^{i\theta}) = e^{i\Phi(\theta)}\,. \tag{7.4}$$

In this context, it is common to speak of the *degree* of the mapping $\widehat{\mathcal{L}}$ which produces $\widehat{\mathrm{L}}$, rather than of the "winding number" of $\widehat{\mathrm{L}}$ (or L). The single real function $\Phi(\theta)$ completely describes the mapping $\widehat{\mathcal{L}}$, and [7.7] shows how we can immediately read off the degree of $\widehat{\mathcal{L}}$ (i.e. ν) from the graph of $\Phi(\theta)$. Make sure you are comfortable with the meaning of such a graph. For example, if z moves at unit speed round C, what does the slope (including the sign) of the graph represent?

7.2.3 *The Explanation**

The archetypal mapping of degree ν is $\widehat{\mathcal{J}_\nu}\,(z) = z^\nu$, for which $\Phi(\theta) = \nu\theta$. Its straight-line graph is shown in [7.7]. As z travels once round C at unit speed, $\widehat{w}$ travels once round $\widehat{\mathrm{J}_\nu}$ with speed $|\nu|$, completing $|\nu|$ circuits of the unit circle [counterclockwise if $\nu > 0$; clockwise if $\nu < 0$].

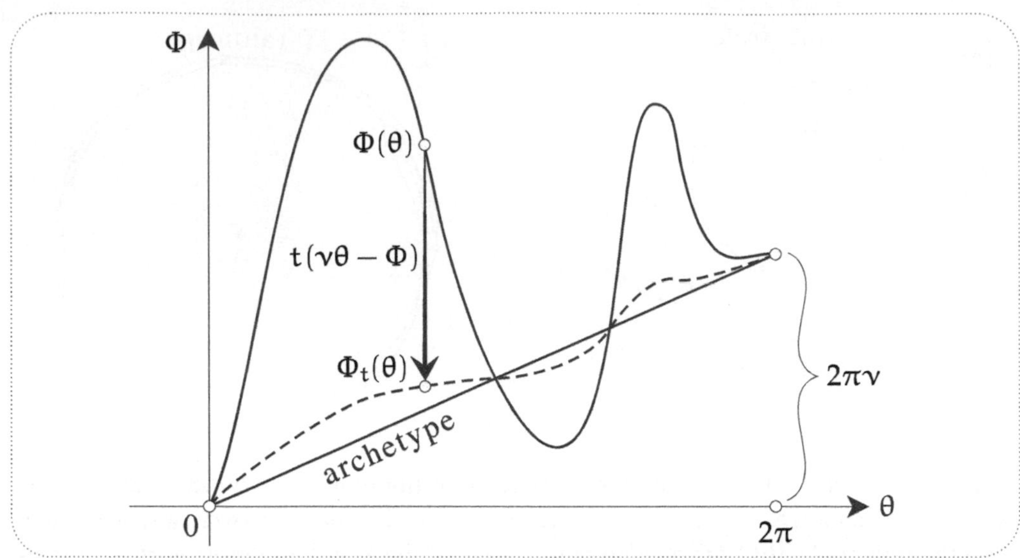

[7.7] Gradual evolution of a loop into the archetypal loop of the same winding number. *If we imagine* $\widehat{L}$ *in the previous figure to be an elastic string wrapped around a cylinder, the slack will automatically be taken up, and the actual mapping* $\widehat{\mathcal{L}}$ *of degree* ν *will evolve into the archetypal mapping of degree* ν*, namely,* $\widehat{\mathcal{J}}_\nu(z) = z^\nu$*, for which* $\Phi(\theta) = \nu\theta$*. This can be achieved explicitly by following the illustrated evolution of* $\Phi_t(\theta) = \Phi(\theta) + t\,[\nu\theta - \Phi(\theta)]$ *from* $t = 0$ *to* $t = 1$.

To see how a typical standardized loop $\widehat{L}$ of winding number ν is related to the archetypal loop $\widehat{J_\nu}$, we return to the example in [7.6b], for which $\nu = 2$. Thinking of the unit circle as the boundary of a solid cylinder, and recalling that the loop is an elastic band that wishes to contract, what will happen if we release $\widehat{L}$? The slack will be taken up, and $\widehat{L}$ will automatically contract itself into the archetypal loop $\widehat{J_2}$. This convincing mental image of the rubber band contracting into $\widehat{J_\nu}$ can be formalized to show that

> *Any* $\widehat{L}$ *of winding number* ν *can be continuously deformed into the archetypal loop* $\widehat{J_\nu}$*, and vice versa.*

The process of "taking up the slack" can be explicitly described in terms of the graph of the Φ that describes $\widehat{L}$. As t varies from 0 to 1, the graph of

$$\Phi_t(\theta) = \Phi(\theta) + t\,[\nu\theta - \Phi(\theta)]$$

continuously and reversibly evolves from the graph of the general Φ into the straight-line graph of the archetype. The dashed curve in [7.7] is the graph of Φ_t for a value of t close to one. Defining

$$\widehat{\mathcal{L}}_t(e^{i\theta}) = e^{i\Phi_t(\theta)}, \tag{7.5}$$

$\widehat{\mathcal{L}}_t(C)$ therefore evolves continuously and reversibly from $\widehat{L}$ into $\widehat{J_\nu}$, as t varies from 0 to 1.

The explicit two-stage deformation given above [(7.3) followed by (7.5)] allows us to deform any loop of winding number ν into the archetypal loop $\widehat{J_\nu}$, and without the origin ever being crossed. Conversely, by reversing these steps, $\widehat{J_\nu}$ may be deformed into any loop of winding number ν. This demonstrates (7.2), for if K and L both have winding number ν, we may first deform K into $\widehat{J_\nu}$, and then deform $\widehat{J_\nu}$ into L.

7.3 Polynomials and the Argument Principle

Let A, B, and C be the complex numbers from the fixed points a, b, and c to the variable point z. Figure [7.8] shows a circle Γ and its image $f(\Gamma)$ under the cubic mapping

$$f(z) = (z-a)\,(z-b)\,(z-c) = ABC\,.$$

Notice that Γ encircles two zeros of the mapping, while $f(\Gamma)$ has a winding number of 2 about zero. This is no accident. Since angles add when we multiply complex numbers, the number of revolutions executed by ABC is just the sum of the revolutions executed separately by each of A, B, and C. But as z goes round Γ once, A and B both execute a complete revolution, while the direction of C merely oscillates. Thus $\nu[f(\Gamma), 0] = 2$.

If we enlarged Γ so that it encircled c, then C would also execute a complete revolution, and the winding number would increase to 3. Once again, the number of points inside Γ that are mapped to 0 is the winding number of the image about that point.

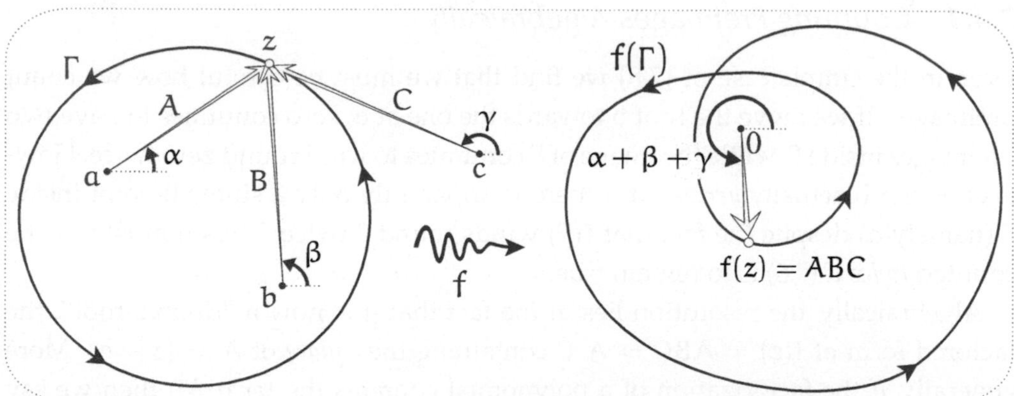

[7.8] If a simple loop Γ winds once around m roots of a polynomial $P(z)$, then $\nu[P(\Gamma), 0] = m$. *In the illustrated cubic example, the number of revolutions executed by $f(z) = ABC$ is just the sum of the revolutions executed separately by each of A, B, and C. But as z goes round Γ once, A and B both execute a complete revolution, while the direction of C merely oscillates, so $f(z)$ winds around 0 twice.*

It is clear that this result is independent of the circularity of Γ, and that it generalizes to the case of a polynomial $P(z)$ of arbitrary degree: *If a simple loop* Γ *winds once around* m *roots of* $P(z)$, *then* $\nu[P(\Gamma), 0] = m$.

Roots are simply preimages of 0, and from the geometric viewpoint there is nothing special about this particular image point. Consequently, in future we will look at the preimages of a general point p and we will call these preimages p-*points* of the mapping.

The *Argument Principle* is a tremendous extension of the above result. Not only does it apply to general analytic mappings but it also contains the converse statement that the winding number tells us the number of preimages:

$$\text{If } f(z) \text{ is analytic inside and on a simple loop } \Gamma, \text{ and } N \text{ is the number of } p\text{-points [counted with their multiplicities] inside } \Gamma, \text{ then } N = \nu[f(\Gamma), p]. \tag{7.6}$$

The meaning of the expression "counted with their multiplicities" will be explained in the next section.

We wish to stress that this result is only peripherally connected with the conformality that has been so central to all our previous thinking. In fact the Argument Principle is a consequence of a still more general *topological* fact concerning mappings that are merely continuous. Our main effort will therefore be directed towards understanding the general result (due to Poincaré), of which the Argument Principle is merely a special case.

7.4 A Topological Argument Principle*

7.4.1 *Counting Preimages Algebraically*

Even in the simple case of [7.8] we find that we must be careful how we count preimages. If we move the root b towards the one at a, zero continues to have two preimages inside Γ, while the image of Γ continues to wind round zero twice. However, when b actually *arrives* at a there is apparently only a single 0-point inside Γ (namely a) despite the fact that $f(\Gamma)$ winds round 0 twice. Thus a must now be counted *twice* if (7.6) is to remain true.

Algebraically, the resolution lies in the fact that a is now a "double-root", the factored form of $f(z) = ABC = A^2C$ containing the *square* of $A = (z - a)$. More generally, if the factorization of a polynomial contains the term A^n then we say that the root a is a 0-point of *algebraic multiplicity* n, and we must count it with this multiplicity in (7.6).

When $n > 1$ there is a further significance to the point a—it is a critical point of the polynomial. In the cubic mapping $f(z) = ABC$, let a, b, and c be real, so that $f(z)$ is real-valued on the real axis. The far left of [7.9] shows the ordinary graph

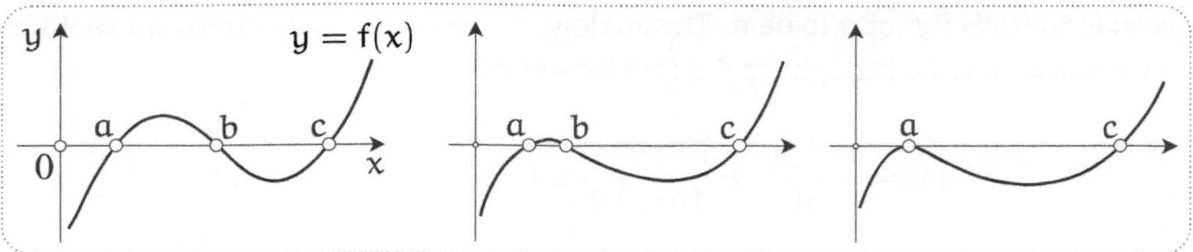

[7.9] When roots coalesce, the derivative must vanish. *The same applies in $\mathbb{C}$, for if $f'(a) \neq 0$ then an infinitesimal disc centred at a is amplitwisted to an infinitesimal disc centred at 0, so that points close to a cannot map to 0.*

of f in this case. As b moves towards a, the slope at a is forced to decrease, finally vanishing at the moment of b's arrival.

In general this vanishing of the derivative (now amplitwist) must occur wherever two or more roots of a polynomial coalesce. Look again at [7.9]. If $f'(a) \neq 0$ then the graph is not flat at a and so neighbouring points cannot map to zero; but this is precisely what we insist on when we merge b into a. Essentially the same thing happens when we return to $\mathbb{C}$, for if $f'(a) \neq 0$ then an infinitesimal disc centred at a is amplitwisted to an infinitesimal disc centred at 0, so that points close to a cannot map to 0.

This conclusion can be refined. If the root a of a polynomial $P(z)$ has multiplicity n then P may be factorized as $A^n\,\Omega(z)$, where $\Omega(a) \neq 0$. It follows by simple calculation [exercise] that the first $(n-1)$ derivatives of P vanish at a, so that a is a critical point of order at least $(n-1)$. We shall see in a moment that the order actually is $(n-1)$.

We next seek to extend the idea of counting preimages "with algebraic multiplicities" to analytic mappings in general. Suppose that a is a p-point of an analytic mapping $f(z)$, i.e., it is a 0-point of $f(z)-p$. What should the algebraic multiplicity of this root be? It is only possible to answer this question because of the remarkable fact that an analytic f can always be represented as a convergent Taylor series in the neighbourhood of a non-singular point. Thus if $\Delta = (z-a)$ is the small complex number from a to a nearby point z, we may write

$$f(z) - p = f(a+\Delta) - f(a) = \frac{f'(a)}{1!}\Delta + \frac{f''(a)}{2!}\Delta^2 + \cdots.$$

The first nonzero term on the right is the one that dominates the local behaviour of $f(z)-p$ and decides what the multiplicity of a should be. Typically a will not be a critical point $[f'(a) \neq 0]$ and so this local behaviour is like Δ to the first power; we say that a is a *simple* root with multiplicity $+1$.

Now consider the rarer case in which a is a critical point. If the order of the critical point is $(n-1)$, so that $f^{(n)}$ is the first nonvanishing derivative at a, then the dominant first term is proportional to Δ^n, and we correspondingly define the *algebraic multiplicity*[2] of a to be n. The analogy between this definition and that for polynomials may be brought to the fore by setting

$$\Omega(z) = \frac{f^{(n)}(a)}{n!} + \frac{f^{(n+1)}(a)}{(n+1)!}\Delta + \frac{f^{(n+2)}(a)}{(n+2)!}\Delta^2 + \cdots,$$

where $f^{(n)}(a)$ is the first nonvanishing derivative. The previous equation can now be written in "factorized" form as

$$f(z) - p = \Omega(z)\,\Delta^n, \tag{7.7}$$

where $\Omega(a) \neq 0$. From this point of view, the only difference between a general analytic mapping and a polynomial is that the latter has a single, "once and for all" factorization, while the former generally requires a different factorization of type (7.7) in the neighbourhood of each p-point.

7.4.2 Counting Preimages Geometrically

Recall that we wish to explain (7.6) as a special case of a more general result dealing with mappings that are merely continuous. But since the very notion of algebraic multiplicity is meaningless for such general mappings, how can we even frame a proposition of type (7.6)?

What is needed is a geometric way of counting preimages that will agree with the previous algebraic definition if we specialize to analytic mappings. To discover the appropriate definition we should therefore return to analytic mappings and ask, "What is the *geometric fingerprint* of a p-point of given algebraic multiplicity?"

Consider the effect of an analytic f on an infinitesimal circle C_a centred at a *simple* p-point a. Since $f'(a) \neq 0$, C_a is amplitwisted to an infinitesimal circle centred at p. We see that the winding number (+1) of this image round p is the same as the algebraic multiplicity of a. In fact, quite generally, if the algebraic multiplicity is n then the winding number of the image will also be n. This is the sought-after geometric fingerprint.

To verify this statement, remember that the local behaviour of f near to a is given by (7.7):

$$f(z) = f(a+\Delta) = p + \Omega(z)\,\Delta^n,$$

with $\Omega(a) \neq 0$. Thus the basic explanation is that as Δ revolves round C_a once, Δ^n rotates n-times as fast, and therefore $f(z)$ completes n revolutions round p. If

[2] Also known as *order* or *valence*.

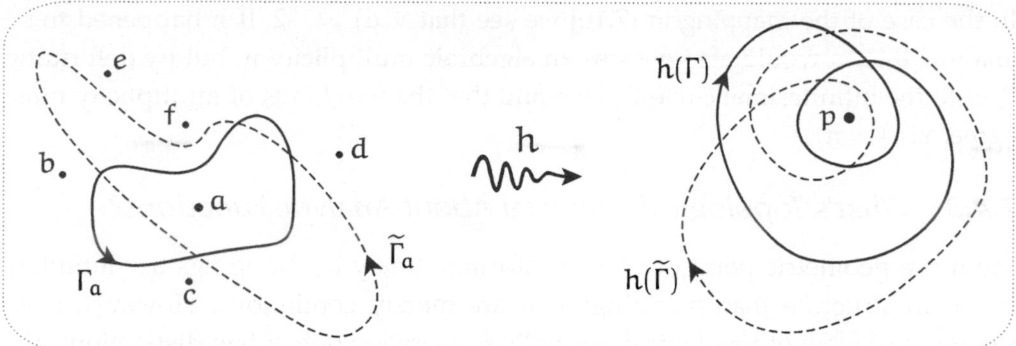

[7.10] Topological Multiplicity as generalization of Algebraic Multiplicity. *On the left are* p-points *of the continuous mapping* h, *i.e., points that are mapped to* p *by* h, *and* Γ_a *is a loop around* a *(and only* a*). If we continuously deform* Γ_a *into another such loop* $\widetilde{\Gamma}_a$ *without crossing* a *(or any other* p*-point) then* $h(\Gamma_a)$ *will continuously deform, but* its winding number around p will not change. *Thus* $\nu(a) \equiv \nu[h(\Gamma_a), p]$ *is really a topological invariant associated with the point* a, *and it is called its* topological multiplicity. *If we specialize* h *to be a complex polynomial, then the topological multiplicity is equal to the* algebraic multiplicity *of* a *as a repeated root of* $h(z) = p$.

Ω were a constant then this argument would be beyond reproach, and we can now do a little calculation to show that a variable Ω does not disturb the conclusion:

$$\begin{aligned} \nu[f(C_a), p] &= \nu[f(C_a) - p, 0] = \nu[\Delta^n \, \Omega(C_a), 0] \\ &= n \, \nu[\Delta, 0] + \nu[\Omega(C_a), 0] \, . \end{aligned} \tag{7.8}$$

As we shrink C_a down towards a, $\Omega(C_a)$ will shrink down towards $\Omega(a)$, but since $\Omega(a) \neq 0$ this implies that the image of a sufficiently small C_a will not wind round 0: $\nu[\Omega(C_a), 0] = 0$. Since $\nu[\Delta, 0] = 1$, we conclude that $\nu[f(C_a), p] = n$, as claimed.

We may now broaden our horizon and use the above idea to define "multiplicity" for a mapping $h(z)$ that is merely continuous. Let Γ_a be any simple loop round a that does not contain other p-points. Figure [7.10] shows such a loop as well as some other p-points b, c, etc. If we continuously deform Γ_a into another such loop $\widetilde{\Gamma}_a$ without crossing a (or any other p-point) then $h(\Gamma_a)$ will continuously deform into $h(\widetilde{\Gamma}_a)$ without ever crossing p, and so

$$\nu[h(\widetilde{\Gamma}_a), p] = \nu[h(\Gamma_a), p] \, .$$

Thus, without specifying Γ_a further, we may unambiguously define the *topological multiplicity*[3] of a to be

$$\nu(a) \equiv \nu[h(\Gamma_a), p] \, .$$

[3] Also known as the *local degree* of h at a.

In the case of the mapping in [7.10] we see that $\nu(a) = -2$. If h happened to be analytic then a would also possess an algebraic multiplicity n, but by deforming Γ_a into the infinitesimal circle C_a, we find that the two kinds of multiplicity must agree: $\nu(a) = n$.

7.4.3 What's Topologically Special About Analytic Functions?

From the geometric point of view, conformal (analytic) mappings are infinitely richer in structure than mappings that are merely continuous. However, from the point of view of topological multiplicity there are only a few distinctions, the following being one of the most striking:

> $\nu(a)$ *is always positive for analytic functions, while it can be negative for nonanalytic functions.*

For example, the mapping in [7.10] cannot possibly be analytic. The positivity of $\nu(a)$ for analytic functions has already been established, so we need only look more closely at the possibility of negative multiplicities for nonanalytic functions.

Since general continuous mappings can actually behave in rather wild ways, let us restrict ourselves to nonanalytic mappings that are at least differentiable in the real sense. For example, consider $h(z) = \overline{z}$. The unique preimage of p is $a = \overline{p}$, and any simple loop Γ_a round this point is reflected by h into a loop that goes once round p in the opposite direction. Thus $\nu(a) = -1$.

More generally, recall [see p. 236] that the local effect of such a mapping at a p-point a is (after translation to p) a stretch by some factor ξ_a in one direction, another stretch by some factor η_a in the perpendicular direction, and finally a rotation through some angle ϕ_a. For example, conjugation has (taking the first expansion to be horizontal) $\xi_a = +1$, $\eta_a = -1$, $\phi_a = 0$. Of course these values of ξ_a, η_a, and ϕ_a are only independent of a because $h(x+iy) = x-iy$ depends linearly on x and y; most mappings have values that *do* depend on the point a.

An infinitesimal circle C_a centred at a is generally distorted into an infinitesimal ellipse E_p centred at p, and if the two expansion factors have the same sign then the mapping *preserves orientation* so that E_p circulates in the same sense as C_a and $\nu(a) = +1$. However, if ξ_a and η_a have opposite signs then the mapping is *orientation reversing*: it turns C_a inside out, so that E_p goes round p in the opposite direction and $\nu(a) = -1$. Our previous example of conjugation was of this type. In summary, we have

$$\nu(a) = \text{the sign of } (\xi_a \eta_a).$$

The local linear transformation at a is encoded by the Jacobian matrix $J(a)$, and we can use its determinant $\det[J(a)]$ to give a more practical formula for the topological multiplicity. We know from linear algebra that the determinant of a constant 2×2 matrix is the factor (including a sign for orientation) by which the area of a figure is expanded. Likewise, $\det[J(a)]$ measures the *local* expansion factor for area at a, and this is just $(\xi_a \eta_a)$. Thus

$$\nu(a) = \text{the sign of } \det[J(a)] . \tag{7.9}$$

Of course this formula is vacuous if $\det[J(a)] = 0$. Geometrically this means that the transformation is locally crushing at a; just as for analytic mappings, such a place is called a *critical point*. However, while the local crushing at a critical point of an analytic mapping is perfectly symmetrical in all directions, this is not true of the more general mappings presently under consideration. For example, if $f(x+iy) = x - iy^3$ then [exercise] $\det[J] = -3y^2$, so although f leaves the horizontal separation of points alone, all the points on the real axis are critical points as a result of crushing in the vertical direction.

This example also serves to illustrate another difference:

> *The critical points of an analytic mapping can be distinguished purely on the basis of topological multiplicity; those of a nonanalytic mapping cannot.*

For analytic functions we have seen that $\nu(a) = +1$ if and only if a is not critical. In the nonanalytic case $\nu(a) = \pm 1$ if a is not critical, but it is also possible for a *critical point* to have one of these multiplicities. Indeed, you can check [exercise] that the above example yields $\nu(a) = -1$ for noncritical and critical points alike.

One final difference:

> $\nu(a)$ *is never zero for analytic mappings, but it can vanish for nonanalytic mappings.*

In the next section we will provide a simple example of a nonanalytic mapping possessing such p-points of vanishing topological multiplicity. Can you think of an example for yourself?

7.4.4 A Topological Argument Principle

Let Γ be a simple loop, and let $h(z)$ be a continuous mapping such that only a finite number of its p-points lie inside Γ. We will show that

> *The total number of* p*-points inside* Γ *(counted with their topological multiplicities) is equal to the winding number of* $h(\Gamma)$ *round* p. (7.10)

If h is analytic this reduces to (7.6). The rest of the chapter will be devoted to mining and extending this simple yet profound result[4].

[4] When interpreted in terms of vector fields (as we shall do in Chapter 10) this is the key to a very surprising and beautiful fact called the *Poincaré–Hopf Theorem*.

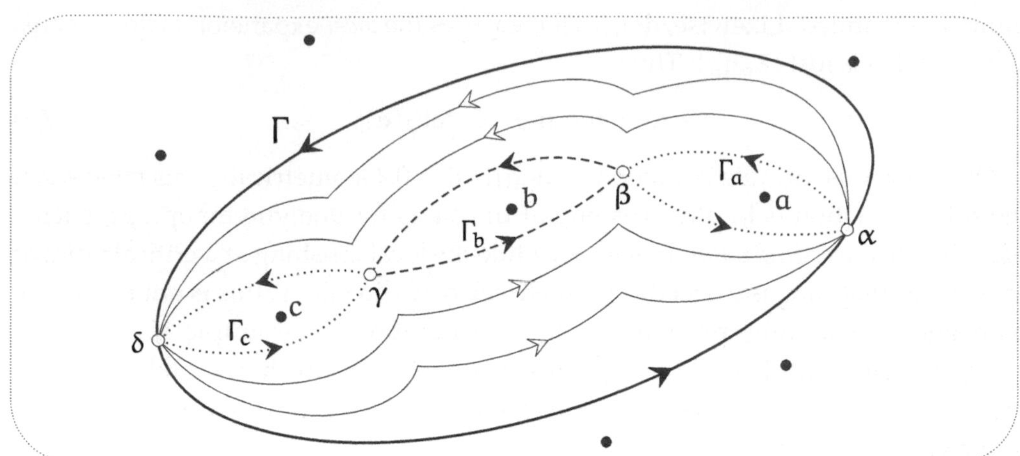

[7.11] The winding number $\nu[h(\Gamma), p]$ equals the sum of the topological multiplicities of the p-points inside Γ. *For this winding number will not change value if we deform Γ as shown into a sum of loops around the individual p-points within.*

Before explaining (7.10), let us describe one of its immediate consequences. As in [7.2], $h(\Gamma)$ will generally partition the plane into a number of sets, and the above result then says that every point in D_j has the same number of preimages lying inside Γ, namely, ν_j. For example, if $h(\Gamma)$ is a simple loop then it partitions the plane into just two sets, namely, its interior and its exterior. If p is in the interior then the result says that the total number of p-points inside Γ is 1. But if h is analytic then these p-points must have strictly positive multiplicities, and so there is exactly one preimage for each point inside $h(\Gamma)$. In other words, we have shown that

> *If an analytic function* h *maps* Γ *onto* $h(\Gamma)$ *in one-to-one fashion, then it also maps the interior of* Γ *onto the interior of* $h(\Gamma)$ *in one-to-one fashion.*

This is *Darboux's Theorem*.

To explain (7.10), consider [7.11]. This shows three p-points a, b, and c lying inside Γ while others lie scattered outside. The essential idea is that we can gradually deform Γ (as shown) into the doubly-pinched loop $\alpha\beta\gamma\delta\gamma\beta\alpha$, which we will call $\widetilde{\Gamma}$. Since no p-points were crossed in the deformation process, $h(\widetilde{\Gamma})$ will wind round p the same number of times as $h(\Gamma)$. The rest is almost obvious: $\widetilde{\Gamma}$ is made up of $\Gamma_a = \alpha\beta\alpha$, $\Gamma_b = \beta\gamma\beta$, $\Gamma_c = \gamma\delta\gamma$, and the winding numbers of their images round p are, by definition, the topological multiplicities of a, b, and c.

We will spell this out in perhaps unnecessary detail. Let K be a path that is not necessarily closed, and define $\mathcal{R}(K)$ to be the net rotation of $h(z)$ round p as z traverses K. For example, if K is closed then $\mathcal{R}(K) = 2\pi\nu[h(K), p]$. Then,

$$\begin{aligned}
2\pi\,\nu[h(\Gamma),p] &= 2\pi\,\nu[h(\widetilde{\Gamma}),p] \\
&= \mathcal{R}(\alpha\beta\gamma\delta\gamma\beta\alpha) \\
&= \mathcal{R}(\alpha\beta)+\mathcal{R}(\beta\gamma)+\mathcal{R}(\gamma\delta)+\mathcal{R}(\delta\gamma)+\mathcal{R}(\gamma\beta)+\mathcal{R}(\beta\alpha) \\
&= \mathcal{R}(\alpha\beta\alpha)+\mathcal{R}(\beta\gamma\beta)+\mathcal{R}(\gamma\delta\gamma) \\
&= \mathcal{R}(\Gamma_a)+\mathcal{R}(\Gamma_b)+\mathcal{R}(\Gamma_c) \\
&= 2\pi[\nu(a)+\nu(b)+\nu(c)]\,.
\end{aligned}$$

Clearly this idea extends to any number of p-points a_1, a_2, etc. lying inside Γ:

$$\nu[h(\Gamma),p] = \sum_{a_j \text{ inside } \Gamma} \nu(a_j)\,. \tag{7.11}$$

7.4.5 Two Examples

Let us immediately illustrate the result with a concrete mapping:

$$h(x+iy) = x + i|y|\,.$$

In terms of our pastry analogy [p. 234] this corresponds to making a crease along the real axis and folding the bottom half of the plane onto the top half; in fact, since no stretching is required in this example, folding a simple piece of paper will do! If $\mathrm{Im}(p) > 0$ then p has two preimages, $a_1 = p$ and $a_2 = \overline{p}$. Figure [7.12] shows that $\nu(a_1) = +1$ and $\nu(a_2) = -1$, and in accord with (7.11), it also shows that if Γ contains a_1 and a_2 then $\nu[h(\Gamma),p] = 0$.

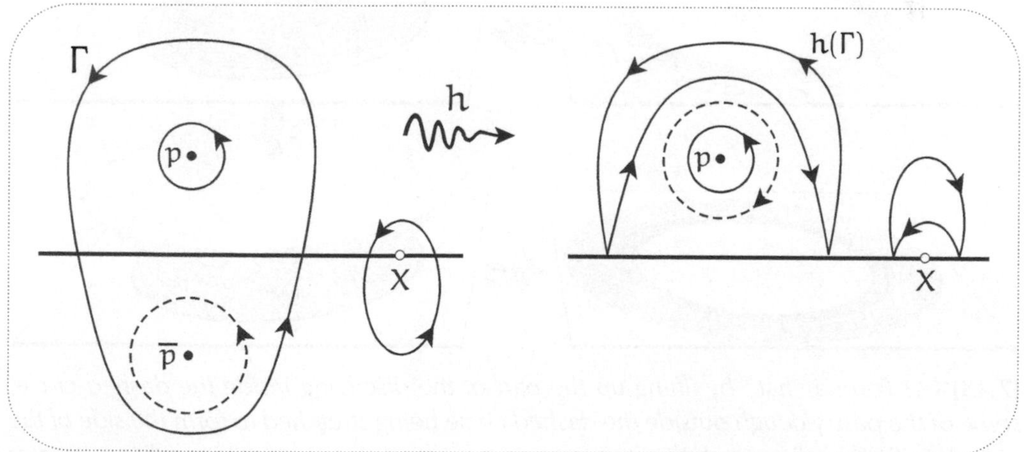

[7.12] Folding a piece of paper across the x-axis *is a continuous mapping* h*, explicitly given by* $h(x+iy) = x + i|y|$*. If* $\mathrm{Im}(p) > 0$*, as shown, then* p *has two preimages,* $a_1 = p$ *and* $a_2 = \overline{p}$*, so* $\nu(a_1) = +1$ *and* $\nu(a_2) = -1$*. Therefore, (7.11) predicts that* $\nu[h(\Gamma),p] = \nu(a_1)+\nu(a_2) = (+1)+(-1) = 0$*, which is indeed true!*

In general, note that $\nu[h(\Gamma), p] = 0$ merely implies that *either* there are no preimages inside Γ *or* the preimages have cancelling topological multiplicities, as above. However, if f is analytic and $\nu[f(\Gamma), p] = 0$ then the conclusion is quite definite: there are no preimages inside Γ. Later we shall return to this important point.

Returning to the example, observe that if $p = X$ is real then there is only one preimage, namely X, and $\nu(X) = 0$. We can look at this in a nice way: as we move p towards X, the two preimages a_1 and a_2 also move towards X, and when they finally coalesce at X their opposite multiplicities annihilate. As we previously pointed out, such points of vanishing multiplicity can *only* exist for nonanalytic mappings.

Figure [7.13] shows a second more elaborate example, in which we subject a unit disc of pastry to a three-stage transformation H that leaves the boundary Γ (the unit circle) fixed: $H(\Gamma) = \Gamma$. Here are the three stages: (A) form a "hat" by lifting up the part of the disc lying inside the dashed circle, some of the pastry dough outside the dashed circle being stretched to form the side of the cylinder. (B) radially stretch the disc forming the top of this "hat" till its radius is greater than one. (C) press down flat, i.e., project each point vertically down onto the plane.

If we pick a point p from the image set [bottom left] then the number of preimages (counted naively) lying in the original disc is the number of layers of pastry lying over p. In the final picture [bottom left] we have used the degree of shading to indicate the number of these layers: *one* over the lightly shaded inner disc; *two*

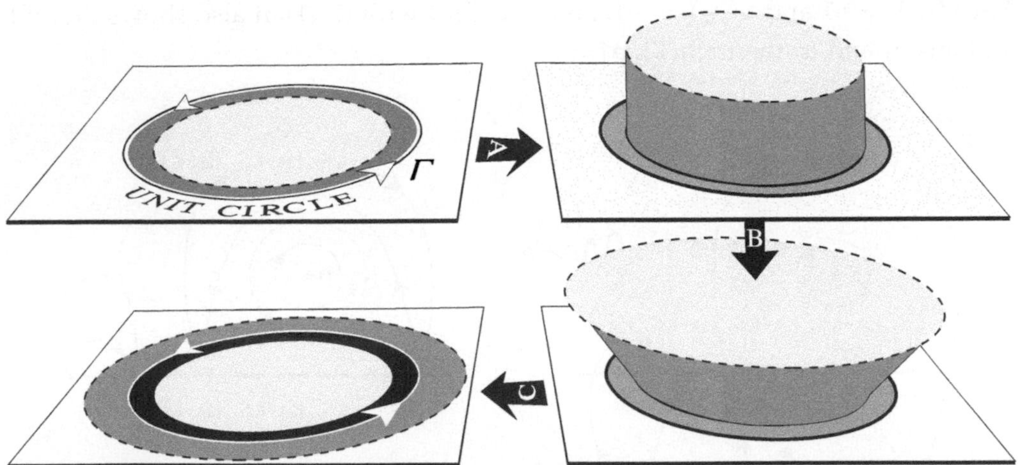

[7.13] *(A) Form a "hat" by lifting up the part of the disc lying inside the dashed circle, some of the pastry dough outside the dashed circle being stretched to form the side of the cylinder. (B) Radially stretch the disc forming the top of this "hat" till its radius is greater than one. (C) Press down flat, i.e., project each point vertically down onto the plane. The final* number *of layers [bottom left] is indicated by the shading:* one *layer over the lightly shaded inner disc;* two *over the darkly shaded outer ring; and* three *over the black ring. You can now geometrically verify that (7.11) works in all three regions.*

over the darkly shaded outer ring; *three* over the black ring. Make sure you can see this.

We can now check (7.11). For example, if p lies in the darkly shaded outer ring, $\nu[H(\Gamma), p] = \nu[\Gamma, p] = 0$ and so the multiplicities of the preimages of such points should sum to zero. By following the effect of the transformation on little loops round each of the two preimages, we confirm this prediction: one preimage has multiplicity $+1$ while the other has multiplicity-1.

Check for yourself that (7.11) continues to work if p instead lies in the inner disc or in the black ring.

7.5 Rouché's Theorem

7.5.1 The Result

Imagine walking a dog round and round a tree in a park, both you and the dog finally returning to your starting points. Further imagine that the dog is on one of those leashes of variable length, similar to a spring-loaded tape measure. On one such walk you keep the leash short, so that the dog stays at your heel. It is then clear that the dog is forced to walk round the tree the same number of times that you do. On another walk, though, you decide to let out the leash somewhat so that the dog may scamper about, perhaps even running circles around you. Nevertheless, provided that you *keep the leash short enough so that the dog cannot reach the tree*, then again the dog must circle the tree the same number of times as you.

Let the tree be the origin of $\mathbb{C}$, and let your walk be the image path traced by $f(z)$ as z traverses a simple loop Γ. Also, let the complex number from you to the dog be $g(z)$, so that the dog's position is $f(z) + g(z)$. The requirement that the leash not stretch to the tree is therefore

$$|g(z)| < |f(z)| \text{ on } \Gamma.$$

Under these circumstances, the previous paragraph states that

$$\nu[(f+g)(\Gamma), 0] = \nu[f(\Gamma), 0].$$

But the Argument Principle then informs us that

> *If* $|g(z)| < |f(z)|$ *on* Γ, *then* $(f+g)$ *must have the same number of zeros inside* Γ *as* f.

This is *Rouché's Theorem.*

Note that while $|g(z)| < |f(z)|$ is a sufficient condition for $(f+g)$ to have the same number of roots as f, it is not a necessary condition. For example, consider $g(z) = 2f(z)$.

7.5.2 *The Fundamental Theorem of Algebra*

A classic illustration of Rouché's Theorem is the Fundamental Theorem of Algebra, which states that a polynomial

$$P(z) = z^n + A\,z^{n-1} + B\,z^{n-2} + \cdots + E$$

of degree n always has n roots. The basic explanation is simple: if $|z|$ is large, the first term dominates the behaviour of $P(z)$ and the image of a sufficiently large origin-centred circle C will therefore wind n times round 0; the Argument Principle then says that $P(z)$ must have n roots inside C.

Rouché's Theorem merely allows us to make the above idea more precise. Let $f(z) = z^n$ be the first term of $P(z)$ and let $g(z)$ be the sum of all the rest, so that $f + g = P$. Now let C be the circle $|z| = 1 + |A| + |B| + \cdots + |E|$. Using the fact that $|z| > 1$ on C, it is not hard to show [exercise] that $|g(z)| < |f(z)|$ on C, and since f has n roots inside C (all at the origin), Rouché says that P must too.

Notice that we have not only confirmed the existence of the n roots, but have also narrowed down their location: they must all lie inside C. In the exercises you will see how Rouché's Theorem can often be used to obtain more precise information on the location of the roots of an equation.

7.5.3 *Brouwer's Fixed Point Theorem**

Sprinkle talcum powder on a cup of coffee and give it a stir. The little white specks will swirl around and eventually come to rest, the speck that was originally at z finally ending up at $g(z)$. If we stir it in a nice symmetrical way then the speck in the centre will remain motionless and its final position will be identical with its starting position. Such a place, for which $g(z) = z$, is called a *fixed point* of g.

Now stir the coffee in a really complicated way and let it again come to rest. Incredible as it may seem, at least one speck will have ended up exactly where it started! This is an example of *Brouwer's Fixed Point Theorem,* which asserts that any continuous mapping of the disc to itself will have a fixed point. Exercise 15 shows this to be true, but for the moment we wish to demonstrate a slightly different result: there must be a fixed point if the disc is mapped into its *interior* and there are at most a finite number of fixed points.

Let the disc D be $|z| \leqslant 1$; the condition that g map D into the interior of D is then $|g(z)| < 1$ for all z in D. Let $m(z)$ be the movement of z under the mapping, i.e., the connecting complex number from z to its destination $g(z)$:

$$m(z) = g(z) - z\,.$$

A fixed point then corresponds to no movement: $m(z) = 0$. Now let $f(z) = -z$. On the boundary of D (the unit circle) we have

$$|g(z)| < 1 = |f(z)|$$

and so Rouché's Theorem says that $m(z) = g(z) + f(z)$ has the same number of roots inside D as f has, namely, one.

If g is merely continuous then there can actually be several fixed points, some of which will necessarily have negative multiplicities, while if g is analytic then there can literally only be one.

7.6 Maxima and Minima

7.6.1 *Maximum-Modulus Theorem*

Take another look at the nonanalytic mapping H of [7.13], and note how the image of the disc "spills over" the image of its boundary: points inside Γ end up in the darkly shaded ring *outside* H(Γ). The central observation of this section is that such spilling over is quite impossible in the case of an analytic mapping:

If f is analytic inside and on a simple loop Γ then no point outside f(Γ) can have a preimage inside Γ. (7.12)

Let's see why. The Argument Principle tells us that the sum of the multiplicities $\nu(a_j)$ of the p-points inside Γ is $\nu[f(\Gamma), p]$, but if p is outside f(Γ) then (by definition) this is zero. Since $\nu(a_j)$ is strictly positive for an analytic function, we conclude that points outside f(Γ) have no preimages inside Γ. On the other hand, if p lies inside f(Γ) then $\nu[f(\Gamma), p] \neq 0$, and so there must be at least one preimage inside Γ. [Unlike (7.12), this is also true of nonanalytic mappings.]

Figure [7.14] illustrates an analytic f sending the shaded interior of Γ strictly to the shaded interior of f(Γ). Since f(Γ) winds round the darker region twice, its points have two preimages in Γ; we can think of this as arising from the overlap of two lightly shaded regions, one preimage per lightly shaded point.

One aspect of the "overspill" produced by H in [7.13] is that the points z which end up furthest from the origin (i.e., for which the modulus $|H(z)|$ is maximum) lie inside Γ. Conversely, the absence of overspill for an analytic f means that

The maximum of $|f(z)|$ on a region where f is analytic is always achieved by points on the boundary, never ones inside.

This is called the *Maximum-Modulus Theorem* and it is illustrated in [7.14]: the maximum of $|f(z)|$ is $|T| = |f(t)|$, where t lies on Γ.

The only exception to this result is the trivial analytic mapping $z \mapsto \text{const.}$, which sends every point to a single image point. To put this more positively, if we know that $|f|$ achieves its maximum at an interior point then $f(z) = \text{const}$.

As an elementary example, consider this problem. Let F(z) be the product of the distances from z to the vertices a, b, c, d of a square. If z lies inside or on the edge of the square, where does the maximum value of F occur? It is certainly tempting

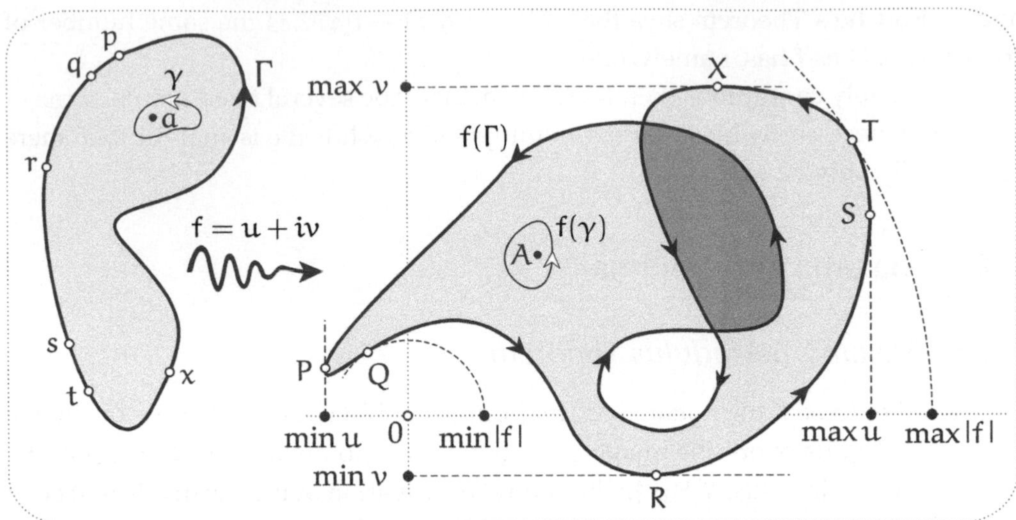

[7.14] Maximum-Modulus Theorem: *The maximum of $|f(z)|$ on a region where f is analytic is always achieved by points on the boundary, never ones inside. For the Argument Principle tells us that the sum of the multiplicities $\nu(a_j)$ of the p-points inside Γ is $\nu[f(\Gamma), p]$, but if p is outside $f(\Gamma)$ then (by definition) this is zero. Since $\nu(a_j)$ is strictly positive for an analytic function, we conclude that points outside $f(\Gamma)$ have no preimages inside Γ. On the other hand, if p lies inside $f(\Gamma)$ then $\nu[f(\Gamma), p] \neq 0$, and so there must be at least one preimage inside Γ.*

to guess that the maximum will occur at the centre of the square, but this is wrong. Since $F(z) = |(z-a)(z-b)(z-c)(z-d)|$ is the modulus of an analytic mapping, the maximum must in fact occur somewhere on the edge of the square. The exact location can now be found [exercise] using nothing more than ordinary calculus.

Returning to matters of theory, recall from Chapter 2 that the "modular surface" of f is the surface obtained by lifting each point z vertically to a height $|f(z)|$ above the complex plane. If we look at the portion of this surface lying above Γ and its interior, the result says that the highest point always lies on the edge, never inside.

Although the absolute maximum of the height always occurs on the edge, could there perhaps be a *local* maximum of $|f|$ at an interior point a, so that the surface would have a peak above a? No! For if we cut out the piece of the surface lying above the interior of any small loop γ round a, the highest point will fail to lie on the edge. Thus a modular surface has no peaks. Further aspects of the modular surface are investigated in the exercises.

This absence of local maxima is re-explained in [7.14]. The Argument Principle says that since γ contains a, $f(\gamma)$ must wind round $A = f(a)$ at least once. This makes it clear that there are always points on γ which have images lying further from the origin than A. More formally (cf. [7.4]), any ray emanating from A must intersect $f(\gamma)$, and by choosing the ray to point directly away from the origin, the intersection point is guaranteed to lie further from the origin than A. Thus $|f|$ cannot have a local maximum.

7.6.2 Related Results

As [7.14] illustrates, the Maximum-Modulus result is only one of several that follow from (7.12). For example, unless there is a 0-point inside Γ, at which $|f(z)| = 0$, the point Q closest to the origin (for which $|f(z)|$ is minimum) must also be the image of a point q lying on the boundary Γ. Naturally enough, this is called the *Minimum-Modulus Theorem*.

Thus if we cut out the piece of the modular surface lying above Γ and its interior, the lowest point will always lie on the edge, unless, that is, the surface actually hits the complex plane at an interior 0-point of f. By the same token, there can be no pits in the surface [local minima of $|f|$] except at 0-points.

As before, the only exception to all this is the mapping $z \mapsto \text{const.}$, for which every point yields the smallest (and only) value of $|f|$. Thus if we know that $|f|$ achieves a positive minimum at an interior point then $f(z) = \text{const.}$

If $f = u + iv$ is analytic then [cf. Ex. 2, p. 293] u and v are automatically "harmonic". As we shall see in Chapter 11, this means that these functions are intimately connected with numerous physical phenomena: heat flow, electrostatics, hydrodynamics, to name but a few. It is therefore of significance that [7.14] shows that u *and v are also subject to the principle that their maxima and (nonzero) minima can only occur on* Γ, *never inside* Γ. As before, if a maximum or minimum occurs at an interior point, the harmonic function must be constant.

7.7 The Schwarz–Pick Lemma*

7.7.1 Schwarz's Lemma

Thinking of the unit disc as the Beltrami–Poincaré model for non-Euclidean geometry, we saw in Chapter 6 that a special role was played by the Möbius transformations of the form

$$M_a^{\phi}(z) = e^{i\phi}\left(\frac{z-a}{\overline{a}z-1}\right) = e^{i\phi}\, M_a(z), \tag{7.13}$$

where a lies inside the disc. These one-to-one mappings of the disc to itself act as rigid motions, for they preserve non-Euclidean distance.

Apart from a digression on Liouville's Theorem, this section continues the work (begun in Chapter 6) of exhibiting the beautiful pre-existing harmony that exists between non-Euclidean geometry and the theory of conformal mappings. Our first new piece of evidence that these two disciplines somehow "know" about each other is the following:

> *Rigid motions of the hyperbolic plane are the* ***only*** *one-to-one analytic mappings of the disc to itself.* (7.14)

There are of course many other kinds of analytic mapping of the disc to itself, but according to (7.14) they must all fail to be one-to-one. For example, $z \mapsto z^3$ maps the disc to itself, but it is *three*-to-one.

Observe that this result establishes a claim we previously made [see p. 204] in connection with Riemann's Mapping Theorem. There we explained that there are as many mappings of one region to another as there are automorphisms of the disc. We already knew that these automorphisms included 3-parameter's worth of Möbius mappings, and (7.14) now tells us that there are no more.

To verify (7.14) we will first establish a lemma (of great interest in itself) due to Schwarz:

> *If an analytic mapping of the disc to itself leaves the centre fixed, then either every interior point moves nearer to the centre, or else the transformation is a simple rotation.*

The example $f(z) = z^2$ shows that the mapping need not be a rotation in order for boundary points to keep their distance from the centre. However, at an interior point we have $|z| < 1$, and so $|f(z)| = |z|^2 < |z|$, in accord with the result.

Let f be any analytic mapping of the disc to itself leaving the centre fixed, so that $|f(z)| \leqslant 1$ on the disc, and $f(0) = 0$. We wish to show that either $|f(z)| < |z|$ at interior points, or else $f(z) = e^{i\phi} z$. To this end, consider the ratio F of image to preimage:

$$F(z) \equiv \frac{f(z)}{z}.$$

At first sight this may look undefined at 0, but a moment's thought shows that as z approaches the origin, $F(z)$ approaches $f'(0)$.

From the previous section we know that the maximum modulus of an analytic function on the disc can only occur at an interior point if the function is constant, otherwise it's on the boundary circle $|z| = 1$. Thus if p is an interior point and z varies over the unit circle C, then

$$|F(p)| \leqslant (\max |F(z)| \text{ on } C) = (\max |f(z)| \text{ on } C) \leqslant 1.$$

Thus it is certainly true that no interior point can end up *further* from the centre. But if even a single interior point q remains at the *same* distance from the centre then $|F(q)| = 1$, which means that F has achieved its maximum modulus at an interior point. In this case F must map the entire disc to a single point of unit modulus, say $e^{i\phi}$, so that $f(z) = e^{i\phi} z$ is a rotation. Done.

The result is illustrated in [7.15]. If f is not a rotation then every point z on a circle such as K is mapped to a point $w = f(z)$ lying strictly inside K, and the shaded region is compressed as shown. If we shrink K down towards the origin then f will amplitwist it to another infinitesimal circle centred at the origin, but having a

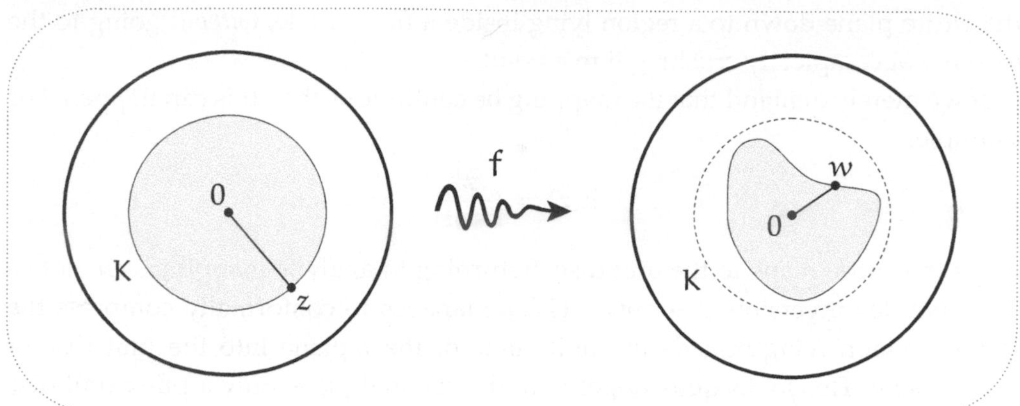

[7.15] Schwarz's Lemma: *If an analytic mapping* $f(z)$ *of the disc to itself leaves the centre fixed, then either every interior point moves nearer to the centre, or else the transformation is a simple rotation. As explained in the text, this follows from applying the Maximum-Modulus Theorem to* $F(z) \equiv f(z)/z$.

smaller radius. Thus the amplification of f at the origin must be less than one. We can only have $|f'(0)|$ *equal* to one in the case of a rotation.

We can now return to (7.14). As in Schwarz's Lemma, first suppose that the mapping f leaves the centre fixed, but now take f to be *one-to-one*, so that it has a well-defined analytic inverse f^{-1} which also maps the disc to itself and leaves the centre fixed. By Schwarz's Lemma, f sends an interior point p to a point q that is no further from centre than p. But f^{-1} is *also* subject to Schwarz's Lemma, and so $p = f^{-1}(q)$ must be no further from the centre than q. These two statements are only compatible if $|q| = |f(p)| = |p|$. Thus f must be a rotation, which is indeed a rigid motion of type (7.13).

Finally, suppose that the one-to-one mapping f does not leave the centre fixed, but instead sends it to c. We can now compose f with the rigid motion M_c which sends c back to 0. We thereby obtain a one-to-one mapping $(M_c \circ f)$ of the disc to itself which does leave the centre fixed, and which must therefore be a rotation M_0^{ϕ}. But this means [exercise] that

$$f = M_c \circ M_0^{\phi}$$

is the composition of two rigid motions, and so is itself a rigid motion. Done.

7.7.2 Liouville's Theorem

The constant mapping $f(z) = c$ crushes the entire plane down to the single image point c. We now ask whether it is possible for an analytic mapping to compress the entire plane down to a region lying inside a finite circle, *without* going to the extreme of completely crushing it to a point.

If we merely demand that the mapping be continuous then this can happen. For example,

$$h(z) = \frac{z}{1+|z|}$$

maps the entire plane to the unit disc. Returning to analytic mappings, we notice that complex inversion, $z \mapsto w = (1/z)$, manages to conformally compress the infinite region lying outside the unit circle of the z-plane into the unit disc of the w-plane. This looks quite hopeful: of the original plane only a puny unit disc remains to be mapped.

To think like this is to completely forget the rigidity of analytic mappings. Having decided to use complex inversion to map the region outside the unit circle, we cannot change the rules when it comes to mapping the remaining disc: the mapping $z \mapsto (1/z)$ acting in the exterior can only be analytically continued to the interior in one way, namely as $z \mapsto (1/z)$. The requirement of analyticity thereby forces the "puny" disc to explode, producing an image of infinite size.

We will now show that

> *An analytic mapping cannot compress the entire plane into a region lying inside a disc of finite radius without crushing it all the way down to a point.*

This is *Liouville's Theorem*. To understand this we must generalize Schwarz's Lemma slightly. Suppose that an analytic function $w = f(z)$ leaves the origin fixed and compresses the disc $|z| \leqslant N$ to a region lying inside the disc $|w| \leqslant M$. By the same reasoning as before, we find that if p lies inside the original disc (boundary circle K) then

$$|F(p)| \leqslant [\max |F(z)| \text{ on } K] = \max \left(\frac{|f(z)|}{N}\right) \text{ on } K \leqslant \frac{M}{N}.$$

Hence,

$$|f(p)| \leqslant \frac{M\,|p|}{N}.$$

But if f compresses the *whole* plane to a region lying inside the disc of radius M, then the above result will continue to hold true no matter how large we make N. Therefore $f(p) = 0$ for all p, and we are done.

Finally, if f does not leave the origin fixed, but instead sends it to c, we may apply the previous argument to the function $[f(z) - c]$. This is the composition of f with the translation which sends c back to 0. Since the image of the plane under f lies inside the disc $|w| \leqslant M$, the translation of $-c$ will produce a region lying inside the disc $|w| \leqslant 2M$. The previous inequality then becomes

$$|f(p) - c| \leqslant \frac{2M\,|p|}{N}.$$

Once again letting N tend to infinity, we conclude that $f(p) = c$ for all p. Done.

7.7.3 Pick's Result

We now turn to a second, rather beautiful piece of evidence that non-Euclidean geometry is intimately connected with the theory of conformal mapping. Reconsider [7.15]. Schwarz's Lemma informs us that (with the exception of rotations) the distance between interior points and the origin is decreased. This result has two blemishes, both related to an exaggerated emphasis on the origin: (i) we require that the mapping leave the origin fixed; (ii) only distances from the origin are shown to decrease.

Consider (ii) first, and for the moment let us simply put up with (i) by continuing to assume that our analytic mapping leaves the origin fixed. Although we did not demonstrate it, perhaps a more symmetrical result holds true—with the exception of a rotation, *will the mapping automatically decrease the distance between* **any** *pair of interior points?*

Sadly, no. Consider the effect of $f(z) = z^2$ (which leaves the origin fixed) on the two interior points $a = (3/4)$ and $b = (1/2)$. The original separation is $|a-b| = 0.25$, while the separation of the images is $|f(a) - f(b)| = 0.3125$. The distance between the pair of points has *increased*.

But now consider the effect of exactly the same mapping on exactly the same two points from the point of view of the *Beltrami–Poincarites*[5]. When *they* measure the distance between a and b, it is found [see (6.46), p. 366] to equal $\mathcal{H}\{a, b\} = 0.8473$, while the separation of the images is $\mathcal{H}\{f(a), f(b)\} = 0.7621$. *The hyperbolic distance has decreased!* Choose any other pair of points for yourself and examine the effect of $z \mapsto z^2$ on their hyperbolic separation.

Pick's splendid discovery[6] was that even if we drop the requirement that the origin be a fixed point, this decrease in *hyperbolic* distance is a universal phenomenon:

Unless an analytic mapping of the disc to itself is a hyperbolic rigid motion, the hyperbolic separation of every pair of interior points decreases. (7.15)

[5] Recall from Chapter 6 that this is the race of beings who inhabit the Beltrami–Poincaré model of the hyperbolic plane.

[6] Georg Alexander Pick [1859–1942] is perhaps best known for his wonderfully simple but completely unexpected *Pick's Theorem,* which determines the area $\mathcal{A}$ of a lattice polygon in terms of the number of lattice points Inside it (I), and on its Boundary (B), namely, $\mathcal{A} = I + (B/2) - 1$. However, his greatest contribution to science may have been the fact that he fought to appoint Einstein at the German University of Prague in 1911, became close friends with him there—playing violin together in a musical quartet—and finally introduced Einstein to the tensor calculus of Ricci and Levi-Civita, without which Einstein could not have formulated General Relativity in 1915! [This is documented in Goodstein (2018, p. 90), though, strangely, Einstein himself only seems to have publicly credited his friend Marcel Grossmann for this critical piece of guidance.] Pick died at the hands of the Nazis at Theresienstadt concentration camp on the 26th of July, 1942, at the age of 82.

Because this result contains Schwarz's Lemma as a special case [we shall clarify this shortly] it is often called the *Schwarz–Pick Lemma*. Despite the startling nature of the result, we can actually understand its essence very simply; we need only ask the question, "How do the *Beltrami–Poincarites* view [7.15]?"

Because their concept of angle is identical to ours, it follows that their concept of an analytic function is also the same as ours—f appears conformal both to us and to them. In addition, we both agree that rays emanating from the origin are straight lines along which we may measure distance. Consequently, the Beltrami–Poincarites willingly concede that w is closer to 0 than z is, although they violently disagree with our quantitative determination of exactly *how much* closer it is. Now recall that there is a small (psychological) flaw in the Beltrami–Poincaré model: 0 appears special to us because it is the centre of the disc, but to the Beltrami–Poincarites who inhabit an infinite, homogeneous plane 0 *is utterly indistinguishable from any other point of their world.*

The above explanation is formalized in [7.16]. In the top left figure we see that the Beltrami–Poincarites have marked an arbitrary point a, drawn a few concentric circles centred there, and on the outermost of these they have marked a second point b. They (and we) now consider the effect of an analytic mapping f of their world to itself. The point a is sent to some image point $A = f(a)$ [top right] and likewise b is sent to $B = f(b)$. In order to compare the separation of A and B with that of a and b, the Beltrami–Poincarites perform a rigid motion ($M_A \circ M_a$ would do) that moves the circles centred at a to circles (of equal hyperbolic size) centred at A. Consequently, the hyperbolic separation of a and b will have been decreased [increased] by f according as B lies inside [outside] the outermost of these circles. In anticipation of Pick's result we have drawn it inside, corresponding to a decrease in hyperbolic separation. However, observe that as in our previous numerical example, to us Euclideans it looks as though the separation has been *increased.*

In order to show us poor blind Euclideans that the circles centred at a and A really are concentric and of equal sizes (so enabling us to see that B really has gotten closer) the Beltrami–Poincarites perform the illustrated rigid motions M_a and M_A. These respectively move a and A to the origin [bottom left and bottom right figures], yielding circles that are as concentric to us as they always were to them. M_a moves b to $z = M_a(b)$, while M_A moves B to

$$w = M_A(B) = (M_A \circ f)(b) = (M_A \circ f \circ M_a)(z).$$

We shall abbreviate $(M_A \circ f \circ M_a)$ to F, so that $w = F(z)$.

We can now see that the following are all equivalent:

$$\mathcal{H}\{A, B\} < \mathcal{H}\{a, b\} \iff \mathcal{H}\{0, w\} < \mathcal{H}\{0, z\} \iff |w| = |F(z)| < |z|.$$

But F is an analytic mapping of the disc to itself which leaves the origin fixed, and so it is subject to Schwarz's Lemma. Thus unless F is a rotation—in which case

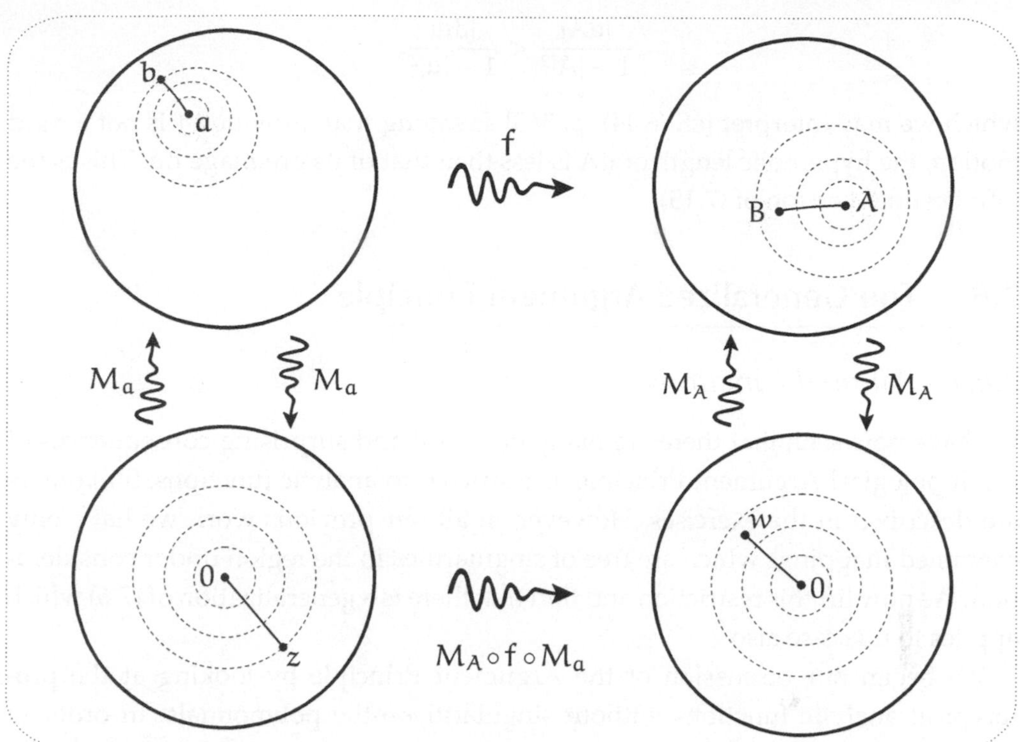

[7.16] The Schwarz–Pick Lemma *is Pick's brilliant generalization of Schwarz's Lemma, stating that the* hyperbolic *separation of every pair of points must decrease. The Beltrami–Poincarites (our imaginary inhabitants of the Beltrami–Poincaré disc) have no way of distinguishing the bottom half of this figure [aka the* already-established *Schwarz's Lemma] from the top half, which is Pick's result.* Q.E.D.!

$f = (M_A \circ F \circ M_a)$ is a rigid motion—we must have $|w| = |F(z)| < |z|$, as depicted. Done.

Finally, let us express the Schwarz–Pick Lemma in symbolic form. If f is not a rigid motion then $|F(z)| < |z|$, which may be written out more explicitly as

$$|(M_A \circ f \circ M_a)(z)| < |z|,$$

which in turn can be written as

$$|(M_A \circ f)(b)| < |M_a(b)|.$$

Thus

$$\left|\frac{B - A}{\overline{A}B - 1}\right| < \left|\frac{b - a}{\overline{a}b - 1}\right|.$$

If we move b closer and closer to a, then $da = (b - a)$ becomes an infinitesimal vector emanating from a whose image under f is an infinitesimal vector $dA = (B - A)$ emanating from A. The above inequality now becomes

$$\frac{|dA|}{1-|A|^2} < \frac{|da|}{1-|a|^2},$$

which we may interpret [cf. (6.44), p. 363] as saying that, provided f is not a rigid motion, the hyperbolic length of dA is less than that of its preimage da. This is the infinitesimal version of (7.15).

7.8 The Generalized Argument Principle

7.8.1 Rational Functions

We have now seen that there are many powerful and surprising consequences of the Topological Argument Principle as restricted to analytic functions. Still others are described in the exercises. However, in all our previous work we have only examined mappings which are free of singularities in the region under consideration. We now lift this restriction and find that there is a generalization of (7.6) which applies to this case also.

We began our discussion of the Argument Principle by looking at the prototypical analytic functions without singularities—the polynomials. In order to understand the generalization to analytic functions *with* singularities, we should correspondingly begin with rational functions.

As in [7.8], let A, B, and C be the complex numbers from the fixed points a, b, and c to the variable point z. The left-hand side of [7.17] shows an expanding circle Γ at three successive stages of its growth: Γ_1, Γ_2, and Γ_3. The right-hand side shows the evolution of the image of Γ under the rational mapping

$$f(z) = \frac{(z-a)\,(z-b)}{(z-c)} = A\,B \cdot \frac{1}{C}. \tag{7.16}$$

By the time Γ has grown into Γ_1 it has enclosed a, and $\nu\,[f(\Gamma_1), 0] = 1$, in accord with the ordinary Argument Principle. As Γ continues to grow it crosses the other 0-point at b, and the winding number of $f(\Gamma)$ correspondingly increases to $\nu\,[f(\Gamma_2), 0] = 2$. Now comes the new phenomenon. As Γ crosses the singularity at c the winding number of its image *decreases* by one so that $\nu\,[f(\Gamma_3), 0] = 1$.

The explanation is simple. As z traverses Γ_3, the winding number of f(z) is the sum of the revolutions executed separately by A, B, and (1/C). The first two go round once, but as C rotates counterclockwise, (1/C) rotates in the opposite direction, finally executing one complete *negative* revolution. By the same token, if the denominator of f instead contained C^m then $(1/C^m)$ would execute $-m$ revolutions, and the winding number would become

$$\nu\,[f(\Gamma_3), 0] = 2 - m.$$

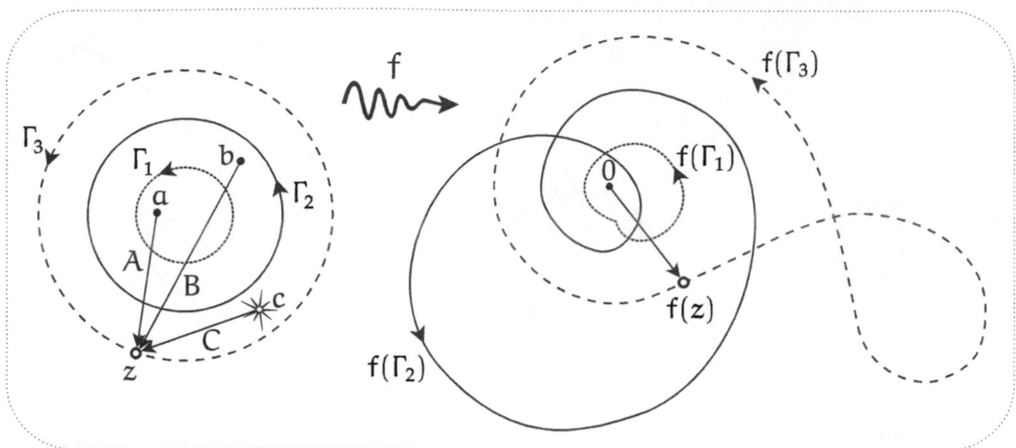

[7.17] The Generalized Argument Principle. *Consider* $f(z) = (z-a)(z-b) \cdot \frac{1}{(z-c)} = A\,B \cdot \frac{1}{C}$. *Traversing* Γ_1, A *executes a complete revolution, while* B *and* C *merely oscillate, so* $\nu[f(\Gamma_1), 0] = 1$. *Expanding* Γ *to enclose* b, *both* A *and* B *now execute complete revolutions, while* C *merely oscillates, so* $\nu[f(\Gamma_2), 0] = 2$. *But when* Γ *expands still further to* Γ_3, *enclosing the pole at* c *(marked by the explosion!) then* $(1/C)$ *executes* a negative *revolution,* decreasing *the winding number by one:* $\nu[f(\Gamma_3), 0] = 1$. *In general, if* N *and* M *are the number of interior* p*-points and poles, both counted with their multiplicities, then* $\nu[f(\Gamma), p] = N - M$.

As with counting zeros, we could say in this case that c was a singularity [or *pole*, as we shall now call such places] of multiplicity m.

The previous equation is an example of the Generalized Argument Principle:

> *Let* f *be analytic on a simple loop* Γ *and analytic inside except for a finite number of poles. If* N *and* M *are the number of interior* p*-points and poles, both counted with their multiplicities, then* $\nu\,[f(\Gamma), p] = N - M$. (7.17)

Simply by allowing an arbitrary number of factors on the top and bottom of (7.16) we see that this result is certainly true when f is any rational function.

Before explaining why it works in general, let us develop a more vivid understanding of how it works in the case of our example (7.16). We have certainly shown that as Γ crosses c the winding number drops from 2 to 1, but exactly *how* does this unwinding occur?

If we look at the image plane just as Γ crosses c then $f(\Gamma)$ undergoes a sudden and violent change of shape as it leaps to infinity and then returns, but this leaves us none the wiser. However, if we instead watch its evolution on the *Riemann sphere* then we gain a new and delightful insight into the process.

Figure [7.18] (which should be scanned like a comic strip) illustrates this. At the time of the first picture [top left] Γ has already enclosed the two roots, and its image is seen to wind round the origin twice. Now follow the evolution of $f(\Gamma)$

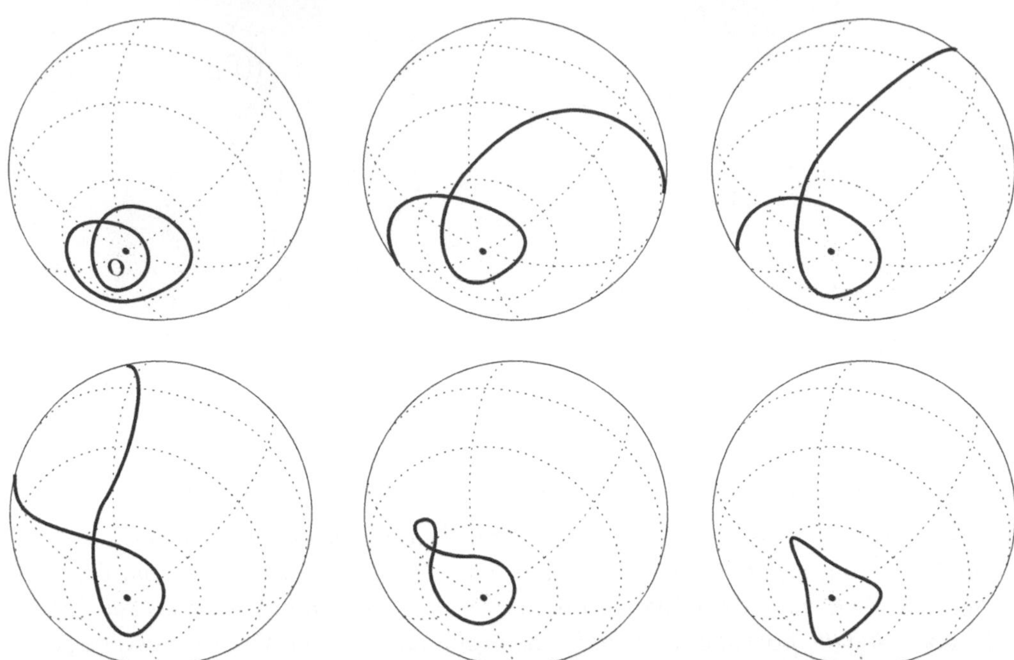

[7.18] Reducing the winding number by passing over the point at infinity. *The Riemann sphere provides a vivid explanation of the previous figure. As* Γ *expands from* Γ_2 *to* Γ_3*, crossing the pole at* c*, the loop on the Riemann sphere simply slides over the north pole, unwinding the loop and* reducing *its winding number from* 2 *to* 1.

through the remaining pictures. As Γ crosses c [top right] there is no longer any excitement—$f(\Gamma)$ merely slides across the north pole, and *this* is how the unwinding is achieved. Try using a computer to animate the evolution of the image $f(\Gamma)$ on the Riemann sphere as Γ expands through the roots and poles of a rational function f of your choosing.

7.8.2 Poles and Essential Singularities

In generalizing the ordinary Argument Principle we had to ask ourselves how we should count p-points of a general analytic function. The factorization (7.7) brought out the analogy with polynomials and gave us a satisfactory definition of the algebraic (and topological) multiplicity of a p-point.

The method of extending (7.17) from rational functions to analytic functions with singularities is essentially the same. The only complication is that there are actually *two* possible kinds of singularity for an otherwise analytic function.

The first kind of singularity is called a *pole*. It is by far the most commonly encountered type in applications of complex analysis, and it is the only type to which (7.17) applies. Here's the definition. If $f(z)$ approaches ∞ as z approaches a *from any direction* then a is a pole of f. We can understand the terminology by thinking of the modular surface of f, for there will be an infinitely high spike or "pole" above the point a. Figure [2.14] on p. 74 is an example of this.

Since f is analytic, it follows that $F(z) \equiv [1/f(z)]$ is also analytic and has a root at a. If this root has multiplicity m then the factorization (7.7) of F is

$$F(z) = (z-a)^m \, \Omega(z), \tag{7.18}$$

where Ω is analytic and nonzero at a; in fact we know that $\Omega(a) = F^{(m)}(a)/m!$. The local behaviour of f near a is therefore given by

$$f(z) = \frac{\widetilde{\Omega}(z)}{(z-a)^m}, \tag{7.19}$$

where $\widetilde{\Omega}(z) = [1/\Omega(z)]$ is analytic and nonzero at a. This expression brings out the analogy with rational functions and enables us to identify m as the algebraic multiplicity or *order* of the pole at a. We call a pole *simple, double, triple,* etc., according as $m = 1, 2, 3$, etc.

Note that we have also found a way of calculating the order of a pole, namely, as the order of the first nonvanishing derivative of $(1/f)$. Once you have identified the locations of the poles, you may use this method [exercise] to find the orders of the poles of the following functions:

$$P(z) = \frac{1}{\sin z}; \quad Q(z) = \frac{\cos z}{z^2}; \quad R(z) = \frac{1}{(e^z-1)^3}.$$

You should have found that P has a simple pole at each multiple of π; Q has a double pole at 0; and R has a triple pole at each multiple of $2\pi i$.

One more piece of terminology. If the only singularities in some region of an otherwise analytic function are poles, the function is called *meromorphic* in that region.

In addition to poles, it is also possible for an otherwise analytic function to possess what are called *essential singularities*. We shall postpone detailed discussion of such places to a later chapter, but it is clear that the behaviour of a function f in the vicinity of an essential singularity s must be very strange and wild. If f were bounded in the vicinity of s then s would not be a singularity at all, but on the other hand $f(z)$ cannot approach ∞ as z approaches s from all directions, for then s would only be a pole.

Consider the standard example $g(z) = e^{1/z}$, which clearly has a singularity of some type at the origin. If we write $z = r\,e^{i\theta}$ then

$$|g(z)| = e^{\frac{\cos\theta}{r}}.$$

Figure [7.19] depicts the modular surface. If z approaches 0 along the imaginary axis then $|g(z)| = 1$. But if the approach is instead made along a path lying to the left of the imaginary axis (where $\cos\theta < 0$) then $g(z)$ tends to 0. Finally, if the

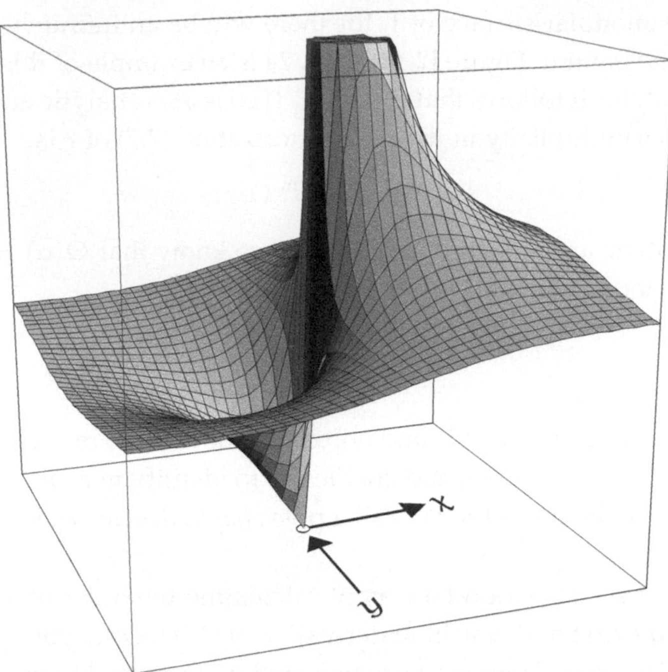

[7.19] The modular surface of $g(z) = e^{1/z}$ reveals its *essential singularity* at the origin. *Since $|g(z)| = e^{\frac{\cos\theta}{r}}$, the behaviour at 0 depends drastically upon the* direction *in which we approach it. If $\cos\theta < 0$ as we approach, then the height $|g(z)|$ drops to zero, and the surface hits $\mathbb{C}$. If we approach* precisely *along the tightrope of the imaginary axis (for which $\cos\theta = 0$) then our height holds precariously steady at 1. But if $\cos\theta > 0$ as we approach, then the height* explodes *to infinity faster than any polynomial!*

approach path lies to the right of the imaginary axis then $g(z)$ tends to ∞. In fact, not only will $|g(z)|$ become infinite in this case, but the *rate* at which it zooms off to ∞ is quite beyond the ken of any pole.

To see this, reconsider (7.19). The greater the order m, the faster the growth of f as the pole at a is approached. However, no matter how great the order happens to be, we know that $(z-a)^m$ dies away fast enough to kill this growth, in the sense that its product with f remains bounded. Indeed, the order of a pole can be *defined* as the smallest power of $(z-a)$ which will curb the growth of f in this way.

Compare this with the growth of $g(z)$ as its essential singularity is approached, say along the positive real x-axis. To confirm that g grows faster than any meromorphic function, we need only recall from ordinary calculus that

$$\lim_{x\to 0} x^m\, e^{1/x} = \lim_{\lambda\to\infty} \frac{e^{\lambda}}{\lambda^m} = \infty,$$

no matter how great the value of m.

7.8.3 *The Explanation**

In order to explain (7.17) let us return to the interpretation of (7.19). If we think of f as mapping into the Riemann sphere Σ, then the north pole (∞) is an image point like any other, and the poles of f are simply its preimages, ∞-points if you will. As we now explain, this means that the topological multiplicity of an ∞-point can be defined in exactly the same way as that of any other p-point, namely, as the number of times that the image of a small loop round a winds round $f(a)$.

Reconsider the mapping F in (7.18). By virtue of (7.8), we know that a sufficiently small circle C_a centred at a will be mapped to a small loop $F(C_a)$ winding round the origin m times. On Σ, the stereographic projection of $F(C_a)$ therefore winds round the south pole m times, counterclockwise as seen from *inside* Σ. Because complex inversion (which sends $F(C_a)$ to $f(C_a) = 1/[F(C_a)]$) rotates Σ about the real axis by π, thereby swapping 0 and ∞, this means that $f(C_a)$ will be a small loop winding m times round ∞. Since it winds counterclockwise as seen from inside Σ, its stereographic projection in the plane is therefore a very large loop winding m times *clockwise* around the origin, i.e., with winding number $-m$.

As an aside, observe that it now makes sense to rewrite (7.19) as

$$f(z) = (z-a)^{-m}\,\widetilde{\Omega}(z),$$

and to correspondingly think of a pole of order m as being a *root of negative multiplicity* $-m$.

Shifting our attention away from the origin, we now consider winding numbers around an arbitrary (finite) point p. By making C_a sufficiently small, we can be certain that $f(C_a)$ will be so large that it will wind $-m$ times round p. But if we expand C_a into any simple loop Γ_a without crossing any p-points or other poles, then the winding number of the image round p cannot change. In other words,

If a is a pole of order m and Γ_a is any simple loop containing a but no p-points and no other poles, then $\nu\,[f(\Gamma_a), p] = -m$. (7.20)

Finally, reconsider figure [7.11]. You may now easily convince yourself that the argument leading to (7.11) remains valid if some of the a_j's are poles instead of p-points. Let's call these singular points s_j. Thus

$$\nu[f(\Gamma), p] = \sum_{\text{p-points}} \nu\,[f(\Gamma_{a_j}), p] \;+ \sum_{\text{poles}} \nu\,[f(\Gamma_{s_j}), p]\,.$$

Using (7.20), this implies that

$$\nu[f(\Gamma), p] = [\text{number of p-points inside } \Gamma] - [\text{number of poles inside } \Gamma],$$

as was to be shown.

7.9 Exercises

1 A "simple" loop can get very complicated (see diagram). However, if we imagine creating this complicated loop by gradually deforming a circle, it is clear that it will wind round its interior points precisely once. Let $N(p)$ be the number of intersection points of the simple loop with a ray emanating from p (cf. [7.4]). What distinguishes the possible values of N(interior point) from those of N(exterior point)? In place of the crossing rule (7.1), you now possess (for simple loops) a much more rapid method of determining whether a point is inside or outside.

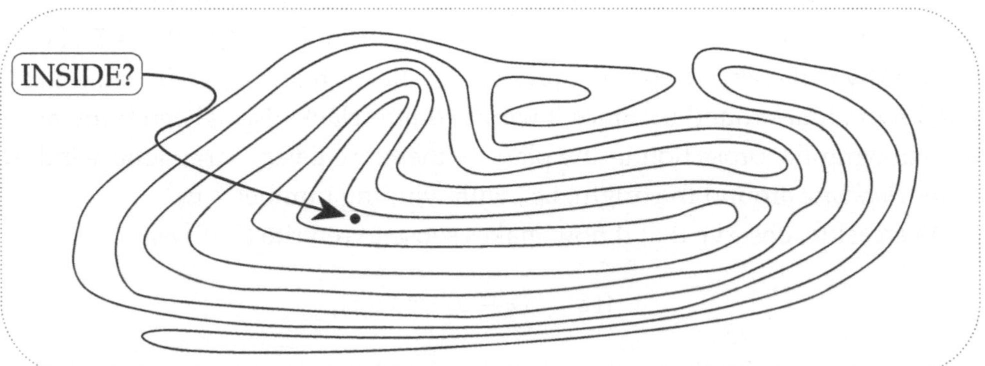

You can use this result to play a trick on a friend F: (1) So that foul play cannot be suspected, get F to draw a very convoluted simple loop for himself; (2) choose a random point in the thick of things and ask F if it's inside or not, i.e., starting at this point, can one escape through the maze to the outside?; (3) after F has been forced to recognize the time and effort required to answer the question, get him to choose a point for you; (4) choosing a ray in your mind's eye, scan along it and count the intersection points. Amaze F with your virtually instantaneous answer!

2 Reconsider the mapping $\widehat{\mathcal{L}}$ in (7.4) of the unit circle to itself, and the associated graph of $\Phi(\theta)$ in [7.7]. If $\Phi'(a) > 0$ then the graph is rising above the point $\theta = a$, and small movement of z will produce a small movement of the w having the *same sense*. We say that Φ is orientation-preserving at a and that the topological multiplicity $\nu(a)$ of $z = e^{ia}$ as a preimage of $w = e^{i\Phi(a)}$ is $+1$. Similarly, if $\Phi'(a) < 0$ then the mapping is orientation-reversing and $\nu(a) = -1$. In other words,

$$\nu(a) = \text{the sign of } \Phi'(a).$$

Compare this with the 2-dimensional formula (7.9).

(i) In [7.7], explain how the complete set of preimages of $w = e^{iA}$ can be found by drawing the family of horizontal lines $\Phi = A, A \pm 2\pi, A \pm 4\pi$, etc.

(ii) If the set of preimages is typical in the sense that $\Phi' \neq 0$ at any of them, what do we obtain if we sum their topological multiplicities? Thus to say that the degree of $\widehat{\mathcal{L}}$ (the winding number of L) is ν is essentially to say that $\widehat{\mathcal{L}}$ is ν-to-one. [*Hint*: In [7.4], consider a ray as describing the location of w.]

3 For each of the following functions $f(z)$, find all the p-points lying inside the specified disc, determine their multiplicities, and by using a computer to draw the image of the boundary circle, verify the Argument Principle.

(i) $f(z) = e^{3\pi z}$ and $p = i$, for the disc $|z| \leqslant (4/3)$.

(ii) $f(z) = \cos z$ and $p = 1$, for the disc $|z| \leqslant 5$.

(iii) $f(z) = \sin z^4$ and $p = 0$, for the disc $|z| \leqslant 2$.

4 Reconsider [7.8].

(i) Use a computer to draw the image under a cubic mapping

$$f(z) = (z - a)(z - b)(z - c)$$

of an expanding circle Γ, and observe the *manner* in which the winding number increases as Γ passes through the roots a, b, and c. In particular, observe that the shape that marks the birth of a new loop is this: $\prec$.

(ii) If $f'(p) \neq 0$ then a little piece of Γ passing through p is merely amplitwisted to another almost straight piece of curve through $f(p)$. Deduce that $\prec$ shapes can only occur when Γ hits a critical point. Explain why the particular shape $\prec$ is consistent with a critical point of order 1.

(iii) Observe that there are only two point in the evolution of Γ at which a $\prec$ shape is produced. Explain this algebraically in terms of the degree of f'.

(iv) Let T be the triangle with vertices a, b, and c. There are many ellipses which can be inscribed in T so as to touch all three sides, but show that there is only one (call it $\mathcal{E}$) that touches T at the *midpoints* of the sides. Hard Show that the two critical points of f are the *foci* of $\mathcal{E}$!

5 As in the text, let ξ_a, η_a, ϕ_a denote the two perpendicular expansion factors and the rotation angle used to describe the SVD decomposition of the local linear transformation at a produced by a mapping. By considering the case of a rotation by $(\pi/4)$, for which J is constant, show that ξ and η are generally *not* the eigenvalues λ_1 and λ_2 of the Jacobian J. However, confirm for this example that $\det(J) = \lambda_1\lambda_2 = \xi\eta$.

6 Even in three or more dimensions the local linear transformation induced by a mapping f at a point a can still be represented by the Jacobian matrix $J(a)$, and if a is not a critical point then its topological multiplicity $\nu(a)$ as a preimage of $f(a)$ is still given by (7.9). If n is the number of real negative eigenvalues of $J(a)$, counted with their algebraic multiplicities, show that

$$\nu(a) = (-1)^n.$$

[*Hint*: Since the characteristic equation $\det[J(a) - \lambda I] = 0$ has real coefficients, any complex eigenvalues must occur in conjugate pairs.]

7 Consider the nonanalytic mapping $h(z) = |z|^2 - i\bar{z}$.

(i) Find the roots of h.

(ii) Calculate the Jacobian J, and hence find $\det(J)$.

(iii) Use (7.9) to calculate the multiplicities of the roots in (i).

(iv) Find the image curve traced by $h(z)$ as $z = 2e^{i\theta}$ traverses the circle $|z| = 2$, and confirm the prediction of the Topological Argument Principle.

(v) Gain a better understanding of the above facts by observing that $h(z) = \bar{z}(z - i)$, and then mimicking the analysis of [7.8].

(vi) Use the insight of the previous part to find $\nu(i/2)$, which cannot be done with (7.9).

8 Let $Q(t)$ be a real function of time t, subject to the differential equation

$$c_n \frac{d^nQ}{dt^n} + c_{n-1}\frac{d^{n-1}Q}{dt^{n-1}} + \cdots + c_1\frac{dQ}{dt} + c_0 Q = 0.$$

Recall that one solves this equation by taking a linear superposition of special solutions of the form $Q_j(t) = e^{s_j t}$. Substitution into the previous equation shows that the s_j's are the roots of the polynomial

$$F(s) \equiv c_n s^n + c_{n-1} s^{n-1} + \cdots + c_1 s + c_0.$$

Note that $Q_j(t)$ will decay with time if s_j has a negative real part. The issue of whether or not the general solution of the differential equation decays away with time therefore reduces to the problem of determining whether or not all n roots of $F(s)$ lie in the half-plane $\text{Re}(s) < 0$. Let $\mathcal{R}$ be the net rotation of $F(s)$ as s traverses the imaginary axis from bottom to top. Explain the following result: *The general solution of the differential equation will die away if and only if*

$$\mathcal{R} = n\pi.$$

This is called the *Nyquist Stability Criterion*, and an F that satisfies this condition is called a *Hurwitz polynomial*. [*Hints*: Apply the Argument Principle to the loop consisting of the segment of the imaginary axis from $-iR$ to $+iR$, followed by one of the two semicircles having this segment as diameter. Now let R tend to infinity.]

9 Referring to the previous exercise, consider the differential equation

$$\frac{d^3Q}{dt^3} - Q = 0.$$

(i) Find $\mathcal{R}$ for this equation. Does it satisfy the Nyquist Stability Criterion?

(ii) Confirm your conclusion by explicitly solving the differential equation.

10 If a is real and greater than 1, use Rouché's Theorem to show that the equation

$$z^n\, e^a = e^z$$

has n solutions inside the unit circle.

11 (i) Applying Rouché's Theorem to $f(z) = 2z^5$ and $g(z) = 8z - 1$, show that all five solutions of the equation $2z^5 + 8z - 1 = 0$ lie in the disc $|z| < 2$.

(ii) By reversing the roles of f and g, show that there is only one root in the unit disc. Deduce that there are four roots in the ring $1 < |z| < 2$.

12 We can formalize, and slightly generalize, our explanation of Rouché's Theorem as follows:

(i) If $p(z)$ and $q(z)$ are nonzero on a simple curve Γ, and $\widetilde{\Gamma}$ is the image curve under $z \mapsto p(z)\, q(z)$, show that

$$\nu\,[\widetilde{\Gamma}, 0] = \nu\,[p(\Gamma), 0] + \nu\,[q(\Gamma), 0].$$

(ii) Write

$$f(z) + g(z) = f(z)\,\left[1 + \frac{g(z)}{f(z)}\right] = f(z)\,H(z).$$

If $|g(z)| < |f(z)|$ on Γ, sketch a typical $H(\Gamma)$. Deduce that

$$\nu\,[H(\Gamma), 0] = 0.$$

Using the previous part, obtain Rouché's Theorem.

(iii) If we only stipulate that $|g(z)| \leqslant |f(z)|$ on Γ, then parts of $H(\Gamma)$ could actually coincide with the circle $|z-1| = 1$, rather than lying strictly inside it, and $\nu\,[H(\Gamma), 0]$ might not be well-defined. However, show that if we further stipulate $f + g \neq 0$ on Γ, then $\nu\,[H(\Gamma), 0] = 0$ as before. Deduce that $\nu[(f + g)(\Gamma), 0] = \nu[f(\Gamma), 0]$.

13 Let $w = f(z)$ be analytic inside and on a simple loop Γ, and suppose that $f(\Gamma)$ is an origin-centred circle.

(i) If Δ is an infinitesimal movement of z along Γ and ϕ is the correspondingly infinitesimal rotation of w, show geometrically that

$$\frac{f'\,\Delta}{f} = i\phi.$$

(ii) As z traverses Γ, explain why $\nu[\Delta, 0] = 1$ and $\nu[i\phi, 0] = 0$.

(iii) Referring to (i) of the previous exercise, show that

$$\nu[f(\Gamma), 0] = \nu[f'(\Gamma), 0] + 1.$$

(iv) Deduce from the Argument Principle that f has one more root inside Γ than f' has. This is sometimes called *Macdonald's Theorem*, though I believe its essence goes back as far as Riemann.

(v) From this we deduce, in particular, that f has at least one root inside Γ. Derive this fact directly by considering the portion of the modular surface lying above Γ and its interior.

14 In contrast to analytic mappings, it is perfectly possible for a continuous non-analytic mapping to completely crush pieces of curve or even areas without crushing the rest of its domain. Let us give a concrete example to show that the Topological Argument Principle does not apply to this case. With $r = |z|$, the mapping $h(z) = \phi(r)\,z$ will be a continuous function of z if $\phi(r)$ is a continuous function of r. Consider the continuous mapping h of the unit disc to itself corresponding to

$$\phi(r) = 0, \qquad 0 \leqslant r < (1/2);$$
$$\phi(r) = 2r - 1, \qquad (1/2) \leqslant r \leqslant 1.$$

(i) Describe this mapping in visually vivid terms.

(ii) Taking Γ to be the circle $|z| = (3/4)$ and letting $p = 0$, try (and fail) to make sense of (7.11).

15 The version of Brouwer's Fixed Point Theorem established in the text fell short of the full result in two ways: (A) we assumed that $|g| < 1$ on D rather than $|g| \leqslant 1$; (B) we essentially used the Topological Argument Principle, which the previous exercise shows to be useless in the general case of a continuous mapping having infinitely many p-points in a finite region. Let's remove these blemishes. Once again let $m(z) = g(z) - z$ be the movement of z, and suppose that Brouwer's result is *false*, so that $m \neq 0$ throughout the disc $|z| \leqslant 1$. Obtain the desired contradiction as follows:

(i) By assumption, $m(z) = g(z) - z = g(z) + f(z)$ does not vanish on the unit circle C. Use Ex. 12(iii) to show that if $|g| \leqslant 1$ then $\nu[m(C), 0] = 1$.

(ii) Let C_r be the circle $|z| = r$, so that $C_1 = C$. By considering the evolution of $\nu[m(C_r), 0]$ as r increases from 0 to 1, obtain a contradiction with (i).

The key fact $\nu[m(C), 0] = 1$ can be obtained more intuitively. Draw a typical movement vector $m(z)$ emanating from z and note that it makes an *acute angle* with the inward unit normal vector $(-z)$ to C, also drawn emanating from z.

But clearly this normal vector undergoes one positive revolution as z traces C. Deduce that the vector m is also dragged round one revolution.

16 Let $f(z)$ be an *odd* power of z, and consider its effect on the unit circle C. Note two facts: (1) if p is on C then $f(-p)$ points in the opposite direction to $f(p)$; (2) $\nu[f(C), 0] = \text{odd}$, in particular it cannot vanish. This is only one example of a general result. Show that (1) always implies (2), even if f is merely continuous. [*Hints*: If f is subject to (1), what can we deduce about the net rotation $\mathcal{R}$ of $f(z)$ as z traverses the semicircle from p to $-p$? How is the rotation produced by the remaining semicircle related to $\mathcal{R}$?]

17 Consider a spherical balloon S resting on a plane. If we gradually deflate S, each point will end up on the plane so that we have a continuous mapping H of S into the plane. Observe that the north and south poles, which are antipodal, have the same image. The *Borsuk-Ulam Theorem* says that *any* continuous mapping H of S into the plane will map some pair of antipodal points to the same image. Consider the mapping $F(p) = H(p) - H(p^{\star})$, where $p^{\star}$ is antipodal to p. The theorem then amounts to showing that F has a root somewhere on S. Prove this. [*Hints*: It is sufficient to examine the effect of F on just the northern hemisphere. By taking the boundary of this hemisphere (the equator) to be the circle C of the previous exercise, deduce that $\nu[H(C), 0] \neq 0$.]

18 Let f be analytic on a simple loop Γ, and let p be a preimage of a point on $f(\Gamma)$ at which $|f|$ is maximum. If ξ is a tangent complex number to Γ at p, and in the same counterclockwise sense as Γ, show geometrically that $\xi f'(p)$ points in the same direction as $if(p)$. What is the analogous result at a positive minimum of $|f|$?

19 (i) If p is not a critical point of an analytic function f, show geometrically that the modulus of f increases most rapidly in the direction $[f(p)/f'(p)]$.

(ii) In terms of the modular surface above p, this direction lies directly beneath the tangent line to the surface having the greatest "slope" (i.e., tan of the angle it makes with the complex plane). Show that the slope of this steepest tangent plane is $|f'(p)|$, and note the analogy with the slope of the ordinary graph of a real function.

(iii) What does the modular surface look like at a root of order n?

(iv) What does the modular surface look like above a critical point of order m? Using the case $m = 1$, explain why such places are called *saddle points*.

(v) Rephrase Macdonald's Theorem [Ex. 13] in terms of the number P of pits and the number S of saddle points in the portion $\mathcal{A}$ of the modular surface lying above Γ and its interior.

(vi) It is a beautiful fact that, expressed in this form, Macdonald's Theorem can be explained almost purely topologically. The following explanation is adapted from Pólya (1954), though I believe the basic idea goes back to Maxwell and Cayley. Since $|f|$ is constant on Γ, the boundary of $\mathcal{A}$ is a horizontal curve K, and since f is analytic, K is higher than the rest of $\mathcal{A}$. Also, recall that there are no peaks. Suppose for simplicity that f and f' have only simple roots (P and S in number) so that the pits are cone-like, and the saddle points really look like saddles or (more geographically) like mountain passes.

Now imagine a persistent rain falling on the surface $\mathcal{A}$. The pits gradually fill with water and so become P lakes, the depths of which we shall imagine are always equal to each other. What happens to the number of lakes as the water successively rises past each of the S passes? How many lakes are left by the time that the water has finally risen to the level of K? As required, deduce that

$$P = S + 1.$$

(vii) Generalize the above argument to roots and critical points that are not simple.

20 Let $f(z)$ and $g(z)$ be analytic inside and on a simple loop Γ. By applying the Maximum Modulus Theorem to $(f - g)$, show that if $f = g$ on Γ then $f = g$ throughout the interior.

21 Let $\mathcal{R}(L)$ be the net rotation of $f(z)$ round p as z traverses a loop L. For example, if L does not contain p then

$$\nu\,[L, p] = \frac{1}{2\pi}\,\mathcal{R}(L).$$

By taking this formula to be the definition of ν, make sense of the statement that the Generalized Argument Principle (7.17) remains valid even if there are some poles and p-points *on* Γ, provided that we count these points with *half* their multiplicities.

22 In [7.11] we used the idea of deformation to derive the argument principle. The figure below shows another method. The interior of Γ, containing various p-points and poles, has been crudely partitioned into cells C_j in such a way that each one contains no more than a single p-point or a single pole. Now think of each cell as being a loop traversed in the conventional direction; this sense is indicated for two adjacent cells in the figure.

(i) What is the value of $\nu\,[f(C_j), p]$ if C_j (1) is empty; (2) contains a p-point of order m; (3) contains a pole of order n?

(ii) Obtain the Argument Principle by showing that

$$\sum_j \nu\,[f(C_j), p] = \nu\,[f(\Gamma), p].$$

[*Hint*: If an edge of a cell does not form part of Γ then it is also an edge of an adjacent cell, *but traversed in the opposite direction*. What is the net rotation of $f(z)$ round p as z traverses this edge once in one direction, then in the opposite direction?]

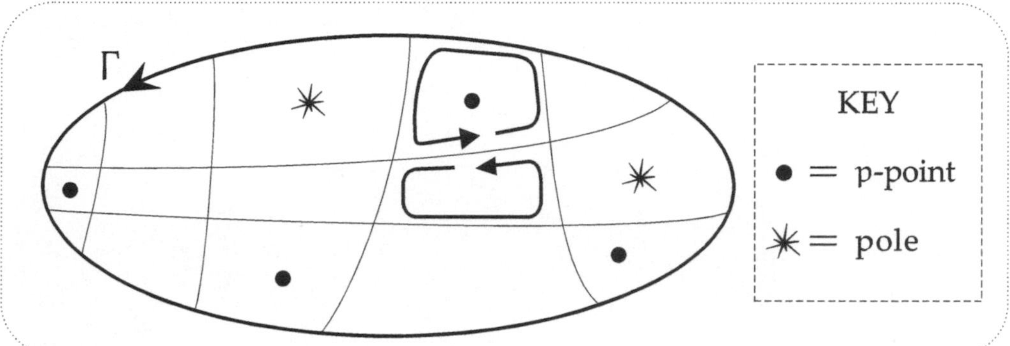

CHAPTER 8

Complex Integration: Cauchy's Theorem

8.1 Introduction

In the last few chapters our efforts to extend the idea of differentiation to complex mappings have been amply rewarded. By innocently attempting to generalize the real derivative we were quickly led to the amplitwist concept, and the subject then came to life with a character all its own. While many of the results cast familiar shadows onto the world of the reals, many did not, and striking indeed was the flavour of the arguments used to grasp them. The ability of z to freely roam the plane unleashed in us a degree of visual imagination that had to remain dormant so long as we could only watch the real number x forlornly pacing its one-dimensional prison.

This little hymn to the glory of the complex plane can be sung again in the context of integration, only louder. If differentiation breathed life into the subject, then integration could be said to give it its soul. Only after we have understood this soul will we be able to demonstrate such fundamental facts as the infinite differentiability of analytic mappings[1].

In ordinary calculus the symbol $\int_a^b$ has a clear meaning. However, if we wish to generalize this to $\mathbb{C}$ then the need for new ideas is immediately apparent, for how are we to *get* from a to b? In $\mathbb{R}$ there was only one way, but a and b are now points in the plane, so we must specify some connecting path (called a *contour*) "along which to integrate". It is then natural to ask whether the value of the integral depends upon the choice of this contour.

In general the value of the integral *will* depend on the route chosen. For example, we will shortly meet an integral of a complex mapping that yields, when evaluated for a closed contour, the area enclosed by the contour—a flagrant dependence of

[1] Since the 1960's it has actually become possible to do such things without integration, thanks to pioneering work by G. T. Whyburn, and others. Nevertheless, integration still appears to provide the simplest approach.

Visual Complex Analysis. 25th Anniversary Edition. Tristan Needham, Oxford University Press.
 DOI: 10.1093/oso/9780192868916.003.0008

value on contour. It should be made clear from the outset that while differentiation only made sense for the strictly limited set of analytic functions, this is *not* the case for integration. Indeed, the example just cited involves the integration of a non-analytic function.

The principal aim of this chapter (beyond the mere construction of an integral calculus) will be the discovery of conditions under which the value of an integral does *not* depend on the choice of contour. One such result is an analogue of the Fundamental Theorem of real analysis, and in deference to that subject it bears the same name. However, in the complex realm this is actually a misnomer, for there exists a still deeper result which has no counterpart in the world of the reals. It is called *Cauchy's Theorem.*

As we have said, it is not only possible, but sometimes useful to integrate non-analytic functions. However, it should come as no surprise to learn that new phenomena arise if we concentrate on the integrals of mappings that *are* analytic. Cauchy's Theorem is the essence of these new phenomena. Essentially it says that any two integrals from a to b will agree, provided that the mapping is *analytic everywhere in the region lying between the two contours*. Almost all the fundamental results of the subject (including some already stated) flow from this single horn of plenty.

8.2 The Real Integral

8.2.1 The Riemann Sum

As we did with differentiation, we begin by re-examining the more familiar idea of integrating a real function. The historical origin of this process, and still the principal means of visualizing it, is the problem of evaluating the area under the graph of a function.

We first approximate the sought-after area with rectangles. See [8.1]. Dividing the interval of integration into n line-segments Δ_j (the bases of the rectangles), we randomly select one point x_j from each segment, and take the height of the corresponding rectangle to be the height of the curve above the point, namely, $f(x_j)$. The area of each rectangle is then $f(x_j)\,\Delta_j$, and thus the total rectangular approximation to the area under f is

$$R \equiv \sum_{i=1}^{n} f(x_j)\,\Delta_j\,. \tag{8.1}$$

The quantity R is called a *Riemann sum*. Finally, by simultaneously letting n tend to infinity while each Δ_j shrinks to nothing, R will tend to the desired area.

In [8.1] we could afford to be indifferent to the precise choice of x_j within each Δ_j because we had our eye on this final limiting process. As each Δ_j shrinks, the

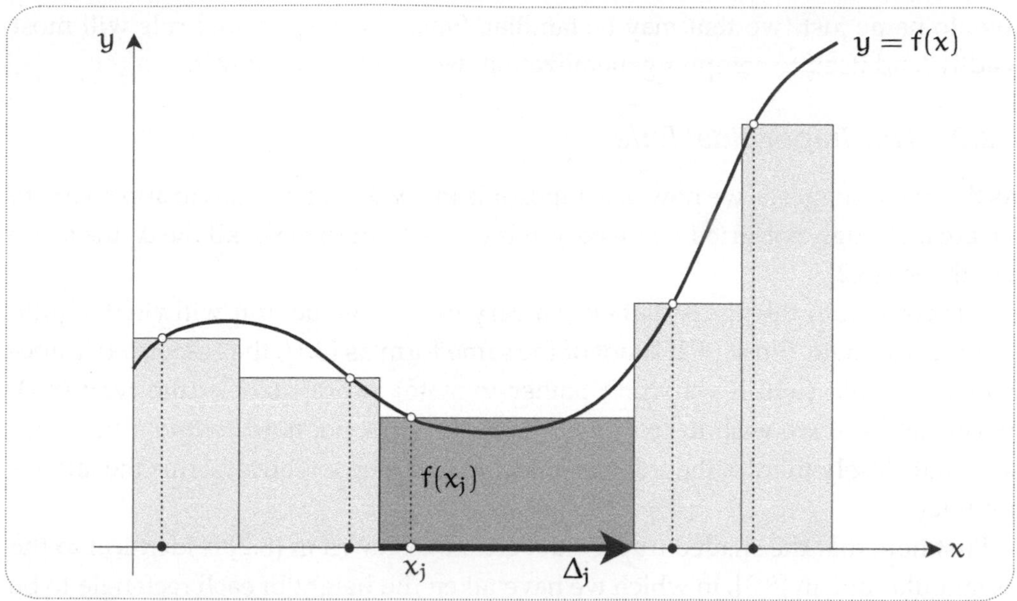

[8.1] A Riemann Sum $\mathbf{R} \equiv \sum_{j=1}^{n} f(x_j)\,\Delta_j$ *approximates the area under a curve with rectangles, the exact answer being obtained in the limit that* n *goes to infinity and each* Δ_j *goes to zero.*

freedom in the choice of x_j becomes more and more limited, and the influence of the choice on the area of the rectangle likewise diminishes. However, if we are unwilling or unable to actually carry out the limiting process, then, as we shall now see, we can ill afford to be so blasé in our choice of x_j.

You probably dimly remember some professor showing you (8.1) before, and perhaps you even evaluated a couple of examples by means of it. However, this was no doubt quickly forgotten once you set eyes upon the Fundamental Theorem of Calculus. In order to integrate x^4, why bother with taking the limit of some complicated series when we know that the answer must be that function which differentiates to x^4, namely, $\frac{1}{5}x^5$?

The Fundamental Theorem *is* a wonderful thing, but one must remember that many quite ordinary functions simply do not possess an antiderivative that is expressible in terms of elementary functions. To take a simple example, the Normal Distribution of statistics requires a knowledge of the area under the curve e^{-x^2}, and this can only be computed *numerically*, perhaps via a Riemann sum.

When doing a numerical calculation with (8.1), it would require an infinite amount of time to find even a single area with perfect precision. It is therefore important to be able to obtain good approximations to $\lim_{n\to\infty} R$ while using only a finite value of n. Several such methods exist: Simpson's rule and the Trapezoidal rule, to name just two that may be familiar. Since the Trapezoidal rule will most readily lend itself to complex generalization, we will now review it.

8.2.2 The Trapezoidal Rule

As the name suggests, we now use trapezoids instead of rectangles to approximate the area. Though not strictly necessary, it is convenient to make all the Δ_j the same length. See [8.2].

It is clear from the figure that even a very modest value of n will yield a quite accurate estimate. Since [8.2] is not of the same form as [8.1], the associated Trapezoidal Formula (which we won't bother to state) is not quite of the type (8.1). Nevertheless, if we wish to continue to use (8.1), it is not hard to find a Riemann sum that closely mimics the trapezoidal sum, and hence which retains the latter's accuracy.

First note that the shaded trapezoidal estimate shown in [8.2] is identical to the rectangular one in [8.3], in which we have taken the height of each rectangle to be the height of the chord at the midpoint of Δ_j. Finally, to recover a Riemann sum, we can replace the height of the chord by the height of the curve at that point. See [8.4].

In other words, the Riemann sum (8.1) will yield an accurate approximation to the integral, using only a modest value of n, provided that we *choose each* x_j *to be at the midpoint of its* Δ_j. We will call this the *Midpoint Riemann Sum*, and write it as R_M.

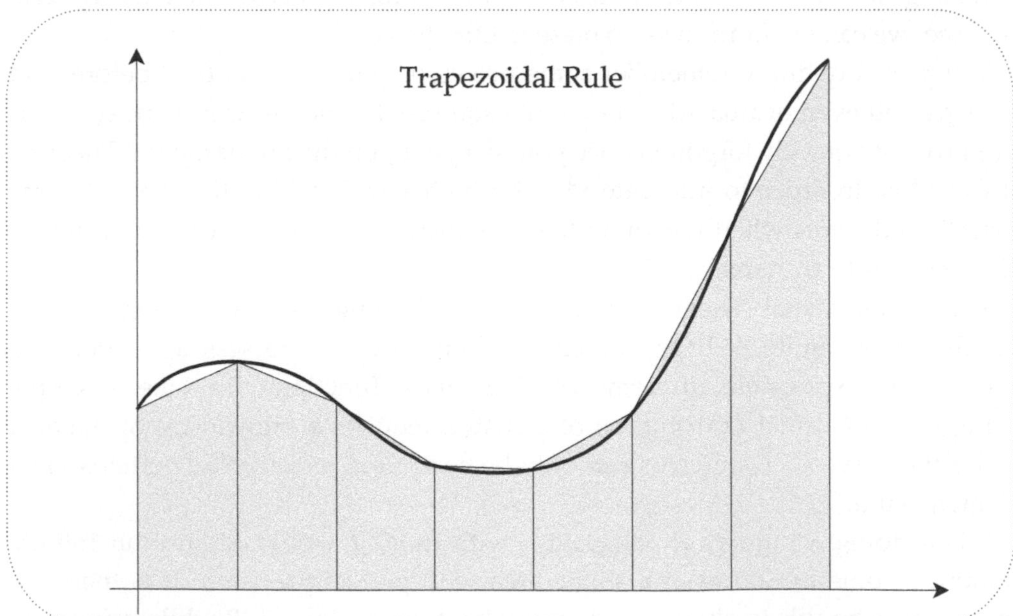

[8.2] The Trapezoidal Rule *replaces the rectangles with trapezoids, and clearly yields an accurate approximation to the integral while using only a modest value of* n.

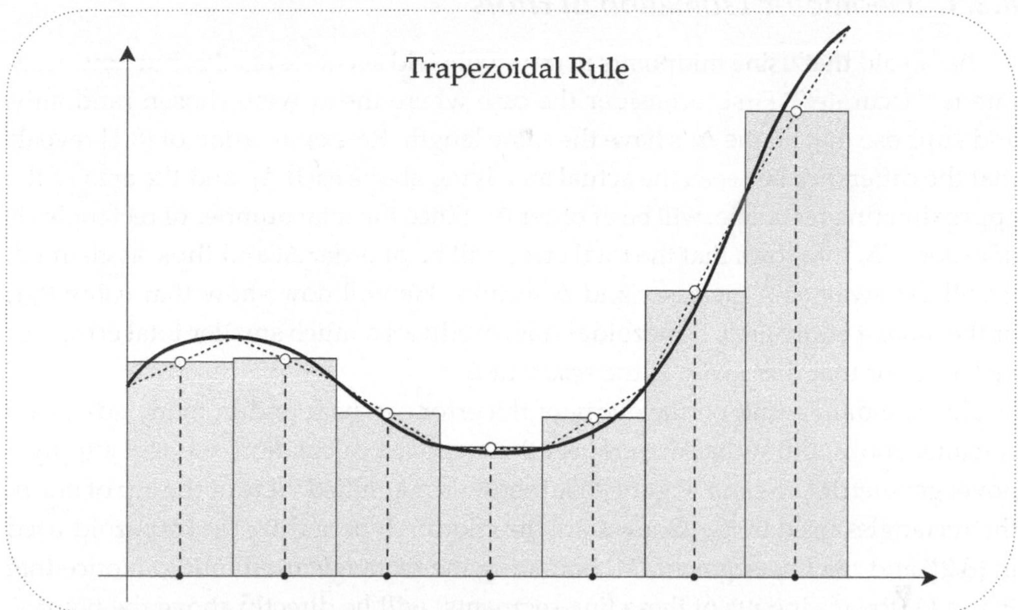

[8.3] *The exact same area as the previous figure, but now represented as rectangles whose heights are given by the midpoints of the illustrated chords.*

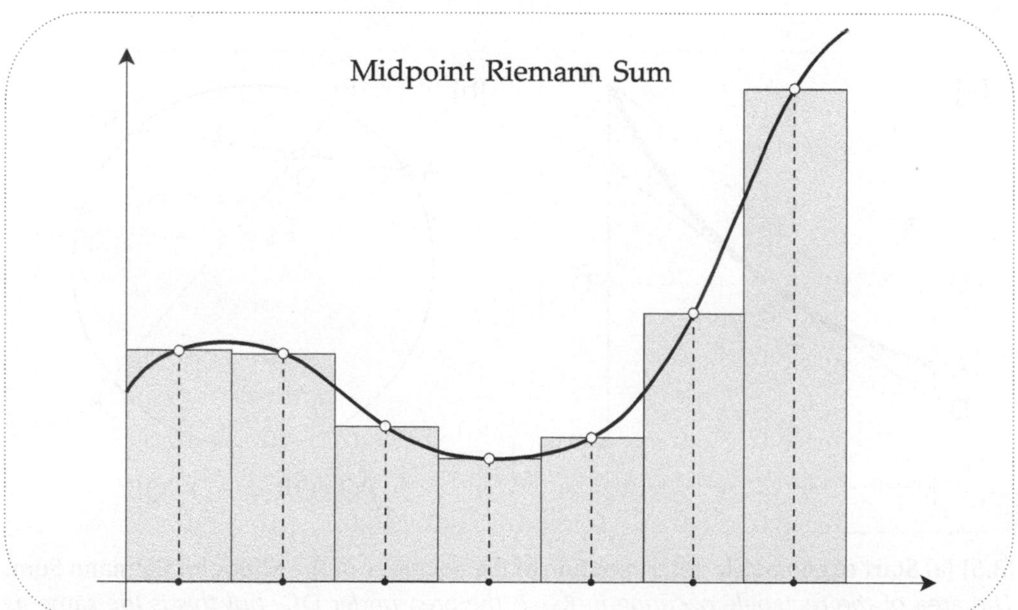

[8.4] The Midpoint Riemann Sum, R_M, *mimics the previous figure, but now uses the height of the graph at each midpoint, instead of the height of the chord. As an approximation to the integral, a typical Riemann sum that uses the height at a random point has a total error that dies away as Δ, but we will show that the* Midpoint *Riemann Sum is* much *more accurate, with an error that dies away as Δ^2.*

8.2.3 Geometric Estimation of Errors

We have said that using midpoints in (8.1) will yield accurate results, but how accurate is "accurate"? First reconsider the case where the x_j were chosen randomly, and suppose that all the Δ_j's have the same length. Re-examination of [8.1] reveals that the difference between the actual area lying above each Δ_j, and the area of the approximating rectangle, will be of order Δ^2. Since the total number of rectangles is of order $1/\Delta$, it follows that the total error will be of order Δ, and thus, as claimed, it will die away as n increases and Δ shrinks. We will now show that using R_M, or the almost equivalent Trapezoidal rule, produces a much smaller total error—in fact an error that dies away as the *square* of Δ.

This standard result on the decay of the error can be found in many advanced calculus books, but rather than repeat the standard calculation, we will supply a novel geometric[2] account. Figure [8.5a] shows a magnified view of the top of one of the rectangles used in R_M. Shown are the chord AB bounding the trapezoid used in [8.2], and the line-segment $\tilde{D}\tilde{C}$ bounding the rectangle used in R_M. Notice that P and Q (the midpoints of these line-segments) will lie directly above the point x_j that is being used in R_M.

Visually, it is easy to compare the area under AB [the Trapezoidal rule] with the actual area under the curve, but the same cannot be said of the area under

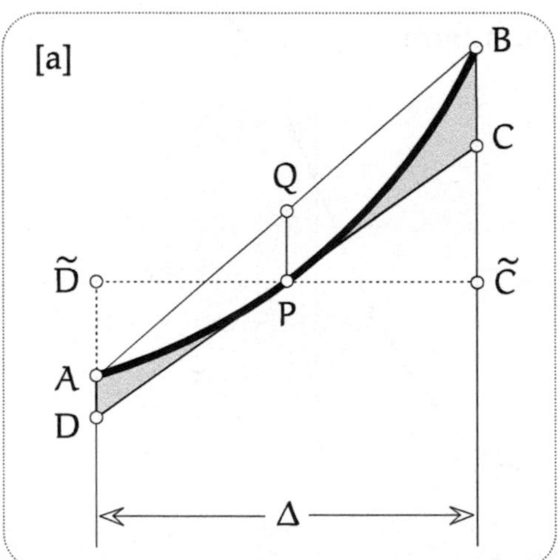

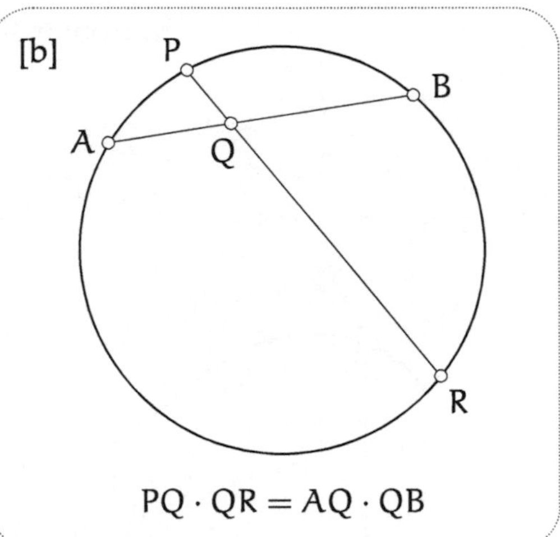

[8.5] [a] Start of geometric determination of the accuracy of the Midpoint Riemann Sum. *The area of the rectangle occuring in R_M is the area under $\tilde{D}\tilde{C}$, but this is the same as the area under the tangent, DC, so the error is the shaded region between DC and the curve. This error is certainly less than the area of ABCD, which is $(PQ) \cdot \Delta$.* **[b]** *As the chord PQR revolves about the fixed point Q, the product $PQ \cdot QR$ remains constant.*

[2] While many of the arguments in this book were merely inspired by Newton's mode of thought in the *Principia*, we have here an example that is very close to his actual methods.

$\widetilde{\mathrm{D}}\widetilde{\mathrm{C}}$ [the R_M rule]. However, note that if we rotate $\widetilde{\mathrm{D}}\widetilde{\mathrm{C}}$ about P (keeping the ends glued to the verticals) until it becomes *tangent* at DC, the area beneath it will remain constant [why?]. Thus we are instead free to visualize each term of R_M as being the area lying beneath a tangent such as DC. It is now clear that the actual area lies between the two values furnished by AB and DC, and that the error induced by using either rule cannot exceed the area of the small quadrilateral ABCD, namely [exercise], $(\mathrm{PQ}) \cdot \Delta$. In order to find this area we will use the elementary property of circles that is illustrated in [8.5b]: *As the chord* PQR *revolves about the fixed point* Q, *the product* $\mathrm{PQ} \cdot \mathrm{QR}$ *remains constant.*

Over a sufficiently tiny distance we can consider any segment of curve to be interchangeable with its tangent. However, over somewhat larger distances (or if we simply require greater accuracy) we must instead replace it by a segment of its *circle of curvature*, that is, the circle whose curvature κ agrees with that of the curve at the point in question. In [8.6] we have drawn this circle for the segment at P. The above result now informs us that

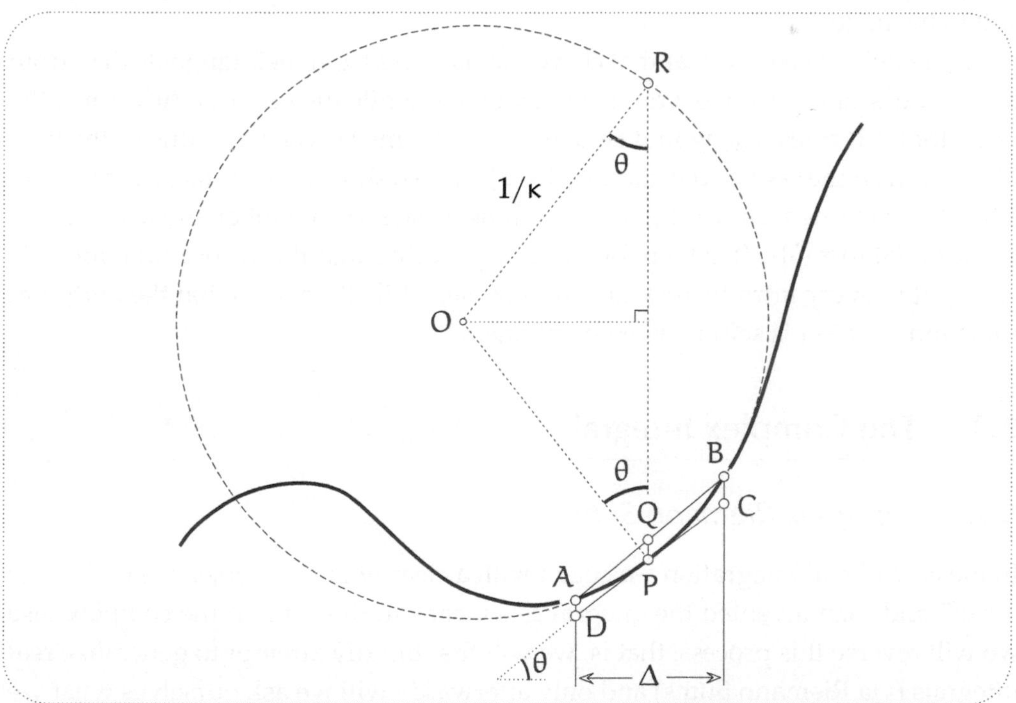

[8.6] Conclusion of geometric determination of the accuracy of the Midpoint Riemann Sum. *Let us approximate the curve in the vicinity of* P *with its circle of curvature, of radius* $(1/\kappa)$. *By virtue of the previous figure,* $\mathrm{PQ} \cdot \mathrm{QR} = (\mathrm{AQ})^2$. *But in the limit of vanishing* Δ, *we have* $\mathrm{AQ} \asymp \mathrm{DP}$ *and* $\mathrm{QR} \asymp \mathrm{PR}$, *so* $\mathrm{PQ} = (\mathrm{AQ})^2/\mathrm{QR} \asymp (\mathrm{DP})^2/\mathrm{PR}$. *If* θ *is the angle of the tangent at* P, *then* $\mathrm{DP} = \frac{1}{2}\Delta \sec\theta$, *and* $\mathrm{PR} = (2/\kappa)\cos\theta$. *Therefore,* area $(\mathrm{ABCD}) = \mathrm{PQ} \cdot \Delta \asymp \left(\frac{1}{8}\kappa \sec^3\theta\right)\Delta^3$.

$$PQ \cdot QR = (AQ)^2. \tag{8.2}$$

As Δ shrinks, both (AQ/DP) and (QR/PR) tend to unity, so in this limit we may substitute DP for AQ and PR for QR. But if the tangent at P makes an angle θ with the horizontal (in which case OP makes angle θ with the vertical) then $DP = \frac{1}{2}\Delta \sec\theta$, and $PR = (2/\kappa)\cos\theta$. Substituting these into (8.2), we obtain the result

$$\text{area (ABCD)} = PQ \cdot \Delta \asymp \left(\tfrac{1}{8}\kappa \sec^3\theta\right)\Delta^3. \tag{8.3}$$

If M denotes the maximum of $\left(\frac{1}{8}\kappa\sec^3\theta\right)$ over the integration range (which we take to be of length L) then each such error will be less than $M\,\Delta^3$. Since the number of these error terms is (L/Δ), we conclude that

$$\text{total error} < (LM)\,\Delta^2,$$

and this indeed dies away in the manner originally claimed. [At this point you may care to look at Ex. 1]

Because the order of the induced error is the same for both R_M and the Trapezoidal rule, we will tend not to distinguish between them when it comes to their complex generalizations. This said, there remains one curious pedagogical point still to be made.

Figure [8.5a] makes it clear that curves deviate less from their tangents than from their chords, and thus one would anticipate that while the order of the error is the same for both rules, R_M would actually yield the more accurate value of the two. This is indeed the case, and in fact [see Ex. 2] one can show that it is *twice* as accurate. In addition to this accuracy, R_M is, if anything, easier to remember and use than the Trapezoidal formula. It is therefore doubly puzzling that the Trapezoidal formula is taught in every introductory calculus course, while it appears that the midpoint Riemann sum R_M is seldom even mentioned.

8.3 The Complex Integral

8.3.1 Complex Riemann Sums

In the case of real integration we began with a clear geometric objective ("Find the area!") and then invented the integral as a means to this end. In the complex case we will reverse this process, that is, we will first blindly attempt to generalize real integrals (via Riemann Sums) and only afterwards will we ask ourselves what we have created. First, in this chapter, we will find one way of picturing an integral as a single complex number; then, in Chapter 11, we will use an entirely different point of view to see that, separately, the real and the imaginary parts of an integral each possess a vivid geometric (and physical) significance. But to guess the relevant geometric entities in advance, and then to invent the complex integral as the appropriate tool with which to find them, would require a prodigious leap of

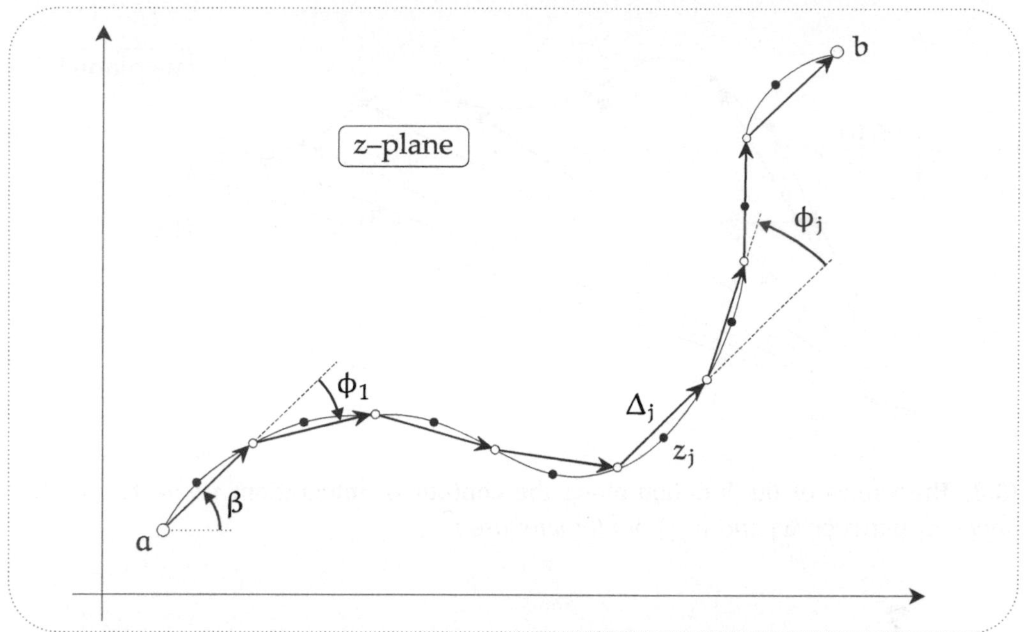

[8.7] Complex Riemann Sums. *To approximate $\int_K f(z)\,dz$, we approximate K with many small complex chords, Δ_j, evaluate $f(z)$ at an arbitrary point z_j within each segment of K, then evaluate the sum, $\sum f(z_j)\,\Delta_j$. Given the established superior accuracy of the Midpoint Riemann Sum (R_M) in the real case, here we have immediately chosen each z_j to be the midpoint of each segment. As we shall explain, the Riemann sum can be visualized more easily if we* choose all the Δ_j's to have the same length. *[The turning angles ϕ_j illustrated here are for later use.]*

imagination—one that historically never took place. A moment's thought reveals that this is similar to the case of differentiation, for there we began with the slope concept, and through an initially blind process of extrapolation we arrived at the very different (but no less intuitive) idea of the amplitwist.

Consider [8.7]. In order to integrate a complex mapping $f(z)$ between the points a and b, we have specified a connecting curve (*contour*) along which to perform the integration. This curve (call it K) now plays the role of the interval of integration, and just as in [8.1], we break it down into small steps Δ_j, which we may conveniently choose to be of equal length. The difference between this and [8.1] is that now the steps are not all in the same direction.

In order to construct a Riemann sum, we randomly pick one point z_j from each little segment of K, and then we form the sum of the products $f(z_j)\,\Delta_j$. Finally, as we increase their number and decrease their lengths, the Δ_j will follow K ever more perfectly, and the Riemann sum will tend to a limiting value (provided only that the mapping is continuous) that serves as our definition of the complex integral, written

$$\int_K f(z)\,dz\,.$$

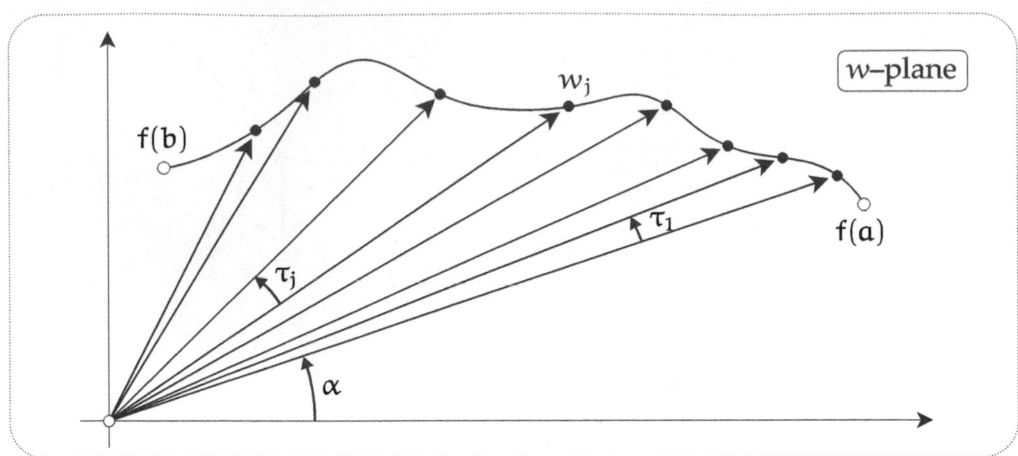

[8.8] The values of the function along the contour of integration: $w_j = f(z_j)$. *[The angles τ_j between w_j and w_{j+1} are for later use.]*

Just as in the real case, we may obtain an accurate estimate of the integral without passing to the limit, simply by choosing the z_j to be at the *midpoints* of the segments of K, rather than at random points. In fact this is the choice that we have illustrated in [8.7]. Once again, this especially accurate Riemann sum will be denoted R_M.

To begin to understand the geometry of R_M, consider [8.8]. This shows the image of K under the mapping $z \mapsto w = f(z)$, and in particular the image w_j of the z_j that was singled out in [8.7]. The corresponding term of R_M is then $\widetilde{\Delta}_j \equiv w_j \Delta_j$, and we will choose to think of this as the arrow that results when w_j "acts on" Δ_j, expanding it by $|w_j|$ and rotating it by $\arg(w_j)$.

Having obtained each $\widetilde{\Delta}_j$ in this manner, we go on to join all these little arrows together (tail to tip), as in [8.9]. The value of R_M, and hence the approximate value of the integral, is then the connecting complex number between the start and the finish[3]. Notice that since the answer is a *connecting* arrow, the point at which we begin drawing R_M is irrelevant.

While [8.9] is intended primarily to convey the general idea, it is in fact a faithful evaluation of the specific R_M corresponding to [8.7] and [8.8], and you may now begin to convince yourself of this. This is perhaps most easily achieved by concentrating on the lengths of the $\widetilde{\Delta}_j$ separately from their angles. As w traces out the image curve in [8.8], its length diminishes, and this produces a corresponding shrinking of the $\widetilde{\Delta}_j$ in [8.9]. Likewise, the increasing angle of w results in progressively greater rotations of the Δ_j.

[3] The great physicist Richard Feynman used a similar kind of picture to explain his quantum-mechanical "path integrals", which are also complex, though they differ from contour integrals. See Feynman (1985).

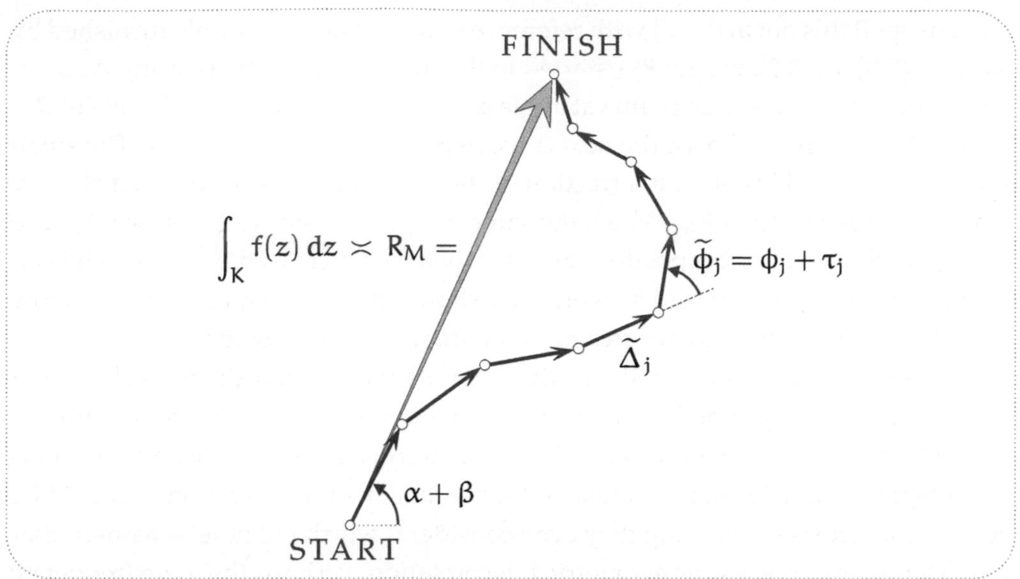

[8.9] Visualizing the Midpoint Riemann Sum, R_m. *The angle of each term $\widetilde{\Delta}_j \equiv w_j\Delta_j$ in $R_M = \sum \widetilde{\Delta}_j$ is the sum of the angles of w_j and Δ_j, so the illustrated turning angle $\widetilde{\phi}_j$ between successive steps along the Riemann sum is therefore the sum of the turning angles ϕ_j and τ_j in the previous two figures: $\widetilde{\phi}_j = \phi_j + \tau_j$.*

8.3.2 *A Visual Technique*

While it is not strictly necessary to choose the same length for all the Δ_j, the benefit of this choice is probably clear: the lengths of the $\widetilde{\Delta}_j$ are simply proportional to $|w_j|$, and it is therefore easy to follow the evolution of $\left|\widetilde{\Delta}_j\right|$ by eye. On the other hand, it is not so easy to visually follow the evolution of the *angle* of $\widetilde{\Delta}_j$.

As we travel along the Δ_j in [8.7], we pass through a sequence of sharp bends. The turning angle at a typical bend is drawn in [8.7], and is denoted ϕ_j. What will be the turning angle $\widetilde{\phi}_j$ at the corresponding bend of the Riemann sum? If, for example, w_{i+1} pointed in the same direction as w_j, then Δ_{i+1} and Δ_j would both suffer the same rotation, and the turning angle $\widetilde{\phi}_j$ of the Riemann sum would equal the original turning angle ϕ_j. More generally, if the angle of w increases by τ_j (see [8.8]), then the turning angle will also increase by τ_j. Thus,

$$\widetilde{\phi}_j = \phi_j + \tau_j. \tag{8.4}$$

This simple observation helps to reduce the difficulty of visualizing R_M. It is no longer necessary to look at the angle of each w_j (which may be large and hard to gauge by eye) and to try and imagine the direction of the rotated Δ. In fact we need now only do this *once*, to find $\widetilde{\Delta}_1$, thereby ensuring that R_M heads off in the correct initial direction. Thereafter, each successive $\widetilde{\Delta}$ is laid down at an angle $\widetilde{\phi}$ to its predecessor, and these $\widetilde{\phi}_j$ may be readily estimated by eye, using (8.4).

Let us spell this out in detail with reference to the concrete example furnished by [8.7] and [8.8]. In [8.9], we get R_M started in the right direction by rotating Δ_1 by α, thereby obtaining $\widetilde{\Delta}_1$ which points at angle $\alpha + \beta$. We can now draw the rest of R_M using only (8.4). To lay down the next $\widetilde{\Delta}$ we need to know $\widetilde{\phi}_1 = \phi_1 + \tau_1$. The small positive τ_1 clearly kills off just a fraction of the negative ϕ_1, resulting in a slightly smaller negative bend in R_M. Much the same happens when we lay down $\widetilde{\Delta}_2$. The angle ϕ_3 at the next bend is positive, and it is therefore increased by τ_3, which itself is about twice as big as τ_1 and τ_2 were. You should now be in a position to follow the rest of R_M's progress in far greater detail than you could before.

Although the above idea will shortly prove its worth on a theoretical level, it is clearly not terribly practical. However, in Chapter 11 we will use an entirely different approach to obtain a second, less strenuous, means of visualizing complex integrals. We will thereby make a double fallacy of an assertion that is to be found in most texts—assuming they even consider it worthy of note!—namely, that complex integrals possess *no* geometric interpretation. Perhaps the mere frequency with which this myth has been reiterated goes some way to explaining how it has acquired the status of fact.

8.3.3 A Useful Inequality

In figure [8.9] it is clear that if we were to straighten out all the bends in R_M then it would get longer. Furthermore, the length of the straightened version would just be the sum of the $\left|\widetilde{\Delta}_j\right|$. Thus

$$|R_M| \leqslant \sum |w_j| \cdot |\Delta_j| ,$$

with equality if[4] and only if all $\widetilde{\phi}_j = 0$. If M denotes the maximum distance from the origin to the image curve in [8.8], it follows that

$$|R_M| \leqslant M \sum |\Delta_j| .$$

But the sum on the right is just the length of the polygonal approximation to K, and hence it cannot exceed the actual length of K. Passing to the limit where R_M becomes the integral, we deduce that

$$\left| \int_K f(z)\, dz \right| \leqslant M \cdot (\text{length of K}) . \tag{8.5}$$

For example, if $f(z) = (1/\overline{z})^2$ and K is the circle $|z| = r$, then (8.5) implies that $\left|\int_K f(z)\, dz\right| \leqslant (2\pi/r)$. In particular this implies that $\lim_{r\to\infty} \int_K f(z)\, dz = 0$. This is a typical (albeit simplistic) application of (8.5): quite often one wishes to demonstrate

[4] The method of visualizing complex integrals in Chapter 11 will enable us to express this condition for equality in a particularly simple form. See Ex. 6, p. 574.

the ultimate vanishing of an integral as K evolves through some family, such as circles of increasing radius. Without knowing the exact value of any of the integrals, (8.5) shows that it is sufficient to demonstrate that the maximum size of $f(z)$ on K dies away faster than the length of K grows.

8.3.4 *Rules of Integration*

Because the complex integral has been defined in complete analogy with the real one, it follows that the former will inherit many of the properties of the latter. We now list some of these shared properties:

$$\begin{aligned}\int_K c\,f(z)\,dz &= c\int_K f(z)\,dz\\ \int_K [f(z)+g(z)]\,dz &= \int_K f(z)\,dz + \int_K g(z)\,dz\\ \int_{K+L} f(z)\,dz &= \int_K f(z)\,dz + \int_L f(z)\,dz\\ \int_{-K} f(z)\,dz &= -\int_K f(z)\,dz.\end{aligned}$$

The meaning of the first two equations is self-evident, but the last two require some clarification.

If L begins where K left off (see [8.10a]), then to integrate along $K + L$ means to integrate along K and then to continue integrating along L, and the resulting integral is then just the sum of the two separate integrals. Notice that the contour is allowed to have a kink in it. In fact the definition of "contour" merely requires that the number of such kinks not be infinite.

Lastly, the fourth rule is analogous to swapping the limits on a real integral, for $-K$ is defined to be the same as K, but traversed in the opposite direction (see [8.10b]).

However familiar you may be with these rules in real calculus, and however readily they may lend themselves to complex generalization, we would nevertheless urge you to make a new and separate peace with each of these results, preferably in terms of pictures such as [8.7], [8.8], and [8.9].

Recall from the Introduction that our main objective is the discovery of conditions under which an integral between two points in the plane does not depend on the connecting route chosen. The last two rules above may be used to recast this problem into a neater form. Suppose that the two paths K and $\widetilde{K}$ in [8.10c] both yield the same value for the integral between a and b. It follows that

$$0 = \int_K f(z)\,dz - \int_{\widetilde{K}} f(z)\,dz = \int_K f(z)\,dz + \int_{-\widetilde{K}} f(z)\,dz = \int_{K-\widetilde{K}} f(z)\,dz.$$

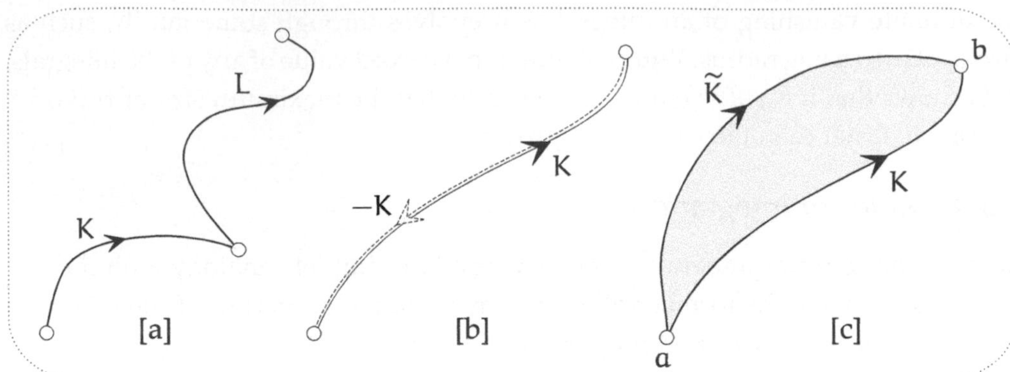

[8.10] [a] *If* L *begins where* K *left off, then to integrate along* K + L *means to integrate along* K *and then to continue integrating along* L*, and the resulting integral is then just the sum of the two separate integrals.* **[b]** *Changing* K *to* −K *reverses the sign of the integral, for* −K *is defined to be the same curve as* K*, but traversed in the opposite direction.* **[c]** *If the integral from* a *to* b *along* K *equals the integral along an alternative route* $\widetilde{K}$*, then the integral along the closed loop* $(K - \widetilde{K}) = (K$ *followed by* $-\widetilde{K})$ *must be zero. Conversely, if the integral vanishes for all closed loops then all curves between* a *and* b *will yield the same value for the integral. In brief:* **Path independence is equivalent to vanishing loop integrals.**

Thus equality of the two integrals is equivalent to the vanishing of the integral taken along the closed loop $(K - \widetilde{K}) = (K \text{ followed by } -\widetilde{K})$. Conversely, if the integral vanishes for all closed loops then all curves between a and b will yield the same value for the integral. In brief: *path independence is equivalent to vanishing loop integrals*. The centrepiece of complex analysis is the link between this phenomenon and analyticity. Cauchy's Theorem consists in recognizing that the vanishing of loop integrals is the nonlocal manifestation of a local property of the mapping, namely, that it is an amplitwist everywhere inside the loop.

8.4 Complex Inversion

8.4.1 A Circular Arc

Probably the single most important integral in all of complex analysis is that of the complex inversion mapping, $z \mapsto 1/z$. While the truth of this assertion will only emerge gradually, this is the reason for the great attention we will now lavish on this particular example.

We begin with the simplest case, namely, where the path of integration K is an arc of the origin-centred circle of radius A (see [8.11a]). As in [8.7], we divide this path into small (ultimately vanishing) steps of equal length. The turning angles ϕ_j clearly all have the same value, say ϕ. Since the angle that each Δ subtends at the

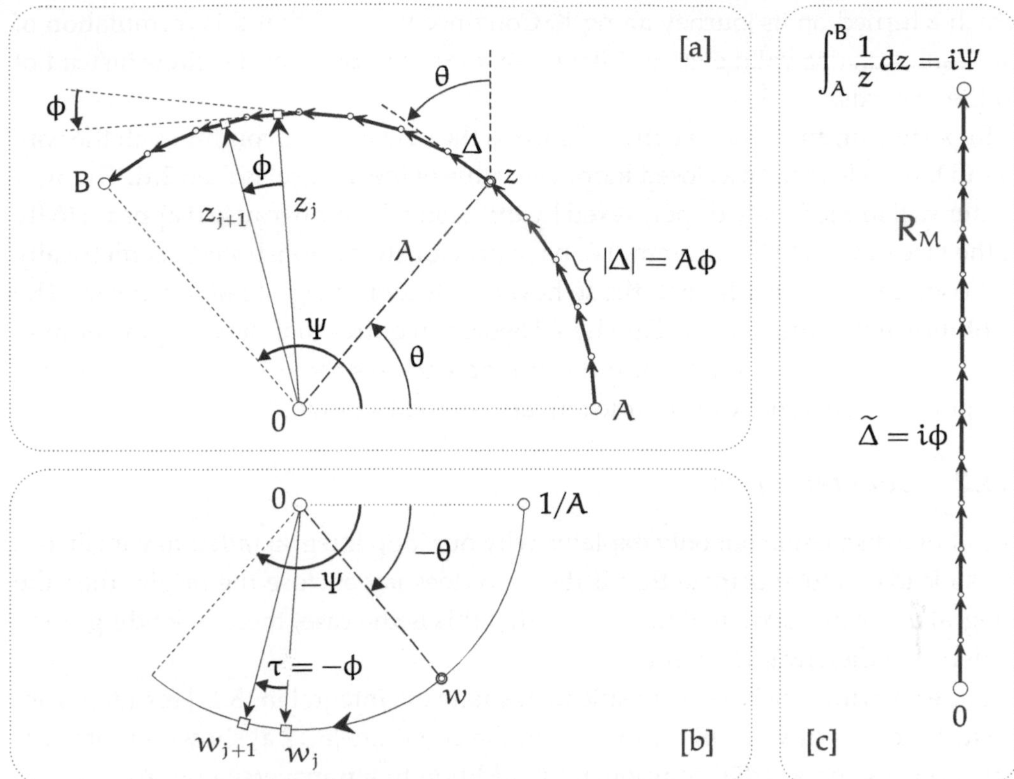

[8.11] The integral of $(1/z)$ along a circle. *Since Δ_1 and w_1 are ultimately vertical and horizontal respectively, it follows that R_M (in* **[c]***) initially heads off in the vertical direction. But since the turning angle τ of $w_j = (1/z_j)$ in* **[b]** *is the opposite of the turning angle ϕ of Δ in* **[a]***, $\widetilde{\phi} = \phi + \tau = 0$. In other words, R_M has no bends, and so it continues on in the imaginary direction. We wanted to illustrate the turning-angle approach for later use, but here is a simpler explanation. Since the Δ_j located at angle θ on the circle itself points at θ to the vertical (as illustrated) and has length $A\phi$, we see that multiplying it by $w = (1/A)e^{-i\theta}$ rotates it by $-\theta$, producing a vertical $\widetilde{\Delta} = w\Delta = i\phi$.*

origin is also given by ϕ, it follows that $|\Delta| = A\phi$. As z travels round the circle its image $w = 1/z$ travels round a circle of radius $1/A$ in the opposite direction (see [8.11b]), and thus w shrinks each Δ to produce a $\widetilde{\Delta}$ of length ϕ.

Since Δ_1 and w_1 are ultimately vertical and horizontal respectively, it follows that R_M (which we choose to begin drawing at the origin in [8.11c]) initially heads off in a vertical direction. But now we observe that $\tau = -\phi$, and consequently that $\widetilde{\phi} = 0$. In other words R_M *has no bends*, and so it continues on in the imaginary direction[5]. Thus, irrespective of the radius, the integral equals i times the total angle Ψ through

[5] We have used the turning angle idea in order to make the subsequent generalization to other powers of z straightforward, but there is actually no need for it in the present case. Since the Δ_j located at angle θ on the circle itself points at θ to the vertical and has length $A\phi$, we see that multiplying it by $w = (1/A)e^{-i\theta}$ rotates it by $-\theta$, producing a vertical $\widetilde{\Delta} = w\Delta$ of length ϕ.

which z turned on its journey along K. Convince yourself that this formulation of the result remains valid even if K begins at a random point of the circle instead of on the real axis.

In particular, and of crucial importance, is the case where z continues all the way round the circle to form a closed loop. The value of the integral is then $2\pi i$. The alert reader will immediately be perplexed by this result. Why? Because it appears to fly in the face of Cauchy's Theorem. We have previously demonstrated geometrically that complex inversion is analytic, so how can its loop integral fail to vanish?! The resolution lies in the fact that Cauchy's Theorem requires that the mapping be analytic *everywhere* inside the loop. But our loop encloses the origin, and just at this one point the analyticity of complex inversion breaks down.

8.4.2 General Loops

The above discussion not only explains why our loop integral failed to vanish, but it also leads us to anticipate that if the loop does *not* enclose the origin, then the integral *will* vanish. We will now show that this is the case, thereby lending some credence to Cauchy's Theorem.

The ease with which we were able to evaluate the integral in [8.11] resulted from the fact that the Δ_j were all orthogonal to the z_j. Figure [8.12 a] shows a more typical Δ possessing a radial component in addition to a transverse one. As you see, Δ can be decomposed into a transverse component $r\,d\theta$ making an angle θ with the vertical, and an orthogonal radial component dr. To obtain the corresponding piece $\widetilde{\Delta}$ of R (see [8.12b]) we multiply by w, thereby rotating these components into

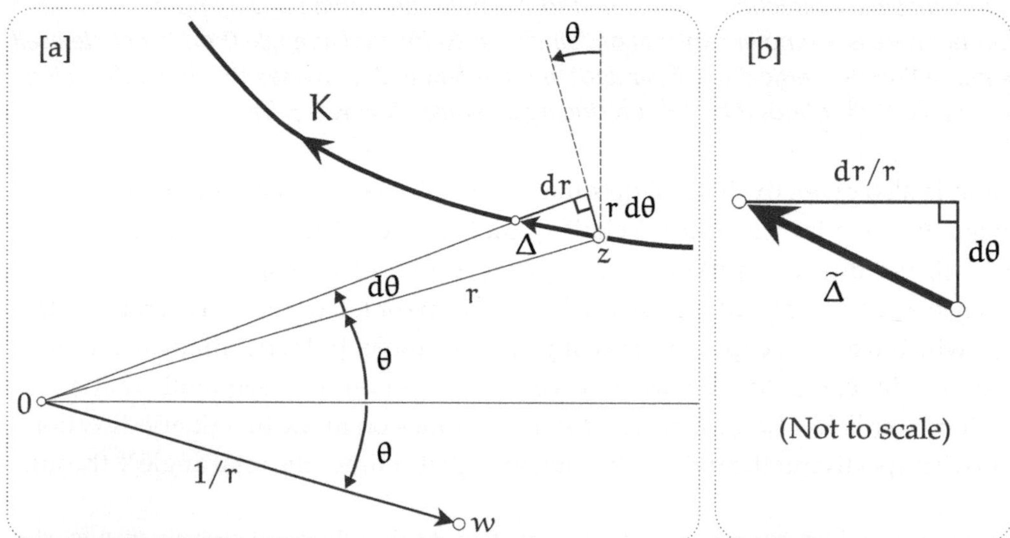

[8.12] The integral of $(1/z)$ along a general contour. [a] *The movement Δ along a general contour can be broken down into a radial component dr and an orthogonal component $r d\theta$ making angle θ with the vertical.* **[b]** *Multiplication by $w = (1/z)$ therefore yields $\widetilde{\Delta} = w\Delta = (dr/r) + i\,d\theta$.*

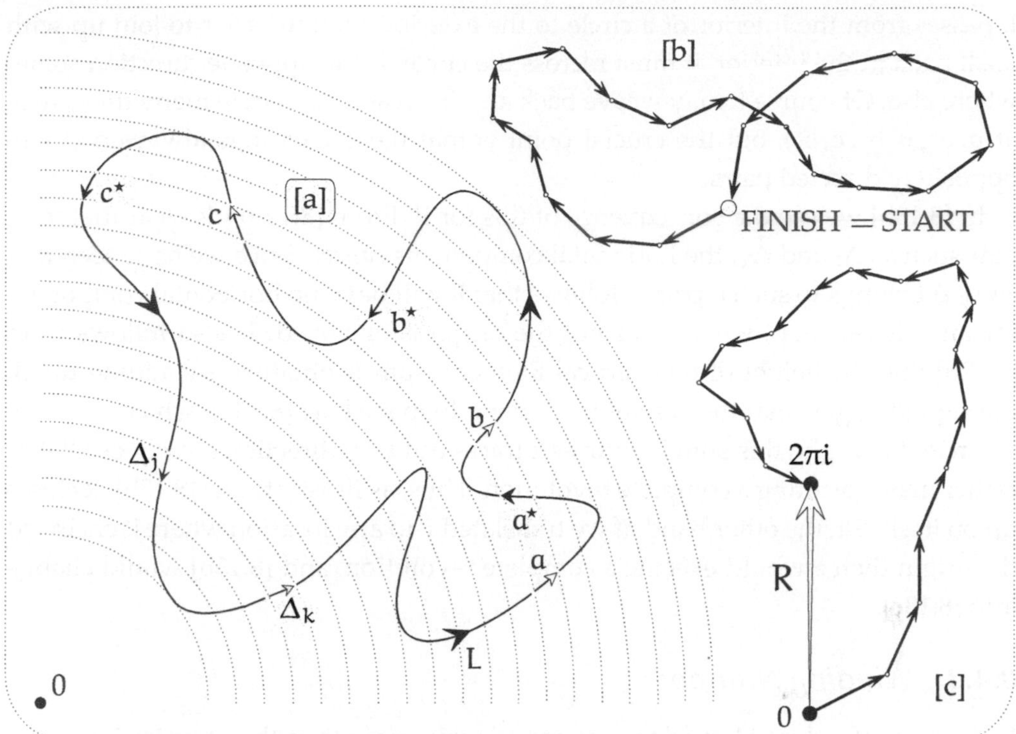

[8.13] The integral of $(1/z)$ along a closed loop. [a] *Instead of choosing Δ_j's of equal length, divide the path up, as shown, using closely spaced concentric circles centred at the origin. Consider the pair Δ_j and Δ_k cut off by the same two circles. From the previous figure, we deduce that real components of $\widetilde{\Delta}_j$ and $\widetilde{\Delta}_k$ in the Riemann sum will cancel. But* every Δ *belongs to such a pair (or pairs), because if the* closed loop L *passes from the interior of a circle to the exterior, then in order to join up with itself back in the interior, it must recross the circle in the opposite direction somewhere else.* **Thus the integral of $(1/z)$ along a closed loop is always imaginary.** *The previous figure also implies that* **this imaginary answer = i *(net angle of rotation as the loop is traversed)*.** **[b]** *The illustrated loop* L *does not enclose the origin, so the integral vanishes.* **[c]** *If the loop does wind around 0 once, then the integral is $2\pi i$.*

the vertical and horizontal directions, as well as shrinking their lengths to $d\theta$ and (dr/r), respectively.

Let us now see what happens if we stick all these $\widetilde{\Delta}_j$ together for a closed loop such as L (see [8.13a]) that does not encircle the origin. In order to accomplish this we will forsake our previous choice of equal lengths for all the Δ_j, and instead divide the path up, as shown, using closely spaced concentric circles centred at the origin[6]. Consider the illustrated pair Δ_j, Δ_k lying between adjacent circles. That the Δ's always *do* occur in such pairs is a consequence of L being a loop. For if

[6] This only fails if part of L coincides with such a circle, but in that event we already know that the contribution to the integral is $i\Psi$.

L passes from the interior of a circle to the exterior, then in order to join up with itself back in the interior, it must recross the circle in the opposite direction somewhere else. Of course L may weave back and forth across a circle many times (e.g. at $a, a^\star, b, b^\star, c, c^\star$), but the crucial point is that these crossings always occur in oppositely directed pairs.

In [8.13b] we see the consequence of this for R. From [8.12b] it's clear that for a pair such as $\widetilde{\Delta}_j$ and $\widetilde{\Delta}_k$, the horizontal components *cancel*. Since we have seen that *every* Δ belongs to such a pair, it follows that R will have no horizontal component for any closed loop, whether or not the origin is encircled. It also follows from [8.12b] that the height of this vertical Riemann sum is obtained by adding up all the signed angles that the Δ's subtend. For a loop such as [8.13a], which does not encircle the origin, this sum is zero: as z traces out L its direction merely oscillates, rather than executing a complete revolution. Thus, as illustrated in [8.13b], R closes up on itself. On the other hand, if we translated L to any location where it encircled the origin then z would execute a complete revolution, and [8.13b] would change into [8.13c].

8.4.3 *Winding Number*

Let's recap. If a closed loop does not encircle the origin then the complex inversion mapping is analytic everywhere inside it, and in accord with Cauchy's Theorem the integral dutifully vanishes. If the origin *is* encircled, then the integral is no longer required to vanish by the theorem: the enclosed region now contains a point at which the mapping is not analytic. Indeed we found that for an origin-centred circle the answer was not zero, but $2\pi i$. Furthermore, the general investigation revealed that we would have obtained exactly the same answer if we had instead used an elliptical loop, or even a square loop. For if we distort the circle into one of these more general shapes, then all that happens to R is that it meanders about (illustrated in [8.13c]) on its net vertical journey to $2\pi i$, instead of marching straight there as it did in [8.11c].

We see that what really matters is not the shape of the loop, but rather its winding number about the origin. Thus we may summarize our findings tidily as follows: If L is *any* closed loop, then

$$\oint_L \frac{1}{z}\, dz = 2\pi i\, \nu(L, 0), \tag{8.6}$$

where the integral sign with a circle through it (which is a standard symbol) serves to remind us that we are integrating around a closed contour. Figure [8.14] shows various loops and the corresponding value of the integral of $(1/z)$ round each of them. Finally, note that (8.6) can easily be generalized [exercise] to

$$\oint_L \frac{1}{z-p}\, dz = 2\pi i\, \nu(L, p). \tag{8.7}$$

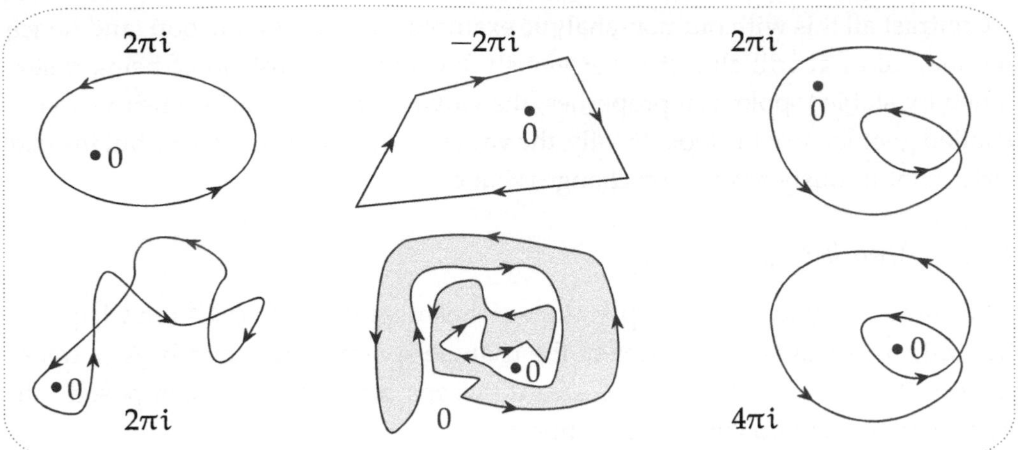

[8.14] *The previous figure proves that* $\oint_L (1/z)\, dz = 2\pi i\, \nu(L, 0)$.

8.5 Conjugation

8.5.1 *Introduction*

In the introduction we stressed that integration makes sense for any continuous complex mapping, regardless of whether or not it is analytic. However, the relatively lawless non-analytic functions give rise to integrals that behave less predictably than their analytic counterparts. In particular, Cauchy's Theorem has no jurisdiction here, and we therefore have no reason to anticipate path independence or, equivalently, vanishing loop integrals. As an example of this type of behaviour, we will show presently that the loop integral of the non-analytic conjugation mapping $z \mapsto \bar{z}$ yields the area enclosed by the loop. Assuming this result for the moment, let us use the examples $\bar{z}$ and $(1/z)$ to spell out more clearly the differences between the non-analytic and analytic cases.

In the analytic case, provided that the special point $z = 0$ was not enclosed, the loop integral vanished. Even when the integral of $(1/z)$ did *not* vanish, its possible values were still neatly *quantized* in units of $2\pi i$; one unit for each time the special point $z = 0$ was enclosed by the loop. As we will see later, this behaviour is typical, although a more general mapping may well possess *several* special points (at which analyticity breaks down) dotted about in the plane. Once again, the integral is not sensitive to the precise shape of the loop. Provided that none of the special points are enclosed by the loop, then the integral vanishes. However, if some of the points are enclosed, then each one makes its own distinctive contribution (generally not $2\pi i$) to the integral, one unit for each time it's encircled. The value of the integral is just the sum of these discrete contributions.

Contrast all this with our non-analytic example. The area of the loop (and hence the integral of $\overline{z}$) will almost never vanish. Furthermore, instead of being determined by stable topological properties, the value of the integral is sensitive to the detailed geometry of the loop. Finally, the value is not neatly quantized, but instead varies continuously as the loop changes shape.

8.5.2 Area Interpretation

Let us now verify the area interpretation of the integral of $\overline{z}$. Recall from Chapter 1 that $\text{Im}(a\overline{b})$ is just twice the area of the triangle spanned by a and b. As z traces the loop L in [8.15a], think of the area it sweeps out as being decomposed into triangular elements, as illustrated. Thus

$$2\,(\text{element of area}) = \text{Im}[(z+\Delta)\overline{z}] = \text{Im}[\overline{z}\Delta].$$

Adding these elements together, we obtain the imaginary part of the Riemann sum corresponding to the integral of $\overline{z}$. Thus we conclude that

$$\text{Im} \oint_L \overline{z}\,dz = 2\,(\text{area enclosed})\,.$$

This result can be further simplified by noticing that $\overline{z}$ and $(1/z)$ both point in the same direction. It follows that we could draw a picture very similar to [8.12], the only difference being that to obtain $\widetilde{\Delta}$ we would multiply by r instead of dividing by

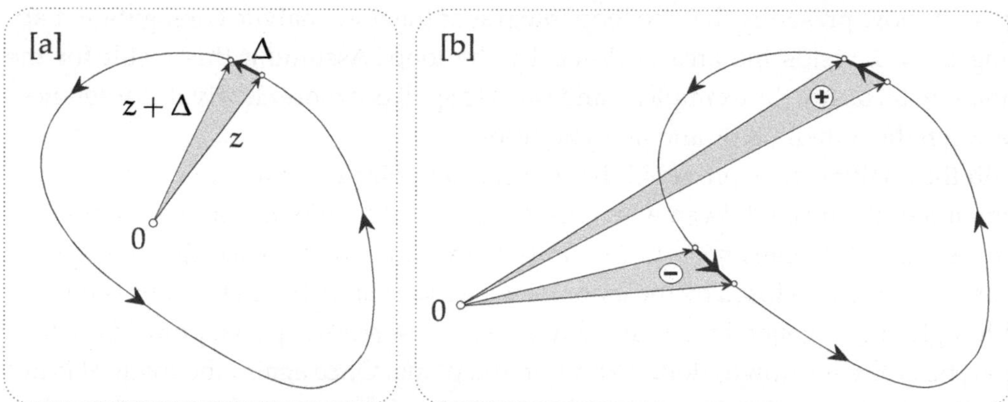

[8.15] $\oint_L \overline{z}\,dz = 2i$ **(area enclosed by L). [a]** *Using (1.20) on page 32, we see that* $2\,(\textit{element of area}) = \text{Im}[(z+\Delta)\overline{z}] = \text{Im}[\overline{z}\Delta]$*, so* $\text{Im} \oint_L \overline{z}\,dz = 2\,(\textit{area})$*. But* $\overline{z}$ *has the same direction as* $(1/z)$*, and the argument given in [8.12] easily generalizes, so* **the integral is purely imaginary***, thereby proving the result.* **[b]** *The geometric meaning of the integral is unaltered if 0 lies outside L, for although the top triangle includes area that lies outside L, this extra area is cancelled by the negative area of triangles like the one at the bottom. Compare this to [1.22b] on page 33.*

it. The argument that followed from [8.12] therefore remains valid, and we deduce that the integral of $\overline{z}$ around a closed loop is purely imaginary. Thus

$$\oint_L \overline{z}\, dz = 2i\,(\text{area enclosed})\,. \tag{8.8}$$

Next we ask how this formula would change if the origin were *outside* the loop. Figure [8.15b] shows that the pleasing answer is, "Not at all!" The point is that the integral adds up the *signed* areas subtended by the Δ's at the origin. On the far side, Δ carries z counterclockwise, yielding a positive element of area; but on the near side z is moving clockwise, yielding a negative element of area. When these are added, the unwanted area lying outside the contour simply cancels, leaving behind just the area enclosed.

As a simple example, consider a circle C of radius r centred at a, the equation of which is $r^2 = |z-a|^2 = (z-a)(\overline{z}-\overline{a})$. Solving this for $\overline{z}$, and using (8.7), we find that

$$\begin{aligned}\oint_C \overline{z}\, dz &= \overline{a}\oint_C dz + r^2 \oint_C \frac{1}{z-a}\, dz \\ &= 0 + r^2\, 2\pi i \\ &= 2i\,(\text{area enclosed}).\end{aligned}$$

From what we have done so far you might be inclined to think that the integral of $\overline{z}$ could never vanish for a nontrivial loop. That this is false can be seen from the figure eight loop in [8.16a]. This may be thought of as the union of two separate loops. The top one is traversed in a positive sense and correspondingly yields its ordinary area $\mathcal{A}_1$; but the bottom one is traversed in a negative sense and yields the

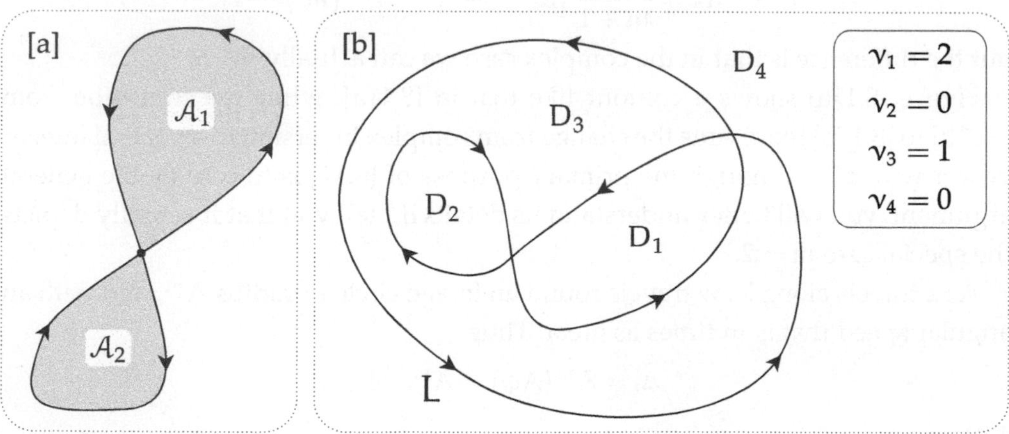

[8.16] [a] *If* L *is a figure eight, then* $\oint_L \overline{z}\, dz = 2i(\mathcal{A}_1 - \mathcal{A}_2)$. *Note that if* L *were symmetrical, then the integral would vanish.* **[b]** *In general,* $\oint_L \overline{z}\, dz = 2i\sum_{\text{inside}} \nu_j\, \mathcal{A}_j = 2i\,[2\mathcal{A}_1 + \mathcal{A}_3]$ *in this example.*

negative ($-\mathcal{A}_2$) of its ordinary area. Thus the integral is $2i(\mathcal{A}_1 - \mathcal{A}_2)$, and if the loop were symmetrical then this would vanish.

8.5.3 General Loops

To finish off this example we will explain how the winding number concept can be used to evaluate the integral for more complicated loops. A typical loop such as [8.16b] will partition the plane into a number of sets D_j, and in the last chapter we defined the "inside" to consist of those D_j for which the corresponding winding number $\nu_j \neq 0$; the remaining D_j constitute the "outside". We can now state the general result and leave you to ponder its truth:

$$\oint_L \bar{z}\,dz = 2i \sum_{\text{inside}} \nu_j \mathcal{A}_j, \tag{8.9}$$

where $\mathcal{A}_j$ denotes the area of D_j. For example, in the case of [8.16b] we obtain $\oint_L \bar{z}\,dz = 2i\,[2\mathcal{A}_1 + \mathcal{A}_3]$. We hope you can already see why this example is correct, but a full explanation of the general formula (8.9) will be provided later in this chapter.

8.6 Power Functions

8.6.1 Integration along a Circular Arc

Having understood the integral of $(1/z)$ it is easy to understand the integrals of other powers. Once again let us begin by integrating along the circular arc K that was used in [8.11]. The result we will obtain is formally identical to the real result

$$\int_A^B x^m\,dx = \frac{1}{m+1}\left[B^{m+1} - A^{m+1}\right] \quad (m \neq -1),$$

but the difference is that in the complex case we can actually *see it!*

Figure [8.17a] shows a contour like that in [8.11a], while the transition from [8.11b] to [8.17b] represents the change from complex inversion to a general integer power $w = z^m$. Although the primary purpose of [8.17] is to convey the general argument, you will better understand its details if I tell you that it actually depicts the special case $m = 2$.

As z travels along K, w travels round an image circle of radius A^m, and with an angular speed that is m times as great. Thus

$$|\widetilde{\Delta}| = A^m(A\phi) = A^{m+1}\phi\,,$$

and

$$\widetilde{\phi} = \tau + \phi = m\phi + \phi = (m+1)\,\phi\,.$$

Since all the $\widetilde{\Delta}$'s have the same length and the same turning angle, it follows that R_M is a polygonal approximation to an arc of a circle, the centre of which we have

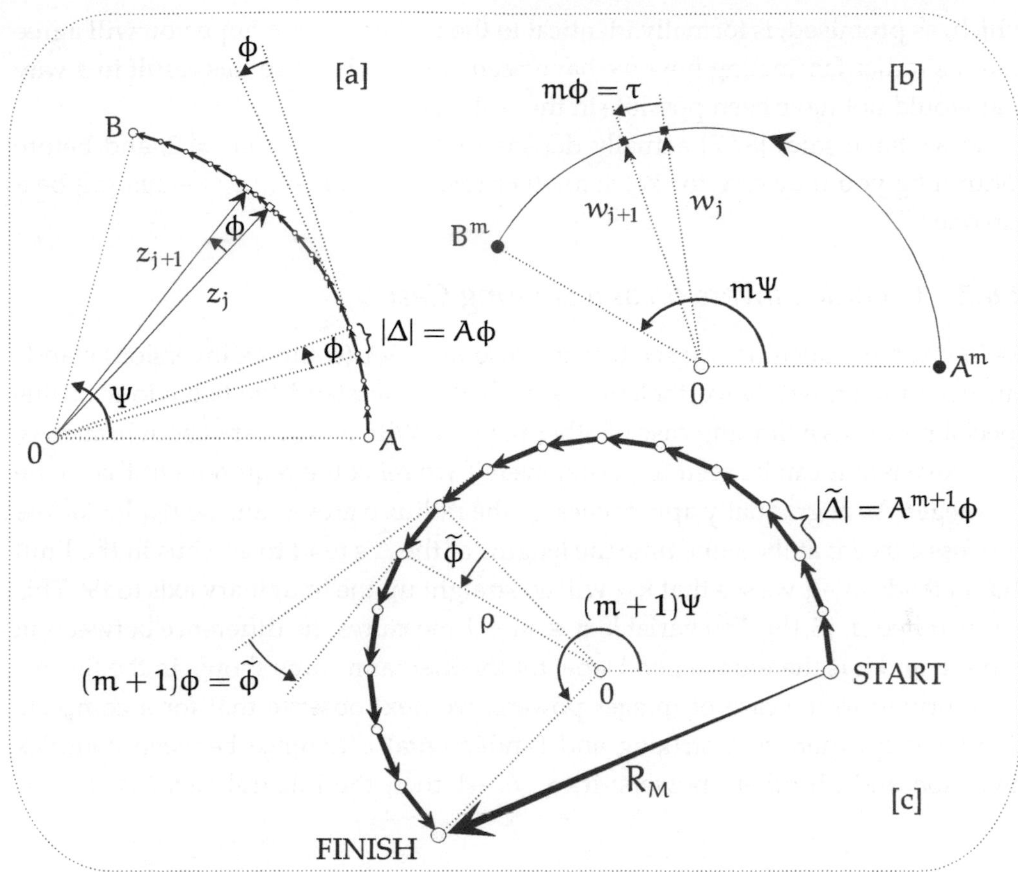

[8.17] $\int_A^B z^m\,dz = \frac{1}{m+1}\left[B^{m+1} - A^{m+1}\right]$, *because* **[a]** $|\Delta| = A\phi$, **[b]** $\tau = m\phi$ *and* $|\widetilde{\Delta}| = A^{m+1}\phi$, *so* **[c]** $\widetilde{\phi} = \tau + \phi = (m+1)\,\phi$, *and therefore* $\rho = |\widetilde{\Delta}|/\widetilde{\phi} = A^{m+1}/(m+1)$.

chosen to place at the origin in [8.17c]. We will now determine the angle subtended by this arc, and also its radius.

The angle that each $\widetilde{\Delta}$ subtends at the origin is the same as the turning angle $\widetilde{\phi}$, namely, $(m+1)$ times the angle subtended by each Δ. Thus

$$\text{angle of FINISH} = (m+1)\Psi.$$

Also, if ρ is the radius, we see from the figure that

$$\rho\,\widetilde{\phi} = |\widetilde{\Delta}| \quad\Longrightarrow\quad \rho = \frac{A^{m+1}}{m+1}.$$

We therefore conclude that if $m \neq -1$ then

$$\begin{aligned} R_M &= \text{FINISH} - \text{START} \\ &= \frac{1}{m+1}\left[A^{m+1}\,e^{i(m+1)\Psi} - A^{m+1}\right] \\ &= \frac{1}{m+1}\left[B^{m+1} - A^{m+1}\right] \end{aligned} \tag{8.10}$$

which, as promised, is formally identical to the real result. We hope you will agree that it's rather fascinating how we have been able to visualize this result in a way that would not have been possible in the real case.

As we have said, [8.17] actually depicts the concrete case $m = 2$, and before continuing you may care to sketch another case for yourself; $m = -2$ might be a fun one.

8.6.2 Complex Inversion as a Limiting Case*

As in ordinary calculus, we see that the case $m = -1$ (complex inversion) stands out from the crowd. Nevertheless, we can still understand the behaviour of this special power as a limiting case of other powers. With a little care about branches, the above result can be seen to persist even if we relax the requirement that m be an integer. As m gradually approaches -1 the radius ρ grows, and so R_M looks less and less curved; at the same time the lengths of the $\widetilde{\Delta}$'s tend to ϕ. Thus in the limit that m tends to -1 we see that R_M will go straight up the imaginary axis to $i\Psi$. This is illustrated in [8.18]. The variable $n \equiv m + 1$ measures the difference between m and -1, and it is therefore a good label for the Riemann sums shown in the figure.

Returning to the case of integer powers, we next observe that for a complete circular loop, there is a striking and fundamental difference between complex inversion and all others powers: if $m \neq -1$ then the integral *vanishes*. This is

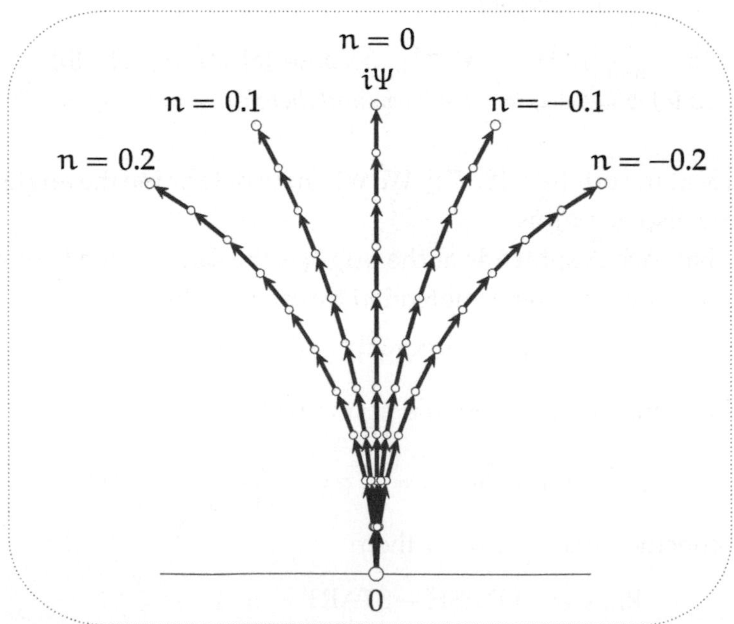

[8.18] The integral of $(1/z)$ as a limiting case of the integral of z^m. *Here, $n \equiv m - (-1)$ measures the "distance" of z^m from the ultimate target of z^{-1}. But there is also a fundamental* difference *between $(1/z)$ and all other powers: its integral around a* complete *circle is $2\pi i$, whereas the integral for all other powers is zero!*

because R_M will now go round in a complete circle $|n|$ times [clockwise if $n < 0$; counterclockwise if $n > 0$], thereby returning to its beginning.

8.6.3 General Contours and the Deformation Theorem

Thus far we have only established (8.10) for the case where A and B are connected by a simple arc, but in fact it is true for almost any contour. Take the case $n > 0$ first. Since z^m is then analytic throughout the plane, it follows directly from Cauchy's Theorem that all contours will yield the same value. However, when $n < 0$ the situation is a little bit more subtle.

Just as complex inversion suffers a breakdown of analyticity at the origin, so too do all the other negative powers of z. Therefore Cauchy's Theorem only guarantees that two connecting paths yield the same integral *provided* that together they do not enclose the origin. For a loop that does enclose the origin, the integral is not required to vanish, and indeed in the case of z^{-1} it equals $2\pi i$.

Nevertheless, our direct evaluation reveals that for all other negative powers the integral round a circular loop *does* vanish, despite not being required to[7]. We will now derive a new form of Cauchy's Theorem that enables us to show that the vanishing of the integral is not a fluke resulting from the special circular shape of the loop.

Consider [8.19a]. The two loops J and L both encircle a singularity of some mapping, and so neither integral is required to vanish by Cauchy's Theorem. However, if the mapping is analytic in the shaded region lying between the loops, then we will now show that the two integrals must be equal. First consider the contribution

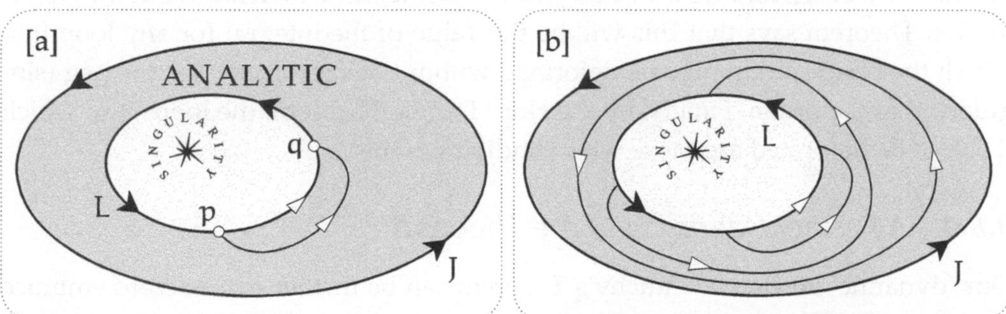

[8.19] Deformation Theorem. [a] *Deform* L *by pushing the segment* pq *outward to form the illustrated bump. Since the mapping is analytic between the two paths connecting* p *and* q*, it follows from Cauchy's Theorem that both integrals are equal, and since the rest of* L *hasn't changed, it follows that* **integral with bump = integral without bump**. **[b]** *We can continue to deform* L *until it becomes* J*, and the value of the integral will not change.*

[7] In Chapter 11 we will give a physical explanation for this difference between complex inversion and the rest of the negative powers.

to the integral round L that comes from the piece between p and q. Suppose that we deform L slightly by replacing this segment by the bump in the figure. Since the mapping is analytic between the two paths connecting p and q, it follows from Cauchy's Theorem that both integrals are equal. Also, since the rest of L hasn't changed, it follows that *integral with bump = integral without bump*. All we need do now, to obtain the stated result, is to let the bump grow and change shape (see [8.19b]) until L has evolved into J.

The crucial idea is this

> ***Deformation Theorem.*** *If a contour sweeps only through analytic points as it is deformed, the value of the integral does not change.* (8.11)

Thus, if you imagine the contour to be a rubber band, and the singularity to be a peg sticking out of the plane (thereby obstructing motion past it), the integral has the same value for all shapes into which the rubber band can be deformed.

We can immediately apply this Deformation Theorem to our problem. For if the mapping is a negative power of z other than z^{-1} then the established fact that the integral vanishes for a circular loop implies it continues to vanish for any loop into which the circle can be deformed without crossing the singularity at the origin. Thus formula (8.10) is path independent even for negative powers.

The Deformation Theorem also provides us with a much simpler derivation of the result (8.6) governing the general loop integral of the complex inversion mapping. Imagine taking a length of elastic string and winding it around an origin-centred circle ν times, finally joining the ends together to form a closed loop. From our earlier work, it follows that the value of the integral is then $2\pi i\nu$. But the Deformation Theorem says that this will be the value of the integral for any loop into which the elastic string may be deformed without being forced over the peg (singularity) at the origin. Finally, by the Hopf Degree Theorem, the loops into which it can so be deformed are those with winding number ν.

8.6.4 A Further Extension of the Theorem

Our 'dynamic' version of Cauchy's Theorem can be further extended to embrace mappings that have *several* singularities. Consider a loop L (see [8.20a]) encircling two singularities (pegs) of some mapping; the generalization to more singularities will be obvious. If we deform L without forcing it over a peg then we know that the integral will remain constant. The process [8.20a]→[8.20b]→[8.20c] is an example of such a deformation. The situation in [8.20c] is now rather interesting. The contour has become pinched together at q, and the value of the integral can be thought of as the sum of the two separate integrals taken round the touching circles. But now, by the same reasoning as usual, we may separately distort these circles so that [8.20c]→[8.20d]. Thus we conclude that

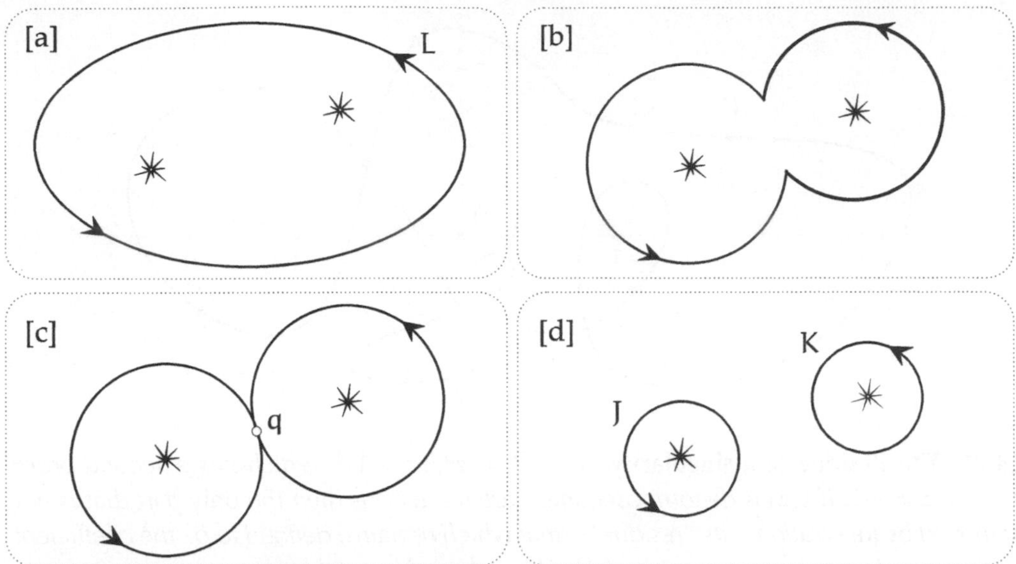

[8.20] Deformation Theorem with multiple singularities. *We can deform* L *into* J + K *without changing the value of the integral.*

$$\oint_L f(z)\,dz = \oint_J f(z)\,dz + \oint_K f(z)\,dz. \tag{8.12}$$

To illustrate (8.12), consider $f(z) = 2/(z^2+1)$ which has singularities at $z = \pm i$. We can evaluate the integral round any loop C by noting this alternative expression:

$$f(z) = \frac{i}{z+i} - \frac{i}{z-i}.$$

Applying (8.7) therefore yields

$$\oint_C f(z)\,dz = 2\pi[\nu(C, i) - \nu(C, -i)].$$

Assuming (as in [8.20a]) that L encloses both singularities, use this formula to verify (8.12) for this particular function.

8.6.5 Residues

Since we now possess a fairly complete understanding of the loop integrals of power functions, it is relatively easy to integrate simple rational functions: we need only find the decomposition into so-called *partial fractions*, and then integrate term by term. Indeed, this is precisely what we did in the example of the last paragraph.

Here is a slightly more complicated example: the integral of $f(z) = z^5/(z+1)^2$ taken round the contour K in [8.21]. This has a second-order pole at $z = -1$, and by writing the numerator as $[(z+1)-1]^5$, we quickly find that

$$f(z) = -\frac{1}{(z+1)^2} + 5\left[\frac{1}{z+1}\right] - 10 + 10\,(z+1) - 5\,(z+1)^2 + (z+1)^3.$$

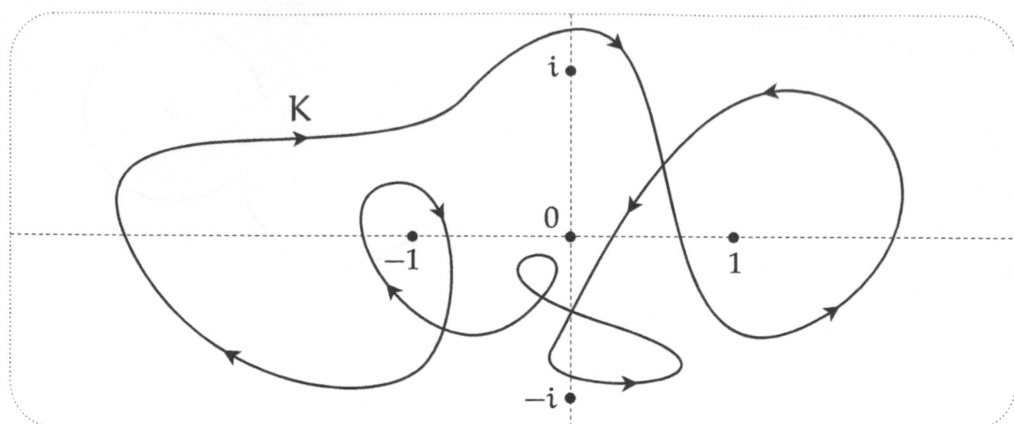

[8.21] The Residue of a singularity. *Let* $f(z) = z^5/(z+1)^2$*, which has a second-order pole at* $z = -1$*. If* $f(z)$ *is decomposed into partial fractions then the* only *part that is not removed by integration is its "residue"—that which remains, defined to be the coefficient of the complex inversion term,* $1/(z+1)$*. This is the* only *part that does not integrate away into thin air. The value of the integral therefore has only two ingredients: the residue and the winding number, and the result is* $2\pi i$ *times their product, which turns out to be* $-20\pi i$ *for this particular function and contour.*

But we know that the loop integral of powers other than −1 is zero, and so *only* the complex inversion term [in square brackets] can contribute. In detail,

$$\oint_K f(z)\, dz = 5 \cdot 2\pi i\, \nu(K, -1) = -20\pi i.$$

Thus the value of the integral has been determined by just two factors: the winding number of the loop, and the amount (i.e., coefficient) of complex inversion contained in the decomposition of the mapping. Because this latter number is the only part of the function that remains after we integrate, it is called the *residue* of the function at the singularity. Quite generally, the residue of $f(z)$ at a singularity s is denoted $\text{Res}\,[f(z), s]$. Thus in the above example, $\text{Res}\,[z^5/(z+1)^2, -1] = 5$.

In fact the residue concept has a significance that extends far beyond simple rational functions, as our next example will illustrate. We have previously [page 259] alluded to the remarkable fact that analytic functions are infinitely differentiable, or equivalently, that they can always be represented by a power series (Taylor's) in the vicinity of a nonsingular point. For example, the Taylor series centred at the origin for $\sin z$ is

$$\sin z = z - \frac{1}{3!}z^3 + \frac{1}{5!}z^5 - \frac{1}{7!}z^7 + \cdots .$$

Clearly no such expansion can be possible at a singular point of a mapping. Nevertheless, we may recover an analogous result near singularities simply by broadening our notion of a power series to include *negative* powers. Such a series is called a *Laurent series*.

Consider $(\sin z)/z^6$. This is singular at the origin, but by simple division of the above Taylor series we obtain the following Laurent series in the vicinity of the singularity:

$$\frac{\sin z}{z^6} = \frac{1}{z^5} - \frac{1}{3!\,z^3} + \frac{1}{5!}\left[\frac{1}{z}\right] - \frac{1}{7!}z + \frac{1}{9!}z^3 - \cdots.$$

Once again, the residue of the function is defined to be the coefficient of the complex inversion term: Res $[(\sin z)/z^6, 0] = 1/5!$ in this case. If a power series converges at every point on a contour, then we may accept for the moment that it makes sense to integrate the series term by term. Once again we see that for a closed loop the sole contribution to the integral comes from the residue. For example, if K is the contour in [8.21] then

$$\oint_K \frac{\sin z}{z^6}\,dz = \frac{1}{5!}\,2\pi i\,\nu(K,0) = -\frac{2\pi i}{5!}.$$

The above examples of evaluating loop integrals in terms of residues are instances of *Cauchy's Residue Theorem*. We will return to these matters at the end of this chapter, and, in greater detail, in the following chapter. For the moment we simply remark that if a mapping possesses several singularities, then a residue can be attributed to each one.

8.7 The Exponential Mapping

In the case of the exponential mapping the easiest contour along which to integrate is a vertical line-segment, say L (see [8.22a]). Once again we will find the result to be formally identical to its real counterpart:

$$\int_L e^z\,dz = e^B - e^A. \tag{8.13}$$

As z travels from A up to B in [8.22a], its image under $w = e^z$ will travel round the arc shown in [8.22b]. In order to verify (8.13), we will now show that (provided we choose to begin drawing R_M at e^A) *this arc is also the precise path taken by the Riemann sum.*

First note that since Δ_1 and w_1 are effectively vertical and horizontal, respectively, it follows that R_M will head off in the required vertical direction. Also observe that all the $\widetilde{\Delta}$'s have the same length, namely, $|\widetilde{\Delta}| = e^A\,|\Delta|$. Finally, since L has no bends (i.e., $\phi = 0$), $\widetilde{\phi} = \tau = |\Delta|$. Because the $\widetilde{\Delta}$'s all have the same length and turning angle, R_M will follow an arc of a circle. It only remains to show that if we begin drawing it at e^A, then this is the *same* arc as in [8.22b].

The next section will reveal the simplest way of seeing this, but the following direct argument is quite straightforward. We first verify that the two arcs have the

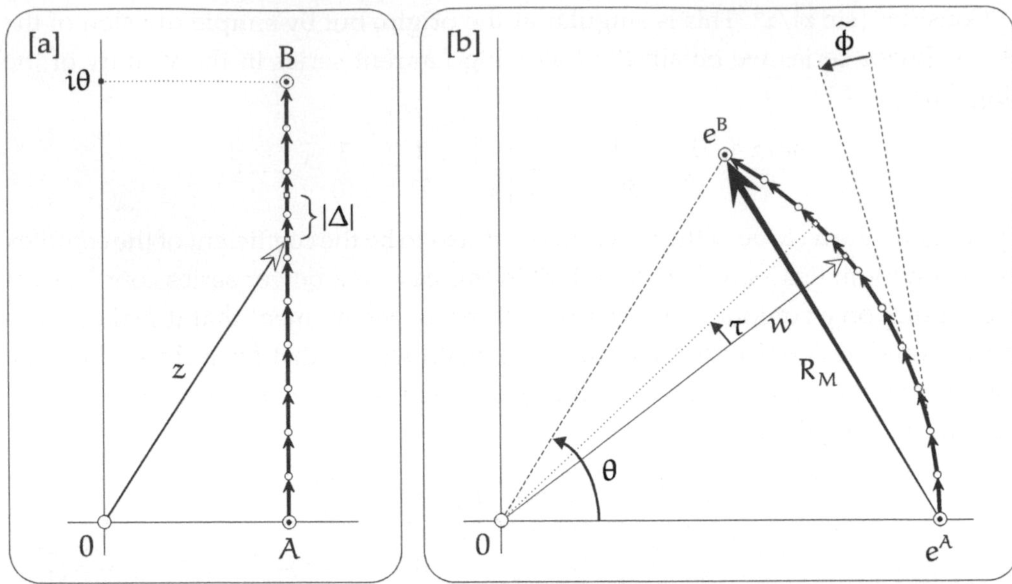

[8.22] Geometric evaluation of $\int_L e^z\,dz$. [a] *Let* L *be the illustrated vertical segment from* A *to* $B = A + i\theta$. **[b]** *If we choose to begin drawing* R_M *at* e^A *then the illustrated arc traced by* $w = e^z$ *is also the precise path taken by the Riemann sum, and therefore* $\int_L e^z\,dz = e^B - e^A$. *First, all the* $\widetilde{\Delta}$*'s have the same length, namely,* $|\widetilde{\Delta}| = e^A\,|\Delta|$. *Since* L *has no bends (i.e.,* $\phi = 0$*),* $\widetilde{\phi} = \tau = |\Delta|$. *Because the* $\widetilde{\Delta}$*'s all have the same length and turning angle,* R_M *will follow an arc of a circle, and it only remains to verify its radius. Try this yourself, or see the text for details.*

same radius. The angle that each $\widetilde{\Delta}$ subtends at the centre of its circle will be the same as its turning angle $\widetilde{\phi}$. Therefore

$$\text{radius} = \frac{\text{arc}}{\text{angle}} = \frac{|\widetilde{\Delta}|}{\widetilde{\phi}} = e^A,$$

as required. Lastly, the total angle subtended by the arc at its centre is just the sum of all the $\widetilde{\phi}_j = |\Delta_j|$, namely θ. The identity of the two arcs is thus established.

Since e^z is singularity-free, Cauchy's Theorem assures us that its loop integral always vanishes. Thus (8.13) must in fact be valid for *any* path from A to B.

8.8 The Fundamental Theorem

8.8.1 Introduction

Through specific geometric constructions, combined with the use of Cauchy's Theorem, we have already learnt a good deal about the integrals of some of the most important functions. However, there are two immediate problems still to be resolved: one is pragmatic, while the other is aesthetic.

The pragmatic one is that the formulae (8.10) and (8.13) are only known to hold for certain special configurations of the points A and B: the derivation of (8.10) assumes that they are equidistant from the origin; while for (8.13) they are assumed to be vertically separated. To be sure, our various forms of Cauchy's Theorem guarantee us that the integrals in question will continue to be path-independent, no matter what the locations of A and B. But the problem is that we haven't yet established that these path independent values will continue to be given by the same formulae as before. In this section we shall see that they are.

The aesthetic concern lies in the manner in which we derived path-independence for negative powers of z. Recall that we were only able to apply Cauchy's Theorem after having explicitly produced an example (a circle) of a loop integral that vanishes *in spite* of enclosing the singularity. Although this was neat enough in itself, one is left with the feeling that Cauchy's Theorem cannot be the most direct way of understanding a loop integral that continues to vanish in the presence of singularities.

A resolution of both these problems is provided by the so-called *Fundamental Theorem of Contour Integration*—a result that is formally identical to its similarly named counterpart in ordinary calculus. The naming of this theorem is not entirely appropriate, at least in the context of complex analysis. After all, so far we have managed quite well *without it*, suggesting that if this theorem is "Fundamental", then Cauchy's must be "Super-Fundamental"!

8.8.2 An Example

As our first example of this theorem, let us return to the exponential mapping of the last section in order to discover why (8.13) is valid for *any* pair of points, not just ones that are vertically separated. As so often happens in mathematics, all that is required is a very slight shift in viewpoint.

Figure [8.23] depicts a curve K (connecting a pair of typical points A and B) being mapped by e^z to the curve $\widetilde{K}$ connecting e^A and e^B. Now let us forget (for a moment) all about integration and Riemann sums, and instead look at the figure from the point of view of *differentiation*.

All the little arrows emanating from a point on K will be mapped to images emanating from a point on $\widetilde{K}$. In particular, if the arrow Δ is a little chord of K [tangent, in the limit that it shrinks], then its image $\widetilde{\Delta}$ will likewise be a little directed chord of $\widetilde{K}$. But for an *analytic* mapping, such as we are now considering, the original arrows are sent to their images by a simple amplitwist:

$$\widetilde{\Delta} = (\text{amplitwist of } e^z) \cdot \Delta = e^z \, \Delta. \tag{8.14}$$

If we now add up all these vector chords of $\widetilde{K}$ then we obtain the connecting vector V between its start and its finish. But (8.14) tells us that this vector V may also be

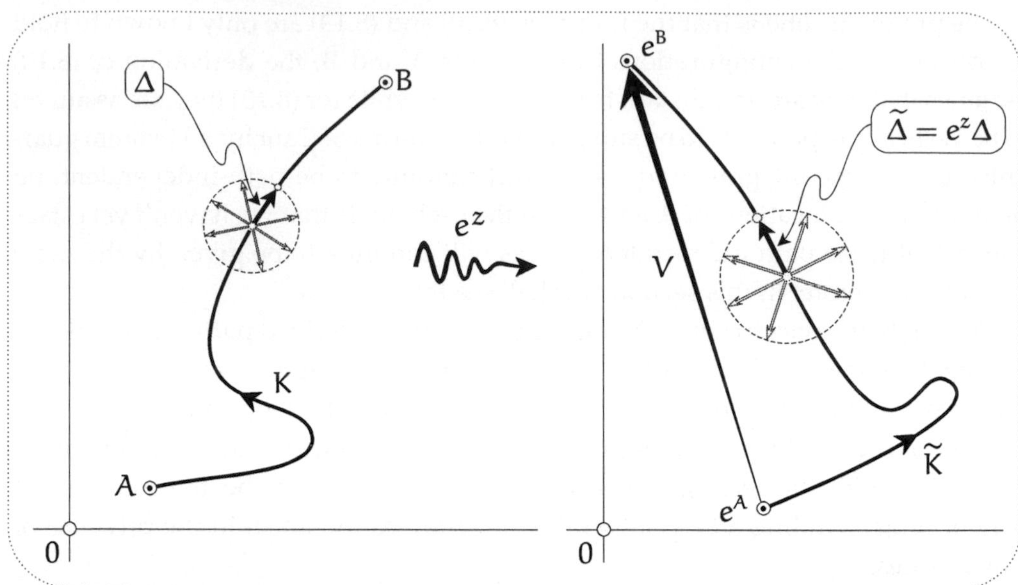

[8.23] The Fundamental Theorem in the case of e^z. *Under $z \mapsto \widetilde{z} = e^z$, the contour K connecting A to B maps to the curve $\widetilde{K}$ connecting e^A to e^B, the connecting complex number from start to finish being $V = e^B - e^A$. But each Δ along K is amplitwisted to $\widetilde{\Delta} = (\text{amplitwist of } e^z) \cdot \Delta = e^z \Delta$ along $\widetilde{K}$, so $\widetilde{K}$ simply* is *the Riemann sum, and therefore $\int_K e^z \, dz = V = e^B - e^A$.*

interpreted as the Riemann sum corresponding to the integral of e^z along K. We have thus established the continued validity of (8.13) for all positions of A and B:

$$\int_K e^z \, dz = V = e^B - e^A .$$

To emphasize the path-independence of the construction, imagine choosing a different contour from A to B. The image curve (i.e., the new Riemann sum) will then take a different route from e^A to e^B, but of course the vector V will be quite unaffected.

8.8.3 The Fundamental Theorem

The Fundamental Theorem amounts to a restatement of the above idea in general terms. Suppose that we wish to evaluate $\int_K f(z) \, dz$ by the method above. We must seek an *analytic* mapping $F(z)$ whose amplitwist $F'(z)$ is given by $f(z)$. Assuming that such an F has been found [whether this animal even *exists* will be discussed shortly], we may then draw [8.24], which depicts the image curve $\widetilde{K}$ under the mapping F. With the same terminology as before, (8.14) now becomes

$$\widetilde{\Delta} = [\text{amplitwist of } F(z)] \cdot \Delta = f(z) \, \Delta .$$

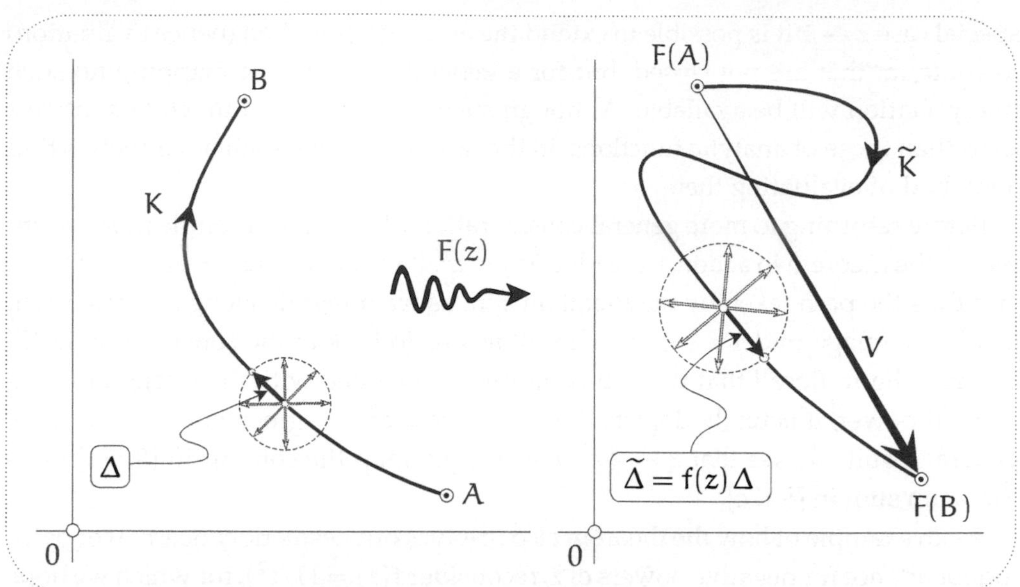

[8.24] The Fundamental Theorem in the general case. *If* F *maps* K *to* $\widetilde{K}$ *and* $F' = f$ *then* Δ *is amplitwisted to* $\widetilde{\Delta} =$ [amplitwist of F(z)] $\cdot \Delta = f(z)\,\Delta$, *and therefore* $\widetilde{K}$ *simply* is *the Riemann sum for the integral of* f. *Therefore,* $\int_K f(z)\,dz = V = F(B) - F(A)$.

Just as before, we conclude that $\widetilde{K}$ is actually the path taken by the Riemann sum of f, and that the vector V is once again the path-independent value of the integral:

$$\int_K f(z)\,dz = V = F(B) - F(A)\,. \tag{8.15}$$

As in ordinary calculus, the function F cannot be unique, for $\widehat{F} \equiv F + \text{const.}$ shares the same amplitwist. In terms of a figure like [8.24] this corresponds to the fact that the effect of $\widehat{F}$ only differs from that of F by a translation; but this has no effect on the connecting arrow V.

In ordinary calculus, a real continuous function f(x) *always* possesses an anti-derivative F(x) for which $F' = f$. Of course it may not be easy to find, and it may not even be expressible in terms of elementary functions (e.g. $f(x) = e^{-x^2}$), but at least it exists. In the complex realm, on the other hand, we know that analytic functions are very special, and so we should not be surprised if the existence of such a function is no longer assured. Remember that when such an F exists, the integral of f is path-independent. It follows, for example, that no such function can exist for the non-analytic mapping $f(z) = \overline{z}$. Indeed, we may fall back on the still unproven result concerning infinite differentiability to see that, quite generally, analyticity of f is a necessary condition for the existence of F. For if F is analytic, then so too is its derivative F', namely f.

When presented with the integral of a non-analytic function, it is therefore hopeless to seek an anti-derivative for use in (8.15)—no such function can exist. For the special case $z \mapsto \bar{z}$ it is possible to extend the area interpretation (hence evaluation) to contours that are not closed, but for a general non-analytic mapping no such interpretation will be available. Although such integrals are of much less interest to us than those of analytic functions, in the next section we shall nevertheless find a method of evaluating them.

Before returning to more general considerations, let us give a couple more examples of the theorem in action. Consider $f(z) = z^2$. If we define $F(z) = \frac{1}{3}z^3$ then $F' = f$, and thus the path taken by the Riemann sum as we integrate along a contour will just be its image under $z \mapsto \frac{1}{3}z^3$. This allows us to look at the construction [8.17] in a new light. Recall that while this figure is concerned with the integration of a general power, it actually depicts the special case z^2. In agreement with our new general result, we see that $z \mapsto \frac{1}{3}z^3$ does indeed map the contour in [8.17a] to its Riemann sum in [8.17c].

As an example of how the theorem also resolves our aesthetic concern over path-independence for negative powers of z, reconsider $f(z) = (1/z^2)$, for which we hope you did actually draw the analogue of [8.17], as suggested. Without appealing to Cauchy's Theorem (thus avoiding the attendant anxiety over the singularity at the origin), we see that since $(-1/z)' = (1/z^2)$, all contours[8] between A and B will yield the same value for the integral:

$$\int_A^B \frac{1}{z^2}\, dz = \frac{1}{A} - \frac{1}{B}\,.$$

Notice, incidentally, that path-independence has allowed us to reinstate the familiar symbol $\int_A^B$ without fear of ambiguity.

Instead of having to use Cauchy's Theorem to extrapolate from the vanishing of the integral for a circle to its vanishing for more general loops, the conclusion is now immediate: since $B = A$ for a closed loop, the above expression vanishes.

8.8.4 The Integral as Antiderivative

We have seen that the existence of an *antiderivative* F (defined by $F' = f$) implies path-independence for the integral of f. We will now show, conversely, that path-independence implies the existence of F.

Let us first give another simple example of the Fundamental Theorem. Since $(\sin z)' = \cos z$, the integral of $\cos z$ will be path-independent, and if we integrate from the origin, for example, to a *variable* endpoint Z, then we obtain a well-defined function of Z:

$$F(Z) = \int_0^Z \cos z\, dz = \sin Z\,.$$

[8] We exclude contours that actually pass *through* the singularity.

We note, without surprise, that this function is the antiderivative of cos Z. If we began our integration at an arbitrary point, instead of at the origin, then the result would only differ by a constant, and so it would still be a perfectly good antiderivative.

In order to establish the claim of the first paragraph, it is only necessary for us to show that the above example is typical. If the integral of a mapping f is known to be path-independent, and A is an arbitrary fixed starting point, then we will show that

$$F(Z) \equiv \int_A^Z f(z)\,dz \tag{8.16}$$

is the antiderivative whose existence is sought. That is, we will verify that infinitesimal arrows emanating from a point P are merely amplitwisted to produce their images under this mapping F, and that the amplitwist at P is just f(P) i.e., $F'(P) = f(P)$.

First we shall need a simple observation on differences of integrals. See [8.25a]. Two paths L and M are shown connecting the point A to the distinct points P and Q. We know that for any function f(z),

$$\int_M f(z)\,dz - \int_L f(z)\,dz = \int_{-L+M} f(z)\,dz\,.$$

The path $(-L+M)$ for the right-hand integral is the round-about route from P to Q shown in [8.25b]. But if the integral is known to be path-independent then we may replace this path with the straight one S. Returning to the notation of (8.16), we thus have

$$F(Q) - F(P) = \int_S f(z)\,dz\,.$$

In the limit that Q coalesces with P, S becomes (with a minor abuse of terminology) an infinitesimal 'vector' Δ emanating from P, and its image $\widetilde{\Delta}$ under F will be given by the left-hand side of the above equation. Thus

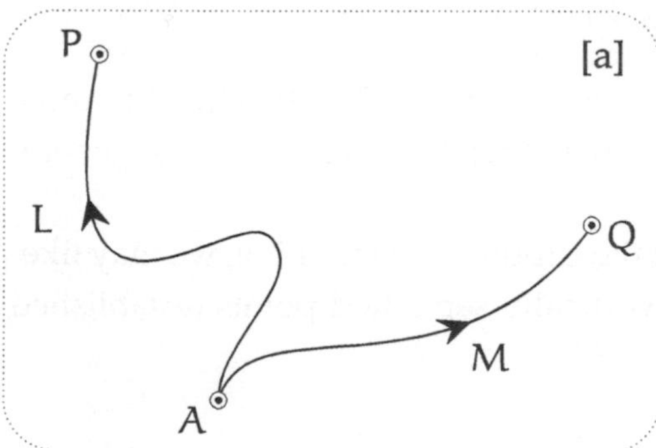

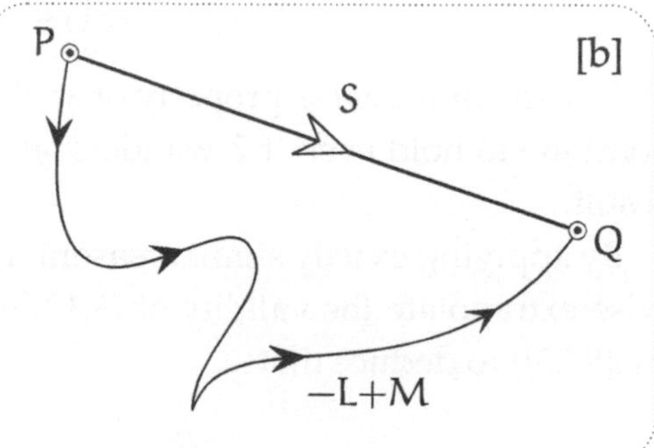

[8.25] **[a]** *Given a fixed point A, suppose we evaluate* $\int_M f(z)\,dz$ *and* $\int_L f(z)\,dz$. **[b]** *Then* $\int_M f(z)\,dz - \int_L f(z)\,dz = \int_{-L+M} f(z)\,dz$ *is an integral along an indirect route from P to Q. If the integral is path-independent, we can replace the indirect route with the direct route S. Defining* $F(Z) \equiv \int_A^Z f(z)\,dz$*, we deduce that* $F(Q) - F(P) = \int_S f(z)\,dz$.

$$F : \Delta \mapsto \widetilde{\Delta} = \int_{\Delta} f(z)\, dz .$$

But if Δ is infinitesimal then the above integral equals $f(P)\,\Delta$, thereby establishing the original claim:

$$F : \Delta \mapsto \widetilde{\Delta} = f(P)\,\Delta .$$

We may now appeal to Cauchy's Theorem to relate the required path-independence to analyticity. If f is analytic throughout some region then its integral is sure to be path-independent, and therefore F will exist. In other words, *every analytic mapping must itself be the derivative of another analytic mapping.*

We conclude this subsection with an interesting application of the analyticity of F. In the previous subsection we completely ignored our previous geometric constructions, and instead appealed to the Fundamental Theorem to show, for example, that

$$\int_A^B z^2\, dz = \frac{1}{3}(B^3 - A^3),$$

for *all* positions of A and B. This was apparently a clear improvement on [8.17] where such formulae were merely established in the special case that A and B were equidistant from the origin. However, we will now see that analyticity makes it possible, paradoxically, for this special case to contain the general case.

Consider these two functions:

$$F(Z) = \int_A^Z z^2\, dz \qquad G(Z) = \frac{1}{3}(Z^3 - A^3),$$

both of which we now recognize as being analytic. From [8.17] we know that if Z moves along the origin-centred circle passing through A then

$$F(Z) = G(Z) .$$

But by the uniqueness property of analytic functions [page 284] this identity must continue to hold even if Z wanders *off* the circle, thereby establishing the general result.

By applying exactly similar reasoning to the exponential mapping, we may likewise extrapolate the validity of (8.13) for vertically separated points (established by [8.22]) to deduce that

$$\int_A^Z e^z\, dz = e^Z - e^A ,$$

even if Z wanders off the vertical line through A.

8.8.5 Logarithm as Integral

In the light of the Fundamental Theorem, we are inclined to jump to the conclusion that because $(\log z)' = (1/z)$,

$$\int_1^Z \frac{1}{z}\, dz = \log Z, \tag{8.17}$$

just as in real analysis. In a sense, this *is* correct, but a little care is required.

The subtlety is, of course, that the singularity at the origin causes the integral of $(1/z)$ not to be single-valued. Thus we must specify the contour K from 1 to Z before the integral in (8.17) becomes well defined. On the other hand, until we choose one of the infinitely many values $\theta(Z)$ for the angle of Z, the RHS of (8.17) is *also* not well defined. These two difficulties now cancel each other out in the following way.

In [8.26] we have drawn three different contours for the specific case $Z = 1+i\sqrt{3}$. If we let $\theta_K(Z)$ stand for the net rotation as we follow K, then

$$\begin{aligned}\theta_{K_0}(Z) &= (\pi/3)\\ \theta_{K_1}(Z) &= (\pi/3) + 2\pi\\ \theta_{K_2}(Z) &= (\pi/3) + 4\pi.\end{aligned}$$

In a sense, including the contour in the definition of angle has rendered it single-valued. Notice that this definition does not depend on the precise shape of K, but only on how many times the origin is encircled.

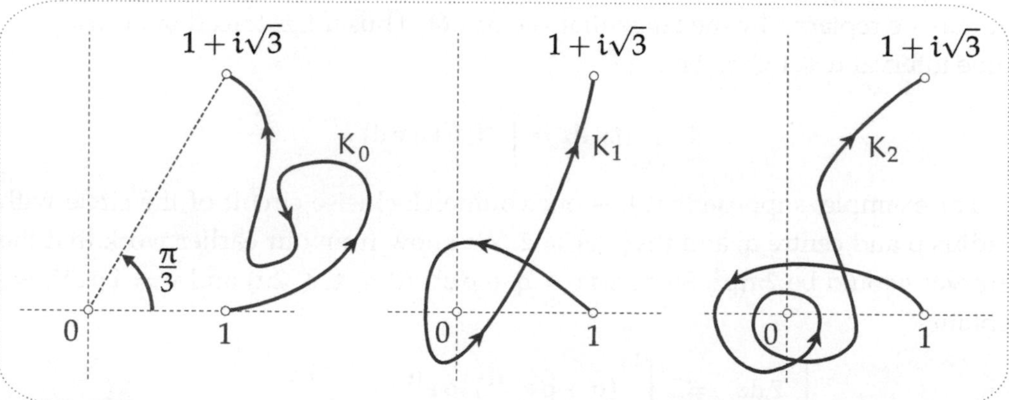

[8.26] Why $\int_1^Z (1/z)\, dz = \log Z$ is a multifunction. *From [8.12] we deduce that* $\int_1^Z (1/z)\, dz = \log_K(Z) = \ln|Z| + i\theta_K(Z)$, *where* $\theta_K(Z)$ *denotes the net change in the angle as z traverses* K. *So the infinitely many values of* $\log Z$ *arise from the number of times* K *loops around* 0 *before finally arriving at* Z, *each extra revolution adding* $2\pi i$ *to the value of* $\log Z$.

In this way, we may absorb the means of reaching Z into the definition of log(Z) in order to obtain a single-valued answer:

$$\log_K(Z) = \ln|Z| + i\,\theta_K(Z)\,.$$

The unambiguously correct version of (8.17) then reads,

$$\int_K \frac{1}{z}\,dz = \log_K(Z)\,.$$

Of course, the multiple-valued nature of log has merely been disguised, not done away with. Nevertheless, by pursuing the above idea one is led to consider so-called *Riemann surfaces*, whereby multifunctions *can* be rendered single-valued. But that is a story for another day.

8.9 Parametric Evaluation

When more elegant means are not available it is nevertheless possible (in principle) to evaluate a contour integral by expressing it in terms of ordinary real integrals. We shall now briefly describe and illustrate this method.

The basic idea is to think of the contour L as being traced by a moving particle whose position at time t is $z(t)$. Next, instead of building the Riemann sum (hence the integral) from very small vectors that are chords of L, we may equally well use very small vectors that are *tangent* to L. This is done using the tangential complex velocity $v = \frac{dz}{dt}$: the chord representing the movement during the instant of time δt may be replaced by the tangential vector $v\,\delta t$. Thus if L is traced out during the time interval $a \leqslant t \leqslant b$, then

$$\int_L f[z]\,dz = \int_a^b f[z(t)]\,v\,dt\,.$$

For example, suppose that L is one counterclockwise circuit of the circle with radius ρ and centre q, and that $f[z] = \overline{z}$. We know from our earlier work that the answer should be $2\pi i\rho^2$. Since $z(t) = q + \rho\,e^{it}$ $(0 \leqslant t \leqslant 2\pi)$ and $v = i\,\rho\,e^{it}$, we obtain

$$\begin{aligned}\int_L \overline{z}\,dz &= \int_0^{2\pi} (\overline{q} + \rho\,e^{-it})\,i\rho\,e^{it}\,dt \\ &= i\rho\overline{q}\int_0^{2\pi} (\cos t + i\,\sin t)\,dt + i\rho^2\int_0^{2\pi} dt \\ &= 2\pi i\rho^2,\end{aligned}$$

as anticipated.

Naturally, the point of this method is not to confirm previously known results, but rather to evaluate integrals that we couldn't do before. For example, with the same contour, but with $f[z] = \bar{z}^2$, the answer can no longer be guessed. However, you should now find it easy to discover that the answer is $4\pi i\bar{q}\rho^2$.

By way of contrast with the non-analytic examples above, and as further practice with this method, confirm (using the same contour L) that $\int_L z^2\,dz = 0$, as predicted by either Cauchy's Theorem or the Fundamental Theorem. Likewise, confirm that $\int_E \bar{z}\,dz = 0$, where E is an origin-centred ellipse. [*Hint:* recall that $z(t) = p\,e^{it} + q\,e^{-it}$ moves on such an ellipse.]

For our last examples, take the contour to be a section of the parabola $y = x^2$ between 0 and $1+i$; in temporal terms this can be represented as $z(t) = t + it^2$ $(0 \leqslant t \leqslant 1)$. Integrate z along this contour, first using the Fundamental Theorem, then parametrically. Likewise, use the Fundamental Theorem to evaluate the integral for e^z. By equating the imaginary part of your answer with the imaginary part of the parametric evaluation, deduce that

$$\int_0^1 (2t\,\cos t^2 + \sin t^2)\, e^t\, dt = e\,\sin 1\,.$$

This result can be verified easily [exercise] without using complex numbers. Later, though, we shall meet real integrals that cannot readily be evaluated by such ordinary means, but which suddenly do become easy when viewed as arising from a *complex* integral.

8.10 Cauchy's Theorem

8.10.1 Some Preliminaries

Having repeatedly witnessed the utility of Cauchy's Theorem in this chapter, it is perhaps time that we checked that it is *true!* We begin with the case where the contour C is a "simple" closed curve, i.e., without self-intersections. See [8.27].

We have filled the interior of C with a grid of small squares, of side length ϵ, aligned with the real and imaginary axes. We have then shaded all those squares that lie wholly within C, and taken the contour K to be the boundary of this shaded region, traversed counterclockwise. Because we have drawn relatively large squares (in order to make the picture clear), K is presently only a crude approximation to C. However, as we let ϵ shrink, the shaded region fills the interior of C ever more completely, and K follows C ever more precisely. Thus, in order to see whether or not the integral of a mapping f along C vanishes, it is sufficient to instead investigate the behaviour of the integral of f along K, as ϵ shrinks to zero. [This is justified in greater detail in Ex. 20.]

Next we seek to relate this integral along K to the behaviour of the mapping *inside* the shaded region that it bounds. Consider the sum of all the integrals of f taken counterclockwise round each of the infinitesimal shaded squares. This

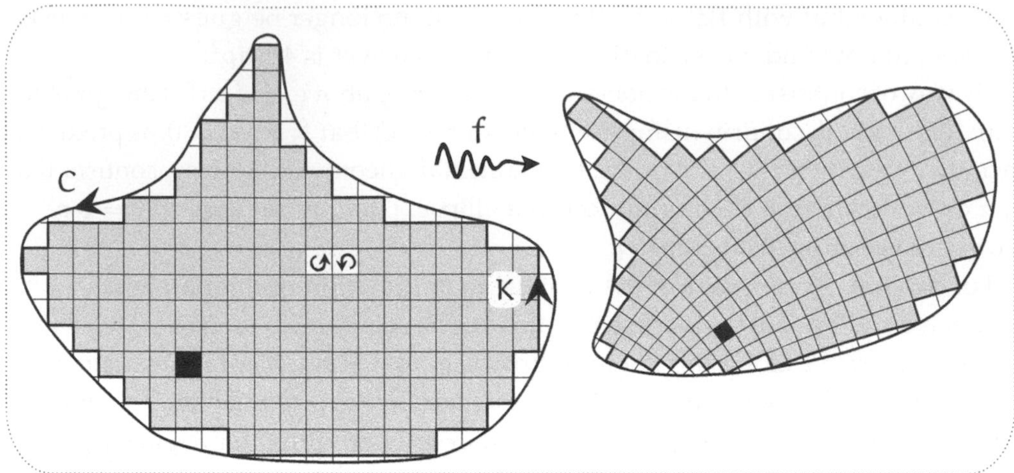

[8.27] Cauchy's Theorem (step one): *Fill the interior of* C *with a grid of small (ultimately vanishing) squares, of side length* ϵ*, aligned with the real and imaginary axes. The sum of the integrals around all these squares yields the integral around the boundary,* K*, for every* interior *edge is matched with an oppositely directed abutting edge that cancels it.*

counterclockwise sense of integration is illustrated in [8.27] for two adjacent squares. When we add the integrals from these two squares, their common edge is traversed *twice*, once in each direction, and hence the integrals along it *cancel*. But this is true of every edge that lies in the shaded region, so that when we sum the integrals for all the shaded squares, the only edges that do not self-destruct in this manner are those that make up K:

$$\oint_K f(z)\,dz = \sum_{\text{shaded squares}} \oint_\square f(z)\,dz\,. \tag{8.18}$$

The investigation of the integral of f along C has thus been reduced to the study of the local effect of f on infinitesimal squares in the interior region.

It should be stressed that the discussion thus far is equally applicable to non-analytic and analytic mappings. For example, with $f(z) = \overline{z}$, (8.18) simply says [see (8.8)] that the area inside K is the sum of the areas of the shaded squares. In order to understand Cauchy's Theorem, we must specialize to the case (illustrated in [8.27]) where the local effect of f is an amplitwist throughout the interior of C. First, though, let us try to guess how the magnitude of a typical integral in the above summation will depend on ϵ (as the squares shrink) for a *general* mapping.

Experience with real integration, as well as the inequality (8.5), might lead one to guess that the integral round an infinitesimal square would die away at the same rate as its perimeter, that is, as ϵ. This is false. The fact that the square is a *closed* contour, together with the fact that complex integration is a type of *vectorial* summation, implies that the integral must decay much faster than this. In the above example of conjugation, we know that the exact value of each term is $2i\epsilon^2$, and this leads us to the correct guess, namely, that the terms die away as the *square* of ϵ. We shall verify this in detail shortly, but for the moment the following rough argument will suffice.

We know that for a general mapping, the integral round K—hence the summation in (8.18)—will be nonzero and finite. This leads us to believe that each term must die away with the reciprocal dependence on ϵ as governs the growth of the number of terms in the series. But the number of terms grows as *(fixed area inside* C*, divided by the area of each square)*, that is as $(1/\epsilon^2)$. Thus the magnitude of each term is expected to die away as ϵ^2. If our original guess had been correct, the order of the sum in (8.18) would have been $\epsilon\,(1/\epsilon^2)$, yielding an infinite result as the squares shrunk. Conversely, any contributions to the terms involving powers of ϵ *greater* than two, cannot have any influence on the final result.

8.10.2 The Explanation

Let us return to [8.27] and to the explanation of Cauchy's Theorem. The analytic mapping f amplitwists the infinitesimal shaded squares on the left to the infinitesimal squares on the right, and [8.28] shows a magnified view of a typical such square and its image (the black ones in [8.27]). According to our especially accurate midpoint Riemann sum (R_M), the integral along the bottom edge of this square can be approximated by the single term $A\,\epsilon$: the image of the midpoint a, times the number along this edge. This conforms to our first, wrong guess concerning the dependence on ϵ of the complete integral round the square. But if we now add this to the integral along the *opposite* edge, the answer is

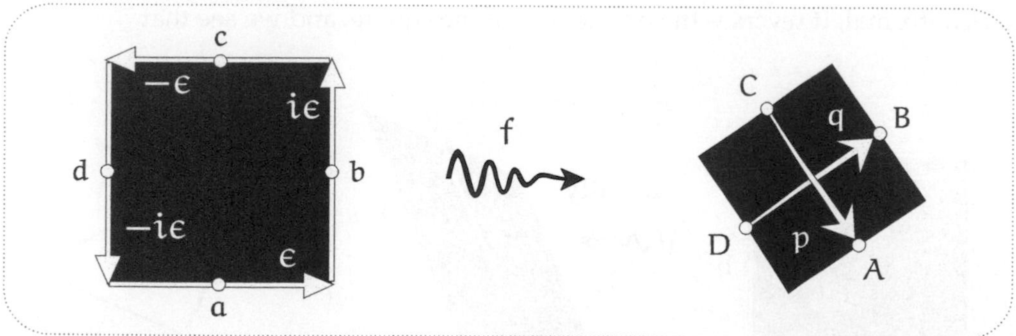

[8.28] Cauchy's Theorem explained. *According to our especially accurate midpoint Riemann sum, R_M, the integral along the bottom edge of this square can be approximated by the single term $A\epsilon$: the image of the midpoint a, times the number along this edge. If we add this to the integral along the* opposite edge, *the answer is* $A\,\epsilon + C\,(-\epsilon) = (A - C)\,\epsilon = p\epsilon$. *Likewise, the contribution from the remaining two edges is also of order ϵ^2, namely,* $(B - D)\,i\epsilon = iq\epsilon$. *But since* f *is locally* an amplitwist, *the image is a* square, *and so* $iq = (q$ rotated through a right angle$) = -p$. *Therefore,* $\oint_{\square} f(z)\,dz = \epsilon\,(p + iq) = 0$.

$$A\,\epsilon + C\,(-\epsilon) = (A - C)\,\epsilon = p\epsilon\,.$$

Even if f is merely differentiable in the real sense, rather than locally an amplitwist, $|p|$ will still be proportional to ϵ, and the magnitude of $p\epsilon$ will therefore be proportional to ϵ^2, as anticipated. Likewise, the contribution from the remaining two edges is also of order ϵ^2, namely, $(B - D)\ i\epsilon = iq\ \epsilon$.

Perhaps you have already seen the light: if f is locally an amplitwist, the image is a *square*, and so

$$iq = q \textit{ rotated through a right angle } = -p$$

$$\Longrightarrow \quad \oint_{\square} f(z)\,dz = \epsilon\,(p + iq) = 0\,. \tag{8.19}$$

We conclude from (8.18) that the vanishing of loop integrals for analytic mappings is indeed the nonlocal manifestation of their local amplitwist property!

Contrast this with non-analytic mappings. See [8.29]. Provided that a mapping is differentiable in the real sense, we know [see page 236] that its local effect is expansion (by different factors) in two perpendicular directions, followed by a twist. Thus the image of an infinitesimal square will generally be a parallelogram; p and q will not have equal length, nor will they be orthogonal. As we see, p and iq no longer cancel, and $\epsilon\,(p + iq)$ is of order ϵ^2. When we add up the terms of (8.18), of order $(1/\epsilon^2)$ in number, the answer will therefore be nonzero and finite.

Conjugation provides a particularly striking example of this noncancellation for non-analytic mappings. See [8.30]. In the terms of the previous paragraph we could say that its expansion factors are everywhere 1 and -1 (in the horizontal and vertical directions), and that its twist is zero. The image of the square is *again* a square, but there is a crucial difference between [8.28] and [8.30]. Because this mapping is *anti*conformal, it reverses the orientation of the square, and we see that

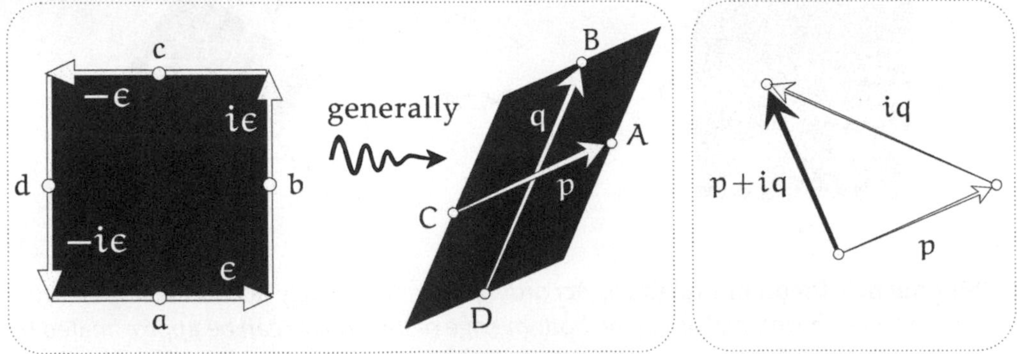

[8.29] The non-analytic case. *The shrinking square is ultimately mapped to a (non-square) parallelogram. As we see,* p *and* iq *no longer cancel, and* $\epsilon(p + iq)$ *is of order* ϵ^2*. When we add up the terms of (8.18), of order* $(1/\epsilon^2)$ *in number, the answer will therefore be nonzero and finite.*

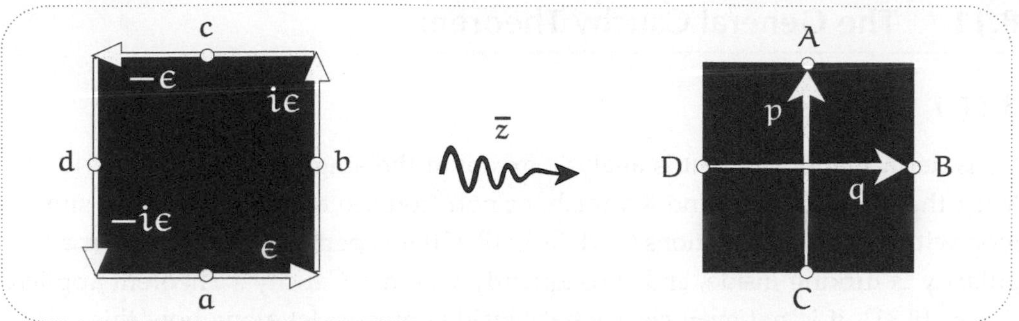

[8.30] $\oint_{\square} \bar{z}\, dz = 2i(\textit{area of } \square)$, *because* $\epsilon(p + iq) = \epsilon(i\epsilon + i\epsilon) = 2i\,\epsilon^2$.

$$\oint_{\square} \bar{z}\, dz = \epsilon\,(p + iq) = \epsilon\,(i\epsilon + i\epsilon) = 2i\,\epsilon^2 = 2i\,(\text{area of } \square).$$

Returning to questions of analyticity, and comparing (8.19) with [8.29], we obtain a converse to Cauchy's Theorem. If all the loop integrals of f are known to vanish, then, in particular, they will vanish for infinitesimal squares, such as the one on the left of [8.29]. Thus, $p + iq = 0$. But it is clear that this can *only* happen if the image is another infinitesimal square with the same orientation as the original (cf. [8.30]). Thus the local effect of f must be an amplitwist. This converse is called *Morera's Theorem*.

As with other new ideas in this book, we have not attempted to present the arguments in rigorous form; "insight", not "proof", is ever our watchword. For example, consider these objections (ascending in severity) to the geometric argument of [8.27] and [8.28]: no matter how small the square, the sides of the image will not be *perfectly* straight (though they will meet in perfect right angles); the midpoint a will not be mapped to the *exact* midpoint of the image; and despite the undoubted accuracy of R_M for a very small contour, it will not yield the *exact* value of the integral.

Nevertheless, it seems plausible that [8.28] and its associated reasoning remain unimpeached when it comes to the evaluation of the *dominant* ϵ^2 contribution. Indeed, the example of [8.30] lends at least some credence to the irrelevance of the above objections, for in that case we know the answer is correct to this order of ϵ. [In fact it comes out exactly right, but that is a fluke.] More generally, recall that parametric evaluation revealed that the real and imaginary parts of any contour integral can be expressed as ordinary real integrals. This means that we may carry over to the complex realm our previous determination (8.3) of the error induced by R_M in real analysis. Thus each of the integrals along the four edges of the square will differ from their R_M-values by an amount that dies away at least as fast as ϵ *cubed* [cf. Ex. 21 and Ex. 22]. But as we have previously argued, as ϵ shrinks to zero, such contributions can have no effect on the sum in (8.18). Although we shall not dwell on them, other objections can be treated in a similar way.

8.11 The General Cauchy Theorem

8.11.1 The Result

Consider a mapping f that is analytic except at the singularity marked in [8.31]. Must the integral of f round K vanish, or not? You see the problem. For a simple loop without self-intersections (such as in [8.27]) it is perfectly clear whether a singularity is lurking inside, and consequently whether Cauchy's Theorem applies. But in [8.31], it is not even clear what "inside" means, let alone how this might relate to Cauchy's Theorem.

Recall that we encountered such problems before when trying to integrate $\overline{z}$ round complicated loops [see (8.9), as well as the discussion on p. 387]. Our solution was to define the "inside" to be all the points for which the winding number does not vanish, and conversely, the "outside" to be all the points for which it does vanish.

With these definitions in place, the completely general version of Cauchy's Theorem is stunning in its simplicity:

> *If an analytic mapping has no singularities "inside" a loop, its integral round the loop vanishes.* (8.20)

This section is devoted to understanding this beautiful result.

First let us answer our opening question. In [8.31], K does not wind around the singularity, and therefore (according to the theorem) the integral should vanish. In

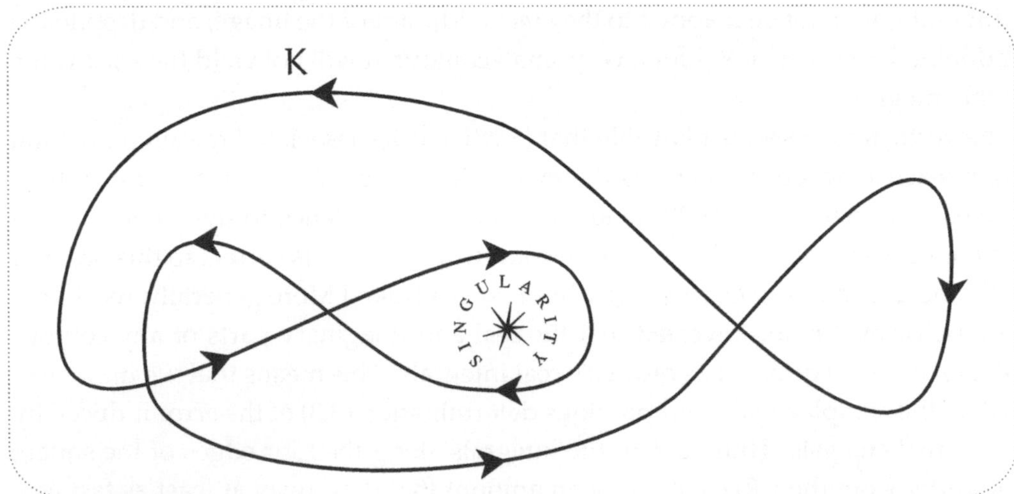

[8.31] *As in Section 7.1.2, we define the "inside" of a closed loop to be the set of points for which the winding number is not zero. Then if an analytic mapping has no singularities inside a loop, its integral round the loop vanishes, as it does in this example.*

the process of understanding this particular instance of the theorem we shall be led to a completely general argument for its validity.

8.11.2 The Explanation

As in [8.16b], the contour K in [8.31] partitions the plane into a number of disjoint regions; in particular, the inside of K is made up of D_1, D_2, and D_3. See [8.32]. Let C_j be the boundary of D_j, traversed counterclockwise. So as not to clutter up the picture, instead of actually drawing these contours in [8.32], we have merely indicated (with ellipses) their common counterclockwise sense. Also shown (in boxes) are the winding numbers of K around each of the regions D_j. Since there are no singularities inside the D_j that make up the inside of K, our basic version of Cauchy's Theorem applies to each of the simple contours C_j, and we have

$$\oint_{C_j} f(z)\, dz = 0\,. \tag{8.21}$$

Now comes the crucial observation. The integral round K can be expressed as a linear combination of the integrals round the C_j's that bound the interior D_j's. In the case of [8.32],

$$\oint_K f(z)\, dz = \oint_{C_1} f(z)\, dz - \oint_{C_2} f(z)\, dz + 2 \oint_{C_3} f(z)\, dz\,. \tag{8.22}$$

Consider, for example, the contour C_3, for which the counterclockwise sense happens to agree with the direction of K. On the other hand, C_1 traverses this portion of K in the *opposite* direction. Consequently, in (8.22), we end up integrating twice in the correct direction, and once in the opposite direction; the net result is to integrate

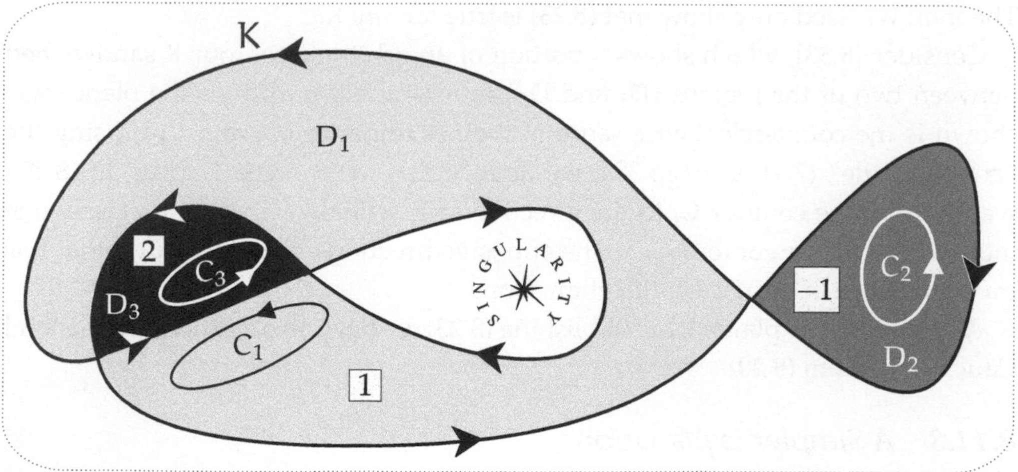

[8.32] $K = \sum_j \nu_j C_j = C_1 - C_2 + 2C_3$ *in this example. [The explanation is in the next figure.]*

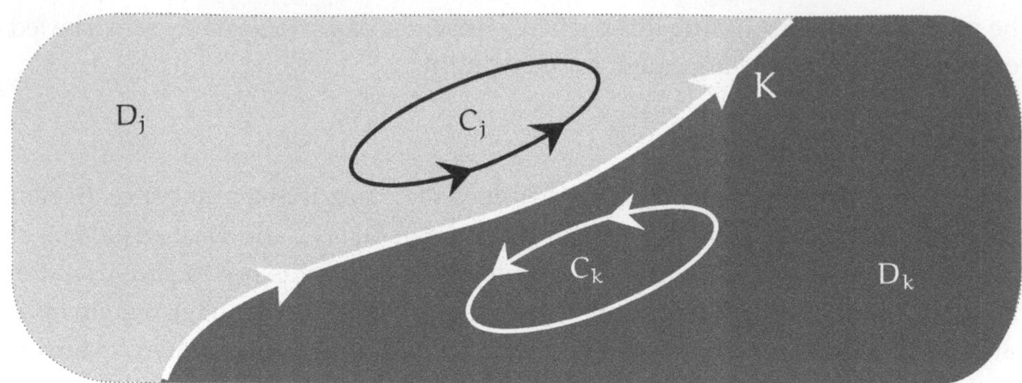

[8.33] Proof that $K = \sum_j \nu_j C_j$. *Using the "crossing rule" (7.1) on page 388, we see that $\nu_j = \nu_k + 1$. Thus the contour C_j in the direction of K will always occur precisely one more time than the contour C_k in the opposite direction—the net result is that K is traversed once in the correct direction.*

along this part of K once in the correct direction. You should check for yourself that all of K is correctly accounted for in this way. Substituting (8.21) into (8.22), we have confirmed the prediction of the general theorem for this particular contour.

Since (8.22) is clearly true for *any* function f, we may abstract it away and write the equation as

$$K = C_1 - C_2 + 2\,C_3 .$$

Notice that the coefficient of C_j in this sum is none other than the *winding number* ν_j of K about D_j, and that we may therefore rewrite the previous equation as

$$K = \sum_j \nu_j\, C_j . \tag{8.23}$$

From the above example, it is clear that to prove the general version of Cauchy's Theorem we need only show that (8.23) is true for *any* K.

Consider [8.33], which shows a portion of an arbitrary contour K sandwiched between two of the regions (D_j and D_k) into which it partitions the plane; also shown is the counterclockwise sense of their boundaries (C_j and C_k). Using the "crossing rule" (7.1) on page 388, we deduce that $\nu_j = \nu_k + 1$. Thus, in (8.23), we find that the contour C_j in the direction of K will always occur precisely one more time than the contour C_k in the opposite direction—the net result is that K is traversed once in the correct direction. Done.

As previously explained, in establishing (8.23) we have also deduced the General Cauchy Theorem (8.20).

8.11.3 A Simpler Explanation

The Deformation Theorem (8.11) was also deduced from the basic Cauchy theorem [unproven at that time], and we will now use it to give a simpler and more intuitive explanation of the General Cauchy Theorem.

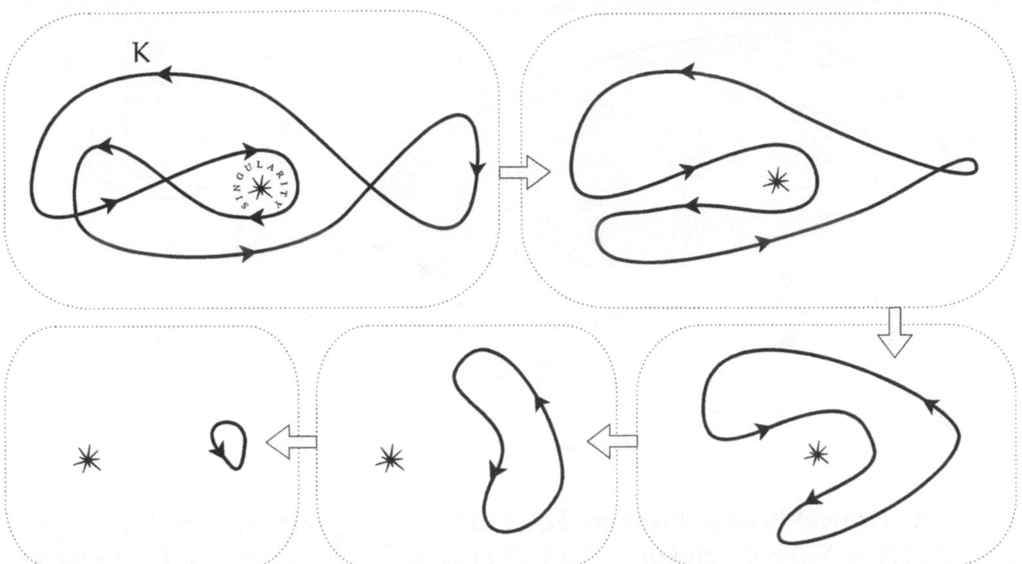

[8.34] If a closed contour can be shrunk to a point without crossing a singularity, the integral round it vanishes. *Hopf's Degree Theorem tells us that this shrinking process is possible if and only if the contour does not wind around any singularities.*

Suppose that a contour can be deformed and shrunk down to a point without ever crossing a singularity of an otherwise analytic function. By inequality (8.5), the value of the integral will be zero at the end of this shrinking process. But by the Deformation Theorem, the value of the integral remains constant throughout this process. In other words,

If a closed contour can be shrunk to a point without crossing a singularity, the integral round it vanishes. (8.24)

To wrap this up, we clearly need a way of recognizing when this shrinking process is possible. For example, is it possible for the contour K in [8.31]? Figure [8.34] shows that it is. Therefore (8.24) implies that the integral along K vanishes, in agreement with the general theorem.

The two theorems are clearly very closely related. In fact we can now deduce the General Cauchy Theorem from (8.24) by observing that the winding number of the final shrunken loop vanishes; Hopf's Degree Theorem then tells us that

The shrinking process in (8.24) is possible if and only if the contour does not wind around any singularities.

8.12 The General Formula of Contour Integration

Consider the general problem of evaluating $\oint_K f(z)\,dz$, where K is a general (possibly self-intersecting) loop, and where f possesses several singularities s_1, s_2, etc., inside K. Figure [8.35] illustrates such a situation.

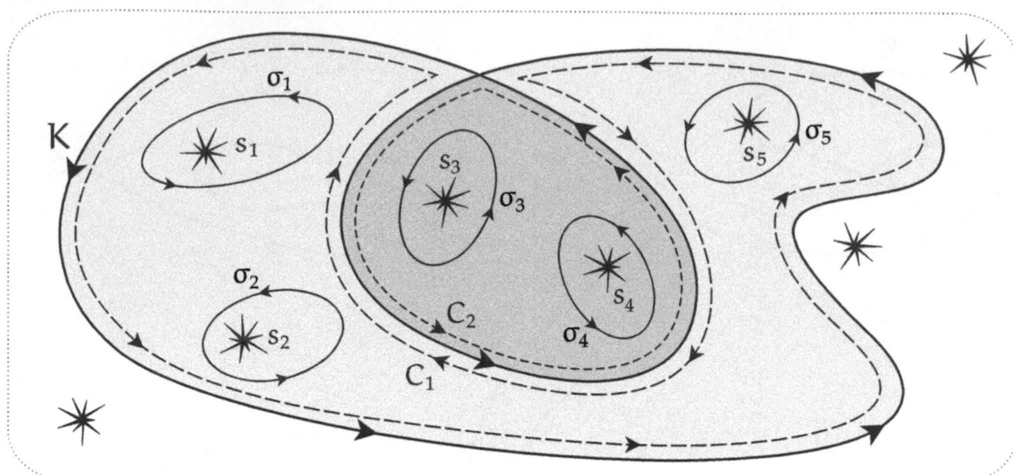

[8.35] The General Residue Theorem: $\oint_K f(z)\,dz = 2\pi i \sum_j \nu(K, s_j)$ Res $[f(z), s_j]$. *We know that* $K = \sum_j \nu_j C_j$, *and therefore* $\oint_K f(z)\,dz = \sum_j \nu_j \oint_{C_j} f(z)\,dz$. *The Deformation Theorem then tells us that each* $\oint_{C_j} f(z)\,dz$ *can be expressed as a sum of integrals around the illustrated loops* σ_j *encircling the singularities within each* C_j. *Finally, each such integral* $\oint_{\sigma_k} f(z)\,dz = 2\pi i$ Res $[f(z), s_k]$.

Here the inside of K consists of two simply connected regions, D_1 [lightly shaded] and D_2 [darkly shaded], with boundaries C_1 and C_2. Since K winds once round points in the lightly shaded region ($\nu_1 = 1$) and twice round points in the darkly shaded region ($\nu_2 = 2$), the general result (8.23) correctly predicts that

$$K = \sum_j \nu_j C_j = 1 \cdot C_1 + 2 \cdot C_2. \tag{8.25}$$

As illustrated, let σ_j be a simple (counterclockwise) contour containing s_j but no other singularities of f, and let us define

$$I_j \equiv \oint_{\sigma_j} f(z)\,dz.$$

By virtue of the Deformation Theorem (8.11), we know that the integral I_j has a characteristic value that does not depend on the size or shape of σ_j. Furthermore, as we saw in [8.20], if a simple loop contains several singularities, then the integral round that loop is the sum of I-values of the singularities it contains. In our example,

$$\oint_{C_1} f(z)\,dz = I_1 + I_2 + I_5 \quad \text{and} \quad \oint_{C_2} f(z)\,dz = I_3 + I_4.$$

Finally, using (8.25), we deduce that

$$\oint_K f(z)\,dz = 1 \cdot [I_1 + I_2 + I_5] + 2 \cdot [I_3 + I_4],$$

in which *each* I*-value has been multiplied by the number of times* K *winds round the corresponding singularity.* Since (8.23) is valid for arbitrary loops it follows that this conclusion is too. As the grand finale to this chapter, we have thus obtained the following completely general formula for the loop integral of an analytic function:

$$\oint_K f(z)\, dz = \sum_j \nu(K, s_j)\, I_j\, .$$

The final icing on the cake is an efficient method of computing the I_j's. In the next chapter we will verify our previous claim that in the neighbourhood of each s_j there exists a unique Laurent series [see page 456], the coefficient of the complex inversion term being (by definition) the residue $\text{Res}\,[f(z), s_j]$. Granted this, we see that $I_j = 2\pi i\, \text{Res}\,[f(z), s_j]$. Thus

$$\oint_K f(z)\, dz = 2\pi i \sum_j \nu(K, s_j)\, \text{Res}\,[f(z), s_j]. \tag{8.26}$$

This is the *General Residue Theorem.* Note that it contains the General Cauchy Theorem as the special case in which each $\nu(K, s_j) = 0$.

We will also see in the next chapter that it is possible to find the residues in this formula directly, without going to the trouble of finding the whole Laurent series. Thus, even before exemplifying its use, it should be clear that in (8.26) we have a result of great practical and theoretical power.

8.13 Exercises

1 Thinking of x as representing time, $z(x) = x + if(x)$ is a parametric description of the ordinary graph, $y = f(x)$.

(i) Show that the complex velocity is $v = 1 + i\tan\theta$, where θ is the angle between the horizontal and the tangent to the graph. Also, show that complex acceleration is $a = if''$.

(ii) Recall from Ex. 20 on page 297 that the curvature of the orbit is $\kappa = [\text{Im}\,(a\overline{v})]/|v|^3$. Deduce from (i) that

$$\kappa \sec^3\theta = f''(x).$$

(iii) From (ii), deduce that the error equation (8.3) can be written as

$$\text{area (ABCD)} = \frac{1}{8} f''(x)\,\Delta^3.$$

2 In [8.6], show that

$$\lim_{\Delta\to 0}\left(\frac{\text{area between the chord AB and the curve}}{\text{area between the tangent CD and the curve}}\right) = 2.$$

In other words, R_M is twice as accurate as the Trapezoidal formula.

3 In the integration of an ordinary real function $f(x)$, let L denote the length of the integration range, and let M denote the maximum size of $f''(x)$ in this range. From the previous two exercises, deduce the standard result,

$$\text{total Trapezoidal error} < \tfrac{1}{12} LM\Delta^2.$$

Likewise, deduce the somewhat less familiar result,

$$\text{total } R_M \text{ error} < \tfrac{1}{24} LM\Delta^2.$$

4 Write down the values of $\oint_C (1/z)\,dz$ for each of the following choices of C, then confirm the answers the hard way, using parametric evaluation.

(i) $|z| = 1$.

(ii) $|z - 2| = 1$.

(iii) $|z - 1| = 2$.

5 Evaluate parametrically the integral of $(1/z)$ round the square with vertices $\pm 1 \pm i$, and confirm that the answer is indeed $2\pi i$.

6 Confirm by parametric evaluation that the integral of z^m round an origin-centred circle vanishes, except when $m = -1$.

7 Hold a coin (of radius A) down on a flat surface and roll another one (of radius B) round it. The path traced by a point on the rim of the rolling coin is called an *epicycloid*, and it is a closed curve if $A = nB$, where n is an integer.

(i) With the centre of the fixed coin at the origin, show that the epicycloid can be represented parametrically as

$$z(t) = B\left[(n+1)\,e^{it} - e^{i(n+1)t}\right].$$

(ii) By evaluating the integral in (8.8) parametrically, show that

$$\text{area of epicycloid} = \pi B^2\,(n+1)\,(n+2)\,.$$

8 The figure below shows four simple loops, and in each case we have indicated how much shaded area is enclosed. Use parametric evaluation to verify equation (8.8) for each of the four loops.

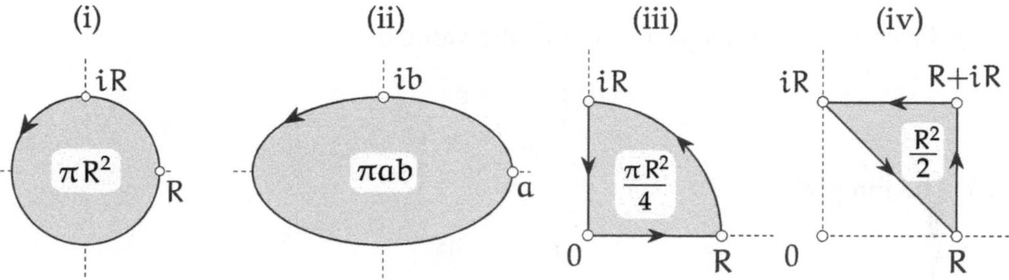

9 What is the generalization of (8.8) to the case where the contour is not closed?

10 Use (8.23) to verify (8.9).

11 The perfect symmetry of figure [8.18] results from integration round the *unit* circle. Roughly how would this figure look if we instead used a somewhat larger circle?

12 Let K be the contour in [8.21].

(i) Evaluate the following integral by factoring the denominator and putting the integrand into partial fractions:

$$\oint_K \left(\frac{z}{z^2 - iz - 1 - i}\right) dz\,.$$

(ii) Write down the Laurent series (centred at the origin) for $(\cos z/z^{11})$. Hence find

$$\oint_K \left(\frac{\cos z}{z^{11}}\right) dz\,.$$

13 This exercise illustrates how one type of difficult real integral may be evaluated easily using a complex integral.

Let L be the straight contour along the real axis from $-R$ to $+R$, and let J be the semi-circular contour (in the upper half plane) back from $+R$ to $-R$. The complete contour $L + J$ is thus a closed loop.

(i) Using the partial fraction idea of the previous exercise, show that the integral

$$\oint_{L+J} \frac{dz}{(z^4+1)}$$

vanishes if $R < 1$, and find its value if $R > 1$.

(ii) Using the fact that z^4+1 is the complex number from -1 to z^4, write down the minimum value of $|z^4+1|$ as z travels round J. Now think of R as large, and use inequality (8.5) to show that the integral round J dies away to zero as R grows to infinity.

(iii) From the previous parts, deduce the value of

$$\int_{-\infty}^{+\infty} \frac{dx}{(x^4+1)}.$$

14 (i) The integral

$$\int_{-\infty}^{+\infty} \frac{dx}{(x^2+1)}$$

is easily found by ordinary means, but evaluate it instead by the method of the previous exercise.

(ii) Likewise, evaluate

$$\int_{-\infty}^{+\infty} \frac{dx}{(x^2+1)^2}$$

by ordinary means and then by contour integration. [*Hint*: The quickest way to find the partial fraction decomposition for this function is to square the decomposition of $1/(z^2+1)$.]

15 (i) Use the Fundamental Theorem to write down the value of

$$\int_{0}^{a+ib} e^z\, dz.$$

(ii) Equate the answer with the one obtained by parametric evaluation along the straight contour from 0 to $(a+ib)$, and deduce that

$$\int_0^1 e^{ax} \cos bx\, dx = \frac{a\,(e^a \cos b - 1) + b\, e^a \sin b}{a^2+b^2},$$

and

$$\int_0^1 e^{ax} \sin bx\, dx = \frac{b\,(1 - e^a \cos b) + a\, e^a \sin b}{a^2+b^2}.$$

(iii) Prove the results in (ii) by ordinary methods.

16 (i) Show that when integrating a product of analytic functions, we may use the ordinary method of "integration by parts".

(ii) Let L be a contour from the real number $-\theta$ to $+\theta$. Show that

$$\int_L z\,e^{iz}\,dz = 2i\,(\sin\theta - \theta\,\cos\theta),$$

and verify this by taking L to be a line-segment and integrating parametrically.

17 Let

$$f(z) = \frac{1}{z}\left(z + \frac{1}{z}\right)^n,$$

where n is a positive integer.

(i) Use the Binomial Theorem to find the residue of f at the origin when n is even and when n is odd.

(ii) If n is odd, what is the value of the integral of f round any loop?

(iii) If $n = 2m$ is even and C is a simple loop winding once round the origin, deduce from part (i) that

$$\oint_C f(z)\,dz = 2\pi i\,\frac{(2m)!}{(m!)^2}.$$

(iv) By taking C to be the unit circle, deduce the following result due to Wallis:

$$\int_0^{2\pi} \cos^{2m}\theta\,d\theta = \frac{(2m)!}{2^{2m-1}\,(m!)^2}\,\pi.$$

(v) Similarly, by considering functions of the form $z^k f(z)$ where k is an integer, evaluate

$$\int_0^{2\pi} \cos^n\theta \cdot \cos k\theta\,d\theta \quad \text{and} \quad \int_0^{2\pi} \cos^n\theta \cdot \sin k\theta\,d\theta$$

18 Let E be the elliptical orbit $z(t) = a\cos t + ib\sin t$, where a and b are positive and t varies from 0 to 2π. By considering the integral of $(1/z)$ round E, show that

$$\int_0^{2\pi} \frac{dt}{a^2\cos^2 t + b^2\sin^2 t} = \frac{2\pi}{ab}.$$

19 Let us verify the claim of Ex. 19, p. 297, that if a function has vanishing Schwarzian derivative, then it must be a Möbius transformation. Following Beardon (1984), suppose that $\{f(z), z\} = 0$, and define $F \equiv (f''/f')$.

(i) Show that $1/F(z) - 1/F(w) = -(z-w)/2$.

(ii) Deduce that $\frac{d}{dz}\log f'(z) = -2/(z-a)$, for some constant a.

(iii) Perform two further integrations to conclude that f(z) is a Möbius transformation.

20 In [8.27], consider the white fragments of squares sandwiched between K and C.

(i) Show that the sum of the integrals round these fragments equals the difference between the integrals round C and K.

(ii) As ϵ shrinks, what is the approximate size of each term in the above series?

(iii) Roughly how many terms are there in the series?

(iv) From the previous parts, what do you conclude about the difference between the integrals round C and K, as ϵ shrinks to nothing?

21 Let K be the straight contour from $a - (\epsilon/2)$ to $a + (\epsilon/2)$, where ϵ is a short complex number in an arbitrary direction.

(i) Use the Fundamental Theorem to integrate z^2 along K, and then write down the value obtained by using a single term in R_M. Show that the error induced by R_M is $\frac{1}{12}\,\epsilon^3$.

(ii) As in part (i), find both the exact value and the R_M value for the integral of e^z along K. By expanding $e^{\epsilon/2}$ as a power series, deduce that the error in this case is roughly $\frac{1}{24}\,e^a\,\epsilon^3$.

(iii) Repeat the error analysis of the previous parts for the non-analytic function $\bar{z}^2$. [You will need to use parametric evaluation to find the exact value of the integral.]

22 Let K be the short contour of the previous exercise. Suppose that $f(z)$ possesses a Taylor series centred at a that converges at points of K:

$$f(a+h) = f(a) + \frac{f'(a)}{1!}\,h + \frac{f''(a)}{2!}\,h^2 + \frac{f'''(a)}{3!}\,h^3 + \cdots .$$

[The existence of such a series for any analytic function is derived in the next chapter.]

(i) By integrating this series along K, show that the difference between the exact integral and the R_M value is roughly $\frac{1}{24}\,f''(a)\,\epsilon^3$. Verify that the results of the first two parts of the previous exercise are in accord with this finding.

(ii) Use the series to show that the complex number from the image of the midpoint of K to the midpoint of the images of the ends of K is roughly $\frac{1}{4}\,f''(a)\,\epsilon^2$. As ϵ shrinks, are these two types of midpoint distinguishable under the magnifying lens that produces figure [8.28]?

(iii) From the Fundamental Theorem, deduce that the existence of such a series implies the vanishing of the integral of f round loops within the disc of convergence.

23 Let $f(z)$ be analytic throughout a region which contains a triangle with vertices a, b, c, and hence with edges $A \equiv (c-b)$, $B \equiv (a-c)$, $C \equiv (b-a)$. Given a pair of point p and q, let us define w_{pq} as a kind of average of $f(z)$ along the line-segment pq:

$$w_{pq} \equiv \frac{1}{(q-p)} \int_p^q f(z)\, dz.$$

Show that this complex average mapping sends the sides of the triangle abc to the vertices w_{ab}, w_{bc}, w_{ca} of a *similar* triangle! We merely rediscovered this result, which is apparently due to Echols (1923).
[*Hint:* Show that $Aw_{bc} + Bw_{ca} + Cw_{ab} = 0$, and use $A + B + C = 0$.]

24 Let K be a closed contour, and let ν be its winding number about the point a. Show that

$$\oint_K \left(\frac{e^z}{z-a}\right) dz = 2\pi i\, \nu\, e^a .$$

[*Hint*: Write e^z as $e^a\, e^{(z-a)}$, and expand $e^{(z-a)}$ as a power series.] This is a special case of *Cauchy's Integral Formula* (explained in the next chapter), which states that if f is analytic inside K, then

$$\oint_K \frac{f(z)}{(z-a)}\, dz = 2\pi i\, \nu\, f(a) .$$

25 Consider the image of the disc $|z| \leqslant R$ under the mapping $z \mapsto k z^m$. As the radius sweeps round the disc once, its image sweeps m times round the image disc of radius $|k|\, R^m$. Thus we may sensibly define the area of the image to be $m\pi\, (|k|\, R^m)^2$. With this understanding, show that if a mapping has a convergent power series

$$f(z) = a + b\, z + c\, z^2 + d\, z^3 + \cdots ,$$

then the area of the image is just the sum of the areas of the images under each of the separate terms of the series:

$$\text{area of image} = \pi\, (|b|^2 R^2 + 2\, |c|^2 R^4 + 3\, |d|^2 R^6 + \cdots) .$$

This is *Bieberbach's Area Theorem*.
Hint: Recall that the local area expansion factor is $|f'|^2$, so the image area is

$$\iint_{|z|\leqslant R} |f'|^2\, dx\, dy = \int_0^R \left[\int_0^{2\pi} f'(r\, e^{i\theta})\, \overline{f'(r\, e^{i\theta})}\, d\theta \right] r\, dr.$$

26 (i) Show that if f is an analytic function without singularities or p -points on a loop L, then

$$\nu\, [f(L), p] = \frac{1}{2\pi i} \oint_L \frac{f'(z)}{f(z)-p}\, dz.$$

(ii) Now let

$$f(z) = \frac{(z-a_1)^{A_1}(z-a_2)^{A_2}\cdots(z-a_n)^{A_n}}{(z-b_1)^{B_1}(z-b_2)^{B_2}\cdots(z-b_m)^{B_m}},$$

and by considering $(\log f)'$, find (f'/f).

(iii) In part (i) put $p = 0$ and take L to be a simple loop containing the roots a_1 to a_r and containing the poles b_1 to b_s. Thereby obtain a calculation proof of the Generalized Argument Principle in the case of rational functions:

$$\begin{aligned} \nu\,[f(L), 0] &= \sum_{j=1}^{r} A_j - \sum_{j=1}^{s} B_j \\ &= \text{(number of interior roots)} - \text{(number of interior poles)}. \end{aligned}$$

CHAPTER 9

Cauchy's Formula and Its Applications

9.1 Cauchy's Formula

9.1.1 *Introduction*

One of the principal objectives of this brief chapter is to tie up various loose ends from previous chapters. In particular, we have previously claimed (but have not yet explained) three important properties of an analytic function $f(z)$:

- We can differentiate $f(z)$ as many times as we please—it is "infinitely differentiable".
- In the vicinity of an ordinary point, $f(z)$ can be expressed as a Taylor series.
- In the vicinity of a singularity, $f(z)$ can be expressed as a Laurent series.

The classical explanation[1] of these facts hinges on the following result. If $f(z)$ is analytic on and inside a simple loop L, and if a is a point inside L, then

$$\frac{1}{2\pi i}\oint_{L} \frac{f(z)}{z-a}\,dz = f(a). \qquad (9.1)$$

This is called *Cauchy's Formula*—it constitutes the precise statement of the "rigidity" of analytic functions that we depicted in [5.3], p. 249. That is, the formula says that the values of f *on* L rigidly determine its values everywhere *inside* L.

We will give two explanations of (9.1), both of which are firmly rooted in Cauchy's Theorem.

[1] In the late 1950s a new approach was developed using topological ideas like those in Chapter 7, and it was our original intention to employ that approach here. However, having lacked both the time and the imagination to reduce the idea to its visual essentials, we have reluctantly fallen back on an integral-based approach. For more on the topological approach, see Whyburn (1955), Whyburn (2015) and Beardon (1979).

Visual Complex Analysis. 25th Anniversary Edition. Tristan Needham, Oxford University Press.
 DOI: 10.1093/oso/9780192868916.003.0009

9.1.2 *First Explanation*

Since $f(z)$ is assumed analytic inside L, the function $[f(z)/(z-a)]$ is also analytic there, *except* that it has a single singularity at $z = a$. Thus it follows from (8.11), p. 454, that the value of the integral in (9.1) will not change if L is deformed into its interior without crossing a.

Let C_r be a circle of radius r, centred at a, and lying strictly inside L. Referring to [9.1a], we may deform L into such a circle without crossing a, and hence without altering the value of the integral:

$$\frac{1}{2\pi i}\oint_L \frac{f(z)}{z-a}\,dz = \frac{1}{2\pi i}\oint_{C_r} \frac{f(z)}{z-a}\,dz. \tag{9.2}$$

The virtue of this transformation is that the integral round C_r turns out to have a simple and helpful interpretation.

First recall that the *average* value $\langle f\rangle_{C_r}$ of $f(z_\theta)$ as $z_\theta = a + r\,e^{i\theta}$ travels round the circle C_r is defined by

$$\langle f\rangle_{C_r} \equiv \frac{1}{2\pi}\int_0^{2\pi} f(z_\theta)\,d\theta.$$

In the previous chapter we saw geometrically that if θ increases by $d\theta$, causing z_θ to move dz along the circle, then $dz/(z-a) = i\,d\theta$. Substituting this into (9.2),

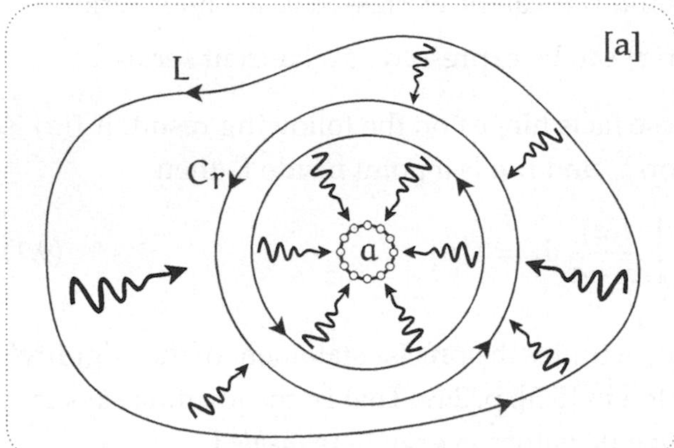

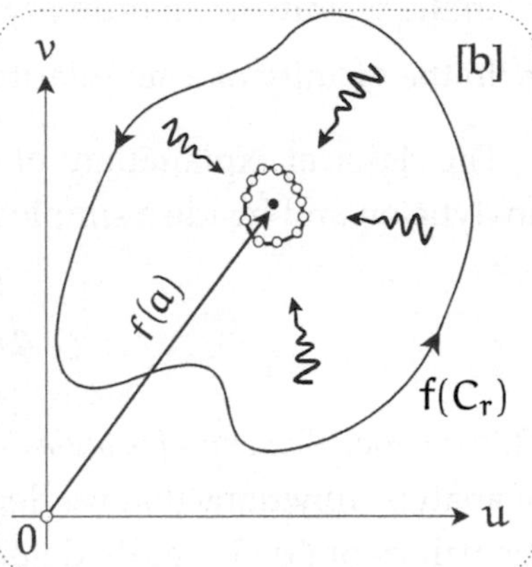

[9.1] First explanation of Cauchy's Formula. [a] *If $f(z)$ is analytic inside L then so is $f(z)/(z-a)$, except at $z = a$, so the Deformation Theorem implies that we can deform L into the circle C_r of radius r without changing the value of the integral. But on C_r we can interpret the integral as the* average value $\langle f\rangle_{C_r}$ *of $f(z)$ on C_r, and, by the Deformation Theorem again, this average value is* independent of the radius. *If z_j are n equally spaced points around C_r then $\langle f\rangle_{C_r}$ can be visualized as the limit as $n \to \infty$ of the centroid of the image points $f(z_j)$.* **[b]** *As C_r shrinks towards a, these image points cluster more and more tightly around $f(a)$, so $\frac{1}{2\pi i}\oint_L \frac{f(z)}{z-a}\,dz = \langle f\rangle_{C_r} = f(a)$.*

we find that the original integral round L may be interpreted as the average value of f on any of the circles C_r:

$$\frac{1}{2\pi i}\oint_L \frac{f(z)}{z-a}\,dz = \langle f\rangle_{C_r}.$$

Note in particular that $\langle f\rangle_{C_r}$ is independent of the radius r of the circle. To complete the derivation of (9.1), it therefore only remains to show that this radius-independent average is the value of f at the centre a.

To better grasp the meaning of the average value $\langle f\rangle_{C_r}$, imagine n equally spaced points $z_1, z_2, \ldots, z_n$ on C_r, and let $w_1, w_2, \ldots, w_n$ be their images under $z \mapsto w = f(z)$. The ordinary average $W_n \equiv \frac{1}{n}\sum_{j=1}^{n} w_j$ of these image points is their *centroid*, and $\langle f\rangle_{C_r}$ is the limiting position of W_n as n tends to infinity. [For a more detailed discussion of averages and centroids, consult the final section of Chapter 2.]

Now shrink the circle C_r towards its centre a, as illustrated in [9.1a]. Even if f is merely *continuous* (rather than analytic), $f(C_r)$ will shrink to $f(a)$, as indicated in [9.1b]. Since the images $w_1, w_2, \ldots, w_n$ of any n points on C_r will all converge to $f(a)$, so will their centroid W_n. Thus

$$\lim_{r\to 0}\langle f\rangle_{C_r} = f(a),$$

and this completes our first explanation of Cauchy's Formula.

9.1.3 *Gauss's Mean Value Theorem*

In the course of the above investigation we have also picked up an interesting bonus result:

> *If* $f(z)$ *is analytic on and inside a circle* C *centred at* a, *then the average value of* f *on* C *is its value at the centre:* $\langle f\rangle_C = f(a)$.

If we go on to split f into real and imaginary parts as $f = u+iv$, then we immediately deduce that $\langle u\rangle_C + i\,\langle v\rangle_C = u(a) + iv(a)$, and so

$$\langle u\rangle_C = u(a) \quad \text{and} \quad \langle v\rangle_C = v(a).$$

Thus if a real function Φ is either the real or the imaginary part of an analytic complex function, then its average on a circle is its value at the centre.

But if we are *given* a function Φ, how can we tell whether there exists an analytic function whose real or imaginary part is equal to Φ? In Ex. 2, p. 293 you showed that a necessary condition is that Φ be *harmonic*, i.e., that it satisfy Laplace's equation,

$$\Delta\Phi \equiv (\partial_x^2 + \partial_y^2)\Phi = 0.$$

In fact in Chapter 12 we will see that this is also a *sufficient* condition, yielding *Gauss's Mean Value Theorem*:

> *The average value of a harmonic function on a circle is equal to the value of the function at the centre of the circle.*

9.1.4 A Second Explanation and the General Cauchy Formula

What will happen to Cauchy's Formula if the loop L is *not* required to be simple? As in the previous chapter, it is now important to carefully define the "inside" of L as the set of points about which L has non-vanishing winding number:

$$\text{"inside"} = \{p \mid \nu[L, p] \neq 0\}.$$

Suppose in [9.2] that f has no singularities "inside" L. Then the only interior singularity of $[f(z)/(z-a)]$ will be the one at $z = a$. Here, L winds round a twice, and it is clear that L may be deformed into a small circle centred at a and traversed $\nu[L, a] = 2$ times. By virtue of Cauchy's Formula for simple loops, we deduce that $\frac{1}{2\pi i}\oint_L \frac{f(z)}{z-a}\,dz = 2\,f(a)$.

More generally, this line of reasoning suggests the following *General Cauchy Formula*: *If* $f(z)$ *is analytic on and "inside" a general loop* L, *then*

$$\frac{1}{2\pi i}\oint_L \frac{f(z)}{z-a}\,dz = \nu[L, a]\,f(a). \tag{9.3}$$

That this is always true is not quite clear from the above line of reasoning. Certainly Hopf's Theorem [(7.2), p. 389] guarantees that without crossing the singularity at a, L may be deformed into a circle centred at a and traversed $\nu[L, a]$ times. But the singularities of f may be scattered in the midst of L, although (by assumption) none lie "inside" L. So is it clear that this deformation can always be performed without crossing any of these singularities?

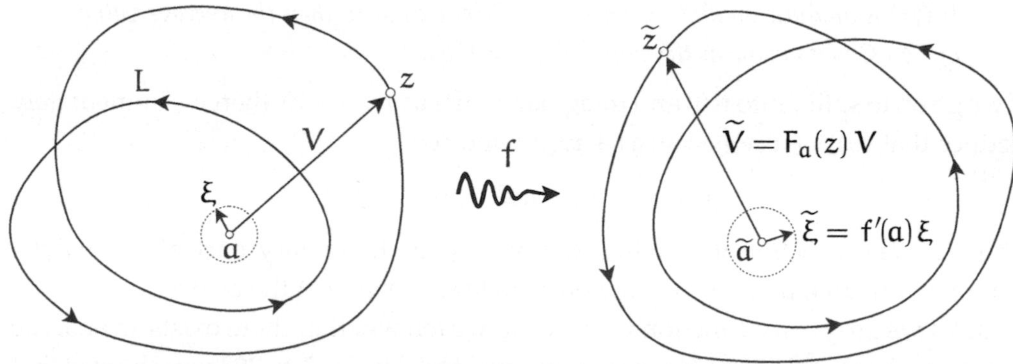

[9.2] Second explanation and the General Cauchy Formula: $\frac{1}{2\pi i}\oint_L \frac{f(z)}{z-a}dz = \nu[L, a]\,f(a)$. *Define* $F_a(z)$ *to be the non-infinitesimal analogue of the amplitwist, expanding and rotating* V *to its "image"* $\tilde{V}$, *so that* $\tilde{V} = F_a(z)\,V$. *Indeed, as* V *vanishes, the analogy becomes exact:* $F_a(a) \equiv \lim_{z\to a} F_a(z) = f'(a)$. *Since* $F_a(z) = [f(z) - f(a)]/[z-a]$ *is analytic inside* L, *its integral vanishes, by Cauchy's Theorem, and therefore* $0 = \frac{1}{2\pi i}\oint_L \frac{f(z)}{z-a}\,dz - f(a)\,\frac{1}{2\pi i}\oint_L \frac{dz}{z-a} = \frac{1}{2\pi i}\oint_L \frac{f(z)}{z-a}\,dz - \nu[L, a]\,f(a)$, *proving the result.*

We encourage you to pursue this idea, but we shall now present a different approach which yields (9.3) cleanly and directly. Consider the mapping $z \mapsto \widetilde{z} = f(z)$ in [9.2], and let us define

$$F_a(z) \equiv \frac{f(z) - f(a)}{z - a} = \frac{\widetilde{z} - \widetilde{a}}{z - a}.$$

If $V \equiv (z - a)$ is pictured as a vector emanating from a, and $\widetilde{V} \equiv (\widetilde{z} - \widetilde{a})$ is pictured as its image emanating from $\widetilde{a}$, then $F_a(z)$ describes the amount of rotation and expansion that carries V into $\widetilde{V} = F_a(z)\, V$. Thus $F_a(z)$ is the non-infinitesimal analogue of the amplitwist $f'(a)$ that carries an infinitesimal vector ξ into its image $\widetilde{\xi} = f'(a)\, \xi$, and

$$F_a(a) \equiv \lim_{z \to a} F_a(z) = f'(a).$$

Since $f(z)$ is assumed to be analytic and to have no singularities "inside" L, it follows that the same is true of $F_a(z)$. Thus the General Cauchy Theorem [(8.20), p. 472] implies that

$$\frac{1}{2\pi i} \oint_L F_a(z)\, dz = 0.$$

In other words,

$$\begin{aligned} 0 &= \frac{1}{2\pi i} \oint_L \frac{f(z)}{z - a}\, dz - f(a) \frac{1}{2\pi i} \oint_L \frac{dz}{z - a} \\ &= \frac{1}{2\pi i} \oint_L \frac{f(z)}{z - a}\, dz - \nu[L, a]\, f(a), \end{aligned}$$

as was to be shown.

9.2 Infinite Differentiability and Taylor Series

9.2.1 *Infinite Differentiability*

Returning to the case where L is a simple loop, let us show that if $f(z)$ is analytic inside L then so is $f'(z)$. From this it will follow by induction that $f(z)$ is *infinitely* differentiable.

What we must show is that if f is conformal, then so is f'. In other words, if f' is thought of as a mapping $z \mapsto \widetilde{z} = f'(z)$, then each infinitesimal vector ξ emanating from a must be rotated and expanded *the same amount* to obtain the image vector $\widetilde{\xi}$ emanating from $\widetilde{a}$. That is, there is a single complex number $f''(a)$ (the amplitwist of f') such that $\widetilde{\xi} = f''(a)\, \xi$.

Our first step is to obtain a neat expression for $f'(a)$ in terms of the values of $f(z)$ on L. Applying Cauchy's Formula to the analytic function $F_a(z)$, we deduce that

$$\begin{aligned} f'(a) = F_a(a) &= \frac{1}{2\pi i}\oint_L \frac{F_a(z)}{z-a}\,dz \\ &= \frac{1}{2\pi i}\oint_L \frac{f(z)}{(z-a)^2}\,dz - \frac{f(a)}{2\pi i}\oint_L \frac{dz}{(z-a)^2}. \end{aligned}$$

Since the second integral vanishes,

$$f'(a) = \frac{1}{2\pi i}\oint_L \frac{f(z)}{(z-a)^2}\,dz. \tag{9.4}$$

Now let's use this to find the image $\widetilde{\xi}$ under $z \mapsto \widetilde{z} = f'(z)$ of a short vector ξ emanating from a. Ignoring a term proportional to ξ^2, we find [exercise] that

$$\widetilde{\xi} \equiv f'(a+\xi) - f'(a) = \left[\frac{2}{2\pi i}\oint_L \frac{f(z)}{[z-(a+\xi)]^2(z-a)}\,dz\right]\xi.$$

Allowing ξ to become infinitesimal, we deduce the desired result: every infinitesimal ξ emanating from a is amplitwisted to $\widetilde{\xi} = f''(a)\,\xi$, where

$$f''(a) = \frac{2}{2\pi i}\oint_L \frac{f(z)}{(z-a)^3}\,dz. \tag{9.5}$$

Observe that since

$$\frac{d}{da}\left[\frac{1}{z-a}\right] = \frac{1}{(z-a)^2} \quad \text{and} \quad \frac{d^2}{da^2}\left[\frac{1}{z-a}\right] = \frac{2}{(z-a)^3},$$

both (9.4) and (9.5) are precisely what we would get if we simply differentiated the formula

$$f(a) = \frac{1}{2\pi i}\oint_L \frac{f(z)}{z-a}\,dz$$

with respect to a. Continuing in this way, we are led to conjecture that the nth derivative $f^{(n)}$ may be represented as

$$f^{(n)}(a) = \frac{n!}{2\pi i}\oint_L \frac{f(z)}{(z-a)^{n+1}}\,dz. \tag{9.6}$$

This is indeed true, as we shall see in a moment.

9.2.2 *Taylor Series*

Now let us show that if $f(z)$ is analytic on and inside an origin-centred circle C of radius R, then $f(z)$ may be expressed as a power series that converges inside this disc:

$$f(z) = c_0 + c_1 z + c_2 z^2 + c_3 z^3 + \cdots .$$

As we saw in Chapter 5, such a power series is infinitely differentiable within its disc of convergence. Thus the existence of the power series expansion will provide a second proof of the infinite differentiability of analytic functions. It also follows that the coefficients c_n may be expressed as

$$c_n = \frac{f^{(n)}(0)}{n!}, \tag{9.7}$$

so the power series is actually a Taylor series, and the coefficients do not depend on R:

$$f(z) = f(0) + f'(0)\, z + \frac{f''(0)}{2!}\, z^2 + \frac{f^{(3)}(0)}{3!}\, z^3 + \cdots .$$

To establish the existence of this series, we return to Cauchy's Formula (9.1). With a change of notation, this may be rewritten as

$$f(z) = \frac{1}{2\pi i} \oint_C \frac{f(Z)}{Z - z}\, dZ = \frac{1}{2\pi i} \oint_C \frac{f(Z)}{Z} \left[\frac{1}{1 - (z/Z)}\right] dZ.$$

See [9.3]. Since z is inside the circle on which Z lies, $|z| < |Z| = R$, and $|(z/Z)| < 1$. Thus $\frac{1}{1-(z/Z)}$ may be viewed as the sum of an infinite geometric series, and

$$f(z) = \frac{1}{2\pi i} \oint_C \frac{f(Z)}{Z} \left[1 + (z/Z) + (z/Z)^2 + (z/Z)^3 + \cdots\right] dZ.$$

Provided it makes sense to integrate this infinite series term by term, we deduce that $f(z)$ can indeed be expressed as a power series:

$$f(z) = \sum_{n=0}^{\infty} c_n z^n \quad \text{where} \quad c_n = \frac{1}{2\pi i} \oint_C \frac{f(Z)}{Z^{n+1}}\, dZ. \tag{9.8}$$

Furthermore, comparing this formula with (9.7), we also deduce (9.6).

To verify that this term-by-term integration is legitimate, consider the sum $f_N(z) \equiv \sum_{n=0}^{N-1} c_n z^n$ of the first N terms of the series (9.8). The result will be established if we can show that $f_N(z)$ tends to $f(z)$ as N tends to infinity.

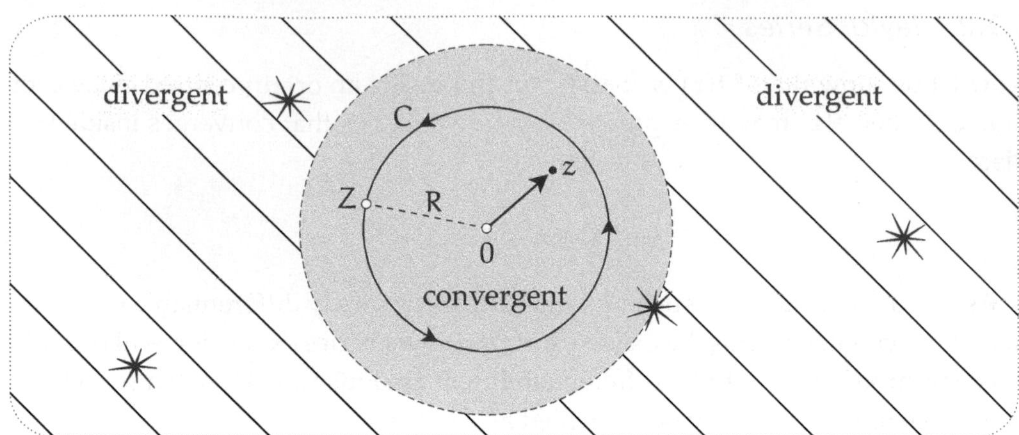

[9.3] If $f(z)$ is analytic inside C, then it has a Taylor expansion there: $f(z)=\sum_{n=0}^{\infty} c_n z^n$, where $c_n=\frac{1}{2\pi i}\oint_C \frac{f(Z)}{Z^{n+1}}\,dZ$. *Cauchy's Formula can be written as $f(z) = \frac{1}{2\pi i}\oint_C \frac{f(Z)}{Z}\left[\frac{1}{1-(z/Z)}\right] dZ$, and then $\left[\frac{1}{1-(z/Z)}\right]$ can be expanded as a geometric series. But the coefficients can also be obtained by differentiation, in the usual way, and therefore* **we obtain a formula for the nth derivative as an *integral*: $f^{(n)}(a) = n!\,c_n=\frac{n!}{2\pi i}\oint_C \frac{f(Z)}{Z^{n+1}}\,dZ$.**

Since

$$\frac{1}{1-(z/Z)} - [1 + (z/Z) + (z/Z)^2 + \cdots + (z/Z)^{N-1}] = \frac{(z/Z)^N}{1-(z/Z)},$$

it follows that

$$f(z) - f_N(z) = \frac{1}{2\pi i}\oint_C \frac{(z/Z)^N f(Z)}{(Z-z)}\,dZ.$$

Finally recall [see (8.5), p. 440] that the modulus of an integral cannot exceed the product of the length of the path and the maximum modulus of the integrand at points on the path. If M stands for the maximum value of $|f(Z)/(Z-z)|$ on C, then it follows that

$$|f(z) - f_N(z)| \leqslant RM|(z/Z)|^N.$$

Thus $\lim_{N\to\infty} f_N(z) = f(z)$, as was to be shown.

What is the radius of convergence of the series we have obtained? We know that if f is analytic inside C then the series (9.8) converges to $f(z)$ in that disc. Thus, referring to [9.3], C may be expanded up to the dashed circle, where it first encounters a singularity of f. More generally, $f(z)$ may be a single-valued branch of a multifunction, and as we learnt in Chapter 2, branch points then act as obstacles just as much as singularities. Thus *the radius of convergence is the distance from the centre of the expansion to the nearest singularity or branch point*.

One final point. We chose the origin as the centre of the expansion in order to avoid algebraic clutter, but this choice really involves no loss of generality. For suppose we instead choose the centre to be at a, meaning that we wish to expand $f(z)$ in powers of $\xi \equiv (z-a)$. If $f(z)$ is analytic at a then $F(\xi) \equiv f(a+\xi) = f(z)$ is analytic at the origin of the ξ-plane, and so it possesses an origin-centred Taylor expansion,

$$F(\xi) = \sum_{n=0}^{\infty} \frac{F^{(n)}(0)}{n!}\,\xi^n \quad \Longrightarrow \quad f(z) = \sum_{n=0}^{\infty} \frac{f^{(n)}(a)}{n!}\,(z-a)^n.$$

Alternatively [exercise], the existence of this series may be deduced by directly generalizing the argument leading to the origin-centred series. Either way, we conclude that

> *If $f(z)$ is analytic, and a is neither a singularity nor a branch point, then $f(z)$ may be expressed as the following power series, which converges to $f(z)$ within the disc whose radius is the distance from a to the nearest singularity or branch point:*

$$f(z) = \sum_{n=0}^{\infty} c_n\,(z-a)^n, \quad \text{where} \quad \frac{f^{(n)}(a)}{n!} = c_n = \frac{1}{2\pi i}\oint_L \frac{f(z)}{(z-a)^{n+1}}\,dz.$$

9.3 Calculus of Residues

9.3.1 *Laurent Series Centred at a Pole*

Suppose that a is a pole of an analytic function $f(z)$, i.e., $\lim_{z\to a} f(z) = \infty$. In Chapter 7 we investigated poles by assuming the existence of Taylor series (which we have just proven), and we found [see (7.19), p. 417] that near a we could express $f(z)$ as

$$f(z) = \frac{\phi(z)}{(z-a)^m},$$

where $\phi(z)$ is analytic, and $\phi(a) \neq 0$. Recall that the positive integer m is called the "order" of the pole, and that the greater the order of the pole, the faster $f(z)$ approaches ∞ as z approaches a.

We know that $\phi(z)$ can be expressed as a Taylor series centred at a:

$$\phi(z) = \sum_{n=0}^{\infty} c_n\,(z-a)^n, \quad \text{where} \quad c_n = \frac{\phi^{(n)}(a)}{n!}.$$

Hence we deduce that

If an analytic function $f(z)$ *has a pole of order* m *at* a*, then in the vicinity of this pole,* $f(z)$ *possesses a Laurent series of the form*

$$f(z) = \frac{c_0}{(z-a)^m} + \frac{c_1}{(z-a)^{m-1}} + \cdots + \frac{c_{m-1}}{(z-a)} + c_m + c_{m+1}(z-a) + \cdots.$$

Recall that the coefficient of $1/(z-a)$ is called the "residue" of $f(z)$ at a, denoted $\operatorname{Res}[f, a]$. Also recall the crucial significance of the residue in evaluating integrals: if L is a simple loop containing a but no other singularities of f, then

$$\oint_L f(z)\, dz = 2\pi i \operatorname{Res}[f, a].$$

More generally, suppose that L is not required to be simple, and that $f(z)$ has several poles, at a_1, a_2, etc. The existence of the Laurent series was the missing ingredient in our discussion of this situation in the previous chapter. Having established that f does indeed possess a Laurent expansion in the vicinity of each of its poles, we have also verified the General Residue Theorem [(8.26), p. 477]:

$$\oint_L f(z)\, dz = 2\pi i \sum_n \nu[L, a_n] \operatorname{Res}[f, a_n]. \tag{9.9}$$

9.3.2 A Formula for Calculating Residues

It is easy enough to find an explicit formula for the residue at a pole. Looking at the derivation of the Laurent series above, we see that

$$\operatorname{Res}[f, a] = c_{m-1} = \frac{\phi^{(m-1)}(a)}{(m-1)!}.$$

Since $\phi(z) = (z-a)^m f(z)$, we deduce that

If a *is an* m^{th} *order pole of* $f(z)$*, then*

$$\operatorname{Res}[f(z), a] = \frac{1}{(m-1)!}\left[\frac{d}{dz}\right]^{m-1} [(z-a)^m f(z)]\Bigg|_{z=a}. \tag{9.10}$$

From this general result one can derive other results that speed up the calculation of residues in commonly encountered special cases. For example, suppose that $f = (P/Q)$ has a "simple" (i.e., order 1) pole at a as a result of Q having a simple root at that point. In that case,

$$\operatorname{Res}[f(z), a] = \lim_{z\to a}(z-a)f(z) = \lim_{z\to a} \frac{P(z)}{\left[\frac{Q(z)-Q(a)}{z-a}\right]}.$$

Thus, *If* $f(z) = \frac{P(z)}{Q(z)}$*, and* a *is a simple root of* Q*, then*

$$\operatorname{Res}[f(z), a] = \frac{P(a)}{Q'(a)}. \tag{9.11}$$

For example, consider $f(z) = e^z/(z^4 - 1)$, which has simple poles at $z = \pm 1, \pm i$. If L is the circle $|z - 1| = 1$ then $z = 1$ is the only pole inside L, so (9.11) yields

$$\oint_L \frac{e^z}{z^4 - 1}\, dz = 2\pi i \operatorname{Res}[f, 1] = 2\pi i \left.\frac{e^z}{4z^3}\right|_{z=1} = \tfrac{1}{2}\pi i e.$$

We can actually check this using Cauchy's Formula. Since

$$(z^4 - 1) = (z - 1)(1 + z + z^2 + z^3),$$

we may write $f(z) = F(z)/(z - 1)$, where $F(z) \equiv e^z/(1 + z + z^2 + z^3)$. Since $F(z)$ is analytic inside L,

$$\oint_L \frac{e^z}{z^4 - 1}\, dz = \oint_L \frac{F(z)}{z - 1}\, dz = 2\pi i\, F(1) = \tfrac{1}{2}\pi i e,$$

just as before.

9.3.3 Application to Real Integrals

In the exercises of the previous chapter we saw how certain kinds of real integrals could be expressed in terms of complex contour integrals. According to (9.9), the evaluation of contour integrals amounts to calculating residues, and we have just seen that this is straightforward. Thus the Residue Theorem leads to a powerful method of evaluating real integrals.

Historically, Cauchy's success in evaluating previously intractable real integrals was one of the first tangible signs of the power of his discoveries. Many modern texts (e.g., Marsden et al. (1999)) continue to celebrate this success with very detailed discussions of how the Residue Theorem may be applied to real integrals. However, there can be little doubt that this application is less important than it used to be. Today,[2] when faced with a tricky integral, a physicist, engineer, or mathematician is less likely to start calculating residues, and is more likely to reach for a computer. We will therefore only do a couple of illustrative examples, though further examples may be found in the exercises.

In Ex. 14, p. 480 we evaluated $\int_{-\infty}^{+\infty} (x^2 + 1)^{-2}\, dx$ using partial fractions. To redo this problem using residues, we integrate $f(z) = 1/(z^2 + 1)^2$ along the simple loop $(L + J)$ shown in [9.4a]. Here L is the segment of the real axis from $-R$ to $+R$, and J is the semi-circular contour (in the upper half plane) back from $+R$ to $-R$. Rewriting $f(z)$ as $f(z) = 1/(z + i)^2(z - i)^2$, we see that the only singularities are the second order poles at $z = \pm i$. Thus if $R > 1$ (as illustrated) then (9.10) yields

[2] Here, "today" refers to 1997! Today, in 2022, my claim is only *more* true than it was then.

$$\oint_{L+J} f(z)\,dz = 2\pi i \,\mathrm{Res}\,[f, i] = 2\pi i \left.\frac{d}{dz}\frac{1}{(z+i)^2}\right|_{z=i} = 2\pi i \frac{-2}{(2i)^3} = \frac{\pi}{2}.$$

But

$$\oint_{L+J} f(z)\,dz = \int_{-R}^{+R} \frac{dx}{(x^2+1)^2} + \int_J f(z)\,dz,$$

and, as you showed in the original exercise, the integral along J tends to zero as R tends to infinity. Thus

$$\int_{-\infty}^{+\infty} \frac{dx}{(x^2+1)^2} = \frac{\pi}{2}.$$

The famous physicist Richard Feynman once bet[3] his colleagues, "I can do by other methods any integral anybody else needs contour integration to do." It is a tribute to complex analysis that Feynman lost this bet. Nevertheless, we can check the above integral using a trick that frequently *did* enable Feynman to dispense with residues: differentiation of a simpler integral with respect to a parameter.

Consider the elementary result,

$$\int_{-\infty}^{+\infty} \frac{dx}{x^2+a^2} = \left[\frac{1}{a}\tan^{-1}\left(\frac{x}{a}\right)\right]_{-\infty}^{+\infty} = \frac{\pi}{a}.$$

Differentiating this with respect to a yields

$$\int_{-\infty}^{+\infty} \frac{2a}{(x^2+a^2)^2}\,dx = \frac{\pi}{a^2},$$

and substituting $a = 1$ then confirms our residue calculation.

For our second example, we will evaluate

$$I \equiv \int_0^{2\pi} \frac{d\theta}{\cos\theta + a}, \quad a > 1,$$

by rewriting it as a contour integral round the unit circle C. See [9.4b]. As illustrated, $\cos\theta$ is the midpoint of z and $(1/z)$, and dz is perpendicular to z and has length $d\theta$: in symbols, $\cos\theta = \frac{1}{2}[z + (1/z)]$ and $dz = iz\,d\theta$. Substituting into I,

$$I = \oint_C \frac{(dz/iz)}{\frac{1}{2}[z+(1/z)] + a} = -2i \oint_C \frac{dz}{z^2 + 2az + 1}.$$

Since the singularities p and q of the integrand satisfy $pq = 1$, only one of them lies inside C—in fact p and q are geometric inverses. Thus [exercise],

$$I = 4\pi\,\mathrm{Res}\left[\frac{1}{(z-p)(z-q)}, q\right] = \frac{4\pi}{(q-p)} = \frac{2\pi}{\sqrt{a^2-1}}.$$

[3] See Feynman (1997).

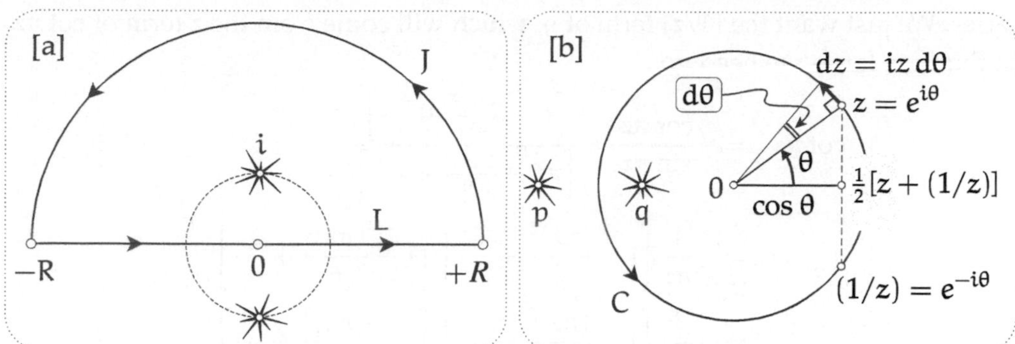

[9.4] Using residues to evaluate real integrals. [a] *To evaluate* $\int_{-\infty}^{+\infty}(x^2+1)^{-2}\,dx$, *consider the integral of* $f(z) = (z^2+1)^{-2}$. *Then,* $\oint_{L+J} f(z)\,dz = 2\pi i\,\text{Res}\,[f, i] = \frac{\pi}{2}$. *But as* $R \to \infty$, $\oint_J f(z)\,dz \to 0$, *and therefore* $\int_{-\infty}^{+\infty}(x^2+1)^{-2}\,dx = \frac{\pi}{2}$. **[b]** *To evaluate* $I \equiv \int_0^{2\pi}\frac{d\theta}{\cos\theta + a}$, *where* $a > 1$, *let* $z = e^{i\theta}$, *so that* $\cos\theta = \frac{1}{2}[z+(1/z)]$ *and* $dz = iz d\theta$. *Then* $I = -2i\oint_C \frac{dz}{z^2+2az+1}$. *Factorizing the denominator into* $(z-p)(z-q)$, *we obtain* $I = 4\pi\,\text{Res}\left[\frac{1}{(z-p)(z-q)}, q\right] = \frac{4\pi}{(q-p)} = \frac{2\pi}{\sqrt{a^2-1}}$.

9.3.4 Calculating Residues using Taylor Series

In order to calculate a residue using (9.10), one must first know the order m of the pole. If $f(z)$ is built out of simple functions whose Taylor series are known, then the quickest method of finding m is by manipulating these series. Furthermore, this approach may be used to calculate the residue itself, often more easily than via formula (9.10). A few examples should suffice to explain the method.

For our first example, let $f(z) = (\sin^2 z/z^5)$, which clearly has a singularity of some kind at the origin. For small values of z, $\sin z \approx z$, so $f(z) \approx (1/z^3)$, and the order of the pole is therefore $m = 3$. By taking more terms of the Taylor series for $\sin z$ we can find more terms in the Laurent expansion of $f(z)$, and hence find the residue:

$$\begin{aligned} f(z) &= \frac{1}{z^5}\left[z - \frac{z^3}{6} + \cdots\right]^2 = \frac{1}{z^5}\left[z^2 - 2z\left(\frac{z^3}{6}\right) + \cdots\right] \\ &= \frac{1}{z^3} - \frac{1}{3z} + \cdots \\ &\Longrightarrow \quad \text{Res}\,[f, 0] = -\frac{1}{3}. \end{aligned}$$

In order to appreciate how efficient this is, try checking the result using formula (9.10) instead.

Our next example will have valuable consequences. Let $g(z) = (1/z^2)\cot(\pi z)$, which is clearly singular at the origin. To find the order of this pole, and its residue, we begin by calculating the Laurent series of $\cot(\pi z)$. When doing such a calculation, it is important to remember that we are not trying to find the whole Laurent series. We just want the $(1/z)$ term of g, which will come from the z term of $\cot \pi z$, so that's as far as we need go:

$$\begin{aligned}
\cot \pi z &= \frac{\cos \pi z}{\sin \pi z} = \frac{\left[1 - \frac{(\pi z)^2}{2!} + \cdots\right]}{\left[\pi z - \frac{(\pi z)^3}{3!} + \cdots\right]} \\
&= \frac{1}{\pi z}\left[1 - \frac{(\pi z)^2}{2} + \cdots\right]\left[1 - \frac{(\pi z)^2}{6} + \cdots\right]^{-1} \\
&= \frac{1}{\pi z}\left[1 - \frac{(\pi z)^2}{2} + \cdots\right]\left[1 + \frac{(\pi z)^2}{6} + \cdots\right] \\
&= \frac{1}{\pi z} - \frac{\pi z}{3} + \cdots .
\end{aligned}$$

In particular, note for future use that $\text{Res}\,[\cot(\pi z), 0] = (1/\pi)$.

Returning to the original function g, we find that

$$g(z) = \frac{1}{\pi z^3} - \frac{\pi}{3z} + \cdots ,$$

and so the origin is a triple pole with $\text{Res}\,[g, 0] = -(\pi/3)$. Again, try checking this using formula (9.10) instead.

Continuing with this example, it's clear that $g(z)$ also has a singularity at each integer n. To find the residue at n, we could write $z = n + \xi$ and expand g as a Laurent series in powers of ξ. However, this is unnecessary. Since $(1/z^2)$ is non-singular at n, and since $\cot[\pi(n + \xi)] = \cot \pi\xi$,

$$\begin{aligned}
\text{Res}\,[(1/z^2)\cot(\pi z), n] &= (1/n^2)\,\text{Res}\,[\cot(\pi z), n] \\
&= (1/n^2)\,\text{Res}\,[\cot(\pi z), 0] \\
&= 1/(\pi n^2).
\end{aligned}$$

More generally, note that if $f(z)$ is any analytic function that is non-singular at n, then

$$\text{Res}\,[f(z)\cot(\pi z), n] = \frac{1}{\pi} f(n). \tag{9.12}$$

This may also be verified [exercise] using (9.11).

9.3.5 *Application to Summation of Series*

Historically, $1 + \frac{1}{2^2} + \frac{1}{3^2} + \frac{1}{4^2} + \cdots$ was the first simple-looking series that mathematicians were unable to sum using elementary algebraic methods. After the Bernoulli family had tried and failed, Euler finally cracked the problem in 1734 by means of a brilliantly unorthodox argument[4]. The answer he found was as unexpected as his methods:

$$\sum_{n=1}^{\infty} \frac{1}{n^2} = \frac{\pi^2}{6}.$$

[4] See Ex. 13 and Stillwell (2010).

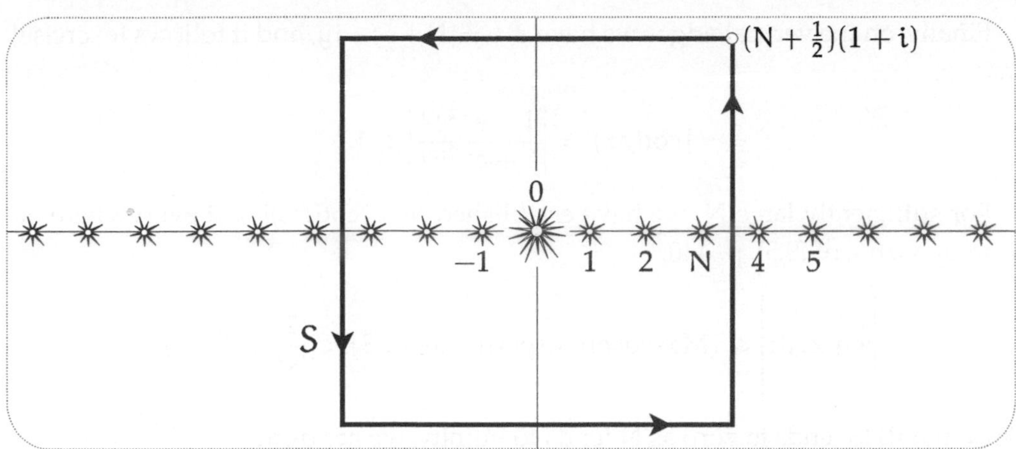

[9.5] Evaluation of $\sum_{-\infty}^{\infty} f(n)$ using Res $[f(z)\cot(\pi z), n] = \frac{1}{\pi}f(n)$. *If we take $f(z) = 1/z^2$, then Res $\left[\frac{\cot(\pi z)}{z^2}, n\right] = \frac{1}{\pi n^2}$. Also, one finds [see text] that Res $\left[\frac{\cot(\pi z)}{z^2}, 0\right] = -\frac{\pi}{3}$. Therefore, $\frac{1}{2\pi i}\oint_S \frac{\cot(\pi z)}{z^2}\,dz = -\frac{\pi}{3} + \frac{2}{\pi}\sum_{n=1}^{N}\frac{1}{n^2}$. But as $N \to \infty$, one finds [see text] that this integral goes to zero, and we thereby obtain Euler's extraordinary discovery of 1734: $\sum_{n=1}^{\infty}\frac{1}{n^2} = \frac{\pi^2}{6}$.*

Today such results can be derived in a systematic way using residues. Reconsider the function $g(z) = (1/z^2)\cot(\pi z)$ above. With N a positive integer, let S be the origin-centred square with vertices $(N+\frac{1}{2})(\pm 1 \pm i)$ shown in [9.5]. Adding up the residues of the illustrated singularities inside S,

$$\frac{1}{2\pi i}\oint_S g(z)\,dz = \text{Res}\,[g(z),0] + \sum_{n=-N}^{-1}\text{Res}\,[g(z),n] + \sum_{n=1}^{N}\text{Res}\,[g(z),n]$$

$$= -\frac{\pi}{3} + \frac{2}{\pi}\sum_{n=1}^{N}\frac{1}{n^2}.$$

As we will now see, the integral on the LHS tends to zero as N tends to infinity, and from this fact we immediately deduce Euler's result.

To show that the integral of $g(z) = (1/z^2)\cot(\pi z)$ does indeed tend to zero as S expands, we must show that the size of the integrand dies away faster than the perimeter $(8N+4)$ of S grows. First the easy part: $|g(z)| = |1/z^2| \cdot |\cot(\pi z)|$, and on S we clearly have $|z| > N$, so $|1/z^2| < (1/N^2)$.

Next we must examine the size of

$$|\cot(\pi z)| = \left|\frac{e^{i\pi z} + e^{-i\pi z}}{e^{i\pi z} - e^{-i\pi z}}\right|$$

on the four edges of S. We begin with the horizontal edges, $y = \pm(N+\frac{1}{2})$. Since $|e^{\pm i\pi z}| = e^{\mp\pi y}$, it is not hard to see [exercise] that if N is reasonably large then $|\cot(\pi z)|$ is very close to 1. Thus for sufficiently large N, $|\cot(\pi z)|$ will certainly be less than 2, for example.

Finally, on the vertical edges we have $z = \pm(N+\frac{1}{2})+iy$, and it follows [exercise] that

$$|\cot(\pi z)| = \left|\frac{1-e^{-2\pi y}}{1+e^{-2\pi y}}\right| \leqslant 1.$$

For sufficiently large N, we have established that $|\cot(\pi z)| < 2$ everywhere on S, so by virtue of (8.5), p. 440,

$$\left|\oint_S g(z)\, dz\right| \leqslant (\text{Max } |g| \text{ on } S)(\text{perimeter of } S) < \frac{2}{N^2}(8N+4).$$

Since the RHS tends to zero as N tends to infinity, we are done.

More generally, let $f(z)$ be an analytic function such that $|f(z)| < (\text{const.})/|z|^2$ for sufficiently large $|z|$. Then it is clear that the above argument applies equally well to the integral of $f(z)\cot(\pi z)$:

$$\begin{aligned} 0 &= \lim_{N\to\infty} \frac{1}{2\pi i}\oint_S f(z)\cot(\pi z)\, dz \\ &= \sum_{\text{all poles}} \text{Res}\,[f(z)\cot(\pi z)] \\ &= \sum_{n=-\infty}^{\infty} \text{Res}\,[f(z)\cot(\pi z), n] + \sum_{\text{poles of } f(z)} \text{Res}\,[f(z)\cot(\pi z)] \\ &= \frac{1}{\pi}\sum_{n=-\infty}^{\infty} f(n) + \sum_{\text{poles of } f(z)} \text{Res}\,[f(z)\cot(\pi z)], \end{aligned}$$

where the last equality follows from (9.12).

Thus

If $f(z)$ is an analytic function such that $|f(z)| < (\text{const.})/|z|^2$ for sufficiently large $|z|$, then

$$\sum_{n=-\infty}^{\infty} f(n) = -\pi \sum_{\text{poles of } f(z)} \text{Res}\,[f(z)\cot(\pi z)]. \tag{9.13}$$

Of course if any of the poles of $f(z)$ happen to be integers, then these values of n are understood to be excluded from the LHS of (9.13).

Note that while symmetry enables us to calculate sums like $\sum_{n=1}^{\infty}(1/n^2)$ and $\sum_{n=1}^{\infty}(1/n^4)$ using (9.13), we *cannot* use (9.13) to calculate a sum like $\sum_{n=1}^{\infty}(1/n^3)$. What, you might ask, is the sum of this last series? The answer is that *nobody knows!*[5]

[5] This was written in 1997, but, 25 years later, it *remains* a mystery.

As a further interesting illustration of (9.13), consider $f(z) = 1/(z-w)^2$, where w is an arbitrary (non-integer) complex number. Geometrically, $|z-w|$ is the distance w to z, and this makes it easy to see that $|f(z)|$ satisfies the requirement of the theorem. Since the only singularity of $f(z)$ is a double pole at $z = w$,

$$\sum_{n=-\infty}^{\infty} \frac{1}{(n-w)^2} = -\pi \operatorname{Res}\left[\frac{\cot(\pi z)}{(z-w)^2}, w\right].$$

Using formula (9.10),

$$\operatorname{Res}\left[\frac{\cot(\pi z)}{(z-w)^2}, w\right] = \frac{d}{dz}\cot(\pi z)\bigg|_{z=w} = -\frac{\pi}{\sin^2(\pi w)}.$$

Thus we obtain the remarkable result,

$$\frac{\pi^2}{\sin^2(\pi w)} = \cdots + \frac{1}{(2+w)^2} + \frac{1}{(1+w)^2} + \frac{1}{w^2} + \frac{1}{(1-w)^2} + \frac{1}{(2-w)^2} + \cdots.$$

Such series were first discovered by Euler in 1748. What is remarkable about such a formula is that the periodicity of the function on the LHS is *explicitly* exhibited by the series on the RHS. That is, if you change w to $(w+1)$, the series is clearly unaltered.

9.4 Annular Laurent Series

9.4.1 An Example

We have seen that the Laurent series is the natural generalization of the Taylor series when the centre of the expansion is a pole rather than a non-singular point. However, this is by no means the *only* situation in which Laurent series are needed.

For example, consider

$$F(z) = \frac{1}{(1-z)(2-z)},$$

whose simple poles are illustrated in [9.6a]. Since F is analytic within the unit disc, it possesses a Taylor series in powers of z. This may be found most easily by splitting F into partial fractions:

$$\begin{aligned} F(z) &= \frac{1}{(1-z)} - \frac{1}{(2-z)} \\ &= \frac{1}{(1-z)} - \frac{1}{2[1-(z/2)]} = \overbrace{\sum_{n=0}^{\infty} z^n}^{\text{for } |z|<1} - \overbrace{\tfrac{1}{2}\sum_{n=0}^{\infty} (z/2)^n}^{\text{for } |z|<2} \\ &= \tfrac{1}{2} + \tfrac{3}{4}z + \tfrac{7}{8}z^2 + \cdots + \left[1-(1/2)^{n+1}\right] z^n + \cdots, \quad \text{for } |z| < 1. \end{aligned}$$

The pole at $z = 1$ means that outside the unit disc F cannot be expressed as a power series in z. However, in the shaded annulus $1 < |z| < 2$ it can be expressed as a *Laurent* series in z:

$$\begin{aligned}
F(z) &= -\frac{1}{z[1-(1/z)]} - \frac{1}{2[1-(z/2)]} = -\overbrace{\sum_{n=0}^{\infty}(1/z)^{n+1}}^{\text{for } |z|>1} - \tfrac{1}{2}\overbrace{\sum_{n=0}^{\infty}(z/2)^{n}}^{\text{for } |z|<2} \\
&= \cdots - \frac{1}{z^3} - \frac{1}{z^2} - \frac{1}{z} - \frac{1}{2} - \frac{z}{4} - \frac{z^2}{8} - \frac{z^3}{16} - \cdots, \quad \text{for } 1 < |z| < 2.
\end{aligned}$$

Finally, in the region $|z| > 2$ beyond the annulus we obtain [exercise] a *different* Laurent series:

$$F(z) = \frac{1}{z^2} + \frac{3}{z^3} + \cdots + \frac{(2^{n-1}-1)}{z^n} + \cdots, \quad \text{for } |z| > 2.$$

9.4.2 *Laurent's Theorem*

What we have just seen is an illustration of a general phenomenon. See [9.6b].

> *If* $f(z)$ *is analytic everywhere within an annulus* A *centred at* a*, then* $f(z)$ *can be expressed as a Laurent series within* A*. In fact, if* K *is any simple loop lying within* A *and winding once round* a,

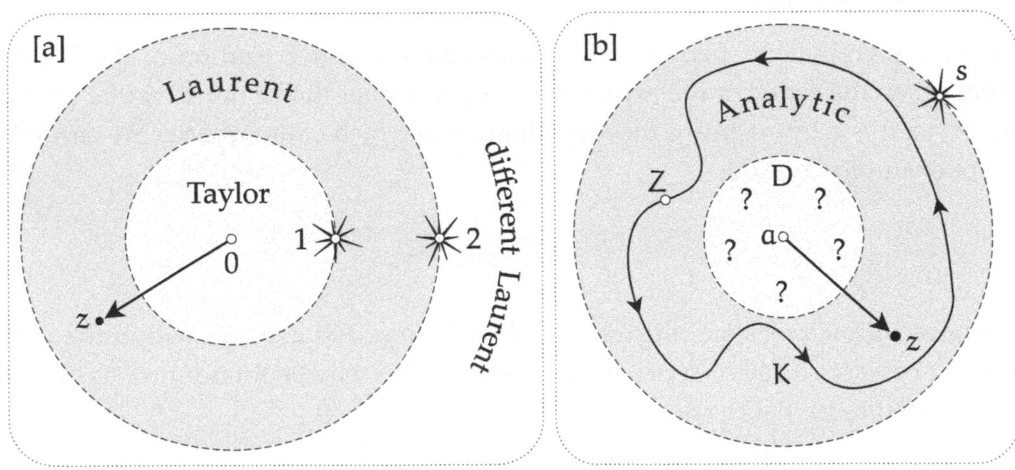

[9.6] Laurent Series. [a] *Let* $F(z) = \frac{1}{(1-z)(2-z)}$. *Then* $F(z)$ *has a Taylor series within the unit disc, but the singularity at* $z = 1$ *means that no power series can exist beyond. However, in the shaded annulus,* $1 < |z| < 2$*, it is possible to express* $F(z)$ *as a so-called* **Laurent series that has positive *and* negative powers of** z: $F(z) = \cdots - \frac{1}{z^3} - \frac{1}{z^2} - \frac{1}{z} - \frac{1}{2} - \frac{z}{4} - \frac{z^2}{8} - \frac{z^3}{16} - \cdots$. *Moving outside the annulus, a Taylor series is again impossible because of the two poles closer in, but, here too, a Laurent series exists. However, it is now a* different *Laurent series:* $F(z) = \frac{1}{z^2} + \frac{3}{z^3} + \cdots + \frac{(2^{n-1}-1)}{z^n} + \cdots$. **[b] Laurent's Theorem.** *If* $f(z)$ *is analytic everywhere within the annulus, then* $f(z) = \sum_{n=-\infty}^{\infty} c_n (z-a)^n$, *where* $c_n = \frac{1}{2\pi i}\oint_K \frac{f(Z)}{(Z-a)^{n+1}}\,dZ$.

$$f(z) = \sum_{n=-\infty}^{\infty} c_n (z-a)^n, \quad \text{where} \quad c_n = \frac{1}{2\pi i}\oint_K \frac{f(Z)}{(Z-a)^{n+1}}\, dZ. \tag{9.14}$$

Before establishing this result, which is called *Laurent's Theorem*, we make the following observations regarding its significance:

- The surprising thing about the result is the *existence* of a Laurent series, not the fact that it converges in an annulus. Since we know that a power series in $(z-a)$ will converge inside a disc centred at a, it follows [exercise] that a power series in $1/(z-a)$ will converge *outside* a disc centred at a. Since a Laurent series is (by definition) the sum of a power series in $(z-a)$ and a power series in $1/(z-a)$, it follows that it will converge in an annulus.
- Previously we were able to deduce the existence of a Laurent series only in the vicinity of a pole. The present result is much more powerful: as indicated by the question marks in [9.6b], we make no assumptions at all concerning the behaviour of $f(z)$ in the disc D bounded by the inner edge of the annulus. In practice, the outer edge of the annulus may be expanded until it hits a singularity s of $f(z)$, and the inner edge may likewise be contracted until it hits the outermost singularity lying in D.
- If there are *no* singularities in D, then the inner edge of the annulus may be completely collapsed, thereby transforming the annulus into a disc. In this case, (9.14) does not contain any negative powers. For if n is negative then $f(z)/(z-a)^{n+1}$ is analytic everywhere inside K, and so $c_n = 0$. In this way we recover the existence of Taylor's series as a special case of Laurent's Theorem.
- Suppose that a is a singularity and that for a sufficiently small value of ϵ there are no other singularities within a distance ϵ of a. In this case one says that a is an *isolated singularity* of $f(z)$. Applying Laurent's Theorem to the annulus $0 < |z-a| < \epsilon$, we find that there are just two fundamentally different possibilities: the principal part of the Laurent series either has finitely many terms, or infinitely many terms. Recall that in the latter case we have (by definition) an "essential singularity". See p. 417, where we considered the example

  $$e^{1/z} = 1 + \frac{1}{1!\,z} + \frac{1}{2!\,z^2} + \frac{1}{3!\,z^3} + \cdots.$$

  To sum up,

  > *An isolated singularity of an analytic function is either a pole or an essential singularity.*

Now let us establish (9.14). In order to simplify the calculations, we will only treat the case $a = 0$, illustrated in [9.7a]. Here, z is a general point in the annulus, $\mathcal{C}$ and $\mathcal{D}$ are counterclockwise circles such that z lies between them, and $\mathcal{L}$ is a simple loop round z, lying within the annulus.

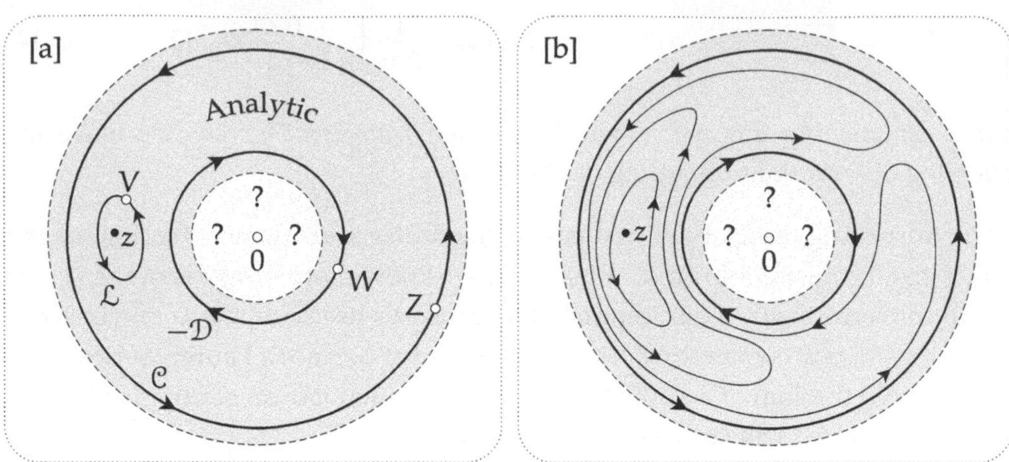

[9.7] Proof of Laurent's Theorem. [a] *The initial key step mimics the rewriting of Cauchy's Formula that was used to prove Taylor's Theorem:* $f(z) = \frac{1}{2\pi i}\oint_{\mathcal{L}} \frac{f(V)}{V-z}\,dV = \frac{1}{2\pi i}\oint_{\mathcal{C}} \frac{f(Z)}{Z}\left[\frac{1}{1-(z/Z)}\right] dZ + \frac{1}{2\pi i}\oint_{\mathcal{D}} \frac{f(W)}{z}\left[\frac{1}{1-(W/z)}\right] dW$*. The theorem then follows quickly by expanding both square brackets into geometric series, in* z *and* $(1/z)$*, respectively.* **[b]** *This picture justifies the previous equation by showing that* $\mathcal{L}$ *can indeed be deformed within the annulus into* $(\mathcal{C}) + (-\mathcal{D})$*, without changing the value of the integral.*

First, by Cauchy's Formula,

$$f(z) = \frac{1}{2\pi i}\oint_{\mathcal{L}} \frac{f(V)}{V-z}\,dV = \frac{1}{2\pi i}\oint_{\mathcal{C}} \frac{f(Z)}{Z-z}\,dZ - \frac{1}{2\pi i}\oint_{\mathcal{D}} \frac{f(W)}{W-z}\,dW,$$

where the second equality follows from the fact that $\mathcal{L}$ may be deformed within the annulus into $(\mathcal{C}) + (-\mathcal{D})$, as indicated in [9.7b].

Next, we rewrite the above equation as

$$f(z) = \frac{1}{2\pi i}\oint_{\mathcal{C}} \frac{f(Z)}{Z}\left[\frac{1}{1-(z/Z)}\right] dZ + \frac{1}{2\pi i}\oint_{\mathcal{D}} \frac{f(W)}{z}\left[\frac{1}{1-(W/z)}\right] dW.$$

The significance of this is that $|(z/Z)| < 1$ and $|(W/z)| < 1$, so both integrands on the RHS can be expanded into geometric series, very much as we did in the example of [9.6a].

Referring back to the derivation of the Taylor series (9.8), the integral round $\mathcal{C}$ can be expressed as

$$\frac{1}{2\pi i}\oint_{\mathcal{C}} \frac{f(Z)}{Z}\left[\frac{1}{1-(z/Z)}\right] dZ = \sum_{n=0}^{\infty}\left[\frac{1}{2\pi i}\oint_{\mathcal{C}} \frac{f(Z)}{Z^{n+1}}\,dZ\right] z^n.$$

Essentially identical reasoning [exercise] also justifies term-by-term integration in the case of the integral round $\mathcal{D}$:

$$\frac{1}{2\pi i}\oint_{\mathcal{D}} \frac{f(W)}{z}\left[\frac{1}{1-(W/z)}\right] dW = \sum_{n=1}^{\infty}\left[\frac{1}{2\pi i}\oint_{\mathcal{D}} W^{n-1}f(W)\,dW\right]\left(\frac{1}{z}\right)^n.$$

Thus the existence of the Laurent series is established:

$$f(z) = \cdots + \frac{d_3}{z^3} + \frac{d_2}{z^2} + \frac{d_1}{z} + c_0 + c_1 z + c_2 z^2 + \cdots,$$

where

$$d_m = \frac{1}{2\pi i}\oint_{\mathcal{D}} W^{m-1} f(W)\,dW \quad \text{and} \quad c_n = \frac{1}{2\pi i}\oint_{\mathcal{C}} \frac{f(Z)}{Z^{n+1}}\,dZ.$$

Finally, the following two observations enable us to tidy up the result. First, by the Deformation Theorem [p. 454], the integrals defining d_m and c_n do not change their values if we allow $\mathcal{C}$ to contract and $\mathcal{D}$ to expand till they coalesce into the *same* circle. Indeed, we may replace both $\mathcal{C}$ and $\mathcal{D}$ with any simple loop K contained in the annulus and winding round it once. Second, if we write $m = -n$ then the integral defining the coefficient d_{-n} of z^n has integrand $W^{-n-1}f(W) = f(W)/W^{n+1}$, which is the same as the integrand for the c_n's. Thus, as was to be shown, the Laurent series may be expressed in the compact form of (9.14):

$$f(z) = \sum_{n=-\infty}^{\infty} c_n z^n, \quad \text{where} \quad c_n = \frac{1}{2\pi i}\oint_K \frac{f(Z)}{Z^{n+1}}\,dZ.$$

9.5 Exercises

1 If C is the unit circle, show that

$$\int_0^{2\pi} \frac{dt}{1+a^2-2a\cos t} = \oint_C \frac{i\,dz}{(z-a)(az-1)}.$$

Use Cauchy's Formula to deduce that if $0 < a < 1$, then

$$\int_0^{2\pi} \frac{dt}{1+a^2-2a\cos t} = \frac{2\pi}{1-a^2}.$$

2 Let $f(z)$ be analytic on and inside a circle K defined by $|z-a| = \rho$, and let M be the maximum of $|f(z)|$ on K.

(i) Use (9.6) to show that

$$\left|f^{(n)}(a)\right| \leqslant \frac{n!M}{\rho^n}.$$

(ii) Suppose that $|f(z)| \leqslant M$ for all z, where M is some constant. By putting $n = 1$ in the above inequality, rederive Liouville's Theorem [p. 410].

(iii) Suppose $|f(z)| \leqslant M|z|^n$ for all z, where n is some positive integer. Show that $f^{(n+1)}(z) \equiv 0$, and deduce that $f(z)$ must be a polynomial whose degree does not exceed n.

3 (i) Show that if C is any simple loop round the origin, then

$$\binom{n}{r} = \frac{1}{2\pi i}\oint_C \frac{(1+z)^n}{z^{r+1}}\,dz.$$

(ii) By taking C to be the unit circle, deduce that

$$\binom{2n}{n} \leqslant 4^n.$$

For other interesting applications of complex analysis to problems involving binomial coefficients, see Bak and Newman (2010).

4 The *Legendre polynomials* $P_n(z)$ are defined by

$$P_n(z) = \frac{1}{2^n n!}\frac{d^n}{dz^n}\left[(z^2-1)^n\right].$$

These polynomials are important in many physical problems, including the quantum mechanical description of the hydrogen atom.

(i) Calculate $P_1(z)$ and $P_2(z)$, and explain why P_n has degree n.

(ii) Use (9.6) to show that

$$P_n(z) = \frac{1}{2\pi i}\oint_K \frac{(Z^2-1)^n}{2^n(Z-z)^{n+1}}\,dZ,$$

where K is any simple loop round z.

(iii) By taking K to be a circle of radius $\sqrt{|z^2-1|}$ centred at z, deduce that

$$P_n(z) = \frac{1}{\pi}\int_0^{\pi} (z+\sqrt{z^2-1}\cos\theta)^n \, d\theta.$$

(iv) Check that this last formula yields the same $P_1(z)$ and $P_2(z)$ as you obtained in part (i).

5 If C denotes the unit circle, show that

$$\int_0^{2\pi} \frac{\sin^2\theta}{5-4\cos\theta}\, d\theta = -\frac{i}{4}\oint_C \frac{(z^2-1)^2}{z^2(z-2)(2z-1)}\, dz = \frac{\pi}{4}.$$

6 Let f(z) be an analytic function with no poles on the real axis, and such that $|f(z)| < (\text{const.})/|z|^2$ for sufficiently large $|z|$. By integrating $f(z)\, e^{iz}$ along the contour (L + J) shown in [9.4a], deduce that

$$\int_{-\infty}^{+\infty} f(x)\cos x\, dx + i\int_{-\infty}^{+\infty} f(x)\sin x\, dx = 2\pi i \sum_{\text{upper half-plane}} \text{Res}\,[f(z)\, e^{iz}].$$

[*Hint:* First show that if $y > 0$, then $|e^{iz}| < 1$.]

7 Use the result of the previous exercise to do the following problems, in which we assume that $a > 0$.

(i) Show that

$$\int_{-\infty}^{+\infty} \frac{\cos x}{x^2+a^2}\, dx = \frac{\pi}{a} e^{-a}.$$

(ii) Evaluate

$$\int_{-\infty}^{+\infty} \frac{x\sin x}{(x^2+a^2)^2}\, dx.$$

8 Let $F_n(z) = 1/(1+z^n)$, where n is even.

(i) Use (9.11) to show that if p is a pole of F_n then $\text{Res}\,[F_n, p] = -(p/n)$.

(ii) With the help of part (i), show that the sum of the residues of F_n in the upper half-plane is a geometric series with sum $1/[in\sin(\pi/n)]$.

(iii) By applying the Residue Theorem to the contour (L + J) shown in [9.4a], deduce that

$$\int_0^{\infty} \frac{dx}{1+x^n} = \frac{\pi}{n\sin(\pi/n)}. \tag{9.15}$$

(iv) Although the above derivation breaks down when n is odd [why?], use a computer to verify that (9.15) is nevertheless still true.

9 Continuing from the previous question, consider the wedge-shaped contour K shown below.

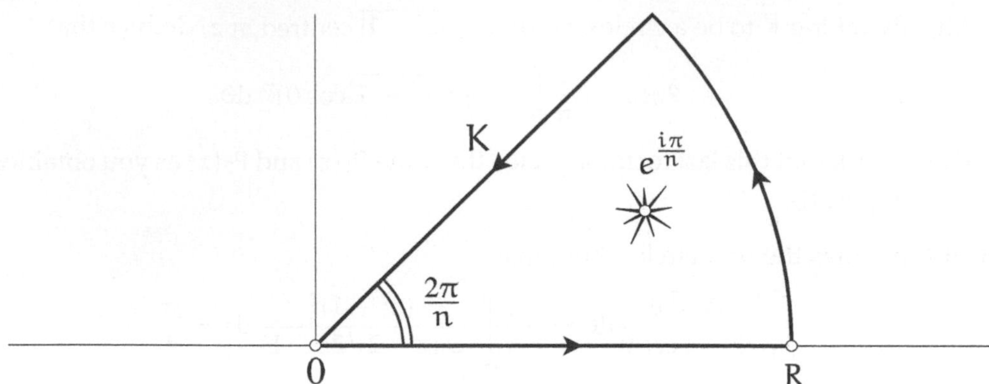

(i) Use the Residue Theorem to show that if $n = 2, 3, 4, \ldots$, and $R > 1$ (as illustrated), then

$$\oint_K \frac{dz}{1+z^n} = -\frac{2\pi i}{n} e^{i(\pi/n)}.$$

(ii) Show that

$$\lim_{R\to\infty} \oint_K \frac{dz}{1+z^n} = \left[1 - e^{i(2\pi/n)}\right] \int_0^\infty \frac{dx}{1+x^n}.$$

(iii) Deduce that (9.15) is indeed valid for odd n as well as even n.

10 Use (9.13) to show that $\sum_{n=1}^{\infty}(1/n^4) = (\pi^4/90)$.

11 Show that if $f(z)$ is an analytic function such that $|f(z)| < (\text{const.})/|z|^2$ for sufficiently large $|z|$, then

$$\sum_{n=-\infty}^{\infty} (-1)^n f(n) = -\pi \sum_{\text{poles of } f(z)} \text{Res}\,[f(z)\,\text{cosec}(\pi z)].$$

In this formula, it is understood that if any of the poles of $f(z)$ happen to be integers, then these values of n are excluded from the LHS.

12 Use the result of the previous question to do the following:

(i) Show that

$$1 - \frac{1}{4} + \frac{1}{9} - \frac{1}{16} + \cdots = \frac{\pi^2}{12}.$$

(ii) Find the sum of the series

$$\frac{1}{2} - \frac{1}{5} + \frac{1}{10} - \frac{1}{17} + \cdots + \frac{(-1)^{n+1}}{(n^2+1)} + \cdots.$$

13 (i) Show that

$$\sum_{n=-\infty}^{\infty} \frac{1}{z^2 - n^2} = \frac{\pi \cot \pi z}{z}.$$

(ii) Show that the previous equation can be rewritten as

$$\cot z = \frac{1}{z} + \sum_{n=1}^{\infty} \frac{2z}{z^2 - n^2\pi^2}.$$

(iii) Show that the previous equation can be rewritten as

$$\frac{d}{dz}\left[\ln(\sin z/z)\right] = \sum_{n=1}^{\infty} \frac{d}{dz} \ln(z^2 - n^2\pi^2).$$

(iv) By integrating along any path from 0 to z that avoids integers, and then exponentiating both sides of the resulting equation, deduce that

$$\sin z = z\left(1 - \frac{z^2}{\pi^2}\right)\left(1 - \frac{z^2}{2^2\pi^2}\right)\left(1 - \frac{z^2}{3^2\pi^2}\right)\cdots.$$

[*Hint:* Recall that $\lim_{z\to 0}(\sin z/z) = 1$.]
This famous formula was discovered by Euler in 1734, in a much simpler way, via an extraordinary leap of imagination; see Stillwell (2010) for the details.

(vi) Multiplying out the brackets on the right, we see that the z^3 terms arise from taking a z^2 term from a single bracket, and 1's from all the others. But, on the left hand side, the power series for $\sin z$ is known. Thus,

$$z - \frac{1}{3!}z^3 + \cdots = z - \left[\frac{1}{\pi^2} + \frac{1}{2^2\pi^2} + \frac{1}{3^2\pi^2} + \frac{1}{4^2\pi^2} + \cdots\right] z^3 + \cdots.$$

Euler was thereby able to sum a series that had baffled the greatest mathematicians that had come before him, making the wonderful and startling discovery that

$$1 + \frac{1}{2^2} + \frac{1}{3^2} + \frac{1}{4^2} + \cdots = \frac{\pi^2}{6}.$$

CHAPTER 10

Vector Fields: Physics and Topology

10.1 Vector Fields

10.1.1 Complex Functions as Vector Fields

Throughout the course of this book we have relied on a *single* means of visualizing a complex function, namely, as a mapping of points in one complex plane to points in another. This idea has proved to be extremely powerful, for in terms of it the complex derivative is nothing more complicated than a local amplitwist. Despite its many virtues, in this chapter we shall abandon the mapping paradigm and introduce a completely new one in its place, thereby gaining a host of fresh insights into the subject and revealing surprising connections with physics.

The new picture of a complex function $f(z)$ involves only a single complex plane. As before, the variable z is thought of as a point in this plane, but now comes the new idea: *the value of* $f(z)$ *is pictured as a vector emanating from* z. The resulting diagram of points with attached vectors is called the *vector field* of f. Figures [10.1a] and [10.1b] illustrate the vector fields of z^2 and $(1/z)$, respectively; before reading further you should study them carefully and convince yourself of their correctness. Try doing a sketch of the vector fields of some other powers, then compare them with accurate ones done by your computer. Also use the computer to examine the vector fields of e^z, $\log z$, and $\sin z$.

The vector field concept remedies a significant defect in the mapping point of view. Although we can learn a lot about a mapping by looking at the images of specific shapes, we do not get a feel for its overall behaviour. But if we let our eyes roam over the vector field of a complex function we *do* get such a view, in much the same way as we can survey the behaviour of a real function by scanning its graph.

Just as a complex mapping determines a vector field, so a vector field determines a mapping—the two concepts are equivalent. More explicitly, given the vector V issuing from the point z, we translate the tail of V to the origin and define the image of z to be the point at the tip.

Visual Complex Analysis. 25th Anniversary Edition. Tristan Needham, Oxford University Press.
 DOI: 10.1093/oso/9780192868916.003.0010

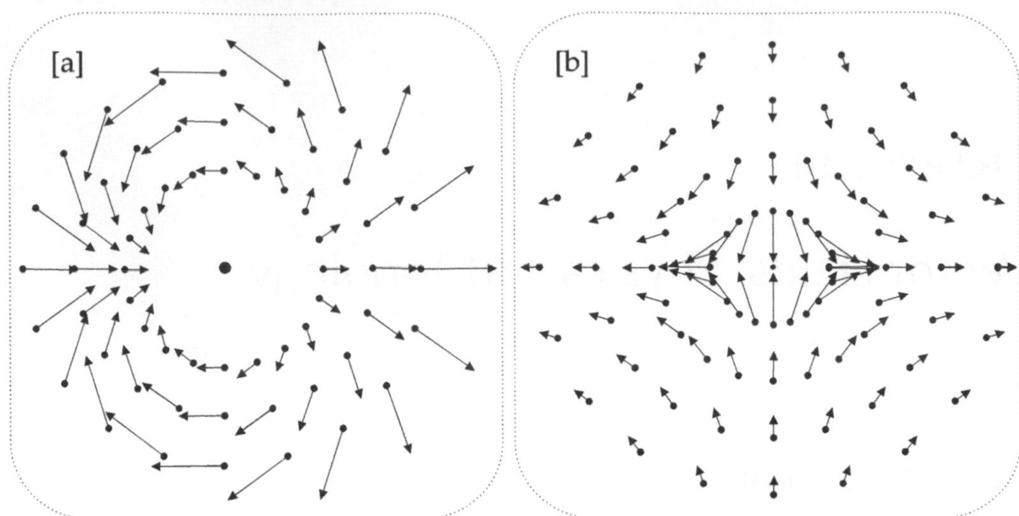

[10.1] Vector field representation of $f(z)$: *Draw the complex number $f(z)$ emanating from z.* **[a]** *The vector field of $f(z) = z^2$.* **[b]** *The vector field of $f(z) = (1/z)$.*

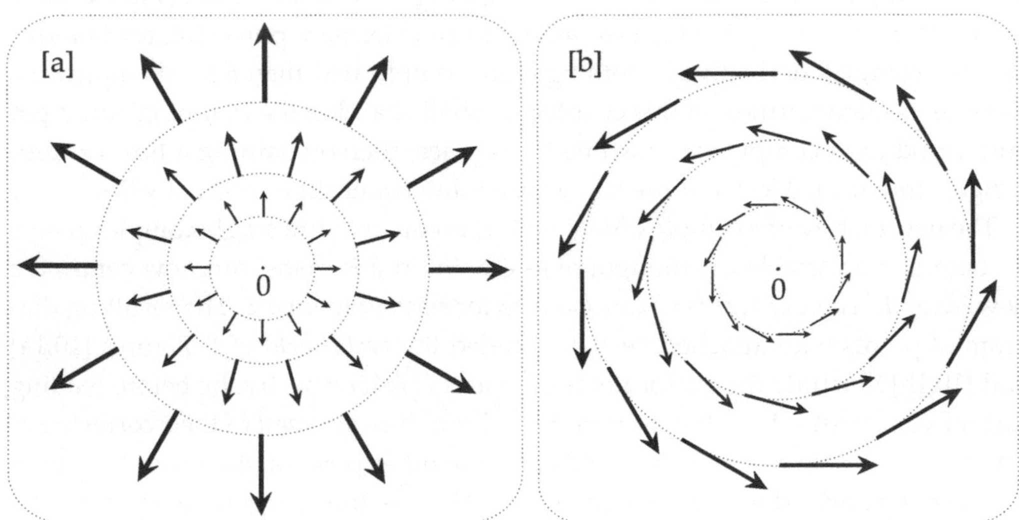

[10.2] A vector field defines a mapping. [a] *This vector field corresponds to the mapping $z \mapsto (1/2)z$.* **[b]** *This vector field corresponds to the mapping $z \mapsto (1/2)iz$.*

Consider the examples in [10.2a] and [10.2b]. If z lies on a circle of radius r then the vector field in [10.2a] is radial and has length $(r/2)$; in [10.2b] the vector field has the same length but is tangential instead of radial. Check that when viewed as mappings, [10.2a] corresponds to an expansion of the plane by $(1/2)$, while [10.2b] corresponds to the same expansion followed by (or preceded by) a rotation through a right angle.

If the vectors in [10.2a] were instead directed inwards, what would the corresponding mapping be? If the vectors in [10.2b] were flowing clockwise, what would the corresponding mapping be?

10.1.2 Physical Vector Fields

Since a vast range of physical phenomena find their most natural description as vector fields, the potential utility of the new way of looking at complex mappings should be obvious.

For example, consider the astonishingly complex array of electromagnetic disturbances zipping through the space around you. The visible light carrying these words to your retina, the totality of television and radio programs simultaneously being broadcast to your home—all this constitutes only a small part of the frenzied activity. But it is a remarkable fact that this great tangle of signals is in fact completely described by just two vector fields! At each instant of time t there is an electric vector $\mathbf{E}(p, t)$ and a magnetic vector $\mathbf{B}(p, t)$ emanating from each point p in space, and these two vector fields constitute the complete description of the electromagnetic field.

If we are to describe such physical vector fields with complex mappings, two problems immediately present themselves. A television set is fixed in space, and the way in which it produces its picture is by monitoring how the electromagnetic vector fields at its location *vary in time*. But a complex mapping is a timeless thing—it assigns the vector $f(z)$ to the point z *once and for all*. This is the first of our two problems. Thus if we are not to radically alter our conception of a complex mapping, the only types of physical vector fields we can describe in this manner are those that do *not* vary with time. We shall call such vector fields *steady*.

Fortunately, steady vector fields are both common and important in physics. For example, the unwavering character of the orbits of the planets reflects the fact that the gravitational field of the sun does not vary with time. In fact Newton informs us that this time-independent force on a particle of unit mass located at a point p in space may be represented by a vector emanating from p, directed to the centre c of the sun, and with a length equal to $M/[cp]^2$, where M is the mass of the sun. Drawing these vectors throughout space we obtain a steady vector field.

The above examples of the electromagnetic and gravitational vector fields illustrate our second problem—they exist in *three*-dimensional space, whereas the complex plane can only accommodate a *two*-dimensional vector field. There is no getting around this problem, but once again it is fortunate that there are certain important types of physical phenomena which are intrinsically two-dimensional in nature, and which can therefore be described in the complex plane. Let us begin with the flow of electricity within a sheet of conducting material.

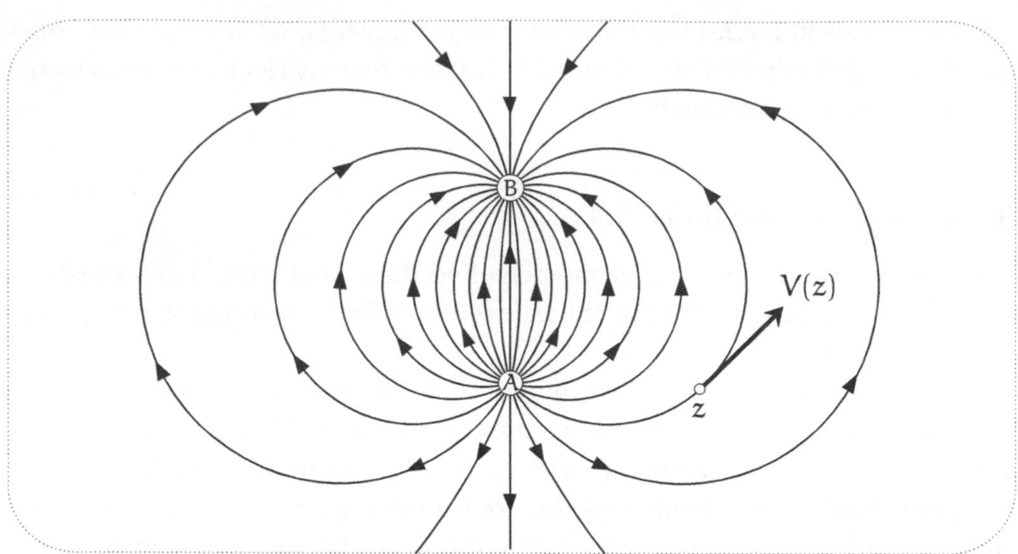

[10.3] The Phase Portrait of $\mathbf{V}(z)$ *is the set of streamlines along which V flows: mathematically, they are the curves that have V as their tangent vectors. Given such a picture, one can immediately deduce the* direction *of* $V(z)$ *at* z*, but its* magnitude *is invisible. However, in many physical examples it is possible to draw the phase portrait in a special way, such that* the closer together the streamlines, the stronger the flow. *This is the case here, where electrodes are attached to a metal plate at A and B, and electricity flows between them—you see that the spacing is smallest, and therefore the flow greatest, along* AB. *Note that the streamlines are arcs of* circles; *this is proved geometrically in the next figure.*

Take two wires and connect them to a battery, then touch the ends to two points A and B of a thin copper plate. Almost instantly a steady flow of electric current from one electrode to the other will be set up in the plate. See [10.3]. At each point z of the plate we now represent this flowing current by a time-independent vector in the direction of the flow, and with a length equal to the strength of the current there. Picturing the plate as a portion of $\mathbb{C}$, the flow is thus expressed as a complex function $V(z)$.

Rather than drawing the actual vector field in [10.3] we have instead shown the paths along which the electricity flows. Such a picture is called the *phase portrait* of the vector field, and the directed curves along which the flow occurs are called the *integral curves* or *streamlines* of the vector field. As illustrated, the streamlines of this example are in fact arcs of circles connecting the two electrodes. We shall justify this shortly.

Phase portraits are easy to take in visually and are thus a common way of representing vector fields. By definition the vector field is everywhere tangent to the streamlines, and thus its direction can be recovered from the phase portrait. On the other hand, it would seem that a phase portrait would necessarily fail to include the information about the *lengths* of the vectors. This is true in general, but for many vector fields that arise in physics it will be shown that there exists a special way of drawing the phase portrait so that the strength of the flow is manifested as the crowding together of the streamlines: *the closer together the streamlines, the stronger the flow*[1]. Later we shall explain this idea in detail, but for the moment we remark that [10.3] has actually been drawn in this special way. For example, as we approach the line-segment connecting the electrodes the streamlines become more and more crowded together, corresponding to a stronger and stronger current.

10.1.3 Flows and Force Fields

One and the same vector field (or phase portrait) can represent many quite different physical phenomena. For example, reconsider the copper plate in [10.3], and imagine that it is now sandwiched between two layers of material which do not conduct heat. Remove the electrodes and instead of supplying electricity at a constant rate at A, let us supply heat. Likewise, let us remove heat at the same constant rate at B. After a short time a steady pattern of heat flow from A to B will be established within the copper plate. In this steady state we may assign to each point a vector in the direction that the heat is flowing there, and having a length equal to the intensity of the heat flow.

Remarkably, in this steady state the physical laws governing the behaviour of the heat are identical to those which previously described the electricity, and thus the phase portrait [10.3] for the electric current is *also* the phase portrait for the new heat flow.

Here is yet another interpretation of [10.3]. In attempting to understand the flow of real liquids, such as water, it is helpful to consider an idealized fluid with the properties of being frictionless, incompressible, and "irrotational"—the precise meaning of the last term will be explained later. Imagine a thin layer of such an ideal fluid sandwiched between two horizontal plates, one of which has two small holes, A and B. If we now connect the holes with a fine tube that passes through a pump, a steady flow will be set up in the layer of fluid, and at each point we may draw its velocity vector. The phase portrait of this steady vector field is *once again* given by [10.3]!

Although there are certainly important differences between these three interpretations of [10.3], we may nevertheless lump them together in one class, for they are all flows of something. Whether it is electricity, heat, or liquid, in each case the vector field can be thought of as the velocity of flowing "stuff", and the streamlines are the paths along which this stuff flows.

A physically quite distinct class is comprised of *force fields*. For example, although we previously discussed how the gravitational field of the sun could be represented as a steady vector field, the vector at a point in space is no longer the

[1] Faraday was the first to conceive of vector fields in this way; Maxwell then rendered the idea mathematically precise and exploited it to the hilt.

velocity of some flowing substance, rather, it represents the *force* experienced by a unit mass placed there. In the context of force fields, integral curves are called *lines of force* rather than streamlines. Here the lines of force are rays coming out of, or rather entering, the centre of the sun. Although this force field is three-dimensional, spherical symmetry[2] means that it will be the same on any plane drawn through the centre of the sun. It can therefore be completely described by a complex function.

Although there is nothing actually flowing along the lines of force, we can switch back to the flow point of view by *pretending* that there is, thereby interpreting a force field as the velocity field of a flowing substance. This is not mere sophistry: it is a remarkable fact that for the most common and important force fields (e.g., gravitational and electrostatic) *this imaginary flowing substance behaves exactly like our previously considered ideal fluid.*

To illustrate this, we turn to an example in electrostatics: equal and opposite charges (per unit length) are induced on two long wires which are then held parallel to each other in empty space. To each point in space we now attach the force vector that a unit electric charge would experience there; this force field is (by definition) the electric field E, and its phase portrait is the same on each plane drawn perpendicular to the wires. Taking [10.3] to be such a plane, with the wires piercing through at A and B, the phase portrait of this force field is exactly the one shown there for the flow of ideal fluid.

10.1.4 Sources and Sinks

In order to make a quantitative analysis of [10.3], we introduce the concepts of (two-dimensional) sources and sinks. Thinking in terms of our layer of ideal fluid, a *source* of *strength* S is a point at which we pump in S units of fluid per unit of time. Figure [10.4a] illustrates the symmetric velocity vector field $V(z)$ of an isolated source at the origin.

Given a curve (open or closed) in a general flow, the amount of fluid flowing across it in each unit of time is called the *flux*. Clearly the flux across an element of the curve is just its length times the component of the velocity perpendicular to the curve. The total flux across the curve is then the sum (i.e., integral) of these elementary fluxes. Returning to the specific case of [10.4a], our assumption of incompressibility says that the flux across any simple loop round 0 must be the same as the amount of fluid S being pumped in at 0. Since the flow is orthogonal to the origin-centred circle C of radius r, we deduce that

$$2\pi r\,|V| = S.$$

Writing $z = r\,e^{i\theta}$, we find that the vector field of the source is therefore

[2] This is an idealization—like the earth, the sun is somewhat flattened at the poles.

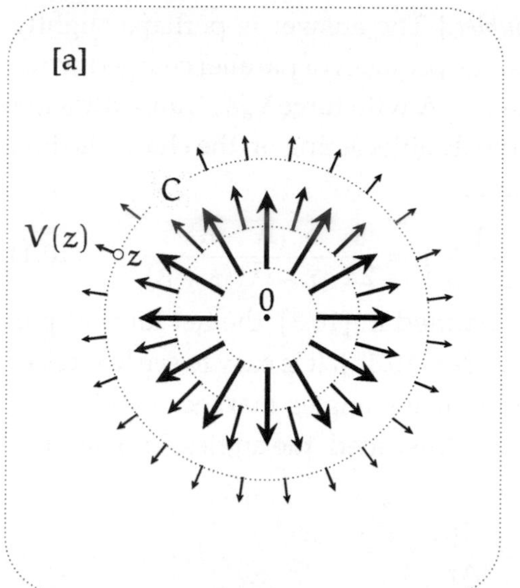

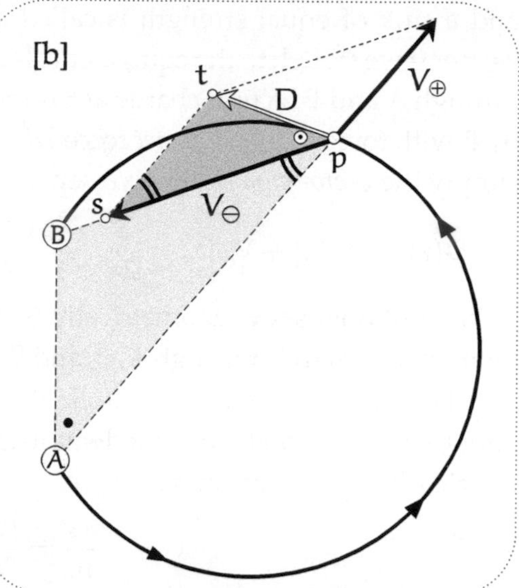

[10.4] [a] A Source. *Fluid is pumped in at a constant rate at 0 and travels outward symmetrically. Its velocity at* $z = r\,e^{i\theta}$ *must be proportional to* $(1/r)$*, for the total amount of fluid crossing a circle of radius* r *per unit of time (the* **flux***) will be this velocity multiplied by* $2\pi r$*.* **[b] The streamlines of a doublet are arcs of circles that pass through the source and the sink.** *To prove that* $D = V_{\oplus} + V_{\ominus}$ *is tangent to the illustrated arc of the circle* ApB*, we must prove that* $\bullet = \odot$*. The angles* ApB *and* pst *are clearly equal, but we also have* $(ts/ps) = |V_{\oplus}|/|V_{\ominus}| = Bp/Ap$*. So, the two shaded triangles are similar, and therefore* $\bullet = \odot$*. Done.*

$$V(z) = |V|\,e^{i\theta} = \frac{S}{2\pi}\left(\frac{e^{i\theta}}{r}\right) = \frac{S}{2\pi}\left(\frac{1}{\overline{z}}\right).$$

[We note without proof that this is also the electric field on a plane at right angles to a very long wire carrying a uniform charge of S per unit length.] The source in [10.3] is at A instead of at the origin, and so it is described by

$$V_{\oplus}(z) = \frac{S}{2\pi}\left(\frac{1}{\overline{z} - \overline{A}}\right).$$

A *sink* may be thought of as a source with a negative strength: it is a place where fluid is pumped out rather than in. In each of the flow experiments which [10.3] purports to describe, the sink at B has the same strength as the source at A, and so its vector field is

$$V_{\ominus}(z) = -\frac{S}{2\pi}\left(\frac{1}{\overline{z} - \overline{B}}\right).$$

We now know the vector fields $V_{\oplus}(z)$ and $V_{\ominus}(z)$ which would be produced by the source or the sink in [10.3] if each were present on its own, but what is the flow when they are both present together? [Incidentally, this combination of a source and a sink of equal strength is called a *doublet*.] The answer is perhaps slightly clearer if we switch to the equivalent electrostatic problem of parallel charged wires through A and B. A unit charge at z is repelled by A with force $V_{\oplus}(z)$, and attracted by B with force $V_{\ominus}(z)$. The *net* force D(z) of the doublet acting on the charge is then simply the *vector sum* of the two separate forces:

$$D(z) = V_{\oplus}(z) + V_{\ominus}(z) = \frac{S}{2\pi}\left(\frac{1}{\overline{z}-\overline{A}} - \frac{1}{\overline{z}-\overline{B}}\right) = \frac{S}{2\pi}\,\frac{(\overline{A}-\overline{B})}{(\overline{z}-\overline{A})(\overline{z}-\overline{B})}. \tag{10.1}$$

We will now show geometrically that, as claimed in [10.3], the net force at p is tangent to the circle through A, p, and B. Consider [10.4b]. It is easy to see [exercise] that D will be tangent to the circle if and only if the angles marked • and ⊙ are equal, so this is what we must demonstrate. As illustrated, the angles ApB and pst are clearly equal. But we also have

$$\frac{ts}{ps} = \frac{|V_{\oplus}|}{|V_{\ominus}|} = \frac{Bp}{Ap}.$$

Thus the two shaded triangles are similar, and therefore • = ⊙, as was to be shown.

10.2 Winding Numbers and Vector Fields*

10.2.1 The Index of a Singular Point

Let us confine all our discussions to vector fields for which the direction is well-defined and continuous at all but a finite number of points. These exceptional places, where the vector field vanishes or becomes infinite, are called *singular points*[3]. They are easy to spot in a phase portrait, usually as the intersection points of distinct streamlines. Figure [10.5] shows the phase portraits in the vicinity of some simple types of singular points, together with their names and their "indices"—a term which we must now explain.

Figure [10.6] is a magnified view of the simple crosspoint (also called a *saddle point*) shown in [10.5]. Round this singular point s we have drawn a simple loop Γ_s, and at some of its points we have also drawn the vectors V. Since Γ_s does not pass through any singular points, the direction of V is well-defined and continuous everywhere on it. Thus we can count the net number of revolutions of V(z) as z traverses Γ_s. We call this number the *index* $\mathcal{I}_V[\Gamma_s]$ of the loop Γ_s with respect to the vector field V. When it is clear which vector field is being considered, we may simplify this notation to $\mathcal{I}[\Gamma_s]$. For example, in [10.6] we see that $\mathcal{I}[\Gamma_s] = -1$. Note that we have drawn the vectors on Γ_s only to make it easier to see the value of the index; actually, since only the *directions* of the vectors are required, the phase portrait is sufficient on its own.

[3] Otherwise known as *critical points* or *singularities*—terms to which we have already attached different meanings.

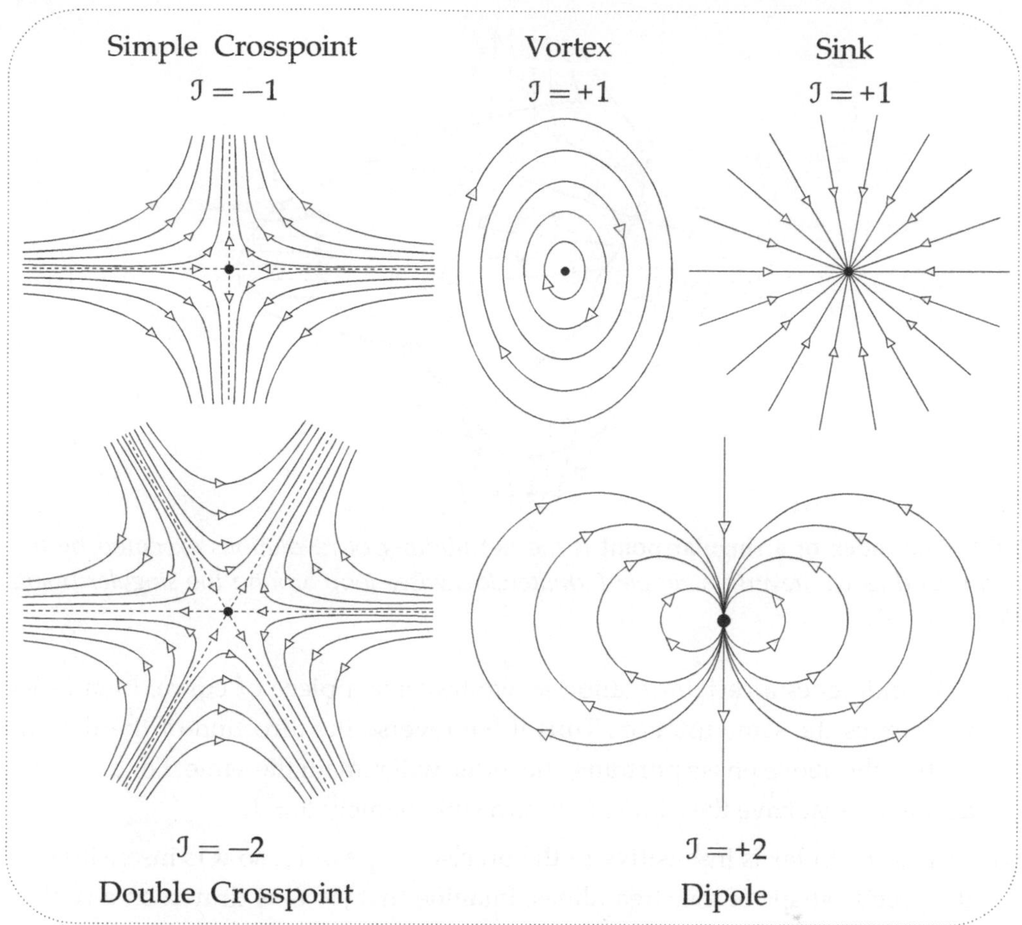

[10.5] The values of the index ($\mathcal{I}$) for various singular points. *The geometric definition of $\mathcal{I}$ is explained in the next figure.*

If we continuously deform Γ_s without crossing s (or any other singular point) then the value of $\mathcal{I}[\Gamma_s]$ will also vary continuously, and since it's an integer, it will therefore remain constant. Thus we may unambiguously define the *index of a singular point* s to be the index of any loop that winds round s once, but does not wind round any other singular points. It should not cause any confusion if we abuse our notation slightly and call this index $\mathcal{I}(s)$. Applying this definition to loops of your choosing, you may now verify each of the given values of $\mathcal{I}$ in [10.5].

Before moving on, we observe three properties of the index:

(i) There is nothing to stop us applying the above definition to a *non*-singular point, but in this case the index must vanish. Choosing Γ_s to be a very small loop, the non-singular nature of s implies that all the vectors on Γ_s will point in roughly the same direction, and so $\mathcal{I}(s) = 0$.

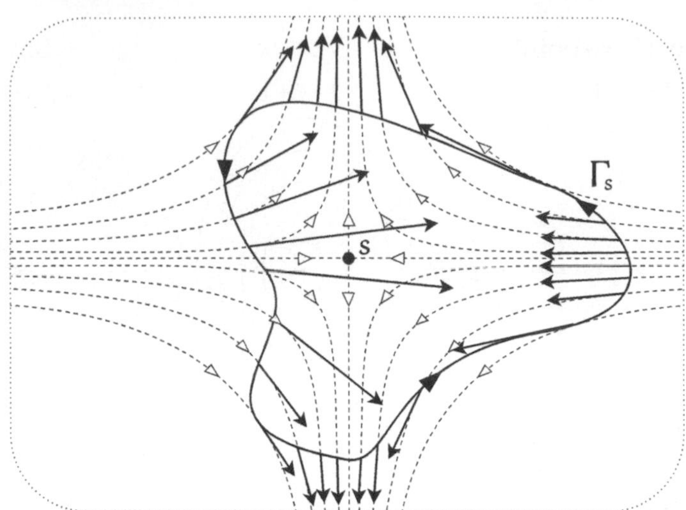

[10.6] The index of a singular point *is the net number of revolutions executed by the vector field as we traverse a simple (counterclockwise) loop around the singular point. Here,* $\mathcal{I} = -1$.

(ii) If V undergoes a certain rotation as we traverse a piece of curve, then $(-V)$ undergoes the same rotation. Thus if we reverse the direction of the flow in each of the above phase portraits, the index will remain the same. For example, a source must have the same index as a sink, namely, $\mathcal{I} = 1$.

(iii) Just as the index is insensitive to the precise shape of Γ_s, so it is insensitive to the precise shape of the streamlines. Imagine that [10.6] is drawn on a rubber sheet which we gradually stretch, so producing a new distorted phase portrait. The direction of V at each point of Γ_s will undergo a continuous change, and so its net revolutions upon traversing Γ_s will likewise vary continuously. The index must therefore remain constant.

Clearly, our new concept of "index" is related to our old concept of "winding number", but how? If we instead think of V as a mapping, sending the points of Γ_s to those of a new loop $V(\Gamma_s)$, then a moment of thought reveals that the index of Γ_s is just a new interpretation of the winding number of its image loop:

$$\mathcal{I}_V[\Gamma_s] = \nu[V(\Gamma_s), 0]. \tag{10.2}$$

This makes it clear [see p. 397] that the index $\mathcal{I}(s)$ of a point s is the same thing as its *topological multiplicity* $\nu(s)$ as a preimage of 0. In particular, if V is analytic then

$$\mathcal{I}\,(\text{root of order } n) = n \quad \text{and} \quad \mathcal{I}\,(\text{pole of order } m) = -m.$$

Check this for the examples in [10.1] .

If you have not done so yet, we urge you to use a computer to draw the vector fields of some simple polynomials and rational functions. Notice how roots and poles show up just as vividly as the corresponding x-intercepts and vertical asymptotes occurring in the graph of a real function. Notice how easy it is to zoom in on the vector field to find their precise locations. But now we can do much better than the real case, for we can also immediately read off the *precise nature* of a root or singularity simply looking at its index!

The following example illustrates the fact that a vector field contains *more* information than an ordinary graph. If we sketch the graphs of

$$F(x) = \frac{(x-1)^2}{(x+2)^3} \quad \text{and} \quad G(x) = \frac{(x-1)^4}{(x+2)^7}$$

the results will be qualitatively the same: both look something like a parabola near $x = 1$; both have branches going to opposite ends of the vertical asymptote at $x = -2$; both look something like $(1/x)$ for large x.

Now use the computer to draw the corresponding vector fields when x is replaced by z. Striking indeed are the differences! As we traverse a small loop around the root at $z = 1$, F makes two positive revolutions while G makes four; doing the same at the pole $z = -2$, F makes three negative revolutions while G makes seven; and on a very large origin-centred circle, F makes one negative revolution while G makes three.

Returning to the general significance of (10.2), consider the ordinary winding number $\nu[L, 0]$ of a loop L. This can now be viewed as the index of L with respect to the vector field of the identity mapping:

$$\nu[L, 0] = \mathcal{I}_z[L].$$

Figure [10.7] illustrates this result with $\mathcal{I}_z[L] = 1$. The winding number of L around a general point a is likewise just its index with respect to the vector field $(z - a)$:

$$\nu[L, a] = \mathcal{I}_{(z-a)}[L].$$

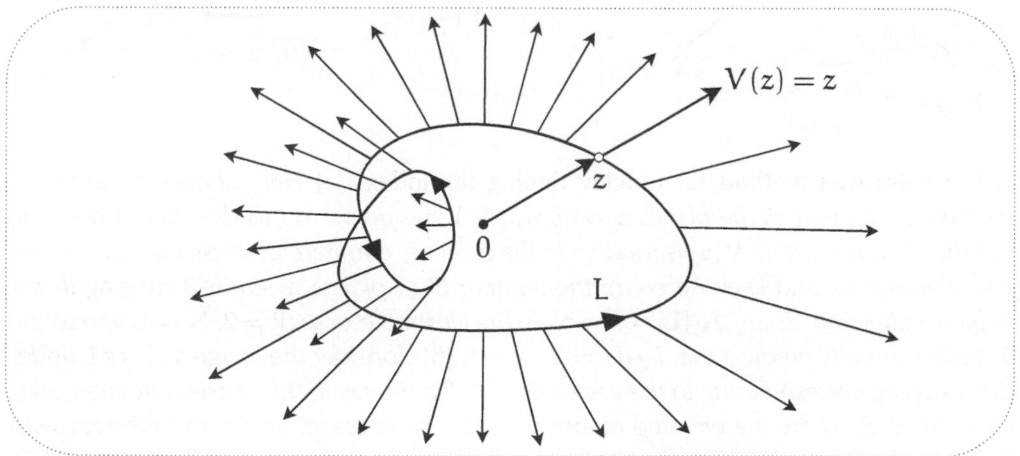

[10.7] Index of the vector field $V(z) = z$ is the winding number of the loop: $\nu[L, 0] = \mathcal{I}_z[L]$.

10.2.2 The Index According to Poincaré

Figure [10.8a] shows a loop L and a vector field V evaluated on it. Let us use this simple example (for which it is obvious that $\mathcal{I}_V[L] = 1$) to explain a quick method (due to Poincaré) of finding the index in more complicated cases.

Consider all the places on L (a, b, c in our case) where V points in one arbitrarily chosen direction. Let P be the number of these places at which V(z) rotates in the positive sense as z passes through it, and let N be the number at which it rotates in the negative sense. Even in relatively complicated cases, P and N are usually quick and easy to find. We now obtain the index as the difference of these two numbers:

$$\mathcal{I}_V[L] = P - N. \tag{10.3}$$

In our case $P = 2$ because of the positive rotation at a and c, and $N = 1$ because of the negative rotation at b. Thus $\mathcal{I}_V[L] = 1$, as it should.

To see just how fast and efficient (10.3) is to use, try it out on each of the examples in [10.5], choosing L to be a counterclockwise circle centred at s, and choosing the direction to be vertically upward. For extra practice, try it again with a different choice of direction.

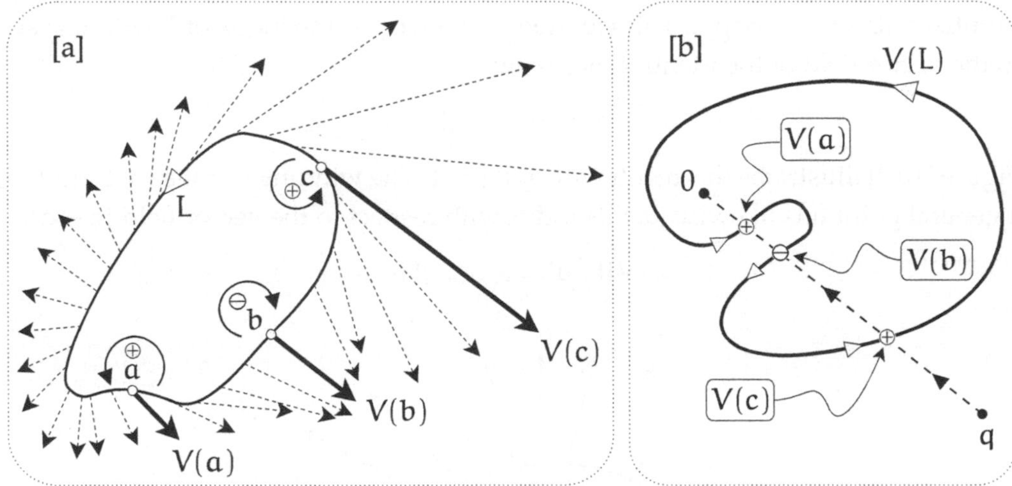

[10.8] Poincaré's method for quickly finding the index. [a] *First, choose an arbitrary direction, then find all the places z_j on L where $V(z_j)$ points in this direction. Count the number P of z_j's where V(z) is rotating in the positive, counterclockwise direction as we pass through z_j, and likewise count the number N of places where it is rotating in the negative direction. Then,* $\mathbf{\mathcal{I}_V[L] = P - N}$. *In the illustrated case, $P = 2$, $N = 1$, and so the formula correctly predicts that $\mathcal{I}_V[L] = 2 - 1 = 1$.* **[b]** *Consider the image V(L) of L under the mapping corresponding to the vector field V. On the ray in the chosen direction, start far away at q, where the winding number $\nu = 0$. The intersection points of the ray with V(L) are the images $V(z_j)$ of the z_j's, and a positive rotation of V at z_j corresponds to a left-to-right crossing, so the* crossing rule *(7.1) increases ν by one. Likewise, a negative rotation of V reduces ν by one. Thus we have proved Poincaré's formula.*

Although the truth of (10.3) is probably clear at an intuitive level, it is nevertheless instructive to deduce it from the "crossing rule" (7.1), p. 388, for computing winding numbers.

Figure [10.8b] shows the image $V(L)$ of L when V is viewed as a mapping. [Check that it really *is* the image!] In these terms, the required index is just $\nu[V(L), 0]$. Draw the ray from 0 in the previously chosen direction, and let the point q travel along it (starting far away), ending up at the origin. In its journey, q will thus cross $V(L)$ at the points $V(c)$, $V(b)$, and $V(a)$. In the vector field picture [10.8a], the positive rotation of V at c now implies (in [10.8b]) that q sees $V(L)$ directed from left to right as it approaches the first crossing at $V(c)$. Oppositely, the negative rotation at b implies that $V(L)$ is directed from right to left as q approaches $V(b)$. But as we previously argued in Chapter 7, $\nu[V(L), 0]$ is the number of points P at which $V(L)$ is directed from left to right (as seen by q as it approaches), minus the number of points N at which it is directed from right to left. Done.

10.2.3 The Index Theorem

With the connection between indices and winding numbers established, the Topological Argument Principle can be reinterpreted in terms of vector fields:

The index of a simple loop is the sum of the indices of the singular points it contains.

Using a neater argument than the one given in Chapter 7, we can now extend this theorem to *multiply connected* regions. As illustrated in [10.9], recall that this means that the region has holes in it; two in our case.

The shaded region consists of the points which are inside C and outside B_1 and B_2. In general there could be more holes, say g of them, with counterclockwise boundary curves $B_1, B_2, \ldots, B_g$. As illustrated, suppose that we have a vector field on such a region, and let $s_1, s_2, \ldots, s_n$ be the singular points within the region. In our case there are only two: s_1 is a dipole, and s_2 is a saddle point. The generalization of the Argument Principle is this:

$$\mathcal{I}[C] - \sum^{g} \mathcal{I}[B_j] = \sum^{n} \mathcal{I}[s_j].$$

It is called the *Index Theorem*.

Perhaps using (10.3), for practice, verify that in our example, $\mathcal{I}[C] = 2$, $\mathcal{I}[B_1] = 0$, and $\mathcal{I}[B_2] = 1$, so that the LHS of the Index Theorem equals 1. But the RHS is

$$\mathcal{I}(\text{dipole}) + \mathcal{I}(\text{saddle point}) = 2 + (-1) = 1,$$

so confirming the prediction of the theorem in this case.

To understand this result, consider [10.10]. Using the dashed curves, break the region into curvilinear polygons in such a way that each one contains at most one

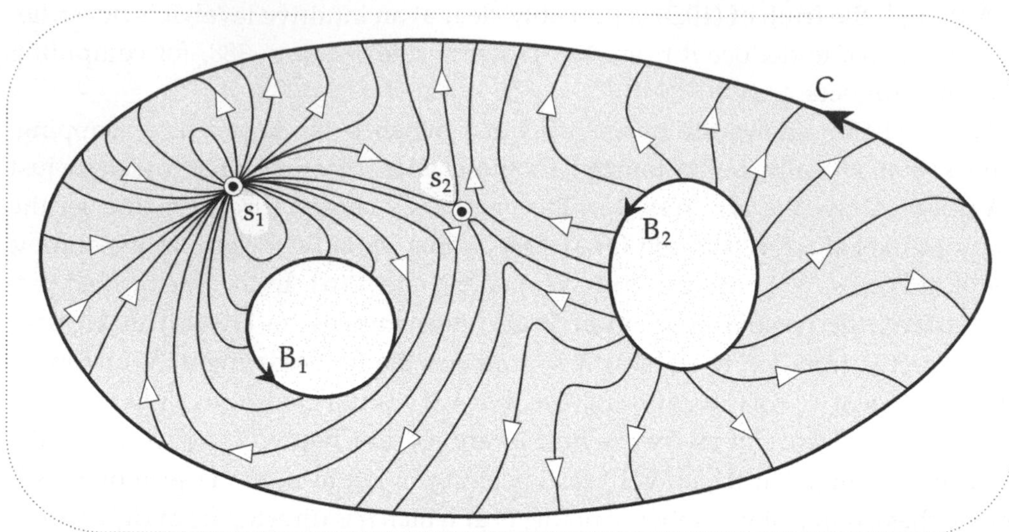

[10.9] The Index Theorem: $\mathfrak{I}[C] - \sum^{g} \mathfrak{I}[B_j] = \sum^{n} \mathfrak{I}[s_j]$. *This is certainly true in the illustrated example, for $\mathfrak{I}[C] - \mathfrak{I}[B_1] - \mathfrak{I}[B_2] = 2 - 0 - 1 = 1$, and $\mathfrak{I}[s_1] + \mathfrak{I}[s_2] = 2 + (-1) = 1$.*

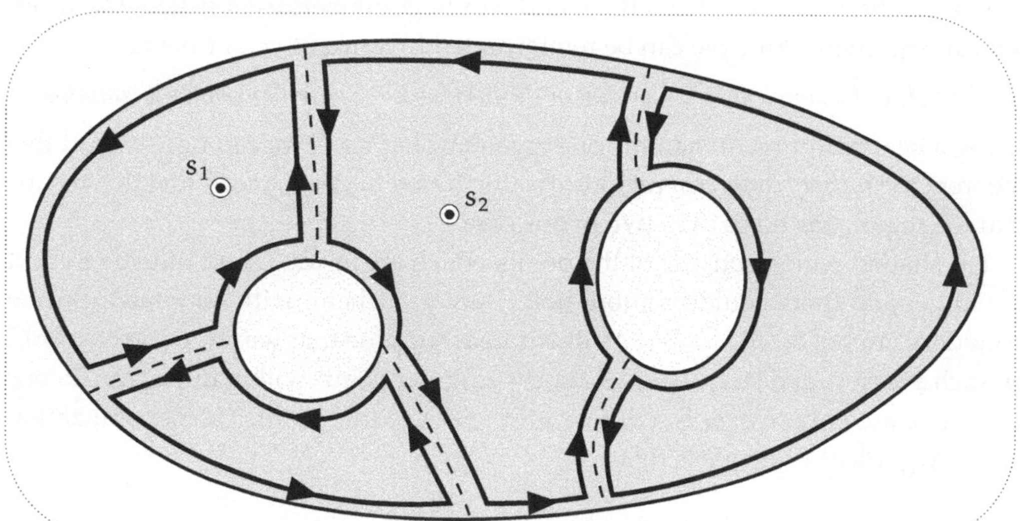

[10.10] Proof of the Index Theorem. *Using the dashed curves, break the region into curvilinear polygons in such a way that each one contains at most one singular point, and let their counterclockwise boundaries be K_j. If we sum the indices of all the K_j's then we obtain the RHS of the Index Theorem. On the other hand, the index of a single K_j is obtained by looking at how much the vector field rotates as one travels along each edge of K_j, then adding up these net rotation angles. But when we sum the indices of all the K_j's, each interior edge [dashed] is traversed twice, once in each direction, and the associated angles of rotation therefore* cancel. *The remaining edges of the K_j's together make up C and $-B_1$, $-B_2$, etc. Summing the associated angles of rotation (divided by 2π) therefore yields the LHS of the Index Theorem. Done.*

singular point, and let their counterclockwise boundaries be K_j. If we sum the indices of all the K_j's then we obtain the RHS of the Index Theorem. For if K_j does not contain a singular point then its index vanishes, while if it does contain one then its index is (by definition) the index of that singular point.

On the other hand, the index of a single K_j is obtained by looking at how much the vector field rotates as one travels along each edge of K_j, then adding up these net rotation angles. But when we sum the indices of all the K_j's, each interior edge [dashed] is traversed twice, once in each direction, and the associated angles of rotation therefore *cancel*. The remaining edges of the K_j's together make up C and $-B_1$, $-B_2$, etc. Summing the associated angles of rotation (divided by 2π) therefore yields the LHS of the Index Theorem. Done.

10.3 Flows on Closed Surfaces*

10.3.1 Formulation of the Poincaré–Hopf Theorem

If a curved surface S in space is "smooth" in the sense that there exists a tangent plane at each of its points, then it makes sense to speak of a vector field that is everywhere tangent to S. Intuitively, we may picture such a vector field as the velocity of a fluid that is flowing over S.

Figure [10.11] shows the streamlines of two such flows on the sphere. Notice that both possess singular points: [10.11a] has two vortices, while [10.11b] has a dipole. In fact there can be *no* vector field on the sphere that is free of singular points. This is

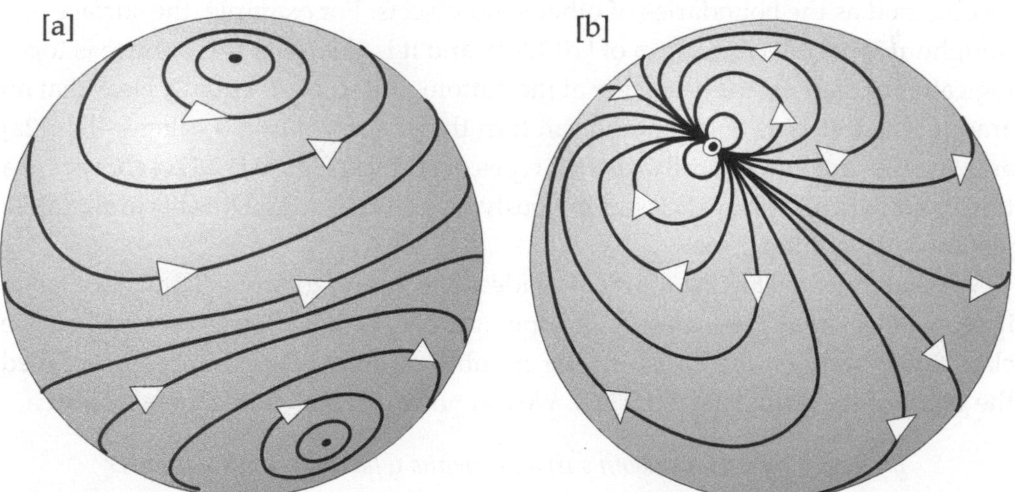

[10.11] Poincaré–Hopf Theorem: If a vector field on a smooth surface of genus g has only a finite number of singular points, then the sum of their indices is $\chi = (2-2g)$. *Two examples:* **[a]** $\mathcal{J}\,(vortex) + \mathcal{J}\,(vortex) = 1 + 1 = 2 = \chi(sphere)$. **[b]** $\mathcal{J}\,(dipole) = 2 = \chi(sphere)$.

one consequence of an extremely beautiful result called the *Poincaré–Hopf Theorem*, the formulation of which we will now sketch.

It is not immediately obvious how to give a precise definition of the "index" of a singular point on a curved surface, but for the moment let us accept that this integer exists, and that its value is the same as for an analogous singular point in the plane. Thus if we sum all the indices in [10.11a] we obtain

$$\mathcal{I}\,(\text{vortex}) + \mathcal{I}\,(\text{vortex}) = 1 + 1 = 2,$$

while if we do the same for [10.11b] we obtain

$$\mathcal{I}\,(\text{dipole}) = 2.$$

Try drawing your own streamlines on an orange, then sum the indices of the singular points. Is this a coincidence??

There *are* no coincidences in mathematics! In the case of the sphere, the Poincaré–Hopf Theorem states that if we sum the indices of *any* vector field on its surface, we will always get 2 for the answer. Indeed, it says that we will get this answer for any surface that is *topologically* a sphere, that is to say, any surface into which the sphere may be changed by a continuous and invertible transformation. If we imagine the sphere to be made of rubber, examples of such transformations and surfaces are given by stretching without tearing. The surfaces of the plum and the wineglass in [10.12a] are two examples of such topological spheres.

The sphere is the boundary of a solid ball, and other closed surfaces may likewise be obtained as the boundaries of other solid objects. For example, the surface of a doughnut is called a torus (top of [10.12b]), and it is clear that this surface is topologically the same as the beach toy at the bottom. But it seems equally clear that no amount of stretching and bending can turn these surfaces into a sphere—[10.12a] and [10.12b] are topologically distinct types of surface. Figure [10.12c] shows yet a third topologically distinct class. Obviously we could continue this list indefinitely just by adding more holes.

We shall not develop the topological ideas[4] necessary to prove it, but once again it seems clear that these classes of topologically distinct closed surfaces can be classified purely on the basis of their number of holes. This number g is called the *genus* of the surface (see [10.12]). We can now formulate the general result:

> *If a vector field on a smooth surface of genus* g *has only a finite number of singular points, then the sum of their indices is* $\chi \equiv (2 - 2g)$. (10.4)

[4] See "Further Reading", at the end of this chapter.

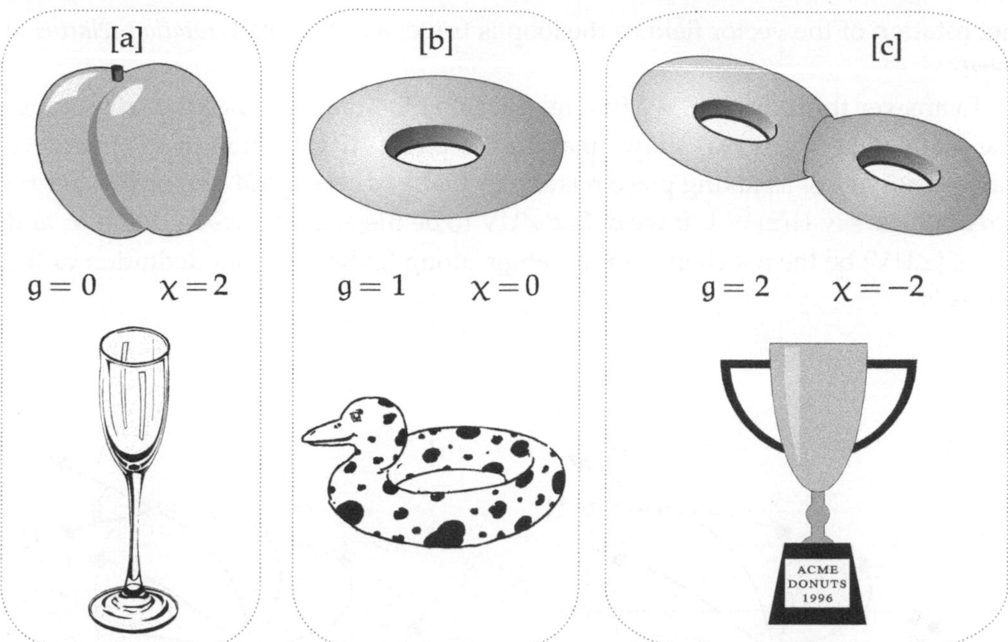

[10.12] The genus (g) and the Euler characteristic (χ) of a closed surface. *The genus counts the number of holes in a closed surface. The Euler characteristic* $\chi \equiv (2-2g)$ *turns out to be ubiquitous in topology, and it is therefore a more natural way of classifying surfaces. Furthermore,* χ *in fact has its own, independent geometrical definition, and it is then a* theorem *that* $\chi = (2-2g)$. *See Ex. 12.*

The number $\chi \equiv (2-2g)$ occurring in this theorem is called the *Euler characteristic* of the surface, and it turns out to be ubiquitous in topology. It is therefore more natural to classify our surfaces using χ rather than g. See [10.12]. Furthermore, χ in fact has its own, independent geometrical definition, and it is then a *theorem* that $\chi = (2-2g)$. See Ex. 12.

An immediate consequence of (10.4) is that a vector field without any singular points can exist only on surfaces of vanishing Euler characteristic, i.e., the topological doughnuts. Even then, the theorem does not actually guarantee that such a vector field exists, it merely says that if there are singular points then their indices must cancel. However, you can see for yourself [draw it!] that on a doughnut there *do* exist at least two topologically distinct vector fields without any singular points: one that circulates around its axis of symmetry, and one that circulates through the hole.

10.3.2 Defining the Index on a Surface

In order to give a precise definition of the "index" of one of the singular points in [10.11], we should presumably draw a loop round it on the surface, then find the net rotation of the vector field as the loop is traversed. But wait, *rotation relative to what?*

To answer this question, we first re-examine the familiar concept of rotation in the plane. Figure [10.13a] shows that (in the plane) the rotation of $V(z)$ along L can be thought of as taking place relative to a *fiducial vector field* having horizontal streamlines, say $U(z) = 1$. If we define $\angle UV$ to be the angle between U and V, and let $\delta_L(\angle UV)$ be the net change in this angle along L, then our old definition of the index is

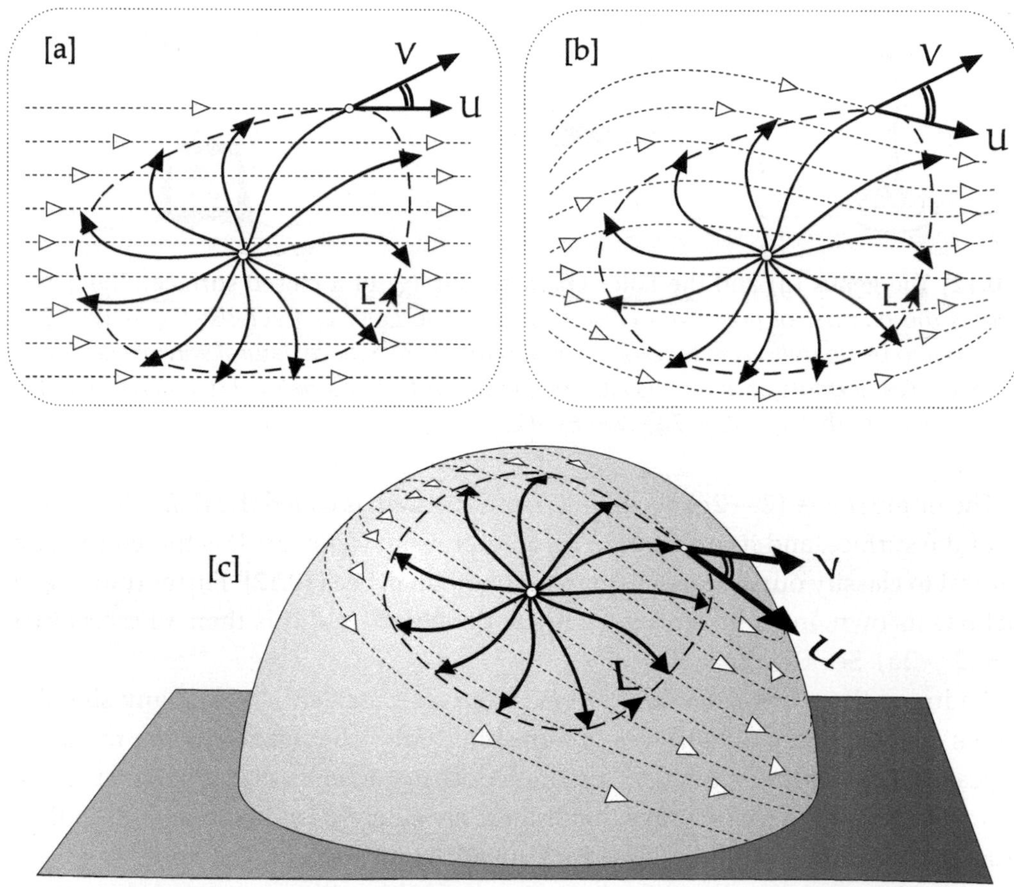

[10.13] Definition of the index of a singular point on a curved surface. [a] *The original definition of the index in the plane can be thought of as the net rotation of V relative to a horizontal* fiducial vector field U: $\mathcal{I}_V[L] = \frac{1}{2\pi}\,\delta_L(\angle UV)$. **[b]** *The definition and value of the index does not change if we replace U with* any *vector field that is nonsingular on and inside L.* **[c]** *Now imagine that [b] is drawn on a rubber sheet. If we continuously stretch it into the form of the curved surface then not only will* $\frac{1}{2\pi}\,\delta_L(\angle UV)$ *remain well-defined, but its value will not change.*

$$\mathcal{I}_V[L] = \frac{1}{2\pi}\,\delta_L(\angle UV). \tag{10.5}$$

If we continuously deform the horizontal streamlines of U in [10.13a] to produce those in [10.13b] then, by the usual reasoning, the RHS of (10.5) will not change. Thus we conclude that this formula yields the correct value of the index if we replace U with *any* vector field that is nonsingular on and inside L.

Now imagine that [10.13b] is drawn on a rubber sheet. If we continuously stretch it into the form of the curved surface in [10.13c] then not only will the RHS of (10.5) remain well-defined, but its value will not change. To summarize: if s is a singular point of a vector field V on a surface S, we define its index as follows. Draw any nonsingular vector field U on a patch of S that covers s but no other singular points; on this patch, draw a simple loop L going round s; finally, apply (10.5), that is count the net revolutions of V relative to U as we traverse L.

10.3.3 An Explanation of the Poincaré–Hopf Theorem

We can now give a very elegant derivation of theorem (10.4), due to Hopf (1956) himself. The argument proceeds in two steps. First, we show that on a surface of given genus, all vector fields yield the same value for the sum of their indices; second, we produce a concrete example of a vector field for which the sum equals the Euler characteristic. This proves the result.

Suppose that V and W are two different vector fields on a given closed surface S. See [10.14]. If v_j are the singular points of V (marked •) and w_j are those of W (marked ⊙), we must show that

$$\sum \mathcal{I}_V[v_j] = \sum \mathcal{I}_W[w_j].$$

Much as we did in [10.10], we divide up S into curvilinear polygons (dashed) such that each one contains at most one v_j and one w_j. taken counterclockwise as viewed from outside S. To find the indices of V and W along boundary of one of the polygons K_j, draw any nonsingular vector field U on the polygon and then use (10.5). The difference of these indices is then

$$\begin{aligned}\mathcal{I}_W[K_j] - \mathcal{I}_V[K_j] &= \frac{1}{2\pi}\left[\delta_{K_j}(\angle UW) - \delta_{K_j}(\angle UV)\right]\\ &= \frac{1}{2\pi}\,\delta_{K_j}(\angle VW),\end{aligned}$$

which is explicitly independent of the local vector field U.

From this we deduce that

$$\begin{aligned}\sum \mathcal{I}_V[v_j] - \sum \mathcal{I}_W[w_j] &= \sum_{\text{all polygons}} (\mathcal{I}_V[K_j] - \mathcal{I}_W[K_j])\\ &= \frac{1}{2\pi}\sum_{\text{all polygons}} \delta_{K_j}(\angle VW)\\ &= 0,\end{aligned}$$

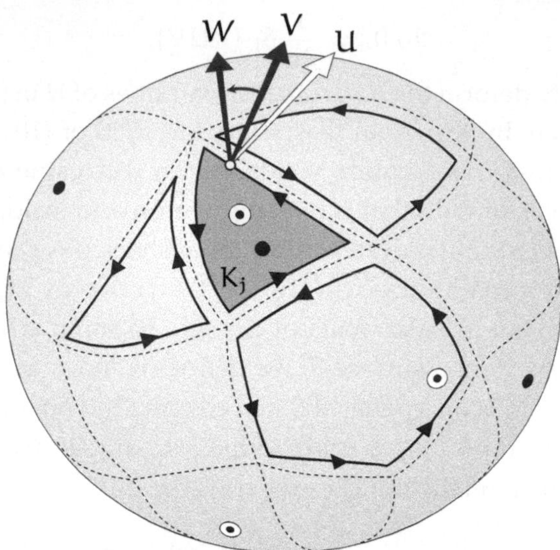

[10.14] Proof that the sum of the indices is the same for all vector fields. *Let* V *(with singular points* v_j *marked* •*) and* W *(with singular points* w_j *marked* ⊙*) be two different vector fields on the surface* S*. Divide up* S *into curvilinear polygons (dashed) such that each one contains at most one* v_j *and one* w_j*. Now consider one of these polygons and its boundary* K_j*. Draw any nonsingular vector field* U *on the polygon. Then* $\mathcal{I}_W[K_j] - \mathcal{I}_V[K_j] = \frac{1}{2\pi}\left[\delta_{K_j}(\angle UW) - \delta_{K_j}(\angle UV)\right] = \frac{1}{2\pi}\delta_{K_j}(\angle VW)$*, which is explicitly independent of the local vector field* U*. If we now sum this over all polygons, each edge is traced twice, in opposite directions, so* $\sum \delta_{K_j}(\angle VW) = 0$*, and therefore* $\sum \mathcal{I}_V[K_j] = \sum \mathcal{I}_W[K_j]$*. Done.*

because every edge of every polygon is traversed once in each direction, producing equal and opposite changes in $\angle VW$. We have thus completed the first step: the sum of the indices is independent of the vector field.

Since the index sum for the example in [10.11a] is 2, we now know that this is the value for any vector field on a topological sphere. The second step of the general argument is likewise to produce an example on a surface of arbitrary genus g, such that the sum is $\chi = (2-2g)$. Figure [10.15] is such an example for $g = 3$, the generalization to higher genus being obvious. Here we imagine that honey is being poured onto the surface at the top—it then flows over the surface, finally streaming off at the very bottom. As the figure explains, and as was required, the sum of the indices is indeed equal to χ.

Further Reading. These topological ideas—in combination with ideas in the next two chapters—open the door to the important subject of *Riemann surfaces*. In particular, we hope you will find it easier to read Klein (1881), which champions

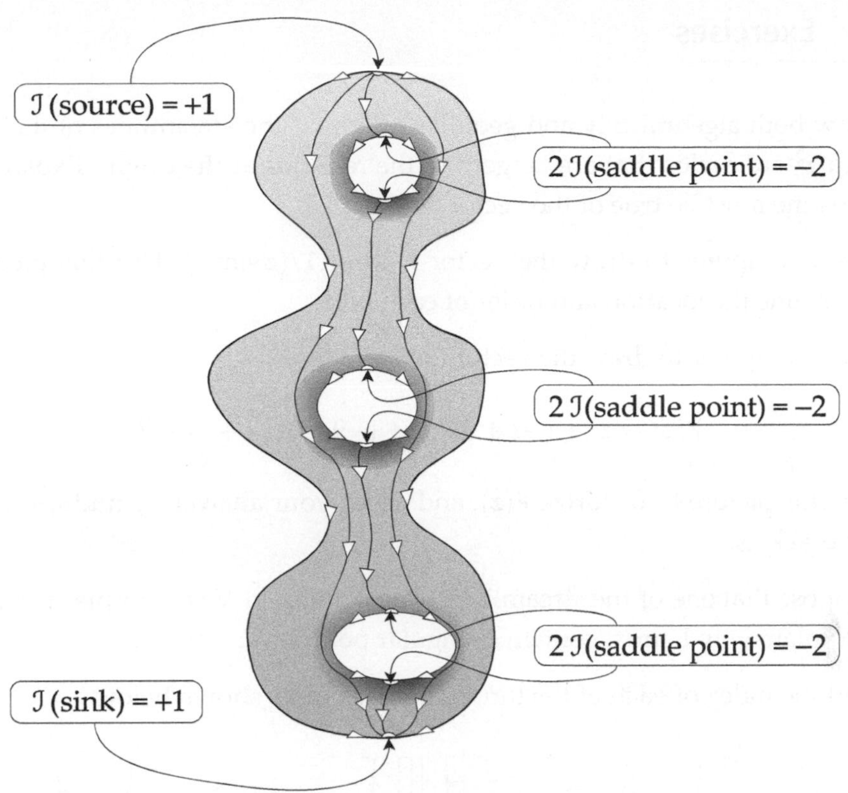

[10.15] A specific vector field on a surface of genus g for which the sum of the indices is χ. *We imagine that honey is being poured onto the surface at the top—it then flows over the surface, finally streaming off at the very bottom. As the figure explains, and as was required, the sum of the indices is indeed* χ. *But we know that this sum is the same for* all *vector fields on the surface, and we have therefore shown that the sum of their indices must* always *equal* χ, *completing the proof of the Poincaré–Hopf Theorem.*

Riemann's original approach to multifunctions in terms of fluid flowing over a surface in space. See also Springer (1981, Ch. 1), which essentially reproduces Klein's 1881 monograph, but with additional helpful commentary. For a good introduction to the more abstract, modern view of Riemann surfaces, see Jones and Singerman (1987). Finally, for more on topology itself, we recommend Hopf (1956), Prasolov (1995), Stillwell (2010), Fulton (1995), and Earl (2019) .

10.4 Exercises

1 Show both algebraically and geometrically that the streamlines of the vector field z^2 are circles that are tangent to the real axis at the origin. Explain why the same must be true of the vector field $1/\overline{z}^2$.

2 Use a computer to draw the vector field of $1/(z \sin^2 z)$. Use this picture to determine the location and order of each pole.

3 Use a computer to draw the vector field of

$$P(z) = z^3 + (-1+5i)\,z^2 + (-9-2i)\,z + (1-7i).$$

Use this picture to factorize $P(z)$, and check your answer by multiplying out the brackets.

4 Suppose that one of the streamlines of a vector field V is a simple closed loop L. Explain why L must contain a singular point of V.

5 Find the index of each of the three singular points shown below.

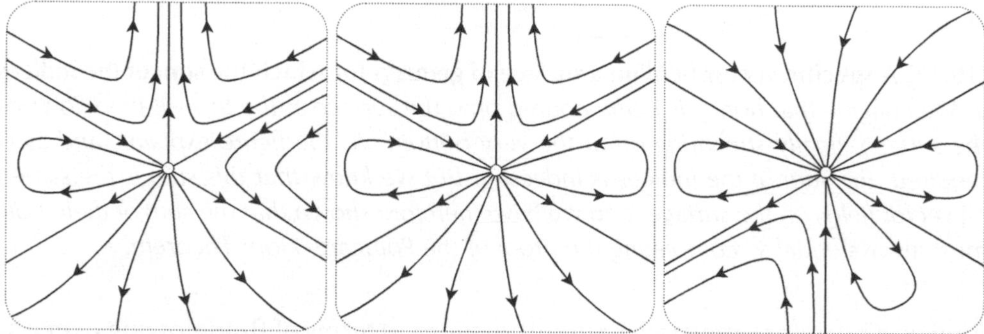

6 Observe that the neighbourhood of every singular point we have examined in this chapter is made up of sectors of one of the three types shown below, called *elliptic*, *parabolic*, and *hyperbolic*. Let e, p, and h denote the number of each type of sector surrounding a singular point.

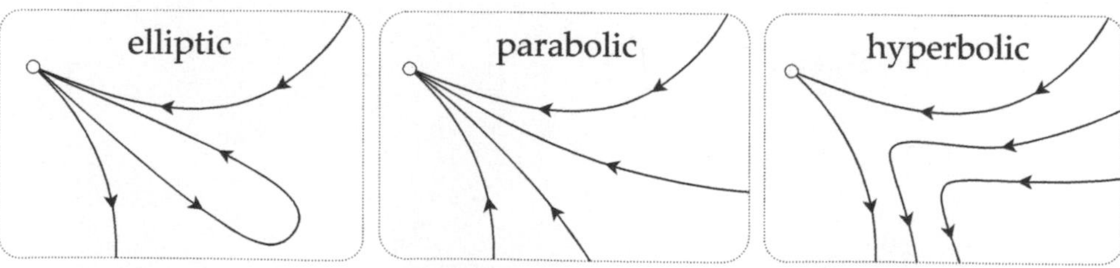

(i) Verify that the index of each of the three singular points in the previous question is correctly (and painlessly) predicted by *Bendixson's Formula*:

$$\mathcal{J} = 1 + \tfrac{1}{2}(e - h).$$

(ii) Explain this formula.

7 Given a vector field V, defined on a circle C, let a vector field W be constructed on C in the manner illustrated below. If $\mathcal{J}_V[C] = n$, find $\mathcal{J}_W[C]$. [This problem is taken from Prasolov (1995, Ch. 6), and the answer may be found there, too.]

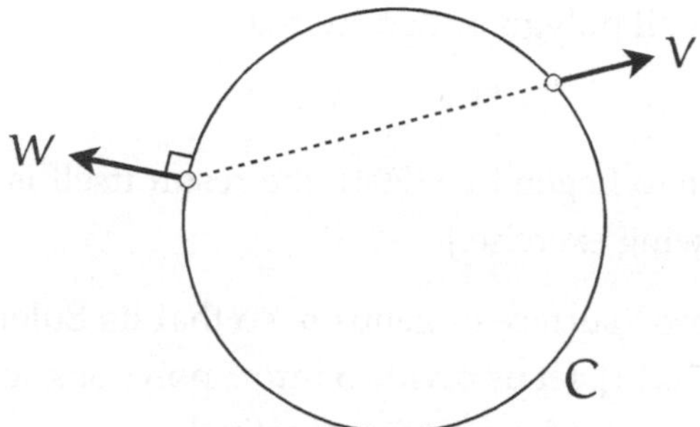

8 If f and g are continuous, one-to-one mappings of the sphere S to itself, then their composition $f \circ g$ is too. Let us prove that *at least one of these three mappings must possess a fixed point.* We proceed by the method of contradiction.

(i) Show if the result is false then, for each point p of S, the points p, f(p), and $[f \circ g](p)$ must be distinct.

(ii) In this case, deduce that there is a unique, directed circle C_p on S passing through these three points in the stated order.

(iii) Imagine a particle orbiting on C_p at unit speed, and let $\mathbf{V}(p)$ be its velocity vector as it passes through p. Since p was arbitrary, **V** is a vector field on S.

(iv) By appealing to the Poincaré–Hopf Theorem, obtain the desired contradiction.

9 Continuing from the previous exercise, apply the result as follows:

(i) By taking $g = f$, deduce that $(f \circ f)$ has a fixed point.

(ii) By taking $g =$ (antipodal mapping), deduce that either f has a fixed point, or f maps some point to its antipodal point.

10 Arbitrarily choose a collection of points $s_1, s_2, \ldots, s_n$ on a closed, smooth surface S. By attempting to draw examples on the surface of a suitable fruit or vegetable, investigate the following claim: There exists a flow on S whose only singular points are $s_1, s_2, \ldots, s_n$, and the type of singular behaviour (dipole, vortex, etc.) at all but one of these points may be chosen arbitrarily.

11 Imagine the surface of the unit sphere divided up into F polygons, the edges all being "straight lines on the sphere", i.e., great circles. Let E and V be the total number of edges and vertices that result from dividing up the sphere in this way.

(i) Let $\mathcal{P}_n$ be an n-gon on the unit sphere. Use (6.9), p. 317, to show that

$$\mathcal{A}(\mathcal{P}_n) = [\text{angle sum of } \mathcal{P}_n\,] - (n-2)\pi.$$

[*Hint:* Join the vertices of $\mathcal{P}_n$ to a point in its interior, thereby dividing it into n triangles.]

(ii) By summing over all polygons, deduce that

$$F - E + V = 2.$$

[This argument is due to Legendre (1794); the result itself is a special case of the result in the following exercise.]

12 Let S be a smooth closed surface of genus g, so that its Euler characteristic is $\chi(S) = 2-2g$. As in [10.14], let us divide S into F polygons, and let E and V be the total number of edges and vertices, respectively.

(i) Draw a simple example on the surface of an orange and convince yourself (by drawing it) that we may obtain a consistent flow over the entire surface whose only singular points are (1) a source inside each of the F polygons; (2) a simple saddle point on each of the E edges; (3) a sink at each of the V vertices. We call this a *Stiefel vector field*, in honour of Hopf's student, Eduard Stiefel [1909–1978]. See Frankel (2012, §16.2b). For an example of a Stiefel vector field, see the front cover of VDGF, Needham (2021).

(ii) By applying the Poincaré–Hopf Theorem to such a flow on the general surface S, deduce the following remarkable result, which we call the *Euler-Lhuilier Formula*:

$$F - E + V = \chi(S).$$

NOTES: (1) This is actually the normal *definition* of χ, and it is then a *theorem* that $\chi(S) = 2-2g$. (2) Euler discovered the original version of this formula in 1750 for topological spheres, with $\chi = 2$; this is called *Euler's Polyhedral Formula*. For a masterful, mathematically accurate, yet riveting account of this history and the connected mathematical ideas, see Richeson (2008).

(iii) Verify this result for your example in (i), then try it out on a doughnut.

13 The figure below shows all the normals that may drawn from the point p to the smooth surface S. Let R(q) be the distance from p to a point q of S, and let us say that q is a *critical point* of R if the rate of change of R vanishes as q begins to move within S; we need not specify the direction in which q begins to move because we are assuming that S has a tangent plane at q.

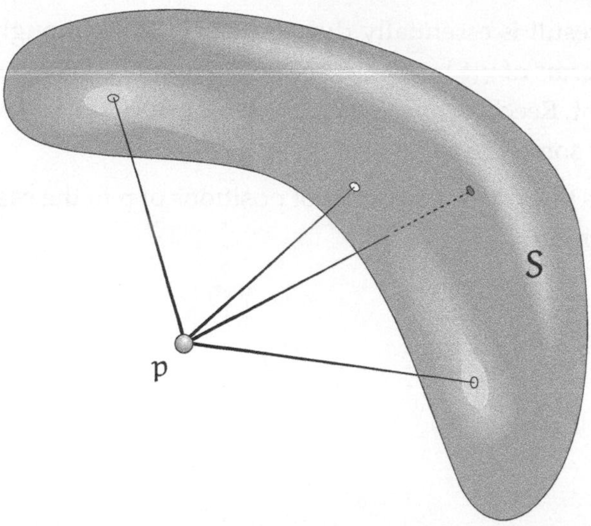

(i) Explain why pq is normal to S if and only if q is a critical point of R.

(ii) The level curves of R on S are the intersections of S with the "onion" of concentric spheres centred at p. Sketch these level curves in the vicinity of the illustrated critical points of R. Notice the distinction between points where R has a local maximum or minimum, versus points where R may increase or decrease depending on the direction (within S) in which one moves away from the critical point.

(iii) Imagine that p generates an attractive force field, so that every point particle in space experiences a force $\mathbf{F}$ directed towards p. For example, we could imagine that p is the centre of the Earth, and that $\mathbf{F}$ is the Earth's gravitational field. If a particle is constrained to move on S, then the only part of $\mathbf{F}$ to which it can respond is the projection $\mathbf{F}_S$ of $\mathbf{F}$ onto S. Sketch the streamlines of $\mathbf{F}_S$. How are they related to the level curves of R in (ii)?

(iv) You have just seen that the critical points of R are the singular points of $\mathbf{F}_S$. How does the index $\mathcal{I}(q)$ of a singular point of $\mathbf{F}_S$ distinguish between the types of critical point discussed in (ii)?

(v) Let us define the *multiplicity* of a normal pq to be this index $\mathcal{I}(q)$. Use the Poincaré–Hopf Theorem to deduce that

> *The total number of normals (counted with their multiplicities) that may be drawn to S from any point* p *is independent of both the location of* p *and the precise shape of S, and is equal to* $\chi(S)$.

[This lovely result is essentially due to Reech (1858), though he did not express it in terms of $\chi(S)$, nor did he use an argument like the one above. With hindsight, Reech's work is a clear harbinger of Morse Theory, which it predates by some 70 years.]

(vi) Verify Reech's theorem for a couple of positions of p in the case where S is a torus (doughnut).

CHAPTER 11

Vector Fields and Complex Integration

11.1 Flux and Work

We promised long ago that there was a more vivid way of understanding complex integrals than the geometric Riemann sum of Chapter 8. In this section we lay the foundations for this elegant new approach. If you are already familiar with vector calculus then you can skip this section and go directly to Section 11.2.

11.1.1 Flux

In order to define the flux a little more carefully than before, consider [11.1]. At each point of the directed path K we introduce a unit tangent vector **T** in the direction of the path, and a unit normal vector **N** pointing to our *right* as we travel along K. In terms of the corresponding complex numbers, this convention amounts to

$$T = iN.$$

The figure also shows how a vector field **X** (which we will first think of as the velocity of a fluid flowing over the plane) can be decomposed into tangential and normal components:

$$\mathbf{X} = (\mathbf{X}\cdot\mathbf{T})\,\mathbf{T} + (\mathbf{X}\cdot\mathbf{N})\,\mathbf{N}.$$

Only the second of these components carries fluid across K, and the amount flowing across an infinitesimal segment ds of the path in unit time (i.e. its flux) is thus $(\mathbf{X}\cdot\mathbf{N})\,ds$. This is a refinement over our previous definition in that the flux now has a *sign*: it is positive or negative according as the flow is from left to right or from right to left. The total flux $\mathfrak{F}[\mathbf{X}, K]$ of **X** across K is then the integral of the fluxes across its elements:

$$\mathfrak{F}[\mathbf{X}, K] = \int_K (\mathbf{X}\cdot\mathbf{N})\,ds.$$

Check for yourself that the flux satisfies

$$\mathfrak{F}[-\mathbf{X}, K] = \mathfrak{F}[\mathbf{X}, -K] = -\mathfrak{F}[\mathbf{X}, K].$$

Visual Complex Analysis. 25th Anniversary Edition. Tristan Needham, Oxford University Press.
 DOI: 10.1093/oso/9780192868916.003.0011

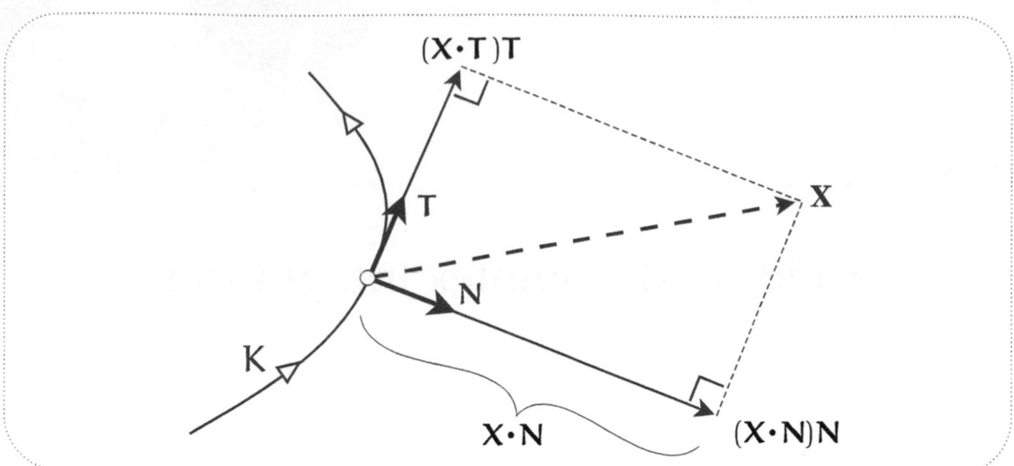

[11.1] Decomposition of a vector into its tangential and normal components. *Note that the normal is defined to go to the* right *of the direction of travel, so that* T = iN.

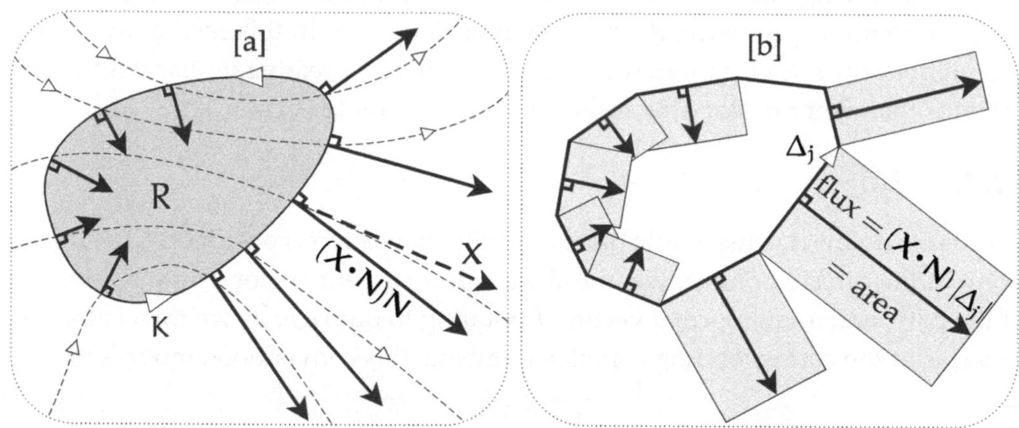

[11.2] If we picture X as the velocity of a fluid, then the flux, $\mathfrak{F}$ [X, K], of X across K represents the net flow of fluid out of R per unit time. [a] *Mathematically, the flux is the integral of the outward normal component of* **X**. **[b]** *The flux can be visualized as the limit of the sum of the illustrated signed areas: positive if the flow is outward, and negative if it is inward.*

The flux concept is further illustrated in [11.2] for the case where K is a simple closed loop bounding the shaded region R. Figure [11.2a] shows the normal components of **X**, the signed magnitudes of which we must integrate to obtain $\mathfrak{F}$ [**X**, K]. Figure [11.2b] shows how we might make an estimate of this flux. We replace K by a polygonal approximation with directed edges Δ_j, and at the midpoint of each one we draw the normal component of **X**. The flux is then approximately given by the algebraic sum of the signed areas of the shaded rectangles. In this case there is clearly more positive area than negative, so the flux is positive. As the Δ_j's become shorter and more numerous, the approximation of course gets better and better.

In the case of the simple loop K in [11.2a] there is another interesting way of looking at the flux:

$$\mathfrak{F}[\mathbf{X}, \mathsf{K}] = [\text{fluid leaving R per unit time}] - [\text{fluid entering R per unit time}].$$

Henceforth we will always take our fluid to be *incompressible*. Thus, provided there are no sources or sinks in R, what flows into R must also flow out of R:

$$\mathfrak{F}[\mathbf{X}, \mathsf{K}] = 0.$$

Indeed, we may turn this around and define a flow to be *sourceless* in a region if *all* simple loops in that region have vanishing flux. The simplest example of such a flow without any (finite) sources or sinks is $\mathbf{X} = \mathtt{const}$. If the loop does contain a source, for example, then incompressibility says that the flux equals the strength of the source.

Although we will only concern ourselves with two-dimensional flows, we should at least mention the concept of flux in three dimensions. If a fluid is flowing through ordinary space, it no longer makes sense to speak of the flux across a curve, but it does make sense to speak of the rate at which the fluid crosses a *surface*. If $\mathbf{N}$ now stands for the normal to this surface, then the flux across an infinitesimal element of *area* $d\mathcal{A}$ is once again given by $(\mathbf{X}\cdot\mathbf{N})\, d\mathcal{A}$. The total flux is then obtained by integrating this quantity over the whole surface. Just as in two dimensions, the incompressibility of a three-dimensional flow is equivalent to the statement that all closed surfaces (that do not contain sources or sinks) have vanishing flux.

Lastly, we should point out that although the word "flux" is Latin for "flow", it is standard practice to retain this terminology when applying our mathematical definition to *any* vector field $\mathbf{X}$, regardless of whether it actually is the velocity of a flowing substance. For example, the electric field represents a force, but one of the four fundamental laws of electromagnetism says that we can think of it as an incompressible flow in which positive and negative electric charges act as sources and sinks, so that its flux through a closed surface in space equals the net charge enclosed.

11.1.2 *Work*

So far we have only studied the normal component of $\mathbf{X}$; we turn next to its tangential component. To do so, let us now imagine that $\mathbf{X}$ is a *force field* rather than a flow.

If a particle on which a force acts is displaced infinitesimally then we know from elementary physics that the work done by the field (i.e. the energy it expends) is the component of the force in the direction of displacement, times the distance moved. Thus if the particle moves an amount ds along K then the work done by $\mathbf{X}$ is $(\mathbf{X}\cdot\mathbf{T})\, ds$. As with flux, this definition contains a *sign*, the physical significance of which we will explain shortly. If the particle is moved along the entire length of K, the total work done by the field is then

$$\mathfrak{W}[\mathbf{X}, \mathsf{K}] = \int_{\mathsf{K}} (\mathbf{X}\cdot\mathbf{T})\, ds.$$

Figure [11.3a] illustrates the tangential components of $\mathbf{X}$ on K, the signed magnitudes of which we must add up to obtain $\mathfrak{W}$.

Just as for $\mathfrak{F}$, check that $\mathfrak{W}$ satisfies

$$\mathfrak{W}[-\mathbf{X}, \mathsf{K}] = \mathfrak{W}[\mathbf{X}, -\mathsf{K}] = -\mathfrak{W}[\mathbf{X}, \mathsf{K}].$$

Note that, unlike $\mathfrak{F}$, no modification of $\mathfrak{W}$ is needed if we wish to extend the idea to three-dimensional force fields: it still makes perfectly good sense to consider the work done by the field as a particle is moved along a curve in space, and the formula is as before.

Figure [11.3b] illustrates a simple thought-experiment for interpreting both the magnitude and the sign of $\mathfrak{W}$. We imagine that the plane in which the force field acts is made of ice on which a very small ice-puck of mass m can slide without friction. We now construct a narrow frictionless channel in the shape of K, just wide enough to accommodate the puck which we fire into it with speed v_{in}. On the initial leg of the journey we see that the force opposes the motion, and thus if v_{in} is not sufficiently great, the puck will slow, stop, and return whence it came. Clearly, though, if we fire the puck with sufficient speed it will overcome all resistance and emerge at the end of the channel, say with speed v_{out}. Let the initial and final kinetic energies of the puck be $\mathcal{E}_{in}$ and $\mathcal{E}_{out}$, so that

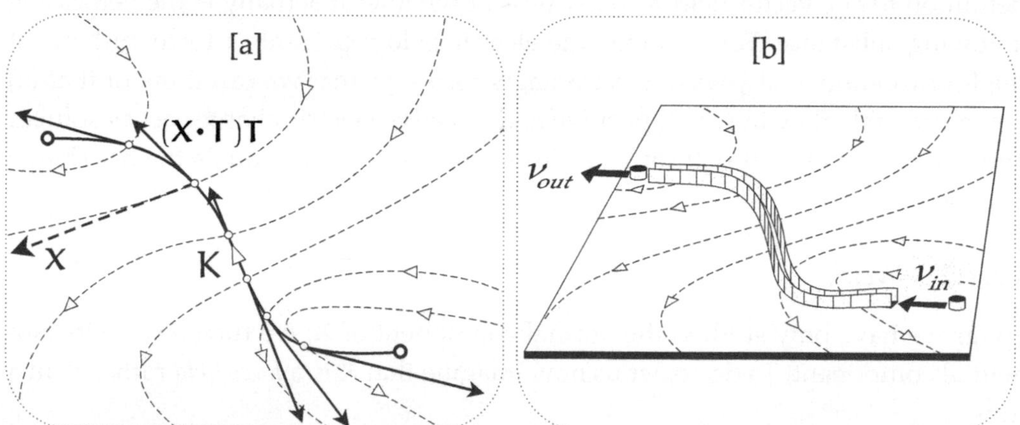

[11.3] [a] If we picture X as a force field, then the integral of the tangential component is the work, $\mathfrak{W}[\mathbf{X}, \mathsf{K}]$, done in moving a particle along K. [b] *If a puck of mass m is fired with speed v_{in} into a frictionless channel following K, and it emerges with speed v_{out}, then the conservation of energy implies that the work is the change in the kinetic energy:* $\mathfrak{W} = \frac{1}{2}mv_{out}^2 - \frac{1}{2}mv_{in}^2$.

$$\mathcal{E}_{\text{in}} = \tfrac{1}{2}m v_{\text{in}}^2 \quad \text{and} \quad \mathcal{E}_{\text{out}} = \tfrac{1}{2}m v_{\text{out}}^2.$$

One of the most sacred principles of physics is the "conservation of energy", which states that energy can never be created or destroyed, only transformed from one kind into another. Thus the energy $\mathfrak{W}$ expended by the force field on the puck does not disappear but instead is transformed into the change in the puck's kinetic energy:

$$\begin{aligned}\mathfrak{W}[\mathbf{X}, \mathsf{K}] &= \mathcal{E}_{\text{out}} - \mathcal{E}_{\text{in}} \\ &= \left[\frac{m\,v_{\text{out}} + m\,v_{\text{in}}}{2}\right](v_{\text{out}} - v_{\text{in}}) \\ &= [\text{average momentum}]\ (\text{change in speed}).\end{aligned}$$

This formula also gives clear meaning to the sign of the work: it is the sign of the change in speed. Thus if $\mathfrak{W}$ is positive the field expends energy speeding up the puck and increasing its kinetic energy, while if $\mathfrak{W}$ is negative then the puck has to give up some of its kinetic energy in doing work against the field.

Next, imagine that we bend K round so that the ends almost join to form a closed loop. When the puck travels along the corresponding channel it will therefore emerge at essentially the same place that it entered. Suppose it were to emerge with greater speed than it entered. Joining the ends of the channel together, the puck would therefore go round the loop faster and faster, gaining energy with each circuit—energy that could be harnessed to solve the world's energy crisis!

Although we may construct mathematical examples for which this happens, if no energy is supplied from outside the puck/field system then a physical force field will not behave in this way; it will conserve energy so that the puck returns to its starting point with exactly the same speed with which it was launched[1]. Such a field is called *conservative*. Mathematically, $\mathbf{X}$ is conservative if and only if

$$\mathfrak{W}[\mathbf{X}, \text{any closed loop}] = 0. \tag{11.1}$$

Just as we applied the concept of flux to vector fields that were not flows, so we may apply the concept of work to vector fields that do not represent force. However, in this general setting it is standard practice to call $\mathfrak{W}[\mathbf{X}, \mathsf{K}]$ the *circulation* of $\mathbf{X}$ along K rather than the work. As with "flux", this terminology originates from thinking of $\mathbf{X}$ as representing a flow. To see why, take K to be a closed loop and consider the following thought-experiment of Feynman (1963). Imagine that the fluid flowing over the plane with velocity $\mathbf{X}$ is instantaneously frozen everywhere

[1] If energy can be supplied from outside the system, then the work need not vanish for a closed loop. In fact the operation of all electrical machines depends on the ability of a moving magnet to create an electric field that *can* speed up our puck. However, there is still no violation of energy-conservation since work is being done to move the magnet. See Feynman (1963).

except within the narrow strip where our channel used to be. The "circulation" is then [exercise] the speed with which the unfrozen fluid flows (or circulates) round K, times the length of K.

If this circulation vanishes for every closed loop then the flow is said to be *irrotational*. Just as "circulation" means $\mathfrak{W}[\mathbf{X}, \mathsf{K}]$, irrespective of the physical nature of $\mathbf{X}$, so with equal generality "irrotational" is short for the mathematical statement (11.1). Thus a conservative force field could also be described as irrotational.

11.1.3 Local Flux and Local Work

At present our definition of a *sourceless and irrotational* vector field $\mathbf{X}$ is that

$$\mathfrak{F}[\mathbf{X}, \text{any closed loop}] = 0 \quad \text{and} \quad \mathfrak{W}[\mathbf{X}, \text{any closed loop}] = 0. \tag{11.2}$$

Our next objective is to show that there are two very simple *local* properties of $\mathbf{X}$ that are equivalent to the non-local ones above.

To do this we must calculate limiting behaviour of the flux and the work for a small loop that shrinks to nothing, i.e., for an "infinitesimal loop". Though it is not entirely obvious, later we will show that this limiting behaviour is independent of the *shape* of the infinitesimal loop. We are thus free to simplify the calculations by choosing the loop to be a small square centred at the point of interest, say z, and having horizontal and vertical edges of length ϵ. See [11.4].

Accurate estimates of $\mathfrak{F}$ and $\mathfrak{W}$ can now be found by evaluating $\mathbf{X}$ at the midpoints (a, b, c, d) of the sides, then summing the appropriate components. In the limit that ϵ shrinks to nothing, this approximation becomes exact, as does the following equation, which we will need in a moment:

$$P(a) - P(c) \asymp \epsilon\, \partial_x P(z),$$

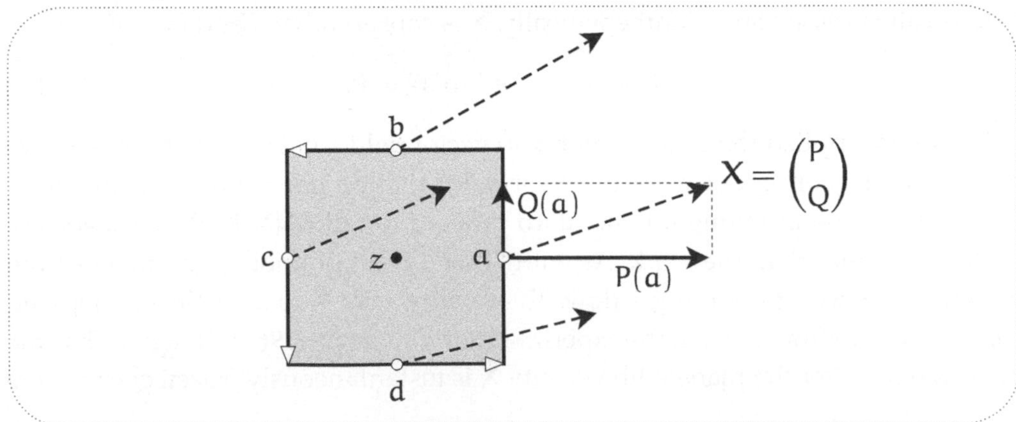

[11.4] Local flux and local work. *Accurate, ultimately exact estimates of* $\mathfrak{F}$ *and* $\mathfrak{W}$ *around a small, ultimately vanishing square can be found by evaluating* $\mathbf{X}$ *at the* midpoints (a, b, c, d) *of the sides, then summing the appropriate components. Writing* $X = P + iQ$, *we find that* $\mathfrak{F}[\mathbf{X}, \square] \asymp (\partial_x P + \partial_y Q)(\text{area of } \square)$, *and* $\mathfrak{W}[\mathbf{X}, \square] \asymp (\partial_x Q - \partial_y P)(\text{area of } \square)$.

where $\partial_x P(z)$ means $\partial_x P$ evaluated at z.

For the flux we find

$$\begin{aligned}\mathfrak{F}[\mathbf{X}, \square] &\asymp \epsilon P(a) + \epsilon Q(b) - \epsilon P(c) - \epsilon Q(d) \\ &\asymp \epsilon\,[\{P(a) - P(c)\} + \{Q(b) - Q(d)\}] \\ &\asymp \epsilon^2\,[\partial_x P(z) + \partial_y Q(z)].\end{aligned}$$

This expression can be simplified by considering the formal dot product of the *gradient operator* $\boldsymbol{\nabla}$ with the vector field:

$$\boldsymbol{\nabla}\cdot\mathbf{X} = \begin{pmatrix}\partial_x \\ \partial_y\end{pmatrix}\cdot\begin{pmatrix}P \\ Q\end{pmatrix} = \partial_x P + \partial_y Q.$$

This quantity $\boldsymbol{\nabla}\cdot\mathbf{X}$ is called the *divergence* of $\mathbf{X}$, and in terms of it we have

$$\mathfrak{F}[\mathbf{X}, \square] \asymp [\boldsymbol{\nabla}\cdot\mathbf{X}(z)]\,(\text{area of } \square). \tag{11.3}$$

In the next section we will see that (11.3) is true if $\square$ is replaced by an infinitesimal loop of arbitrary shape. This important result explains the term "divergence", for it says that $\boldsymbol{\nabla}\cdot\mathbf{X}$ is the local flux per unit area flowing *away* from z, i.e., diverging from z. In future we will abbreviate "local flux per unit area" to "flux density".

Repeating the above analysis for the work, we find [exercise]

$$\mathfrak{W}[\mathbf{X}, \square] \asymp [\boldsymbol{\nabla}\times\mathbf{X}(z)]\,(\text{area of } \square), \tag{11.4}$$

where the formal cross product is defined by

$$\boldsymbol{\nabla}\times\mathbf{X} = \begin{pmatrix}\partial_x \\ \partial_y\end{pmatrix}\times\begin{pmatrix}P \\ Q\end{pmatrix} = \partial_x Q - \partial_y P.$$

The quantity $\boldsymbol{\nabla}\times\mathbf{X}$ is called the *curl* of $\mathbf{X}$. Geometrically, it measures the extent to which $\mathbf{X}$ 'curls around' the point z. Physically, in terms of force fields, the above result says that the curl is the local work per unit area, or work density. There is also a vivid interpretation in terms of flows. If we drop a small disc of paper onto the surface of the flowing liquid at z, in general it will not only start to move (translate) along the streamline through z with speed $|\mathbf{X}(z)|$, but it will also rotate about its centre with some angular speed $\omega(z)$. It can be shown that the aspect of $\mathbf{X}$ which determines the rate of rotation ω is none other than the curl:

$$\omega(z) = \tfrac{1}{2}\,[\boldsymbol{\nabla}\times\mathbf{X}(z)].$$

For this reason "curl" is sometimes denoted "rot", which is short for "rotation".

*11.1.4 Divergence and Curl in Geometric Form**

The above expression for the divergence was obtained by considering the flux out of a shape having no connection with the flow. Greater insight is gained by

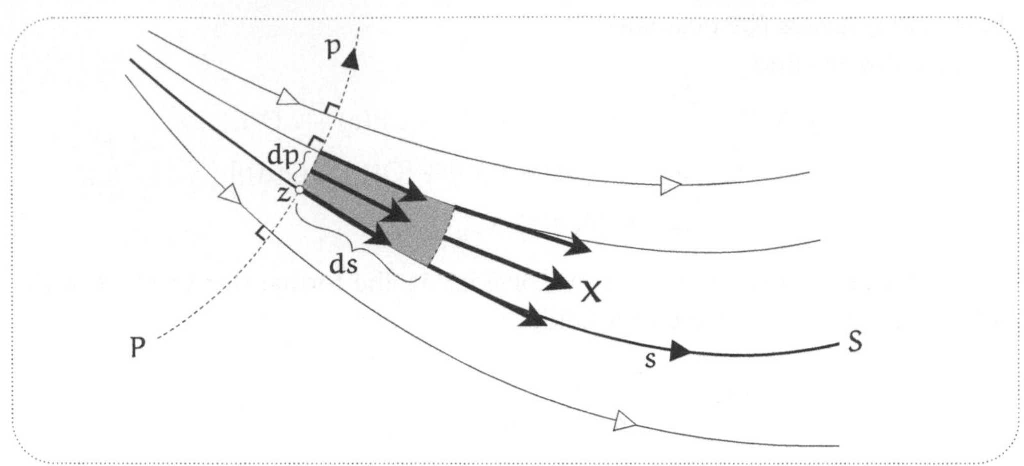

[11.5] Divergence and curl in geometric form. *Let* P *be the illustrated orthogonal trajectory of the streamlines, let* p *measure distance along it, and let* s *denote distance along the streamline* S*. Also, let* κ_P *and* κ_S *denote the curvatures of* P *and* S*, respectively. Then* $\nabla \cdot \mathbf{X} = \partial_s|X| + \kappa_P|X|$ and $\nabla \times \mathbf{X} = -\partial_p|X| + \kappa_S|X|$.

considering the flux out of an infinitesimal "rectangle" R, two sides of which are segments of streamlines of **X**, while the other two sides are segments of orthogonal trajectories through the streamlines. See [11.5].

Here z is the point down to which R will ultimately be collapsed in order to find the divergence there, S and P are the streamline and orthogonal trajectory through z, and s and p are arc-length along S and P, the direction of increasing p being chosen to make a positive right angle with **X**.

The net flux out of R is the difference between the fluxes entering and leaving. The flux entering is $|X|\,dp$, while the flux leaving is the same expression evaluated on the opposite side of R. It is now clear that two factors contribute to more fluid leaving than entering: (1) greater fluid speed $|X|$ as the fluid exits; (2) greater separation dp of the streamlines as the fluid exits.

The second factor is clearly governed by how much the direction of **X** changes along dp, in other words, by the *curvature* κ_P of P at z. More precisely, if δ denotes the increase in a quantity as we move ds along the streamline, then [exercise] $\delta(dp) = \kappa_P\, ds\, dp$. Thus

$$\begin{aligned}
(\text{Net flux out of R}) &= \delta\{|X|\,dp\} \\
&= (\delta|X|)\,dp + |X|\,\delta(dp) \\
&= (\partial_s|X| + \kappa_P|X|)\,(\text{area of R}).
\end{aligned}$$

The flux density is therefore

$$\nabla \cdot \mathbf{X} = \partial_s|X| + \kappa_P|X|. \tag{11.5}$$

In fact [exercise] this formula is still true for a three-dimensional vector field, provided that there exists[2] a *surface* P orthogonal to the streamlines, and κ_P is taken to be the sum of its principal curvatures.

Turning to the circulation round R, identical reasoning yields [exercise] an equally neat formula for the curl:

$$\nabla \times \mathbf{X} = -\partial_p |X| + \kappa_S |X|, \tag{11.6}$$

where κ_S is the curvature at z of the streamline S.

Although we suspect that (11.5) and (11.6) must have been known to the likes of Maxwell, Kelvin, or Stokes, we have not found any reference to these formulae in modern literature.

11.1.5 Divergence-Free and Curl-Free Vector Fields

From the definition (11.2) and the results (11.3) and (11.4) it follows that if $\mathbf{X}$ is sourceless and irrotational throughout some region R, then at each point of R we have

$$\nabla \cdot \mathbf{X} = 0 \quad \text{and} \quad \nabla \times \mathbf{X} = 0.$$

The vector field is then said to be *divergence-free* and *curl-free* in R.

For example, consider the vector field of a point source with strength S:

$$X(z) = \frac{S}{2\pi\overline{z}} \iff \mathbf{X} = \frac{S}{2\pi}\begin{pmatrix} x/(x^2+y^2) \\ y/(x^2+y^2) \end{pmatrix}.$$

This should have zero flux density (i.e. divergence) everywhere except at the origin, where it should be undefined. Check that this is so. Recall that we previously claimed that this was also the electrostatic field of a long, uniformly charged wire. We can now see that this makes physical sense in that the field is locally conservative. Thus if we fire our puck (which must now carry electric charge in order to experience the force) round an infinitesimal loop, it will return to its starting point with its kinetic energy unchanged. To verify this statement you need only check that the field is curl-free.

We have seen that a sourceless and irrotational field is divergence-free and curl-free. To end this section we wish to establish the converse result: if the divergence and curl vanish throughout a region, the flux and work vanish for all simple loops in that region. We will then have,

A vector field is sourceless and irrotational in a simply connected region if and only if it is divergence-free and curl-free there.

To understand this converse, consider [11.6] which essentially reproduces part of [8.27], p. 468. Let us now recycle the line of reasoning associated with that figure.

[2] The condition for existence is that the curl either vanish or be orthogonal to the vector field.

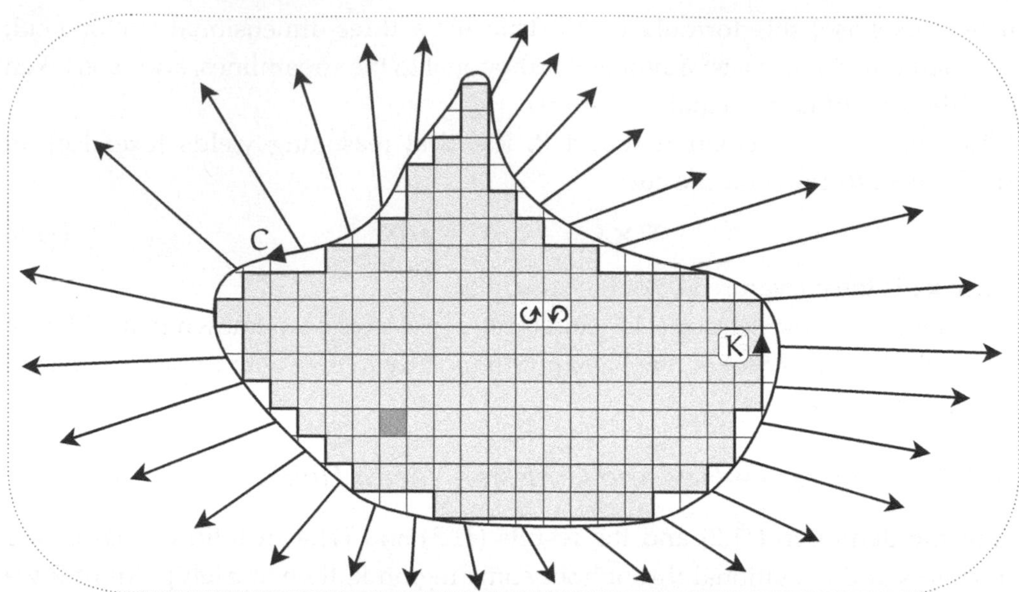

[11.6] Gauss's Theorem and Stokes's Theorem. *Using the same logic as in the proof of Cauchy's Theorem, summing integrals of the normal or the tangential component of* $\mathbf{X}$ *around the boundaries of the shaded squares yields the integral along the outer boundary* K. *But then the flux and work around each square can be expressed in terms of the divergence and curl of* $\mathbf{X}$, *via (11.3) and (11.4). This immediately yields Gauss's Theorem (11.7) and Stokes's Theorem (11.8).*

We begin by noting that as the grid gets finer and finer, the flux or work for K becomes the flux or work for C. Next we relate these quantities to the divergence and curl inside K. Check for yourself that exactly the same mathematical reasoning which previously yielded

$$\oint_{K} f(z)\,dz = \sum_{\text{shaded squares}} \oint_{\square} f(z)\,dz,$$

now yields

$$\mathfrak{F}[\mathbf{X}, K] = \sum_{\text{shaded squares}} \mathfrak{F}[\mathbf{X}, \square]$$

and

$$\mathfrak{W}[\mathbf{X}, K] = \sum_{\text{shaded squares}} \mathfrak{W}[\mathbf{X}, \square].$$

However, in the present context these results become accessible to physical intuition. The first says that the total amount of fluid flowing out of K is the sum of fluxes out of the interior squares. What does the second one say?

Now let the squares of the grid shrink so as to completely fill the interior R of C. Using (11.3) and (11.4) and replacing the sum over squares by a double integral over infinitesimal areas $d\mathcal{A}$, we obtain *Gauss's Theorem*,

$$\mathfrak{F}[\mathbf{X}, C] = \iint_R [\nabla \cdot \mathbf{X}] \, d\mathcal{A}, \tag{11.7}$$

and *Stokes's Theorem*,

$$\mathfrak{W}[\mathbf{X}, C] = \iint_R [\nabla \times \mathbf{X}] \, d\mathcal{A}. \tag{11.8}$$

From these we see that if the divergence and curl vanish everywhere in R then the flux and work for C also vanish, as was required.

Again following the logic in Chapter 8, consider what happens to the flux and work as we continuously deform a closed contour, or an open contour with fixed end points. You should be able to see that (11.7) and (11.8) imply two deformation theorems:

If the contour sweeps only through points at which the divergence vanishes, the flux does not change. (11.9)

If the contour sweeps only through points at which the curl vanishes, the work does not change. (11.10)

11.2 Complex Integration in Terms of Vector Fields

11.2.1 The Pólya Vector Field

Consider

$$\int_K H(z) \, dz$$

from the vector field point of view. See [11.7]. In forming a Riemann sum with terms $H\,dz$ we now have the minor advantage that $H = |H|\,e^{i\beta}$ and $dz = e^{i\alpha}\,ds$ are not drawn in separate planes, as they were in Chapter 8. However, we still face the problem that $H\,dz = |H|\,e^{i(\alpha+\beta)}\,ds$ involves the *addition* of angles, which is not easy to visualize. Just as it is more natural to subtract vectors [yielding connecting vectors] than to add them, so it is also more natural to subtract angles, for this yields the angle contained between two directions.

The simple and elegant solution to our problem is to consider a new vector field: instead of drawing $H(z)$ at z we draw its *conjugate* $\overline{H(z)} = |H|\,e^{-i\beta}$. We shall call this the *Pólya vector field* of H. Before showing how this solves our problem, let us offer (i) a caution and (ii) a reassurance:

(i) The Pólya vector field of H is *not* obtained by reflecting the picture of the ordinary vector field for H in the real axis, for this would attach $\overline{H(z)}$ to $\overline{z}$ instead of z. This will become very clear if you (or your computer) draw the Pólya vector fields of z and z^2, for example. Comparison with [10.1], p. 512, reveals that the resulting phase portraits (not the vector fields themselves) are identical to those of $(1/z)$ and $(1/z^2)$. This is because $\overline{z^n}$ points in the same direction as $(1/z^n)$.

(ii) As we will see in a moment, much is gained by representing H by its Pólya vector field, but we also wish to stress that nothing is lost: the new field contains exactly the same information as the old one. For example, it is clear that the index of a loop L merely changes sign when we switch to the Pólya vector field:

$$\mathcal{I}_{\overline{H}}[L] = -\mathcal{I}_{H}[L].$$

Thus an nth order root of an analytic H still shows up clearly in its Pólya vector field as a singular point, but now with index $-n$ instead of n. Likewise, a pole of order m produces a singular point of index m instead of $-m$.

Returning to integration, the great advantage of the Pólya vector field is that the angle θ that it makes with the contour (see [11.7]) is given by $\theta = \alpha - (-\beta)$, and this is precisely the angle we were trying to visualize—the angle of the term $H\,dz$ in the Riemann sum. Better still, we find that

$$\begin{aligned} H\,dz &= |\overline{H}|\,e^{i\theta}\,ds \\ &= \left[|\overline{H}|\,\cos\theta + i\,|\overline{H}|\,\sin\theta\right]\,ds \\ &= \left[\overline{\mathbf{H}}\boldsymbol{\cdot}\mathbf{T} + i\,\overline{\mathbf{H}}\boldsymbol{\cdot}\mathbf{N}\right]\,ds. \end{aligned}$$

Thus the real and imaginary parts of each term in the Riemann sum are the work and flux of the Pólya vector field for the corresponding element of the contour. We have thus discovered a vivid interpretation (due to Pólya[3] and first championed

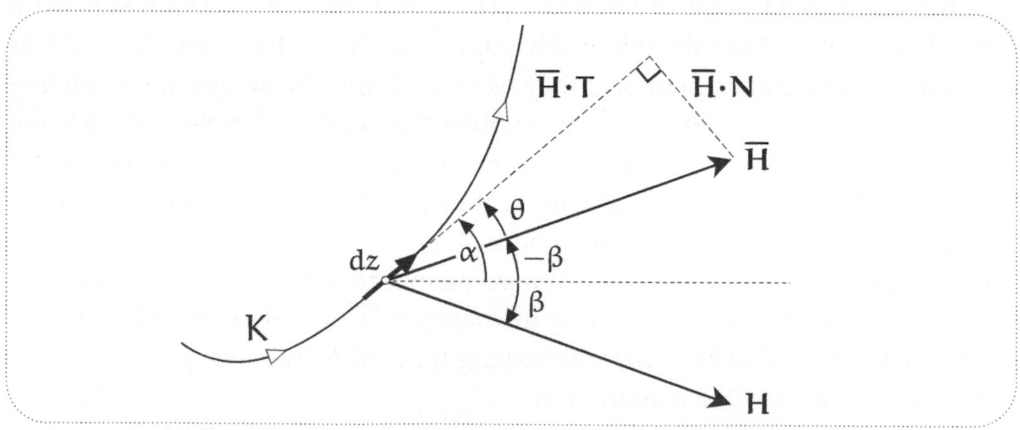

[11.7] How the Pólya vector field $\overline{H}$ makes it possible to visualize $\int_K H(z)\,dz$. *We can see that* $H\,dz = \left[\overline{\mathbf{H}}\boldsymbol{\cdot}\mathbf{T} + i\,\overline{\mathbf{H}}\boldsymbol{\cdot}\mathbf{N}\right]ds$, *so this immediately yields the extremely important and useful physical and geometrical interpretation of the complex contour integral:* $\int_K H(z)\,dz = \mathfrak{W}[\overline{H}, K] + i\,\mathfrak{F}[\overline{H}, K]$.

[3] See Pólya and Latta (1974).

by Bart Braden[4].) of the complex integral of H in terms of the work and flux of its Pólya vector field along the contour:

$$\boxed{\int_K H(z)\,dz = \mathfrak{W}[\overline{H}, K] + i\,\mathfrak{F}[\overline{H}, K].} \tag{11.11}$$

This interpretation is rendered particularly useful by the fact that a computer can instantly draw the Pólya vector field of any function you wish to integrate. You can then quickly get a feel for the value of the integral by looking at how much the field flows along and across the contour. For example, the integral of $(\overline{z}^2 z)$ along the line-segment from $1-i$ to $1+i$ is clearly a positive multiple of i. Why? For more on the nitty-gritty of estimating integrals with (11.11), see Braden (1987).

Our interest in (11.11) will be less in this practical aspect, and more in its theoretical import: ideas about flows and force fields can shed light on complex integration, and vice versa. In what follows we shall give examples in both directions.

11.2.2 Cauchy's Theorem

Given a picture of the vector field of a complex mapping $H(z) = u+iv$, how can we tell whether or not H is *analytic*? To my knowledge there is no satisfactory answer to this question as posed. However, there is an answer if we instead look at the Pólya vector field, and it is an answer that exhibits a beautiful connection between physics and complex analysis:

The Pólya vector field of H *is divergence-free and curl-free if and only if* H *is analytic.* (11.12)

The verification is a simple calculation:

$$\nabla \cdot \overline{H} = \begin{pmatrix} \partial_x \\ \partial_y \end{pmatrix} \cdot \begin{pmatrix} u \\ -v \end{pmatrix} = \partial_x u - \partial_y v,$$

and

$$\nabla \times \overline{H} = \begin{pmatrix} \partial_x \\ \partial_y \end{pmatrix} \times \begin{pmatrix} u \\ -v \end{pmatrix} = -(\partial_x v + \partial_y u).$$

Thus the divergence and curl of $\overline{H}$ will both vanish if and only if the Cauchy–Riemann equations are satisfied. Note for future use that these two equations are really two aspects of a single complex equation,

$$i\,\partial_x H - \partial_y H = \nabla \times \overline{H} + i\,\nabla \cdot \overline{H}, \tag{11.13}$$

the vanishing of the LHS being the compact form of the CR equations, (5.1) on page 246.

[4] See Braden (1985), Braden (1987)—which won the MAA's *Carl B. Allendoerfer Award* in 1988—and Braden (1991).

With this connection established, we now have a second, *physical* explanation of Cauchy's Theorem which is scarcely less intuitive than the geometric one in Chapter 8. For if H is analytic everywhere inside a simple loop K bounding a region R, its Pólya vector field in R will have (as a flow) zero flux density and (as a force field) zero work density. This means that there is no net flux of fluid out of R, and that a puck fired round K returns with its kinetic energy unchanged. From (11.11) we see that the integral of H round K must vanish.

A more mathematical version of this physical explanation was given at the end of the last section in terms of the theorems of Gauss and Stokes. Restating that argument in the present context, for a simple loop K bounding a region R, substitution of (11.7) and (11.8) into (11.11) yields

$$\oint_K H(z)\,dz = \iint_R [\nabla\times\overline{H}]\,d\mathcal{A} + i\iint_R [\nabla\cdot\overline{H}]\,d\mathcal{A}, \qquad (11.14)$$

which vanishes if $\overline{H}$ is curl-free and divergence-free in R.

11.2.3 *Example: Area as Flux*

As a fun and instructive example, let us reconsider the result

$$\oint_K \overline{z}\,dz = 2i\mathcal{A} \qquad (11.15)$$

in the light of the physically intuitive theorems of Gauss and Stokes.

Observe that the Pólya vector field of $H(z) = \overline{z}$ is $\overline{H(z)} = z$, which flows radially outwards from the origin, like a source. However, unlike a source, here the speed of the flow *increases* with distance, making it clear that this flow cannot be divergence-free. Indeed, calculating its flux density, we find that

$$\nabla\cdot\overline{H} = \begin{pmatrix}\partial_x\\ \partial_y\end{pmatrix}\cdot\begin{pmatrix}x\\ y\end{pmatrix} = 2.$$

In other words, in each unit of time, 2 units of fluid are being pumped into each unit of area. The flux of fluid out of K is therefore $2\mathcal{A}$. On the other hand the flow *is* curl-free:

$$\nabla\times\overline{H} = \begin{pmatrix}\partial_x\\ \partial_y\end{pmatrix}\times\begin{pmatrix}x\\ y\end{pmatrix} = 0,$$

so there is no circulation round K. Inserting these facts into (11.14) we obtain (11.15).

Figure [11.8] is a concrete example of this new way of looking at (11.15), the shape of K having been chosen so as to make the values of the circulation and flux obvious.

Clearly $\overline{H(z)} = z$ has no circulation along either of the arcs, and it has equal and opposite circulations along the line-segments. The total circulation round K

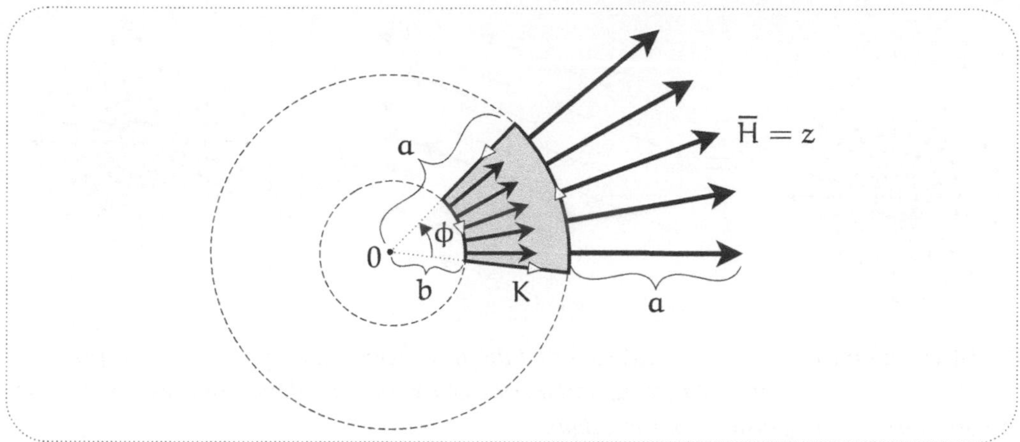

[11.8] Area as one half the flux of $\overline{H} = z$. *Clearly, $\overline{H(z)} = z$ has no circulation along either of the arcs, and it has equal and opposite circulations along the line-segments. The total circulation round K therefore vanishes. Equally clearly, there is no flux across the line-segments, but there is across the arcs. The larger arc has length $a\phi$ and the speed of the fluid crossing it is a, so the flux across it is $a^2\phi$; similarly, for the smaller arc it is $b^2\phi$. Thus, $\mathfrak{F}[z, K]$ = (fluid out) − (fluid in) = 2 (shaded area). Therefore,* $\oint_K \overline{z}\, dz = \mathfrak{W}[z, K] + i\mathfrak{F}[z, K] = 2i\mathcal{A}$.

therefore vanishes. Equally clearly, there is no flux across the line-segments, but there is across the arcs. The larger arc has length $a\phi$ and the speed of the fluid crossing it is a, so the flux across it is $a^2\phi$; similarly, for the smaller arc it is $b^2\phi$. Thus,

$$\mathfrak{F}[z, K] = (\text{fluid out}) - (\text{fluid in}) = 2\left[\tfrac{1}{2}\, a^2\, \phi - \tfrac{1}{2}\, b^2\phi\right] = 2\,(\text{shaded area}).$$

Before moving on, let us clear up a paradoxical feature of the vector field z: fluid is being pumped in *uniformly* throughout the plane, and yet the flow appears to radiate from one special place, namely, the origin. The resolution (see [11.9]) lies in the trivial identity $z = z_0 + (z - z_0)$, which says that the flow from the origin is the superposition of the sourceless, irrotational field z_0 and a copy of the original flow, but now centred on the arbitrary point z_0 instead of the origin.

11.2.4 Example: Winding Number as Flux

Next, let us see how the Pólya vector field also breathes fresh meaning into the fundamentally important formula

$$\oint_L \frac{1}{z}\, dz = 2\pi i\, \nu\, [L, 0]. \tag{11.16}$$

According to (11.11),

$$\oint_L \frac{1}{z}\, dz = \mathfrak{W}\,[(1/\overline{z}), L] + i\,\mathfrak{F}\,[(1/\overline{z}), L].$$

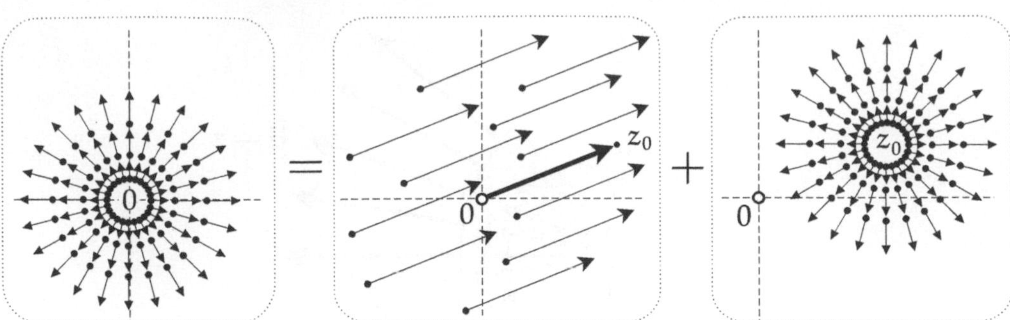

[11.9] *The identity $z = z_0 + (z - z_0)$ says that the flow from the origin is the superposition of the sourceless, irrotational field z_0 and a copy of the original flow, but now centred on the arbitrary point z_0 instead of the origin.*

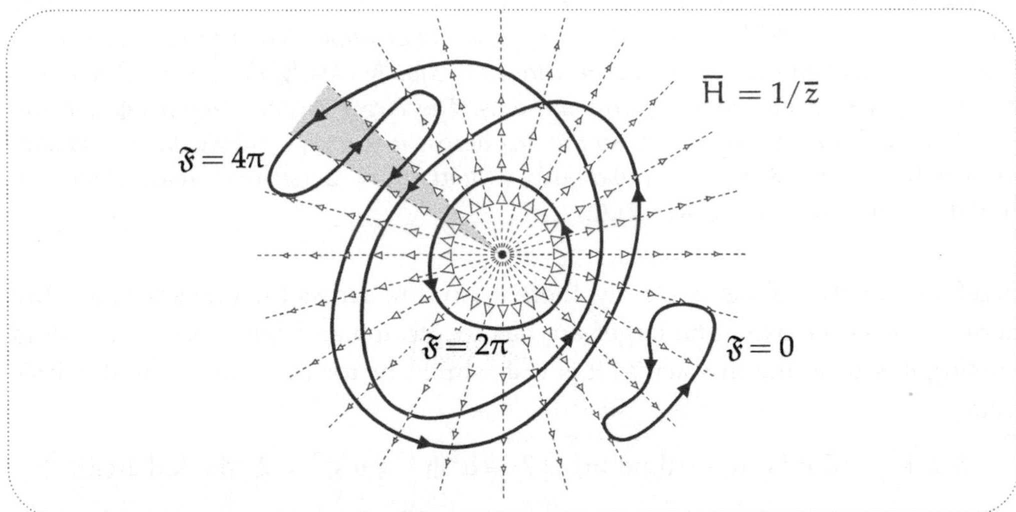

[11.10] Winding Number interpreted as the flux of a source of strength $(1/2\pi)$. *As we discussed in the new Preface, the Pólya vector field of $(1/z)$, namely $(1/\overline{z})$, represents the illustrated source of strength 2π at the origin. Choosing K to be an origin-centred circle K of radius r, traced counterclockwise, we obtain an* immediate visual and physical explanation *of the iconic fact that $\oint_K \frac{1}{z} dz = 2\pi i$. For clearly there is no flow along K, and since P flows orthogonally across K (from left to right) with speed $(1/r)$, its flux across K is $(2\pi r)(1/r) = 2\pi$. Furthermore, this physical interpretation explains why the value of the integral will* remain $2\pi i$ *if K is continuously deformed into a general loop encircling the source at the origin, so long as K does not* cross *that point as it is deformed. And if the loop encircles the source ν times, then the intercepted flux is clearly $2\pi\nu$. Thus if the strength of the source is reduced to $(1/2\pi)$, then ν simply* is *the flux.*

But the Pólya vector field $(1/\overline{z})$ is an old friend—it is a source of strength 2π located at the origin.

Figure [11.10] illustrates the intuitive nature of the result from the new point of view. If a loop does not enclose the source, just as much fluid flows out as in; if a simple loop does enclose the source, it intercepts the full 2π of fluid being pumped in at the origin; more generally, a loop will accrue 2π of flux each time it encircles the source.

To finish the explanation of (11.16) we must show that a source is *pure* flux, i.e. every loop has vanishing work or circulation. Since a source is curl-free except at the origin, Stokes' Theorem guarantees vanishing work for simple loops that do not contain 0. If the loop does contain 0 then it's not so obvious. However, it *is* obvious for an origin-centred circle. You can now finish the argument for yourself by appealing to the Deformation Theorem (11.10).

In connection with another matter, consider the shaded sector in [11.10]. The same amount of fluid will cross each segment of a contour which passes through it, but the *sign* of the flux will depend on the direction of the contour. Try meditating on the connection between this fact and the crossing rule for winding numbers [(7.1), p. 388].

*11.2.5 Local Behaviour of Vector Fields**

We previously showed that $\nabla\cdot\overline{H}$ and $\nabla\times\overline{H}$ represent the flux density and work density of $\overline{H}$ for infinitesimal squares. However, in order for the formulae (11.7) and (11.8) to really make sense it is necessary that these interpretations persist for infinitesimal loops of *arbitrary* shape. Let us now place (11.7) and (11.8) on firmer ground by verifying the shape-independent significance of the divergence and curl. To do so we will first analyse the local behaviour of a general Pólya vector field $\overline{H}$ in the neighbourhood of the origin. The generalization to points other than the origin will be obvious.

As the point $z = x + iy$ moves towards the origin, $H(z)$ will ultimately equal the following formula, in which the partial derivatives are evaluated at 0:

$$\begin{aligned} H(z) - H(0) &\asymp x\,\partial_x H + y\,\partial_y H = \tfrac{1}{2}(z+\overline{z})\,\partial_x H - \tfrac{i}{2}(z-\overline{z})\,\partial_y H \\ &= \tfrac{1}{2}\,[\partial_x H - i\,\partial_y H]\,z + \tfrac{1}{2}\,[\partial_x H + i\,\partial_y H]\,\overline{z}. \end{aligned}$$

This will become exact in the limit that $|z|$ shrinks to nothing.

Turning to the Pólya vector field itself, and substituting (11.13), we find

$$\overline{H(z)} \asymp \overline{H(0)} + (\nabla\cdot\overline{H})\frac{z}{2} + (\nabla\times\overline{H})\frac{iz}{2} + C\,\overline{z}, \tag{11.17}$$

where $\overline{C} = \frac{1}{2}\,[\partial_x H - i\,\partial_y H]$. Note that if H is analytic, in which case $\overline{H}$ is sourceless and irrotational, then (11.17) correctly reduces to the first two terms of Taylor's series: $H(z) = H(0) + H'(0)z + \cdots$.

The meaning of the decomposition (11.17) is illustrated in [11.11]. Unless $H(0) = 0$, the constant first term dominates: vectors near the origin differ little from the vector *at* the origin. The remaining three terms correct this crude approximation. The second term describes a vector field (cf. figure [11.8]) that is

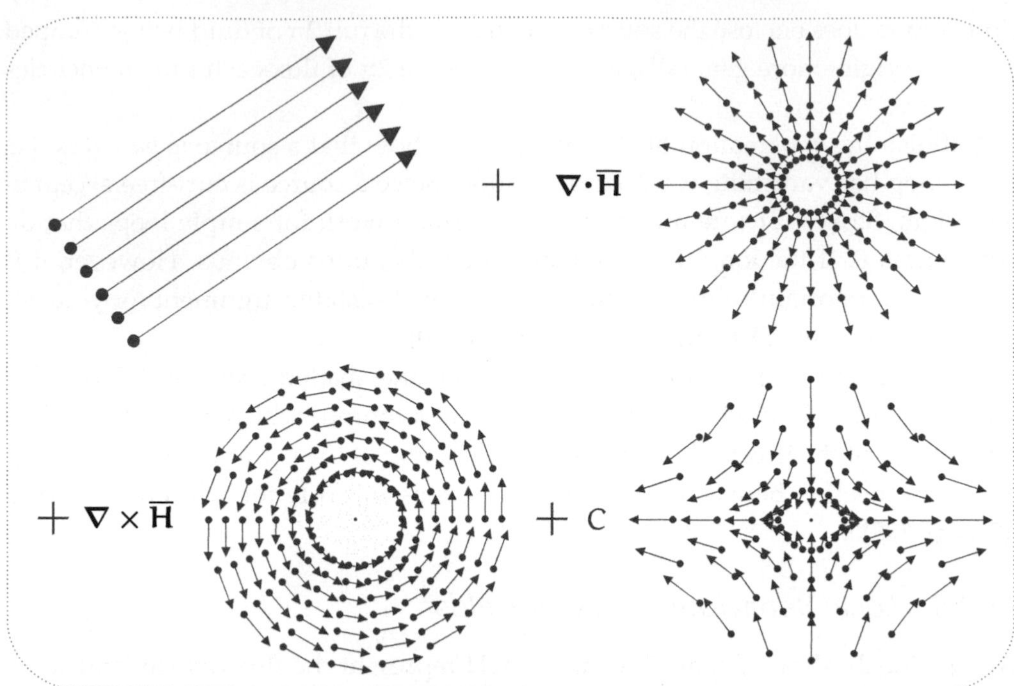

[11.11] Local behaviour of a general vector field. *In the limit that* $z \to 0$*, the general Pólya vector field is given by this ultimate equality:* $\overline{H(z)} \asymp \overline{H(0)} + \frac{1}{2}(\nabla \cdot \overline{H})z + \frac{1}{2}(\nabla \times \overline{H})iz + C\,\bar{z}$*. Unless* $H(0) = 0$*, the constant first term dominates: vectors near the origin differ little from the vector at the origin. The remaining three terms correct this crude approximation. The second term describes a vector field (cf. [11.8]) that is irrotational and has constant divergence, equal to that of* $\overline{H}$ *at the origin. The third term describes a vector field that is sourceless and has constant curl, equal to that of* $\overline{H}$ *at the origin. The final term is both irrotational and sourceless.*

irrotational and has constant divergence, equal to that of $\overline{H}$ at the origin. The third term describes a vector field that is sourceless and has constant curl, equal to that of $\overline{H}$ at the origin. The final term is both irrotational and sourceless.

Note that this decomposition is geometrically meaningful because the appearance of each of the component vector fields is qualitatively unaffected by the value of its coefficient[5]. We hope these observations make the formula (11.17) both plausible and meaningful.

Now let us return to the original problem. Let K be a small simple loop of arbitrary shape round the origin, and let $\mathcal{A}$ be the area it encloses. We wish to show that the divergence and curl of $\overline{H}$ at 0 are the limiting values of the flux per unit area and work per unit area as K shrinks to the origin. Using (11.17) in (11.11) we find

[5] This is obvious for the source and vortex terms, but not for the last term; see Ex. 10.

$$
\begin{aligned}
&\mathfrak{W}[\overline{H}, K] + i\,\mathfrak{F}[\overline{H}, K] \\
&= \oint_K H(z)\,dz \\
&= H(0)\oint_K dz + \tfrac{1}{2}\left[\nabla\cdot\overline{H} - i\,\nabla\times\overline{H}\right]\oint_K \overline{z}\,dz + \overline{C}\oint_K z\,dz.
\end{aligned}
$$

This becomes exact as K shrinks to nothing. But even if K is not small, we know that the exact values of these three integrals are

$$\oint_K dz = 0, \quad \oint_K \overline{z}\,dz = 2i\mathcal{A}, \quad \oint_K z\,dz = 0.$$

Thus

$$\mathfrak{W}[\overline{H}, K] + i\,\mathfrak{F}[\overline{H}, K] = \left[\nabla\times\overline{H} + i\,\nabla\cdot\overline{H}\right]\mathcal{A}.$$

Equating real and imaginary parts, we obtain the desired results.

11.2.6 Cauchy's Formula

The Pólya vector field also allows us to cast the mathematical explanation of Cauchy's Formula into a form that is more accessible to physical intuition.

Consider the function

$$H(z) = \frac{f(z)}{(z-p)},$$

where $f(z)$ is analytic. Since H is analytic except at p, its Pólya vector field $\overline{H}$ will have vanishing flux and circulation densities except at p. Thus if C is a simple loop round p, all of its flux and circulation must have originated at p. To find $\mathfrak{W}[\overline{H}, C]$ and $\mathfrak{F}[\overline{H}, C]$ we should therefore examine $\overline{H}$ in the immediate vicinity of p.

If $f(p) = A + iB$, then very close to p the Pólya vector field $\overline{H}$ will be indistinguishable from

$$\frac{A - iB}{\overline{z} - \overline{p}} = A\left[\frac{1}{\overline{z} - \overline{p}}\right] - B\left[\frac{i}{\overline{z} - \overline{p}}\right].$$

Figure [11.12] illustrates this field for positive A and B, as well as showing the geometric significance of the algebraic decomposition above.

The first term is familiar as a source at p of strength $2\pi A$, a negative value for A corresponding to a sink. The second term is a multiple of the less familiar field $i/(\overline{z} - \overline{p})$ which represents a *vortex*[6] at p. It is easy to see that the circulation round one of its circular streamlines is 2π, so this will also be its value for any simple loop round p—we say that the vortex has strength 2π. On the other hand its flux vanishes for all loops. While a source is pure flux, a vortex is pure circulation.

[6] We are now using this term in a narrow sense—previously "vortex" referred to all vector fields of this topological form.

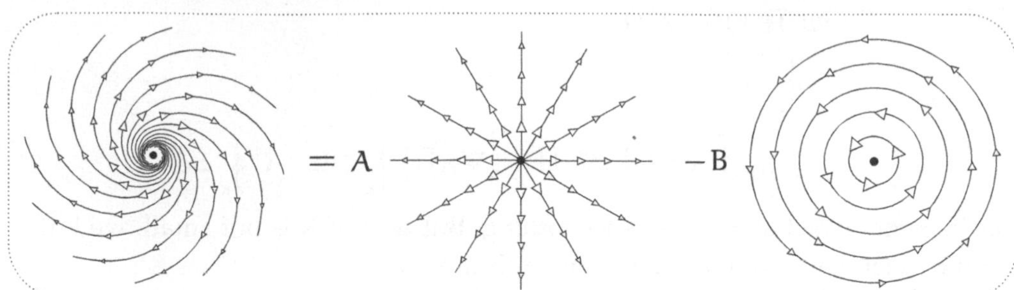

[11.12] A more physical view of Cauchy's Formula. *If* $f(p) = A + iB$, *then as* $z \to p$, *the local behaviour of* $H(z) \equiv f(z)/(z-p)$ *is* $H(z) \asymp A[1/(\bar{z}-\bar{p})] - B[i/(\bar{z}-\bar{p})]$, *as illustrated. The first term is a source (pure flux) of strength* $2\pi A$. *The second term represents a* vortex *(pure circulation) such that the circulation round one of its circular streamlines is* $-2\pi B$. *If* C *is a simple loop around* p, *then* $\oint_C \frac{f(z)}{(z-p)}\, dz = \mathfrak{W}[\overline{H}, C] + i\,\mathfrak{F}[\overline{H}, C] = -2\pi B + i\,2\pi A = 2\pi i\, f(p)$.

These observations give us a more physical view of Cauchy's Formula:

$$\begin{aligned}\oint_C \frac{f(z)}{(z-p)}\, dz &= \mathfrak{W}[\overline{H}, C] + i\,\mathfrak{F}[\overline{H}, C] \\ &= -2\pi B + i\,2\pi A \\ &= 2\pi i\, f(p).\end{aligned}$$

11.2.7 Positive Powers

If n is a positive integer then z^n is analytic everywhere and its Pólya vector field $\bar{z}^n$ is correspondingly divergence-free and curl-free. Our physical version of Cauchy's Theorem therefore gives

$$\oint_C z^n\, dz = \mathfrak{W}[\bar{z}^{\,n}, C] + i\,\mathfrak{F}[\bar{z}^{\,n}, C] = 0.$$

At least in the case of an origin-centred circle we can make this much more vivid[7]. Figure [11.13] illustrates the behaviour of $\bar{z}$ and $\bar{z}^2$ on such a circle. It now seems clear that as much fluid flows into each shaded disc as flows out, so that $\mathfrak{F} = 0$, and also (when viewed as force fields) that no net work is done in transporting a particle round the boundary of each disc, so that $\mathfrak{W} = 0$.

We can make this idea precise. First note that for any vector field on the circle, the work and flux will not change their values if we perform an arbitrary rotation of the diagram about the centre of the circle. Next, let us exploit the attractive symmetries of these particular vector fields. Rotating the picture of $\bar{z}$ through $(\pi/2)$ clearly

[7] In the particular case of z^2 this has also been observed by Braden (1991), though he did not supply the general argument which follows.

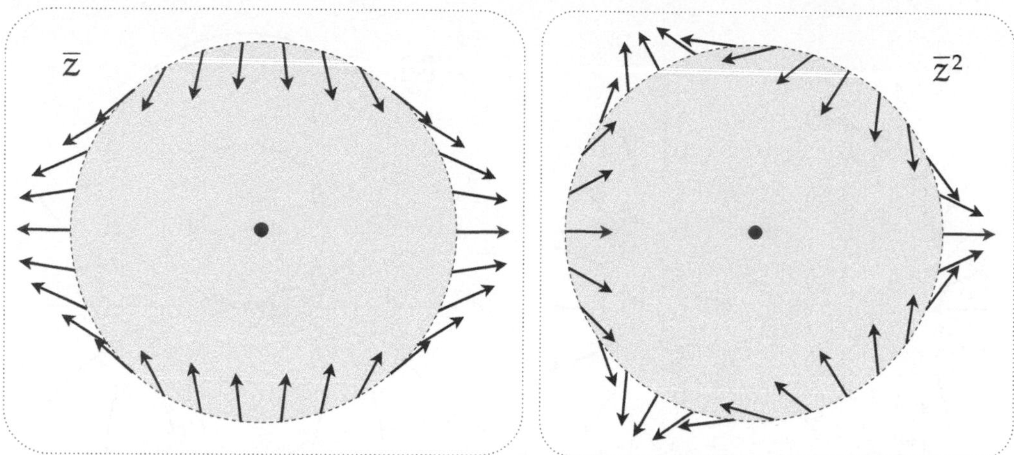

[11.13] Pólya vector field explanation of $\oint_C z^n\,dz = \mathfrak{W}[\overline{z}^n, C] + i\mathfrak{F}[\overline{z}^n, C] = 0$. *Draw the Pólya vector field $\overline{z}^n$ of z^n on an origin-centred circle. Rotating this picture about the origin through any angle cannot change either the work or the flux. But rotating the picture by $\pi/(n+1)$ yields the* negative *of the original field, changing the sign of both the work and the flux. Since the work and the flux must remain unchanged* and *reverse sign, they must both vanish.*

yields the negative of the original field and, correspondingly, the negative of the original work and flux. Since $\mathfrak{W}$ and $\mathfrak{F}$ are simultaneously required to remain the same *and* to reverse sign, they must both vanish.

The same argument applies to $\overline{z}^2$ under a rotation of $(\pi/3)$, and to $\overline{z}^n$ under a rotation of $\pi/(n+1)$. Use your computer to check this for $n = 3$. To understand this symmetry better, consult Ex. 10.

11.2.8 Negative Powers and Multipoles

Consider the negative power functions $(1/z^m)$, where m is a positive integer. Their Pólya vector fields $(1/\overline{z^m})$ will be divergence-free and curl-free except at the singularity at the origin. Thus if a simple loop C does not enclose the origin, its circulation and flux will vanish. However, since we know from the case $m = 1$ that singularities *are* capable of generating flux and circulation, it remains something of a mystery that (except for $m = 1$) $\mathfrak{W}$ and $\mathfrak{F}$ also vanish if C *does* enclose the singularity.

In the case of an origin-centred circle we can visualize this result exactly as for positive powers. Figure [11.14a] illustrates this for the so-called *dipole* field $(1/\overline{z^2})$. The argument is also the same as before: this vector field is reversed under a rotation of π, and for $(1/\overline{z^m})$ it is reversed under a rotation of $\pi/(m-1)$. Knowing that $\mathfrak{W}$ and $\mathfrak{F}$ vanish for the circle tells us [see (11.9), (11.10)] that they will continue to vanish for any loop into which we may deform the circle without crossing the origin.

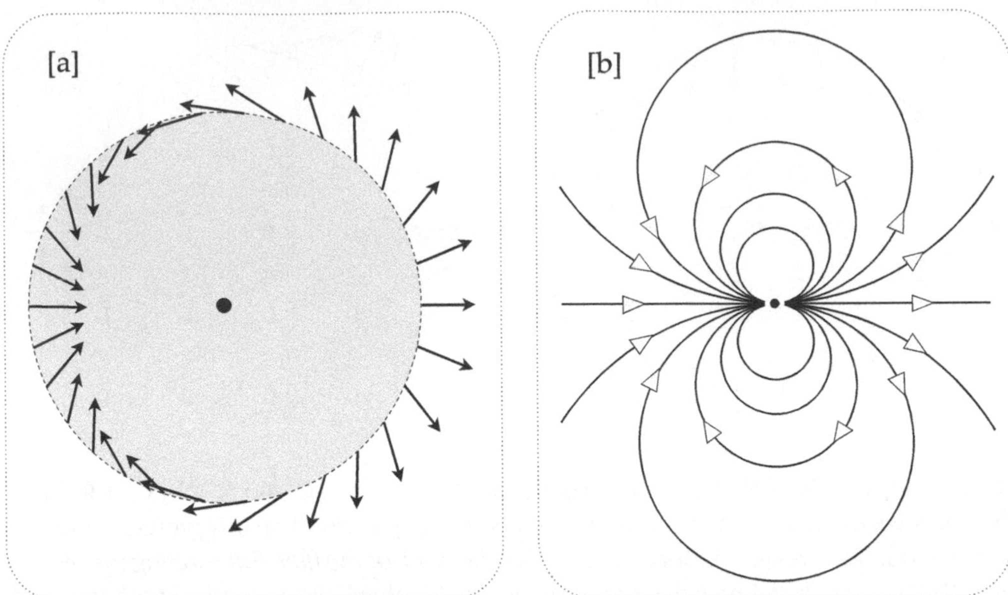

[11.14] The Dipole Pólya vector field $(1/\overline{z}^2)$. [a] *If we draw the field on a circle, then a rotation of π reverses the field, so by the same reasoning as the previous figure, $\oint_C (1/z^2)\,dz = \mathfrak{W}[(1/\overline{z}^2), C] + i\mathfrak{F}[(1/\overline{z}^2), C] = 0$. In general, $(1/\overline{z}^m)$ is reversed under a rotation of $\pi/(m-1)$, so $\oint_C (1/z^m)\,dz = 0$.* **[b]** *The full phase portrait of the dipole field. Note that the streamlines are perfect circles.*

Let us now go beyond this geometric explanation in search of a compelling *physical* explanation. Figure [11.14b] shows the phase portrait of the dipole $(1/\overline{z^2})$, the streamlines of which are apparently circular; a simple geometric argument [exercise] confirms their perfect circularity. Where have we seen something like this before? Answer: the doublet field consisting of a source and sink of equal strength S (see [10.3], p. 514). It therefore looks as though we can obtain the dipole simply by coalescing the source and sink. This solves our mystery in a surprising and elegant fashion: neither the source nor the sink generate circulation, and a loop enclosing both receives equal and opposite fluxes.

This explanation is essentially correct. However, as the sink and source move closer and closer together, a greater and greater proportion of the fluid from the source is swallowed up by the sink before it can go anywhere, and at the moment of coalescence the source and sink annihilate each other, leaving no field at all. Let us investigate this algebraically using (10.1), p. 518.

Suppose that the source and sink approach the origin along a fixed line L making an angle ϕ with the real axis. This streamline of symmetry L is called the *axis* of the doublet. Putting $A = \epsilon\, e^{i\phi} = -B$, the doublet field (10.1) becomes

$$D(z) = \left[\frac{2\epsilon S\, e^{-i\phi}}{2\pi}\right] \frac{1}{(\overline{z}^2 - \epsilon^2\, e^{-i2\phi})}, \tag{11.18}$$

which dies away as the source/sink separation 2ϵ tends to zero. The solution is to increase the strength S in inverse proportion to the separation 2ϵ, so that $2\epsilon S$ remains constant. If we call this real constant $2\pi k$, the limiting doublet field (as $\epsilon \to 0$) is

$$D(z) = \frac{k\, e^{-i\phi}}{\overline{z}^2},$$

i.e. the general dipole field obtained by rotating [11.14] by $+\phi$ and scaling up the speed of the flow by k, which we may think of as the "strength" of the dipole. Thus the Pólya vector field of (d/z^2) is a dipole whose axis points in the direction of d, and whose strength is $|d|$. The complex number d is called the *dipole moment*.

We created the dipole by coalescing equal and opposite sources, increasing their strength so as to avoid mutual annihilation. Continuing this game, we ask, "What will happen if we coalesce equal and opposite *dipoles*, increasing their strength so as to avoid mutual annihilation?" Figure [11.15] reveals the pleasing answer. Figure [11.15a] represents a pair of equal and opposite dipoles located at $\pm\epsilon$ and having real dipole moments $\pm d$, while [11.15b] is the Pólya vector field of $(1/z^3)$. The resemblance is striking, and we can show algebraically that [11.15b], which is called a *quadrupole*, is indeed the appropriate limiting case of [11.15a].

The field for [11.15a] is

$$Q(z) = d\left[\frac{1}{(\overline{z}-\epsilon)^2} - \frac{1}{(\overline{z}+\epsilon)^2}\right] = 4d\epsilon\,\frac{\overline{z}}{(\overline{z}^2-\epsilon^2)^2}. \tag{11.19}$$

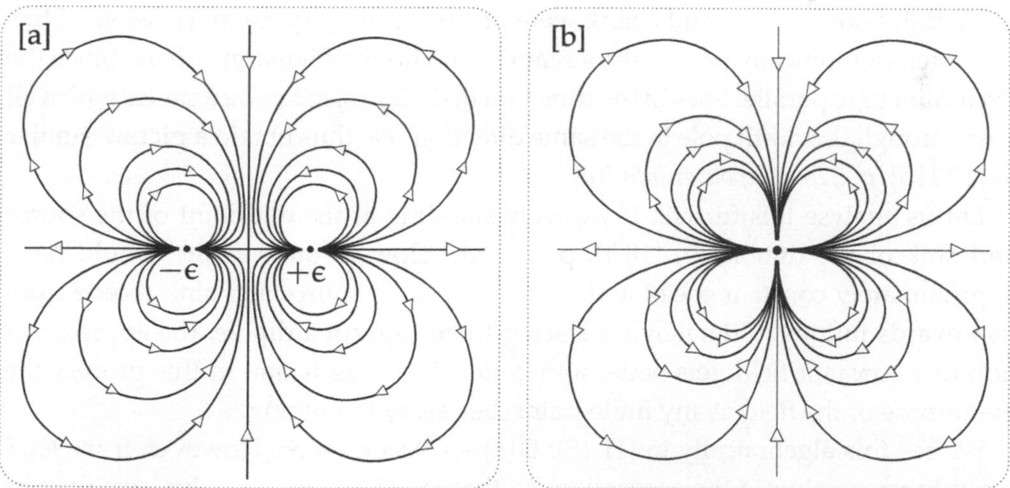

[11.15] The Quadrupole Pólya vector field ($1/\overline{z}^3$). [a] *Dipoles of strength $\pm d$ at $\pm\epsilon$ are merged by letting $\epsilon \to 0$. But in order to form a quadrupole, the dipole moments d must grow in inverse proportion to the separation ϵ, so that $d\epsilon$ remains constant, otherwise the two opposite dipoles would simply annihilate each other, leaving no field.* **[b]** *The full phase portrait of the resulting quadrupole.*

Once again letting the strength d grow in inverse proportion to the separation, so that $k=4d\epsilon$ remains constant, the coalescence of the dipoles yields the quadrupole:

$$Q(z) = \frac{k}{\overline{z}^3}.$$

In general, the Pólya vector field of (q/z^3) is called a quadrupole with *quadrupole moment* q.

We have thus explained the vanishing circulation and flux of $(1/\overline{z^3})$: each of the dipoles in [11.15a] is known not to generate any circulation or flux, so the quadrupole in [11.15b] won't either. You are invited to continue this line of thought by showing (geometrically and algebraically) that the fusion of two quadrupoles yields the so-called *octupole* field, $(1/\overline{z^4})$, and so on.

Dipoles, quadrupoles, octupoles, etc., are collectively known as *multipoles*. Similarly, dipole moments, quadrupole moments, etc., are collectively known as *multipole moments*.

11.2.9 Multipoles at Infinity

Although there is no mystery surrounding the vanishing circulation and flux for positive powers, it would still be nice to find a physical explanation analogous to the one for negative powers. To see how this can be done, we begin by considering the constant function $f(z) = a$, the Pólya vector field of which is a flow of constant speed $|a|$ in the direction $\overline{a}$.

Standing in the midst of this flow, the fluid seems to originate far over the horizon in the direction $-\overline{a}$ and to disappear over the horizon in the opposite direction, as though both a source and a sink were present at infinity. To make sense of this idea, stereographically project the streamlines onto the Riemann sphere. Since the streamlines are parallel lines in the direction $\overline{a}$, their projections are circles which all pass through the north pole in the same direction. We thus obtain a picture similar to [10.11b], p. 525: a *dipole at infinity!*

Let us analyse this further. If we were standing at the midpoint of the source and sink of the doublet in [10.3], p. 514, the flow in our vicinity would have approximately constant speed and direction. As the source and sink recede from us towards infinity, ultimately coalescing there to form a dipole, the approximation to a constant field gets better and better. The snag is that in this process the magnitude of the field at any finite point dies away to nothing.

We see this algebraically in (11.18): $D(z) \to 0$ as $\epsilon \to \infty$. However, if we let S grow in proportion to the separation, so that $(S/\epsilon) = \text{const.} = k\pi$, say, then as $\epsilon \to \infty$ the doublet field tends to the constant field $D(z) = -k\,e^{i\phi}$.

Given that z^0 yields a dipole at infinity, what might the Pólya vector field of z^1 correspond to? Use your computer to see that it is a quadrupole at infinity. Verify this algebraically using (11.19). Continuing in this fashion, one finds [exercise] that z^2 corresponds to an octupole at infinity, and so on.

11.2.10 Laurent's Series as a Multipole Expansion

The above ideas shed new light on the Laurent series and the Residue Theorem. Suppose that an otherwise analytic function $f(z)$ has a triple pole at the origin. We know from Chapter 9 that $f(z)$ will have a Laurent series of the form

$$f(z) = \underbrace{\frac{q}{z^3} + \frac{d}{z^2} + \frac{\rho}{z}}_{P(z)} + a + bz + cz^2 + \cdots . \tag{11.20}$$

In the vicinity of the singularity, the behaviour of f is governed by its principal part P, the Pólya vector field of which is

$$\overline{P(z)} = \frac{\overline{q}}{\overline{z}^3} + \frac{\overline{d}}{\overline{z}^2} + \frac{\overline{\rho}}{\overline{z}}.$$

This we now recognize to be the superposition of a quadrupole, a dipole, and a source-vortex combination of the type shown in [11.12]. Thus the principal part of the Laurent series amounts to what a physicist would call a *multipole expansion*.

To visually grasp the meaning of such an expansion, consider [11.16] which illustrates a typical $\overline{P}$. Very close to the singularity, the field is completely dominated by the quadrupole with its characteristic four loops, but as we move slightly further away the quadrupole's influence wanes relative to the dipole. Indeed, at this intermediate range we clearly see the characteristic two loops of a dipole. Finally, at still greater distances, both the quadrupole and the dipole become insignificant relative to the source-vortex, the precise form of which is determined solely by the residue ρ. Compare with [11.12], in which $\rho = A + iB$.

Continuing our outward journey, now well beyond the unit circle, the entire principal part becomes negligible relative to the remaining terms of (11.20). First a becomes important, then bz takes over, and so on. Thus as we approach infinity the field at first resembles a dipole, then a quadrupole, and so on. However, unlike the approach to the pole, on the journey to infinity we may experience multipoles of greater and greater order, without end.

Of course in general f may possess other singularities and (11.20) will cease to be meaningful when $|z|$ increases to the distance of the nearest one. Nevertheless, in the region where it is valid, we may still think of the non-negative powers as representing multipoles at infinity.

To recap, Laurent's series and the Residue Theorem may be conceived of physically as follows. The only term capable of generating circulation and flux is $\overline{(\rho/z)}$, which may itself be decomposed into a vortex of strength $\mathfrak{W} = -2\pi\,\mathrm{Im}(\rho)$ and a source of strength $\mathfrak{F} = 2\pi\,\mathrm{Re}(\rho)$. All the other terms correspond to multipoles which generate neither circulation nor flux; a finite collection of these reside at the pole, while the rest are at infinity.

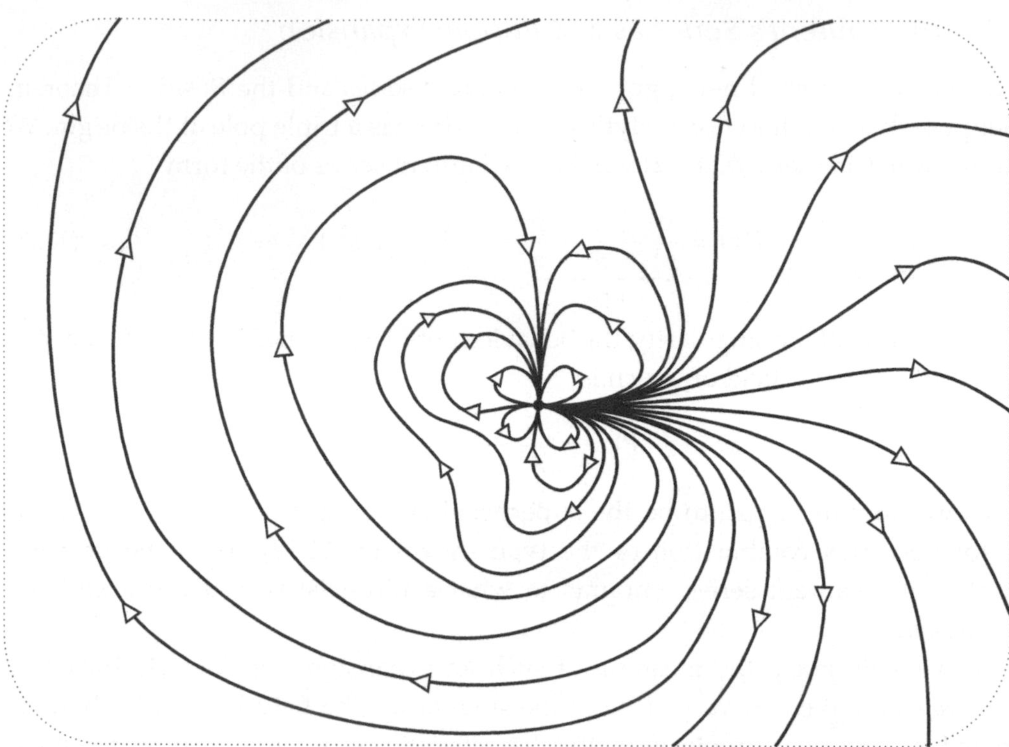

[11.16] Multipole expansions and the Residue Theorem. *Consider the Pólya vector field* $\overline{f(z)}$ *of an analytic function in the vicinity of a triple pole. The dominant behaviour near the singularity is given by the conjugate of the principal part of the Laurent series:* $\overline{P(z)} = \frac{\overline{q}}{\overline{z}^3} + \frac{\overline{d}}{\overline{z}^2} + \frac{\overline{\rho}}{\overline{z}}$, *which we recognize to be the superposition of a quadrupole, a dipole, and a source–vortex combination of the type shown in [11.12]. Thus the principal part of the Laurent series amounts to what a physicist would call* a multipole expansion. *Very close to the singularity, the field is completely dominated by the quadrupole with its characteristic four loops, but as we move slightly further away the quadrupole's influence wanes and we see the characteristic two loops of a dipole. Finally, at still greater distances, both the quadrupole and the dipole become insignificant relative to the source–vortex, the precise form of which is determined solely by the residue* ρ. *Neither the quadrupole nor the dipole generates any circulation or flux, so if* L *encircles the pole, the circulation and flux is entirely due to the source–vortex:* $\oint_L f(z)\,dz = 2\pi i\rho$.

11.3 The Complex Potential

11.3.1 Introduction

Phase portraits are so convenient that it is easy to forget that in general they cannot represent the *lengths* of the vectors. In this section we shall see that if a vector field is either sourceless or irrotational (or both) then there exists a special way of drawing the phase portrait so that the lengths *are* represented.

Although we shall ultimately be concerned with the Pólya vector fields of analytic functions, which are both sourceless *and* irrotational, it is more instructive to analyse the implications for sourcelessness and irrotationality separately. Nevertheless, in view of the final objective, we shall continue to write the vector field as $\overline{H}$.

11.3.2 The Stream Function

First let $\overline{H}$ be a sourceless flow of fluid. The Deformation Theorem (11.9) tells us that the flux across a curve connecting two given points is independent of the choice of the curve. Thus if K is any contour from an arbitrary fixed point a to a variable point z, the flux across it, namely

$$\Psi(z) \equiv \mathfrak{F}[\overline{H}, K],$$

will be a well-defined function of z, called the *stream function*. If we choose a different point a then the new stream function will only differ from the old one by an additive constant.

Suppose that z lies anywhere on the streamline through a. See [11.17]. Choosing K to be the portion of the streamline from a to z, we see that $\Psi(z) = 0$. Similarly, suppose that q lies anywhere on a streamline through another point p. Taking K to be a path from a to p, followed by the section of the streamline from p to q, we see that $\Psi(q) = \Psi(p)$. In other words,

The streamlines are the level curves of the stream function Ψ.

Instead of constructing the phase portrait by drawing random streamlines, suppose we do it as follows: *choose a number* k *and draw just those streamlines for which* $\Psi = 0, \pm k, \pm 2k, \pm 3k, \ldots$ See [11.17]. Having drawn the phase portrait in this special way, the speed of the flow is represented by the crowding together of the streamlines. Let's justify this claim and make it more precise.

Since no fluid crosses the streamlines, we may think of the region lying between two adjacent ones as a tube down which fluid flows. Any curve connecting the two sides will have the same flux, namely k. Adapting the language of Faraday and Maxwell, we may thus describe the tube more quantitatively as a k-*flux tube*.

The shaded area in [11.17] is part of one such tube, the initial and final cross-sections (lengths ϵ_1 and ϵ_2) having been drawn perpendicular to the flow. If k is chosen small, the speed $v = |\overline{H}|$ of the flow will be approximately constant across these ends, say v_1 and v_2. Thus the fluxes into and out of the shaded region (which must both equal k) are approximately $\epsilon_1 v_1$ and $\epsilon_2 v_2$. As k is chosen smaller and smaller, these expressions become more and more accurate:

$$v_1 = \frac{k}{\epsilon_1} \quad \text{and} \quad v_2 = \frac{k}{\epsilon_2}. \tag{11.21}$$

In order to maintain a constant flux k, the speed must decrease as the tube widens.

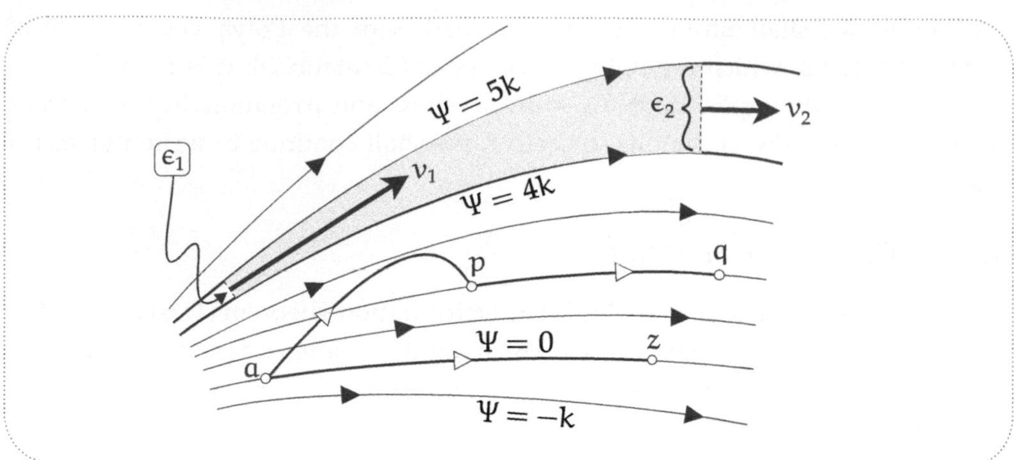

[11.17] The Stream Function Ψ and the k-flux tubes of a sourceless flow. *If* $\overline{H}$ *is sourceless, and* K *is any contour connecting an arbitrary fixed point* a *to a variable point* z*, the flux across it is independent of* K*, yielding the well-defined* stream function, $\Psi(z) \equiv \mathfrak{F}[\overline{H}, K]$. *Now choose a number* k *and draw just those streamlines for which* $\Psi = 0, \pm k, \pm 2k, \pm 3k, \ldots$ *Any cross section of a channel between neighbouring streamlines therefore intercepts the same flux* k*, and so we call it a* k-flux tube. *This special phase portrait (pioneered by Faraday and Maxwell) has the great advantage that the speed of the flow is now* visible via $|\overline{H}| \asymp (k/\epsilon)$: *the denser the streamlines, the faster the flow.*

To summarize:

Let the phase portrait of a sourceless vector field be constructed using k-*flux tubes. If* k *is chosen small, the speed of the flow at any point will be approximately given by* k *divided by the width of the tubes in the vicinity of the point. For infinitesimal* k*, the result is exact.* (11.22)

However, since the number of k-flux tubes passing through a given region will vary inversely with k, our phase portrait will get very cluttered if k is chosen too small. In practice (cf. [10.3], p. 514) we get a good feel for the speed of the flow with relatively few streamlines.

Let's apply these ideas to the simple (non-analytic) example $H(z) = i\bar{z}$. The Pólya vector field is then

$$\overline{H} = -iz \iff \overline{H} = \begin{pmatrix} y \\ -x \end{pmatrix},$$

the streamlines of which are clockwise circles round the origin, the speed of the flow round each one being equal to its radius. See [11.18].

Although this vector field is not irrotational [$\nabla \times \overline{H} = -2$], it is sourceless [$\nabla \cdot \overline{H} = 0$], and thus it possesses a stream function. For convenience's sake, let's choose $a = 0$. We already know that the streamlines are origin-centred circles, so to find the value of Ψ on the streamline of radius R we must find the flux for any

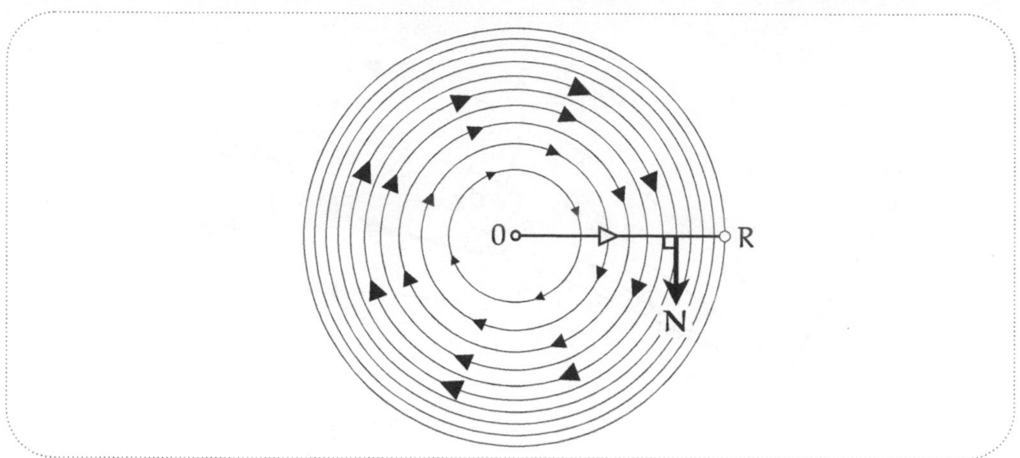

[11.18] The Pólya vector field $\overline{H} = -i\,z$, drawn using k-flux tubes.

path from the origin to any point on this circle. Choosing the path to be the portion of the positive real axis from 0 to R, we see that

$$ds = dx \quad \text{and} \quad \mathbf{N} = \begin{pmatrix} 0 \\ -1 \end{pmatrix}.$$

Thus

$$\Psi = \int (\overline{H} \cdot \mathbf{N})\, ds = \int_0^R x\, dx = \tfrac{1}{2} R^2.$$

Knowing the stream function we are now in a position to draw the special phase portrait. Choosing $k = (1/2)$ we find that the radii of the streamlines are $\sqrt{1}, \sqrt{2}, \sqrt{3}, \ldots$ Figure [11.18] illustrates these streamlines, and qualitatively confirms the prediction of (11.22). As we move outward from the origin the streamlines become more crowded together, reflecting the increasing speed of the flow.

11.3.3 The Gradient Field

We have seen in geometrical terms how it is possible to reconstruct a sourceless vector field $\overline{H}$ from a knowledge of its stream function Ψ. In order to find a simple *formula* for $\overline{H}$ in terms of Ψ, we need the concept of the *gradient field* $\nabla\Psi$. This is defined to be the vector field

$$\nabla\Psi = \begin{pmatrix} \partial_x \\ \partial_y \end{pmatrix} \Psi = \begin{pmatrix} \partial_x \Psi \\ \partial_y \Psi \end{pmatrix} \quad \Longleftrightarrow \quad \nabla\Psi = \partial_x \Psi + i\,\partial_y \Psi.$$

The gradient field $\nabla\Psi$ has a simple geometric interpretation in terms of the streamlines of [11.17]. To see this, we express the infinitesimal change $d\Psi$ resulting from an infinitesimal movement $dz = dx + i\,dy$ as a dot product:

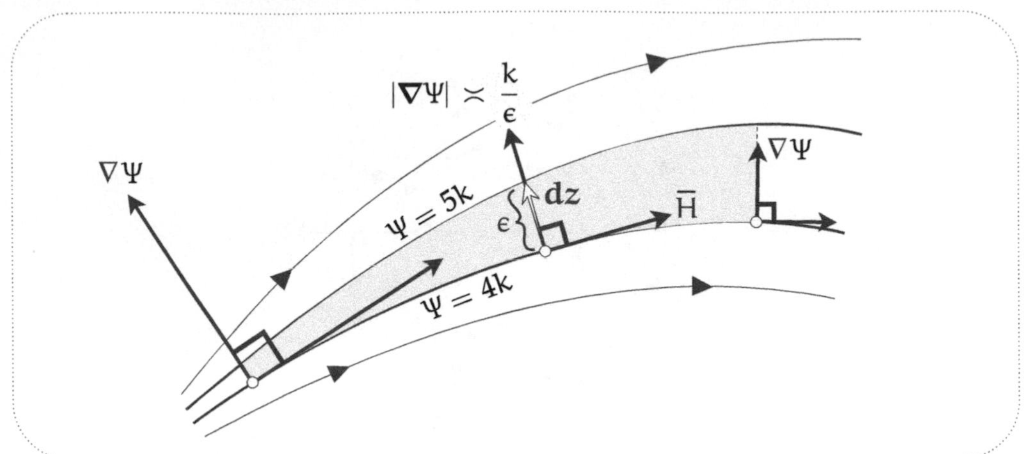

[11.19] The field in terms of the stream function: $\overline{H} = -i\,\nabla\Psi$. *First, the direction of $\nabla\Psi$ is the one that is orthogonal to the streamlines and along which Ψ increases. Thus $-i\,\nabla\Psi$ points in the direction of $\overline{H}$. Second, $|\nabla\Psi| \asymp (k/\epsilon) \asymp |\overline{H}|$. Done.*

$$d\Psi = (\partial_x\Psi)\,dx + (\partial_y\Psi)\,dy = \begin{pmatrix} \partial_x\Psi \\ \partial_y\Psi \end{pmatrix} \cdot \begin{pmatrix} dx \\ dy \end{pmatrix} = \boldsymbol{\nabla}\Psi \cdot \mathbf{dz}.$$

If dz is tangent to a streamline then $d\Psi = 0$, so $\nabla\Psi$ has vanishing dot product with this direction. Also, Ψ increases when dz makes an acute angle with $\nabla\Psi$. Thus

The direction of $\boldsymbol{\nabla}\Psi$ is the one that is orthogonal to the streamlines and along which Ψ increases. Thus $-i\,\nabla\Psi$ points in the direction of $\overline{H}$. (11.23)

See [11.19].

So much for the direction of $\boldsymbol{\nabla}\Psi$; what about its magnitude? In [11.19] (which is basically a copy of [11.17]) we imagine that k is infinitesimal. Choosing $dz = e^{i\theta}\,ds$ in the direction of $\nabla\Psi$, we find $d\Psi = |\nabla\Psi|\,ds$. In particular, if we let ds equal ϵ (the width of the k-flux tube) then $d\Psi$ will equal k. Thus

$$|\nabla\Psi| \asymp (k/\epsilon).$$

But this is precisely the formula we previously obtained for the speed $v = |\overline{H}|$ of the flow! Thus $|\nabla\Psi| = |\overline{H}|$.

Combining this result with (11.23) we obtain the following simple formula for $\overline{H}$ in terms of Ψ:

$$\overline{H} = -i\,\nabla\Psi \quad \Longleftrightarrow \quad \overline{\mathbf{H}} = \begin{pmatrix} \partial_y\Psi \\ -\partial_x\Psi \end{pmatrix}. \tag{11.24}$$

Try this out on our previous example $H(z) = i\,\overline{z}$, the Pólya vector field of which had stream function $\Psi = (x^2 + y^2)/2$.

Now consider the question, "What additional condition must be satisfied by Ψ if $\overline{H}$ is also required to be *irrotational*?" The answer is that it must satisfy *Laplace's equation*:

$$\Delta\Psi \equiv \partial_x^2\,\Psi + \partial_y^2\,\Psi = 0.$$

Solutions of this equation are called *harmonic*, so we may restate the result as follows:

A sourceless field is irrotational if and only if its stream function is harmonic.

The verification is a simple calculation:

$$\nabla\times\overline{H} = \begin{pmatrix}\partial_x\\ \partial_y\end{pmatrix} \times \begin{pmatrix}\partial_y\Psi\\ -\partial_x\Psi\end{pmatrix} = -\Delta\Psi.$$

11.3.4 The Potential Function

Next suppose that $\overline{H}$ is a force field which is known to be conservative (irrotational). In this case it is the work rather than the flux which must be path-independent. Thus if K is any contour from an arbitrary fixed point a to a variable point z, the work done by the field in moving the particle along K is a well-defined function of z,

$$\Phi(z) = \mathfrak{W}\,[\overline{H}, K].$$

This is called the *potential function*, though there are several pseudonyms depending on the context: e.g., in electrostatics it is called the "electrostatic potential", in hydrodynamics it is called the "velocity potential", and in the case of flowing heat it is already familiar as the temperature. As with the stream function, changing the choice of a merely changes Φ by an additive constant.

Let's investigate Φ as we did Ψ. The level curves $\Phi = \text{const}$. are called *equipotentials*; what is their geometric significance? As [11.20] illustrates, the answer is that

The equipotentials are the orthogonal trajectories through the lines of force. (11.25)

The reason should be clear. A certain amount of work $\Phi(p)$ is done in moving the particle from a to p, but then no additional energy is expended in moving it to q along the orthogonal trajectory through p. Thus $\Phi(q) = \Phi(p)$.

Instead of illustrating random equipotentials, [11.20] mimics the special construction used in [11.17]: *we draw just those equipotentials for which* $\Phi = 0, \pm l, \pm 2l, \pm 3l, \ldots$. In this picture the same amount of work l is required to move the particle from each equipotential to the next. Let us therefore call the region lying between two such adjacent equipotentials an l*-work tube.*

Suppose that l is chosen small, and consider the work done in moving a particle along the correspondingly short cross-section δ in [11.20]. In the limit of vanishing l we find that

$$|\overline{H}| = \frac{l}{\delta}. \tag{11.26}$$

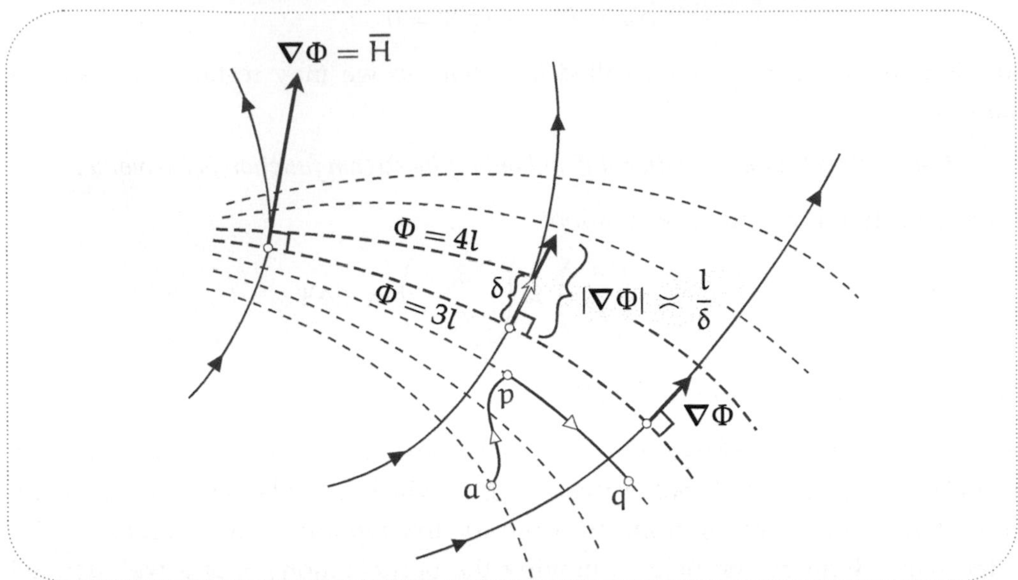

[11.20] The Potential Function Φ of a conservative force field. *If* K *is any contour connecting an arbitrary fixed point* a *to a variable point* z*, the work done by the conservative (irrotational) field in moving the particle along* K *is a well-defined function of* z*, independent of* K*, called the* potential function, $\Phi(z) = \mathfrak{W}[\overline{H}, K]$. *Just as we did with the stream function, we draw just those equipotentials for which* $\Phi = 0, \pm l, \pm 2l, \pm 3l, \ldots$ *In this picture the same amount of work* l *is required to move the particle from each equipotential to the next, so we call the region lying between two such adjacent equipotentials an* l-work tube. *The equipotentials are the orthogonal trajectories through the lines of force. Also,* $|\nabla\Phi| \asymp (l/\delta) \asymp |\overline{H}|$, *so* $\overline{H} = \nabla\Phi$.

Thus the magnitude of the force is represented by the crowding together of the equipotentials:

> *Let the equipotentials of a conservative force field be constructed using* l*-work tubes. If* l *is chosen small, the magnitude of the force at any point will be approximately given by* l *divided by the width of the tubes in the vicinity of the point. For infinitesimal* l*, the result is exact.* (11.27)

Since the gradient field $\nabla\Phi$ is automatically orthogonal to the equipotentials and has magnitude (l/δ), we may combine (11.25) and (11.27) into the simple formula

$$\overline{H} = \nabla\Phi \iff \overline{H} = \begin{pmatrix} \partial_x \Phi \\ \partial_y \Phi \end{pmatrix}. \tag{11.28}$$

Lastly, suppose that $\overline{H}$ is required to be sourceless. Since

$$\boldsymbol{\nabla}\cdot\overline{H} = \begin{pmatrix} \partial_x \\ \partial_y \end{pmatrix} \cdot \begin{pmatrix} \partial_x \Phi \\ \partial_y \Phi \end{pmatrix} = \Delta\Phi, \tag{11.29}$$

we see that

> *A conservative force field is sourceless if and only if its potential function is harmonic.* (11.30)

11.3.5 The Complex Potential

We now know two things about a vector field $\overline{H}$ that is irrotational *and* sourceless: (i) both Φ and Ψ exist; (ii) it is the Pólya vector field of an analytic function. In this section we shall attempt to illuminate the connections between these two facts.

Since Φ and Ψ both exist, we may superimpose pictures of types [11.17] and [11.20], thereby simultaneously dividing the flow into mutually orthogonal k-flux tubes and l-work tubes. Before drawing this picture let us *choose the increment of work to be numerically equal to the increment of flux*: $l = k$.

Let us call the intersection of a k-flux tube with a k-work tube a k-*cell*. We already know that the sides of each k-cell meet at right angles, so for small k they will be approximately rectangles. The sides of such a rectangle will be the previously considered widths ϵ and δ of the two kinds of tube. But combining the results (11.21) and (11.26) we see that

$$\frac{k}{\epsilon} = |\overline{H}| = \frac{k}{\delta} \quad \Longrightarrow \quad \delta = \epsilon.$$

Thus

> *In the limit of vanishing* k, *the* k-*cells are squares.* (11.31)

The LHS of [11.21] illustrates such a division into approximately square k-cells. We have labelled $\Phi = 11k$ and $\Psi = 3k$, but we have left it to you [exercise] to label the remaining streamlines and equipotentials; this can only be done in one way.

Note that once such a special phase portrait (including the equipotentials) has been drawn with a small value of k, the value of $\int_L H\,dz$ is easy to find. For if L crosses m equipotentials and n streamlines, an accurate estimate of the integral will be $k(m + in)$. If L crosses an equipotential or streamline more than once, how should m and n be counted?

We mention in passing that there is an interesting physical interpretation of the k-cells which is due to Maxwell (1881). Suppose that the vector field represents the flow of a fluid having unit mass per unit area. In the limit of vanishing k, the speed v will be constant throughout any particular cell, and the kinetic energy of the fluid in that cell will be

$$\text{kinetic energy} = \tfrac{1}{2}(\text{area})\, v^2 = \tfrac{1}{2}\epsilon^2 \left(\frac{k}{\epsilon}\right)^2 = \tfrac{1}{2}k^2.$$

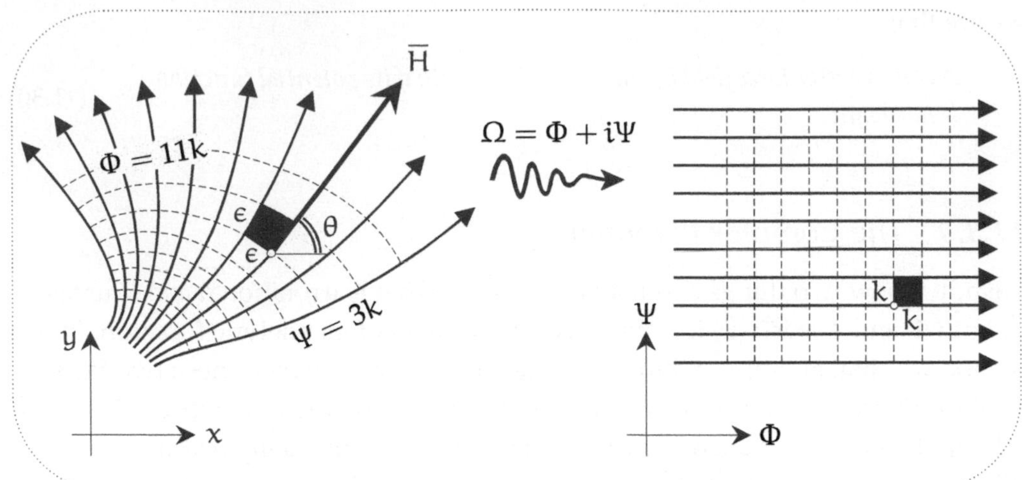

[11.21] The Complex Potential, $\Omega = \Phi + i\Psi$. *If* $H(z)$ *is analytic, then* $\overline{H(z)}$ *is both irrotational and sourceless, so we may simultaneously define and superimpose* l*-work tubes orthogonal to the streamlines with* k*-flux tubes along the streamlines, thereby dividing the region of the flow into small "rectangles" (as* l *and* k *tend to zero). We now go one step further and choose* $l = k$*, dividing the region into squares bounded by equipotentials and streamlines. This makes it vividly clear that if we combine the potential and stream functions into the single* complex potential $\Omega = \Phi + i\Psi$*, then the mapping* Ω *is locally an amplitwist, i.e.,* Ω' *exists. Indeed, we will show that* $\mathbf{H} = \mathbf{\Omega'}$.

Thus

> *Each* k*-cell contains the same amount of energy, and the total energy in a region is thus obtained by counting the number of* k*-cells contained within it.*

If we reinterpret the vector field as an electrostatic field, and correspondingly reinterpret "energy" as electrostatic energy, the result is still valid; this was the context in which Maxwell discovered it.

The result (11.31) is intimately connected with ideas of complex analysis. To see this, let us combine the potential and stream functions into a single complex function Ω called the *complex potential*:

$$\Omega(z) = \Phi(z) + i\,\Psi(z).$$

Returning to the dominant point of view of this book, think of Ω as a *mapping*. The RHS of [11.21] shows the image of the special phase portrait under this mapping:

> *The complex potential maps streamlines to horizontal lines and equipotentials to vertical lines. Furthermore, each square* k*-cell is mapped to a square of side* k*. Thus* Ω *is an analytic mapping.*

We may check this symbolically. Equating (11.24) and (11.28) we obtain

$$\begin{pmatrix} \partial_x \Phi \\ \partial_y \Phi \end{pmatrix} = \begin{pmatrix} \partial_y \Psi \\ -\partial_x \Psi \end{pmatrix},$$

which are the CR equations for Ω.

What is the amplitwist of the complex potential? By considering the effect of Ω on the black k-cell in [11.21] we see that if the streamline through z makes an angle θ with the horizontal, the twist of Ω at z is $-\theta$, which we recognize as the angle of $H(z)$. We also see that the amplification of Ω is (k/ϵ), which we recognize as $|\overline{H}| = |H|$. Thus

$$\Omega' = H.$$

Since H is the derivative of an analytic function, it must itself be analytic. We have thus obtained a second, more geometrical proof that the class of sourceless, irrotational vector fields is the same as the class of Pólya vector fields of analytic functions.

The result $\Omega' = H$ can be checked symbolically. Substituting one of the CR equations for Ω into (11.28), we obtain

$$\overline{H} = \nabla\Phi = \partial_x\Phi + i\,\partial_y\Phi = \partial_x\Phi - i\,\partial_x\Psi = \overline{\partial_x\Omega} = \overline{\Omega'}.$$

When we thought of an analytic function f as a conformal mapping, f' represented its amplitwist. But since any such function may instead be thought of as the complex potential of a flow, we now have another interpretation of differentiation: f' is the conjugate of the velocity of the flow described by f. Correspondingly, we also have a new interpretation of critical points: they are the places where the velocity vanishes. Such places are called *stagnation points* in the flow.

By analysing the implications of sourcelessness and irrotationality separately, we have been able to understand the Pólya vector fields of non-analytic functions that may possess a stream function or a potential function, but not both. If we had instead restricted ourselves from the outset to the Pólya vector fields of analytic functions, the complex potential could have been obtained more rapidly (but less revealingly) as follows.

If L is any contour from an arbitrary fixed point a to a variable point z, we may define

$$\Omega_L(z) = \int_L H(w)\,dw = \mathfrak{W}[\overline{H}, L] + i\,\mathfrak{F}[\overline{H}, L].$$

But, as we saw in Chapter 8, if H is analytic then this integral is independent of L, and the well-defined function

$$\Omega(z) = \int_a^z H(w)\,dw = \Phi(z) + i\Psi(z)$$

is in fact the antiderivative of H. More explicitly, the image $\Omega(L)$ of a contour L from p to q is the path taken by the Riemann sum for the integral of H along L. The value of the integral is then the vector connecting the start of $\Omega(L)$ to its finish, namely, $\Omega(q) - \Omega(p)$.

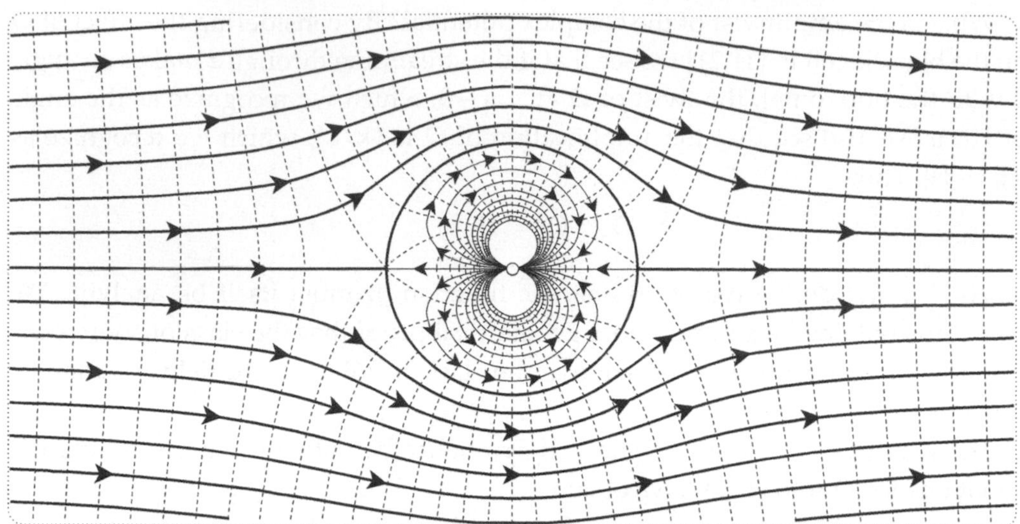

[11.22] $\Omega(z) = z + \frac{1}{z}$ *represents the superposition of a uniform eastward flow and a dipole.*

11.3.6 Examples

(1) We previously claimed that the streamlines of the dipole $\overline{H} = (1/\overline{z}^2)$ in [11.14b] were perfect circles, and we asked you to provide a simple geometric proof. A second demonstration is obtained by finding the complex potential:

$$\overline{H} = \frac{1}{\overline{z}^2} \implies \Omega' = \frac{1}{z^2} \implies \Omega = -\frac{1}{z} + c.$$

The streamlines are the images under $\Omega^{-1}(z) = -1/(z-c)$ of horizontal lines. The result follows from the fact that inversion sends straight lines to circles through the origin.

(2) A uniform eastward flow has complex potential $\Omega = z$. If we insert a dipole of complex potential $\Omega = (1/z)$ into this flow then the new flow will be the superposition of the two individual flows and thus will have complex potential

$$\Omega(z) = z + \frac{1}{z}.$$

Using your computer you may verify that the streamlines and equipotentials are as shown in [11.22]. Note how the streamlines emanating from the dipole are deformed out of perfect circularity by the uniform flow, but that this distortion diminishes as the origin is approached.

(3) A source of strength 2π at the origin has vector field $\overline{H} = (1/\overline{z})$. If we choose to measure work and flux along a path L emanating from $z = 1$ then [see p. 465] the complex potential is

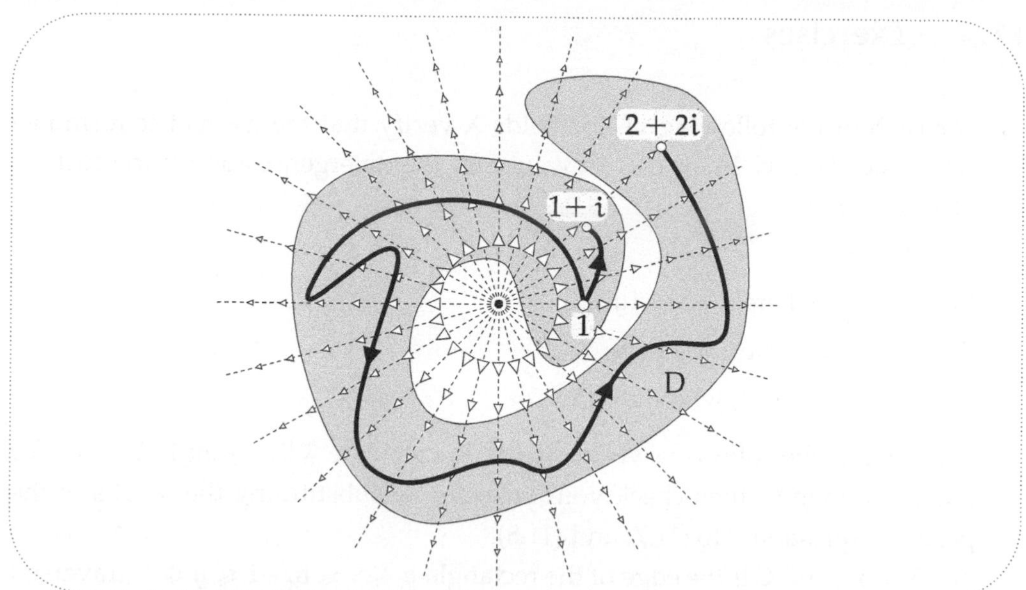

[11.23] The complex potential of the source $(1/\overline{z})$ is a multifunction: *although the work is single-valued, the flux increases by 2π with every extra revolution around the source. But if we restrict attention to a simply connected region D that does not contain the source, then Ω becomes single-valued.*

$$\Omega(z) = \Phi(z) + i\,\Psi(z) = \log_L(z) = \ln|z| + i\,\theta_L(z).$$

While the work Φ is single-valued, the flux Ψ is a multifunction whose values differ from each other by multiples of 2π. This makes perfect sense since each time L encircles the source it intercepts the full 2π of fluid being pumped in there. Note that the single-valued inverse function $\Omega^{-1}(z) = e^z$ does indeed map horizontal and vertical lines to the source's streamlines and equipotentials.

If we wish to obtain a single-valued complex potential we may do so by confining our attention to any simply connected region not containing the source. The shaded region D in [11.23] is an example. Any two paths from 1 to z that lie wholly within D may be deformed into each other without ever leaving D, hence without crossing the source, hence without altering the flux. For example, we see that for the particular choice of D in [11.23], the unique values of Ψ at $(1+i)$ and at $(2+2i)$ are $(\pi/4)$ and $(9\pi/4)$. However, a different choice of D might well yield different values of Ψ at these two points.

More generally, if D is any simply connected region not containing any singularities of an otherwise analytic H, the Pólya vector field $\overline{H}$ will possess a single-valued complex potential in D.

11.4 Exercises

1 For each of the following vector fields X verify that the geometric formulae (11.5) and (11.6) yield the correct values for the divergence and for the curl:

(i) $X = (1/\overline{z})$.

(ii) $X = \overline{z}$.

(iii) $X = x^2$, where $z = x + iy$.

(iv) $X = y^2$, where $z = x + iy$.

(v) $X = i(1/r^2)e^{i\theta}$, where $z = re^{i\theta}$.

2 For each of the following vector fields X, calculate $\mathfrak{F}[X, C]$ and $\mathfrak{W}[X, C]$ for the given loop C, then check your answers by substituting the results of the previous question into (11.7) and (11.8).

(i) $X = x^2$, and C is the edge of the rectangle $a \leqslant x \leqslant b, -1 \leqslant y \leqslant 1$, traversed counterclockwise.

(ii) $X = i(1/r^2)e^{i\theta}$, and C is the edge of the region $a \leqslant r \leqslant b, 0 \leqslant \theta \leqslant \pi$, traversed counterclockwise.

3 Use a computer to draw the Pólya vector field of $f(z) = 1/[z \sin z]$ and thereby identify the locations and orders of the poles of $f(z)$. For each of the following choices of C, numerically estimate $\oint_C f(z)\, dz$ by making on-screen measurements of the vectors, then estimating the flux and circulation round C. In each case check your estimate by calculating the exact answer using residue theory.

(i) Let C be a small circle centred at $-\pi$.

(ii) Let C be a small circle centred at 0.

(iii) Let C be a small circle centred at π.

(iv) Let C be a small circle centred at 2π.

(v) Let C be the boundary of the rectangle $1 \leqslant x \leqslant 7, -1 \leqslant y \leqslant 1$.

4 Repeat parts (i) and (ii) of the previous question using $f(z) = z \operatorname{cosec}^2 z$.

5 Let L be a contour from the real number $-\theta$ to $+\theta$. By choosing L to be a line-segment, and then sketching the Pólya vector field at points along L, show that $\int_L z\, e^{iz}\, dz$ is purely imaginary. Verify this by calculating the exact value of the integral.

6 All complex analysis texts recognize the great utility of the inequality

$$\left| \int_L f(z)\, dz \right| \leqslant \int_L |f(z)| \cdot |dz|, \tag{11.32}$$

but none that we know of have sought to answer the question, "When does equality hold?" This is probably because no elegant answer is forthcoming (cf. our attempt in Chapter 8) without the concept of the Pólya vector field. However, armed with the Pólya vector field, we have what we shall call *Braden's Theorem*[8]:

> *Equality holds in* (11.32) *if and only if the contour* L *cuts the streamlines of the Pólya vector field of* f *at a constant angle.*

Explain Braden's Theorem.

7 Continuing from the previous question, suppose that $f(z) = \overline{z}$.

(i) Show that if L is a segment of the spiral with polar equation $r = e^{\theta}$, then the condition of Braden's Theorem is met.

(ii) Verify by explicit calculation that equality does indeed hold in (11.32), as predicted.

8 Consider the flow created by $(2n+1)$ sources, each of strength 2π, located at

$$0,\ \pm\pi,\ \pm 2\pi,\ \ldots, \pm n\pi.$$

(i) If $\Omega_n(z)$ denotes the complex potential of this flow, show that

$$\Omega_n(z) = \ln\left[z\left(1-\frac{z^2}{\pi^2}\right)\left(1-\frac{z^2}{2^2\pi^2}\right)\cdots\left(1-\frac{z^2}{n^2\pi^2}\right)\right] + \text{const}.$$

(ii) Ignoring the constant, and referring to Ex. 13 on p. 508, deduce that as the number of sources increases without limit, $\Omega_n(z)$ tends to $\Omega(z) = \ln[\sin z]$.

(iii) Check that this answer makes sense by using a computer to draw the velocity vector field, $V = \overline{\Omega'}$.

9 (i) Explain why the derivative of the complex potential of a source yields the complex potential of a dipole.

(ii) Referring to the previous question, draw a sketch predicting the appearance of the flow whose complex potential is $\Omega(z) = \frac{d}{dz}\ln[\sin z]$. Check your answer by getting the computer to draw this flow.

10 Reconsider the term $C\overline{z}$ in the local decomposition (11.17) of a general vector field. See [11.11].

(i) Show that the visual appearance of the vector field $C\overline{z}$ is essentially independent of the value of C. More precisely, show that if $C = e^{i\phi}$ then increasing ϕ merely causes the entire picture of the vector field $C\overline{z}$ to rotate, in fact exactly half as fast as $e^{i\phi}$ rotates.

[8] See Braden (1987). We independently recognized this fact, probably at about the same time as Braden himself.

(ii) To make the result vivid, create a computer animation of the vector field $e^{i\phi}\,\overline{z}$ as ϕ increases from 0 to π.

(iii) More generally, show that if n is an integer and $F(z)$ stands for either $\overline{z^n}$ or z^{-n}, then the vector field of $e^{i\phi}F$ is obtained by rotating the vector field of F through $\phi/(n+1)$. [Note that the $n = -1$ fields (including sources and vortices) are exceptional.]

11 Consider a flow such that the inverse complex potential is $\Omega^{-1}(w) = w + e^w$. Use a computer to draw the streamlines, and verify mathematically that the picture may be interpreted as the flow out of a channel $-\pi \leqslant \text{Im}(z) \leqslant \pi$, $\text{Re}(z) \leqslant -1$.

12 Consider the flow with complex potential

$$\Omega(z) = \frac{1}{2}\left[\frac{e^z+1}{e^z-1}\right].$$

Use a computer to draw the streamlines, and verify mathematically that the picture may be interpreted as the flow that results when the dipole with complex potential $\Omega(z) = (1/z)$ is confined to the channel $-\pi \leqslant \text{Im}(z) \leqslant \pi$.

13 Continuing from the previous question, what would the new complex potential be if fluid were flowing down the channel with speed v prior to the insertion of the dipole? Check your answer by using a computer to draw the streamlines.

14 Suppose that the doublet consisting of a source of strength 2π at $z = 1$ and a sink of equal strength at $z = -1$ is inserted into the uniform flow with real, positive velocity v. Locate the "stagnation points" (singular points of zero velocity) of the net flow, and describe (perhaps with the aid of a computer animation) how they move as v varies from 0 to 3.

15 If two sources are located at opposite corners of a square, and two sinks are located at the other two corners, and all four are of equal strength, then show that the circle through these four points is a streamline. Check this by getting the computer to draw the complete flow.

16 Show that the streamlines produced by two vortices of equal strength are Cassinian curves (figure [2.8b], p. 68) whose foci are the locations of the two vortices. [Note that your reasoning immediately generalizes: *Cassinian curves with* n *foci are the streamlines of* n *equal vortices placed at the foci.*]

17 Show that the streamlines [11.15b] and the equipotentials of a quadrupole are lemniscates (see [2.9], p. 69).

CHAPTER 12

Flows and Harmonic Functions

12.1 Harmonic Duals

12.1.1 *Dual Flows*

As in the previous chapter, let $\overline{H} = \overline{\Omega'}$ be a steady, sourceless, irrotational vector field with complex potential $\Omega = \Phi + i\Psi$. If at each point we rotate $\overline{H}$ through a fixed angle ϑ then we obtain the Pólya vector field of the analytic function $H_\vartheta \equiv e^{-i\vartheta}H$, namely, $\overline{H_\vartheta} = e^{i\vartheta}\overline{H}$. Thus this rotated vector field is automatically sourceless and irrotational, and its complex potential is $\Omega_\vartheta = e^{-i\vartheta}\Omega$. Writing $\Omega_\vartheta \equiv \Phi_\vartheta + i\Psi_\vartheta$, the potential and stream functions are therefore

$$\Phi_\vartheta = (\cos\vartheta)\,\Phi + (\sin\vartheta)\,\Psi \quad \text{and} \quad \Psi_\vartheta = (\cos\vartheta)\,\Psi - (\sin\vartheta)\,\Phi.$$

Henceforth we shall concentrate on the particularly simple and important case in which $\vartheta = +(\pi/2)$. After rotating $\overline{H}$ through this right angle we obtain the Pólya vector field of $H_{\pi/2}$, for which we shall use the special symbol $\widehat{H}$. Thus $\overline{\widehat{H}} \equiv \overline{H_{\pi/2}} = i\overline{H}$. In complex analysis, the standard terminology is to say that $\overline{\widehat{H}}$ is "conjugate" to the original flow $\overline{H}$. However, I know of no mathematical[1] connection between this concept and the familiar one of complex conjugation. Furthermore, our use of Pólya vector fields (involving genuine complex conjugation) brings these two senses of "conjugate" into direct conflict, for the complex conjugate of the original flow is *not* the "conjugate" flow.

Fortunately, in other areas of mathematics (e.g., topology) there is another term that is commonly used to describe this idea. We therefore propose to call $\overline{\widehat{H}}$ the *dual* of $\overline{H}$. Similarly, let us call the potential and stream functions of the dual flow the *dual potential* and the *dual stream function*.

[1] Linguistically, the common origin of both terms is the Latin word "conjugatus", meaning joined together.

Visual Complex Analysis. 25th Anniversary Edition. Tristan Needham, Oxford University Press.
 DOI: 10.1093/oso/9780192868916.003.0012

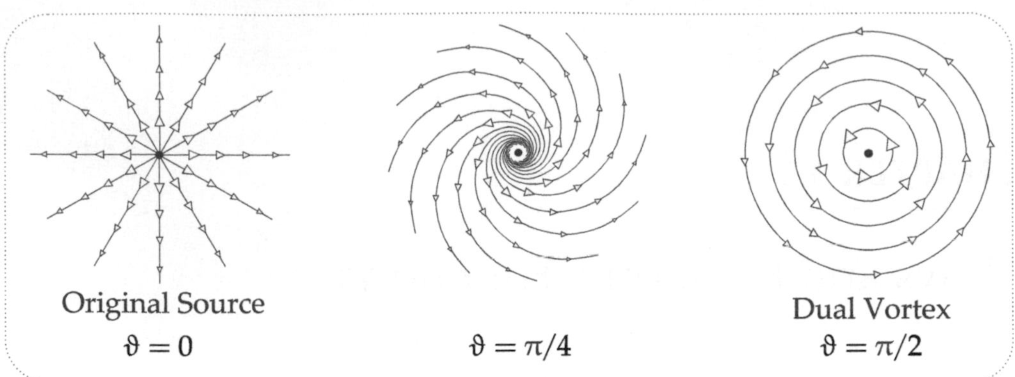

[12.1] Evolution of a source into its dual vortex: *As ϑ goes from 0 to $(\pi/2)$, the field $\overline{H}_{\vartheta} \equiv e^{i\vartheta}\overline{H} = e^{i\vartheta}(1/\overline{z})$ rotates from a pure source into its* dual *pure vortex, $\widehat{\overline{H}} \equiv \overline{H}_{\pi/2} = i\overline{H} = (i/\overline{z})$.*

Later we shall see that the concept of a dual flow is very useful. For example, having found the flow of a fluid round an obstacle, the dual flow represents the electric field which solves an analogous problem in electrostatics.

As interesting examples of dual flows, consider what happens in the vicinity of a singularity. Figure [12.1] illustrates how, as ϑ varies from 0 to $(\pi/2)$, a source gradually evolves into a dual vortex of equal strength. Note (cf. [11.12], p. 556) that the intermediate flow may also be viewed as a superposition of the original flow and its dual. Indeed, this is true quite generally:

$$\overline{H}_{\vartheta} = (\cos\vartheta)\overline{H} + (\sin\vartheta)\widehat{\overline{H}}.$$

Check for yourself that the type of qualitative change of flow exhibited in [12.1] does *not* occur in the case of higher multipoles. For example, the dual of a dipole is just another dipole. As ϑ varies from 0 to $(\pi/2)$, are all the intermediate flows dipoles, as well? See Ex. 10 of the previous chapter.

Observe that in passing from a flow to its dual the roles of the streamlines and equipotentials are interchanged: the streamlines of the dual flow are the equipotentials of the original, while the equipotentials of the dual flow are the streamlines of the original. Symbolically, this interchange of roles is manifested in the fact that the dual potential and stream functions are

$$\widehat{\Phi} = +\Psi \quad \text{and} \quad \widehat{\Psi} = -\Phi.$$

The difference of sign in these two equations is easily understood when we look at [12.2], which depicts a typical flow and its dual. [Streamlines are solid and equipotentials are dashed.] Recall that if we think of these pictures as force fields, work is done by the field when a particle moves along a line of force, so the original and dual potentials increase in the illustrated directions. Similarly, when thought of as a fluid flow, the flux across a directed segment of curve is positive when the

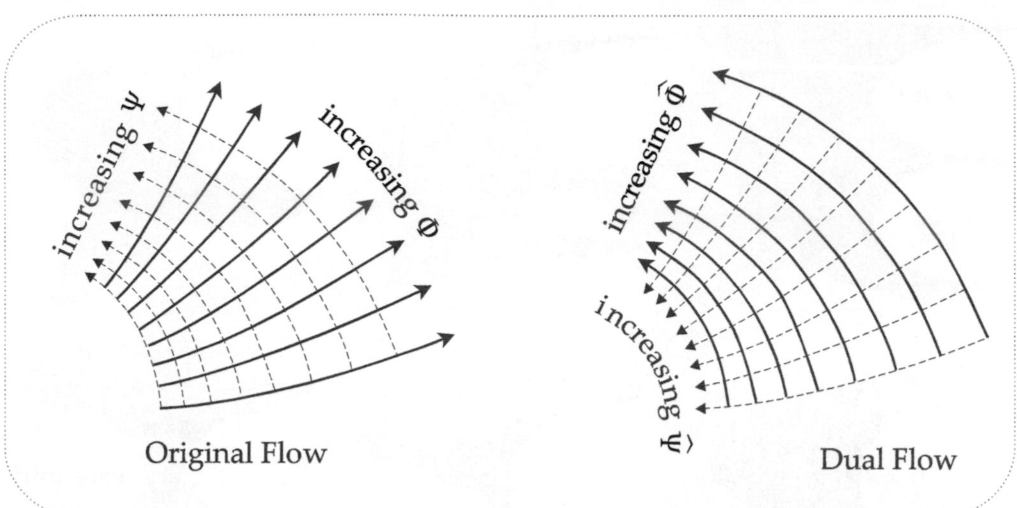

[12.2] In passing from a flow to its dual, the streamlines and equipotentials are interchanged.

fluid crosses it from left to right, so the original and dual stream functions increase in the illustrated directions. We now see clearly that $\widehat{\Phi}$ and Ψ increase in the same direction, while $\widehat{\Psi}$ and Φ increase in opposite directions.

Given a complex potential $\Omega = \Phi + i\Psi$, we may thus think of Ψ as either the stream function, or as the dual of the potential function. Likewise, Φ may be thought of as either the potential function, or as minus the dual of the stream function. Since any analytic function $f = u + iv$ may be thought of as a complex potential, we may extend this language and say that v is dual to u, and that $-u$ is dual to v.

Finally, we cannot resist at least mentioning two miraculous connections between the above ideas and the study of soap films, also known as *minimal surfaces*, characterized by *vanishing mean curvature*.

Imagine yourself standing on the tangent plane to a saddle-shaped surface. In the case of a soap film, as you turn around, the average (or "mean") curvature vanishes: the surface bends away from the tangent plane *equally and oppositely* as you turn around.

First miracle: Each complex analytic function $H(z)$ describes the shape of a minimal surface, and vice versa. **Second miracle:** Varying ϑ causes the minimal surface corresponding to $H_{\vartheta}(z)$ to undergo stretch-free bending: *all these minimal surfaces have identical intrinsic geometry.* For example, if H corresponds to the so-called *helicoid*, then $\widehat{H}$ corresponds to the so-called *catenoid*, and [12.3] illustrates the stretch-free bending of one into the other, each intermediate surface itself being a minimal surface.

For an elementary introduction to the fascinating subject of minimal surfaces, see Hildebrandt and Tromba (1985); for the mathematical details, see Nitsche (1989).

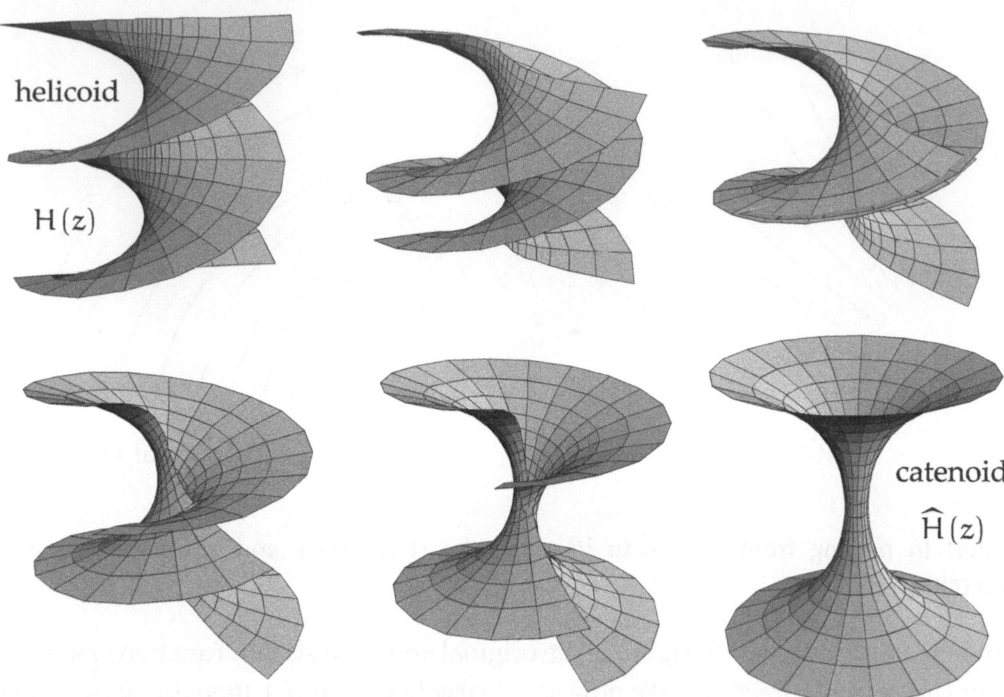

[12.3] Miraculously, complex analytic functions correspond to minimal surfaces. *In particular, the helicoid and catenoid correspond to dual complex functions, and the evolution as* ϑ *goes from* 0 *to* $(\pi/2)$ *is* isometric: *the intrinsic geometry does not change throughout the evolution,* and *all the intermediary surfaces are minimal, too!*

NOTE added in this *25th Anniversary Edition*: An especially good undergraduate introduction to this subject has now appeared in Woodward and Bolton (2019). The discussion includes the 1982 breakthrough discovery by Celso Costa of new minimal surfaces (beyond planes, catenoids, and helicoids) that are closed and without self-intersections. Furthermore, unusually, the authors provide a thorough discussion of the generalization to *surfaces of constant mean curvature* (CMC surfaces), in which the average curvature is the same at every point of the surface, but its value does *not* vanish (as it must do for soap films).

12.1.2 Harmonic Duals

We know that both the real and the imaginary parts of an analytic function are automatically harmonic. It is therefore natural to wonder if, conversely, every harmonic function is the real (or imaginary) part of some analytic function. As we shall see, this is indeed the case. That is, given a harmonic function u we can always find another harmonic function v, the *harmonic dual* of u, such that $f = u + iv$ is analytic. [Again, the standard terminology is that v is the "harmonic *conjugate*" of u.]

We make two remarks before proceeding. First, if v is an harmonic dual then so is $v + \text{const.}$, and consequently v will only be uniquely determined if we impose additional conditions, such as v vanishing at a particular point. Second, the harmonic dual of a single-valued function may itself be a multifunction. Witness the case $u = \ln|z|$, illustrated in [12.1], for which $v = \arg(z)$.

Given an irrotational vector field, we know how to construct a potential function. But, conversely, if we are given a real function $\Phi(z)$ then we may *construct* an irrotational vector field $\overline{H}$ for which Φ is the potential function, namely,

$$\overline{H} = \nabla\Phi.$$

If Φ is harmonic then we know [(11.30), p. 569] that $\overline{H}$ will be sourceless, and so it will possess a stream function Ψ. Since $\overline{H}$ is irrotational, Ψ is harmonic. The complex potential $\Omega = \Phi + i\Psi$ will then be an analytic function having as its real part the given harmonic function Φ. In other words we have shown that

> *The harmonic dual of a given harmonic function Φ is the stream function of the vector field $\nabla\Phi$.*

Alternatively, Ψ is the potential function for the dual of $\nabla\Phi$.

This result means that facts about analytic functions can sometimes be recast as facts about harmonic functions. For example, in Chapter 9 we saw that if $f = u + iv$ is analytic then $\langle f \rangle = f(p)$, where $\langle f \rangle$ denotes the average of f over any circle centred at p. It follows that the harmonic real part of f obeys the law $\langle u \rangle = u(p)$. But we now know that if u is any *given* harmonic function then we may construct an analytic function for which it is the real part. We thus obtain *Gauss's Mean Value Theorem*:

> *The average value of a harmonic function on a circle is equal to the value of the function at the centre of the circle.*

We obtain another example by reconsidering [7.14], p. 406, in which we saw that if $f = u + iv$ is analytic in some region whose boundary is Γ, then the maximum of u occurs on Γ. The existence of harmonic duals therefore implies that

> *If a function is harmonic in some region, its maximum occurs on the boundary of that region.*

The same goes for a (nonzero) minimum of a harmonic function.

Next we give explicit formulae for the construction of harmonic duals. To make Ψ unique, let us demand that it vanish at some point a. Then if K is any path from a to p, we have the flux formula

$$\Psi(p) = \int_K (\boldsymbol{\nabla}\Phi) \cdot \mathbf{N}\, ds.$$

Alternatively, in terms of complex integration, we have

$$\Psi(p) = \operatorname{Im}\left[\int_K (\overline{\nabla\Phi})\, dz\right].$$

As we have seen, if we restrict ourselves to a simply connected region throughout which Φ is harmonic, these integrals are single-valued. However, if the region is not simply connected, or if Φ has singularities, then (in general) Ψ will be a multifunction.

We illustrate these formulae with the example $\Phi = x^3 - 3xy^2$, which is easily seen to be harmonic. Choose $a = 0$, let $p = X + iY$, and choose K to be the line-segment between them, which may be represented parametrically as $z = x + iy = (X + iY)t$, where $0 \leqslant t \leqslant 1$. Since

$$\boldsymbol{\nabla}\Phi = \begin{pmatrix} 3x^2 - 3y^2 \\ -6yx \end{pmatrix} = \begin{pmatrix} 3X^2 - 3Y^2 \\ -6YX \end{pmatrix} t^2, \quad \mathbf{N} = \frac{1}{\sqrt{X^2 + Y^2}} \begin{pmatrix} Y \\ -X \end{pmatrix},$$

and $ds = \sqrt{X^2 + Y^2}\, dt$, the first formula yields [exercise]

$$\Psi = 3X^2Y - Y^3.$$

Alternatively, since

$$\overline{\nabla\Phi} = (3x^2 - 3y^2) + i\,6xy = 3z^2,$$

the second formula yields

$$\Psi = \operatorname{Im}\left[\int_0^{X+iY} 3z^2\, dz\right] = \operatorname{Im}(X + iY)^3 = 3X^2Y - Y^3.$$

The simplicity of the second method depended crucially on our ability to express $\overline{\nabla\Phi}(x, y)$ as a function $\Omega'(z)$ of z, but it is not always so obvious how to do this. However, there does exist a systematic method of doing this in the case where Φ is defined in a region containing a segment of the real axis.

Let $V(x)$ be the vector field evaluated on the real x-axis, i.e., $V(x) = \overline{\nabla\Phi}(x, 0)$. If $\Phi(x, y)$ is an explicit formula in terms of the familiar functions (powers, trigonometric, exponential) that possess complex analytic generalizations, then $V(x)$ is such a formula also. Hence if in the formula for $V(x)$ we now replace the symbol x with the complex variable z then we obtain an analytic function $V(z)$ which agrees with $\Omega'(z)$ when z is real. But, as we saw in Chapter 5, this implies that the two functions must continue to agree when z becomes *complex*.

Thus our recipe for finding $\Omega'(z)$ as an explicit formula in z is to calculate $\overline{\nabla\Phi}(x, y)$, set $y = 0$, then substitute z for x:

$$\Omega'(z) = \overline{\nabla\Phi}(z, 0). \tag{12.1}$$

For example, if $\Phi = \cos[\cos x \sinh y]\, e^{\sin x \cosh y}$ then [exercise]

$$\overline{\nabla\Phi}(x, y) = \cos[\cos x \sinh y]\, e^{\sin x \cosh y} \cos x \cosh y + F,$$

where F stands for three terms which vanish when $y = 0$. Using (12.1) we get $\Omega'(z) = e^{\sin z} \cos z$, and hence $\Psi = \operatorname{Im} e^{\sin z}$.

12.2 Conformal Invariance

12.2.1 Conformal Invariance of Harmonicity

Let $w = f(z)$ be a complex analytic function of z, which we will think of as a conformal mapping (rather than as a vector field) from the z-plane to the w-plane. Using f, any real function $\Phi(z)$ in the z-plane may be copied over (or "transplanted") to a function $\widetilde{\Phi}(w)$ in the w-plane by defining

$$\widetilde{\Phi}[f(z)] \equiv \Phi(z). \tag{12.2}$$

In other words, corresponding points in the two planes are assigned equal function values. We will now show (first symbolically then geometrically) that

Harmonicity is conformally invariant: $\widetilde{\Phi}(w)$ is harmonic if and only if $\Phi(z)$ is harmonic. (12.3)

As before, think of $\widetilde{\Phi}(w)$ as the potential of the vector field $\widetilde{V} \equiv \nabla\widetilde{\Phi}$. If and only if $\widetilde{\Phi}$ is harmonic, $\widetilde{V}$ possesses an analytic complex potential $\widetilde{\Omega}(w) = \widetilde{\Phi}(w) + i\widetilde{\Psi}(w)$, where the stream function $\widetilde{\Psi}$ is the harmonic dual of $\widetilde{\Phi}$. Since f is analytic, so is its composition with $\widetilde{\Omega}$:

$$\Omega(z) \equiv \widetilde{\Omega}[f(z)] = \Phi(z) + i\Psi(z).$$

Thus $\Phi(z)$ is the real part of an analytic function, and so it is harmonic.

There is a very simple geometric idea behind this important result. Figure [12.4] illustrates a visual means of checking whether or not a given real function Φ is harmonic. Once again, think of Φ as the potential of the force field $V = \nabla\Phi$. We know that Φ is harmonic if and only if V admits a complex potential. This we know occurs if and only if the field may be divided into a grid of (infinitesimal) square k-cells.

To check this we should therefore construct a "test grid":

(i) With a small value of k, draw the equipotentials $\Phi = 0, \pm k, \pm 2k, \pm 3k, \ldots$.

(ii) Choose one of the resulting k-work tubes [shaded in the figure] and draw line-segments across it in such a way that the tube is divided into squares.

(iii) Extend these line-segments into lines of force [dashed] of V, i.e., orthogonal trajectories through the equipotentials.

Then Φ *is harmonic if and only if these lines of force divide each* k*-work tube into squares.* Figure [12.4a] illustrates this test for a Φ that is harmonic, while [12.4b] illustrates it for one that is not. The result (12.3) can now be seen as nothing more than a statement of the conformal invariance of this geometric test. Let us spell this out.

Equation (12.2) defines the potential of each point in the z-plane to be the same as its image point (under f) in the w-plane. Thus f maps the k-work tubes of Φ to the k-work tubes of $\widetilde{\Phi}$. See [12.5]. Finally, since f is conformal, the constructed test

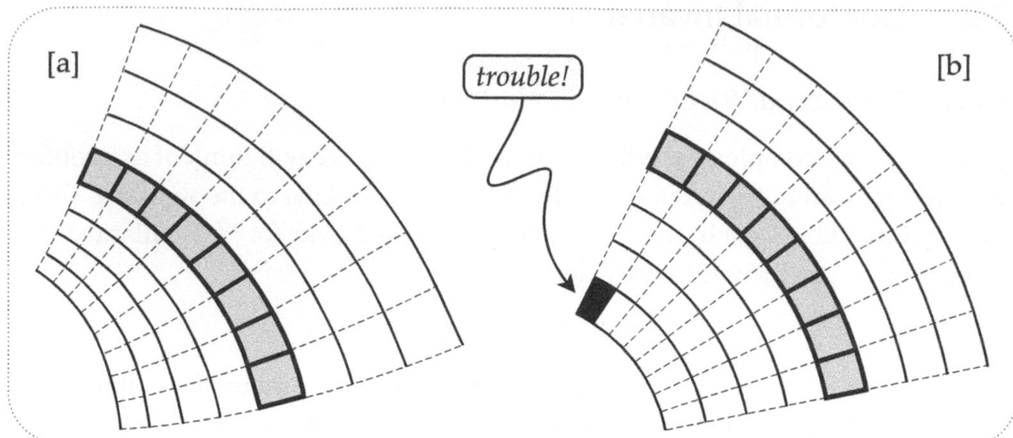

[12.4] Geometric test for harmonicity of Φ. *STEP ONE: With a small value of* k*, draw the equipotentials* $\Phi = 0, \pm k, \pm 2k, \pm 3k, \ldots$ *STEP TWO: Divide a* k*-work tube [shaded] into squares. STEP THREE: Extend the edges of the squares into lines of force. Then* **Φ is harmonic if and only if these lines of force divide each k-work tube into squares.** *So* **[a]** *is harmonic, and* **[b]** *is not.*

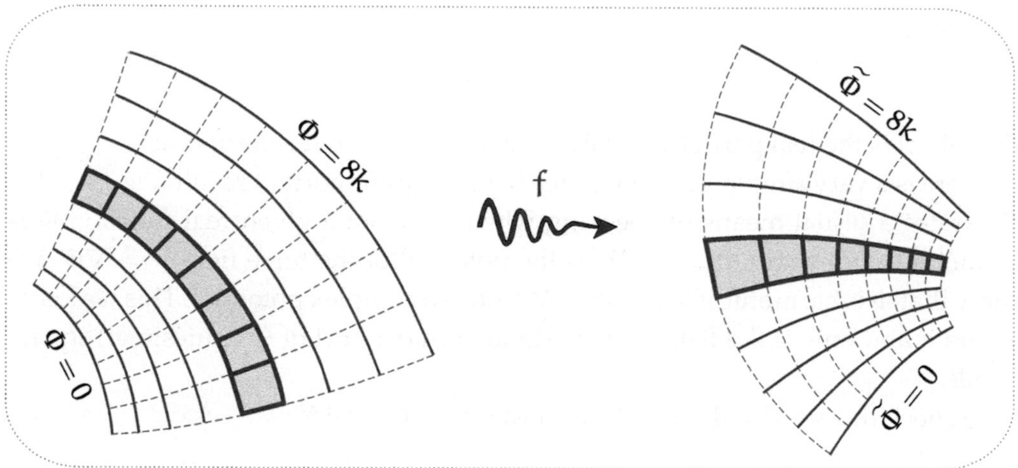

[12.5] Conformal invariance of the harmonicity test for Φ. *If* f *is conformal, then the image test grid will pass the harmonicity test if and only if the original test grid passes the test. Here, the initial grid passes, so its image does, too.*

grid for Φ will be composed of squares if and only if the image grid is composed of squares. Figure [12.5] illustrates the case where the potentials *are* harmonic.

12.2.2 Conformal Invariance of the Laplacian

The result (12.3) is merely a special case of the following more general result on the conformal invariance of the Laplacian operator Δ:

$$\Delta\widetilde{\Phi}(w) = \frac{1}{|f'(z)|^2}\Delta\Phi(z). \tag{12.4}$$

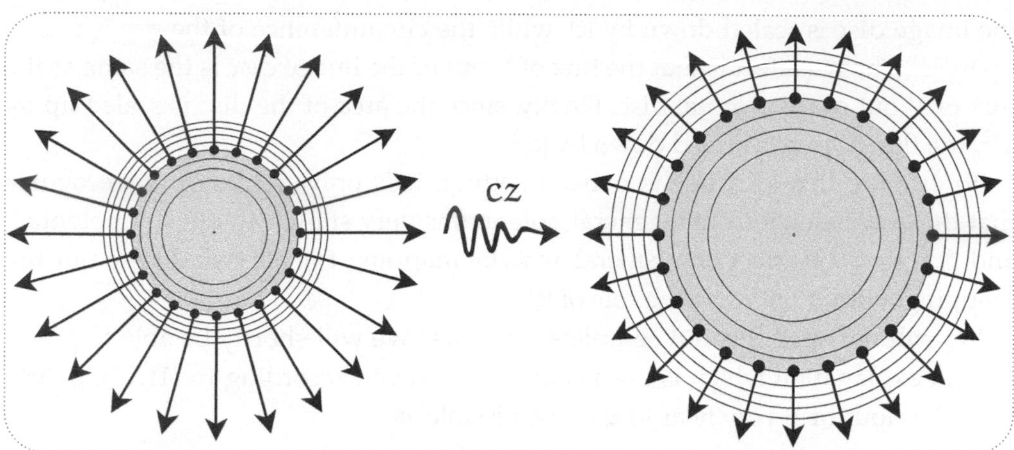

[12.6] Geometric explanation of $\boldsymbol{\nabla}\cdot\widetilde{\mathbf{V}}(w) = [1/|f'(z)|^2]\,\boldsymbol{\nabla}\cdot\mathbf{V}(z)$. *Let us explain this for* $f(z) = cz$; *in the general case, we need only then replace* c *with the local amplitwist,* $f'(z)$. *Compare the flux leaving the shaded disc on the left with the flux leaving the shaded image disc on the right. Since the separation of adjacent equipotentials is scaled up by* $|c|$, *the strength of the field on the rim of the image disc is scaled down by* $|c|$, *while the circumference of the rim is scaled up by* $|c|$. *The net effect is that the flux of* $\widetilde{V}$ *out of the image disc is the same as the flux of* V *out of the original disc. Finally, since the area of the disc is scaled up by* $|c|^2$, *the flux density is scaled down by* $|c|^2$. *Done.*

We will give two explanations of this result.

For the first explanation, we rephrase the result in terms of flux densities. Just as we did for $\widetilde{\Phi}$ in the w-plane, let us construct the vector field $\mathbf{V} \equiv \boldsymbol{\nabla}\Phi$ in the z-plane. We wish to understand this:

$$\boldsymbol{\nabla}\cdot\widetilde{\mathbf{V}}(w) = \frac{1}{|f'(z)|^2}\boldsymbol{\nabla}\cdot\mathbf{V}(z). \tag{12.5}$$

Now consider [12.6], which illustrates a toy model of the phenomenon. The potential $\Phi(z) = (S/4)\,|z|^2$ generates a vector field $V = (S/2)\,z$ of uniform divergence $\boldsymbol{\nabla}\cdot\mathbf{V} = S$. With a small value of k, the LHS of [12.6] shows the special equipotentials $\Phi = 0, \pm k, \pm 2k, \pm 3k, \ldots$, for which the strength of the field is inversely proportional to the separation of the curves. Now apply the mapping $w = f(z) = cz$, which is a rotation and an expansion by $|c|$. By definition, these expanded circles are equal-valued equipotentials of $\widetilde{\Phi}(w)$, so that

$$\widetilde{\Phi}(w) = \Phi(z) = \tfrac{1}{4}S\,|z|^2 = \tfrac{1}{4}\frac{S}{|c|^2}\,|w|^2.$$

The field $\widetilde{V} = \boldsymbol{\nabla}\widetilde{\Phi}$ therefore has uniform flux density $\boldsymbol{\nabla}\cdot\widetilde{\mathbf{V}} = S/|c|^2$, proving (12.5) for this case.

More intuitively still, in [12.6] compare the flux leaving the shaded disc on the left with the flux leaving the shaded image disc on the right. Since the separation of adjacent equipotentials is scaled up by $|c|$, the strength of the field on the rim of the image disc is scaled down by $|c|$, while the circumference of the rim is scaled up by $|c|$. The net effect is that the flux of $\widetilde{V}$ out of the image disc is the same as the flux of V out of the original disc. Finally, since the area of the disc is scaled up by $|c|^2$, the flux *density* is scaled down by $|c|^2$.

To employ this idea in the general setting, it is only necessary to recognize that the *local* behaviour of a general potential is very similar to our toy potential, and that the *local* effect of a general analytic mapping f is very similar to our toy mapping, with $|f'|$ playing the role of $|c|$.

We will not spell this out completely because we will shortly be able to give a second explanation which is even simpler. However, according to (11.17), p. 553, the behaviour of V very near to z_0 is expressible as

$$V(z) = [\nabla \boldsymbol{\cdot} \mathbf{V}(z_0)] \frac{(z - z_0)}{2} + Y(z),$$

where Y is sourceless and, of course, irrotational. Correspondingly, the local behaviour of the potential is[2]

$$\Phi(z) = \tfrac{1}{4}[\nabla \boldsymbol{\cdot} \mathbf{V}(z_0)]\, r^2 + \Upsilon(z), \tag{12.6}$$

where Υ is harmonic, and $r = |z - z_0|$ is the small distance from z_0. Having made explicit the connection with the toy model, we leave the remaining details to the interested reader.

12.2.3 *The Meaning of the Laplacian*

Given a real function $\Phi(z)$ in the z-plane, we have seen that its gradient vector field $\nabla\Phi$ is a geometric quantity, independent of the coordinates used to describe z. We have also seen that the divergence of a vector field measures its flux-density, so it too is geometrically defined. It follows that the Laplacian $\Delta\Phi = \nabla \boldsymbol{\cdot} \nabla\Phi$ must possess a coordinate-independent interpretation.

In order to state this interpretation, recall that if C is a circle centred at p then $\langle\Phi\rangle$ denotes the average value of Φ on C. We will show that

> *The Laplacian of Φ at p measures the amount by which the average value of Φ on an infinitesimal circle centred at p exceeds the value of Φ at p itself. More precisely, if r is the infinitesimal radius of this circle then* (12.7)
>
> $$\langle\Phi\rangle - \Phi(p) = \tfrac{1}{4} r^2\, \Delta\Phi.$$

Note that this result is in accord with Gauss's Mean Value Theorem, which says that if Φ is harmonic then $\langle\Phi\rangle - \Phi(p) = 0$ for circles of any size, not just infinitesimal ones. In fact if you have already convinced yourself of (12.6) then [exercise] you

[2] This may also be derived directly by taking the Taylor series for Φ and rewriting it in a rather unobvious way.

may derive (12.7) by using the fact that the harmonic function Υ obeys Gauss's Mean Value Theorem.

Before giving a more direct derivation of (12.7), let us return to Gauss's Mean Value Theorem itself and rederive it without appealing to complex analysis[3]. Let $\mathbf{V} = \nabla\Phi$ be the vector field of the potential function Φ. The flux of $\mathbf{V}$ out of a (non-infinitesimal) circle C of radius r is then [exercise]

$$\mathfrak{F}\,[\mathbf{V}, C] = 2\pi r\,\partial_r\langle\Phi\rangle.$$

Thus if Φ is harmonic, so that $\mathbf{V}$ is sourceless, then $\mathfrak{F} = 0$, by (11.7), p. 547. Since $\partial_r\langle\Phi\rangle = 0$, we see that $\langle\Phi\rangle$ is independent of the radius of C. Shrinking C down to p, we deduce that this radius-independent value must be $\Phi(p)$. Done.

Now suppose that $\mathbf{V}$ is not sourceless, but that its flux-density $\nabla\cdot\mathbf{V} = \Delta\Phi$ is constant. Gauss's Divergence Theorem [(11.7), p. 547] then yields $\mathfrak{F}\,[\mathbf{V}, C] = \pi r^2\,\Delta\Phi$. Inserting this into the previous result, we find that

$$\partial_r\langle\Phi\rangle = \tfrac{1}{2}\,r\,\Delta\Phi,$$

which may be integrated to yield the formula in (12.7). To complete the explanation of (12.7) it is only necessary to observe that the Laplacian of an arbitrary Φ is constant within an infinitesimal circle.

Knowing the meaning of the Laplacian, it is a simple matter to understand its conformal invariance as expressed in (12.4). The analytic mapping f amplitwists an infinitesimal circle C centred at p to an infinitesimal circle $\widetilde{C}$ centred at $\widetilde{p}$, the new radius being $\widetilde{r} = |f'(p)|\,r$. By definition, $\widetilde{\Phi}(\widetilde{p}) = \Phi(p)$. Likewise, the values of $\widetilde{\Phi}$ at points of $\widetilde{C}$ are the same as those of Φ at the preimages on C, so $\langle\widetilde{\Phi}\rangle$ on $\widetilde{C}$ equals $\langle\Phi\rangle$ on C. Thus (12.7) implies

$$r^2\,\Delta\Phi(p) = \widetilde{r}^2\,\Delta\widetilde{\Phi}(\widetilde{p}) = |f'(p)|^2 r^2\,\Delta\widetilde{\Phi}(\widetilde{p}),$$

from which (12.4) follows immediately.

12.3 A Powerful Computational Tool

A zealot might wish for an ideal world in which calculation would always be relegated to the confirmation of insights provided by geometry. Alas, even this author must confess to occasional lapses in which calculation has preceded understanding! We now describe a powerful computational tool which in many areas of complex analysis provides a considerable saving of labour. In the next section the study of the "complex curvature" [cf. Chapter 5] will provide a good showcase for its simplicity and elegance.

[3] Previously we got it from $\langle f\rangle = f(p)$, which in turn came from Cauchy's formula.

The gradient operator $\boldsymbol{\nabla}$ of vector calculus acts on a real function $R(x, y)$ to produce the gradient vector field

$$\boldsymbol{\nabla}R = \begin{pmatrix} \partial_x \\ \partial_y \end{pmatrix} R = \begin{pmatrix} \partial_x R \\ \partial_y R \end{pmatrix},$$

and we are free (as we have previously done) to think of this as a complex function

$$\nabla R = \partial_x R + i\,\partial_y R.$$

From this we may abstract the *complex gradient operator* ∇, together with the conjugate operator $\overline{\nabla}$:

$$\nabla = \partial_x + i\,\partial_y \quad \text{and} \quad \overline{\nabla} = \partial_x - i\,\partial_y.$$

These two operators open the way to an exciting new method of calculation.

Given a vector field

$$\mathbf{f} = \begin{pmatrix} u \\ v \end{pmatrix},$$

we have seen how the real version of ∇ may be formally dotted or crossed with $\mathbf{f}$ to yield its divergence $\boldsymbol{\nabla}\boldsymbol{\cdot}\mathbf{f}$ or its curl $\boldsymbol{\nabla}\boldsymbol{\times}\mathbf{f}$. The interpretations of these quantities as flux and work densities shows them to be truly geometric, that is to say, coordinate-independent. However, there would seem to be no natural way of applying $\boldsymbol{\nabla}$ directly to $\mathbf{f}$ to obtain a new vector field $\boldsymbol{\nabla}\mathbf{f}$. However, if we replace $\boldsymbol{\nabla}$ by its complex version ∇, and replace the vector field $\mathbf{f}$ by the complex function $f = u + iv$, then there *is* a natural definition:

$$\nabla f = (\partial_x + i\,\partial_y)\,(u + iv) = (\partial_x u - \partial_y v) + i\,(\partial_x v + \partial_y u).$$

The equivalent expression

$$\nabla f = \nabla u + i\,\nabla v = \nabla u + \left(\nabla v \text{ rotated } \tfrac{\pi}{2}\right)$$

helps to see that ∇f is geometrically meaningful (because ∇u and ∇v both are).

The power of the complex gradient derives from the following fundamental result [exercise]:

> *A complex function* f *is analytic if and only if* $\overline{\nabla} f = 0$, *in which case we also have* $\nabla f = 2f'$.

You may easily verify the following useful properties of ∇:

- $\nabla(f + g) = \nabla f + \nabla g$.
- $\nabla(fg) = f\nabla g + g\nabla f$.
- If f is analytic then $\nabla f[g(z)] = f'[g(z)]\,\nabla g$. For example, $\nabla e^{g(z)} = e^{g(z)}\,\nabla g$.

- The concepts of divergence and curl are neatly subsumed by ∇:

$$\overline{\nabla}f = \boldsymbol{\nabla}\cdot\mathbf{f} + i\,\boldsymbol{\nabla}\times\mathbf{f}.$$

Similarly,

$$\nabla f = \boldsymbol{\nabla}\cdot\overline{\mathbf{f}} - i\,\boldsymbol{\nabla}\times\overline{\mathbf{f}},$$

which shows, once again, that a vector field is sourceless and irrotational if and only if it is the Pólya vector field of an analytic function.

- The Laplacian operator Δ can be expressed neatly as $\nabla\overline{\nabla} = \Delta = \overline{\nabla}\nabla$.

In the next section, and in the exercises at the end of the chapter, you will see the strength of the new technique. You may also find that exercises from previous chapters are solved more readily by this method. For the moment, here are just two examples of the use of the complex gradient.

The first is simply to observe how neatly the theorems of Gauss and Stokes [p. 547] may be combined into a single complex result: If C is the boundary curve of a simply connected region R then

$$\oint_C f\,dz = i\iint_R \nabla f\,d\mathcal{A}.$$

If f is analytic ($\nabla f = 0$) we immediately obtain Cauchy's Theorem. This is not a new explanation, of course, merely a mathematically streamlined version of our previous physical one.

Our second example is another derivation of the result (12.4). Let $z = x + iy$ and $w = u + iv$, so that the complex gradient operators in the two planes are $\nabla_z = \partial_x + i\,\partial_y$ and $\nabla_w = \partial_u + i\,\partial_v$. The result we wish to prove can then be expressed (less ambiguously than before) as

$$\nabla_z\overline{\nabla}_z\widetilde{\Phi}[f(z)] = |f'|^2\,\nabla_w\overline{\nabla}_w\widetilde{\Phi}(w).$$

Since $w = f(z)$, a straightforward application of the chain rule yields [exercise]

$$\nabla_z = \overline{f'}\,\nabla_w \quad \text{and} \quad \overline{\nabla}_z = f'\,\overline{\nabla}_w. \tag{12.8}$$

For example,

$$\overline{\nabla}_z f(z) = f'\,\overline{\nabla}_w w = f'\,(\partial_u - i\,\partial_v)(u + iv) = 2f',$$

as it should. Returning to the problem, we easily obtain

$$\begin{aligned}\nabla_z\overline{\nabla}_z\widetilde{\Phi}[f(z)] &= \nabla_z\left\{f'\,\overline{\nabla}_w\widetilde{\Phi}(w)\right\} \\ &= (\nabla_z f')\,\overline{\nabla}_w\widetilde{\Phi}(w) + f'\,\nabla_z\overline{\nabla}_w\widetilde{\Phi}(w) \\ &= |f'|^2\,\nabla_w\overline{\nabla}_w\widetilde{\Phi}(w).\end{aligned}$$

12.4 The Complex Curvature Revisited*

12.4.1 Some Geometry of Harmonic Equipotentials

Given a real function $\Phi(x, y)$, figure [12.4] provided a geometric test for harmonicity. However, the first step of the test is not purely geometric in that it uses the *values* of Φ, not just the geometry of the curves $\Phi = \text{const}$. This leads us to pose the following more subtle problem. *Given a family of curves $\mathcal{E}$ filling a region of the plane, how may we decide whether or not there exists a harmonic function Φ such that $\mathcal{E}$ is its family of equipotentials?*

For example, let $\mathcal{E}$ be the set of origin-centred circles. If we assign potentials to these curves according to the rule $\Phi(z) = |z|$ then our previous test yields figure [12.4b], which shows that this potential is not harmonic. But the question we are now asking is whether these same curves can be assigned potentials according to a different rule that *is* harmonic? In fact they can: let $\Phi(z) = \ln|z|$.

When the family does admit a harmonic assignment of potential in this way, we shall simply extend our use of the word "harmonic" to the family of curves itself. Thus we would say that the family of origin-centred circles is *harmonic*. The opening question may then be rephrased succinctly:

What geometric property of $\mathcal{E}$ determines whether it is harmonic?

One way to answer this question is to generalize the test in [12.4] to the one in [12.7]:

(i) Choose two members of $\mathcal{E}$ that are very close together.

(ii) Draw line-segments across the region between them [shaded] so as to divide it into squares.

(iii) Extend these line-segments into streamlines [dashed], i.e., orthogonal trajectories through $\mathcal{E}$.

(iv) Choose one of the resulting flux tubes [darkly shaded] and draw line-segments across it so as to divide it into squares.

(v) Extend these line-segments into members of $\mathcal{E}$.

Then $\mathcal{E}$ *is harmonic if and only if the resulting grid is composed of squares.*

We already know that a family of concentric circles is harmonic, and [12.7a] shows how symmetry guarantees that it passes the test. Figure [12.7b] shows that a family of similar, concentric ellipses is not harmonic.

12.4.2 The Curvature of Harmonic Equipotentials

We now turn to a second, more elegant answer to our question. Let $\mathcal{S}$ be the family of curves (streamlines) orthogonal to $\mathcal{E}$. When one looks at [12.7] one gets the feeling that in order for a grid of squares to form, the bending of the curves of $\mathcal{E}$

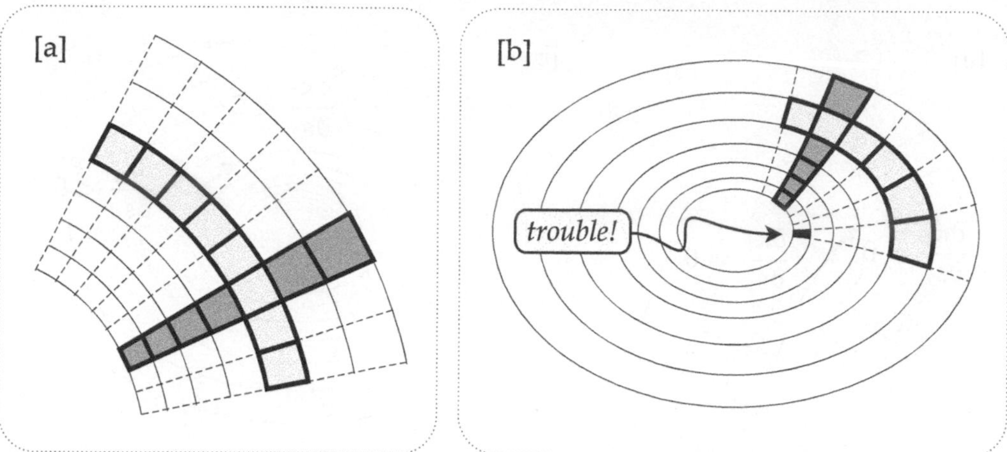

[12.7] Geometric harmonicity test for a family of curves, $\mathcal{E}$. *STEP ONE: Choose two members of $\mathcal{E}$ that are very close together. STEP TWO: Draw line-segments across the region between them [shaded] so as to divide it into squares. STEP THREE: Extend these line-segments into streamlines [dashed], i.e., orthogonal trajectories through $\mathcal{E}$. STEP FOUR: Choose one of the resulting flux tubes [darkly shaded] and draw line-segments across it so as to divide it into squares. STEP FIVE: Extend these line-segments into members of $\mathcal{E}$. Then* **$\mathcal{E}$ is harmonic if and only if the resulting grid is composed of squares. [a]** *Concentric circles are harmonic.* **[b]** *Concentric ellipses are not harmonic.*

must be connected in some special way with the bending of the curves of $\mathcal{S}$. By examining the curvatures of the two types of curve we shall see that this is indeed the case.

First let us attach directions to the curves of $\mathcal{E}$ and $\mathcal{S}$, so that their curvatures will have well-defined signs. Choose the direction for $\mathcal{S}$ arbitrarily, but then define the direction of $\mathcal{E}$ to be that of a tangent vector to $\mathcal{S}$ rotated through a *positive* right angle. Through any given point w_0 there passes one member $\mathcal{C}_1$ of $\mathcal{S}$ and one member $\mathcal{C}_2$ of $\mathcal{E}$. Let the curvatures of $\mathcal{C}_1$ and $\mathcal{C}_2$ be κ_1 and κ_2, and let s_1 and s_2 be the arc lengths along $\mathcal{C}_1$ and $\mathcal{C}_2$. We then have the following striking result:

$\mathcal{E}$ *is harmonic if and only if*

$$\frac{\partial \kappa_1}{\partial s_1} + \frac{\partial \kappa_2}{\partial s_2} = 0. \tag{12.9}$$

In other words, if the rates of change of the curvatures along the two types of curve are exactly equal and opposite. Note that both these rates of change are well-defined even in the absence of a choice of direction for the curves, for reversing such a choice changes the sign of both κ and ∂s.

Figure [12.8] illustrates the new test for the concentric circles and ellipses we considered in [12.7]. Since the circles and lines of [12.8a] have constant curvature,

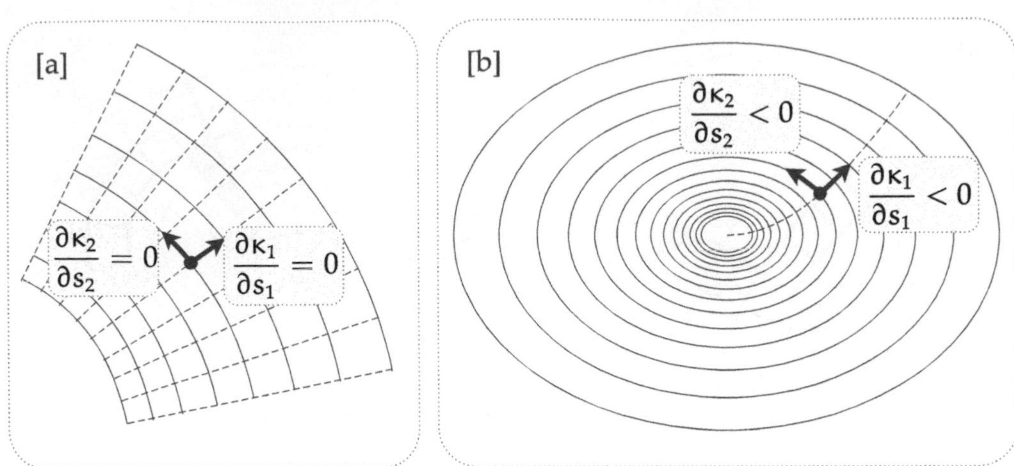

[12.8] Geometric curvature test for harmonicity: $\mathcal{E}$ is harmonic if and only if $\frac{\partial\kappa_1}{\partial s_1} + \frac{\partial\kappa_2}{\partial s_2} = 0$. [a] *Concentric circles are harmonic.* **[b]** *Concentric ellipses are not harmonic.*

(12.9) is trivially satisfied. In [12.8b] it is clear that both curvatures are decreasing at the indicated point, so (12.9) is violated there.

The result (12.9) seems to have been first published by Bivens (1992). We had also hit upon the same result, but as a natural consequence of investigating our *complex curvature* concept, which we introduced in Chapter 5. Here is the pertinent result:

$$\text{\textit{The complex curvature vector field of an analytic mapping } f \text{ \textit{is automatically sourceless, and its stream function is} } \Psi = 1/|f'|. \tag{12.10}$$

To see the connection between the two results, first observe that a family $\mathcal{E}$ in the w-plane is harmonic if and only if it is the image set of vertical lines in the z-plane under an analytic mapping $w = f(z)$. For if $\mathcal{E}$ is harmonic then its potential function $\Phi(w)$ is the real part of a complex potential $z = \Omega(w)$ which maps $\mathcal{E}$ to vertical lines, and so $f(z) \equiv \Omega^{-1}(z)$ has the required property. Conversely, if $\mathcal{E}$ is the image set of vertical lines under an analytic f then it is harmonic, indeed its harmonic potential is the real part of the complex potential $\Omega(w) \equiv f^{-1}(w)$.

Thus if $\mathcal{E}$ is harmonic then the curves $\mathcal{C}_1$ and $\mathcal{C}_2$ through $w_0 = f(z_0)$ are the images under f of the horizontal and vertical lines through z_0. But referring to [5.20], p. 272, we see that curvatures occurring in (12.9) are simply the real and imaginary parts of the complex curvature of f at z_0:

$$\mathcal{K}(z_0) = \kappa_1 + i\kappa_2.$$

Since infinitesimal horizontal and vertical movements dx and dy are amplified by $|f'(z_0)|$ to movements ds_1 and ds_2 along $\mathcal{C}_1$ and $\mathcal{C}_2$, the flux-density of the complex curvature is

$$\nabla \cdot \mathcal{K} = \frac{\partial \kappa_1}{\partial x} + \frac{\partial \kappa_2}{\partial y} = |f'(z_0)| \left[\frac{\partial \kappa_1}{\partial s_1} + \frac{\partial \kappa_2}{\partial s_2} \right].$$

The result (12.10) therefore implies that equation (12.9) is a necessary condition for $\mathcal{E}$ to be harmonic; the question of sufficiency will be addressed shortly.

To prove (12.10) we will use the complex gradient technique of the previous section; later we will give a proper geometric explanation. We must show that $\Psi = 1/|f'|$ is the stream function for the complex curvature

$$\mathcal{K} = \frac{i\,\overline{f''}}{\overline{f'}\,|f'|},$$

in other words [cf. (11.24), p. 566], $\mathcal{K} = -i\,\nabla\Psi$.

Since

$$-i\,\nabla\Psi = -i\,\nabla\left(\frac{1}{|f'|}\right) = \frac{i}{|f'|^2}\nabla|f'|,$$

we need to know $\nabla|f'|$. Because f is analytic, so is f', and this implies $\nabla f' = 0$ and $\overline{\nabla} f' = 2f''$. Hence

$$\begin{aligned} 2|f'|\,\nabla|f'| &= \nabla|f'|^2 = \nabla(f'\overline{f'}) = f'\nabla\overline{f'} + \overline{f'}\,\nabla f' = 2f'\,\overline{f''} \\ \Longrightarrow \nabla|f'| &= \frac{f'\,\overline{f''}}{|f'|}. \end{aligned}$$

Substituting this into the previous equation, we obtain the desired result:

$$-i\,\nabla\Psi = \frac{i\,f'\,\overline{f''}}{|f'|^3} = \frac{i\,\overline{f''}}{\overline{f'}\,|f'|} = \mathcal{K}.$$

The fact that harmonicity implies (12.9) can also be understood without appealing to ideas from complex analysis. Let $\mathcal{S}$ and $\mathcal{E}$ be the streamlines and orthogonal trajectories of a vector field $\mathbf{X}$. With the present notation, the results (11.5) and (11.6) from the previous chapter [see p. 544] become

$$\nabla \cdot \mathbf{X} = \frac{\partial |X|}{\partial s_1} + \kappa_2|X| \quad \text{and} \quad \nabla \times \mathbf{X} = -\frac{\partial |X|}{\partial s_2} + \kappa_1|X|.$$

In order for $\mathcal{E}$ (or $\mathcal{S}$) to be harmonic, $\mathbf{X}$ must be divergence-free and curl-free, so

$$\kappa_1 = \frac{\partial}{\partial s_2}\ln|X| \quad \text{and} \quad \kappa_2 = -\frac{\partial}{\partial s_1}\ln|X|,$$

from which (12.9) immediately follows.

We conclude this section by establishing the converse result that equation (12.9) is a sufficient condition for $\mathcal{E}$ to be harmonic. In the w-plane, let $\Theta(w)$ be the angle that the curve $\mathcal{C}_1$ through w makes with the horizontal. The angle of the curve $\mathcal{C}_2$ through w is therefore $\Theta + (\pi/2)$. Since the curvatures of $\mathcal{C}_1$ and $\mathcal{C}_2$ are the rates of change of these two angles with respect to the distances along the curves, we then have

$$\kappa_1 = \frac{\partial \Theta}{\partial s_1} \quad \text{and} \quad \kappa_2 = \frac{\partial \Theta}{\partial s_2}.$$

Next we calculate the Laplacian of Θ; the reason will be clear in a moment. Because the Laplacian is coordinate-independent, we may choose our coordinate directions tangent to $\mathcal{C}_1$ and $\mathcal{C}_2$, obtaining

$$\Delta\Theta = \frac{\partial}{\partial s_1}\left(\frac{\partial \Theta}{\partial s_1}\right) + \frac{\partial}{\partial s_2}\left(\frac{\partial \Theta}{\partial s_2}\right) = \frac{\partial \kappa_1}{\partial s_1} + \frac{\partial \kappa_2}{\partial s_2}.$$

Thus equation (12.9) implies that $\Theta(w)$ is harmonic, in which case it is the real part of an analytic function, say, $G(w)$. We may now define an analytic function $H(w) = e^{-iG} \propto e^{-i\Theta}$ such that $\overline{H} \propto e^{i\Theta}$ is everywhere tangent to $\mathcal{S}$. Thus $\mathcal{S}$ and $\mathcal{E}$ are the streamlines and equipotentials of the Pólya vector field of an analytic function. Done.

12.4.3 Further Complex Curvature Calculations

Suppose once again that $\mathcal{S}$ and $\mathcal{E}$ are the images in the w-plane of horizontal and vertical lines in the z-plane under an analytic mapping $w = f(z)$. They are then the streamlines and equipotentials of the vector field $\overline{\Omega'}$ in the w-plane, where Ω is the complex potential $z = \Omega(w) = f^{-1}(w)$ mapping $\mathcal{S}$ and $\mathcal{E}$ in the w-plane back to horizontal and vertical lines in the z-plane.

At present, the curvatures κ_1 and κ_2 at a point $w_0 = f(z_0)$ in the w-plane are represented as the components of the complex curvature of f at z_0 in the z-plane: $\mathcal{K}_f(z_0) = \kappa_1 + i\kappa_2$. [We have added the subscript f because we will shortly be considering the complex curvature of more than one mapping.] But suppose we think of the complex potential $\Omega(w)$ as fundamental, not merely the inverse of f; how can we express the curvatures directly in terms of Ω?

We shall answer this question by deriving a remarkably simple relationship between the complex curvature $\mathcal{K}_f(z)$ of f at z, and the complex curvature $\mathcal{K}_\Omega(w)$ of Ω at the image point w. Since $f'(z) = 1/\Omega'(w)$, equation (12.8) yields

$$\begin{aligned} \mathcal{K}_f(z) &= -i\,\nabla_z(1/|f'|) \\ &= -\frac{i}{\overline{\Omega'}}\nabla_w|\Omega'| \\ &= -\frac{|\Omega'|^2}{\overline{\Omega'}}\,[-i\,\nabla_w(1/|\Omega'|)] \\ &= -\Omega'(w)\,\mathcal{K}_\Omega(w). \end{aligned} \tag{12.11}$$

This is interesting. Recall that an infinitesimal complex number ϵ emanating from z is amplitwisted to yield the image complex number $f'(z)\,\epsilon$ emanating from

w. Thus, since (12.11) may also be written as

$$\mathcal{K}_{\Omega}(w) = -f'(z)\,\mathcal{K}_f(z), \tag{12.12}$$

we see that transforming $\mathcal{K}_f(z)$ as if it were an infinitesimal vector yields an image vector at w which is simply the negative of $\mathcal{K}_{\Omega}(w)$.

In the next section we shall shed some geometric light on this result, but for the moment (12.11) yields the desired formula (also known to Bivens (1992)) for the curvature of the streamlines and equipotentials of a sourceless, irrotational vector field in terms of its complex potential:

$$\kappa_1 + i\kappa_2 = -i\,\frac{|\Omega'|\,\overline{\Omega''}}{(\overline{\Omega'})^2}.$$

Next we turn to the curl of $\mathcal{K}$. The complex curvature is the Pólya vector field of the function $(-if''/f'|f'|)$, and the presence of $|f'|$ in the denominator prevents this from being analytic. Thus while $\mathcal{K}$ has been shown to be divergence-free, it cannot also be curl-free. What then is its curl?

To find it, recall that

$$\overline{\nabla}\mathcal{K} = \boldsymbol{\nabla}\boldsymbol{\cdot}\mathcal{K} + i\,\boldsymbol{\nabla}\boldsymbol{\times}\mathcal{K}.$$

Since $\overline{\nabla f'} = 0 = \overline{\nabla f''}$,

$$\overline{\nabla}\mathcal{K} = \overline{\nabla}\left\{\frac{i\,\overline{f''}}{\overline{f'}\,|f'|}\right\} = \frac{\overline{f''}}{\overline{f'}}\left\{i\,\overline{\nabla}\left(\frac{1}{|f'|}\right)\right\} = \frac{\overline{f''}}{\overline{f'}}\,\{\overline{\mathcal{K}}\} = -i\frac{|f''|^2}{|f'|^3}.$$

Thus $\boldsymbol{\nabla}\boldsymbol{\cdot}\mathcal{K} = 0$, which we already knew, and

$$\boldsymbol{\nabla}\boldsymbol{\times}\mathcal{K} = -\frac{|f''|^2}{|f'|^3}.$$

Although we have just differentiated $\mathcal{K}$, note that the result does not depend on any higher derivatives of f than occur in $\mathcal{K}$ itself. Indeed, the result may be re-expressed as

$$\boldsymbol{\nabla}\boldsymbol{\times}\mathcal{K} = -|f'|\,|\mathcal{K}|^2. \tag{12.13}$$

Since $\mathcal{K}$ is geometrically defined by f, and since the curl operator is also geometric, the curl of $\mathcal{K}$ must encode some (presumably simple) geometric information about the mapping f. Unfortunately, we have not yet succeeded in decoding this information.

12.4.4 Further Geometry of the Complex Curvature

The above results were derived by pure calculation in order to illustrate the complex gradient technique. We now revert to form and seek more geometrical explanations, beginning with (12.10).

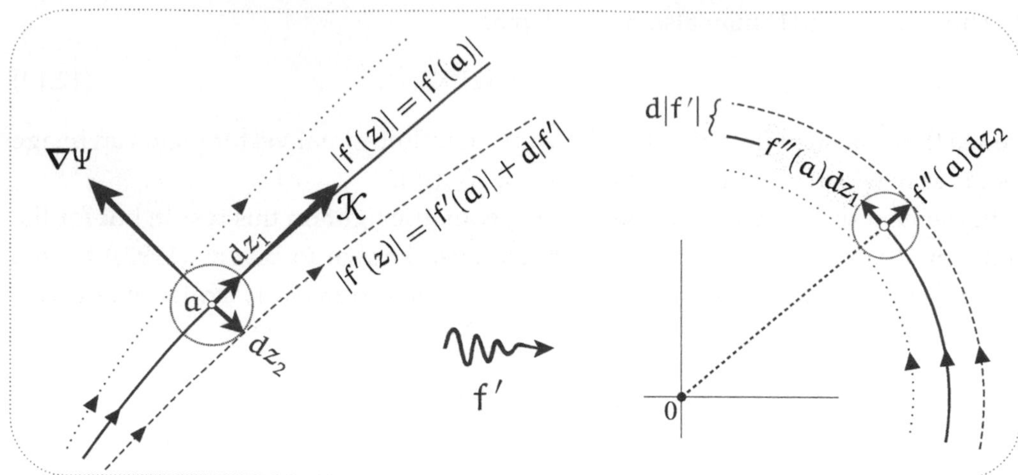

[12.9] Geometric proof that the complex curvature $\mathcal{K}$ has stream function $\Psi = 1/|f'|$. *We know from (5.32) that* the streamlines of the complex curvature are the level curves of the amplification, $|f'|$. *It follows readily that* $-i\nabla\Psi$ *points in the direction of* $\mathcal{K}$. *Next, observe that under the mapping* f', *the movement* dz_1 *along* $\mathcal{K}$ *is mapped to a movement tangent to the circle* $f' = \text{const.}$, *and therefore* dz_2 *is mapped to* a radial *movement of length* $d|f'| = |f''(a)\,dz_2|$. *Finally,* $|\nabla\Psi| = -\frac{d\Psi}{|dz_2|} = \frac{1}{|f'|^2}\frac{d|f'|}{|dz_2|} = \frac{|f''|}{|f'|^2} = |\mathcal{K}|$. *Done.*

The geometrically derived result (5.32), p. 273, says, in part, that *the streamlines of the complex curvature are the level curves of the amplification,* $|f'|$. Figure [12.9] illustrates this: sections of adjacent streamlines are mapped by f' to arcs of origin-centred circles. It also shows how the infinitesimal complex number dz_1 along the streamline through the point a is amplitwisted by $f''(a)$ to a complex number at $f'(a)$ that points counterclockwise along the circle $|f'(z)| = |f'(a)| = \text{const}$. Correspondingly, the orthogonal number $dz_2 = -i\,dz_1$ is amplitwisted to a complex number at $f'(a)$ that points radially outwards.

In order for $\mathcal{K}$ to be sourceless it must be of the form $\mathcal{K} = -i\nabla\Psi$, so that $\nabla\Psi$ (the direction of maximum increase of Ψ) must be directed as shown, and Ψ must be a function of $|f'|$. Since $|f'|$ increases in the direction dz_2, while $\nabla\Psi$ is in the opposite direction, Ψ must be a decreasing function of $|f'|$. Thus if Ψ is any decreasing function of $|f'|$ then $-i\nabla\Psi$ will be a sourceless vector field in the direction of $\mathcal{K}$. It only remains to show that for the particular function $\Psi = 1/|f'|$, the *magnitudes* also agree, i.e., $|\nabla\Psi| = |\mathcal{K}|$.

Let $d|f'|$ and $d\Psi$ be the changes in $|f'|$ and Ψ that result from the movement dz_2. From the picture we see that $d|f'| = |f''(a)\,dz_2|$, so that with $\Psi = 1/|f'|$,

$$|\nabla\Psi| = -\frac{d\Psi}{|dz_2|} = \frac{1}{|f'|^2}\frac{d|f'|}{|dz_2|} = \frac{|f''|}{|f'|^2} = |\mathcal{K}|,$$

as was to be shown.

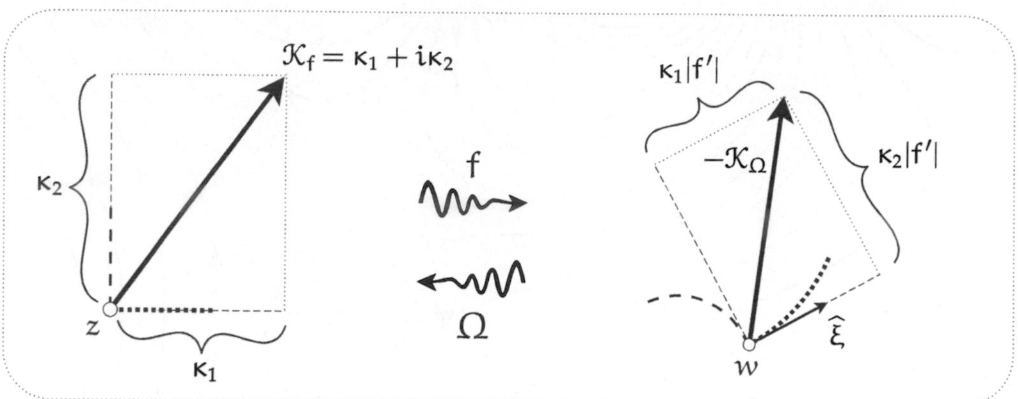

[12.10] Geometric proof that $-\mathcal{K}_\Omega(w) = f'(z)\,\mathcal{K}_f(z)$. *The horizontal line-segment at z is mapped by f to a segment of curvature κ_1, with unit tangent $\widehat{\xi}$. The inverse mapping Ω unbends this piece of curve, so, by the general transformation law of curvature, (5.31), $\mathcal{K}_\Omega \cdot \widehat{\xi} + \kappa_1/|\Omega'| = 0$. Therefore, the component of $-\mathcal{K}_\Omega$ in the direction of $\widehat{\xi}$ is $\kappa_1/|\Omega'| = \kappa_1|f'|$. Likewise, the orthogonal component is $\kappa_2|f'|$. Thus $-\mathcal{K}_\Omega$ is obtained by expanding $\mathcal{K}_f$ by $|f'|$, and rotating it by $\arg(\widehat{\xi})$. But $\arg(\widehat{\xi}) = \arg(f')$, so $-\mathcal{K}_\Omega(w) = f'(z)\,\mathcal{K}_f(z)$. Done.*

Next we give a more geometric derivation of (12.12). Figure [12.10] shows a horizontal line-segment at z being mapped by f to a segment of curve at w whose curvature is κ_1 and whose unit tangent is $\widehat{\xi}$. The inverse mapping Ω unbends this piece of curve and sends it back to the straight line-segment at z. Hence, by the general transformation law of curvature, (5.31), page 272,

$$\mathcal{K}_\Omega \cdot \widehat{\xi} + \kappa_1/|\Omega'| = 0.$$

In other words, as illustrated, the component of $-\mathcal{K}_\Omega$ in the direction of $\widehat{\xi}$ is $\kappa_1/|\Omega'| = \kappa_1|f'|$. Likewise, the orthogonal component is $\kappa_2|f'|$.

We now see that $-\mathcal{K}_\Omega$ is obtained by expanding $\mathcal{K}_f$ by $|f'|$, and rotating it by $\arg(\widehat{\xi})$. But since an infinitesimal real number ϵ at z is amplitwisted to a complex number $\epsilon f'(z)$ at w which points in the direction of $\widehat{\xi}$, we see that $\arg(\widehat{\xi}) = \arg(f')$. Hence

$$-\mathcal{K}_\Omega(w) = f'(z)\,\mathcal{K}_f(z),$$

as was to be shown.

It is also possible to give a more geometric derivation of the result (12.13). However, we shall not bother to do this since we have not yet been able to establish the significance of that result.

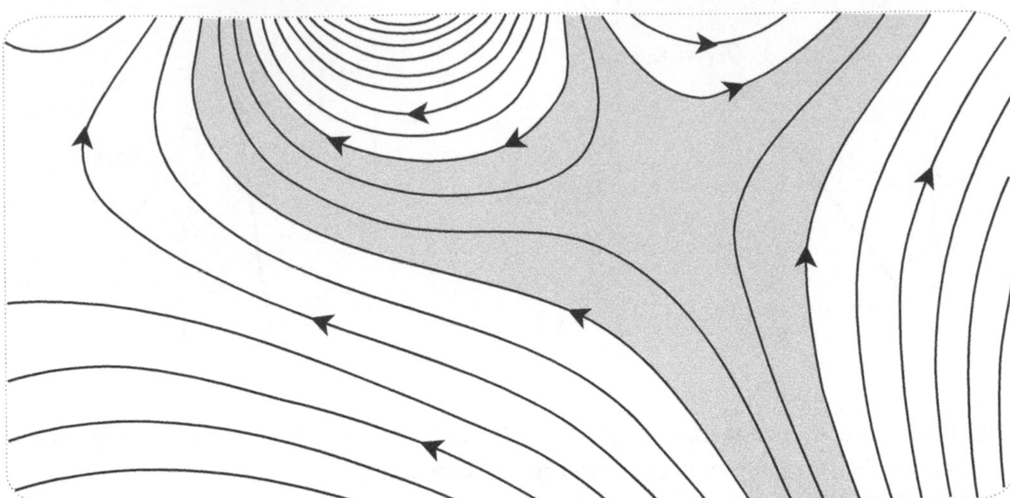

[12.11] Finding the flow around a given obstacle, B. *Given a frictionless flow, imagine that we instantly freeze the shaded region between streamlines, turning it into an obstacle, B, around which the flow continues, undisturbed. The image of the boundary under the conformal complex potential is a horizontal line, and, conversely, the problem of finding the flow around a given B therefore amounts to finding a conformal mapping Ω with this property.*

12.5 Flow Around an Obstacle

12.5.1 Introduction

Consider a typical fluid flow such as [12.11]. From our assumption that the fluid has no viscosity [this is most certainly an idealization] it follows that if we were to suddenly freeze the fluid within a flux tube, such as the shaded region in [12.11], then the unfrozen fluid would continue to flow in exactly the same way as before. The same idea applies if [12.11] instead represents flowing heat: if the shaded region of the metal plate were suddenly replaced with material which did not conduct heat then there would be no disturbance to the flow of heat in the remainder of the plate. This is much less of an idealization than in the case of fluid flow.

Conversely, if we insert an obstacle into a flow then the new disturbed flow must be such that the boundary B of this obstacle is a streamline, or is made up of segments of streamlines. If we think of the complex potential Ω of this disturbed flow as a mapping, this means that Ω maps B to a horizontal line, or to segments of a horizontal line.

The problem of finding flows around a given B therefore amounts to finding conformal mappings Ω with this property. In fact, since the complex potential of a given flow is only defined up to a constant, we may further demand that the horizontal image line be the real axis. Alternatively, this characterization may be

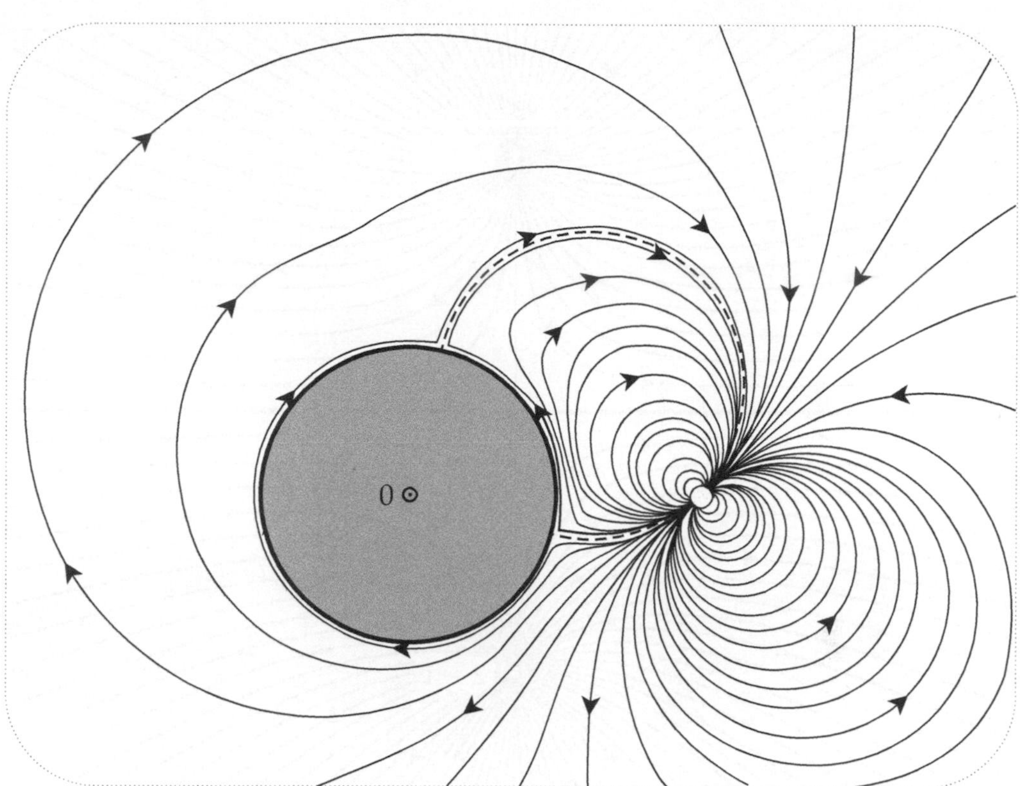

[12.17] The flow resulting from inserting the unit disc into a dipole field.

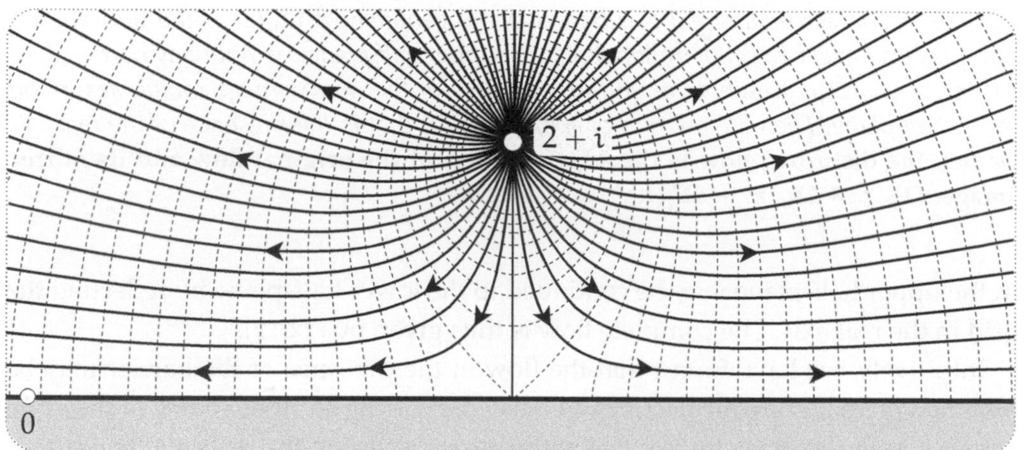

[12.18] The flow of a source in the presence of a barrier along the real axis.

As we saw in Chapter 5, Schwarz's Symmetry Principle (p. 286) says that a function which is analytic on one side of the real line, and which takes real values on that line (e.g., Ω_d), may be analytically continued to the other side by taking conjugate points to have conjugate images. This is vividly clear in [12.18]: the grid of squares

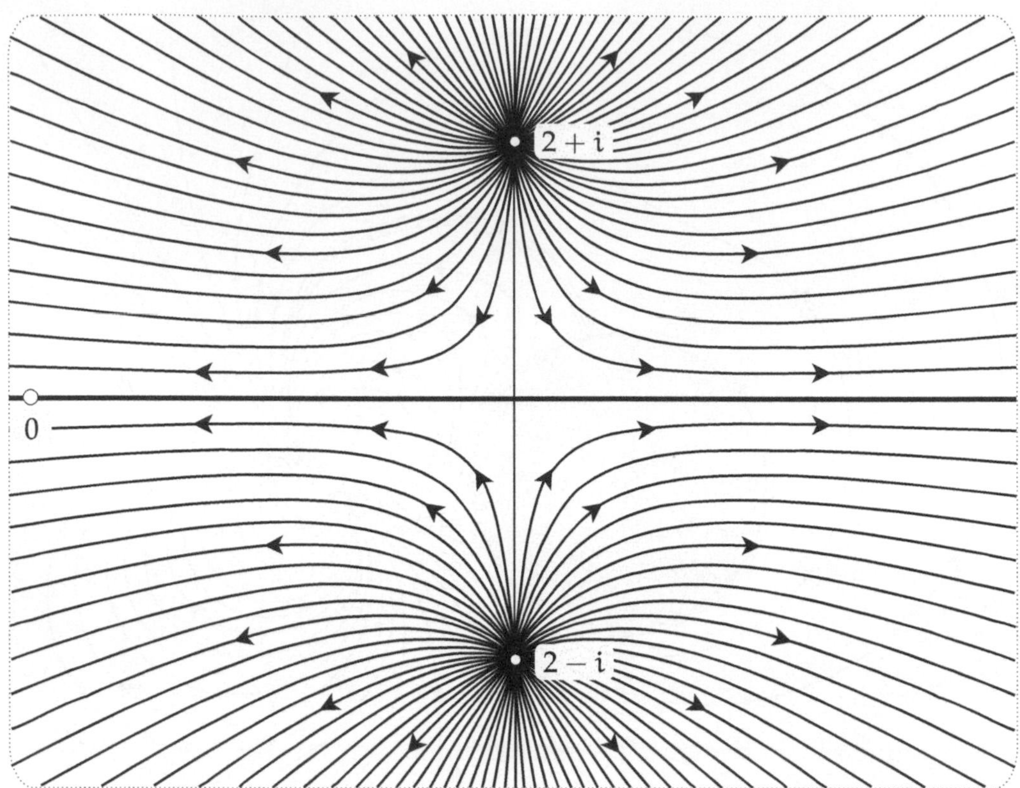

[12.19] The Method of Images with a linear boundary. *To find the flow depicted in the previous figure, we superpose the original source with its image (of equal strength) after reflection across the real axis barrier. More generally, if* $\Omega_u(z)$ *is the undisturbed complex potential of a set of multipoles in the upper half-plane, without any barrier, then* $\Omega^{\star}_u(z) \equiv \overline{\Omega_u(\overline{z})}$ *represents the undisturbed flow of the mirror-image multipoles in the lower half-plane. Then the method of images tells us that in the presence of the real axis barrier,* **the disturbed flow is the superposition of the original flow and its mirror image:** $\boldsymbol{\Omega_d(z) = \Omega_u(z) + \Omega^{\star}_u(z)}$.

in the upper half-plane may be continued to the lower half-plane by reflecting the grid in the real axis. The complete flow is thus given by [12.19].

Intuitively, we have found that the flow in the presence of the barrier may be obtained by removing the barrier and instead inserting another source of the same strength as the original, but located at the mirror image in the real axis. Thus Ω_d is the superposition of these two sources:

$$\Omega_d(z) = \log(z - 2 - i) + \log(z - 2 + i).$$

Use a computer to verify that this formula produces [12.19].

More generally, if Ω_u is the undisturbed complex potential of a superposition of multipoles (sources, vortices, dipoles, etc.) in the upper half-plane, then Ω_d will be the superposition of the undisturbed flow of this set of multipoles and the undisturbed flow of their mirror images. Note that while the mirror image multipoles will be of the same type and strength as the originals, the direction of their multipole moments will be different. For example, a dipole in the direction $(3-2i)$ reflects to one in the conjugate direction $(3+2i)$. More generally, a multipole with multipole moment Q reflects to one with moment $\overline{Q}$.

Now let us turn this method into a formula. In Chapter 5 we showed that from a given analytic mapping $f(z)$ we can produce a new analytic mapping $f^{\star}(z)$ according to the recipe

$$f^{\star}(z) \equiv \overline{f(\overline{z})}. \tag{12.15}$$

The physical significance of this new analytic function is easy to see: if Ω_u again represents the complex potential of a superposition of multipoles in the upper half-plane, then $\Omega_u^{\star}$ will be the undisturbed complex potential of their mirror images in the lower half-plane. The formula for the disturbed complex potential is thus

$$\Omega_d(z) = \Omega_u(z) + \Omega_u^{\star}(z). \tag{12.16}$$

Note that this formula does indeed satisfy Schwarz's Symmetry Principle: $\Omega_d^{\star}(z) = \Omega_d(z)$, so the real axis is a streamline. Naturally, if Ω_u instead represents a collection of multipoles in the lower half-plane, then (12.16) is again the solution in the presence of the barrier.

Essentially the same method may be used to find the disturbed flow in [12.17], in which the barrier is now a circle rather than a line. Reconsider that figure. We have drawn the streamlines $\Psi = \text{const.}$ at random, but had we instead chosen the values of Ψ in arithmetic progression (as we did in [12.12]) then it would have been possible to divide this flow into a grid of infinitesimal squares, with the unit circle being comprised of edges of these squares. As we saw in Chapter 5, it is again possible to extend this grid across the barrier (cf. [12.12]), but to do so we must replace reflection in a line by its analogue for circles, namely, *inversion*.

Performing this inversion in the unit circle we obtain [12.20], the dipole at 2 inverting to another dipole at $(1/2)$. It is now clear that to find Ω_d we should remove the barrier and superpose the undisturbed flows of these two dipoles. But what is the undisturbed complex potential of this new dipole at $(1/2)$?

As we saw in Chapter 5, if reflection in the real line is replaced by inversion in the unit circle then the recipe (12.15) may be modified to generate a new analytic function $f^{\dagger}$ given by

$$f^{\dagger}(z) \equiv \overline{f\left(\frac{1}{\overline{z}}\right)}. \tag{12.17}$$

The physical significance of this new analytic function is much as before: if Ω_u represents the complex potential of a superposition of multipoles outside the unit

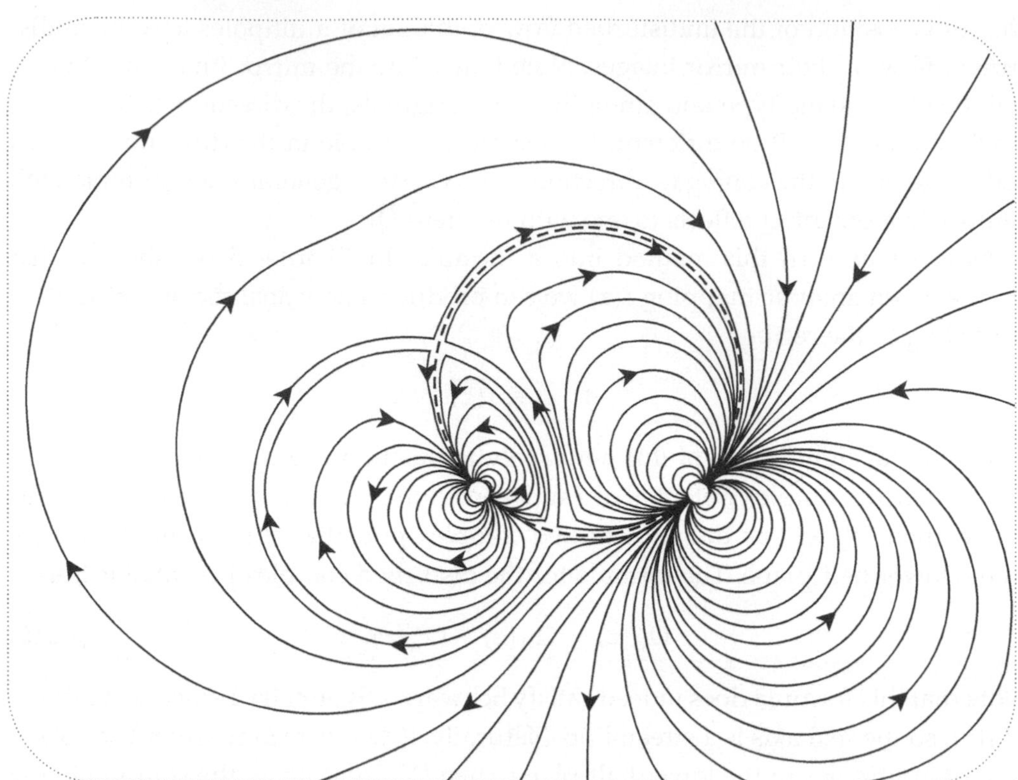

[12.20] The Method of Images with a circular boundary: ***Milne–Thomson Circle Theorem.*** *To find the flow depicted in [12.17], we superpose the original dipole with its image under inversion in the unit circle. More generally, if $\Omega_u(z)$ is the undisturbed complex potential of a set of multipoles outside the unit circle, without any barrier, then $\Omega^\dagger(z) \equiv \overline{\Omega\left(\frac{1}{\bar{z}}\right)}$ represents the undisturbed flow of the mirror-image multipoles inside the unit disc, and vice versa. Then the method of images tells us that in the presence of the unit circle barrier,* **the disturbed flow is the superposition of the original flow and its mirror image: $\Omega_d(z) = \Omega_u(z) + \Omega_u^\dagger(z)$.**

circle, then $\Omega_u^\dagger$ will be the undisturbed complex potential of their images under inversion. The analogue of (12.16) is now

$$\Omega_d(z) = \Omega_u(z) + \Omega_u^\dagger(z), \tag{12.18}$$

which automatically satisfies the symmetry requirement $\Omega_d^\dagger(z) = \Omega_d(z)$, so that the unit circle is a streamline. This result is known as the *Milne–Thomson Circle Theorem.* If Ω_u instead represents a collection of multipoles *all* of which lie inside the unit circle, then this Ω_d is again the solution in the presence of the barrier. The reason for the emphasis on "all" will be explained later.

Let us apply this method to find the disturbed complex potential of [12.20] (and hence of [12.17]). If Ω_u is given by (12.14) then

$$\Omega_u^\dagger(z) = \frac{(1-i)}{(1/z)-2} = \frac{(1-i)z}{1-2z}.$$

That this is indeed a dipole at (1/2) may be seen by rewriting it as

$$\Omega_u^\dagger(z) = -\frac{(1-i)}{4}\left[\frac{1}{z-(1/2)}\right] - \frac{(1-i)}{2}.$$

Because constants have no effect on the flow, this is a dipole at (1/2) with dipole moment $(1-i)/4$. Unlike the case of reflection across a line, note that it is not only the direction of the dipole moment which is affected by the inversion, but also its magnitude: here the strength of the inverted dipole is one quarter the strength of the original. Superposing the two dipoles we obtain

$$\Omega_d(z) = \Omega_u(z) + \Omega_u^\dagger(z) = \frac{(1+i)}{z-2} + \frac{(1-i)z}{1-2z},$$

and you may use a computer to verify that this formula does yield the flow in [12.20].

Like [12.19], figure [12.20] has symmetry, but it is of a more subtle kind than before: by construction, the figure reproduces itself under inversion in the unit circle. This symmetry can be made to leap from the page by projecting [12.20] onto the Riemann sphere. As we learnt in Chapter 3, inverting in the unit circle is equivalent to reflecting the Riemann sphere in its equatorial plane. The flow on the northern and southern hemispheres should therefore be mirror images of each other in this plane. Behold figure [12.21]! Note that on the sphere the strengths of the two dipoles become equal.

We can now see that [12.12] is simply a limiting case of [12.21], for as the southern dipole moves towards the south pole (0), its reflection moves towards the north pole (∞), and a solitary dipole at ∞ projects to a uniform flow in the plane.

In Chapter 5 we saw that both reflection in a line and inversion in a circle were special cases of Schwarzian reflection in a curve; see [5.30], page 290. We now use this fact to generalize the method of images. For example, suppose that we insert an ellipse E into a uniform flow. Clearly the flow will look something like [12.22], and that this is in fact the exact flow is once again apparent from its divisibility into squares. How did we find it?

The generalization of the formulae (12.15) and (12.17) to the case of Schwarzian reflection $\mathfrak{R}_K$ in a curve K is

$$f^\ddagger(z) \equiv \overline{f[\mathfrak{R}_K(z)]}.$$

For example, if K is the real line (so that $\mathfrak{R}_K(z) = \bar{z}$) then $f^\ddagger = f^\star$, while if K is the unit circle (so that $\mathfrak{R}_K(z) = 1/\bar{z}$) then $f^\ddagger = f^\dagger$. Thus if Ω_u is the undisturbed

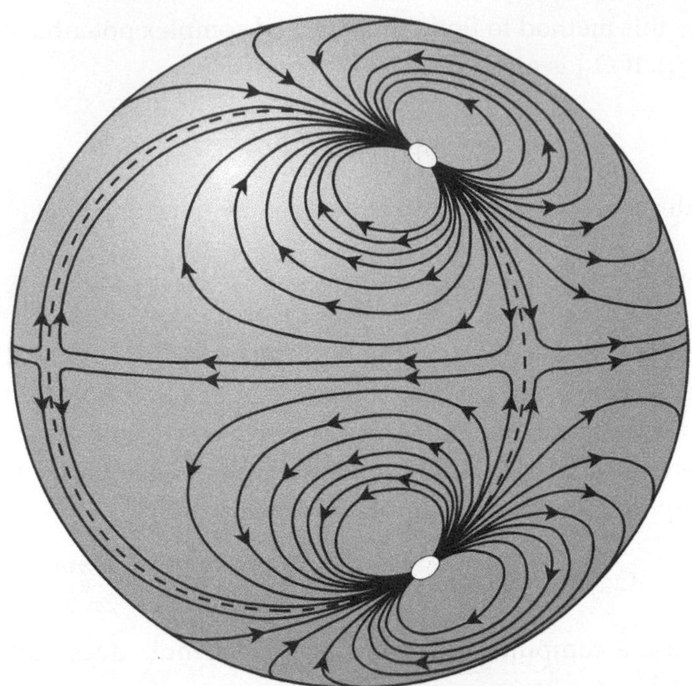

[12.21] Projection of the previous figure onto the Riemann sphere. *By construction, the previous figure is symmetric under inversion in the unit circle, which induces reflection of the Riemann sphere in its equatorial plane. So the flow on the northern and southern hemispheres should therefore be mirror images of each other, and indeed we see that they are! Note that the unequal strengths of the original dipoles in the plane project to dipoles of equal strength on the sphere. Also observe that [12.12] may now be understood as a limiting case of this figure, for as the southern dipole moves towards the south pole (0), its reflection moves towards the north pole (∞), and a solitary dipole at ∞ projects to a uniform flow in the plane.*

complex potential of a collection of multipoles on one side of K then the disturbed complex potential with K as barrier will be

$$\Omega_{\mathrm{d}}(z) = \Omega_{\mathrm{u}}(z) + \Omega_{\mathrm{u}}^{\ddagger}(z).$$

This formula automatically satisfies $\Omega_{\mathrm{d}}^{\ddagger} = \Omega_{\mathrm{d}}$, so that K is a streamline.

For example, take the ellipse E in [12.22] to have equation $(x/2)^2 + y^2 = 1$. In this case we have [see p. 291]

$$\mathfrak{R}_{\mathrm{E}}(z) = \tfrac{1}{3}\left[5\,\overline{z} - 4\sqrt{\overline{z}^2 - 3}\,\right].$$

Thus, with $\Omega_{\mathrm{u}}(z) = z$,

$$\Omega_{\mathrm{d}}(z) = \tfrac{4}{3}\left[2\,z - \sqrt{z^2 - 3}\,\right],$$

and you may use a computer to verify that this formula yields figure [12.22]. Note, however, that everything is not quite as it seems. While Ω_{u} is a uniform

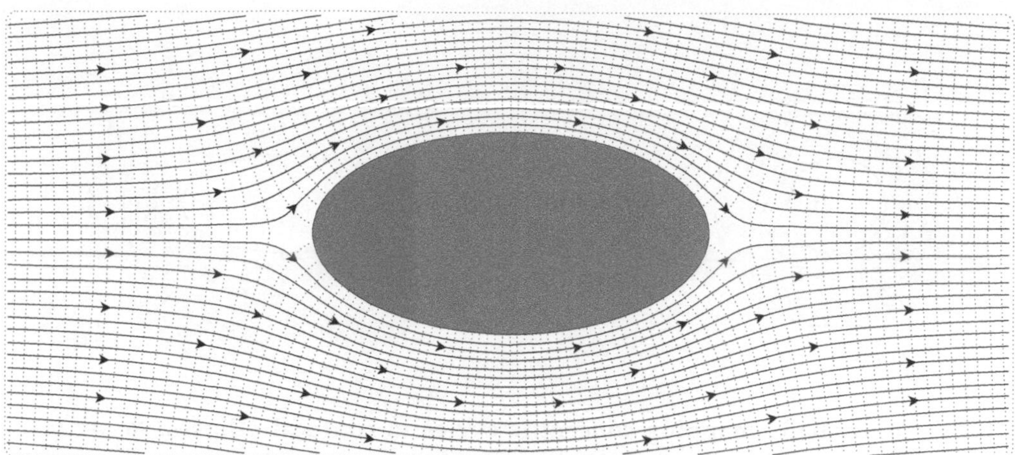

[12.22] The Method of Images with a general boundary via Schwarzian reflection. *To find the illustrated flow around the ellipse, E, we superpose a steady eastward flow with its Schwarzian reflection $\mathfrak{R}_E$ across E; see [5.31]. More generally, if $\Omega_u(z)$ is the undisturbed complex potential of a set of multipoles on one side of a curve K, without any barrier, then $\Omega_u^{\ddagger}(z) \equiv \overline{\Omega_u[\mathfrak{R}_K(z)]}$ represents the undisturbed flow of the Schwarzian reflections of the multipoles on the other side of K, and vice versa. Then the method of images tells us that in the presence of the barrier, K,* **the disturbed flow is the superposition of the original flow and its mirror image: $\Omega_d(z) = \Omega_u(z) + \Omega_u^{\ddagger}(z)$.** *This subsumes both previous methods of images as special cases, for if K is the real axis then $\mathfrak{R}_K(z) = \bar{z}$ and $f^{\ddagger} = f^{\star}$, while if K is the unit circle then $\mathfrak{R}_K(z) = 1/\bar{z}$ and $f^{\ddagger} = f^{\dagger}$.*

flow to the right with speed 1, the behaviour of Ω_d for large values of $|z|$ is a uniform flow to the right with speed $(4/3)$. Of course if we wish the flow to have unit speed far from the ellipse then we need only multiply this Ω_d by $(3/4)$.

12.5.4 Mapping One Flow Onto Another

We previously established the fact that a conformal mapping sends the streamlines and equipotentials of steady, sourceless, irrotational flow to the streamlines and equipotentials of another such flow. This idea has many theoretical and practical uses.

A theoretical benefit is a fresh insight into the very concept of a complex potential. Reconsider figure [11.21], p. 570. Applying any conformal mapping f to the flow on the left yields another steady, sourceless, irrotational flow on the right. The complex potential may now be defined as the special mapping $f = \Omega$ for which the image flow is uniform with velocity 1. For example, what is the complex potential of a uniform flow with velocity 1? From the new point of view it is the conformal mapping which sends this flow to a uniform flow with velocity 1. Thus it is the identity mapping $\Omega(z) = z$, as it should be!

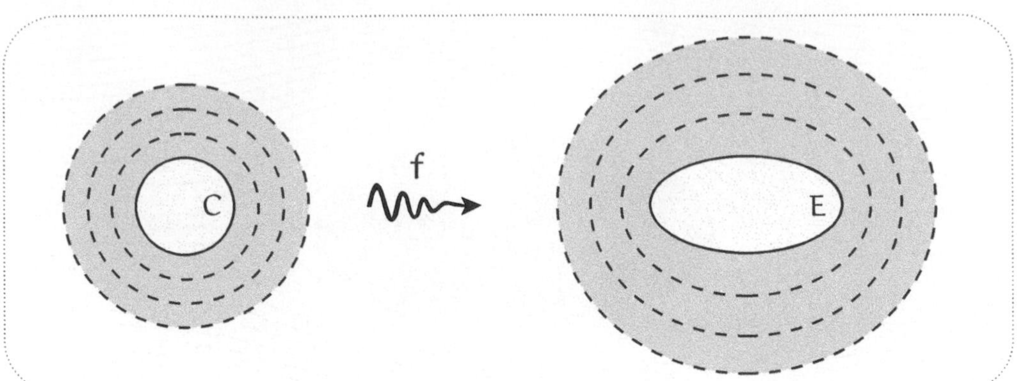

[12.23] Conformally mapping concentric circles to confocal ellipses. *Recall from [5.22], on page 276, that as* $z = e^{it}$ *describes the unit circle,* C*,* $w = p\,e^{it} + q\,e^{-it}$ *describes the ellipse* $(x/a)^2 + (y/b)^2 = 1$*, where* $p = (a+b)/2$ *and* $q = (a-b)/2$*. Thus, in the specific example of* $(x/2)^2 + y^2 = 1$*, the conformal mapping we seek of* C *to* E *is* $z \mapsto w = f(z) = \frac{1}{2}\left(3z + \frac{1}{z}\right)$*. Circles concentric to* C *are mapped to ellipses confocal to* E*. Far from* C*, the mapping behaves like* $\frac{3}{2}z$*.*

To demonstrate the practicality of mapping one flow onto another, let us return to the problem of finding the flow round an obstacle, restricting our attention to the case where the obstacle is inserted into a *uniform* flow. For example, let us rederive the flow round the ellipse $(x/2)^2 + y^2 = 1$ in [12.22]. Suppose that we knew of a one-to-one conformal mapping $w = f(z)$ from the exterior of the unit circle C in the z-plane to the exterior of this ellipse E in the w-plane. We already know the flow round C when it is inserted into a uniform flow of velocity 1, so applying f to these streamlines yields *some* flow round E. If we want the flow round E to be uniform far from E then we must further demand that $f(z)$ behave like a multiple of the identity far from C: $f(z) \approx cz$ if $|z|$ is large. Assigning equal values of Ψ to the original and image streamlines, the image flow far from E will then be uniform, with speed $(1/|c|)$ and direction c [why?].

Recall figure [5.22], p. 276. As $z = e^{it}$ describes C, $w = p\,e^{it} + q\,e^{-it}$ describes the ellipse $(x/a)^2 + (y/b)^2 = 1$, where $p = (a+b)/2$ and $q = (a-b)/2$. Thus in the case of $(x/2)^2 + y^2 = 1$, the mapping we seek is

$$w = f(z) = \frac{1}{2}\left(3z + \frac{1}{z}\right). \tag{12.19}$$

As illustrated in [12.23], C is mapped to E and circles concentric to C are mapped to ellipses confocal to E. Far from C the mapping behaves like $(3/2)z$: you can see that as the circles on the left grow, their images approach circles that are $(3/2)$ as big. Thus f should map the flow round C to the flow that results from inserting E into a uniform flow of velocity $(2/3)$.

Let's go through the details and check that we recover the flow in [12.22]. Since the flow round C has complex potential $\Omega(z) = z + (1/z)$, and since the complex potential $\widetilde{\Omega}(w)$ at the image $w = f(z)$ of z is defined to be the same as Ω at z, we have $\widetilde{\Omega}(w) = z + (1/z)$. To express this as an explicit function of w we solve (12.19) for z and obtain [why the choice of +?]

$$z = \tfrac{1}{3}\left(w + \sqrt{w^2 - 3}\right).$$

Although we could immediately insert this into the formula for $\widetilde{\Omega}(w)$, we may save ourselves a little algebra by first noting that (12.19) implies $(1/z) = 2w - 3z$. Thus

$$\widetilde{\Omega}(w) = z + (1/z) = 2(w - z) = \tfrac{2}{3}\left(2w - \sqrt{w^2 - 3}\right).$$

Apart from the factor of (2/3), signifying that the velocity is (2/3) far from E (as anticipated), this is the same formula we previously obtained by Schwarzian reflection.

As another illustration of this idea, suppose that E were instead inserted into a uniform flow in the direction of $e^{i\phi}$. To find the flow round E we need only find the flow round C when it is inserted into such a flow, and then apply f. Since the undisturbed complex potential is $\Omega_u(z) = e^{-i\phi}z$, the method of images says that the flow round C is

$$\Omega_d(z) = \Omega_u(z) + \Omega_u^{\dagger}(z) = e^{-i\phi}z + \frac{e^{i\phi}}{z}.$$

With $\phi = (\pi/4)$ this flow is illustrated on the left of [12.24]. On the right is the desired flow round E obtained by applying f to the flow round C. Use your computer to verify this figure.

In this manner, we may derive the flow round an infinite variety of obstacles: choose any analytic mapping $w = f(z)$ which is one-to-one outside C and which behaves like cz for large $|z|$, then $\widetilde{\Omega}(w) = f^{-1}(w) + [1/f^{-1}(w)]$ is the flow round an obstacle whose boundary curve B is $f(C)$.

12.6 The Physics of Riemann's Mapping Theorem

12.6.1 *Introduction*

Recall that Riemann's Mapping Theorem [p. 204] asserts that any simply connected region R (other than the entire plane) may be mapped one-to-one and conformally to any other such region S. Granted this, we saw that no loss of generality results from taking S to be the unit disc D, and that there must be as many mappings from R to S as there are automorphisms of D. Later [p. 407] we showed that these automorphisms are the hyperbolic rigid motions, which have three degrees of freedom.

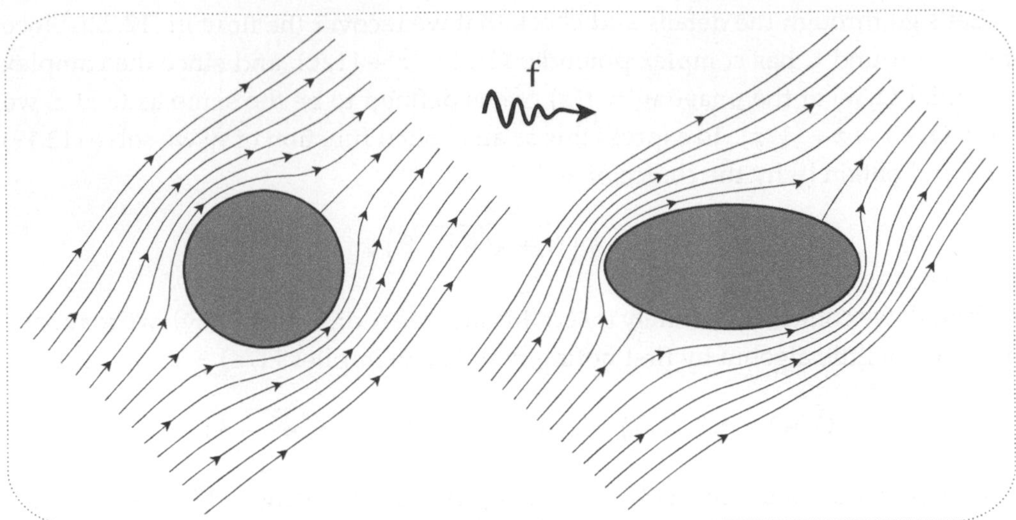

[12.24] Conformally mapping one flow onto another. *Using the mapping obtained in the previous figure, the known flow on the left around the unit disc may be conformally mapped to the flow around the ellipse on the right. [NOTE: In the text, we confirm that this new approach yields the same answer that we formerly obtained via the method of images.] In this manner, we may derive the flow round an infinite variety of obstacles: choose any analytic mapping* $w = f(z)$ *which is one-to-one outside* C *and which behaves like* cz *for large* $|z|$, *so that the flow becomes uniform, then* $\widetilde{\Omega}(w) = f^{-1}(w) + [1/f^{-1}(w)]$ *is the flow round an obstacle whose boundary curve* B *is* $f(C)$.

Thus the complete result we seek to understand is this: *there exists a three-parameter family of one-to-one, conformal mappings between* R *and* D.

There are at least two standard proofs of this fundamental result, and most advanced books on complex analysis include one of these. Despite the fact that one of the arguments (due to Koebe) is constructive in nature—and therefore, in principle, comprehensible—we have not yet found a way to present it in a manner consistent with the aims of this book. However, the interested reader will find an excellent description of the idea underlying Koebe's proof in Hilbert (1952), and a clear description of the technical details in Nehari (1952) or Nevanlinna and Paatero (2007). To our knowledge, the deepest investigation of this idea is that given by Henrici (1991).

On a brighter note, the above ideas on flows *will* enable us to gain considerable insight into both the existence of such mappings and the fact that they have three degrees of freedom. We shall do so by reversing the idea of conformally mapping one flow onto another. That is, by resorting to physical experiment one may obtain flows, and these may then be used to construct conformal mappings.

12.6.2 *Exterior Mappings and Flows Round Obstacles*

Consider [12.25]. The top part depicts an obstacle R that has been inserted into uniform flow having velocity 1. If we do not introduce any circulation round R

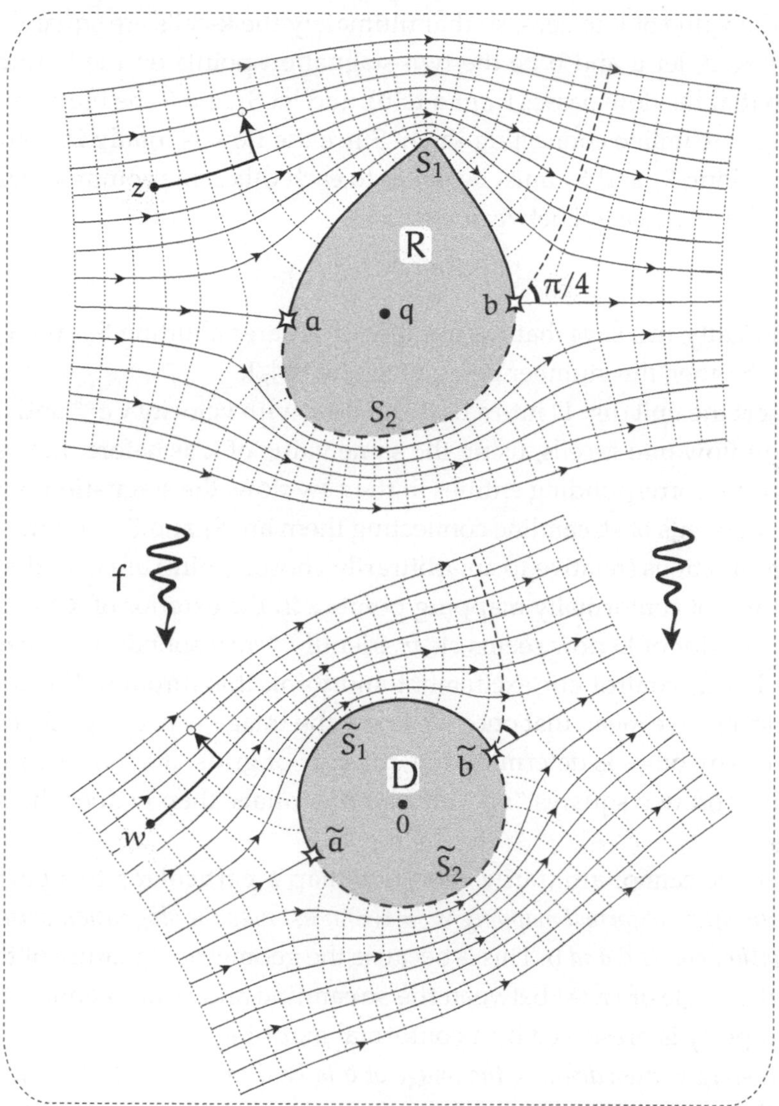

[12.25] Constraint on the mapping between flows. *Divide the flow around* R *and the flow around the unit disc* D *into small, square* k*-cells. If the illustrated point* z *is mapped by* f *to* $w = f(z)$*, then if* z *moves four squares downstream and two squares "up" the equipotentials,* w *must do the same. But the sought-after conformal mapping* $w = f(z)$ *has a constraint: it must map the stagnation points* a *and* b *to the stagnation points* $\tilde{a}$ *and* $\tilde{b}$*. This is because the geometric signature of the critical point* b *is the angle of* $(\pi/4)$ *between the streamline and equipotential through* b*, and this property is preserved by a conformal mapping.*

then, as we have previously remarked, the flow is unique. Next, we have arbitrarily chosen a point [not shown] from which to measure circulation and flux, i.e., a point at which the potential Φ and stream function Ψ both vanish. With a small value of k, we have then constructed the k-flux tubes and k-work tubes, thereby dividing the exterior of R into small, approximately square, k-cells. As usual, we may imagine shrinking the value of k to zero, so that ultimately the k-cells are square.

As illustrated, let a and b be the two stagnation points on the boundary of R (ordered so that the flow passes from a to b), and let S_1 and S_2 be the two segments of boundary streamline connecting them. Since the flow is totally irrotational, the circulations along S_1 and S_2 must be equal to each other, the common value being the potential difference $[\Phi]$ between a and b:

$$[\Phi] = \Phi(b) - \Phi(a).$$

Put geometrically, this says that the number of squares abutting S_1 and S_2 must be equal to each other, this number being given by $[\Phi]/k$.

Now insert the unit disc D into a uniform flow with velocity $v\,e^{i\phi}$ and divide up the resulting flow into k-cells, using the same value of k as before. Let us employ tildes to denote corresponding entities in the new flow: the stagnation points are $\widetilde{a}$ and $\widetilde{b}$, the segments of streamline connecting them are $\widetilde{S}_1$ and $\widetilde{S}_2$, and the potential and stream functions (relative to an arbitrarily chosen point) are $\widetilde{\Phi}$ and $\widetilde{\Psi}$.

The method of conformally mapping points z in the exterior of R to points $w = f(z)$ in the exterior of D now seems clear: identify "corresponding" squares in the two grids! Let us control our excitement and think this through. Presumably, by "corresponding" we mean that once we know the image w of one grid point z then the image of any other is determined by the pair of grids: if z moves four squares downstream and two squares "up" the equipotentials, then w does the same. See [12.25].

However, we cannot arbitrarily choose to map a particular z to a particular w, for *the sought-after conformal mapping* $w = f(z)$ *must map the stagnation points* a *and* b *to the stagnation points* $\widetilde{a}$ *and* $\widetilde{b}$. This is because the geometric signature of the critical point b is the angle of $(\pi/4)$ between the streamline and equipotential through b, and this property is preserved by a conformal mapping.

Now the snag is this: *defining the image of* a *to be* $\widetilde{a}$ *(as we must),* b *will map to* $\widetilde{b}$ *if and only if the number of squares abutting* $\widetilde{S}_1$ *is equal to the number abutting* S_1. Looking closely, we see this is not the case in [12.25]: there are fewer squares along $\widetilde{S}_1$ than S_1. To correct the situation we must slightly increase the speed v of the flow round the disc. More precisely, v must be chosen so that the potential difference $[\widetilde{\Phi}]$ across D equals the potential difference $[\Phi]$ across R. Since we know that $[\widetilde{\Phi}] = 4v$, we deduce that

> *A one-to-one, conformal mapping between the two grids is obtained if and only if the disc is inserted into a flow of speed* $v = [\Phi]/4$.

Although we hope our use of grids has helped to make the mapping vivid (and to show that it is both one-to-one and conformal), we should perhaps point out that the grids are in no way essential to the definition of f. All we are doing is identifying points by means of circulation and flux: the image $w = f(z)$ of z is defined by

$$\widetilde{\Omega}(w) = \Omega(z) + \left\{\widetilde{\Omega}(\,\widetilde{a}\,) - \Omega(a)\right\},$$

the constant having been chosen so as to ensure that $\widetilde{a} = f(a)$.

Since the choice of the zero point of flux and circulation in each of the flows has no bearing on the construction of f, we may choose these two points to our advantage. Henceforth, *we will take the zero point in the image flow to be the image of the zero point in the original flow.* In [12.25], for example, this convention says that if we choose the zero point for the flow round R to be at a (i.e., $\Omega(a) = 0$), then we must take the zero point for the flow round D to be at $\widetilde{a}$ (i.e., $\widetilde{\Omega}(\,\widetilde{a}\,) = 0$). The above equation then simplifies to

$$\widetilde{\Omega}(w) = \Omega(z).$$

In other words, w and z correspond if and only if their circulation and flux are equal.

In this case the mapping may be written as $f = \widetilde{\Omega}^{-1} \circ \Omega$, which we may interpret as follows. Refer to [12.25] and [12.15]. The complex potentials $\Omega(z)$ and $\widetilde{\Omega}(w)$ map points lying strictly outside R and, respectively, D to points in a plane that is slit along the real axis from 0 to $[\Phi]$. Thus f may be thought of as first mapping the exterior of R to the slit plane by means of Ω, then mapping the slit plane to the exterior of D by means of $\widetilde{\Omega}^{-1}$.

We now return to the mappings themselves and ask, how "many" of them do we obtain by means of this construction? We begin by explaining the illusory nature of some of the apparent freedoms in the construction. First, why not insert R into a uniform flow of arbitrary speed? Of course we can, but nothing new results from doing so. For example, if we double this speed then we must also double v, because we must maintain equality between the number of k-cells along S_1 and $\widetilde{S}_1$. But this yields the same mapping f as before. Second, in constructing the grid outside D, why not use a different value of k, say $\widetilde{k}$, from that used outside R? In order to maintain equality between the number of squares along S_1 and $\widetilde{S}_1$ we would then have to change v to $\widetilde{k}[\Phi]/4k$ [why?], and this would produce the same grid and mapping as before.

Clearly, however, we do obtain new mappings by varying the direction ϕ of the flow round D. If F denotes the particular mapping f corresponding to $\phi = 0$, then the mapping F_ϕ corresponding to a general value of ϕ is $F_\phi(z) = e^{i\phi}F(z)$, namely, F followed by a rotation.

It would seem plausible that still other mappings could be obtained by varying the direction—let us call it θ—of the flow into which R is inserted. Not so. This follows from the fact that the flow round R in this case may be obtained by applying F^{-1} to the flow round the disc when *it* is inserted into a uniform flow in the direction θ. Clarify this in your own mind by referring to figures [12.23] and [12.24]. Thus the mapping between the two flows is only sensitive to the angle *between* their directions: inserting R and C into flows with directions θ and ϕ, respectively, yields the mapping $F_{(\phi-\theta)}$.

Since ϕ is the only genuine degree of freedom, the mappings we have constructed belong to a one-parameter family. We are therefore missing two degrees of freedom. Though we will explain this mystery shortly, you may care to think about it on your own before reading further.

Granted the generosity of Nature in providing such flows round obstacles, we now have a physical method of determining f, but we lack a mathematical procedure for doing so. This is a very hard problem. However, there does exist an explicit formula for f in the case that R is bounded by a polygon. This is called the *Schwarz-Christoffel formula*; see Nehari (1952) or Pólya and Latta (1974) for good discussions of the result. We shall not enter into this here, except to say that the advent of high-speed computers has opened the way to approximating f by approximating the boundary of a given R with a polygon; see Trefethen (1986) and Henrici (1991) for this and other algorithmic approaches to the problem.

12.6.3 Interior Mappings and Dipoles

In [12.25], suppose that the speed of the flow round D has been adjusted to $[\Phi]/4$, yielding a conformal mapping between the exteriors of R and D. If we now perform inversion in the boundary of D, we obtain the flow [12.26b] inside the unit circle C. This is the familiar flow of a dipole inserted into the centre of a circular pool of fluid.

Similarly, if we choose the arbitrarily selected interior point q of R shown in the previous figure, then perform an inversion[4] in a circle of unit radius centred there, we obtain [12.26a]. Use a computer, or your Peaucellier linkage [see p. 205], to verify that the boundary of R does indeed invert to the illustrated curve $R^{\dagger}$. In [12.25], the streamlines far from R are parallel lines, and the inversion therefore sends these to small, mutually tangent circles through q. Thus, as illustrated, the inverted flow in [12.26a] represents a dipole which has been inserted at q into the $R^{\dagger}$-shaped pool

[4] Ideas related to those which now follow may be found in Bak and Newman (2010), Siegel (1969, p. 148), and especially Courant (1950). Although Riemann himself employed physical reasoning, the idea of relating his mapping theorem to dipoles seems to have originated with Hilbert (1965, Vol. 3, pp. 73–80).

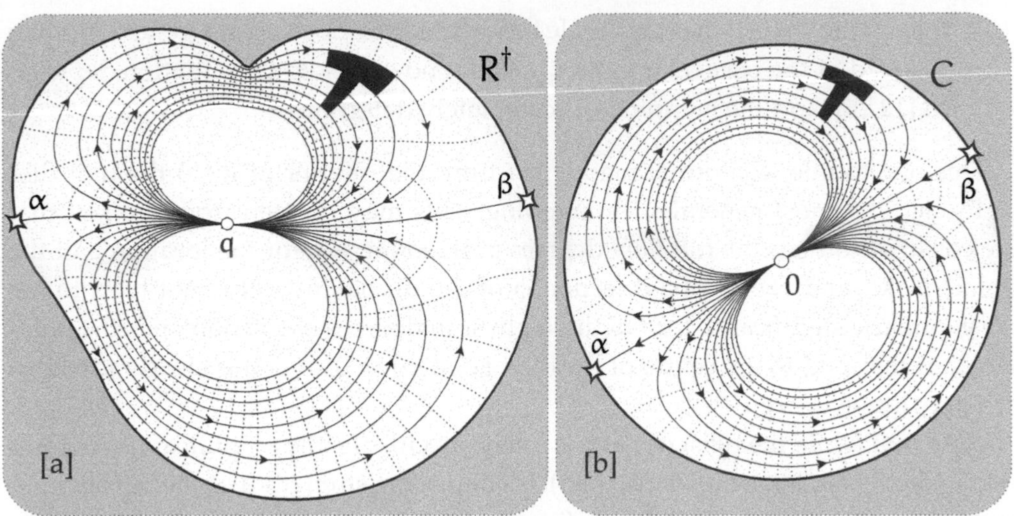

[12.26] Interior mappings and dipoles. *Referring to the previous figure, inversion of* R *in a circle centred at the arbitrarily selected interior point* q *yields* **[a]**, *while inversion of* D *in the unit circle yields* **[b]**. *In both cases, we obtain dipoles. In this way, the one-to-one, conformal mapping between the exteriors of* R *and* C *now yields another such mapping between the interior dipole flows of* $R^\dagger$ *and* C: *for example, the black T-shape is mapped to the black T-shape.*

of fluid. In fact [exercise] since the original uniform flow had unit speed, the dipole at q has unit strength.

In this way, the one-to-one, conformal mapping between the exteriors of R and C now yields another such mapping between the interiors of $R^\dagger$ and C. For example, the stagnation point α maps to the stagnation point $\widetilde{\alpha}$, q maps to 0, and the black T-shape inside $R^\dagger$ maps to the one inside C. Although the speed of the flow grows arbitrarily large as we approach q, there is nothing dramatic about the behaviour of the mapping near q. For example, imagine sliding the T-shape along the streamlines towards q, and consider its image.

We may look at this construction rather differently. Suppose that $R^\dagger$ is a *given*, fixed curve whose interior we wish to map to the unit disc. We may obtain such a mapping f as follows:

(i) Insert a horizontal dipole of unit strength into the interior of $R^\dagger$ at an arbitrary interior point p. Divide the interior into k-cells of the flow, and let N denote the number of them abutting $R^\dagger$.

(ii) At the centre 0 of C, insert a dipole of strength d, and direction ϕ. Divide the interior of C into the k-cells of this flow, and adjust the strength d until the number of them abutting C is equal to N. If $[\Phi]$ again denotes the circulation along a segment of boundary streamline connecting the stagnation points α and β, this condition is met by $d = [\Phi]/4$.

(iii) Except for the fact that the dipole inside $R^\dagger$ is now located at a general point p (instead of the particular point q), we are now back to the situation depicted in [12.26], and the mapping f is completely determined.

This one-to-one, conformal mapping f from the interior of $R^\dagger$ to the unit disc is, in fact, the most general such mapping. This may be seen from the fact that the construction has the full *three* degrees of freedom: *two* for the location p of the dipole in $R^\dagger$, and *one* for the direction ϕ of the dipole in C. It is not hard to see the geometric significance for f of these physical degrees of freedom: p is the point which is mapped to the centre of the disc, i.e., $f(p) = 0$; and ϕ is clearly the twist of f at p, i.e., $\phi = \arg[f'(p)]$.

Note that while the twist of f at p is freely specifiable, the amplification $|f'(p)|$ is not. In fact [exercise, or read on] $|f'(p)|$ is simply the strength d of the dipole in C, and we have seen how the latter is fixed in the course of the above construction. Put differently, $|f'(p)| = [\Phi]/4$.

We can now return to the construction in [12.25] and see why it is missing two degrees of freedom. To obtain the general mapping between the exteriors of R and D, we may take this newly constructed *general* mapping between the interiors of $R^\dagger$ and C, then reverse (i.e., repeat) the inversions which took us from [12.25] to [12.26].

In the above process of generalization the fundamental step was moving the dipole at q to an arbitrary interior point p. Since the flow in [12.26b] did not undergo any fundamental change, inversion returns it to a flow round D like that shown in [12.25]. However, to send $R^\dagger$ back to the boundary of R we must invert in the circle of unit radius centred at q (which we now take to be the origin), and *the flow inside* $R^\dagger$ *of the dipole at* p *inverts to the flow outside* R *of a dipole placed at* $(1/\overline{p})$. Figure [12.17] illustrates such a flow when R is a disc.

The construction in [12.25] is now recognizable as the special $p = q$ in which the dipole outside R has been placed at ∞ rather than at a general point. Here are our two missing degrees of freedom: there was nothing wrong with choosing to place one of the dipoles (the one outside D) at ∞, but we then (unnecessarily) insisted on placing the other dipole there too. In terms of the resulting mapping between the exteriors, this amounted to insisting that ∞ map to ∞, whereas the general construction allows us to map any point outside R to ∞ by placing a dipole there.

12.6.4 *Interior Mappings, Vortices, and Sources*

Having had the inspiration to construct mappings by means of multipoles, it is natural to wonder if we may simplify this dipole method by employing the most primitive of all multipoles, the source. However, if we continue to think in terms of fluid flows then this idea cannot work. For if we place a source inside the region we wish to map, the fluid flowing out of the source and into the region has nowhere to go! The only way out is to also insert a sink of equal strength[5], thereby creating a doublet. But this is really no improvement over our earlier dipole construction, for a dipole is merely a limiting form of a doublet.

However, as we shall explain shortly, the failure of this attempted use of sources is merely a consequence of our thinking in terms of fluid flowing within a strangely-shaped pool; by thinking in terms of electric fields or heat flows, we *can* use sources to construct mappings.

Continuing with the fluid interpretation for the time being, there does exist a simple alternative to the dipole method, but in place of sources we must use *vortices*. Let B be the boundary curve of the simply connected region which we wish to map to the interior of the unit circle C. To construct the mapping, insert a vortex into B at an arbitrary interior point p, and insert another vortex into C at 0. See [12.27]. Dividing each flow into k-cells, as before, conformal mappings leap from the page. Let's examine the details.

Firstly, in defining the correspondence between the grids we are forced to map p to 0. Secondly, having chosen the strength S of the vortex in B, the strength $\widetilde{S}$ of

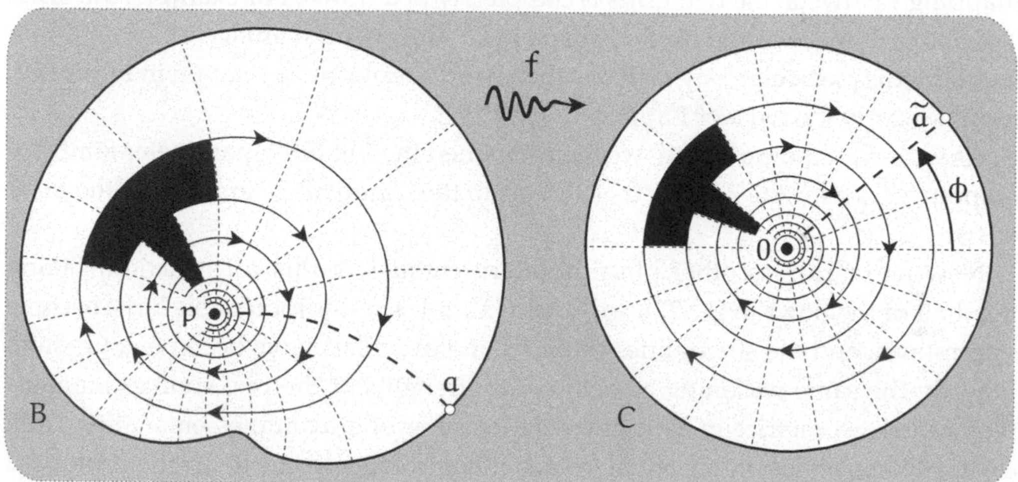

[12.27] Mapping interiors with vortices instead of dipoles. *Place a source of strength* S *at* p *inside* B*, and a source of strength* $\widetilde{S}$ *at the centre of* C*. Dividing each flow into small, square* k*-cells, we readily obtain a conformal mapping from one interior to the other, with* p *mapping to* 0*. But this is only possible if the number* N *of* k*-cells abutting* B *equals the number* $\widetilde{N}$ *abutting* C*. This implies that* **the two vortices must be of equal strength:** $S = Nk = \widetilde{N}k = \widetilde{S}$*. Finally, we may specify which boundary point* $\widetilde{a}$ *of* C *corresponds to the boundary point* a *of* B*. Now the mapping is completely tied down, and, for example, the black T-shape is mapped to the black T-shape.*

[5] Of course we could insert several sinks whose strengths summed to that of the source, but that would be messier still.

the one in C must be chosen so that (as illustrated) the number $\widetilde{N}$ of k-cells abutting C is equal to the number N abutting B. This is the same idea as in the dipole construction, but the answer in this case is much simpler: put $\widetilde{S}$ *equal* to S.

The explanation is as simple as the answer. Concentrate on one of the vortices, say the one in B. Its strength S is, by definition, the circulation along any simple loop round p, and we may conveniently take this to be one of the streamlines. The equipotentials cut this streamline into N segments, each having circulation k. Thus $S = Nk$, and likewise $\widetilde{S} = \widetilde{N}k$. The condition $\widetilde{S} = S$ follows immediately.

Finally, observe that we have not yet pinned down the mapping. In both [12.25] and [12.26] there were stagnation points which we were forced to identify, thereby tying down the mapping. However, here there are none, and we must do the job by hand. A common procedure goes like this. We know that the streamline B maps to the streamline C, and we may now insist that a particular point a on the first maps to a particular point $\widetilde{a}$ on the second.

If we wished to be more definite, we might do the following. Consider the heavily-dashed equipotential in B which exits p travelling due east, and choose a to be its intersection with B. See [12.27]. If we now choose $\widetilde{a} = e^{i\phi}$ then the mapping f between the two grids is completely tied down. For example, the black T-shape in B maps to the black T-shape in C. The only advantage of specifying a and $\widetilde{a}$ in this particular way is that ϕ then has a simple interpretation in terms of f, for it is clearly the twist of f at p: $\phi = \arg[f'(p)]$.

As with the dipole method, we have obtained the full three-parameter family of mappings: *two* for the point p that f maps to the centre of C, and *one* for the twist of f at p.

Now let us turn to other physical interpretations of this construction. Taking the dual of the flows in [12.27], we obtain [12.28]. The equipotentials have become streamlines, and the streamlines (B and C in particular) have become equipotentials. By the same reasoning as before, the strengths of the two sources manifest themselves geometrically as k times the number of streamlines emanating from each, and these must be set equal to each other in order to construct the conformal mapping between the two grids.

Let us digress briefly. Since the strength of a source is measured by the number of k-flux tubes emanating from it, a conformal mapping sends a source to another source of equal strength. Obviously, the same goes for a sink. Returning to [12.26], we can understand why the strength d of the dipole inside C is the amplification of f at p. The dipole of unit strength at p may be thought of as a doublet in which the source and sink are an infinitesimal distance ϵ apart, and each has strength $(1/\epsilon)$. The conformal mapping f sends this to another doublet at 0: the strengths of the source and sink are preserved, but their separation (and hence the strength of dipole) is amplified by $|f'(p)|$, as was to be shown.

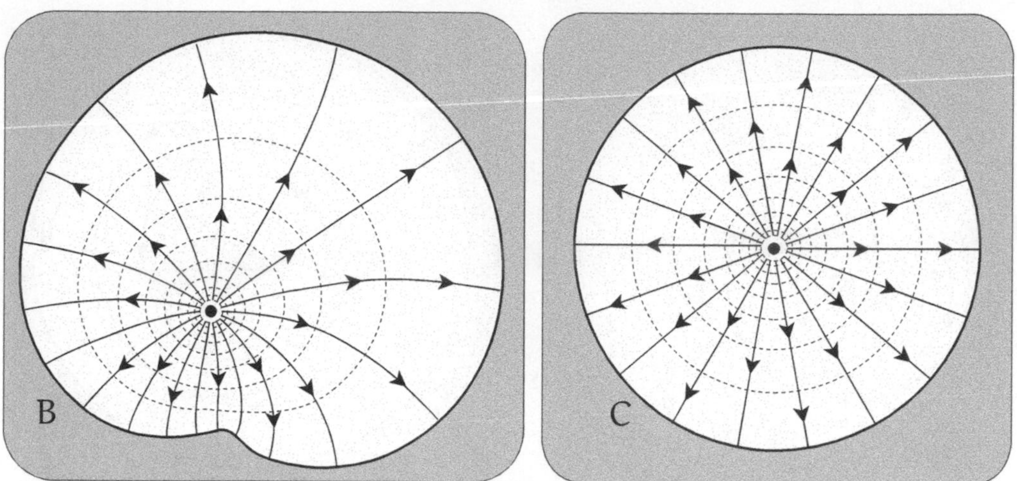

[12.28] The source construction dual to the vortex construction of the previous figure. *In order to make physical sense, here we must replace the fluid flow with an electrostatic field, the sources being electric charges. By the same reasoning as before, the strengths of the two charges manifest themselves geometrically as* k *times the number of electric field lines emanating from each, and these must be set equal to each other in order to construct the one-to-one conformal mapping between the two grids.*

Returning to the physical interpretation of [12.28], we have already observed that such pictures of sources make no sense when thought of as fluid flows. However, they do make perfect sense when thought of as electrostatic fields. Imagine that the dark regions in [12.28] represent cross-sections of blocks of copper through which we have bored "cylindrical" holes with cross-sections given by B and C. Now imagine that p and 0 are the cross-sections of two long, very thin, uniformly (and equally) charged wires running down these holes. The electrostatic fields generated by these wires automatically have B and C as equipotentials, so [12.28] faithfully represents these fields, with the streamlines being the electric field lines.

We may also interpret [12.28] in terms of heat flows. Imagine that the white shapes bounded by B and C have been cut from a sheet of heat-conducting metal, and imagine that the dark regions are filled with ice, thereby maintaining B and C at constant temperature (i.e., potential). If heat is introduced at a steady rate at p and at 0 then the flow of heat will eventually settle down to the one in [12.28], with the dashed equipotentials being the isotherms.

It must be observed, however, that a point source of heat is a much less physical concept than its electrostatic analogue, which may be realized (as we have said) by a very thin charged wire. The reason is that the potential function Φ becomes arbitrarily large as we approach a source. In electrostatics this does not present any difficulties, but in the heat-flow interpretation $-\Phi$ represents temperature,

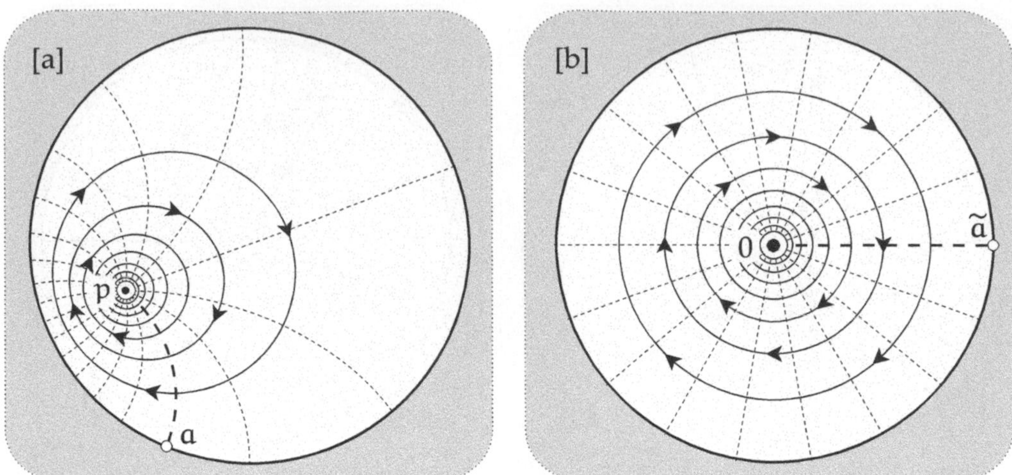

[12.29] Using flows to construct the most general automorphism of the disc. *Returning to [12.27], suppose we choose* B = C *to be the unit circle. Then the vortex flow shown here in* **[a]** *may be viewed as a one-to-one conformal mapping to the flow in* **[b]**, *i.e., it is an automorphism of the disc. Recall that we previously used the Symmetry Principle to find the formula for the most general such automorphism: (3.52), on page 201. We can now rederive this formula by finding the flow in* **[a]**, *which we accomplish by means of the method of images, in the next figure.*

so the metal in the vicinity of such an imagined source of heat would *vaporize*! For an excellent discussion of such physical distinctions, see Maxwell (1881), p. 51 onwards.

While many readers may be unfamiliar with electrostatics, few will be unfamiliar with heat. Thus, ignoring the above objection, we shall persist in expressing ideas mainly in the language of heat rather than electrostatics.

12.6.5 An Example: Automorphisms of the Disc

Let us explicitly carry out the construction in [12.27] for the case B = C, and thereby re-obtain the automorphisms of the unit disc from a fresh point of view.

Figure [12.29] illustrates the construction in this case. Previously we did not specify the strength S of the two vortices because the choice had no effect on the resulting mapping, but let us now choose $S = -2\pi$. Taking the zero of flux and circulation in [12.29b] to be at 1, the complex potential for [12.29b] is then

$$\widetilde{\Omega}(w) = i \log w.$$

Now consider the flow in [12.29a]. We may find its complex potential by the method of images[6]. That is, we superpose the real vortex at p with its reflection

[6] However, note that we are *not* using (12.18). See Ex. 14 to understand why.

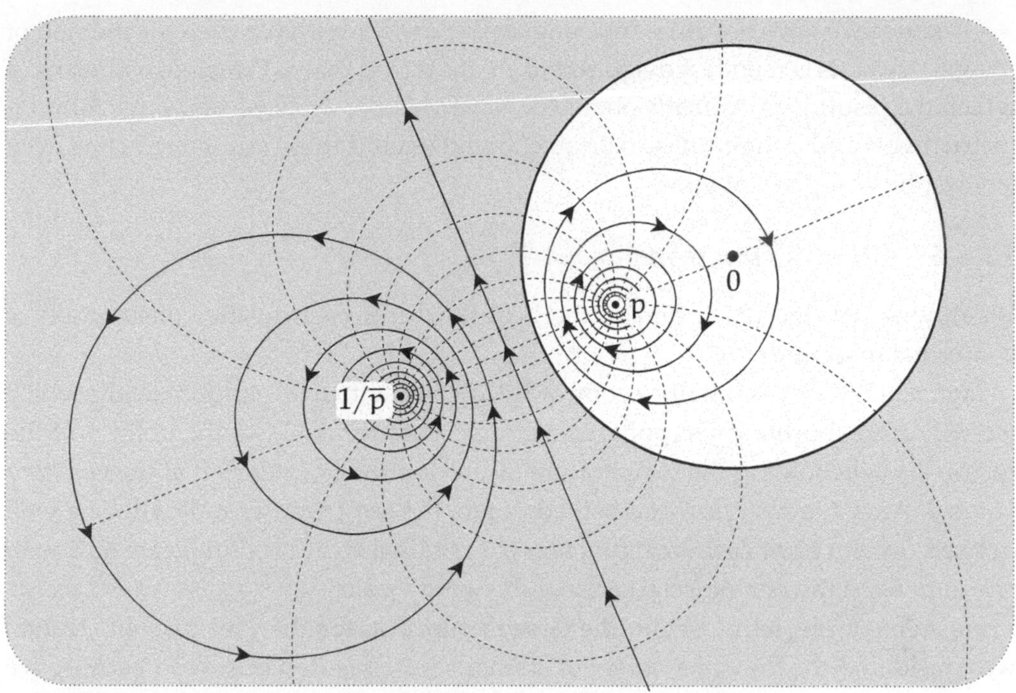

[12.30] Using the method of images to construct the most general automorphism of the disc. *To obtain the flow in [12.29a], we superpose a vortex at* p *with an equal and opposite vortex at its reflection,* $1/\overline{p}$.

in C, namely, a fictitious vortex of equal and opposite strength at $(1/\overline{p})$. See [12.30]. The complex potential of [12.29a] is thus

$$\begin{aligned}\Omega(z) &= i\log(z-p) - i\log(z-1/\overline{p}) - \gamma \\ &= i\log\left[\frac{z-p}{\overline{p}\,z-1}\right] - \delta \qquad (12.20)\end{aligned}$$

where γ and δ are constants.

In [12.29b] we chose to measure circulation and flux from a boundary point, and we now choose to do the same in [12.29a]. In terms of the above equation this is equivalent [exercise] to demanding that the constant δ be an arbitrary *real* number.

To pin down the mapping we must now choose a and $\widetilde{a}$. Instead of doing this as we did in [12.27], let us this time choose them to be the two boundary points from which we have elected to measure circulation and flux in the two flows. As previously discussed, the mapping is then effected by equating circulation and flux:

$$\widetilde{\Omega}(w) = \Omega(z).$$

Solving for w, we do indeed recover the familiar automorphisms of the disc:

$$w = f(z) = e^{i\delta}\left[\frac{z-p}{\overline{p}\,z-1}\right]. \qquad (12.21)$$

Of course we have not done anything really new, for we have used the method of images, and this is merely a disguised form of the Symmetry Principle by means of which the result was *originally* obtained. Nevertheless, we hope you have found it instructive—and delightful!—to be able to understand these automorphisms from a new, physical point of view.

12.6.6 Green's Function

We now return to [12.28] and to the heat-flow approach to the construction of conformal mappings.

Figure [12.31] is essentially a copy of [12.28], but with a few added details which we will need shortly. We supply heat at the constant rate 2π to the point p of the region R while holding the temperature all around the boundary B at the constant value 0. After the heat flow has settled down, the temperature in R will be a well defined (except at p) harmonic function $\mathcal{G}_p(z)$ called the *Green's function* of R with *pole* at p. Note the new, special sense of the word "pole".

As an example, let us obtain the Green's function for the unit disc. In [12.29a] we considered the flow of a vortex of strength -2π inserted at p. And because we chose a point on C as the zero of circulation and flux, the stream function vanished on the boundary streamline. The dual of this flow is therefore precisely what we are after: a source of strength 2π at p, with the boundary at zero potential. By (12.20), the complex potential is

$$\Omega(z) = \Phi(z) + i\,\Psi(z) = \log\left[\frac{z-p}{\overline{p}\,z-1}\right] + i\delta,$$

where δ is a real constant. Thus the temperature in the disc is given by

$$\mathcal{G}_p(z) = -\Phi(z) = -\ln\left|\frac{z-p}{\overline{p}\,z-1}\right|. \tag{12.22}$$

Note that $\mathcal{G}_p(z) = -\ln|f(z)|$, where $f(z)$ is any of the one-parameter family of mappings to the unit disc such that $f(p) = 0$. We shall see that this is true quite generally.

Also note that this Green's function has a very interesting symmetry property:

$$\mathcal{G}_p(q) = \mathcal{G}_q(p).$$

Thus, with the boundary packed with ice, the steady-state temperature at q due to a point source of heat at p is the same as the temperature at p when the source is instead at q. Remarkably, we shall see that this symmetry holds true for the Green's function in a region of *arbitrary* shape!

The Green's function is a powerful tool in several areas of mathematics and physics. For the time being, we will concern ourselves primarily with its relationship to the conformal mappings $w = f(z)$ from the interior of B to the unit disc. In the next section we shall discuss another important application.

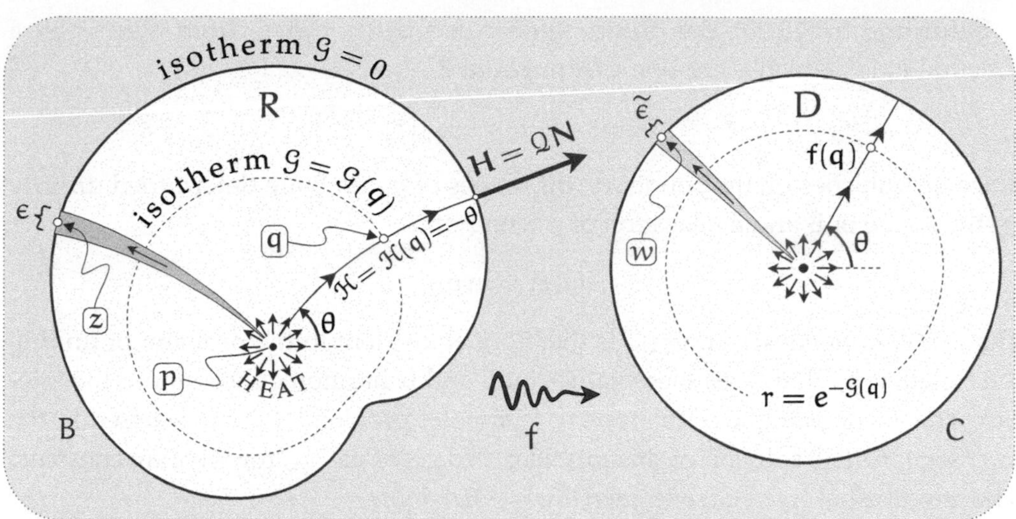

[12.31] The Green's Function $\mathcal{G}_p(z)$ of a region R *is defined to be the harmonic equilibrium temperature distribution when heat is supplied at constant rate* 2π *to the point* p *while holding the temperature all around the boundary* B *at the constant value* 0. *We may then construct a harmonic dual* $\mathcal{H}_p(z)$, *that is, a harmonic function whose level curves are the paths along which the heat flows (orthogonally through the isotherms* $\mathcal{G}_p = \text{const.}$*) from* p *to the boundary. Thus knowledge of* $\mathcal{G}_p$ *is sufficient for the construction of the complete complex potential,* $\Omega(z) = -[\mathcal{G}_p(z) + i\,\mathcal{H}_p(z)]$, *and the conformal mapping of* R *to* D *is then explicitly given by* $w = f(z) = e^{\Omega(z)} = e^{-\mathcal{G}}\,e^{-i\mathcal{H}}$.

Returning to the general case in [12.31], suppose $\mathcal{G}_p$ is known. As previously described, we may then construct a harmonic dual $\mathcal{H}_p(z)$, that is, a harmonic function whose level curves are the paths along which the heat flows (orthogonally through the isotherms $\mathcal{G}_p = \text{const.}$) from p to the boundary. Thus knowledge of $\mathcal{G}_p$ is sufficient for the construction of the complete complex potential, $\Omega(z) = -[\mathcal{G}_p(z) + i\,\mathcal{H}_p(z)]$.

Consider the behaviour of $\mathcal{G}_p$ in the immediate vicinity of p. Physical intuition leads us to expect, irrespective of the temperatures assigned to B, that the flow out of p will almost be like that of an isolated source, so that $\Omega(z) \approx \log(z - p)$. Thus if $z = p + \rho\,e^{i\theta}$ then

$$\mathcal{G}_p(z) \approx -\ln \rho, \tag{12.23}$$

for small values of ρ.

Similarly, $\mathcal{H}_p(z) \approx -(\theta + \phi)$, where ϕ is a constant. The freedom in choosing ϕ is equivalent, as we shall see, to the freedom of choosing which point of B maps to a particular point of C. If we choose $\phi = 0$ then the value of $\mathcal{H}_p$ at a typical point q has a particularly simple interpretation. See [12.31]. Following the flow of heat back from q to p, the angle at which it enters p is $-\mathcal{H}_p(q)$.

Returning to (12.23), the precise version is that $\mathcal{G}_p$ differs from $-\ln(\rho)$ by a function $g_p(z)$ which is harmonic throughout R:

$$\mathcal{G}_p(z) = -\ln \rho + g_p(z).$$

Since $\mathcal{G}_p$ vanishes on the boundary, the values of g_p on B are determined directly by the shape of B and the location of p within it:

$$g_p(z) = \ln \rho.$$

The problem of constructing $\mathcal{G}_p$ is therefore equivalent to the problem of finding the function g_p that takes these values on B and is harmonic throughout the interior. This is an example of the type of "Dirichlet problem" to be discussed in the next section. The solution of this problem also gives us $\mathcal{H}_p$, for we may construct a harmonic dual h_p of g_p, and then $\mathcal{H}_p = -\theta + h_p$.

To construct a conformal mapping $w = f(z)$ to the unit disc D such that $f(p) = 0$, we may (as previously explained) equate the complex potential $\Omega(z)$ of the flow inside R with the complex potential $\widetilde{\Omega}(w) = \log w$ in D of the heat source at 0, with the boundary at zero temperature. Thus

$$w = f(z) = e^{\Omega(z)} = e^{-\mathcal{G}}\, e^{-i\mathcal{H}}. \tag{12.24}$$

As illustrated in [12.31], the dashed isotherm at temperature $\mathcal{G}(q)$ is mapped to the dashed circle of radius $e^{-\mathcal{G}(q)}$, and the streamline $\mathcal{H} = \mathcal{H}(q) = -\theta$ entering p at angle θ maps to the ray entering 0 at angle θ. Thus f has zero twist at p. This is the significance for f of our previously choosing $\phi = 0$; in general $\arg[f'(p)] = \phi$.

Now that we possess the mapping f, any harmonic temperature distribution $T(z)$ on R may be conformally transplanted to a harmonic function $\widetilde{T}(w)$ on D (and vice versa) by assigning equal temperatures to corresponding points of the two regions: $\widetilde{T}[f(z)] \equiv T(z)$. In particular, the values of T on B are transplanted to C.

We can use this to understand the following. If we know a conformal mapping $w = f(z)$ from R to D such that $f(p) = 0$, then the Green's function of R with pole at p is

$$\mathcal{G}_p(z) = -\ln |f(z)|.$$

To see this, consider the temperature distribution $\widetilde{\mathcal{G}}_p(w)$ in D obtained by transplanting $\mathcal{G}_p$ with f. We know that this conformal transplantation preserves the harmonicity of the original, that it sends the source at p to an equal source at $f(p) = 0$, and that the vanishing temperature on B is transplanted to vanishing temperature on C. Thus this temperature distribution $\widetilde{\mathcal{G}}_p(w) = \mathcal{G}_p[f^{-1}(w)]$ in D must be the Green's function with pole at 0, and we already know that this is $-\ln|w|$. Done.

Exactly the same reasoning yields the following generalization. Let $J(z)$ be a one-to-one conformal mapping of R to some other simply connected region S with boundary Y. Then J conformally transplants the Green's function $\mathcal{G}_a(z)$ of R with pole at a to the Green's function of S with pole at J(a). In particular, the streamlines of the flow in R map to the streamlines of the flow in S. In this sense, *the concept of the Green's function is conformally invariant.*

This result immediately yields the following generalization of (12.6.6) to the case where the pole of the Green's function of R is an arbitrary point s, rather than $f^{-1}(0)$. From (12.22), we know the formula for the Green's function in the disc when the pole is moved from 0 to f(s). Conformal transplantation by means of f^{-1} carries this pole to the desired point s of R, so

$$\mathcal{G}_s(z) = -\ln\left|\frac{f(z) - f(s)}{\overline{f(s)}\, f(z) - 1}\right|.$$

As a bonus, notice that this general formula establishes the previously claimed "symmetry property" of the Green's function:

$$\mathcal{G}_s(z) = \mathcal{G}_z(s).$$

For a more common approach to the symmetry property, see Ex. 15.

We end this section with a result which we will need later. The analogue for heat flows of the velocity of a fluid flow is the *heat flow vector* H; in the present case, $H = -\nabla\mathcal{G}$. Let us call its magnitude $\mathcal{Q} \equiv |H|$ the *local heat flux*; this is the analogue of fluid speed, and it represents the heat flux (per unit length) across a short line-segment at right angles to the flow. Since B is an isotherm, H is orthogonal to B, and so the local heat flux at a boundary point z may be expressed as

$$\mathcal{Q}(z) = -\frac{\partial\mathcal{G}}{\partial n}, \tag{12.25}$$

where n measures distance in the direction of N, the outward unit normal vector to B (see [12.31]). We can now state the result:

At a boundary point z, the local heat flux is equal to the amplification of $f : \mathcal{Q}(z) = |f'(z)|$. (12.26)

For example, using (12.21), the result predicts that the local heat flux at the boundary of the unit disc is given by [exercise]

$$\mathcal{Q}(z) = |f'(z)| = \frac{1 - |p|^2}{|z - p|^2} = \frac{1 - |p|^2}{\rho^2}. \tag{12.27}$$

This formula will play a central role in the next section. Of course $\mathcal{Q}(z)$ may instead be calculated directly by substituting (12.22) into (12.25), but this is a little easier said than done [exercise].

The general result (12.26) can be understood very intuitively. See [12.31]. With an infinitesimal value of k, consider the shaded k-flux tube emanating from p which hits B at z, and let its width there be ϵ. Its image under f is a k-flux tube emanating from 0 and hitting C at $w = f(z)$. [Remember, f was originally *defined* to have this property!] Let $\widetilde{\epsilon}$ be the width of this image tube at w. Since the segment of B at z of length ϵ is amplitwisted by $f'(z)$ to the segment of C at w of length $\widetilde{\epsilon}$,

$$|f'(z)| = \frac{\widetilde{\epsilon}}{\epsilon}.$$

Next, recall that the width of a k-flux tube at any given point is equal to k divided by the local heat flux [previously fluid speed] at that point. Since the local heat flux is constant on C, its value at w is simply the ratio of the strength of the source at 0 to the perimeter of C, and by construction this ratio is $(2\pi/2\pi) = 1$. Since the local heat flux at z is $\mathcal{Q}(z)$, we see that

$$\widetilde{\epsilon} = k/1 \quad \text{and} \quad \epsilon = k/\mathcal{Q}.$$

Thus

$$|f'(z)| = \frac{\widetilde{\epsilon}}{\epsilon} = \frac{k}{k/\mathcal{Q}} = \mathcal{Q}(z),$$

as was to be shown.

12.7 Dirichlet's Problem

12.7.1 *Introduction*

Consider a steady heat flow within a metal plate whose faces are insulated. Other than at singularities, the temperature $T(z)$ is then a harmonic function, and the (locally) sourceless heat flow vector field is $H = -\nabla T$.

Let us measure the temperature around the circumference C of a circle of radius R, the interior of which is free of sources and sinks, and the centre of which we may conveniently choose to be the origin. As $z = R\,e^{i\theta}$ moves round C, we may express the measured temperature as a function of the angle: $T = T(\theta)$. We hope that it may seem physically plausible that these values actually determine the temperature at any interior point a. Indeed, if $a = 0$ then we know [from Gauss's Mean Value Theorem] that the temperature at the centre of C is simply the average of the temperatures on C:

$$T(0) = \langle T \rangle = \frac{1}{2\pi}\int_{-\pi}^{\pi} T(\theta)\, d\theta. \tag{12.28}$$

Eventually we will discover that the generalization of this result to the case $a \neq 0$ is given by

$$T(a) = \frac{1}{2\pi}\int_{-\pi}^{\pi} \left[\frac{R^2 - |a|^2}{|z-a|^2}\right] T(\theta)\, d\theta. \tag{12.29}$$

Writing $a = r\,e^{i\alpha}$ $(r < R)$, and appealing to the cosine formula [exercise], this is usually written as

$$T(a) = \frac{1}{2\pi}\int_{-\pi}^{\pi}\left[\frac{R^2 - r^2}{R^2 + r^2 - 2Rr\cos(\theta - \alpha)}\right] T(\theta)\, d\theta.$$

This is called *Poisson's formula,* and the quantity in square brackets is called the *Poisson kernel,* which we shall write as $\mathcal{P}_a(z)$.

Formula (12.29) says that $T(a)$ is a *weighted* average of T on C, the temperature of each element of C contributing to $T(a)$ in proportion to its weight $\mathcal{P}_a(z)$. Notice that $\mathcal{P}_a(z)$ dies away inversely with the square of the distance between a and the element of C, so that if one element is twice as far from a as another, its influence on the temperature at a is only one quarter as great. If $a = 0$ then all parts of C have equal influence (for all are equally far from a) and you can see that we do recover (12.28).

Poisson's formula is connected with the following important and difficult issue. Instead of dealing with a pre-existing harmonic function, *Dirichlet's problem* demands that we arbitrarily (but piecewise continuously) assign values to the boundary of a simply connected region R and then seek a continuous harmonic function in R which takes on these values as the boundary is approached.

In the case of the disc, H. A. Schwarz demonstrated that not only does the solution to Dirichlet's problem exist, but it is explicitly given by (12.29). If we are handed the piecewise continuous values $T(\theta)$ on C then we may construct a function $T(a)$ in the interior according to Poisson's recipe. Schwarz's solution then amounted to showing that $T(a)$ is automatically harmonic, and that as a approaches a boundary point at which $T(\theta)$ is continuous, $T(a)$ approaches the given value $T(\theta)$. Let us begin to explain all this.[7]

12.7.2 Schwarz's Interpretation

In 1890 Schwarz[8] discovered a lovely geometric interpretation of formula (12.29) which deserves to be far better known than it is:

> *To find the temperature at* a, *transplant each temperature on* C *to the point directly opposite to it as seen from* a, *then take the average of the new temperature distribution on* C. (12.30)

Schwarz deduced this from Poisson's formula, itself derived by computation. We shall instead demonstrate his result directly and geometrically, only then producing the Poisson formula as a corollary.

The example in [12.32] illustrates the beauty of (12.30). In [12.32a] half of C is kept at 100 degrees with steam, while the other half is kept at 0 degrees with ice.

[7] Much of the following material previously appeared in Needham (1994).
[8] Schwarz (1972b).

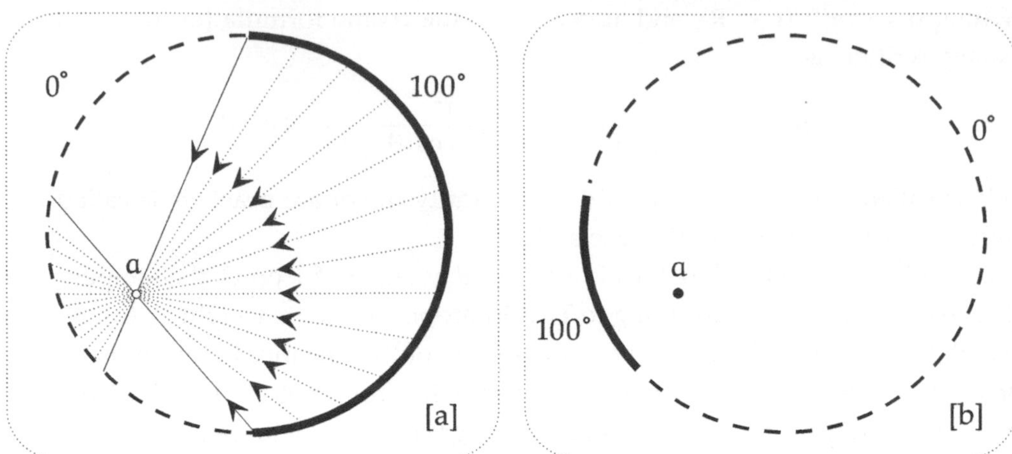

[12.32] In 1890 Schwarz discovered a beautiful geometrical interpretation of Poisson's formula. [a] *Transplant each temperature on C to the point directly opposite to it as seen from a, thereby obtaining the new temperature distribution shown in* **[b]. Then the temperature at a is simply the average of the new temperature distribution on C.**

Being close to the cold side, we would expect a to be cool. Figure [12.32b] shows the new temperature distribution obtained by projection through a. It is now vividly clear how the distant hot semicircle is 'focused' through a onto a much smaller arc, yielding a low average temperature on C and hence a low temperature at a itself.

To begin to establish (12.30), recall the conformal invariance of harmonic functions: if $T(z)$ is any harmonic function and $h(z)$ any conformal mapping, then $T(z^\star)$ is automatically harmonic, with $z^\star = h(z)$.

Suppose now that $h(z)$ maps the disc to *itself*. If $z = R\,e^{i\theta}$ lies on C then so does $z^\star = R\,e^{i\theta^\star}$, and since we suppose that we have measured the temperature all round C, we therefore know the temperature $T(\theta^\star)$ at $z^\star$. Having the values of $T(\theta^\star)$, we may now compute the integral in (12.28) for the harmonic function $T[h(z)]$ to obtain

$$T(0^\star) = \frac{1}{2\pi} \int_{-\pi}^{\pi} T(\theta^\star)\, d\theta\,, \tag{12.31}$$

in which it should be stressed that the averaging is still taking place with respect to the angle of z, not its image $z^\star$.

We may interpret (12.31) as follows: the temperature at $0^\star$ is the average of the new temperature distribution on C obtained by transplanting the temperature measured at each z to the new location $z^\star$. We are now half way to Schwarz's result. To find the temperature at a we must find a conformal mapping of the disc to itself such that 0 is sent to a, then take the average of the new temperature distribution.

But viewing the disc as the Beltrami–Poincaré model of the hyperbolic plane, we are already very familiar with such mappings! Peek at [12.33b], to which we

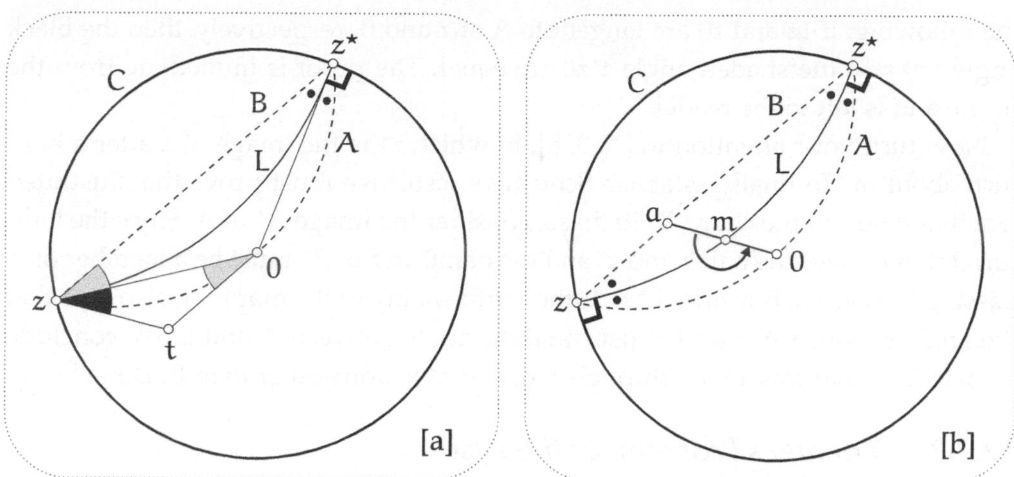

[12.33] Hyperbolic geometry proof of Schwarz's interpretation of Poisson's formula. *According to (12.31), if we can find a conformal mapping of the disc to itself that sends* 0 *to* $\mathfrak{a}$*, sending boundary points* z *to* $z^\star$*, then we may find the temperature at* $\mathfrak{a}$ *by averaging these transplanted temperatures at* $z^\star$*. But* **[b]** *reminds us that we have already met such a transformation in hyperbolic geometry! In the Beltrami–Poincaré disc, let* $\mathfrak{m}$ *be the midpoint (in the hyperbolic sense) of the line-segment* $0\mathfrak{a}$*, then the half-turn* $z \mapsto z^\star = \mathcal{M}_\mathfrak{a}(z)$ *of the hyperbolic plane about* $\mathfrak{m}$ *interchanges* 0 *and* $\mathfrak{a}$: $0^\star = \mathfrak{a}$ *(as we desire) and* $\mathfrak{a}^\star = 0$*. As explained in the text,* **[a]** *allows us to prove that in fact* $z^\star$ *lies at the end of the Euclidean chord through* $\mathfrak{a}$ *and* z*, as illustrated, thereby proving Schwarz's result.*

shall return in a moment. If $\mathfrak{m}$ is the midpoint (in the hyperbolic sense) of the line-segment $0\mathfrak{a}$, then the half-turn[9] $z \mapsto z^\star = \mathcal{M}_\mathfrak{a}(z)$ of the hyperbolic plane about $\mathfrak{m}$ interchanges 0 and $\mathfrak{a}$: $0^\star = \mathfrak{a}$ (as we desire) and $\mathfrak{a}^\star = 0$. Thus to establish (12.30) we need only demonstrate the illustrated fact that *if* z *lies on* C *then* $z^\star$ *lies at the end of the (Euclidean) chord* B *passing from* z *through* $\mathfrak{a}$. In Chapter 3 we derived a formula for $\mathcal{M}_\mathfrak{a}(z)$, so we could easily obtain this result by calculation; however, we prefer a direct geometric approach.

First we need a simple result which is explained in [12.33a]. Consider the family

$$\mathcal{N} \equiv \{\text{circular arcs passing from } z \text{ to } z^\star\},$$

where for the moment $z^\star$ may be thought of as any given point of C. The figure shows three members of $\mathcal{N}$: the arc A through 0, the Euclidean chord B, and the hyperbolic line L. [Recall that in terms of hyperbolic geometry the members of $\mathcal{N}$ consist of the equidistant curves of L.] The result we need is this:

The Euclidean chord B *is the unique member of* $\mathcal{N}$ *such that* L *bisects the angle contained by* A *and* B. (12.32)

Since each member of $\mathcal{N}$ is uniquely determined by the direction in which it emerges from z, and since the radius $z0$ is tangent to L at z, this is equivalent to

[9] See (3.53), p. 201.

the following: if tz and t0 are tangent to A at z and 0, respectively, then the black angle tz0 and the shaded angle $z^{\star}z0$ are equal. The proof is immediate from the figure and is left to the reader.

Now turn your attention to [12.33b], in which $z^{\star}$ is the image of z after a half-turn about m. To finally establish Schwarz's result we must prove the illustrated fact that a lies on the chord B. To do so, consider the image $A^{\star}$ of A. Since the half-turn interchanges the pair z and $z^{\star}$ and the pair 0 and a, $A^{\star}$ must be a member of $\mathcal{N}$ passing through a. But since $L^{\star} = L$, the conformality of the mapping also says that the angle between $A^{\star}$ and L must equal the angle between A and L. We conclude from (12.32) that this arc $A^{\star}$ through z, a, and $z^{\star}$ is none other than B. Done[10].

12.7.3 Dirichlet's Problem for the Disc

Our example in [12.32] was a trifle hasty. For the moment, Schwarz's result merely says how the interior values of a given harmonic function in the disc may be found from the values on C. But in [12.32] we blithely assumed that we could also use it to *construct* such a function in the disc, given arbitrary piecewise continuous boundary values. In other words we assumed Schwarz's solution of Dirichlet's problem for the disc, as outlined in the introduction. We now justify this.

Figure [12.34a] shows a approaching a boundary point z; also shown are the images ($C_1^{\star}$ and $C_2^{\star}$) under projection through a of the two small arcs (C_1 and C_2) adjacent to z. If the given boundary values are continuous at z then T is essentially constant on $C_1 \cup C_2$, and so the new temperature distribution is likewise almost constant on $C_1^{\star} \cup C_2^{\star}$. As required, the constructed function $T(a)$ therefore *does* approach $T(z)$ as a approaches z.

Although Dirichlet's problem makes no demands on the behaviour of $T(a)$ as a approaches a boundary point at which T is *dis*continuous, it is easy to see (though not to calculate!) what actually happens. Suppose that the boundary temperature jumps from T_1 to T_2 as we pass from C_1 to C_2. If a arrives at z while travelling in a direction making an angle $\beta\pi$ with C, then [exercise] $T(a)$ approaches $[\beta T_1 + (1-\beta)T_2]$. This result is relevant to the representation of discontinuous functions by Fourier series.

It now only remains to show that the constructed function is indeed harmonic. First we shall pause to recover Poisson's formula in its classical form. We begin by noting that (12.31) may be re-expressed [why?] as

$$T(a) = \frac{1}{2\pi}\int_{-\pi}^{\pi} T(\theta)\, d\theta^{\star}. \tag{12.33}$$

In order to put this into the same form as (12.29), we now require $d\theta^{\star}$ in terms of $d\theta$.

Consider [12.34b], which shows the movement $R\Delta\theta^{\star}$ of $z^{\star}$ resulting from a movement $R\Delta\theta$ of z. These arcs are ultimately equal to the chords t and s, so

[10] This argument is perhaps conceptually clearer than the more elementary one in Needham (1994).

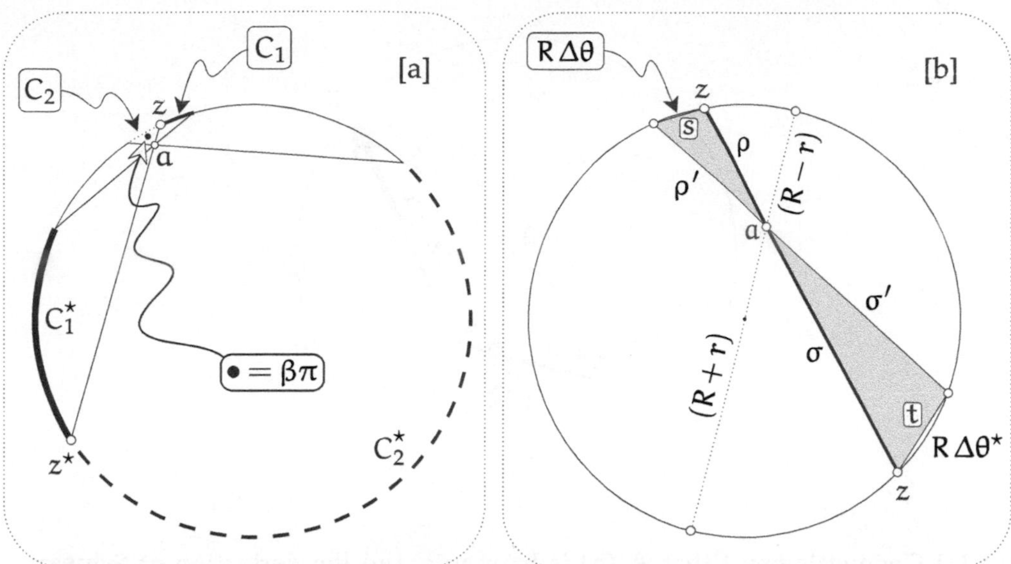

[12.34] Solving Dirichlet's Problem with Schwarz's geometric construction. [a] *Suppose that the boundary temperature jumps from* T_1 *to* T_2 *as we pass from* C_1 *to* C_2*. If* a *arrives at* z *while travelling in a direction making an angle* $\beta\pi$ *with* C*, then* $T(a)$ *approaches* $[\beta\, T_1 + (1-\beta)\, T_2]$. **[b]** *To derive Poisson's formula from (12.33), consider the movement* $R\,\Delta\theta^\star \asymp t$ *of* $z^\star$ *resulting from a movement* $R\,\Delta\theta \asymp s$ *of* z*. So* $(\Delta\theta^\star/\Delta\theta) \asymp (t/s)$*. But* t *and* s *are corresponding sides of two similar triangles [shaded], so* $(t/s) = (\sigma'/\rho)$*. And since* $(\sigma'/\rho) \asymp (\sigma/\rho)$*, we obtain* $\frac{d\theta^\star}{d\theta} = \left[\frac{\sigma}{\rho}\right]$*. Finally,* $\rho\,\sigma = \rho'\,\sigma' = (R+r)(R-r) = (R^2 - r^2)$*, so* $\frac{d\theta^\star}{d\theta} = \left[\frac{\sigma}{\rho}\right] = \left[\frac{R^2-r^2}{\rho^2}\right] = \mathcal{P}_a(z)$*. Done.*

that $(\Delta\theta^\star/\Delta\theta)$ is ultimately equal to (t/s). But t and s are corresponding sides of two similar triangles [shaded], so $(t/s) = (\sigma'/\rho)$. Finally, since (σ'/ρ) is ultimately equal to (σ/ρ), we obtain

$$\frac{d\theta^\star}{d\theta} = \left[\frac{\sigma}{\rho}\right].$$

Thus (12.33) becomes

$$T(a) = \frac{1}{2\pi}\int_{-\pi}^{\pi} \left[\frac{\sigma}{\rho}\right] T(\theta)\, d\theta\,. \tag{12.34}$$

Consequently, to derive Poisson's formula we need only show that $[\sigma/\rho]$ is the Poisson kernel $\mathcal{P}_a(z)$. This was precisely how Schwarz, working in the opposite direction, originally deduced his result *from* Poisson's formula.

Since $\rho\,\sigma = \rho'\,\sigma'$ is constant, we may evaluate it for the dotted diameter through a to obtain $\rho\,\sigma = (R^2 - r^2)$. Thus we do indeed find that

$$\left[\frac{\sigma}{\rho}\right] = \left[\frac{R^2 - r^2}{\rho^2}\right] = \mathcal{P}_a(z).$$

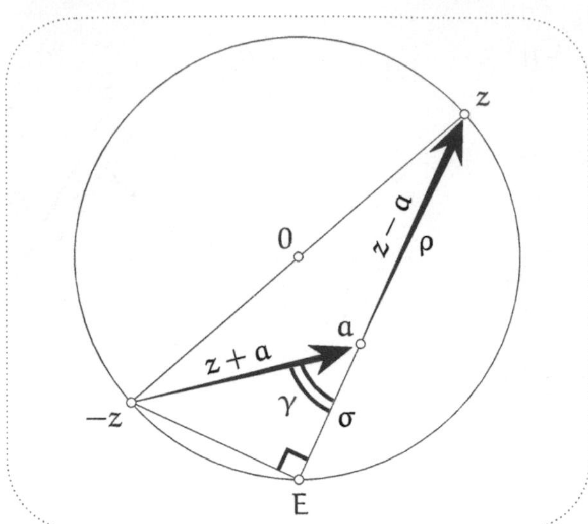

[12.35] Geometric proof that $\mathcal{P}_a(z)$ is harmonic, and the derivation of Schwarz's formula. *Since the angle at* E *is a right angle, we have* $\mathcal{P}_a(z) = \left[\frac{\sigma}{\rho}\right] = \frac{|z+a|\cos\gamma}{|z-a|} =$ Re $\left(\frac{z+a}{z-a}\right)$. *Since* $\mathcal{P}_a(z)$ *is the real part of an analytic function, it is indeed harmonic. Let* S *be a harmonic dual of* T, *so that* $f = T + iS$ *is an analytic function. This function is uniquely defined (up to a constant), and so it must be given by* **Schwarz's formula***:* $f(a) = \frac{1}{2\pi}\int_{-\pi}^{\pi}\left(\frac{z+a}{z-a}\right)T(\theta)\,d\theta$, *for this is analytic and has* $T(a)$ *as its real part. Thus Schwarz was able to resurrect the complete analytic function* f *everywhere* inside C *just from the ashes of its real part* on C*!*

As an interesting consequence of the geometric interpretation of the Poisson kernel, we see that (with z fixed) the level curves of $\mathcal{P}_a$ are the circles which are tangent to C at z (i.e., horocycles), with $\mathcal{P}_a = 0$ being C itself.

Returning to the issue of harmonicity, we see that if we permit ourselves differentiation under the integral sign of (12.34), then it is sufficient to show that $[\sigma/\rho]$ is a harmonic function of a. To see that it is, consider [12.35].Since the angle at E is a right angle, we have

$$\left[\frac{\sigma}{\rho}\right] = \frac{|z+a|\cos\gamma}{|z-a|} = \text{Re}\left(\frac{z+a}{z-a}\right).$$

Because it is the real part of an analytic function of a, $[\sigma/\rho]$ is automatically harmonic, and we are done.

This line of reasoning yields a bonus result. Let S be a harmonic dual of T, so that $f = T + iS = -$(complex potential) is an analytic function. This function is uniquely defined (up to an additive imaginary constant) and so it must be given by

$$f(a) = \frac{1}{2\pi}\int_{-\pi}^{\pi}\left(\frac{z+a}{z-a}\right)T(\theta)\,d\theta,$$

for this is analytic and has $T(a)$ as its real part. This result is called *Schwarz's formula*, and it enables us to resurrect the complete analytic function f everywhere *inside* C just from the ashes of its real part *on* C.

12.7.4 The Interpretations of Neumann and Bôcher

If we specify arbitrary piecewise continuous temperatures $T(x)$ along the edge (the real axis) of the upper half-plane, then there is another formula due to Poisson that yields the temperature at any point $a = X + iY$ $(Y > 0)$:

$$T(a) = \frac{1}{\pi}\int_{-\infty}^{\infty}\left[\frac{Y}{(X-x)^2+Y^2}\right] T(x)\,dx. \tag{12.35}$$

We shall explain this result by reinterpreting (12.33) in terms of elementary hyperbolic geometry. The transition from (12.29) to (12.35) will then be seen as nothing more than a transition between the Beltrami–Poincaré disc and upper half-plane models of the hyperbolic plane. First, however, let us obtain still another geometric interpretation of Poisson's formula.

For simplicity, let us employ the *unit* circle. Consider [12.36]. Let the arc K be heated to unit temperature while the rest of C is kept at zero degrees. By Schwarz's result, the temperature at a is $T(a) = (K^\star/2\pi)$, while the temperature at the centre of the circle is $T(0) = (K/2\pi)$.

Next, imagine yourself standing at a looking out at a vast number of thermometers placed at equal intervals along the circle. As you turn your head through a full revolution—remembering to turn your feet!—let $\langle\!\langle T\rangle\!\rangle_a$ denote the average (over all directions) of the temperatures you see. For example, the average $\langle T\rangle$ occurring in Gauss's Mean Value Theorem is $\langle\!\langle T\rangle\!\rangle_0$.

In our case $\langle\!\langle T\rangle\!\rangle_a = (\lambda/2\pi)$, where λ is the angle subtended by K at a. But we see from the figure that[11]

$$\lambda = \tfrac{1}{2}\,(K^\star + K)\,,$$

so $\langle\!\langle T\rangle\!\rangle_a = \frac{1}{2}\,[T(a) + T(0)]$: *the average of the boundary temperatures as they appear to you is equal to the average of the temperature where you are and the temperature at the centre.* It is then easy to see that this is still true if we instead have many arcs at different temperatures, and ultimately a general piecewise-continuous temperature distribution. Thus Poisson's formula may be re-expressed as what we shall call *Neumann's formula*:

$$T(a) = 2\,\langle\!\langle T\rangle\!\rangle_a - T(0).$$

This result is due to Neumann (1884); we merely rediscovered it, as did Duffin (1957) from another point of view. For an interesting generalization, see Perkins (1928).

[11] Incidentally, this means that the isotherms are the arcs of circles through p and q.

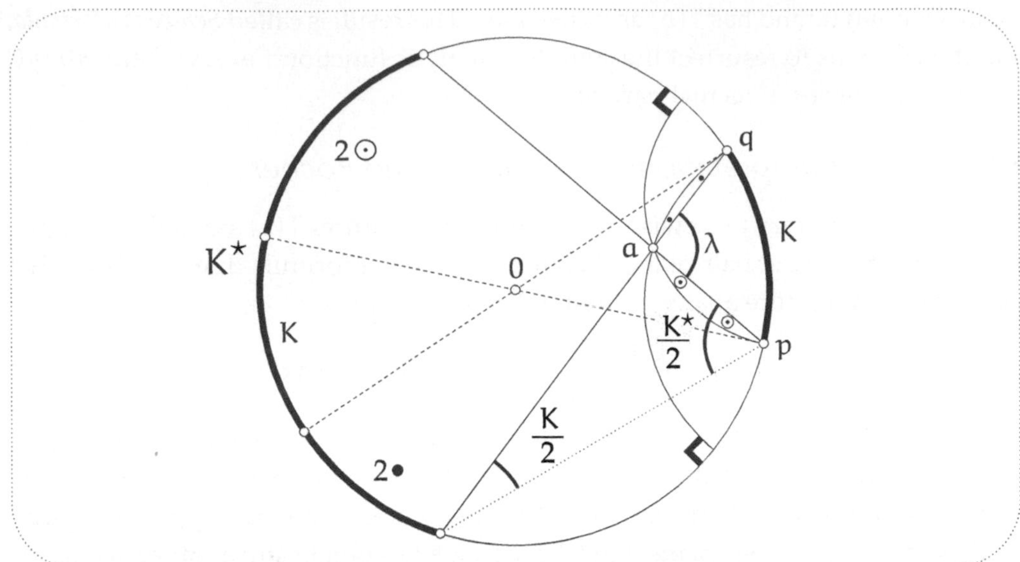

[12.36] Neumann's geometrical interpretation of Poisson's formula. *Let the arc* K *be heated to unit temperature while the rest of* C *is kept at zero degrees. By Schwarz's result, the temperature at* $\mathfrak{a}$ *is* $T(\mathfrak{a}) = (K^{\star}/2\pi)$, *while the temperature at the centre of the circle is* $T(0) = (K/2\pi)$. *Standing at* $\mathfrak{a}$, *turn your head through a full revolution, and let* $\langle\langle T\rangle\rangle_{\mathfrak{a}}$ *denote the average (over all directions of your head) of the temperatures you see. In our case* $\langle\langle T\rangle\rangle_{\mathfrak{a}} = (\lambda/2\pi)$, *where* λ *is the angle subtended by* K *at* $\mathfrak{a}$. *But we see from the figure that* $\lambda = \frac{1}{2}(K^{\star} + K)$, *so* $\langle\langle T\rangle\rangle_{\mathfrak{a}} = \frac{1}{2}[T(\mathfrak{a}) + T(0)]$: **the average of the boundary temperatures as they appear to you is equal to the average of the temperature where you stand and the temperature at the centre.** *This is still true if we instead have many arcs at different temperatures, and ultimately a general piecewise-continuous temperature distribution. Thus Poisson's formula may be re-expressed as what we shall call* **Neumann's formula:** $\mathbf{T(\mathfrak{a}) = 2\,\langle\langle T\rangle\rangle_{\mathfrak{a}} - T(0)}$.

Figure [12.37] is intended to make this result vivid.Turning one's head successively through the same small angle marked • one would see the thermometers located at the white dots on the boundary. The average of their temperatures is then a good approximation [exact as $\bullet \to 0$] to $\langle\langle T\rangle\rangle_{\mathfrak{a}}$, and hence to the average of the temperature where we stand and the temperature at 0. Note how the white dots become crowded together on the part of the boundary nearest us. As anticipated, this part of the boundary therefore has the greatest influence on the temperature where we stand.

To obtain our third and final interpretation of Poisson's formula, imagine that the disc is the Beltrami–Poincaré model of the hyperbolic plane, and that you are once again standing at the point $\mathfrak{a}$ looking out to K, which is now infinitely far away on the horizon. How big does K appear to you in this distorted geometry? To a Godlike observer looking down on this model of the hyperbolic plane, the straight

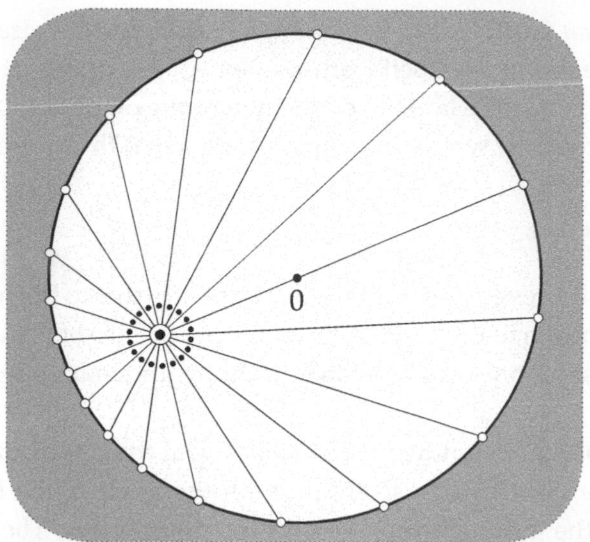

[12.37] Visualizing Neumann's geometrical interpretation of Poisson's formula. *Turning one's head successively through the same small angle marked • one would see the thermometers located at the white dots on the boundary. The average of their temperatures is then a good approximation [exact as • → 0] to $\langle\langle T \rangle\rangle_a$, and hence to the average of the temperature where we stand and the temperature at 0. Note how the white dots become crowded together on the part of the boundary nearest us. As anticipated, this part of the boundary therefore has the greatest influence on the temperature where we stand.*

lines along which light travels to you now appear to be arcs of circles orthogonal to C, and so *you* see the angular size of K as being

$$\text{hyperbolic angle} = \lambda + (\bullet + \odot)\,.$$

But we see in the figure that

$$(\bullet + \odot) = \tfrac{1}{2}\,(K^{\star} - K),$$

and hence we obtain the following remarkable fact:

$$\begin{aligned} \text{hyperbolic angle} &= \tfrac{1}{2}\,(K^{\star} + K) + \tfrac{1}{2}\,(K^{\star} - K) \\ &= K^{\star} \\ &= 2\pi\,T(a). \end{aligned}$$

The temperature where you are is simply proportional to how big K looks! [The result can also be obtained directly by appealing to the conformal and circle-preserving nature of the hyperbolic half-turn $M_a(z)$ considered earlier.]

Reinterpreting (12.33), we now see that $d\theta^{\star}$ is simply the hyperbolic angle subtended at a by the element of C: *the temperature of each element of* C *contributes to the temperature at an interior point in proportion to its hyperbolic size as seen from that point.* Much as we did in the Euclidean case, let $\prec T \succ_a$ denote the average of the temperatures you see on the horizon of the hyperbolic plane as you turn your head through a full revolution while standing at a. We have thus found what we shall call *Bôcher's formula*:

$$T(a) = \prec T \succ_a . \tag{12.36}$$

This result (exceeding even the beauty of Schwarz's) is due to Bôcher (1898), Bôcher (1906). We have chosen to present (12.36) as a consequence of Schwarz's result, but at the end of the section we shall see that it can be understood in a much simpler way.

The analogue of [12.37] is now [12.38]. Standing at the same point as before, and again turning one's head successively through the angle $\bullet$, the figure shows the new locations of the thermometers we see on the boundary. The average of their temperatures is then a good approximation [exact as $\bullet \to 0$] to $\prec T \succ_a$, and hence to the temperature where we stand. Note how the white dots again become crowded together on the part of the boundary nearest us, so that this part of the boundary has the greatest influence on the temperature where we stand.

From the vantage point of (12.36), the distinction between (12.28) and (12.33) evaporates. Every point of the hyperbolic plane is on an equal footing with every other, it is merely that the hyperbolic angle $d\theta^\star$ happens to coincide with the more familiar Euclidean angle $d\theta$ when $a = 0$.

Formulated in this way, we may carry the result over to the upper half-plane model for hyperbolic geometry. [The full justification for this transition will be explained at the end of the section.] The horizon is now the real axis and 'straight lines' are now (for our Godlike observer) semicircles meeting the real axis at right angles. The temperature where you stand is now the average (as $\bullet \to 0$) of the temperatures at the white boundary points in [12.39].

Figure [12.40] analyses this in greater detail. It shows both the hyperbolic angle $\Delta\theta^\star$ and the Euclidean angle $\Delta\theta$ subtended at a by the element Δx of the horizon. Thinking of Δx as sufficiently small that $T(x)$ is essentially constant on it, the contribution to the temperature at a is $(1/2\pi)\, T(x)\, \Delta\theta^\star$. Integrating along the entire horizon we obtain

$$T(a) = \frac{1}{2\pi} \int_{x=-\infty}^{x=\infty} T(x)\, d\theta^\star . \tag{12.37}$$

In order to put this into precisely the same form as (12.35), we need to find $(d\theta^\star/dx)$. We shall do this via an attractive and rather surprising fact: *The non-Euclidean angle $\Delta\theta^\star$ is exactly double the Euclidean one $\Delta\theta$, even if Δx is not small.* To see this, concentrate on the semicircle meeting the axis at p. The angle between the dotted tangent at a and the vertical is clearly double that between the chord ap and the vertical. The result then follows immediately.

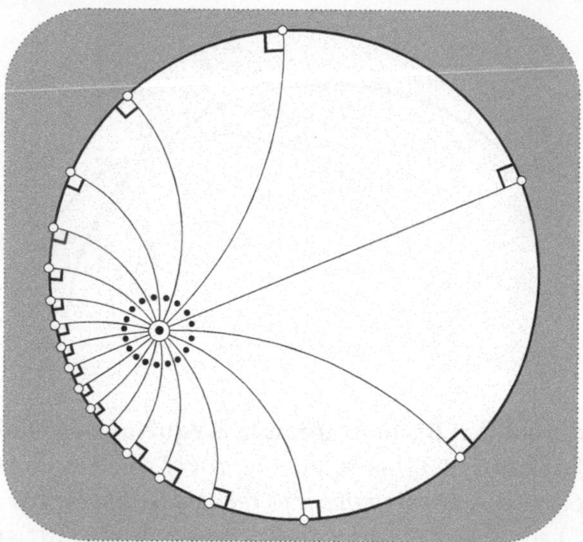

[12.38] Bôcher's hyperbolic interpretation of Poisson's formula. *Let* $\prec T \succ_a$ *denote the average of the temperatures you see on the horizon of the hyperbolic plane as you turn your head through a full revolution while standing at* a*. Then we have what we shall call* **Bôcher's formula:** $T(a) = \prec T \succ_a$. *This figure visualizes this average: turning your head successively through the small angle* $\bullet$*, the white dots show the locations of the thermometers you see on the boundary as light travels along hyperbolic straight lines to your eye. The average of their temperatures is then a good approximation [exact as* $\bullet \to 0$*] to* $\prec T \succ_a$*, and hence to the temperature where you stand. Note how the white dots again become crowded together on the part of the boundary nearest to you, so that this part of the boundary has the greatest influence on the temperature where you stand.*

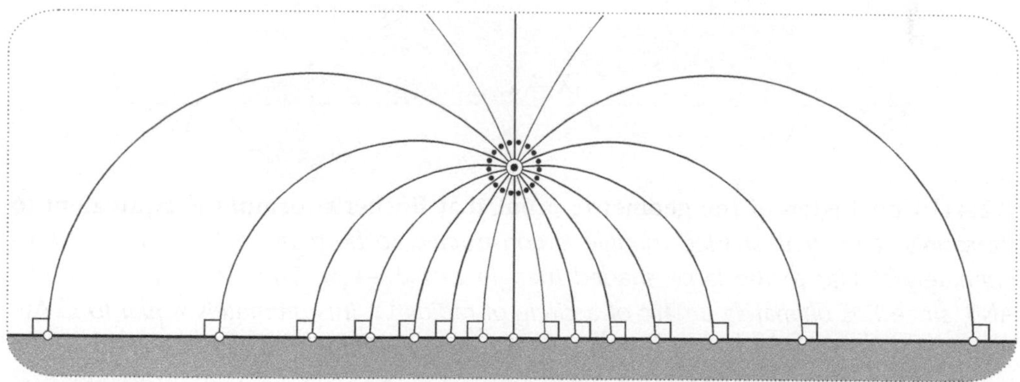

[12.39] Bôcher's hyperbolic interpretation of Poisson's formula in the upper half-plane. *The hyperbolic interpretation of the previous figure carries over without change to the Beltrami–Poincaré upper half-plane model of the hyperbolic plane. The horizon is now the real axis and hyperbolic straight lines are now semicircles meeting the real axis at right angles. The temperature where you stand is now the average (as* $\bullet \to 0$*) of the temperatures at the white boundary points.*

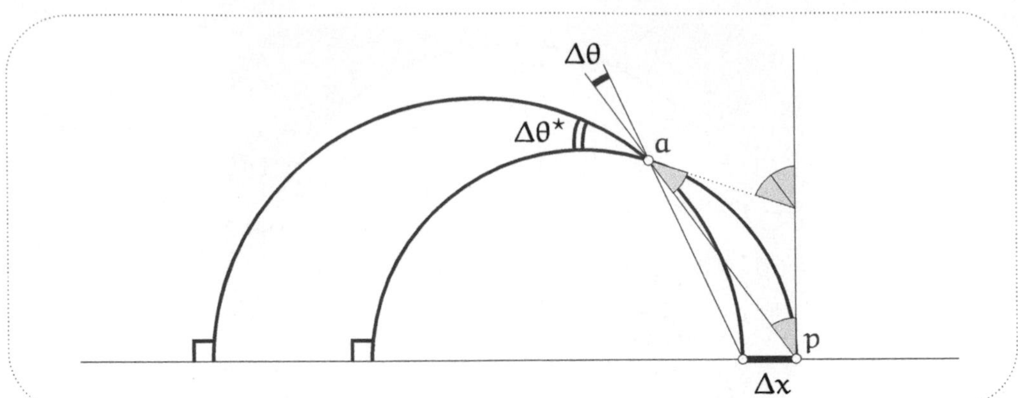

[12.40] Geometric proof that Bôcher's formula is equivalent to Poisson's half-plane formula. *To establish that (12.37) is equivalent to (12.35), we must find* $(d\theta^\star/dx) \asymp (\Delta\theta^\star/\Delta x)$. *The first step in evaluating this is to recognize that even if* Δx *is* not *small (as we shall ultimately need it to be),* **the non-Euclidean angle $\Delta\theta^\star$ subtended at a by Δx is exactly double the Euclidean one, $\Delta\theta$.** *To see this, concentrate on the semicircle meeting the axis at* p. *The angle between the dotted tangent at* a *and the vertical is clearly double that between the chord* ap *and the vertical. The result then follows immediately.*

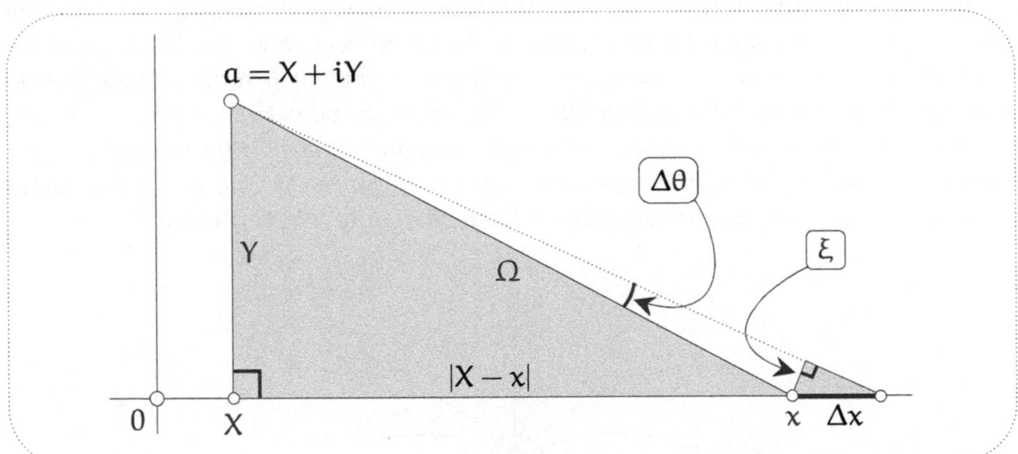

[12.41] Conclusion of the geometric proof that Bôcher's formula is equivalent to Poisson's. *The small shaded triangle is constructed to be right-angled, and it is thus ultimately similar to the large shaded triangle as* $\Delta\theta \to 0$. *Thus* $(\xi/\Delta x) \asymp (Y/\Omega)$. *Also, since* ξ *is ultimately an arc of a circle of radius* Ω, *it is ultimately equal to* $\Omega\,\Delta\theta$. *Thus as* $\Delta\theta \to 0$, $\frac{\Omega\,\Delta\theta}{\Delta x} \asymp \frac{\xi}{\Delta x} \asymp \frac{Y}{\Omega}$. *We can now combine this with the result of the previous figure to obtain* $\frac{d\theta^\star}{dx} = 2\left[\frac{d\theta}{dx}\right] = 2\left[\frac{Y}{\Omega^2}\right] = 2\left[\frac{Y}{(X-x)^2+Y^2}\right]$. *Inserting this into (12.37), we do indeed obtain (12.35).*

Now consider [12.41]. The small shaded triangle is constructed to be right-angled, and it is thus ultimately similar to the large shaded triangle as $\Delta\theta \to 0$. Thus $(\xi/\Delta x)$ is ultimately equal to (Y/Ω). Also, since ξ is ultimately an arc of a circle of radius Ω, it is ultimately equal to $\Omega\,\Delta\theta$. Thus as $\Delta\theta \to 0$,

$$\frac{\Omega\,\Delta\theta}{\Delta x} \asymp \frac{\xi}{\Delta x} \asymp \frac{Y}{\Omega}.$$

We can now combine this with the result of the previous figure to obtain

$$\frac{d\theta^*}{dx} = 2\left[\frac{d\theta}{dx}\right] = 2\left[\frac{Y}{\Omega^2}\right] = 2\left[\frac{Y}{(X-x)^2+Y^2}\right].$$

Inserting this into (12.37), we obtain (12.35).

While the precise form of the above argument may be new, the basic idea of transferring Bôcher's result from the disc to the half plane was given by Osgood (1928). For a different approach to (12.35), see Lange and Walsh (1985). For more on all three of the interpretations thus far obtained, see Perkins (1928).

12.7.5 Green's General Formula

If R is a simply connected region of *arbitrary* shape there exists a generalization of Poisson's formula (due to Green) for finding the temperature at any point a inside R in terms of the values $T(z)$ on the boundary B. As before, let $\mathcal{G}_a(z)$ be the Green's function of the region when the heat source is placed at the point a, so that local heat flux at the boundary point z is given by

$$\mathcal{Q}_a(z) = -\frac{\partial \mathcal{G}_a}{\partial n}.$$

With the aid of $\mathcal{Q}_a$ we may now determine $T(a)$. Here is the remarkable *Green's formula*:

$$T(a) = \frac{1}{2\pi}\oint_B \mathcal{Q}_a(z)\,T(z)\,ds, \tag{12.38}$$

where ds is an element of arc length along B. Thus $\mathcal{Q}_a$ now plays the same role as the Poisson kernel did in (12.29). Indeed, we previously calculated $\mathcal{Q}_a$ for the unit disc, and we now recognize the result (12.27) as the Poisson kernel $\mathcal{P}_a$.

Although formula (12.38) is valuable both in theory and practice, we should point out that it is less explicit than Poisson's formula, for to find $\mathcal{Q}_a$ we must first find the Green's function. But as we previously explained, the problem of finding $\mathcal{G}_a$ is *itself* a Dirichlet problem: to construct

$$\mathcal{G}_a(z) = -\ln\rho + g_a(z),$$

we must find the harmonic function g_a with boundary values $g_a(z) = \ln\rho$. Formula (12.38) says that if we can just solve this particular boundary value problem then we can solve them all.

To begin with, imagine that $T(z)$ is a given harmonic function in R whose value $T(a)$ at an interior a we wish to determine from the boundary values. The idea behind our explanation of (12.38) is very simple[12]. The Green's function $\mathcal{G}_a$ enables

[12] For a beautiful *physical* explanation of (12.38) in terms of electrostatic energy, see Maxwell (1873) or, better still, Maxwell (1881, Ch. III).

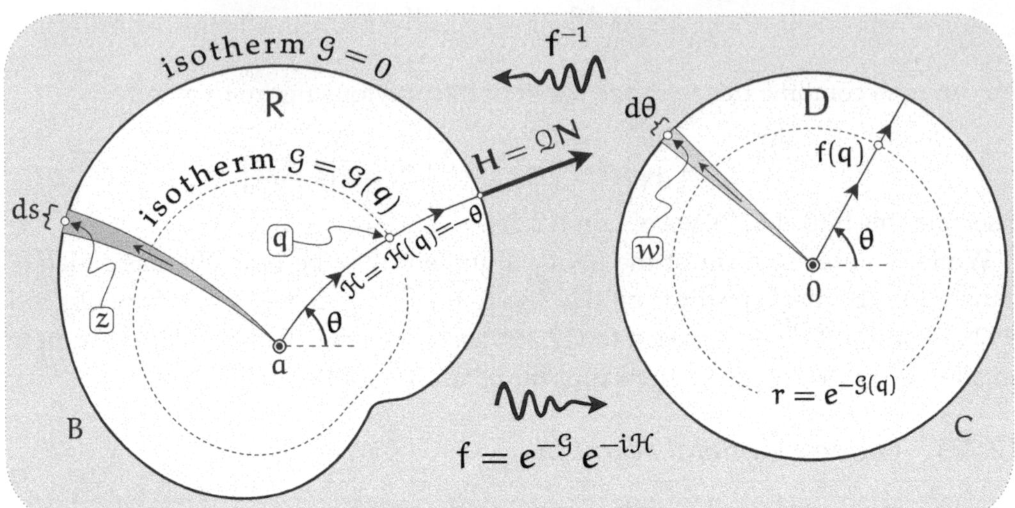

[12.42] Geometric proof of Green's formula. *If we know the Green's function of* R*, then, as previously explained, we can use it to conformally map* R *to the disc* D*:* $z \mapsto w = f(z) = e^{-\mathcal{G}}\, e^{-i\mathcal{H}}$*, and* $f(a) = 0$*. The harmonic function* $T(z)$ *in* R *may be transplanted to the harmonic function* $\widetilde{T}(w) \equiv T[f^{-1}(w)]$ *in* D*, the boundary values on* B *becoming the boundary values on* C*. But the average of these boundary temperatures on* C *is the temperature at the centre, and this is precisely what we want, because* $\widetilde{T}(0) = T(a)$*. Therefore,* $T(a) = \frac{1}{2\pi}\oint_C \widetilde{T}(w)\, d\theta = \frac{1}{2\pi}\oint_B T(z)\,|f'(z)|\, ds$*. Finally, recall the result (12.26): the amplification* $|f'(z)|$ *equals the local heat flux* $\mathcal{Q}_a(z)$*. This concludes the derivation of Green's formula, (12.38).*

us to construct a conformal mapping f (and an inverse mapping f^{-1}) between R and the unit disc D such that a corresponds to 0. Figure [12.42] essentially reproduces [12.31] in which this was explained.

Using the mapping $z \mapsto w = f(z)$, the harmonic function $T(z)$ in R may be transplanted to the harmonic function $\widetilde{T}(w) \equiv T[f^{-1}(w)]$ in D, the boundary values on B becoming the boundary values on C. But the average of these boundary temperatures on C is the temperature at the centre, and this is precisely what we were after, because $\widetilde{T}(0) = T(a)$.

Expressing this idea symbolically, we have

$$T(a) = \frac{1}{2\pi}\oint_C \widetilde{T}(w)\, d\theta.$$

If the element $d\theta$ of C at $w = f(z)$ is the image of the element ds of B at z then

$$d\theta = |f'(z)|\, ds,$$

so

$$T(a) = \frac{1}{2\pi}\oint_B T(z)\,|f'(z)|\, ds.$$

Finally, recall the result (12.26): the amplification $|f'(z)|$ equals the local heat flux $\mathcal{Q}_a(z)$. This concludes the derivation of formula (12.38).

This argument also explains the stronger result that (12.38) solves Dirichlet's problem for R. Using f to conformally transplant the given boundary values from B to C, we know that Poisson's formula allows us to construct the solution to Dirichlet's problem in D. Transferring this solution back from D to R with f^{-1}, we have found the harmonic function T in R, and its value at a must then be given by (12.38). You can now understand why we lavished so much attention on the special case of the disc.

We end this section with the observation that Green's formula (12.38) possesses the beautiful geometric interpretation shown on the LHS of [12.43]. Just as in [12.38], one imagines standing at a and turning one's head successively through the small angle $\bullet$. But now *suppose that light travels along the illustrated streamlines of the heat flow* $H = -\nabla \mathcal{G}_a$ *associated with the Green's function.* We would then see the thermometers at the illustrated points on the boundary. The general formula (12.38) says that the average of these temperatures (as $\bullet \to 0$) is the temperature where one stands! Bôcher's interpretation is clearly just a special case[13].

The explanation essentially reiterates the derivation of (12.26). Let z_θ be the boundary point we see when we look in the direction θ. Green's formula says that the temperature $T(z_\theta)$ of the element ds contributes to the temperature at a in proportion to $\mathcal{Q}_a(z_\theta)\,ds$, which is the flux of H through ds. Now follow the shaded flux tube back to the source at a, and let $d\theta$ be its angular width there. Since 2π of flux emerges symmetrically from a, the flux $\mathcal{Q}_a(z_\theta)\,ds$ emitted into our tube is equal to $d\theta$. Thus (12.38) may be re-expressed as

$$T(a) = \frac{1}{2\pi} \oint_B T(z_\theta)\, d\theta, \qquad (12.39)$$

namely, as the average of the boundary temperatures $T(z_\theta)$ over all directions. Done.

We have presented [12.43] as a geometric interpretation of (12.38), but we may instead use it to simplify and illuminate our derivation of that formula. The key observation is that (even without passing to the limit of vanishing $\bullet$) *the average in [12.43] of the observed temperatures on* B *is conformally invariant.* As before, let $J(z)$ be a one-to-one conformal mapping of R to some other simply connected region S with boundary Y. Just as we did with f, let us choose J so that the directions of curves through a are preserved (i.e., $\arg[J'(a)] = 0$). Let $w_\theta \equiv J(z_\theta)$ be the image on Y of z_θ on B.

By the conformal invariance of the Green's function, the image of the streamline leaving a at angle θ is the streamline leaving $J(a)$ at the same angle. Thus w_θ is not

[13] If we define the distance between two infinitesimally separated points of R to be the hyperbolic distance between their images in D, then R becomes a (non-standard) conformal model of the hyperbolic plane, and the geodesics emanating from a are the streamlines of [12.43]. The two results may then be viewed as identical.

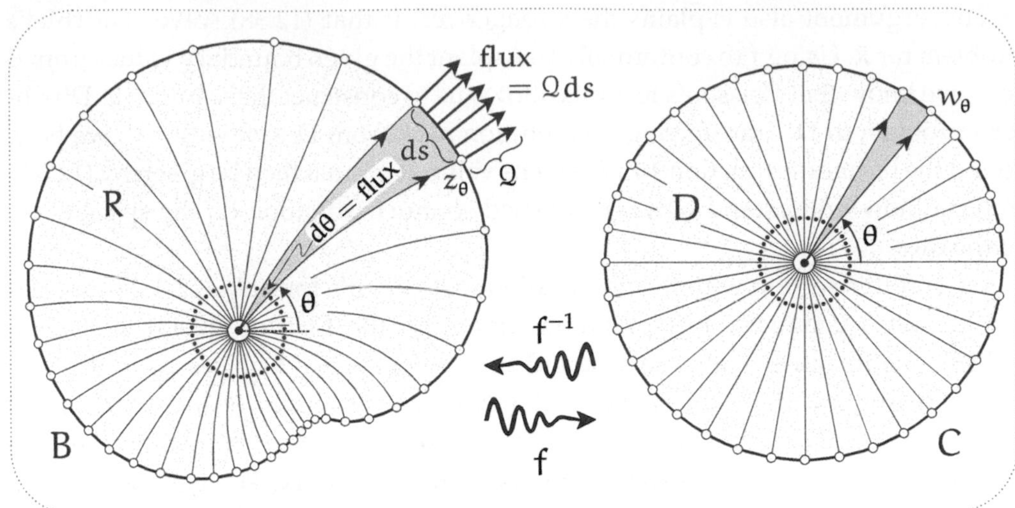

[12.43] The geometric meaning of Green's formula. *Green's formula (12.38) possesses the beautiful geometric interpretation shown on the left. Just as in [12.38], one imagines standing at* a *and turning one's head successively through the small angle •. But now* suppose that light travels along the illustrated streamlines of the heat flow $H = -\nabla \mathcal{G}_a$ associated with the Green's function. *We would then see the thermometers at the illustrated points on the boundary. The general formula (12.38) says that the average of these temperatures (as • → 0) is the temperature where one stands! Bôcher's interpretation is clearly just a special case. Indeed, if we define the distance between two infinitesimally separated points of* R *to be the hyperbolic distance between their images in* D*, then* R *becomes a (non-standard) conformal model of the hyperbolic plane, and the geodesics emanating from* a *are the streamlines. The two results may then be viewed as* identical.

only the image of z_θ, it is also the boundary point which an observer at $J(a)$ sees when he looks in the direction θ. But, by definition, the temperature at each point z_θ on B is transplanted to w_θ on Y, so the observer at $J(a)$ sees exactly the same temperatures on Y as the original observer at a saw on B. Done.

Passing to the limit of vanishing •, the conformal invariance of this average may expressed as

$$\frac{1}{2\pi}\oint_B T(z_\theta)\, d\theta = \frac{1}{2\pi}\oint_Y \widetilde{T}(w_\theta)\, d\theta.$$

Figure [12.43] illustrates the particular case where $J = f$ is the previously constructed function which maps R to D and a to 0. The virtue of this special case is that the conformally invariant average may now be *evaluated*. By Gauss's Mean Value Theorem, the average of the transplanted temperatures $\widetilde{T}(w_\theta) \equiv T(z_\theta)$ on C is the temperature $\widetilde{T}(0) \equiv T(a)$ at the centre:

$$\frac{1}{2\pi}\oint_B T(z_\theta)\, d\theta = \frac{1}{2\pi}\oint_C \widetilde{T}(w_\theta)\, d\theta = T(a).$$

Thus, returning to R and passing to the limit of vanishing •, the average of the observed temperatures is the temperature where you stand. Finally, the argument associated with [12.43] shows that (12.39) is equivalent to (12.38).

As illustrated in [12.44], this idea of a conformally invariant hyperbolic average lends unity to much of what we have done. Top centre is a depiction of Gauss's

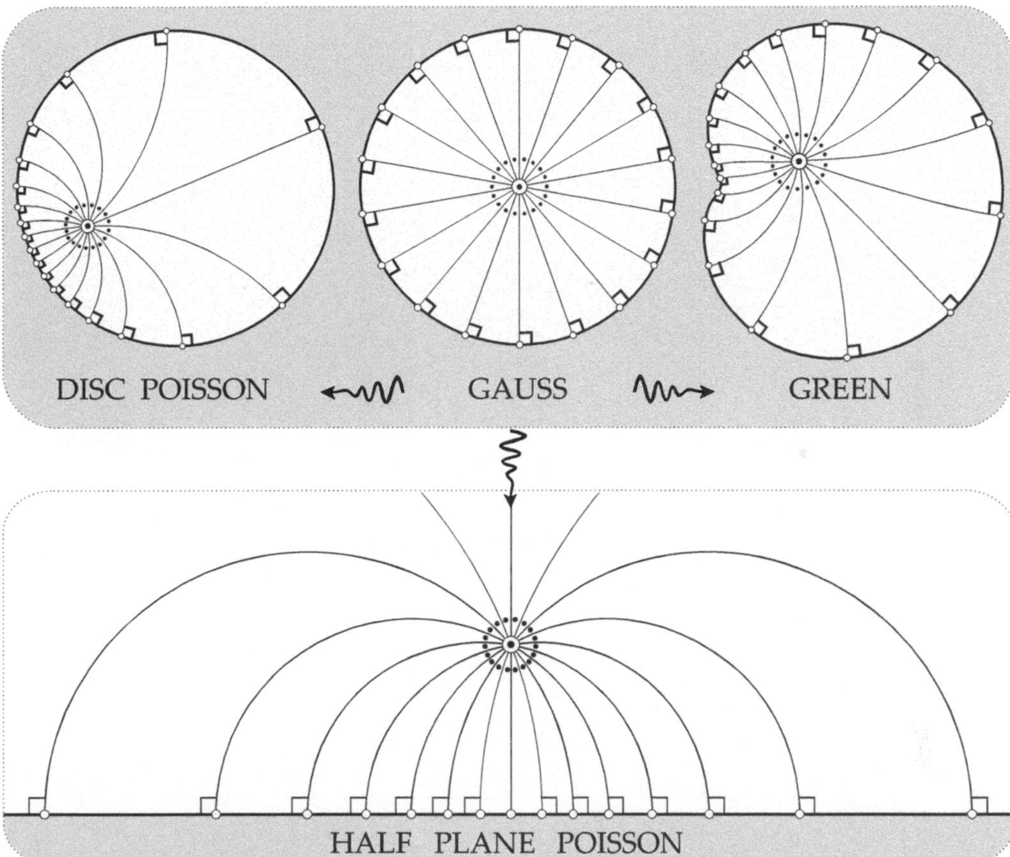

[12.44] Hyperbolic geometry unifies all methods of solving Dirichlet's Problem. *For the Beltrami–Poincarites who inhabit the hyperbolic plane, the conformally invariant average that Gauss calls for at the centre of the disc [see top centre] is utterly indistinguishable from the Poisson disc-formula [top left] at an arbitrary point. In both cases, the Beltrami–Poincarite turns her head successively through the small angle •, and both figures show the locations of the thermometers she sees on the boundary as light travels along hyperbolic straight lines (geodesics) to her eye. The average of their temperatures is then a good approximation [exact as • → 0] to the temperature where she stands. But she could also not detect any change if we transplanted her and the boundary temperatures to the Beltrami–Poincaré upper half-plane: the temperature would* still *be the average over all directions. Finally, if we conformally transplant both the boundary temperatures* and *the hyperbolic metric to a general region* R *[top right], then it becomes a (nonstandard) conformal model of the hyperbolic plane. All four figures may then be viewed* as hyperbolically indistinguishable.

theorem: as $\bullet \rightarrow 0$ the average of the temperatures at the white boundary points is equal to the temperature at the centre of the disc. Applying an automorphism to this picture yields the visual form of Poisson's formula for the disc; applying a Möbius transformation from the Beltrami–Poincaré disc model to the half-plane model yields the visual form of Poisson's formula for the half-plane; and applying a more general conformal mapping yields the visual form of Green's formula.

12.8 Exercises

1 (i) Show that the dual of a dipole is a dipole.

(ii) Think of a dipole as the limiting form of a doublet [a source and sink of equal strength]. Sketch the dual of a doublet, and thereby make geometric sense of part (i).

2 If Φ is a real harmonic function, and $\widehat{\Phi}$ is its harmonic dual, show that $\Phi\widehat{\Phi}$ is also harmonic. [*Hint*: Consider the complex function $(\Phi + i\,\widehat{\Phi}\,)$.]

3 Use Gauss's Mean Value Theorem to show that a harmonic function cannot have a local maximum.

4 Find the generalization of (12.7) in three dimensional space.

5 (i) If f is analytic, show that $\nabla|f| = (f\overline{f'})/|f|$, and explain how this agrees with part (i) of Ex. 19, p. 425.

(ii) Show that if Δ is the Laplacian, then $\Delta|f|^2 = 4\,|f'|^2$. Try deriving this by brute force.

6 Let $f(z)$ be analytic. By applying $\overline{\nabla}$ in turn to $(f + \overline{f})$ and to $(f\,\overline{f})$, show that if either the real part or the modulus of f is constant, then f itself is constant.

7 (i) By thinking of z and $\overline{z}$ as functions of x and y, use the chain rule of partial differentiation to show (at least formally) that

$$\partial_z = \tfrac{1}{2}\,\overline{\nabla} \quad \text{and} \quad \partial_{\overline{z}} = \tfrac{1}{2}\,\nabla.$$

(ii) Deduce (at least formally) that *an analytic function* f *depends on* z *but not on* $\overline{z}$*:*

$$\partial_z f = f' \quad \text{and} \quad \partial_{\overline{z}} f = 0.$$

8 Let J denote the Jacobian of a transformation $z \mapsto w$. Referring to the previous question, show that the determinant is given by $\det(J) = |\partial_z w|^2 - |\partial_{\overline{z}}\, w|^2$.

9 From Ex. 7(i) we see that $\Delta = \overline{\nabla}\nabla = 4\,\partial_z\partial_{\overline{z}}$. Use this fact to solve the following problems:

(i) Show that $\Phi = [1 - (x^2 + y^2)]^{-1}$ satisfies $\Delta\Phi = 4\Phi^2(2\Phi - 1)$.

(ii) Solve $\Delta F = e^z$ for F by formally integrating with respect to z and then with respect to $\overline{z}$. Deduce that $R = \frac{1}{4}e^x(x\cos y + y\sin y)$ is a solution of $\Delta R = e^x \cos y$. Verify this by calculating ΔR explicitly.

(iii) Show that if f is the most general harmonic function in the plane, then $f(z, \overline{z}) = p(z) + q(\overline{z})$, where p and q are arbitrary analytic functions.

10 As usual, let $\widehat{\xi}$ be the unit tangent to a curve of curvature κ, and let $\widetilde{\kappa}$ be the curvature of the image under the analytic mapping $f(z)$. Also let s and $\widetilde{s}$ denote arc length of the original and image curves. Finally, let $\Psi = (1/|f'|)$ be the stream function of the complex curvature, $\mathcal{K} = -i\nabla\Psi$.

(i) Use (5.31), p. 272 to show that $\partial_{\widetilde{s}}\,\widetilde{\kappa} = \Psi\,\partial_s[\kappa\Psi + \mathcal{K}\boldsymbol{\cdot}\widehat{\xi}\,]$.

(ii) Show that $\partial_s\widehat{\xi} = i\kappa\widehat{\xi}$ and $\partial_s\Psi = \widehat{\xi}\boldsymbol{\cdot}(i\mathcal{K})$.

(iii) Deduce that

$$\partial_{\widetilde{s}}\,\widetilde{\kappa} = \Psi^2\partial_s\kappa + \Psi\,(\partial_s\mathcal{K})\boldsymbol{\cdot}\widehat{\xi}\,.$$

[*Hint:* Remember (or prove) that $(i\mathbf{a})\boldsymbol{\cdot}(\mathbf{b}) + (\mathbf{a})\boldsymbol{\cdot}(i\mathbf{b}) = 0$.]

(iv) Recall Ex. 18, p. 296, in which we saw how Newton attempted to define the "angle" Θ between two touching curves as the difference of their curvatures: $\Theta \equiv (\kappa_1 - \kappa_2)$. Although this is not quite conformally invariant, show that $\left[\Theta^2/\partial_s\Theta\right]$ *is* conformally invariant. This geometrically meaningful generalization of the concept of angle is called *Kasner's Invariant*.

11 With the same notation as in the previous exercise, let p measure distance in the direction $i\widehat{\xi}$ perpendicular to the curve. By substituting $\mathcal{K} = -i\nabla\Psi$ into (5.31), p. 272, show that the image curvature is given by the tidy formula,

$$\widetilde{\kappa} = \partial_p\Psi + \kappa\Psi.$$

12 Consider the image under an analytic mapping f of a source of strength S located at p.

(i) Show geometrically, then algebraically, that if p is not a critical point of f (i.e., $f'(p) \neq 0$) then the image is another source of strength S at $f(p)$.

(ii) Show geometrically, then algebraically, that if p is a critical point of f of order $(n-1)$ then the image is again a source, but now of strength (S/n).

13 Repeat the investigation of the previous question in the case of a dipole located at p.

14 Show that applying the Milne-Thomson formula (12.18) to a vortex at a point p inside the unit circle yields *two* new vortices: at $1/\overline{p}$ and at 0. Explain this with the aid of a picture of the flow on the Riemann sphere.

15 (i) Show that if u and v are harmonic then $X \equiv (u\boldsymbol{\nabla}v - v\boldsymbol{\nabla}u)$ is sourceless.

(ii) Prove (12.38) by taking $u = T$ and $v = \mathcal{G}_a$, then equating the flux of X out of B with the flux out of an infinitesimal circle centred at a.

(iii) By taking $u = \mathcal{G}_a$ and $v = \mathcal{G}_b$, prove the symmetry property of the Green's function: $\mathcal{G}_a(b) = \mathcal{G}_b(a)$.

16 Let $T(z)$ be the temperature distribution in the unit disc if the top semicircle ($\text{Im}(z) > 0$) is kept at temperature $+(\pi/2)$ and the bottom semicircle is kept at temperature $-(\pi/2)$. Show that

$$T(z) = \text{Arg}\left[\frac{1+z}{1-z}\right].$$

17 Let $T(z)$ be a *non-negative* temperature distribution in the disc $|z| \leqslant R$. Writing $|a| = r$, use Poisson's formula (12.29) to derive *Harnack's inequality*:

$$\left(\frac{R-r}{R+r}\right) T(0) \leqslant T(a) \leqslant \left(\frac{R+r}{R-r}\right) T(0).$$

18 Use Harnack's inequality [previous exercise] to prove the following analogue of Liouville's Theorem: *If* T *is harmonic in the whole plane and is bounded from above (or below), then* T *is a constant.*

19 Substitute the disc Green's function (12.22) into (12.25), thereby confirming the formula (12.27) for the heat flux at the boundary of the disc.

20 (i) Use the method of images to find the Green's function for the upper half-plane.

(ii) Use this to show that Green's general formula (12.38) does indeed yield the Poisson half-plane formula (12.35).

21 Use the idea behind the method of images to show that if $0 < \text{Re}\,(p) < 1$ then the Green's function of the half-disc $\text{Re}\,(z) \geqslant 0$, $|z| \leqslant 1$ is

$$\mathcal{G}_p(z) = -\ln\left|\frac{z-p}{\overline{p}\,z-1}\right| + \ln\left|\frac{z+\overline{p}}{p\,z+1}\right|.$$

Check this by getting the computer to draw the level curves of $\mathcal{G}_p(z)$.

BIBLIOGRAPHY

Ahlfors, L. V. 1979. *Complex analysis: an introduction to the theory of analytic functions of one complex variable*, International series in pure and applied mathematics, 3d ed edition. New York: McGraw-Hill.

Altmann, S. L. 1989. Hamilton, Rodrigues, and the quaternion scandal. *Mathematics Magazine*, 62(5):291–308.

Arianrhod, R. 2019. *Thomas Harriot: A Life in Science*. Oxford University Press.

Arnol'd, V. I. 1989. *Mathematical Methods of Classical Mechanics*, volume 60 of *Graduate Texts in Mathematics*, 2nd edition. New York: Springer-Verlag.

Arnol'd, V. I. 1990. *Huygens and Barrow, Newton and Hooke*. Basel: Birkhäuser Verlag. Pioneers in mathematical analysis and catastrophe theory from evolvents to quasicrystals, Translated from the Russian by Eric J. F. Primrose.

Bak, J. and D. J. Newman 2010. *Complex Analysis*, Undergraduate texts in mathematics, 3rd ed edition. New York: Springer.

Beardon, A. F. 1979. *Complex Analysis: the Argument Principle in Analysis and Topology*. Chichester: Wiley.

Beardon, A. F. 1984. *A Primer on Riemann Surfaces*, volume 78. Cambridge: Cambridge University Press.

Beardon, A. F. 1987. Curvature, circles, and conformal maps. *The American Mathematical Monthly*, 94(1):48–53.

Beltrami, E. 1868a. Essay on the interpretation of non-Euclidean geometry. In *Stillwell (1996)*. American Mathematical Society.

Beltrami, E. 1868b. Fundamental theory of spaces of constant curvature. In *Stillwell (1996)*. American Mathematical Society.

Bivens, I. C. 1992. When do orthogonal families of curves possess a complex potential? *Mathematics Magazine*, 65(4):226–235.

Bloye, N. and S. Huggett 2011. Newton, the Geometer. *Newsletter of the European Mathematical Society*, (82):19–27.

Boas, R. P. and H. P. Boas 2010. *Invitation to complex analysis*, MAA textbooks, 2nd ed edition. Mathematical Association of America.

Bôcher, M. 1898. Note on poisson's integral. *Bull. Amer. Math. Soc.*, 4(9):424–426.

Bôcher, M. 1906. On harmonic functions in two dimensions. *Proceedings of the American Academy of Arts and Sciences*, 41(26):577–583.

Bohlin, K. 1911. Note sur le problème des deux corps et sur une integration nouvelle dans le problème des trois corps. *Bull. Astr.*, 28:113–119.

Braden, B. 1985. Picturing functions of a complex variable. *The College Mathematics Journal*, 16(1):63–72.

Braden, B. 1987. Polya's geometric picture of complex contour integrals. *Mathematics Magazine*, 60(5):321–327.

Braden, B. 1991. *Visualization in teaching and learning mathematics: a project*, volume no. 19, chapter : The Vector Field Approach in Complex Analysis. Mathematical Association of America.

Brieskorn, E. and H. Knörrer 2012. *Plane Algebraic Curves*, Modern Birkhäuser classics. Basel: Birkhäuser.

Carathéodory, C. 1937. The most general transformations of plane regions which transform circles into circles. *Bulletin of the American Mathematical Society*, 43(8):573–579.

Carathéodory, C. 1964. *Theory of Functions of a Complex Variable*, 2nd english ed edition. New York: Chelsea Pub. Co.

Chandrasekhar, S. 1995. *Newton's Principia for the Common Reader*. Oxford: The Clarendon Press, Oxford University Press.

Courant, R. 1950. *Dirichlet's Principle, Conformal Mapping, and Minimal Surfaces*, volume 3. New York: Interscience Publishers.

Coxeter, H. S. M. 1946. Quaternions and reflections. *The American Mathematical Monthly*, 53(3):136–146.

Coxeter, H. S. M. 1967. *Proceedings of the International Conference on the Theory of Groups*, chapter : The Lorentz Group and the Group of Homographies, Pp. 73–77. New York: Gordon and Breach Science Publishers.

Coxeter, H. S. M. 1969. *Introduction to Geometry*, 2nd edition. New York: Wiley.

Coxeter, H. S. M. and S. L. Greitzer 1967. *Geometry Revisited*, volume 19. Random House.

Davis, P. and H. Pollak 1958. On the analytic continuation of mapping functions. *Transactions of the American Mathematical Society*, 87(1):198–225.

Davis, P. J. 1974. *The Schwarz Function and its Applications*, volume no. 17. Mathematical Association of America.

de Gandt, F. 1995. *Force and Geometry in Newton's Principia*. Princeton, NJ: Princeton University Press. Translated from the French original and with an introduction by Curtis Wilson.

do Carmo, M. P. 1994. *Differential Forms and Applications*, Universitext. Berlin: Springer-Verlag.

Dray, T. 2015. *Differential Forms and the Geometry of General Relativity*. CRC Press, Taylor & Francis Group.

Duffin, R. J. 1957. A note on Poisson's integral. *Quarterly of Applied Math*, 15:109–111.

Earl, R. 2019. *Topology: A Very Short Introduction*, Very short introductions. Oxford University Press.

Echols, W. H. 1923. Some properties of a skewsquare. *The American Mathematical Monthly*, 30(3):120–127.

Eves, H. 1992. *Fundamentals of Modern Elementary Geometry*. Boston: Jones and Bartlett.

Feynman, R. P. 1963. *The Feynman Lectures on Physics*. Reading, Mass.: Addison-Wesley Pub. Co.

Feynman, R. P. 1966. Nobel lecture: "The Development of the Space-Time View of Quantum Field Theory". *Physics Today*, August:31–34.

Feynman, R. P. 1985. *QED: the Strange Theory of Light and Matter*. Princeton, N.J.: Princeton University Press.

Feynman, R. P. 1997. *Surely you're joking, Mr. Feynman!: Adventures of a Curious Character*, Norton paberback edition. New York: W.W. Norton.

Finney, R. L. 1970. Dynamic proofs of Euclidean theorems. *Mathematics Magazine*, 43(4):177–185.

Ford, L. R. 1929. *Automorphic Functions*, 1st edition. New York: McGraw-Hill book company, inc.

Frankel, T. 2012. *The Geometry of Physics: an Introduction*, 3rd ed edition. Cambridge: Cambridge University Press.

Fulton, W. 1995. *Algebraic Topology: A First Course*, volume 153. New York: Springer-Verlag.

Gauss, C. F. 1827. *General Investigations of Curved Surfaces*, Translated from the Latin and German by Adam Hiltebeitel and James Moreh ead, 1965 edition. Raven Press, Hewlett, N.Y.

Goodstein, J. R. 2018. *Einstein's Italian Mathematicians: Ricci, Levi-Civita, and the Birth of General Relativity*. AMS.

Greenberg, M. J. 2008. *Euclidean and non-Euclidean Geometries*, 4th edition. New York: W.H. Freeman.

Guicciardini, N. 1999. *Reading the Principia*. Cambridge: Cambridge University Press. The debate on Newton's mathematical methods for natural philosophy from 1687 to 1736.

Guicciardini, N. 2009. *Isaac Newton on Mathematical Certainty and Method*, Transformations. Cambridge, Mass.: MIT Press.

Henrici, P. 1991. *Applied and Computational Complex Analysis*, volume 3 of *Wiley classics library*, wiley classics library ed edition. New York: Wiley.

Hilbert, D. 1902. *The Foundations of Geometry [Second English Edition of Tenth German Edition, translated by Leo Unger, Revised and Enlarged by Dr. Paul Bernays, published 1971]*. Chicago: The Open Court.

Hilbert, D. 1952. *Geometry and the Imagination*. New York: Chelsea Pub. Co.

Hilbert, D. 1965. *Gesammelte Abhandlungen*. Bronx, N.Y.: Chelsea Pub. Co.

Hildebrandt, S. and A. Tromba 1985. *Mathematics and Optimal Form*. New York: Scientific American Library.

Hoggar, S. G. 1992. *Mathematics for computer graphics*, volume 14. Cambridge: Cambridge University Press.

Hopf, H. 1946–1956. *Differential Geometry in the Large: seminar lectures, New York University, 1946 and Stanford University, 1956 [Second Edition, 1983]*, volume 1000 of *Lecture notes in mathematics*. Berlin: Springer-Verlag.

Horn, B. K. P. 1991. Relative orientation revisited. *J. Opt. Soc. Am. A*, 8(10): 1630–1638.

Jones, G. A. and D. Singerman 1987. *Complex Functions: An Algebraic and Geometric Viewpoint*. Cambridge: Cambridge University Press.

Kasner, E. 1912. Conformal geometry. In *Proceedings of the International Congress of Mathematicians, Vol. 2*.

Kasner, E. 1913. *Differential-geometric Aspects of Dynamics*, American Mathematical Society. American Mathematical Society.

Katok, S. 1992. *Fuchsian Groups*, Chicago lectures in mathematics series. Chicago: University of Chicago Press.

Klein, F. 1881. *On Riemann's Theory of Algebraic Functions and their Integrals: A Supplement to the Usual Treatises ; translated from the German by Frances Hardcastle*. Mineola, NY: 2003, Dover Publications.

Körner, T. W. 1988. *Fourier Analysis*. Cambridge: Cambridge University Press.

Lanczos, C. 1966. *Discourse on Fourier series*. Edinburgh: Oliver and Boyd.

Lange, R. and R. A. Walsh 1985. A heuristic for the poisson integral for the half plane and some caveats. *The American Mathematical Monthly*, 92(5): 356–358.

Ludvigsen, M. 1999. *General Relativity: A Geometric Approach*. Cambridge: Cambridge University Press.

Markushevich, A. 2005. *Theory of Functions of a Complex Variable*, number pt. 11 in AMS Chelsea Publishing Series, 2nd edition. AMS Chelsea Pub.

Marsden, J., U. Marsden, M. Hoffman, U. Hoffman, and T. Marsden 1999. *Basic Complex Analysis*. W. H. Freeman.

Maxwell, J. 1873. *A Treatise on Electricity and Magnetism*, A Treatise on Electricity and Magnetism 2 Volume Paperback Set. 2010, Cambridge University Press.

Maxwell, J. 1881. *An Elementary Treatise on Electricity*, Cambridge Library Collection - Physical Sciences. 2011, Cambridge University Press.

Milnor, J. 1982. Hyperbolic geometry: the first 150 years. *Bull. Amer. Math. Soc. (N.S.)*, 6(1):9–24.

Misner, Thorne, and Wheeler 1973. *Gravitation*. San Francisco: W. H. Freeman [A marvellous new hardback edtition was published by Princeton University Press in 2017. It contains a new introduction by Charles Misner and Kip Thorne, discussing exciting developments in the field since the book's original publication.].

Mughal, A. and D. Weaire 2009. Curvature in conformal mappings of two-dimensional lattices and foam structure. *Proc. R. Soc.*, A(465):219–238.

Needham, T. 1993. Newton and the transmutation of force. *Amer. Math. Monthly*, 100(2):119–137.

Needham, T. 1994. The geometry of harmonic functions. *Mathematics Magazine*, 67(2):92–108.

Needham, T. 2014. Visual Differential Geometry and Beltrami's Hyperbolic Plane. In *The Art of Science: From Perspective Drawing to Quantum Randomness*, R. Lupacchini and A. Angelini, eds., Pp. 71–99. Springer International Publishing.

Needham, T. 2021. *Visual Differential Geometry and Forms: A Mathematical Drama in Five Acts*. Princeton, N.J.: Princeton University Press.

Nehari, Z. 1952. *Conformal mapping*. New York: Dover Publications.

Neumann, C. 1884. *Vorlesungen über Riemann's Theorie der Abel'schen Integrale*. Leipzig : B. G. Teubner.

Nevanlinna, R. H. and V. Paatero 2007. *Introduction to Complex Analysis*, 2nd edition. Providence, R.I.: AMS Chelsea Pub./American Mathematical Society.

Newton, I. 1670. *The Mathematical Papers of Isaac Newton*, volume III. Cambridge: 1981, Cambridge U.P.

Newton, I. 1687. *The Principia: Mathematical Principles of Natural Philosophy*, 1999 edition. Berkeley, CA: University of California Press. A new translation by I. Bernard Cohen and Anne Whitman, assisted by Julia Budenz, Preceded by "A guide to Newton's *Principia*" by Cohen.

Nikulin, V. V. and I. R. Shafarevich 1987. *Geometries and Groups*, Universitext. Berlin: Springer-Verlag.

Nitsche, J. C. C. 1989. *Lectures on Minimal Surfaces*. Cambridge: Cambridge University Press.

Ogilvy, C. S. 1990. *Excursions in Geometry*. New York: Dover Publications.

O'Neill, B. 2006. *Elementary Differential Geometry*, Revised 2nd edition. Amsterdam: Elsevier Academic Press.

Osgood, W. 1928. *Lehrbuch der Funktionentheorie*. Teubner.

Penrose, R. 1978. The geometry of the universe. In *Mathematics today: twelve informal essays*, L. A. Steen, ed., Pp. 83–127. New York: Vintage Books.

Penrose, R. 2005. *The Road to Reality*. New York: Alfred A. Knopf Inc. A complete guide to the laws of the universe.

Penrose, R. and W. Rindler 1984. *Spinors and Space-Time*. Cambridge: Cambridge University Press.

Perkins, F. W. 1928. An intrinsic treatment of Poisson's Integral. *American Journal of Mathematics*, 50(3):389–414.

Poincaré, H. 1985. *Papers on Fuchsian Functions*. New York: Springer-Verlag.

Pólya, G. 1954. *Mathematics and Plausible Reasoning*. Princeton, N.J.: Princeton University Press.

Pólya, G. and G. Latta 1974. *Complex Variables*. New York: Wiley.

Prasolov, V. V. 1995. *Intuitive Topology*, volume v. 4. American Mathematical Society.

Reech 1858. *Journal de l'École Polytechnique*, 27:169–178.

Richeson, D. S. 2008. *Euler's Gem: The Polyhedron Formula and the Birth of Topology*. Princeton, N.J.: Princeton University Press.

Schutz, B. F. 1980. *Geometrical Methods of Mathematical Physics*. Cambridge: Cambridge University Press.

Schutz, B. F. 2022. *A First Course in General Relativity*, 3rd edition. Cambridge: Cambridge University Press.

Schwarz, H. 1972a. *Gesammelte Mathematische Abhandlungen*, volume II, Pp. 144–171. Chelsea Publishing.

Schwarz, H. 1972b. *Gesammelte Mathematische Abhandlungen*, volume II, P. 360. Chelsea Publishing.

Shapiro, H. S. 1992. *The Schwarz Function and its Generalization to Higher Dimensions*, volume v. 9. New York: Wiley.

Shaw, W. T. 2006. *Complex Analysis with Mathematica*. Cambridge, UK: Cambridge University Press.

Siegel, C. L. 1969. *Topics in Complex Function Theory*, volume 25. New York: Wiley-Interscience.

Sommerville, D. 1958. *The Elements of Non-Euclidean Geometry*. Dover.

Springer, G. 1981. *Introduction to Riemann Surfaces*, 2nd ed edition. New York: Chelsea Pub. Co.

Stewart, G. W. 1993. On the early history of the singular value decomposition. *SIAM Review*, 35(4):551–566.

Stewart, I. and D. O. Tall 2018. *Complex Analysis: The Hitch Hiker's Guide to the Plane*, second edition. Cambridge University Press.

Stillwell, J. 1992. *Geometry of Surfaces*, Universitext. New York: Springer-Verlag.

Stillwell, J. 1994. *Elements of Algebra: Geometry, Numbers, Equations*. New York: Springer-Verlag.

Stillwell, J. 1996. *Sources of Hyperbolic Geometry*, volume 10 of *History of Mathematics*. Providence, RI: American Mathematical Society.

Stillwell, J. 2010. *Mathematics and its History*, third edition. New York: Springer.

Thorne, K. S. and R. D. Blandford 2017. *Modern Classical Physics: Optics, Fluids, Plasmas, Elasticity, Relativity, and Statistical Physics*. Princeton University Press.

Thurston, W. P. 1997. *Three-Dimensional Geometry and Topology*, volume 35. Princeton, N.J.: Princeton University Press.

Trefethen, L. N. 1986. *Numerical Conformal Mapping*. Amsterdam: North-Holland.

Vitelli, V. and D. R. Nelson 2006. Nematic textures in spherical shells. *Phys. Rev. E*, 74:021711.

Waerden, B. L. v. d. 1985. *A history of algebra: from al-Khwārizmī to Emmy Noether*. Berlin: Springer-Verlag.

Wald, R. M. 1984. *General Relativity*. Chicago: University of Chicago Press.

Westfall, R. S. 1980. *Never at Rest*. Cambridge: Cambridge University Press. A biography of Isaac Newton.

Whyburn, G. 1955. *Introductory Topological Analysis*, number v. 783, no. 22 in Introductory Topological Analysis. University of Michigan Press.

Whyburn, G. 2015. *Topological Analysis*, Princeton Legacy Library, second edition. Princeton University Press.

Woodward, L. M. and J. Bolton 2019. *A First Course in Differential Geometry: Surfaces in Euclidean Space*. Cambridge University Press.

Yaglom, I. M. 1988. *Felix Klein and Sophus Lie: evolution of the idea of symmetry in the Nineteenth Century*. Boston: Birkhäuser.

INDEX